# NREL/SNL PHOTOVOLTAICS PROGRAM REVIEW

# NREL/SNL PHOTOVOLTAICS PROGRAM REVIEW

Proceedings of the 14th Conference—
A Joint Meeting

Lakewood, Colorado November 1996

EDITORS
C. Edwin Witt
Mowafak Al-Jassim
*National Renewable Energy Laboratory*

James M. Gee
*Sandia National Laboratories*

AIP CONFERENCE
PROCEEDINGS 394

**American Institute of Physics**  Woodbury, New York

Authorization to photocopy items for internal or personal use, beyond the free copying permitted under the 1978 U.S. Copyright Law (see statement below), is granted by the American Institute of Physics for users registered with the Copyright Clearance Center (CCC) Transactional Reporting Service, provided that the base fee of $10.00 per copy is paid directly to CCC, 222 Rosewood Drive, Danvers, MA 01923. For those organizations that have been granted a photocopy license by CCC, a separate system of payment has been arranged. The fee code for users of the Transactional Reporting Service is: 1-56396-687-5/ 97 /$10.00.

© 1997 American Institute of Physics

Individual readers of this volume and nonprofit libraries, acting for them, are permitted to make fair use of the material in it, such as copying an article for use in teaching or research. Permission is granted to quote from this volume in scientific work with the customary acknowledgment of the source. To reprint a figure, table, or other excerpt requires the consent of one of the original authors and notification to AIP. Republication or systematic or multiple reproduction of any material in this volume is permitted only under license from AIP. Address inquiries to Office of Rights and Permissions, 500 Sunnyside Boulevard, Woodbury, NY 11797-2999; phone: 516-576-2268; fax: 516-576-2499; e-mail: rights@aip.org.

L.C. Catalog Card No. 97-72645
ISBN 1-56396-687-5
ISSN 0094-243X
DOE CONF- 961178

Printed in the United States of America

CONTENTS

Preface .................................................................. xiii

## THIN FILMS I

Progress Report on the Amorphous Silicon Teaming Activities............... 3
    B. von Roedern, K. Zweibel, E. Schiff, J. D. Cohen, S. Wagner,
    S. S. Hegedus, and T. Peterson

A Record Setting Amorphous Silicon Alloy Triple-Junction Solar
Cell With 14.6% Initial and 12.8% Stable Efficiencies...................... 13
    J. Yang, A. Banerjee, and S. Guha

Hot Wire Deposited Hydrogenated Amorphous Silicon Solar Cells ............ 27
    A. H. Mahan, E. Iwaniczko, B. P. Nelson, R. C. Reedy Jr., T. Unold,
    R. S. Crandall, S. Guha, and J. Yang

Stability Properties of Amorphous Silicon and Silicon-Germanium Films
and Devices Deposited Using Remote ECR-Techniques ....................... 33
    V. L. Dalal, S. Kaushal, T. Maxson, R. Girvan, and A. Boerner

Deposition of Transparent Conducting Oxides For Solar Cells............... 39
    R. G. Gordon

Progress in Amorphous Silicon Multijunction Research at Solarex ........... 49
    R. R. Arya, D. E. Carlson, L. Yang, L. F. Chen, F. Willing, K. Rajan,
    K. Jansen, C. Poplawski, D. Bradley, and G. Wood

Why is $CuInSe_2$ Tolerant to Defects and What Is the Origin of "Ordered
Defect Structures" ....................................................... 63
    A. Zunger, S. B. Zhang, and S.-H. Wei

## THIN FILMS II

Team-Based Thin-Film CIS Research Activities ............................. 75
    H. S. Ullal

Materials and Processing Issues in Thin-Film $Cu(In,Ga)Se_2$—Based
Solar Cells .............................................................. 83
    J. R. Tuttle, M. A. Contreras, K. R. Ramanathan, S. E. Asher,
    R. Bhattacharya, T. A. Berens, J. Keane, and R. Noufi

Low Cost CIS Device Processing .......................................... 107
    B. M. Basol, V. K. Kapur, C. R. Leidholm, R. Roe, A. Halani,
    and G. Norsworthy

Technologies for Thin Film CIGS-Based Photovoltaics ..................... 115
    J. S. Britt, A. E. Delahoy, and Z. J. Kiss

Fabrication and Characterization of $Cu(InGa)Se_2$ Solar Cells
with Absorber Bandgap from 1.0 to 1.5 eV ................................ 123
    W. N. Shafarman, R. W. Birkmire, M. Marudachalam, B. E. McCandless,
    and J. M. Schultz

Progress in Thin Film CIGS Modules ...................................... 133
    S. Wiedeman, J. Kessler, L. Russell, J. Fogleboch, S. Skibo, and R. Arya

**Progress in CIS-Based Module Development** .............................. 143
    D. E. Tarrant, J. Bauer, R. Dearmore, M. E. Dietrich, G. T. Fernandez,
    O. D. Frausto, C. V. Fredric, C. L. Jensen, A. R. Ramos,
    J. A. Schmitzberger, R. D. Wieting, D. Willet, and R. R. Gay

**CdTe Team Activities** .................................................. 153
    P. V. Meyers

**CdTe Solar Cells: Electronic and Morphological Properties** ................ 162
    J. J. Kester, S. Albright, V. Kaydanov, R. Ribelin, L. M. Woods,
    and J. A. Phillips

**Recent Progress in CdTe Solar Cell Research at SCI** ....................... 171
    R. A. Sasala, R. C. Powell, G. L. Dorer, and N. Reiter

## CRYSTALLINE MATERIALS I

**The Crystalline-Silicon Photovoltaic R&D Project at NREL and SNL** ........ 189
    J. M. Gee and T. F. Ciszek

**Gettering and Passivation of High Efficiency Multicrystalline Silicon Solar Cells** ............................................................. 199
    A. Rohatgi, S. Narasimha, and L. Cai

**Physical and Numerical Modeling of Impurity Gettering in Silicon** .......... 215
    T. Y. Tan, R. Gafiteanu, and U. M. Gösele

**Carrier Recombination in Silicon Materials Used for Photovoltaic Devices** .... 225
    R. K. Ahrenkiel

## CRYSTALLINE MATERIALS II

**Recent Developments in Terrestrial Concentrator Photovoltaics** .............. 247
    S. R. Kurtz and D. Friedman

**High-Efficiency GaAs Solar Cells on mm and sub-mm Grain-Size Polycrystalline Ge Substrates** ........................................... 259
    R. Venkatasubramanian, B. O'Quinn, and E. Siivola

**The "Micromorph" Cell: A New Way to High-Efficiency-Low-Temperature Crystalline Silicon Thin-Film Cell Manufacturing?** ........ 271
    H. Keppner, U. Kroll, P. Torres, J. Meier, R. Platz, D. Fischer, N. Beck,
    S. Dubail, J. A. Anna Selvan, N. Pellaton Vaucher, M. Goerlitzer, Y. Ziegler,
    R. Tscharner, Ch. Hof, M. Goetz, P. Pernet, N. Wyrsch, J. Vuille, J. Cuperus,
    A. Shah, and J. Pohl

## COMPONENT AND SYSTEM EVALUATION AND RELIABILITY

**Photovoltaic Performance and Reliability Workshop Summary** .............. 285
    B. Kroposki

**Service Lifetime Prediction for Encapsulated Photovoltaic Cells/Minimodules** ........................................................... 295
    A. W. Czanderna and G. J. Jorgensen

Can the Staebler-Wronski Effect Account for the Long-Term
Performance of α-Si PV Arrays?........................................ 313
    B. von Roedern and B. Kroposki
PV System Field Experience and Reliability ............................. 323
    S. Durand, A. Rosenthal, and M. Thomas
Inverter Testing at Sandia National Laboratories........................ 335
    J. W. Ginn, R. H. Bonn, and G. Sittler
Photovoltaic Module and Array Performance Characterization Methods
for All System Operating Conditions..................................... 347
    D. L. King
PV Standards Overview .................................................. 369
    R. DeBlasio
Battery Testing for Photovoltaic Applications .......................... 379
    T. Hund
PV Solar Radiometric Measurements ...................................... 395
    D. R. Myers and T. W. Cannon

## PHOTOVOLTAIC MANUFACTURING

Photovoltaic Manufacturing Technology (PVMaT) Project—Latest Results.... 407
    R. L. Mitchell, C. E. Witt, and H. P. Thomas
Improvements in Cast Polycrystalline Silicon PV Manufacturing
Technology.............................................................. 415
    J. H. Wohlgemuth
Progress in High-Throughput Manufacturing of Thin-Film CdTe Modules .... 425
    D. W. Sandwisch
Improvements in Cz Silicon PV Module Manufacturing ..................... 433
    R. R. King, K. W. Mitchell, and T. L. Jester
Large-Area Silicon-Film™ Manufacturing ................................. 445
    D. H. Ford, A. M. Barnett, J. C. Checchi, J. S. Culik, R. B. Hall,
    E. L. Jackson, C. L. Kendall, and J. A. Rand
PVMaT Improvements in Monolithic α-Si Modules on Continuous
Polymer Substrates...................................................... 451
    F. R. Jeffrey, D. P. Grimmer, S. Brayman, B. Scandrett, M. Thomas,
    S. A. Martens, W. Chen, and M. Noak
The SunSine300 AC/PV Module........................................... 463
    M. C. Russell, G. A. Kern, and C. K. P. Handleman
Design, Fabrication and Certification of Advanced Modular PV
Power Systems .......................................................... 471
    G. E. Minyard and T. J. Lambarski

## MARKETS AND APPLICATIONS

Commercialization of Multijunction α-Si Modules......................... 479
    D. E. Carlson, R. R. Arya, L.-F. Chen, R. Oswald, J. Newton, K. Rajan,
    R. Romero, F. Willing, and L. Yang

**Photovoltaics in the Department of Defense**.............................. 491
    R. N. Chapman
**The USAID/DOE Mexico Renewable Energy Program:**
**Using Technology to Build New Markets** ................................. 513
    C. J. Hanley
**The Ramakrishna Mission PV Project—A Cooperation Between**
**India and the United States** ............................................... 521
    J. L. Stone and H. S. Ullal
**Sino-American Cooperation for Rural Electrification in China** .............. 529
    W. L. Wallace and Y. S. Tsuo

## POSTERS

### THIN FILMS (P)

**Potential Fluctuations in Intrinsic Hydrogenated Amorphous Silicon** ......... 537
    S. Dong, J. Liebe, Y. Tang, R. Braunstein, and B. von Roedern
**Transparent Conducting Oxide Contacts for n-i-p and p-i-n Amorphous**
**Silicon Solar Cells.**........................................................ 547
    S. S. Hegedus, W. A. Buchanan, E. Eser, J. E. Phillips, and W. N. Shafarman
**The Potential of Hydrogenated Amorphous Silicon—Chalcogen Alloys**
**for Photovoltaic Applications: The Role of Persistent Photoconductivity** ...... 557
    S. L. Wang, J. M. Viner, P. C. Taylor, T. Itoh, and S. Nitta
**Pulse Duration and Wavelength Effects in Laser Scribing of Thin-Film**
**Polycrystalline PV Materials.**............................................. 567
    A. D. Compaan, I. Matulionis, S. Nakade, and U. Jayamaha
**FT-PL Analysis of CIGS/CdS/ZnO Interfaces** ............................. 573
    J. D. Webb, B. M. Keyes, K. Ramanathan, P. Dippo, D. W. Niles, and R. Noufi
**Reaction Engineering and Precursor Film Deposition for CIS Synthesis** ...... 579
    B. J. Stanbery, A. Davydov, C. H. Chang, and T. J. Anderson
**The Effect of Surface Processing Conditions on the Junction**
**Properties of $CuIn_xGa_{1-x}Se_2$ Solar Cells.**.................................. 589
    S. Zafar, J. D'Amico, S. Karthikeyan, R. Narayanaswamy, P. Panse,
    H. Sankaranarayanan, C. S. Ferekides, and D. L. Morel
**Challenge of Replacing CdS in $CuInSe_2$-Based Solar Cells**................... 597
    L. C. Olsen, F. W. Addis, W. Lei, and H. Aguilar
**Monocrystalline $CuInSe_2$-CdO Cell**....................................... 603
    Z. A. Shukri and C. H. Champness
**Techniques for Increasing Ga Content in $CuIn_{1-x}Ga_xSe_2$ Thin Films**
**Prepared by Two-Stage Selenization Process** .............................. 613
    K. Lynn and N. G. Dhere
**Sodium Dependence of $Cu(In,Ga)Se_2$ Junction Electronics**................... 621
    J. E. Granata, J. R. Sites, and J. R. Tuttle
**Effects of Processing Temperature on the Thickness of CdS**
**and the Performance of CdTe Solar Cells** ................................. 631
    C. S. Ferekides, B. Tetali, D. Marinskiy, S. Marinskaya, and D. Morel

Effect of Cu Doping on the Properties of ZnTe:Cu Thin Films
and CdS/CdTe/ZnTe Solar Cells.................................................... 639
      J. Tang, D. Mao, and J. U. Trefny

Processing Issues for Thin Film CdTe/CdS Solar Cells ...................... 647
      B. E. McCandless, R. W. Birkmire, D. G. Jensen, J. E. Phillips,
      and I. Youm

Raman and RBS Studies of Interdiffusion in RF-Sputtered CdS/CdTe
Solar Cells........................................................................... 655
      A. Fischer, U. N. Jayamaha, E. Bykov, D. Grecu, R. G. Bohn,
      and A. D. Compaan

Tin Oxide Stability Effects—Their Identification, Dependence on
Processing, and Impacts on CdTe/CdS Solar Cell Performance .............. 665
      D. Albin, D. Rose, R. Dhere, D. Niles, A. Swartzlander, A. Mason, D. Levi,
      H. Moutinho, and P. Sheldon

Nanoparticle Colloids as Spray Deposition Precursors to CIGS
Photovoltaic Materials................................................................ 683
      D. L. Schulz, C. J. Curtis, R. A. Flitton, H. Wiesner, J. Keane, R. J. Matson,
      P. A. Parilla, R. Noufi, and D. S. Ginley

CdS/CdTe Thin-Film Devices Using a $Cd_2SnO_4$ Transparent
Conducting Oxide.................................................................... 693
      X. Wu, P. Sheldon, T. J. Coutts, D. H. Rose, W. P. Mulligan,
      and H. R. Moutinho

Polycrystalline MBE-Grown GaAs for Solar Cells .......................... 703
      D. J. Friedman, S. R. Kurtz, A. E. Kibbler, M. Al-Jassim, K. Jones,
      B. Keyes, and R. Matson

O Impurity Chemistry in CdS Thin-Films Grown by Chemical Bath
Deposition: An Investigation with X-ray Photoelectron Spectroscopy ......... 709
      D. W. Niles, G. Herdt, and M. Al-Jassim

## CRYSTALLINE MATERIALS (P)

Thin Film GaAs Solar Cells on Glass Substrates by Epitaxial Liftoff......... 719
      X. Y. Lee, M. Goertemiller, M. Boroditsky, R. Ragan, and E. Yablonovitch

Molecular Dynamics Modeling of Hydrogen in Silicon ...................... 729
      S. K. Estreicher and P. A. Fedders

High Concentration Low Wattage Solar Arrays and Their Applications........ 739
      R. Hoffmann, J. O'Gallagher, and R. Winston

Plasma Processing Applied to Crystalline-Silicon Solar Cells .............. 745
      D. S. Ruby, C. B. Fleddermann, M. Roy, and S. Narayanan

High-Flux Solar Furnace Processing of Crystalline Silicon Solar Cells........ 751
      Y. S. Tsuo, J. R. Pitts, P. Menna, M. D. Landry, J. M. Gee, and T. F. Ciszek

Aluminum Gettering and Transition Metal Precipitates in PV Silicon ........ 759
      H. Hieslmair, S. A. McHugo, and E. R. Weber

Incorporation of Cu and Al in Thin Layer Silicon Grown from Cu-Al-Si ..... 771
      T. H. Wang and T. F. Ciszek

Current Status of HEM Grown Silicon Ingots.................................. 779
      C. P. Khattak and F. Schmid

Distributed Control and Process Monitoring for Photovoltaic Applications .... 787
    M. D. Landry, Y. S. Tsuo, T. F. Ciszek, R. Roze, and D. Hoegh

### COMPONENT AND SYSTEM EVALUATION AND RELIABILITY (P)

Development of New EVA Formulations for Improved Performance
at NREL. .......................................................... 795
    F. J. Pern
A Study of Various Encapsulation Schemes for c-Si Solar Cells
with EVA Encapsulants ............................................... 811
    F. J. Pern and S. H. Glick

### PHOTOVOLTAIC MANUFACTURING (P)

Development of a Low Cost Integrated 15 kW A.C. Solar Tracking
Sub-Array for Grid Connected PV Power Systems Application .............. 827
    M. Stern, R. West, G. Fourer, W. Whalen, M. Van Loo, and G. Duran
PVMaT Improvements for Commercial Production of Thin-Film
CdTe Modules. ....................................................... 835
    J. Phillips and T. Brog
PVMaT Improvements in the Manufacturing of the PVI Powergrid™ ........ 841
    N. Kaminar
Market-Driven Improvements in the Manufacturing of EFG Modules ........ 851
    M. Kardauskas, J. Kalejs, J. Cao, E. Tornstrom, R. Gonsiorawski,
    C. O'Brien, and M. Prince
Advanced Polymer PV System ......................................... 859
    J. I. Hanoka, P. M. Kane, R. G. Chleboski, and M. A. Farber
The AC Photovoltaic Module is Here!. .................................. 867
    S. J. Strong, J. H. Wohlgemuth, and R. H. Wills
Photovoltaic Manufacturing Technology (PVMaT) Improvements
for ENTECH's Fourth-Generation Concentrator Systems. ................. 873
    M. J. O'Neill and A. J. McDanal
PVMaT Manufacturing Improvements for Continuous Roll-to-Roll
Amorphous Silicon Module Production ................................. 881
    M. Izu, H. C. Ovshinsky, and S. R. Ovshinsky

### SMALL BUSINESS INNOVATIVE RESEARCH (P)

Photovoltaic Research in the Small Business Innovative Research
Program. ........................................................... 893
    W. I. Bower and A. Bulawka

## ENVIRONMENT, SAFETY, AND HEALTH IN PHOTOVOLTAICS

**Emerging Photovoltaic Technologies: Environmental and Health Issues Update** .................................................. 903
    V. M. Fthenakis and P. D. Moskowitz

**Author Index** .................................................. 915

## PREFACE

The mission of the U.S. Department of Energy's (DOE) National Photovoltaic Program is to make photovoltaics (PV) a significant part of the domestic economy—as an industry and an energy source. To help achieve this goal, the DOE Photovoltaic Program sponsors a PV research program at the National Renewable Energy Laboratory (NREL) and Sandia National Laboratories (SNL). A review of the laboratories' programs, the "NREL/SNL Photovoltaics Program Review Meeting" was held November 18–22, 1996 in Lakewood, Colorado. This conference was designed as the successor to a series of conferences known as the "NREL Photovoltaic Program Reviews", the last of which was held in May of 1995. The tome documents the proceedings of this first NREL and SNL joint program review and highlights recent progress achieved in photovoltaic technologies.

This conference was a conference of "firsts". It was the first combined NREL and SNL program review, thus providing a more comprehensive and uniform review of the laboratories' programs. Also, during the opening session on November 19, Christine Ervin, Deputy Assistant Secretary of Energy for the Office of Energy Efficiency and Renewable Energy, announced the new DOE laboratory system collaboration designated as the "National Center for Photovoltaics" (NCPV). The new NCPV utilizes NREL and SNL core expertise to guide operations and coordinate support from National Program resources such as the universities, labs, and industrial partners reporting in these proceedings. The Monday, November 19 luncheon was the scene for a third first as NREL Director, Dr. Charles Gay, announced Dr. Lloyd Herwig as the first recipient of NREL's "Paul Rappaport Award". Dr. Herwig was recognized for his early support of solar as a technology of significance and for his long and fruitful career as a program manager in government service. His luncheon presentation, entitled "My 25 Year Solar Energy Odyssey," was appreciated by all.

The Keynote Address on renewable energy and its relationship to restructuring of utilities was given by the Honorable Dan Schaefer from Colorado's 6th District. Later in the Conference, a panel discussion entitled "Financing Photovoltaic Deployment" was also held. Throughout the conference, technology progress papers presented R&D results and promise for technical improvement and cost reductions in the cell, device, module, system, and market areas. Together, these presentations and discussions described many aspects of the process necessary to realize and implement a young technology.

The panel discussion was lively. The questions from the audience were numerous and obviously the result of much interest. Our thanks go to the panelists:

| | |
|---|---|
| Christy Herig | Moderator, "Financing Photovoltaic Deployment" |
| Neil Holstead | Tucson Electric Power Company |
| Clay Aldrich | Solar Energy Industries Association |
| George Cody | Exxon |
| Don Osburn | SMUD |

We would also like to thank the people who helped plan and arrange the fine program. They include Al Czanderna, XiaoNan Li, Harv Mahan, Angelo Mascarenhas, Tom McMahon, Syl Morgan-Smith, Bob Noun, John Pern, Bhushon Sopori, John Thornton, and Simon Tsuo.

Special thanks go to Joan Ross, our conference coordinator, who walked into the middle of this project only a few months before the conference and did an outstanding job of orchestrating the final preparation. Also, thank you to those who assisted her; Heather Bulmer, Wendy Larsen, Cricket Pierce, and Jeri Windschell.

C. Edwin Witt, General Chair<br>
Mowafak Al-Jassim, Co-Chair<br>
James M. Gee, Co-Chair

# THIN FILMS I

# Progress Report on the Amorphous Silicon Teaming Activities

B. von Roedern, K. Zweibel, E. Schiff,[1] J.D. Cohen,[2] S. Wagner,[3] S.S. Hegedus,[4] and T. Peterson[5]

National Renewable Energy Laboratory (NREL), Golden, Co 80401-3393
[1]Department of Physics, Syracuse University, Syracuse, NY 13244-1130
[2]Department of Physics, University of Oregon, Eugene, OR 97403
[3]Department of Electrical Engineering, Princeton University, Princeton, NJ 08544
[4]Institute of Energy Conversion, University of Delaware, Newark, DE 19716
[5]Electric Power Research Institute, 3412 Hillside Ave., Palo Alto, CA 94304

**Abstract.** We review the progress of the teamed amorphous silicon research activities sponsored by NREL Thin-Film Partnership and by EPRI. We summarize the technical progress made since the last published report and attempt to assess the operational efficiency of the teams and the value of teaming.

## INTRODUCTION

Since the last report (1) on the teaming activities, steady progress has been made toward reaching the goals of the teams, culminating in the demonstration of an approximately 12.5%-efficient (active-area) stabilized triple-junction cell by United Solar Systems Corporation (USSC) (2). There were no major changes in the objectives and approaches of each team. The purpose of this paper is to provide recent technical highlights and to summarize the high operational efficiency reached by the teams. The formal merger of the National Renewable Energy Laboratory (NREL)- and Electric Power Research Institute (EPRI)-supported teams, which had previously operated separately, also aided the efficiency of team operation.

## THE VALUE OF TEAMING

In the following, we summarize the value of teaming for the NREL Thin-Film Partnership and EPRI-sponsored amorphous silicon research. This summary incorporates results from questionnaires sent to team leaders by NREL program management in July 1996. The following issues were cited multiple times:

1. The teamed research has led to a much better understanding of the needs of the industrial researchers by the academic (and NREL in-house) research community, and vice versa.

A true team spirit has evolved. Prior to the teams, some industrial groups might have looked upon academic research as helpful to provide certain useful characterization services, but as non-essential in moving device performance forward. An attitude prevailed that progress would come from individual research groups finding a breakthrough that would lead to greater cell efficiencies. Now, it is well accepted that progress in cell efficiencies will be more likely if pursued by the entire team.

2. Device results, rather than material parameters, have become the accepted yardstick to evaluate the potential of new materials and processing concepts as to their potential to reach the 15% stabilized efficiency goal.

This represents a major shift from a former point of view that device making would be too complicated for university researchers. Although the industrial team members are still the most skillful device makers, the device fabrication capabilities of the device makers at Pennsylvania State University, Iowa State University, the Institute of Energy Conversion (University of Delaware), Princeton University, and at NREL have made considerable progress. It is now an accepted mode of operation that these device-making groups must first evaluate a new material or deposition concept and compare it to their "baseline" cell performance. Only approaches that have successfully passed this screening procedure will be taken up by the industry groups that will then attempt to process a champion cell.

Significant assistance and transfer of know-how was provided by the industry groups to advance the device making know-how at the academic device-making groups and at NREL. The fabrication of substrate devices (stainless steel/n-i-p/TCO) solar cells has been significantly advanced among the academic device makers and at NREL, with much support and many helpful hints from the groups at USSC and Energy Conversion Devices (ECD). This device structure is particularly suited for evaluating deposition sequences that call for higher than standard (~250°C) i-layer deposition temperatures, because the higher i-layer temperatures require a substantial reoptimization of the p-layer in glass/(textured)TCO/p-i-n/(TCO)/metal superstrate devices.

For diagnostic purposes, the teams have also prepared Schottky barrier junction devices due to their simplicity in fabrication. Again, USSC provided processing hints and guidance in appropriate characterization and correlations of Schottky barrier device characteristics with the performance in p-i-n or n-i-p junctions.

3. Results obtained by each team member have reached a much higher level of credibility. This enables the team members to engage effectively

in a "lessons learned" approach, especially if the results obtained were unexpected.

The teams have developed a mode of operation where everyone desires to better understand all results obtained and to bring issues to closure. Before the teams were formed, the mode of operation of the research groups of many entities was to run with successful schemes and drop unsuccessful ones, without seeking an explanation for such decisions. The danger of the latter approach is that after a while, unsuccessful research approaches will be revisited, because no fundamental root cause for the original unsatisfactory results was ever determined. The teams realize that to significantly advance the performance of a-Si-based solar cells, high-risk approaches must be pursued. Thus, the probability that a chosen approach may be unsuccessful is high, making it even more important to document limitations.

4. Teaming is fun. The team setting has encouraged collaborations and increased the value of each individual's research.

This is self explanatory. The perceived advantages far outweigh the disadvantages. Teaming does increase the workload for everybody. The mode of interactions in the a-Si teams has evolved. The teams now meet only once every 9 month (compared to 6-month intervals in the beginning).

## HIGH-LEVEL OPERATIONAL EFFICIENCY

Teaming would not have been implemented without the original vision of those who viewed it as of potential value. It is important to note that "natural" teaming that occurs within organizations is different from this kind of teaming, which imposes a structure on participants who would otherwise be competing rather than working together. They recognized a common need for overcoming the problem: that the available amorphous materials and device structures limited amorphous silicon-based PV device efficiency from reaching the 15% mark, especially as impacted by light-induced instability.

Through the effects of teaming, significant changes occurred in the operational characteristics of the various entities, focusing them on high priorities. Participants *voluntarily* agreed to change their own research focus to support the team's consensus. This pursuit of an agreed-upon set of research priorities is one of the clearest positive results of teaming. *Group debate and consensus development built around common needs turned out to be a more effective means of developing focus than command-and-control strategies.*

Once institutions agreed to common priorities, the next stage of change took place. Universities began to approach fundamental research problems from a perspective of the needs of the ultimate product: a PV device. Less-focused materials research was replaced with materials research focused on improved PV devices and processes.

The change of focus of industrial participants was more subtle, but still important. Companies were able to develop collaborations with universities and NREL that allowed them to off-load more fundamental tasks and to take advantage of the capabilities of these other institutions. Sometimes, even the negative results of the universities were useful to the companies, because they showed that certain avenues could be avoided without loss of opportunity. Attending to these approaches had previously cost money and time. Through teaming, the focus of corporate actions broadened, while the focus of the academic and federal institutions narrowed to more pragmatic priorities. *One clear result of the team consensus planning process is that the most important research priorities were recognized and resources were directed to their solution.*

There had always been collaborations within the PV R&D community. The formation of teams increased the priority of forming collaborations. The result was more systematic collaborations, faster and improved sample preparation, more rapid sample exchange, better feedback on sample processing and characterization of sample experiments, and improved and broadened feedback on experimental results. Increased formal and systematic collaborations were stimulated (1) by being an acknowledged priority of the partnership, (2) because team and subteam activities required them, (3) because semiannual subteam and team meetings allowed for greater inter-organizational interactions, and (4) because the team supported the idea of reducing barriers between organizations/competitors. *Increased and more effective collaborations are among the most important positive results of the teaming activity.*

Various important values emerged from the consistent exposure of the research community to the technical state-of-the-art within their area of interest. For example, several university team members were early in the learning curve of cell fabrication. They universally acknowledged that immersion in discussions and collaborations with leaders in PV added tremendously to their expertise and significantly compressed the developmental cycle. For example, the progress of NREL and Iowa State in developing the reverse (substrate) a-Si cell structure for new materials, deposited by using the hot-wire and electron cyclotron resonance (ECR) deposition techniques, would have been much slower without the support from USSC. Beyond these cases of easily measured progress, all participants gained insights into key issues, allowing them to improve their devices and processing. An example of such insights was the awareness of the general value of hydrogen dilution in a-Si device processing to reduce degradation by up to 50%. These and similar details about best practices led to accelerated progress among team members.

An unexpected but valuable aspect of teaming is that resources have been saved by avoiding repetitive failures. Shared awareness of the limits of existing procedures forms a road map of potential resource sinks. This is an unexpected, but valuable, pay-off of the teaming arrangement. On the other hand, many research groups made progress due to being exposed to the expertise of state-of-the-art device makers. Within the context of the team, traditional rivals relaxed and shared much more information than previously.

## TECHNICAL HIGHLIGHTS FROM THE TEAMS

The most significant achievement was USSC's triple-junction cell with a stabilized efficiency of about 12.5% (active-area). This efficiency was achieved by a combination of improvements of the device design, which have been documented in a report published by USSC (3). In conjunction with the recent guidance team meeting (August 1996), the value of numerical solar cell modeling was reviewed. Most researchers believed that modeling is valuable for insight into the details of device operation, but that current models have difficulties quantitatively predicting the best cell performance expected from using specific materials or from a particular device design. The complex numerical models like Pennsylvania State's AMPS model require a very large number of input parameters for each layer that make up a solar cell. Many researchers expressed the need for help and more consistency in managing these parameter sets. Below, we summarize the technical highlights for each of the four teams.

### Wide-Bandgap Team

The NREL/EPRI Wide-Bandgap Team aims to demonstrate a 68 W/m$^2$ solar cell that uses only the blue/green portion of the solar spectrum. The exact spectral region available to this cell is determined by the detailed engineering of a triple-junction, a-Si-based solar cell; the provisional requirement is $\lambda < 530$ nm (hv $>$ 2.4 eV), corresponding to a maximum short-circuit current density ($J_{sc}$) of about 82 A/m$^2$. The targets for the stabilized fill-factor and open-circuit voltage ($V_{OC}$) are 0.75 and 1.10 V, respectively.

Industrial members of the NREL/EPRI Wide-Bandgap Team (United Solar Systems Corp., Energy Conversion Devices, Inc., and Solarex Thin Films Division) have demonstrated solar cells that are fairly close to these wide-bandgap targets. An open-circuit voltage of 1.054 V and a power of 50 W/m$^2$ have been achieved. The most significant factor in obtaining the higher voltages comes from optimizing hydrogen dilution schemes (depositing from $H_2/SiH_4$ gas mixtures with a ratio >10).

The academic members of the team are currently placing great emphasis on understanding and improving the open-circuit voltage for wide-bandgap cells. Several new insights have emerged recently from this work. At Pennsylvania

State University, researchers have done extensive studies on the *intensity-dependence* of $V_{OC}$. For high $V_{OC}$ cells from USSC, the measurements indicate that open-circuit voltages below 1.05 V are limited by the bulk properties of the intrinsic layer. For higher open-circuit voltages, an interface effect appears to dominate; such an effect may originate in low values of the built-in potential or photocarrier recombination in interface states. Similar results were obtained for cells prepared at Penn State using high hydrogen dilution and with $V_{OC}$ = 0.95 V.

These conclusions appear consistent with recent estimates of the built-in potential from Syracuse University; this group uses an electroabsorption technique to estimate the built-in potential $V_{BI}$. For high $V_{OC}$ cells, the group measured $V_{BI}$ = 1.25 V, consistent with models proposed by the Penn State group for the $V_{OC}$ vs. intensity data. Most workers have assumed that $V_{BI}$ is determined by the difference in Fermi energies of the $p^+$ and $n^+$ layers. The Syracuse group proposed that the measured value of $V_{BI}$ is significantly lowered below this limit by *interface dipoles* at the p/i interface; varying dipole effects would then also account for some of the interface effects well known in $V_{OC}$ and $V_{BI}$ measurements.

Two implications for device improvement can be drawn from this characterization research. It appears that intrinsic layers with a somewhat larger bandgap will be required to achieve the $V_{OC}$ target of 1.10 V; the high hydrogen dilution material has a nominal, "Tauc" bandgap of 1.80-1.85 eV. At the same time, further improvements in $p^+$ and/or interfaces will be necessary to fully exploit such a bandgap increase.

The Wide-Bandgap Team seeks to realize these improvements in devices. The critical i/p region in n-i-p cells (amorphous intrinsic layer, microcrystalline p layer) is being studied during deposition using spectroscopic ellipsometry. This work at Penn State has clarified the enormous effects of a hydrogen plasma on this interface; we note that a very high level of hydrogen dilution is an essential component of the deposition of microcrystalline silicon layers. The effects uncovered include:

- bandgap widening of the intrinsic layer up to 25 nm below the interface
- significant roughening of the interface. This roughening proved to be crucial to rapid coalescence of the microcrystalline film and to $V_{OC}$; in the absence of the hydrogen plasma treatment, coalescence was very slow and $V_{OC}$ reached only about 0.4 V.
- Very large optical absorption onsets (2.2-2.4 eV) have been found in microcrystalline layers for the first 10-20 nm of their nucleation; the effect presumably reflects quantum confinement in nanocrystals. Thicker layers have lower onsets, indicating larger grain sizes.

Additional work under way at Penn State includes:

(i) *filament-produced* atomic hydrogen (as an interface treatment and a substitute for the hydrogen plasma). The interface roughening effect is very different than for plasma treatment.

(ii) *optimization* studies of a-SiC:H and high hydrogen dilution a-Si:H films. Promising as-deposited a-SiC:H films with 1.95 bandgap were obtained, but these degraded badly under illumination. High hydrogen dilution a-Si:H have somewhat larger bandgap than conventional a-Si:H; they also have higher stability and strikingly different degradation kinetics than undiluted material.

(iii) *custom cells* incorporating an intrinsic region with two distinct layers, prepared with or without high hydrogen dilution. These cells permit quantification of the role of bulk and p/i interfaces on $V_{OC}$ and degradation kinetics.

## Metastability and Mid-Bandgap Team

Recent progress in the Mid-Band-Gap and Metastability Team can be divided into three general areas. First, there is good evidence that some forms of a-Si:H are significantly more stable, as demonstrated in studies of both films and solar cell devices. The most complete studies have been carried out for glow-discharge material grown under high hydrogen dilution, and several types of studies were reported by Mid-Gap Team members. In one study, glow-discharge n-i-p structures were deposited either using pure $SiH_4$ or a mixture of $H_2$ and $SiH_4$ in a 15:1 ratio. Using capacitance methods to deduce the deep defect density, the University of Oregon group found that the defect density for the H-diluted material remained below $1.2 \times 10^{16}$ cm$^{-3}$ even after 300 hours of high-intensity (3.4 W/cm$^2$) light exposure, compared to $5 \times 10^{16}$ cm$^{-3}$ for the standard (100% $SiH_4$) sample. The Solarex group, long a proponent of $H_2$-diluted a-Si:H for improved device stability, demonstrated that p-i-n devices incorporating this material actually recovered to within 95% of their undegraded efficiencies at a temperature of only 65°C (compared to 80% for the pure $SiH_4$ cells). Researchers at Penn State correlated the degradation of cell efficiencies and the degradation of photoconductivity in companion films. Moreover, the stable steady-state properties of devices *and* films were achieved in less than 100 hours under 1-sun illumination for the H-diluted materials.

In addition to the H-diluted glow-discharge studies, ECR-deposited material grown under H dilution also appears to be quite promising as a new form of a-Si:H with improved stability. In a study at Iowa State University comparing the cell performance for glow-discharge vs. a H-ECR devices, the latter was found to exhibit a fill-factor deterioration of less than 5% after exposure to 1.6 suns for 75 hours. The glow-discharge cell fill-factor deteriorated more than 20% under these conditions. USSC reported improved stability of solar cells where the intrinsic layers were deposited from deuterated $SiD_4/D_2$ mixtures (3). Solarex obtained inconclusive results when substituting deuterium for hydrogen. USSC suggested that the microstructure of the deuterated films may be different, because deuterium evolves at lower temperatures than hydrogen when H(D)-evolution experiments were carried out.

The second general area of progress concerns a-Si:H produced by the hot-wire deposition process. This material has exhibited considerable promise as a stable, lower-bandgap form of a-Si:H. The experimental findings during this past year concern the fundamental materials properties of hot-wire material compared to glow-discharge a-Si:H These results underscore the fact that the hot-wire material is a distinctly different kind of material. Key findings include: (1) nuclear magnetic resonance (NMR) results from North Carolina State indicating a unique new distribution of hydrogen environments compared to glow-discharge material; (2) transport studies at Syracuse University, using time-of-flight, and at the University of California, using photomixing; the Syracuse results indicate an enhanced hole drift mobility by a factor of ~5 compared to a-Si:H prepared by the glow-discharge method. (3) H-diffusion studies at NREL indicating nearly 100-fold higher hydrogen diffusion rates; and (4) X-ray diffraction results from NREL and Colorado School of Mines show a much narrower bond length distribution for hot-wire samples grown at higher temperatures. Such results may help with ongoing efforts to fabricate high-performance cells using hot-wire material.

The third general area concerns progress in diagnostic methods and theoretical calculations. Studies are progressing in understanding the mechanisms of particle formation in the plasma during growth at the University of Colorado, in understanding the role of charged defects near the n-i and p-i interfaces on cell performance through device modeling at NREL, and in using first-principles theory calculations at Iowa State University to examine defect dynamics following their capture of electrons. These latter two types of studies may ultimately offer important insights into the detailed mechanisms for a-Si:H cell degradation.

## Narrow-Bandgap Team

The a-SiGe:H cell designed as the bottom cell of the triple-junction device is apparently close to the 4% efficiency target. However, this observation may be misleading, because the current is more than the target (80 A/m$^2$), whereas the FF and voltage are below the target values. Therefore, more improvement is needed than what would appear from having obtained the present stabilized 3.8% efficiency result on the bottom cell. The continuous improvement in a-SiGe:H alloy quality and stability was demonstrated by photocapacitance and drive-level capacitance measurements on USSC material carried out at the University of Oregon.

The recent progress in triple-junction cell efficiencies at USSC relies on the fact that the three component cells are current *mis*matched such that the top cells are limiting the current density. This leads to higher FF of the triple junction device. Thus, the high currents produced by the bottom cell have been advantageous for the triple-junction device. However, until the FF and $V_{OC}$ values of the bottom cell can be improved also, the 15% stabilized-efficiency triple-junction device may not be reached. If such improvements in the bottom cell can be obtained, this would reduce present design constraints of the top and middle cells.

The team is developing improved narrow-bandgap cells by focusing on synthesizing better a-SiGe:H alloys. The deposition schemes include glow-discharge deposition from halogenated precursor gases at Princeton, ECR deposition at Iowa State University, and hot-wire-deposited alloys at NREL. Some of this work is progressing slower because the groups involved plan to first establish their deposition techniques and prepare baseline devices using unalloyed a-Si:H prepared with the respective deposition methods. In most instances, such layers have bandgaps already lower than glow-discharge deposited a-Si:H layers, which provides a further incentive to the Narrow-Bandgap Team to pursue these techniques.

Due to limited resources, the Narrow-Bandgap Team presently cannot pursue radically different alternatives to the a-SiGe:H material for the bottom cell, such as nanocrystalline Si layers that have a similar bandgap.

## Multijunction Device Team

The focus of the Multijunction Device Team is to help meet the 15% stabilized efficiency goal by generating higher $J_{SC}$ in cells with thinner i-layers. This requires substantial improvements in the theory and application of light-trapping or optical enhancement to minimize parasitic absorption losses. Because it is likely that a high performance solar cell will have TCO contacts to both p and n layers, the team is researching ways to improve the TCO/a-Si interface for both p and n layers.

The NREL optical model now explains why the TCO/metal back reflector is optically better than pure metal. Absorption in the metal is a significant loss, and the placement of a dielectric layer with n~2 between Si and metal reduces transmission into the metal by increasing reflection into the Si. Quantitative agreement between measured and calculated reflection of glass/a-Si/ZnO/Al structures provided by the Institute of Energy Conversion (IEC) has been obtained.

A robust atmospheric pressure chemical vapor deposition (APCVD) process to deposit textured ZnO has been developed at Harvard with much better uniformity and reproducibility than the previous process. This was achieved by developing a new metal-organic Zn precursor. Aside from allowing better process control, the new precursor, unlike diethyl zinc, is not pyrophoric, making the handling less critical. The new ZnO layers were evaluated at Solarex and at IEC in tandem and single-junction devices. Improved values of $J_{SC}$ and $V_{OC}$ were obtained, but the FF was still low compared to cells prepared on $SnO_2$ substrates. Laser-scribing and shunting studies were conducted at Solarex to test how these new ZnO layers affect the fabrication of integrated PV modules.

New structures were developed at IEC to characterize the back reflector (BR) in conjunction with substrate texture. An optically coupled, external BR allows separation of substrate texture and reflection properties on the same device. It was found that substrate texture is more important than BR reflectivity to enhance the red response. Texture also has positive impact on reducing front reflection losses.

Tunnel-junction losses were reduced with a two-step p-layer at ECD. Deposition of a very thin, highly doped layer was followed by deposition of a thicker, lower-doped layer. Higher doping of the surface of the p-layer provides better contact with the subsequent n layer, increasing $V_{oc}$ of the multijunction cell.

Modeling of multijunction devices shows that i-layer defects in all three component cells must be reduced an order of magnitude to meet the 15% goal.

## CONCLUSIONS

The recent progress in amorphous silicon device efficiency suggests that we continue to consider a-Si an important thin-film photovoltaic technology. The NREL Thin Film Partnership and EPRI-sponsored teamed research activities focus on finding materials and device structures that will permit higher stabilized efficiencies, to reach the goal of 15% stabilized device efficiencies. Along the way, processing and device optimization schemes are discovered that allow incremental improvements of products manufactured today. The current teams are addressing these topics in an appropriate and effective manner. We believe that this has significantly increased the return on the dollars expended on research and development of a-Si PV technology.

## REFERENCES

1. Luft, W., Branz, H.M., Dalal, V.L., Hegedus, S.S. and Schiff, E.A., AIP Conference Proceedings **353** *(13th NREL Photovoltaics Program Review)*, pp. 81-100 (1995).
2. Yang, J., Banerjee, A., and Guha, S., this volume.
3. Guha, S., "Amorphous Silicon Research: Phase II," *Annual Technical Progress Report, NREL/SR-520-21964* (1996).

# A Record Setting Amorphous Silicon Alloy Triple-Junction Solar Cell With 14.6% Initial and 12.8% Stable Efficiencies

Jeffrey Yang, Arindam Banerjee, and
Subhendu Guha

*United Solar Systems Corp., Troy, Michigan 48084*

**Abstract.** World record 14.6% initial and 12.8% stable conversion efficiencies have been achieved using amorphous silicon based alloy in a spectrum-splitting, triple-junction structure. This performance exceeds our previous record of 13.2% initial and 11.8% stable efficiencies and establishes a new milestone toward reaching the 15% stable module goal. Key factors leading to this major advance include: (a) Improvement in the low bandgap amorphous silicon-germanium component cell that resulted in enhanced red response and provided desired current mismatching, (b) improvement in the pn tunnel junction between component cells by incorporating microcrystalline p and n layers in a multilayered structure that resulted in reduced optical and electrical losses, and (c) improvement in the top conducting oxide that resulted in reduced absorption and enhanced blue response without increasing the top cell thickness. Details of these advances along with light-soaking data for high efficiency cells will be discussed.

## INTRODUCTION

Amorphous silicon (a-Si) alloy photovoltaic (PV) technology has received a great deal of attention in the last two decades due mainly to its low material cost and ease of manufacturing with good yield (1). One major challenge, however, is to improve the stabilized module efficiency to 15% by the year 2005 (2). A mission-focused integrated team, the Thin Film Partnership, among the a-Si alloy PV industry, universities, and the National Renewable Energy Laboratory (NREL) was formed to foster research integration (3). We are pleased to report in this paper that we have made significant progress in improving cell efficiency under the Thin Film Partnership program. Using a-Si based alloy in a spectrum-splitting, triple-junction structure, a world record 14.6% initial active-area efficiency has been achieved and measured by NREL to be 14.5%, breaking through our previous record of 13.2%.

Light-soaking experiments have been carried out for the high efficiency devices. Similar to our previous observation in the degradation of triple-junction structures, these devices reach substantial saturation after ~300 hours of one-sun exposure. The highest stabilized efficiency after 600 hours is 12.8%, surpassing our previous record of 11.8%, and establishing a new milestone toward reaching the 15% stable module goal.

Three key factors, in addition to our previously reported improvements that led to the 11.8% stable cell efficiency (4), have contributed to the record performance. Recent advances have been made in (a) the low bandgap amorphous silicon-germanium component cells, (b) the $p$-$n$ tunnel junction between the component cells, and (c) the top conducting oxide. We shall discuss the details of these advances in this paper.

## EXPERIMENTAL RESULTS AND DISCUSSION

We shall first describe the triple-junction structure and briefly review our previous result. Figure 1 shows a schematic diagram of a triple-junction structure. The top intrinsic ($i$) layer uses wide bandgap a-Si alloy for absorbing the blue photons, while the middle and bottom $i$ layers incorporate intermediate and narrow bandgap amorphous silicon-germanium (a-SiGe) alloys with different silicon-to-germanium ratios for absorbing the green and red photons, respectively. A textured silver/zinc-oxide back reflector is used to facilitate the desired light trapping effect. The red photons that reach the back surface are scattered back at an oblique angle so as to enhance optical path and absorption. The top contact uses evaporated indium-tin-oxide (ITO) which also serves as an antireflection coating. Finally, a metal grid is deposited on top of ITO for collecting current.

All the three $i$ layers use heavy hydrogen dilution during film growth to obtain superior quality, as reported previously (6), while the middle and bottom a-SiGe $i$ layers employ bandgap profiling for better carrier collection (7). Microcrystalline $p$ layers with high conductivity and low optical absorption are used as the window layer as well as in the tunnel junctions.

Using the above approach, along with an appropriate current mismatching cell design, we recently reported the achievement of an 11.8% stable cell efficiency. Table 1 shows the J-V characteristic of this device in both the initial and degraded states. Figure 2 plots the initial quantum efficiency versus wavelength data for this device. It is noted from Table 1 and Fig. 2 that the total photocurrent generated in the triple stack is ~25 mA/cm$^2$.

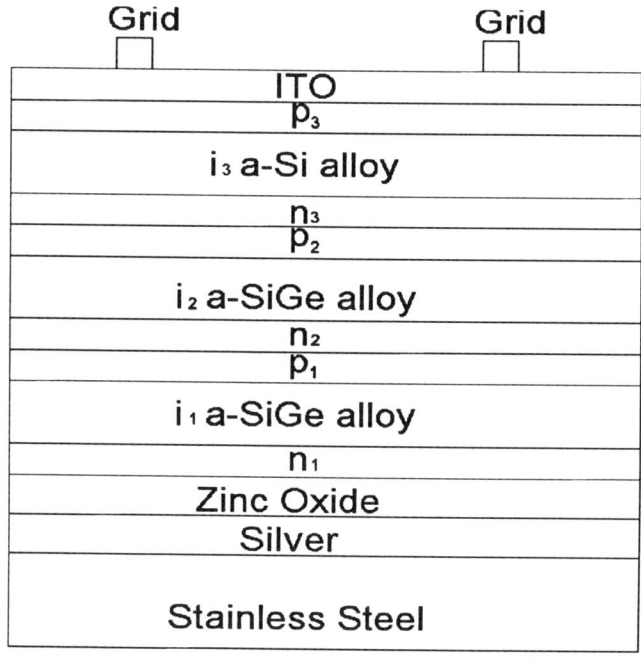

**Figure 1.** Schematic diagram of a triple-junction cell structure.

**TABLE 1.** Characteristics of our Previous Best Triple-junction Cell.

| Smpl. | State | $J_{sc}^{a}$ (mA/cm$^2$) | $V_{oc}$ (V) | FF | Q (mA/cm$^2$) Top | Mdl | Btm | Total (mA/cm$^2$) | $\eta$ (%) |
|---|---|---|---|---|---|---|---|---|---|
| L7907 | Initial | 7.64 | 2.340 | 0.739 | 7.64 | 8.03 | 9.51 | 25.18 | 13.21 |
|  | Degraded (1100 h) | 7.49 | 2.283 | 0.692 | 7.49 | 7.77 | 9.33 | 24.59 | 11.83 (-10.5) |

[a] United Solar uses $J_{sc}$ values as obtained from the quantum efficiency measurement.

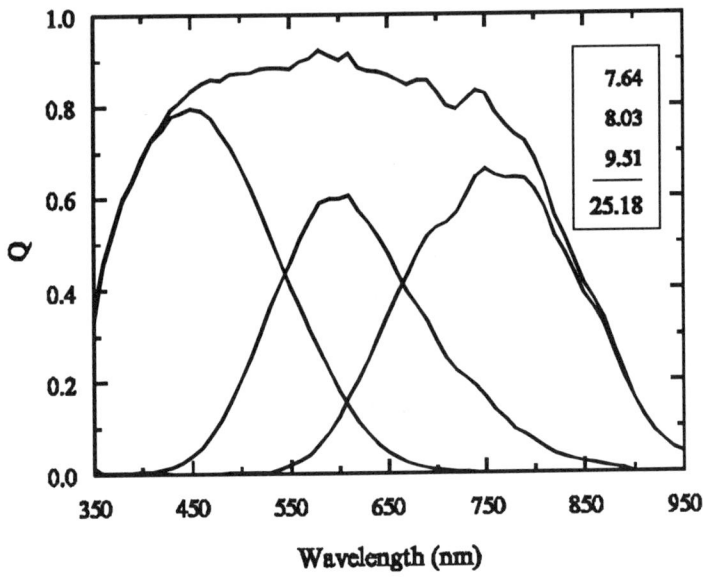

**Figure 2.**  Initial quantum efficiency of the triple cell in Table 1.

A careful analysis of the quantum efficiency data in Fig. 2 reveals that one can further improve the spectrum-splitting feature in three areas: (a) the red response in the long wavelength ($\lambda > 800$ nm) region, (b) the response associated with the absorption of the tunnel junction between the top and the middle cells in $\lambda \sim 500$-$600$ nm region, and (c) the blue response in the short wavelength ($\lambda < 450$ nm) region. We shall discuss the three areas separately.

## The Red Response

One way to enhance the red response, hence increasing the triple-cell current with desired current mismatch, is to increase the Ge content in the bottom cell. This, however, often results in a poorer material quality and deteriorates the cell performance. Using heavy hydrogen dilution during film growth and incorporating proper bandgap profiling, we have improved the bottom component cell with an enhanced red performance. Table 2 compares the performance of the previous and recent bottom cells measured under AM1.5 illumination through a $\lambda > 630$ nm red filter. It is readily observed that the recent cell has a higher red current and output power. The higher $V_{oc}$ and FF demonstrate the benefit of proper hydrogen dilution and bandgap profiling. Since the previous cell was used to obtain the 11.8% stable triple-junction device, the improved cell should set a good foundation for a triple-stack structure. Figure 3 shows the J-V characteristic of the improved bottom cell under AM1.5 illumination, achieving an efficiency of 10.37%. Figure 4 plots the corresponding quantum efficiency versus wavelength for this device. It is noted that at 850 nm, the quantum efficiency is 45%, a significant improvement over the 35% value reported for the previous bottom cell (4).

**TABLE 2.** Initial J-V Characteristics of Previous and Recent Bottom a-SiGe Cells Measured under AM1.5 Illumination with a λ>630 nm Filter.

|  | $J_{sc}$ (mA/cm$^2$) | $V_{oc}$ (V) | FF | $P_{max}$ (mW/cm$^2$) |
|---|---|---|---|---|
| Previous | 11.9 | 0.611 | 0.634 | 4.6 |
| Recent | 12.2 | 0.631 | 0.671 | 5.2 |

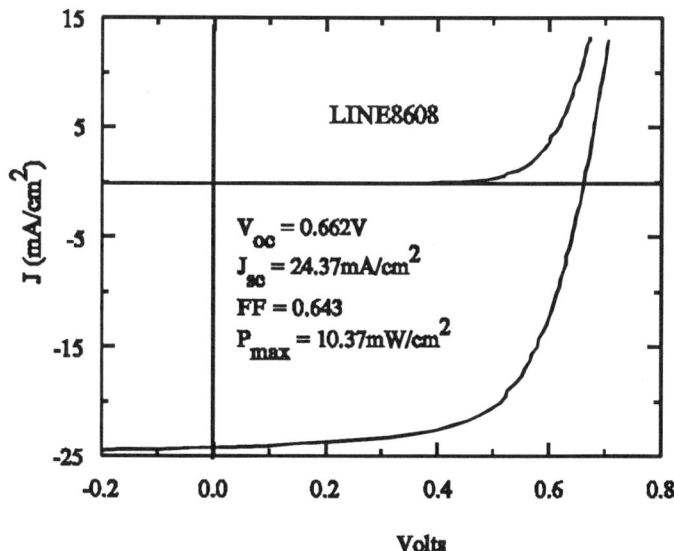

**Figure 3.** Initial J-V characteristic of an a-SiGe alloy bottom cell.

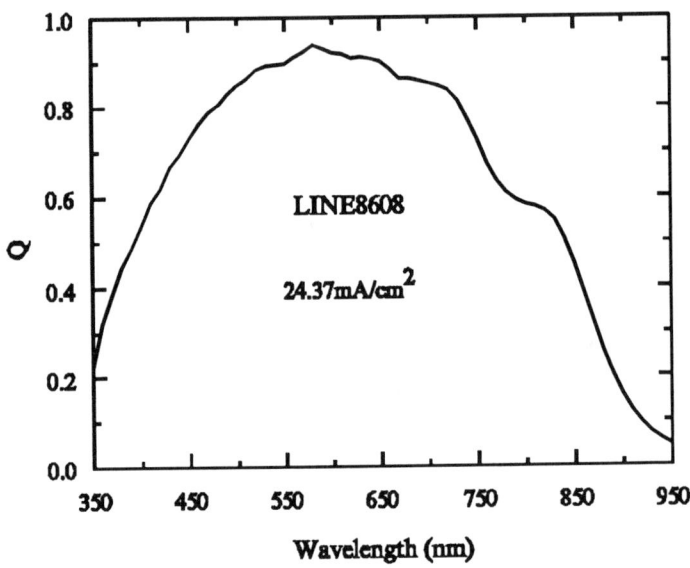

**Figure 4.**   Quantum efficiency of the bottom cell in Fig. 3.

## The Tunnel Junction

Next we discuss the improvement in the tunnel junction. One can see from Fig. 1 that there are two tunnel junctions in a triple structure. One is $p_2n_3$ between the top and the middle cells, and the other is $p_1n_2$ between the middle and the bottom cells. These tunnel junctions, while essential in the triple-cell operation, result in optical and electrical losses. Any reduction in their optical absorption or electrical resistance can give rise to a better cell performance. In our earlier triple-junction cells, we only used microcrystalline $p$ layer but not microcrystalline $n$ layer in the tunnel junctions.

Recently, we carried out a systematic study to evaluate the benefit of using microcrystalline $n$ layer in the tunnel junction. Since the $p_2n_3$ tunnel junction is closer to where light enters the device than the $p_1n_2$ junction, any improvement in the $p_2n_3$ top tunnel junction is expected to result in a larger improvement in the cell performance.

To evaluate the top tunnel junction, we first studied two a-Si/a-Si double-junction cells. The only difference between the two samples is that one uses amorphous $n$ layer while the other uses microcrystalline $n$ layer in the tunnel junction. To our surprise, we find that the $V_{oc}$ value for the tandem cell having a microcrystalline *pn* junction is much lower ($V_{oc}$=1.657 V) than the corresponding cell ($V_{oc}$=1.804 V). We attribute the lowering of $V_{oc}$ to the following two possible reasons: (a) There may be incompatibility between the microcrystalline *n* layer and the adjacent *i* layer. Microcrystalline *n* layer may have much lower bandgap than the amorphous *i* layer, and the band edge discontinuity (9) may cause the lowering of $V_{oc}$. (b) There may be intermixing of dopants between the microcrystalline doped layers. It is well known that the deposition conditions for microcrystalline material generally require high rf power and high $H_2$ dilution (8). These conditions can cause the intermixing of the dopants in the thin doped layers.

We have, therefore, developed appropriate buffer layers and inserted them between the microcrystalline *p* and *n* layers and between the microcrystalline *n* and amorphous *i* layers. We reevaluated the double-junction performance by carrying out a series of experiments on a-Si/a-Si and a-Si/a-SiGe double-junction structures; we compared the cell performance by using a conventional amorphous *n* layer and a microcrystalline *n* layer sandwiched between buffer layers in the tunnel junction. We have taken care to deposit the companion pairs onto the same back reflector. The pairs also have the same ITO deposited onto them. For the a-Si/a-SiGe structure, we have used top cell/middle cell parameters from a typical triple-junction structure. The results are listed in Table 3. It is observed that the microcrystalline *pn* multilayered tunnel junction gives rise to cells with higher $J_{sc}$, $V_{oc}$, total current, and efficiency. The higher total current and lower series resistance demonstrate the improvement due to the new tunnel junction structure.

## The Blue Response

As discussed earlier, the top conducting oxide serves as the top contact (see Fig. 1) as well as an antireflecting coating. If one can reduce the absorption in ITO without sacrificing its conductivity, one can gain in photocurrent. This will benefit the entire triple stack with the top cell receiving the most advantage. We have, therefore, reoptimized the ITO evaporation conditions by adjusting the oxygen partial pressure and obtained a higher ITO transmission. Figure 5 shows J-V of a top cell deposited onto a stainless substrate with the improved ITO layer. The $J_{sc}$ value is ~0.5 mA/cm$^2$ higher than that obtained from a similar cell with previous ITO. The higher current is obtained without increasing the *i* layer thickness which is certainly desirable from the stability point of view. $J_{sc}$ value of ~8.5 mA/cm$^2$ is also suitable for the top cell of a triple structure with the improved bottom component cell.

**TABLE 3.** Characteristics of a-Si/a-Si and a-Si/a-SiGe Double Junctions with Different Tunnel Junction Structures.

| Tunnel Junction Structures | $J_{sc}$ (mA/cm$^2$) | $V_{oc}$ (V) | FF | η (%) | Qtop/Qbtm (mA/cm$^2$) | Qtotal (mA/cm$^2$) | $R_s$ (Ω cm$^2$) |
|---|---|---|---|---|---|---|---|
| a-Si/a-Si microcrystalline p amorphous n | 7.80 | 1.901 | 0.752 | 11.15 | 7.97/7.8 | 15.77 | 15.0 |
| a-Si/a-Si microcrystalline p multilayered n | 8.06 | 1.919 | 0.766 | 11.85 | 8.06/8.28 | 16.34 | 14.3 |
| a-Si/a-SiGe microcrystalline p amorphous n | 8.29 | 1.694 | 0.772 | 10.84 | 8.29/14.33 | 22.72 | 14.1 |
| a-Si/a-SiGe microcrystalline p multilayered n | 8.52 | 1.710 | 0.770 | 11.22 | 8.52/14.56 | 23.08 | 13.1 |

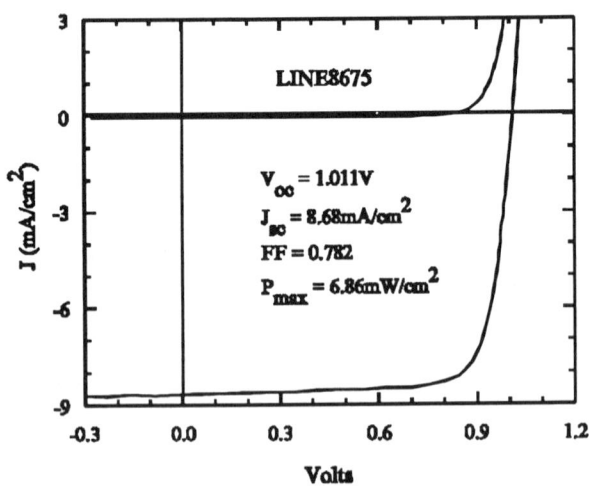

**Figure 5.** Initial J-V characteristic of an a-Si alloy top cell.

# The Triple-junction Cell

Incorporating the improvements described above, we then proceeded to make triple-junction cells with microcrystalline *p* and multilayered *n* structure in both tunnel junctions. Several triple cells with initial efficiencies exceeding 14% were obtained. The highest efficiency achieved is 14.6%, representing a 10% improvement over our previous record of 13.2%, also setting a new world record.

Figure 6 shows the J-V characteristic of the 14.6% triple-junction cell. Compared to our earlier best triple cell shown in Table 1, it is noted that $J_{sc}$ is increased significantly from 7.64 to 8.57 mA/cm$^2$, while $V_{oc}$ and FF have similar values. Quantum efficiency data shown in Fig. 7 reveal that the total photocurrent from the triple stack is 26.88 mA/cm$^2$, a 1.7 mA/cm$^2$ increase over our earlier value of 25.18 mA/cm$^2$ (see Table 1).

Comparing Figs. 2 and 7, one can readily see that (a) the quantum efficiency for $\lambda$>800 nm is increased significantly due to the improvement in the red response; (b) the quantum efficiency for $\lambda$~500-600 nm region is improved due to the incorporation of microcrystalline *p* and multilayered *n* in the tunnel junction; and (c) the overall quantum efficiency is enhanced, especially for the $\lambda$<450 nm region due to the improvement in ITO. We indeed benefit from the three improvements discussed above.

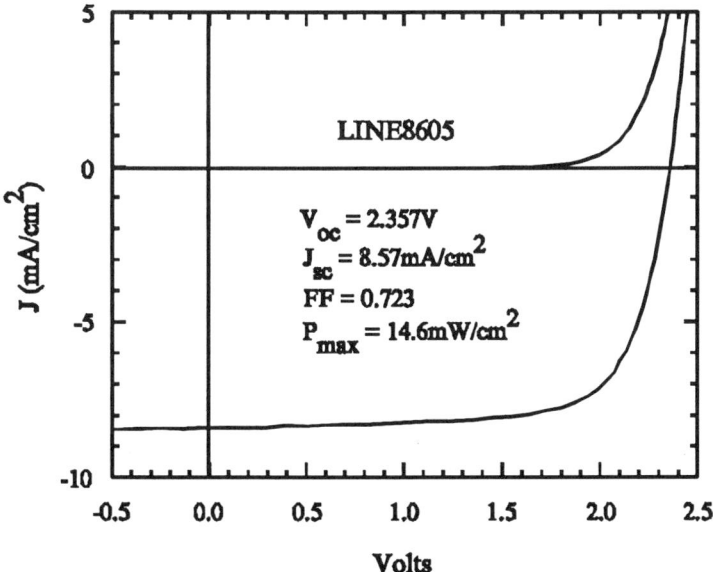

**Figure 6.** J-V characteristic of a triple-junction cell showing a 14.6% initial efficiency.

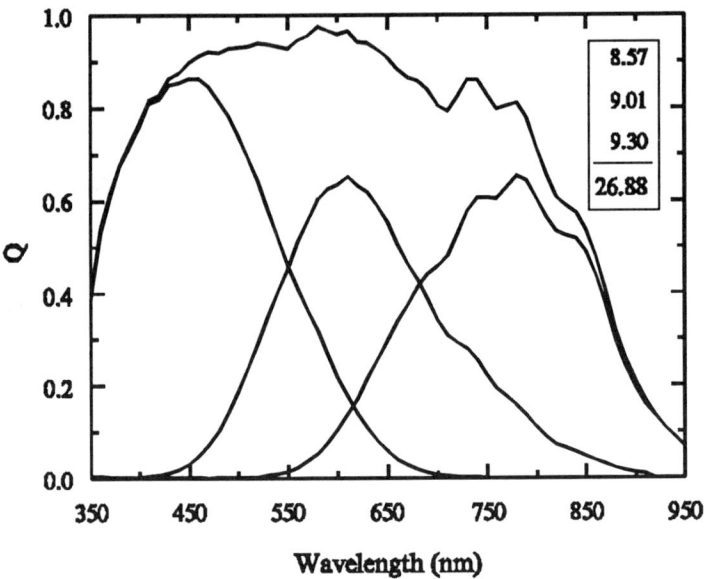

**Figure 7.** Quantum efficiency of the triple cell in Fig. 6.

A total of four devices was sent to NREL for triple-source measurement. Table 4 summarizes results obtained at NREL and United Solar. It should be pointed out that the $J_{sc}$ value at United Solar is based on quantum efficiency measurement. NREL measures total area efficiency only; the active-area current density is obtained by first subtracting the grid coverage from the total area and then calculating its active-area current density. In general, the active-area efficiency values for these four devices are in good agreement, with the highest measured to be 14.5%.

These devices were returned and were subjected to light soaking under one-sun, 50 °C, and open-circuit conditions. After ~300 hours of light soaking, their performance was substantially stabilized. The J-V characteristics of two devices after 600 hours are listed in Table 5 along with their initial characteristics. The best stabilized efficiency is 12.8%. This is the highest stable efficiency for amorphous silicon alloy solar cells, surpassing our previous record of 11.8%. Figure 8 plots progress made in United Solar on stable cell efficiency since the beginning of the Thin Film Partnership program. Further improvement in the

**TABLE 4.** Comparison of Initial Triple-cell Efficiency (area of ~0.25 cm$^2$) as Measured at NREL and United Solar.

| Sample | $V_{oc}$ (V) | $I_{sc}$ (mA) | FF (%) | $J_{sc}$ (total area) (mA/cm$^2$) | $\eta$ (total area) (%) | $J_{sc}$ (active area) (mA/cm$^2$) | $\eta$ (active area) (%) | |
|---|---|---|---|---|---|---|---|---|
| L8605 #42 | 2.357 | 2.104 | 74.39 | 7.721 | 13.5 | 8.28 | 14.5 | NREL |
| L8605 #42 | 2.357 |  | 72.3 |  |  | 8.57 | 14.6 | United Solar |
| L8605 #34 | 2.335 | 2.116 | 73.29 | 7.721 | 13.2 | 8.22 | 14.1 | NREL |
| L8605 #34 | 2.349 |  | 72.1 |  |  | 8.57 | 14.5 | United Solar |
| L8606 #21 | 2.336 | 1.973 | 76.09 | 7.403 | 13.2 | 7.93 | 14.1 | NREL |
| L8606 #21 | 2.344 |  | 74.5 |  |  | 8.23 | 14.4 | United Solar |
| L8606 #34 | 2.340 | 1.955 | 75.23 | 7.265 | 12.8 | 7.77 | 13.7 | NREL |
| L8606 #34 | 2.334 |  | 74.8 |  |  | 8.23 | 14.4 | United Solar |

understanding of plasma chemistry, the film growth, and the degradation mechanism should help in reaching the 15% stable module goal. Also shown in Table 5 are the measurements on the cell showing initial 14.6% efficiency using a single-source simulator. The initial efficiency is 14.3% and the stable efficiency is 13%. The light-soaked devices were sent to NREL for triple-source measurement again; the NREL results, showing slightly lower values, are also listed in Table 5. The cause for this discrepancy in the stabilized values is now being investigated.

**TABLE 5.** Initial and Stable Triple-cell Efficiency (area of ~0.25 cm$^2$) as Measured at NREL and United Solar.

| Sample | $V_{oc}$ (V) | $I_{sc}$ (mA) | FF (%) | $J_{sc}$ (total area) (mA/cm$^2$) | η (total area) (%) | $J_{sc}$ (active area) (mA/cm$^2$) | η (active area) (%) | |
|---|---|---|---|---|---|---|---|---|
| L8605 #42 Initial | 2.357 | 2.104 | 74.39 | 7.721 | 13.5 | 8.28 | 14.5 | NREL |
| Stable (600 hrs) | 2.279 | 1.995 | 68.76 | 7.320 | 11.5 | 7.85 | 12.3 (-15.2) | NREL |
| L8605 #42 Initial | 2.357 | | 72.3 | | | 8.57 | 14.6 | United Solar |
| Stable (600 hrs) | 2.294 | | 68.3 | | | 8.18 | 12.8 (-12.3) | United Solar |
| L8605 #42 Initial[a] | 2.357 | | 72.3 | | | 8.41 | 14.3 | United Solar |
| Stable[a] (600 hrs) | 2.294 | | 68.3 | | | 8.31 | 13.0 (-9.1) | United Solar |
| L8606 #34 Initial | 2.340 | 1.955 | 75.23 | 7.265 | 12.8 | 7.77 | 13.7 | NREL |
| Stable (600 hrs) | 2.265 | 1.907 | 71.46 | 7.087 | 11.5 | 7.58 | 12.3 (-10.2) | NREL |
| L8606 #34 Initial | 2.334 | | 74.8 | | | 8.23 | 14.4 | United Solar |
| Stable (600 hrs) | 2.277 | | 70.7 | | | 7.87 | 12.7 (-11.8) | United Solar |

[a] Single-source solar simulator data

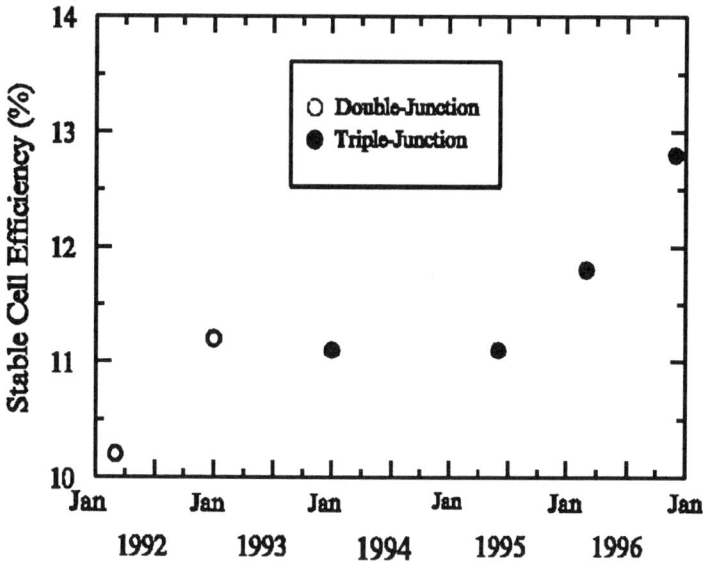

**Figure 8.** Progress of stable multijunction cell efficiency at United Solar.

## CONCLUSION

We have improved the narrow bandgap a-SiGe bottom cell, the tunnel junction, and the top conducting oxide and achieved a new world record initial efficiency of 14.6% and stable efficiency of 12.8% on a spectrum-splitting, triple-junction cell.

## ACKNOWLEDGEMENT

The authors thank E. Chen, G. Hammond, M. Hopson, H. Laarman, J. Noch, and T. Palmer for their assistance in sample preparation and measurements, S. Sugiyama and T. Glatfelter for discussions, and V. Trudeau for help in manuscript preparation. Stimulating discussions and constant encouragement from S. R. Ovshinsky are acknowledged. The work was supported in part by NREL under Subcontract Nos. ZAN-4-13318-02 and ZAF-5-14142-01.

# REFERENCES

1. Guha, S., Yang, J., Banerjee, A., Glatfelter, T., Hoffman, K., Ovshinsky, S. R., Izu, M., Ovshinsky, H. C., and Deng, X., "Amorphous Silicon Alloy Photovoltaic Technology—from R&D to Production,"in *Proceedings of the Mat. Res. Soc. Symp.*, 1994, pp. 645-655.
2. Luft, W., Branz, H. M., Dalal, V. L., Hegedus, S. S., and Schiff, E. A., "Recent Progress in Amorphous Silicon PV Technology," in *Proceedings of the 12th NREL Photovoltaic Program Review*, 1993, pp. 31-45.
3. Luft, W., "NREL/Industry Interaction: Amorphous Silicon Alloy Research Team Formation," in *Proceedings of the 12th NREL Photovoltaic Program Review*, 1993, pp. 46-50.
4. Yang, J., Xu, X., Banerjee, A., and Guha, S., "Progress in Triple-junction Amorphous Silicon Alloy Solar Cells with Improved Current Mismatch in Component Cells," in *Proceedings of the 25th IEEE PVSC*, 1996, pp. 1041-1044.
5. Ross, R., Mohr, R., Fournier, J., and Yang, J., "Status of Fluorinated Amorphous Silicon-Germanium Alloys and Multijunction Devices", in *Proceedings of 19th IEEE PVSC*, 1987, pp. 327-330.
6. Yang, J., Banerjee, A., Glatfelter, T., Hoffman, K., Xu, X., and Guha, S., "Progress in Triple-junction Amorphous Silicon-based Alloy Solar Cells and Modules Using Hydrogen Dilution," in *Proceedings of the First World Conference on Photovoltaic Energy Conversion*, 1994, pp. 380-385.
7. Guha, S., Yang, J., Pawlikiewicz, A., Glatfelter, T., Ross, R., and Ovshinsky, S. R., *Appl. Phys. Lett.* **54**, 2330-2332 (1989).
8. Guha, S., Yang, J., Nath, P., and Hack, M., *Appl. Phys, Lett.* **49**(4), 218-219 (1986).
9. Xu, X., Yang, J., Banerjee, A., Guha, S., Vasanth, K., and Wagner, S., *Appl. Phys. Lett.* **67**, 2323-2325 (1995).

# Hot Wire Deposited Hydrogenated Amorphous Silicon Solar Cells

A. H. Mahan, E. Iwaniczko, B. P. Nelson, R. C. Reedy Jr.,
T. Unold, R. S. Crandall, S. Guha* and J. Yang*

National Renewable Energy Laboratory, Golden, CO 80401
* United Solar, Troy, MI 48084

## ABSTRACT

This paper details the results of a study in which low H content, high deposition rate hot wire (HW) deposited amorphous silicon (a-Si:H) has been incorporated into a substrate solar cell. We find that the treatment of the top surface of the HW i-layer while it is cooled from its high deposition temperature is crucial to device performance. We present data concerning these surface treatments, and correlate these treatments with Schottky device performance. We also present first generation HW n-i-p solar cell data, where a glow discharge (GD) µc-Si(p) layer completes the partial devices. No light trapping layer is used to increase the device $J_{sc}$. Our preliminary results yield efficiencies of up to 6.8% for a cell with a 4000 Å thick HW i-layer, which degrade less than 10% after a 900h AM1 light soak. We suggest areas for further improvement of our devices.

## INTRODUCTION

This paper describes recent device results obtained using the hot wire (HW) deposition technique. Using this technique we previously found, by raising the substrate temperature ($T_S$) to values higher than that commonly used to deposit GD a-Si:H cells and modules, that device quality a-Si:H with bonded H contents as little as 1 at. % could be deposited for the first time [1,2]. We also measured saturated defect densities for these high $T_S$, low H content films, using the constant photocurrent (CPM) technique, which were significantly lower than had been previously reported [3,4]. These HW films were deposited at deposition rates between 4-8 Å/sec. Thus, we have undertaken to incorporate this HW a-Si:H material into a solar cell. We originally incorporated a HW i-layer into a superstrate p-i-n cell [5], but were unable to reach the high $T_S$ needed to deposit this low H content HW a-Si:H material. This paper details preliminary results obtained using the SS/n-i-p/ITO substrate cell approach. To explore incorporation of a high $T_S$ HW i-layer into a substrate solar cell, we first examine the Schottky barrier structure, because this structure must be optimized in any case to optimize an n-i-p solar cell [6]. Then, results on our first generation n-i-p HW "hybrid" cells, deposited with HW n- and i-layers and GD µc-Si(p) layers, are described.

## EXPERIMENTAL

The HW films and devices were deposited in a non-load-locked, single chamber system using deposition conditions described elsewhere [1,2,7]. The polished SS substrates were supplied by United Solar. No light trapping layers were used in this feasibility study. The n-layers were either deposited by GD or HW. In the former case, the GD n-layers were transferred at room temperature in air to the HW reactor, where they were heated to the $T_S$ (360°-400°C) needed for deposition of the HW i-layer. Nominally, these GD n-layers sat at the elevated temperatures for 1 h before the HW i-layer deposition to insure $T_S$ uniformity and adequate chamber outgassing. When the n- and i-layers were deposited by HW in the same chamber, a purging procedure was used to attempt to minimize dopant contamination of the i-layer. After the HW i-layer deposition, the partial device was cooled, using appropriate and varied surface treatments, and was finished by either a Pd top contact or, for the completed devices, a $\mu c$-Si(p)/ITO combination. In the latter case, the partial devices were sent to United Solar for completion. All i-layers were ~ 4000-4500 Å, and were deposited at ~ 8 Å/s. For the completed devices, the light soaking was done under 100 mW/cm$^2$ ELH light at ~40°C, and I-V measurements were done using an XT-10 simulator and a 4 probe geometry.

## RESULTS AND DISCUSSION

In our initial Schottky barrier study, we used GD n-layers for all devices, with an n-layer $T_S$ of ~375°C. This elevated temperature was used to roughly match the HW i-layer $T_S$ that we used. To demonstrate our Schottky device expertise, we first made a 4000 Å thick, all-GD device, using 100% silane and a substrate temperature of 250°C for the i-layer, and demonstrated (white light FF=0.63) that we could obtain device parameters similar to those reported elsewhere [8]. Next, we deliberately introduced a room temperature air break in the middle of the GD i-layer deposition and found (FF=0.61) that this air break was not crucial to device performance, suggesting that the GD i-layer, when exposed to air at room temperature for short times, is not very reactive. We deposited this device structure to test the feasibility of transferring (partially finished) i-layers in a device between different deposition systems which are not connected with a load lock. In this device, the $T_S$ of the GD i-layer deposited after the air break was raised to 350°C to roughly mimic the $T_S$ that would be needed for the HW i-layer. Then, on successive depositions, we progressively substituted the (high $T_S$) HW i-layer for the GD i-layer, while keeping the total device thickness the same. We found that when either a thin or thick HW i-layer is substituted for the GD i-layer, the device white light FF droped from >0.60 to ~0.52 and was independent of the HW i-layer thickness [9].

The constancy of the FF when HW i-layers of different thicknesses were progressively substituted for GD i-layers suggested that the device performance was not limited by the transport properties of the (different) i-layers, but rather by some interface layer. Therefore, we decided to vary the surface treatment of the HW i-layer as we cooled it from its high $T_S$ and to correlate these surface treatments with SIMS H profiles and device performance. In this investigation, HW n-layers were used for all the devices. Stopping the HW i-layer growth completely and cooling the SS/n-i structure in vacuum produces a device with a very low FF (0.28), while various surface treatments during cooling yield a significant FF improvement [10]. Fig. 1 shows SIMS H profile measurements on two selected devices with widely varying surface treatments. Without any special precautions (i.e., stopping the HW growth and cooling in vacuum), a significant amount of H diffuses out of the HW i-layer surface during cooling, producing a device with a very poor FF (0.28), while a surface treatment which inhibits this out-diffusion, and actually puts H back into the surface layer, produces a device with a significantly higher FF (0.60). These results readily suggest that the defects that were created by the H out-diffusion can, to a large extent, be passivated by a surface treatment procedure. Our best HW results to date for our low H content, high $T_S$, high deposition rate HW material yield a FF=0.65 and a $V_{OC}$ =0.54. These values compare favorably to those for our best GD devices, deposited without H dilution, which have a FF=0.63 and a $V_{OC}$=0.51.

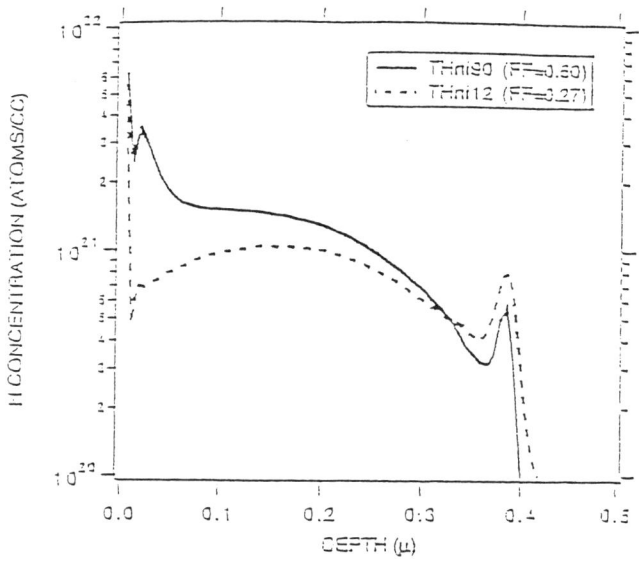

Figure 1. SIMS H profiles for two Schottky devices subjected to very different surface treatments during, and after, the cooling process.

We now report on the device performance, including the results of light soaking, of our first generation n-i-p HW "hybrid" devices, which were made in collaboration with United Solar. In these devices, the n- and i-layers were both deposited by HW in the same chamber, since in our initial study no significant difference was found in devices fabricated with either GD or HW n-layers [11]. Based on published results, a state-of-the-art GD SS/n-i-p/ITO test device with a

| THDni113 | as grown | 887 hr light soak |
|---|---|---|
| white FF | .605 | .545 |
| $V_{OC}$ (V) | 0.88 | 0.86 |
| $J_{SC}$ (mA/cm2) | 11.0 to 12.8 | no change |
| efficiency (%) | 6.81 (best) | 6.13 (best) |
| % change | | ~ 10% |

Table 1. Light soaking data for a typical SS/n(HW)-i(HW)-μc-Si(p)(GD)/ITO device. No measuring mask was used to define the area, and thus determine $J_{SC}$.

4000Å i-layer and no light trapping, deposited using H dilution, would have an initial efficiency of ~7.8 % and would degrade after 600 h light soaking under one sun by ~16%, to an efficiency of ~6.5% [12]. Note that a reduction in all three I-V parameters were reported when this GD test device was light soaked. These results have been corroborated at NREL, using our XT-10 Simulator, on a standard United Solar device kindly sent to NREL for analysis. Table 1 shows the I-V results for a typical HW device, both in its initial state, and after having been light soaked for 887 h. The range in $J_{SC}$ is representative of the spread in the 11 (out of 12) active devices which were measured.

The primary change in the HW I-V parameters, measured after the light soaking, is seen to be in the FF. The $V_{OC}$ changed minimally over the course of the light soaking, and we observed no measurable decrease in the device $J_{SC}$. As a consequence, the overall device efficiency degraded by ~10%. Other HW "hybrid" devices fabricated by the same procedure degraded by a similar amount. This degradation compares quite favorably to changes reported in GD n-i-p devices containing i-layers of similar thickness [12].

The major differences in the initial state between the GD and HW "hybrid" devices are in the values of $V_{OC}$ (0.94 vs. 0.88 V) and FF (0.66 vs. 0.605). Addressing first the differences in $V_{OC}$, we suggest that the lower $V_{OC}$ for the HW "hybrid" device is due primarily to the lower Tauc's bandgap of the high $T_S$ HW i-layers (1.65-1.70 eV), compared with devices using GD i-layers deposited with H dilution (> 1.70 eV). Indeed, the $V_{OC}$ of a HW "hybrid" device with the HW i-layer deposited at a lower $T_S$ (280°C) and thus exhibiting a higher bandgap, exhibits a similarly high $V_{OC}$ (0.93 V). Therefore, we suggest that the feasibility of depositing a $\mu$c-Si(p) layer on a high $T_S$ HW a-Si:H i-layer has been successfully demonstrated, as seen by the high $V_{OC}$ of our "hybrid" device.

We now address possible reasons for the low FF of the HW "hybrid" device compared to its GD counterpart, when both are measured in their initial states. We note that, in addition to the low initial white light FF's observed, both Quantum Efficiency (QE) and high illumination intensity red light FF measurements (using a lens and a > 610 nm bandpass filter) suggest problems with the red response as well. We believe that several indicators point to the need to reexamine the fabrication of the n/i interface, rather than being concerned about and probing the quality of the HW i-layer in the device itself.

First, if the low FF's in our devices are due to (uniformly) high HW i-layer defect densities, then certainly thinner Schottky devices should have higher FF's, and we do not observe this within the i-layer thickness range 2000-8000 Å. Further, we fabricated a HW Schottky device with an i-layer thickness of 1.5 $\mu$. The 600nm red light FF in this device (0.54) is much higher than that predicted (< 0.48) using AMPS modeling and standard AMPS parameters [13]; the only way to explain this high value for such a thick device is to assume a low bulk defect density (< 1 x $10^{16}$ cm$^{-3}$) and a narrower valence bandtail (39 mV). Note that these modeling predictions are consistent both with CPM bulk defect densities and with enhanced hole mobilities recently reported in our (thick) low H content HW devices by time of flight measurements [14].

Second, SIMS P depth profiles, in which the HW n- and i-layers were deposited in the same chamber, show consistent tailing of the dopant into the i-layer. A representative profile is shown in Fig. 2. We observed that wide variations in our gas purging procedure, from a weekend pump to a 5 min purge between deposition of the HW n- and HW i-layers, made little difference in either the SIMS P profiles or the device FF's. We now believe that we had inadequate "burial" of the P on the chamber walls surrounding the sample before the i-layer deposition, due to the geometry of the shutter used for device fabrication. Our next generation HW reactor, a multichamber, load-locked system incorporating a new HW chamber design, will directly address this issue.

Finally, while acknowledging that thin, high defect density i-layers can have a different effect on cell efficiencies when placed at different depth positions within the cell, we believe that attention must now be paid to the sizeable H out-diffusion observed at the back of the HW i-layer (n/i interface). As seen in Fig. 1 and discussed earlier, addressing this out-diffusion at the top surface (the surface most easily accessible to H 'treatments') has enabled us to increase the device FF significantly. However, H out-diffusion at the top surface is the less severe of the two profiles, due to the fact that the back surface of the device sits longer at the high Ts. In addition, preliminary ESR and CPM measurements on thick samples prepared so that the great majority of the H diffusion is at the back of the sample (i-layer/substrate interface) suggest that defects do occur in the (back-surface) region where the H has out-diffused. We are currently exploring ways to vary the H profiles in this region and correlate its effect on device efficiency.

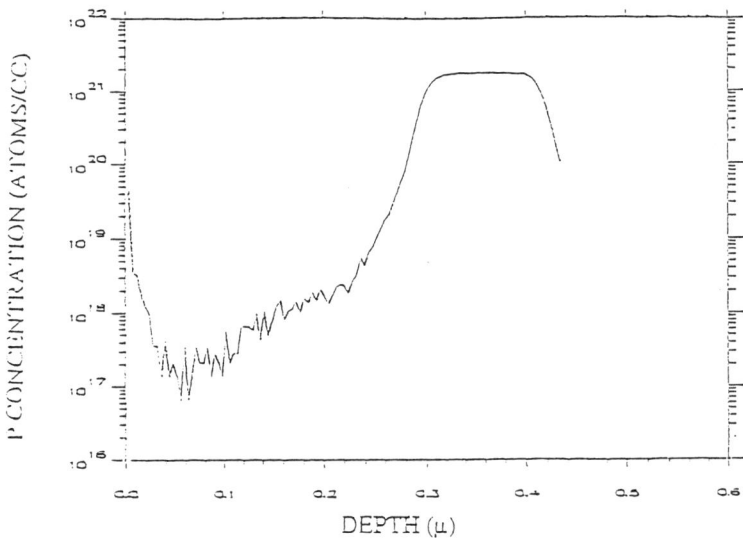

Figure 2. SIMS P depth profiles for a typical HW Schottky device, where the HW n- and i-layers were deposited consecutively in the same deposition reactor.

## CONCLUSIONS

We have presented device data, including the results of light soaking, on our first generation of HW "hybrid" SS/n-i-p/ITO solar cells, where the n- and the i-layers were deposited by HW and the $\mu c$-Si(p) layer by GD. No light trapping layers were used in this initial feasibility study. We have shown that we can fabricate high efficiency substrate solar cells, where the n- and i-layers are deposited by HW at high substrate temperatures and at high deposition rates. A major consideration in obtaining these cell efficiencies is the procedure used to treat the top surface of the HW i-layer while it is cooling from its high deposition temperature. When this is taken into account, initial solar cell efficiencies as high as 6.8% can be achieved, and the amount of device degradation after light soaking compares quite favorably with published reports of GD cells with similar i-layer thicknesses. Finally, we have discussed research avenues which, in our opinion, will result in further improvements in HW "hybrid" devices.

## ACKNOWLEDGEMENTS

We thank Y. Xu for GD device fabrication, V. Dalal and S. Hugedus for QE measurements, and A. C. Gallagher for valuable support and many stimulating discussions during the entire course of this work. The work at NREL was supported by the U. S. Department of Energy under Contract No. DE-AC36-83CH10093, and the work at United Solar was supported in part by NREL under subcontract No. ZAN-4-13318-02.

## REFERENCES

[1]. A. H. Mahan, J. Carapella, B. P. Nelson, R. S. Crandall, and I. Balberg, *J. Appl. Phys.* **69**, 6728 (1991).
[2]. A. H. Mahan, B. P. Nelson, S. Salamon, and R. S. Crandall, *MRS Symp. Proc.* **219**, 673 (1991).
[3]. A. H. Mahan and M. Vanecek, *AIP Conf. Proc.* **234**, 195 (1991).
[4]. R. S. Crandall, A. H. Mahan, B. P. Nelson, M. Vanecek and I. Balberg, *AIP Conf. Proc.* **268**, 81 (1992).
[5]. B. P. Nelson, E. Iwaniczko, R. E. I. Schropp, A. H. Mahan, E. C. Molenbroek, S. Salamon, and R. S. Crandall, *Proc. 12th European PV Solar Energy Conference*, 679 (1994).
[6]. A. H. Mahan, B. P. Nelson, E. Iwaniczko, Q. Wang, E. C. Molenbroek, S. E. Asher, R. C. Reedy Jr., and R. S.Crandall, *AIP Conf. Proc.* **353**, 67 (1995).
[7]. E. C. Molenbroek, E. J. Johnson, and A. C. Gallagher, *Proc. 13th European PV Solar Energy Conference*, 1995, in press.
[8]. X. Deng, S. J. Jones, J. Evans and M. Izu, *MRS Symp. Proc.* **377**, 803 (1995).
[9]. When these initial HW Schottky barrier devices were made, we were in fact, due to our initial shutter design, continuing to grow very slowly during the i-layer cooling, so the FF's of these HW devices were not representative of devices cooled in vacuum (see text).
[10]. A. H. Mahan, B. P. Nelson, E. Iwaniczko, Q. Wang, Y. Xu, R. S. Crandall, S. Guha, and J. Yang, IEEE PV Spec. Conf. Proc. **25**, 1065 (1996).
[11]. We believed at the time that an air break at the n/i interface, necessitated by the fabrication of the GD n- and the HW i-layers in different deposition systems, was more detrimental to device performance than possible trace P contamination in the HW i-layer when both layers were deposited in the same deposition system.
[12]. S. Guha, J. Yang, S. J. Jones, Y. Chen and D. L. Williamson, *Appl. Phys. Lett.* **61**, 1444 (1992).
[13]. J. K. Arch and S. J. Fonash, *Appl. Phys. Lett.* **60**, 757 (1992).
[14]. Q. Gu, E. A. Schiff, R. S. Crandall, E. Iwaniczko, A. H. Mahan and B. P. Nelson, *Proc. 1996 MRS Spring Meeting*, in press.

# Stability Properties of Amorphous Silicon and Silicon-Germanium Films and Devices Deposited Using Remote ECR-Techniques

Vikram L. Dalal, Sanjeev Kaushal, Tim Maxson, Robert Girvan, and Alan Boerner

Iowa State University, Ames, Iowa 50011

## ABSTRACT

We report on the preparation and properties of a-Si:H and a-(Si,Ge):H films and devices prepared using remote, low pressure ECR discharge. We show that deposition chemistry has a profound influence on the stability of both films and devices. In particular, the use of remote H beam to etch the film during growth leads to significant improvements in the stability of both films and devices. The devices produced using the hydrogen beam are more stable than devices produced using either glow discharge or Helium beams. A new model of growth chemistry which can explain these differences in terms of localized differences in growth when H is present is suggested to explain the improved stability of devices produced using hydrogen-ECR process.

## INTRODUCTION

It is now recognized that the localized microstructure of amorphous Si and Silicon-Germanium alloys has a profound influence on the light-induced degradation of both films and devices[1,2]. In particular, the presence of microvoids and other growth defects has a deleterious effect on stability. In recent years, our group at Iowa State University has shown that the use of a remote electron-cyclotron-resonance(ECR) deposition technique, where the sample is subjected to energetic beams of H or He during growth, leads to better stability of films and devices[3,4]. In this paper, we further examine the stability characteristics of films and devices in both a-Si:H and a-(Si,Ge):H prepared using remote ECR techniques.

## EXPERIMENTS

The devices were all of p-i-n type on stainless steel substrates. The substrates were bright polished but not electro-polished. On top of the substrate, successive n, i and p layers were grown. To minimize contamination from the n layer, a thick n layer was first grown in a separate reactor, followed by a thin n layer in the ECR reactor. The n layer was followed by an i layer which was grown using either He or H as the plasma gas. Silane and Germane were introduced downstream near the substrate. Details of the deposition reactor have been published previously[3,4]. The deposition temperatures were 350-380 C, and the deposition pressures were in the range of 10-15 mT. Pressures below 10 mT sometimes led to the onset of microcrystallinity. The typical growth rates for the i layer were in the range of 1.0 to 1.5 A/sec. The thickness of the i layer was in the range of 0.4 $\mu$m. The final p layer was a-(Si,C):H, and was grown at a low temperature( 200 C) so as to avoid CVD deposition of low-gap p-type a-Si:H. The p layer was followed by a semi-transparent 100 A Cr layer, with a central thick Al bussbar for putting down the current probe on the device. The Cr contact allowed about 20%

of the light to pass through to the cell. Fig. 1 shows the schematic diagram of the device.

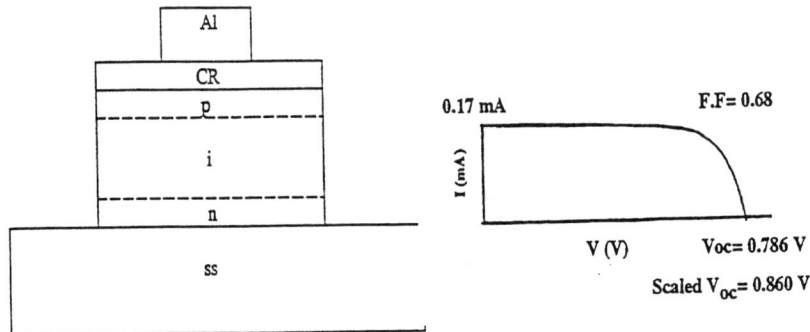

Fig. 1 Device structure for substrate cell

Fig.2 I(V) curve for a typical cell

## RESULTS

The I(V) curves of the devices were measured under AM1.5 illumination. Fig. 2 shows the typical device I(V) curve, with a very good fill factor. The open circuit voltage is in the range of 0.8 V, which increases to about 0.85 V when the cell receives full light intensity. The internal quantum efficiency(QE) was measured under a white light bias, and the results are shown in Fig. 3. QE was measured under both short-circuit and forward bias conditions. The ratio of QE vs. wavelength is shown in Fig. 4, and a low QE low ratio ( about 1.1) indicates good hole collection, i.e. good hole mobility-lifetime product.

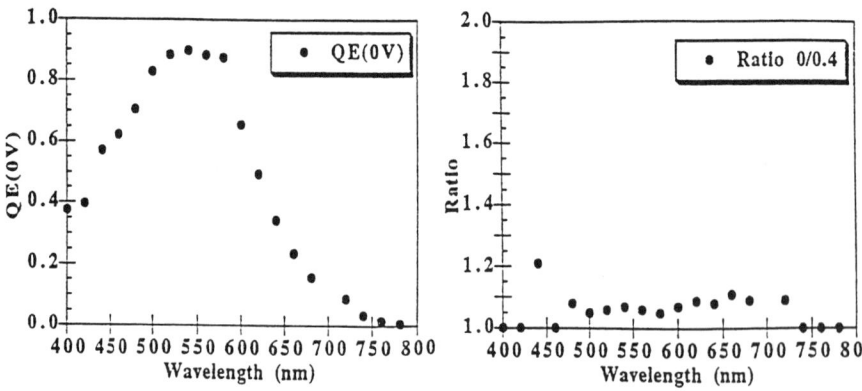

Fig. 3 Quantum Efficiency of the cell

Fig.4 QE ratio vs. wavelength plot

For stability testing, three different types of devices were made. Each was of the p-i-n type on stainless steel substrates. One device had the i layer made using H-ECR, where hydrogen was used as the plasma gas. Another device had the i layer made using He-ECR, where He was used as the plasma gas. The third device was made using a RF triode glow-discharge process. All three devices had similar thicknesses of the i layer, similar initial fill factors, and comparable initial open circuit voltages, indicating comparable built-in voltages and comparable hole collection properties before light soaking.

## RESULTS FROM STABILITY TESTING

In Fig. 5, we show the results of the stability testing of the three devices under 8x sunlight. With the 20% transmission of the Cr layer, about 1.6xsun is actually incident on the cell. Since fill factor is the parameter that is most affected by degradation of the material, we only plot the fill factor as a function of illumination time. The results from Fig. 5 are striking. The cell made using a H-ECR i layer degrades much less than either the cell made using the He-ECR or the cell made using the glow-discharge process. The degradation of the latter two cells is comparable.

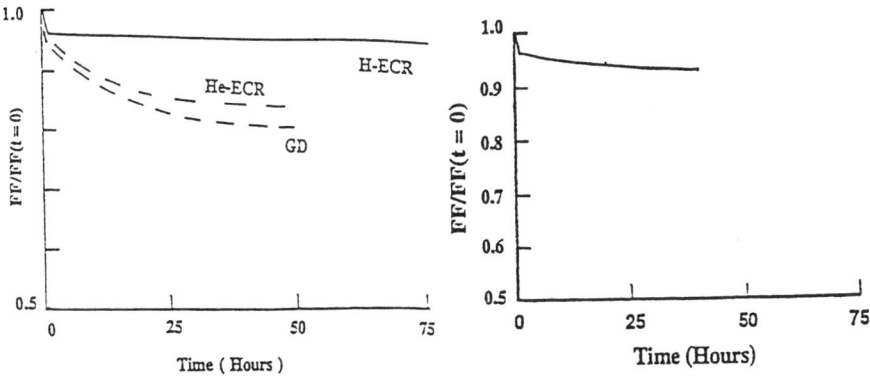

Fig.5  Degradation in fill factors of 3 similar a-Si:H cells.

Fig.6  Degradation in fill factor of an a-Si,Ge:H cell

This result is particularly noteworthy because the cell made using H as the plasma gas had a larger bonded H-content(7-8%) than the cell made using He as the plasma gas(4.5%). Therefore, we conclude that it is NOT the amount of bonded H that is important, but rather, what H does during growth. We address this point more thoroughly in the next section.

We have also started some work on fabricating a-(Si,Ge):H devices, also using H-ECR process, and the preliminary results on stability of these devices are shown in Fig. 6. Once again, the stability of the a-(Si,Ge):H device produced using H-ECR process is remarkable.

## INFLUENCE OF DEPOSITION CHEMISTRY ON STABILITY

The striking results of the H vs. He plasma experiments reported above indicates that there is a fundamental difference in the stability of the a-Si:H film when grown using different approaches. This result can be explained by examining the growth chemistry.

Consider a growing Si surface, as shown in Fig. 7(A). There, we show a Si surface waiting to accept an appropriate Si radical that it can bond with. Plasma generates many different types of radicals from silane, e.g. $SiH_3$, $SiH_2$, SiH etc. In particular, the rates of production of $SiH_3$ and $SiH_2$ radicals are not too different in a silane plasma without H dilution. At most, the rate of production of these two radicals may be different by a factor of 5.[5,6]. Now, if $SiH_2$ inserts in one site, and $SiH_3$ in the next, then the resulting surface looks as shown in Fig. 7(B) where the first site is active, but the second site is passivated with H(i.e. unable to bond with another Si radical). Thus, we have converted a surface with all active sites into a surface with some active and some passive sites. This is a prescription for non-uniform growth,i.e. microvoids, at hence, instability.

However, when H is present so that the mixture is highly dilute, the most likely radical to arrive at the surface is H. Thus, all sites are initially passive. Then, the next H radical abstracts the initial H from the surface, and makes the sites active. Due to the great abundance of H radicals in a highly dilute plasma, the sites are active some fraction of the time, and passive some other fraction. Thus, H serves to homogenize the sites. We call this mechanism site homogenization. See Fig. 8.

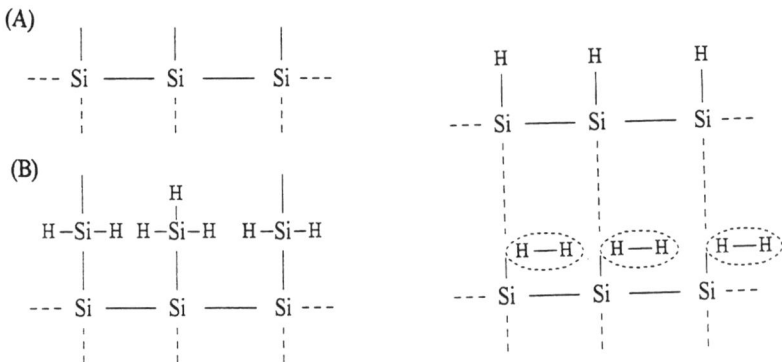

Fig. 7 Schematic diagram showing how different radicals lead to non-homogeneous sites

Fig. 8 Schematic diagram showing how the presence of significant H leads to site homogenization

The second role played by H in promoting better growth is related to enhancing radical selectivity. H will promote the production of $SiH_3$ radicals by the reaction:
$H + SiH_4 = SiH_3 + H_2$. With a high flux of H impinging on the surface, we can enhance the production of $SiH_3$ radicals so that the ratio of $SiH_3$ to $SiH_2$ becomes 20:1 instead of between 1:1 to 5:1. This mechanism, in itself, promotes a more homogeneous growth.

The third major role played by H is one of etching during growth. This mechanism is known to promote a more homogeneous microstructure(micro-crystallinity)by etching away the weakest bonds first. This mechanism will also promote the growth of more homogeneous films.

Therefore, from fundamental arguments of growth chemistry, we can postulate that having a high flux of energetic H impinge on the surface during growth will be beneficial for promoting a more homogeneous microstructure, and therefore, better stability. He, in contrast, should not participate in surface chemistry, but will lead to ion bombardment during growth, which may be beneficial upto a point.

## CONCLUSIONS

In conclusion, we have shown that the use of a high H dilution ECR process can produce a-Si:H devices which are much more stable than comparable devices produced using either glow-discharge or non-hydrogen ECR. We have suggested that H plays three different roles in deposition chemistry, and that each one of these roles, namely site homogenization, enhancement of radical selectivity, and etching-during-growth, can promote a more homogeneous microstructure, and hence, better stability.

## REFERENCES

1. D. Williamson, Proc. of Materials Res. Soc. 377,251(1995)
2. S. Guha, X. Xu and J. Yang, Proc. Mat'l. Res. Soc. 336,675(1994)
3. V.L. Dalal et al, " Growth of a-Si films with significantly improved stability", Appl.Phys.Lett.64,1862(1994)
4. V. L. Dalal et al, " Significant improvements in stability of a-Si:H solar cells using ECR deposition", J.Non-Cryst. Solids, 198-200, 1101(1995)
5. M. Kushner, J. Appl. Phys.,63,2532(1988)
6. J. P. Schmitt, J.Non-Cryst.Solids,59-60,649(1983)

# Deposition Of Transparent Conducting Oxides For Solar Cells

Roy G. Gordon

*Chemical Laboratories, Harvard University
Cambridge, Massachusetts 02138*

**Abstract.** Transparent conductors are needed as the front surface electrodes in all types of solar cells. The electrical and optical performance of a transparent conductor may be rated by a figure of merit defined as the ratio of the electrical conductivity to the optical absorption coefficient of the layer. Fluorine-doped zinc oxide is shown to have the highest figure of merit. ZnO:F films with a sheet resistance of 5 ohms per square can have a visible absorption of less than 3 per cent. This high performance makes zinc oxide a candidate for replacing tin oxide in thin film amorphous silicon solar cells, or for replacing part of the highly-doped silicon layer in crystalline silicon solar cells. A new, cost-effective process is described for the chemical vapor deposition of ZnO:F at atmospheric pressure.

## BACKGROUND ON TRANSPARENT CONDUCTORS

The first transparent conductors to be used were based on tin oxide. During the second world war, tin chloride solutions were sprayed onto hot glass, to produce electrically conductive, chlorine-doped tin oxide coatings on glass. Electricity was passed through these coatings to heat aircraft windows.[1] Doping with antimony[2] and fluorine[3] was found to give tin oxide films with even higher conductivity.

Conductive tin-doped indium oxide (ITO) films were also prepared by spray pyrolysis[4] and sputtering[5]. Sputtering has been the preferred mode for its production. The high cost of ITO precludes its application to photovoltaics.

Chemical vapor deposition (CVD) was later shown to be a more effective process for making fluorine-doped tin oxide films[6]. Transparency, electrical conductivity and thickness uniformity are better for the CVD films than for the sprayed films. Since the 1980's, CVD has been widely adopted in continuous production of glass coated with fluorine-doped tin oxide.[7] By far the largest area of transparent conducting oxide (TCO) films are currently produced in this way. Most of this material is used for energy-conserving ("low-emissivity") windows in buildings, with smaller amounts going into thin-film photovoltaics.

Conductive zinc oxide films were first produced by spray pyrolysis with indium doping[8] and sputtering with doping by aluminum[9], indium[10] or gallium[11]. CVD of conductive zinc oxide films has been achieved with doping by boron[12], aluminum[13], gallium[14], indium[15] and fluorine[16]. Fluorine-doped zinc oxide was found to have higher transparency for a given sheet resistance, than any of the other transparent conductors. The low cost of zinc, compared to that of indium or tin, gives zinc oxide the potential to be the least expensive TCO material, as well as the best performer.

What has been lacking is an inexpensive process for the manufacture of fluorine-doped zinc oxide. Industrial experience with tin oxide has shown that CVD at atmospheric pressure is the most cost-effective method for its production. We report in this paper a new atmospheric pressure CVD process for producing fluorine-doped zinc oxide, which has the potential to make this high-performance TCO at low cost.

# DESIRABLE FEATURES OF A CVD PROCESS

There are many requirements for an effective CVD process. The reactants should be readily available and cheap, and easy to purify to the required level of purity. They should remain pure and stable during shipment and storage, and should not react with air or water. They should not be pyrophoric, flammable, toxic or corrosive.

The reactants should be gases or liquids at room temperature, because their use is easier than that of solid precursors. Gases can be metered into the process with standard flow-controllers. Liquids can be evaporated from bubblers, or pumped by a metering pump into a vaporizer. The viscosity of the liquid should be low enough so that it can be pumped and metered easily. The precursors should have a high enough vapor pressure (over about 1 Torr) at a temperature low enough (under about 200 °C) so that standard components and materials may be used in the vaporizer. The vaporization process should be rapid and reproducible. In order for this to happen, the liquid should not be associated or polymeric. The liquid should remain thermally stable during the vaporization process.

The CVD reaction should take place at a temperature above the vaporization temperature, but below a temperature that would damage the substrate or degrade the properties of the deposited film. If there are two or more reactant vapors, they should remain unreactive to each other at the vaporization temperature of the least volatile reactant, so that the vapors may be mixed to a uniform and homogeneous composition before they reach the CVD chamber. The use of a homogeneous reactant vapor mixture allows films to be deposited with more uniform composition and thickness, than in the case that the reactants react prematurely, so that they must be mixed near the surface of the substrate.

The reaction should be carried out at atmospheric pressure, rather than at low pressure, so that expensive vacuum chambers and pumps are not required. The yield of film should be high, with minimal production of gaseous byproducts or powder. This requirement is important not only for the economical use of the reactants, but also so that the CVD apparatus will require only infrequent cleaning, and so that the treatment of the byproduct gases will be simplified. The growth rate of the film must be high, so that the productivity of the apparatus is high.

Finally, of course, the properties of the deposited film must be excellent, including high electrical conductivity and high transparency, good adhesion to the substrate. The morphology of the films is also important for solar cell applications. A rough, textured surface, on the scale of the wavelength of light, increases the amount of light absorbed by the cell.

# SURVEY OF CVD REACTIONS USED FOR ZINC OXIDE TRANSPARENT CONDUCTORS

In this section, we will review the reactions previously reported for the CVD of zinc oxide films. Then we will report on a new CVD reaction which appears to overcome the main difficulties with the previously used reactions.

One commonly used precursor for zinc oxide films is diethylzinc. It reacts readily with water vapor to deposit zinc oxide films, at temperatures as low as about 150 °C:

$$Zn(C_2H_5)_2 + H_2O \rightarrow ZnO + 2\ C_2H_6 \qquad (1)$$

This reaction is so fast that the reactant vapors must be mixed close to the surface of the substrate. It has been run successfully at atmospheric pressure and at low pressure. Textured films, with good light-trapping properties, are easily obtained from this reaction. The deposited zinc oxide film does contain a significant amount of hydrogen, probably in the form of zinc hydroxide. Because of this impurity, these films tend to desorb water vapor when heated.

Highly transparent zinc oxide films can be prepared by the CVD reaction of diethyl zinc vapor and ethanol vapor:

$$Zn(C_2H_5)_2 + C_2H_5OH \rightarrow ZnO + 2\ C_2H_6 + C_2H_4 \qquad (2)$$

In the absence of other materials, however, this reaction does not begin spontaneously. If a small amount of water vapor is also present, it does begin the fast reaction (1), to deposit an initial small amount of zinc oxide. This initial deposit of zinc oxide then catalyzes on its surface the decomposition of ethanol into water and ethylene:

$$C_2H_5OH \rightarrow H_2O + C_2H_4 \qquad (3)$$

This decomposition reaction becomes rapid at surface temperatures above about 400 °C. The water thus produced then reacts with the diethylzinc according to reaction (1) to form more zinc oxide. The net effect of combining reactions (1) and (3) is the production of zinc oxide by the overall reaction (2). The water vapor acts as a catalyst for reaction (2).

The deposition of zinc oxide may also be initiated in a water-free vapor mixture by an already-deposited layer of zinc oxide, which then grows thicker. Other materials, such as a surface of aluminum oxide, also start the growth of zinc oxide, by catalyzing the decomposition of ethanol according to reaction (3).

An excess of ethanol vapor is necessary to make sure that the deposited zinc oxide is highly transparent. It is preferable to have a molar ratio of ethanol to diethylzinc of at least about 10. Using less ethanol produces zinc oxide films with a brown color due to absorption of visible light, probably from small amounts of carbon impurity in the film.

Dimethylzinc can also be used as a zinc source. The vapor of dimethylzinc, diluted in an inert carrier gas, such as nitrogen, may be mixed with oxygen or air for the CVD of zinc oxide films.[17] At higher vapor concentrations, spontaneous burning produces powdered zinc oxide, rather than a film. The pyrophoric properties of both dimethylzinc and diethylzinc make them somewhat dangerous precursors, requiring special care for their safe handling.

Another zinc precursor is zinc acetylacetonate. This material is stable in dry air, and is not pyrophoric. It does, however, absorb water vapor from humid air. The CVD reaction of zinc acetylacetonate with water vapor deposits zinc oxide at temperatures over about 400 °C, while the reaction with oxygen begins at about 500 °C. A considerable excess of oxygen is require to produce a highly transparent film from this reaction. Smaller amounts of oxygen leave a brown color in the film. The volatility of zinc acetylacetonate is low, so the growth rates from its CVD reaction are also low.

Table 1 summarizes the ratings of these reactions, with respect to the properties of an ideal CVD reaction. In summary, all of these previously-used CVD reactions for zinc oxide are less than ideal.

**TABLE 1.** CVD Reactions For Producing Zinc Oxide.
\*\*\* = excellent, \*\* = good, \* = fair, uns = unsatisfactory

| Zn precursor | $Et_2Zn$ | $Et_2Zn$ | $Me_2Zn$ | $Zn(acac)_2$ | $Zn(acac)_2$ | $Et_2Zn(TEED)$ |
|---|---|---|---|---|---|---|
| O precursor | $H_2O$ | EtOH | $O_2$ | $H_2O$ | $O_2$ | EtOH |
| liquid precursor | *** | *** | *** | * | * | *** |
| inexpensive | *** | *** | * | *** | *** | *** |
| stable storage | *** | *** | *** | *** | *** | *** |
| non-flammable | uns | uns | uns | ** | ** | * |
| non-toxic | *** | *** | *** | *** | *** | *** |
| non-corrosive | *** | *** | *** | *** | *** | *** |
| volatile | *** | *** | *** | * | * | *** |
| fast vaporizing | *** | *** | *** | ** | ** | *** |
| stable vapor | *** | *** | *** | ** | ** | *** |
| low CVD temp. | *** | ** | *** | ** | * | ** |
| deposition rate | *** | ** | *** | ** | ** | ** |
| uniformity | ** | ** | *** | ** | *** | *** |
| film purity | ** | *** | ** | ** | ** | *** |
| conductivity | ** | *** | ** | ** | ** | *** |
| transparency | *** | *** | *** | *** | ** | *** |
| light-trapping | *** | *** | ** | ** | ** | *** |
| byproducts | ** | ** | ** | ** | ** | ** |

We have recently discovered a CVD reaction for zinc oxide which comes much closer to the ideal. In it, the diethylzinc is complexed with a diamine, such as tetraethylethylenediamine (TEED). The resulting liquid compound is not pyrophoric, although it still does react with water vapor. The vapor of this compound may be premixed with ethanol vapor, so that very uniform films can be deposited. Deposition becomes rapid at temperatures above about 400 °C.

Fluorine dopant can be introduced into the films using a variety of fluorocarbon gases. Acetyl fluoride is particularly effective as a source for fluorine doping. Using this new reaction, films of ZnO:F were prepared with sheet resistance as low as 8 ohms per square, along with an absorption of visible light less than 2%.

## COMPARISON OF THE OPTICAL AND ELECTRICAL PERFORMANCE OF TRANSPARENT CONDUCTORS

The best transparent conductor should combine a high electrical conductivity with a low loss of light through absorption. In order to measure the performance of a transparent conductor, we proposed[18] the use of a figure of merit defined by the ratio of the electrical conductivity, $\sigma$, to the visible absorption coefficient, $\alpha$:

$$\sigma/\alpha = -\{R_{\square}\ln(T+R)\}^{-1} \qquad (4)$$

This ratio can be evaluated by measuring the sheet resistance $R_{\square}$, in ohms per square, and using an integrating sphere spectrometer to find the total visible transmission T, and the total visible reflectance R, expressed as fractions. $\sigma/\alpha$ is thus a figure of merit for rating transparent conductors. A larger value of $\sigma/\alpha$ indicates better performance of the transparent conductor. This figure of merit is independent of film thickness, to the extent that the material properties of a film are independent of its thickness.

**TABLE 2.** Figure of Merit for Some Transparent Conductors

| Material | Sheet Resistance | Visible Absorption | Figure of Merit |
|---|---|---|---|
| | (ohms/square) | | (inverse ohms) |
| ZnO:F | 5 | 0.03 | 6.6 |
| ZnO:Al | 3.8 | 0.05 | 5.1 |
| $In_2O_3$:Sn | 6 | 0.04 | 4.1 |
| $SnO_2$:F | 8 | 0.04 | 3.1 |
| ZnO:Ga | 3.2 | 0.12 | 2.5 |
| ZnO:B | 8 | 0.06 | 2.0 |
| $SnO_2$:Sb | 20 | 0.12 | 0.4 |
| ZnO:In | 20 | 0.20 | 0.2 |

Figures of merit for some transparent conductors are given in Table 2. The values are for the best samples that we have prepared in our laboratory by CVD at atmospheric pressure, except for the indium oxide value, which is the best that we have measured for a commercially available film.

These results show that fluorine-doped zinc oxide gives the best figure of merit among the known transparent conductors.

# COMPARISON OF AMORPHOUS SILICON SOLAR CELL PERFORMANCE ON DIFFERENT TCO MATERIALS

Samples of fluorine-doped zinc oxide were deposited on glass substrates at a temperature of about 450 °C by the new APCVD process. Some were sent to the Institute for Energy Conversion, University of Delaware and to the Thin Film Division of Solarex, for evaluation of their use in superstrate amorphous silicon solar cells. Optical measurements at Solarex confirmed the high visible transmission of these films. Visible haze values ranged up to 12%, just at the lower end of the desired range of 12 to 18%. The thicknesses of the films were considered to be quite uniform over their 10 cm by 10 cm areas, with thicknesses around 800 nm. Future samples will be made slightly thicker in order to increase the haze values.

An amorphous silicon solar cell was made on ZnO:F at IEC, Delaware. Figure 1 shows the quantum efficiency for this cell as a function of wavelength. As expected from the optical measurements, the current is higher for zinc oxide than for tin oxide at all wavelengths, with the smallest increases in the red end of the spectrum. When ZnO:F samples with higher haze levels are prepared, even higher currents should be found for the red wavelengths.

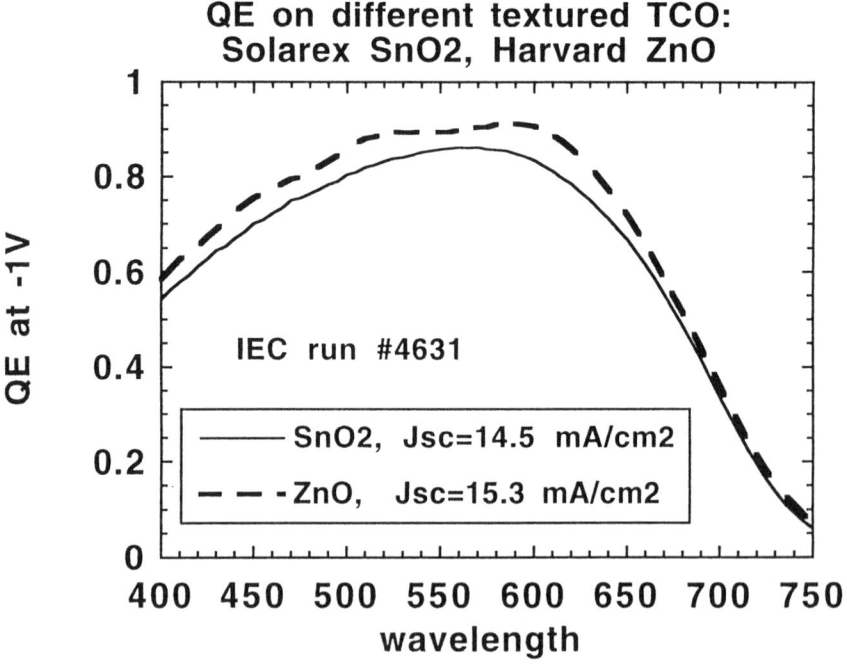

**FIGURE 1.** Quantum efficiency as a function of wavelength (nm) for amorphous silicon solar cells produced at IEC, Delaware on fluorine-doped tin oxide and zinc oxide.

Electrical parameters for the IEC single-junction cell are given in Table 3.

**TABLE 3.** Solar Cell Parameters for IEC Amorphous Silicon Solar Cells on Different TCO Films.

| TCO Type | Voc V | FF % | Jsc mA/cm$^2$ | Effic. % | Roc Wcm$^2$ |
|---|---|---|---|---|---|
| APCVD ZnO:F | 0.856 | 60.5 | 15.3 | 7.9 | 12.5 |
| Solarex SnO$_2$ | 0.875 | 67.0 | 14.5 | 8.5 | 7.0 |

Similar results are given in the table 4 for a superstrate tandem cell made at Solarex, in comparison with two of their TCO materials.

**TABLE 4.** Solar Cell Parameters for Solarex Tandem Amorphous Silicon Solar Cells on Different TCO Films.

| TCO Type | Voc V | FF % | Jsc mA/cm$^2$ | Effic. % | J1 mA | J2 mA |
|---|---|---|---|---|---|---|
| APCVD ZnO:F | 1.45 | 67.3 | 10.22 | 9.97 | 9.72 | 10.3 |
| LPCVD ZnO:B | 1.45 | 70.2 | 9.62 | 9.79 | 9.12 | 10.55 |
| SnO$_2$ | 1.47 | 70.5 | 9.82 | 10.18 | 9.32 | 9.74 |

The voltages of the cells prepared on ZnO are about 20 mV lower than those prepared on SnO$_2$. The fill factors are also significantly lower, by an average of 5%. These deleterious factors negate the advantage of the higher currents found for the ZnO. Both of these deleterious effects relate to an electrical contact between the p-type amorphous silicon and ZnO that is more resistive than that with SnO$_2$. On the other hand, many other groups have reported that amorphous zinc oxide makes low-resistance electrical contact to n-type amorphous silicon.

These observations about contact resistance can be rationalized if (n-type) tin oxide has a higher work function (lower conduction band energy) than that of (n-type) zinc oxide. Thus the conduction band of tin oxide is closer to the valence band energy of silicon, and has a lower electrical barrier in contact with p-type silicon. The conduction band energy of zinc oxide is further above the valence band energy of silicon, and forms a higher-resistance contact.

The opposite situation holds for contacts to n-type silicon. The conduction band of zinc oxide lies closer to the conduction band of n-type silicon, forming a low-barrier, low-resistance contact. The conduction band of tin oxide lies further below the conduction band of silicon, resulting in a higher barrier and larger contact resistance to n-type silicon.

It is planned to prepare transparent conductors with the structure thin SnO$_2$:F/thick ZnO:F/glass. This structure should combine the best features of the two materials for use in superstrate amorphous silicon solar cells. The thick ZnO:F should provide high transparency, and the thin SnO$_2$:F should provide good electrical contact to the silicon.

# PROPOSED USE OF TCO MATERIALS IN CRYSTALLINE OR POLYCRYSTALLINE SOLAR CELLS

Fluorine-doped zinc oxide may also have a beneficial role to play in crystalline and polycrystalline silicon solar cells, because the transparency of ZnO:F is higher than that of the doped silicon layer on the front surface of conventional cells. If the diffused n-silicon layer were made thinner and a layer of fluorine-doped zinc oxide were applied to the front surface, a more efficient cell should result. Figure 2 shows a schematic cross section of this structure. Because the ZnO:F would greatly reduce the sheet resistance of the surface, fewer metal grid lines would be needed, thereby reducing the surface recombination velocity and increasing the voltage. The current should be increased, both because of the reduced shading from fewer grid lines and the reduced absorption in the thinner diffused layer. The textured surface of the polycrystalline ZnO:F will also help provide broadband antireflection to the cell.

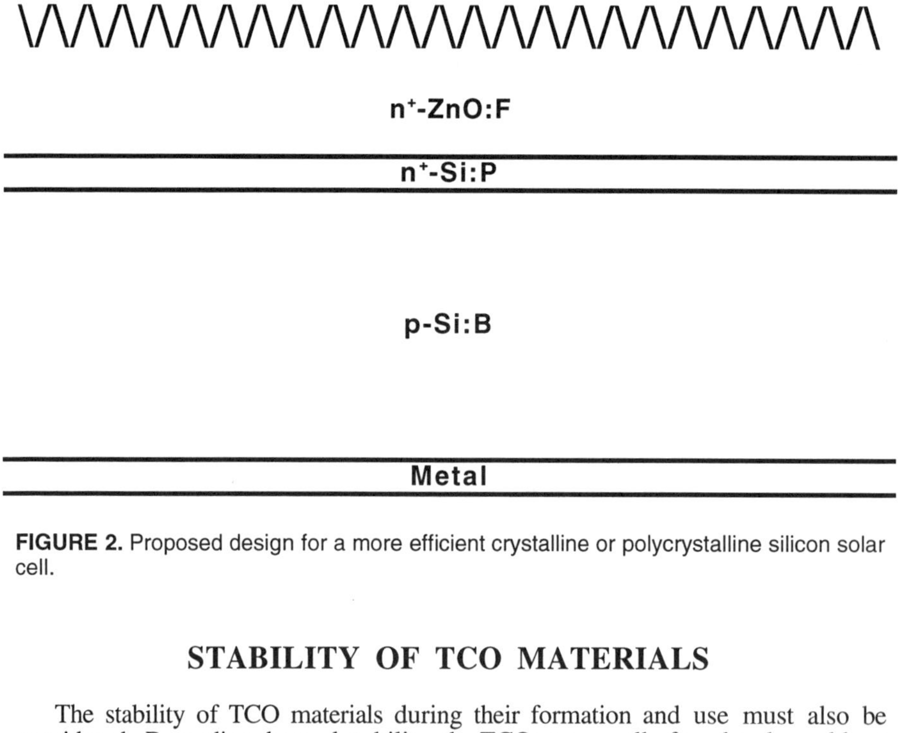

$\bigwedge\bigwedge\bigwedge\bigwedge\bigwedge\bigwedge\bigwedge\bigwedge\bigwedge\bigwedge\bigwedge\bigwedge\bigwedge\bigwedge\bigwedge\bigwedge\bigwedge\bigwedge\bigwedge\bigwedge\bigwedge\bigwedge\bigwedge$

$n^+$-ZnO:F

---

$n^+$-Si:P

---

p-Si:B

---

Metal

**FIGURE 2.** Proposed design for a more efficient crystalline or polycrystalline silicon solar cell.

# STABILITY OF TCO MATERIALS

The stability of TCO materials during their formation and use must also be considered. Regarding thermal stability, the TCOs are usually found to be stable at least up to their deposition temperature. Because tin oxide is usually deposited at temperatures around 550 °C or higher, it is stable at least up to these high temperatures. APCVD fluorine-doped zinc oxide is deposited between 400 to 500 °C, and is usually stable up to about 500 °C. LPCVD boron-doped zinc oxide increases its resistance when heated above its deposition temperature of around 200 °C.[19]

Zinc oxide is much more resistant to hydrogen plasma reduction than is tin oxide.[20]

Sodium in soda-lime substrates is another factor affecting TCO materials. Sodium can diffuse out of a soda-lime substrate into the TCO and increase its resistance. This effect can be important for tin oxide, because the sodium is particularly mobile at the high substrate temperatures used in the CVD of tin oxide. In order to retard the diffusion of sodium out of the glass, a diffusion barrier of silica is usually deposited on the glass before the CVD of the tin oxide. Even though the silica does not entirely prevent the diffusion of sodium, it does reduce the amount of sodium reaching the tin oxide.[21] The silica layer usually serves a second purpose, that of eliminating the interference colors that would otherwise be shown by the tin oxide film.[22]

Because the CVD zinc oxide processes operate at lower glass temperature, under 500 °C, sodium diffusion from the glass is negligible, and the extra step of applying a sodium diffusion barrier is not needed. The low glass temperature also helps to avoid warping the glass substrates during the CVD process.

## CONCLUSIONS

Because of the many requirements for a TCO, many factors must be taken into account in the choice of the most appropriate material. Table 5 summarizes the relative properties of the most widely used TCO materials.

**TABLE 5.** Criteria for the Choice of a Transparent Conductor.
\*\*\*=excellent, \*\*=good, \*=fair.

| MATERIAL | TRANS-PARENCY | STABILITY TO H PLASMA | THERMAL STABILITY | CONTACT RESIST-ANCE | COST | ABUN-DANCE |
|---|---|---|---|---|---|---|
| ZnO:F | *** | *** | ** | ***(n-Si) | *** | *** |
| SnO$_2$:F | * | * | *** | ***(p-Si) | *** | ** |
| In$_2$O$_3$:Sn | ** | ** | * | ** | * | * |

It is clear that no one material is best in all respects. For the highest currents in solar cells, the high transparency of zinc oxide is the most important factor. Low contact resistance is also essential for high voltages and fill factors, so the optimum choice for a transparent electrode over p-type amorphous silicon may require a composite of a thick zinc oxide layer and a thin tin oxide layer in contact with the p-type silicon.

Applying a layer of conductive ZnO:F to the front surface of crystalline or polycrystalline silicon solar cells should allow thinner diffused layers and fewer metal grid lines, increasing both the current and the voltage of the cells.

For large-scale use in photovoltaics, abundance in the Earth's crust will eventually favor the use of zinc, over less abundant tin, which is mined in only a few areas, and indium, a rare metal in limited supply.

# ACKNOWLEDGMENTS

This work was supported in part by the National Renewable Energy Laboratory, under a subcontract from the United States Department of Energy. Steven Hegedus at the Institute for Energy Conversion, University of Delaware, and Liyou Yang at Solarex prepared amorphous silicon solar cells as part of an NREL-sponsored team collaboration.

# REFERENCES

1. McMaster, H. A., U.S. Patent No. 2,429,420 (21 October 1947).
2. Mochel, J. M., U.S. Patent No. 2,564,706 (7 April 1947).
3. Lytle, W. O., and Junge, A. E., U.S. Patent 2,566,346 (4 September 1951).
4. Mochel, J. M., U.S. Patent 2,564,707 (21 August 1951).
5. Holland, L. and Siddall, G., Vacuum III (1953).
6. Gordon, R. G., U.S. Patent No. 4,146,657 (27 March 1979).
7. McCurdy, R. J., to be published in the Proceeding of the Materials Research Society Spring Meeting (1996).
8. Major, S., Banerjee, A., and Chopra, K. L., Thin Solid Films 122, 31 (1984).
9. Minami, T., Nanto, H., and Takata, S., Jpn. J. Appl. Phys. 23, L280 (1984); Minami, T., Sato, H., Nanto, H., and Takata, S., Jpn. J. Appl. Phys. 24, L781 (1985).
10. Qiu, S. N., Qiu, C. X., and Shih, I., Solar Energy Mater. 15, 261 (1987).
11. Choi, B. H., Im, H. B., Song, J. S., and Yoon, K. H., Thin Solid Films 193, 712 (1990).
12. Vijayakumar, P. S., Blaker, K. A., Weiting, R. D., Wong, B., Halani, A. T., and Park, C., U.S. Patent No. 4,751,149 (1988).
13. Hu, J. and Gordon, R. G., J. Appl. Phys. 71, 880 (1992).
14. Hu, J. and Gordon, R. G., J. Appl. Phys. 72, 5381 (1992).
15. Hu, J. and Gordon, R. G., "Electrical and optical properties of indium doped zinc oxide films prepared by atmospheric pressure chemical vapor deposition," in *Mat. Res. Soc. Proc.* 283, 1993, pp. 891-896.
16. Hu, J. and Gordon, R. G., Solar Cells 30, 437 (1991).
17. Hu, J. and Gordon, R. G., "Deposition of highly transparent and conductive fluorine doped zinc oxide films," in *Mat. Res. Soc. Symp. Proc.* 202, 1991, pp. 457-462.
18. Gordon, R. G.,"Preparation and properties of transparent conductors," in *Mat. Res. Soc. Symp. Proc.* 1996 (in press).
19 Hegedus, S., Liang, H., Gordon, R. G., "Tranparent Conducting Oxides for Amorphous Silicon Soclar Cells," presented at the 13th NREL Program Review, May 17-19, 1995; American Institute of Physics Conference Proceedings, vol. 355, p. 465.
20. Wanka, H., Lotter, E., and Shubert, M., in *Amorphous Silicon Technology-1994*, edited by Schiff, E. A., Hack, M., Madan, A., Powell, M., and Matsuda, A., (Mat. Res. Soc. Proc. 336, Pittsburgh, PA, 1994), pp. 657-662.
21. Chapple-Sokol, J., and Gordon, R. G., to be published.
22. Gordon, R. G., U.S. Patent 4,187,336 (5 February 1980); U.S. Patent 4,419,386 (6 December 1983).

# Progress in Amorphous Silicon Multijunction Research at Solarex

R. R. Arya, D. E Carlson, L. Yang, L. F. Chen, F. Willing, K. Rajan, K. Jansen, C. Poplawski, D. Bradley, and G. Wood

*Solarex, a Business Unit of Amoco/Enron Solar*
*826 Newtown-Yardley Road*
*Newtown, Pennsylvania 18940, USA*
*Phone: (215) 860-0902*

**Abstract.** Large strides have been made at Solarex in advancing amorphous silicon multijunction technology to a maturity level where large-area commercial modules are technically and economically viable. Tandem junction modules (4 $ft^2$) have been demonstrated with average stabilized efficiency of 8%. While maintaining stabilized efficiency the a-Si alloy deposition time has been reduced by 28% and the material usage reduced by 38%. Progress has also been made in understanding and improving ZnO front contact and stability.

## INTRODUCTION

Amorphous silicon thin-film technology for photovoltaic applications was started at RCA Laboratories in 1974 [1]. Over the past 20 years this technology has seen rapid progress in the development of new alloys, multijunction solar cells and modules with higher stabilized conversion efficiencies. The technology has found widespread use in consumer applications and is increasingly finding use in terrestrial applications. Large strides have been made at Solarex (a business unit of Amoco/Enron Solar) in advancing the technology over the past 10 years. This has resulted in improvements in a-Si:H based alloys, both intrinsic and doped, and with the use of the multijunction approach has led to the demonstration of stabilized conversion efficiencies in the 8%-9% range for 1 $ft^2$ and 4 $ft^2$ modules [2]. Solarex has advanced the amorphous silicon based multijunction technology to a maturity level where large-area commercial modules are technically and economically viable. Consequently, Solarex is building a manufacturing plant in James City County, Virginia, to produce multijunction large-area modules with a capacity of 10 MW/year. The construction of the plant is nearing completion, and commercial production is slated to start in January 1997.

# RESEARCH PROGRESS

In 1994 Solarex decided to rapidly commercialize the multijunction technology, and since then the commercialization effort has been the guiding principle for research activities at Solarex. The research activities can be broadly classified into two categories, (1) research related to the development of a cost-effective, manufacturable product from the first thin-film plant and (2) research related to product improvement for future thin film plants.

The important research issues addressed for developing a manufacturable tandem junction module were: (a) basic tandem junction module (b) amorphous silicon process robustness, (c) amorphous silicon deposition throughput and material utilization, and (d) stabilized performance.

## Tandem Junction Module

Tandem junction module development involves the scale-up of various thin-film processes, such as the textured tin oxide front contact, the amorphous silicon based alloys and the zinc oxide / aluminum rear contact, first to 1 ft$^2$ and then to 4 ft$^2$, and the optimization of the laser scribing and the segment width to minimize sheet-resistance and interconnection losses.

The basic tandem junction device structure employed is:

glass/TCO/p-i-n/p-i-n/ZnO/Al

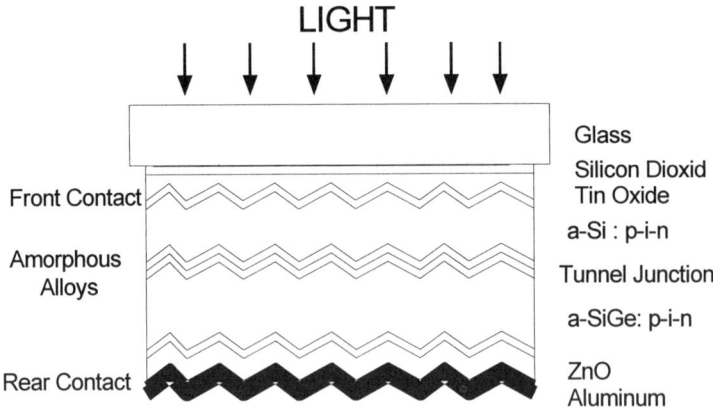

**FIGURE 1.** Schematic of a-Si:H/a-SiGe:H tandem junction cell structure

which is schematically shown in Figure 1. In this device structure the front junction employs an intrinsic a-Si:H absorber layer with a bandgap of ~1.78 eV and the back junction employs an a-SiGe:H alloy with a bandgap ranging from 1.45 eV to 1.55 eV. The front contact is ~600 nm of textured tin oxide, a transparent conductive oxide (TCO) deposited by atmospheric pressure chemical vapor deposition (APCVD). The amorphous silicon based layers are deposited by plasma-enhanced chemical vapor deposition (PECVD). The first layer consists of ~10 nm of p-type a-SiC:H alloy deposited on to the CTO coated glass. The front intrinsic a-Si:H layer is deposited from a mixture of silane and hydrogen. The intrinsic layer is followed by ~10 nm of phosphorus doped microcrystalline silicon. The second a-SiC:H p-layer forms the tunnel junction and part of the second junction which is followed by a bandgap graded a-SiGe:H alloy layer deposited from a mixture of silane, germane and hydrogen. This intrinsic layer is followed by ~20 nm of an amorphous silicon n-layer which completes the p-i-n/p-i-n structure. The rear contact comprises of ~100 nm of ZnO deposited by low-pressure chemical vapor deposition (LPCVD) followed by ~300 nm of aluminum deposited by magnetron sputtering [4].

This device structure resulted in an average initial efficiency of ~9.6% with a standard deviation of ± 0.39% on 1 $ft^2$ modules. The basic module processes were then scaled-up to a 4 $ft^2$ substrate size (16" X 36"), and the pilot production of 4 $ft^2$ tandem junction (a-Si/a-SiGe) modules was started..

## *Baseline 1 Process*

The "baseline 1" Si process initially developed for the 4 $ft^2$ pilot line had a relatively low bandgap (~ 1.45 eV), relatively thick a-SiGe:H i-layer in the back junction which generated a current density of about 10 $mA/cm^2$. The average initial efficiency of this process (at 100% yield) was ~ 9.5% and at 70% yield was 9.7%. The average light-induced degradation of a large number of modules after 600 hours of continuous illumination under one-sun has been established to be ~17% which was confirmed by an independent measurement at NREL. Thus, the process led to modules with average stabilized efficiencies over 8%. Figure 2 shows the distribution of initial efficiencies of a batch of 40 modules produced with the baseline 1 process. We have established a systematic difference in indoor measurements made at Solarex and at NREL. On the average, the conversion efficiency measured at NREL is about 6% higher than that measured at Solarex. Table 1 shows the difference in measurements made at NREL and those made at Solarex on the same 4 $ft^2$ tandem junction modules from the baseline 1 process.

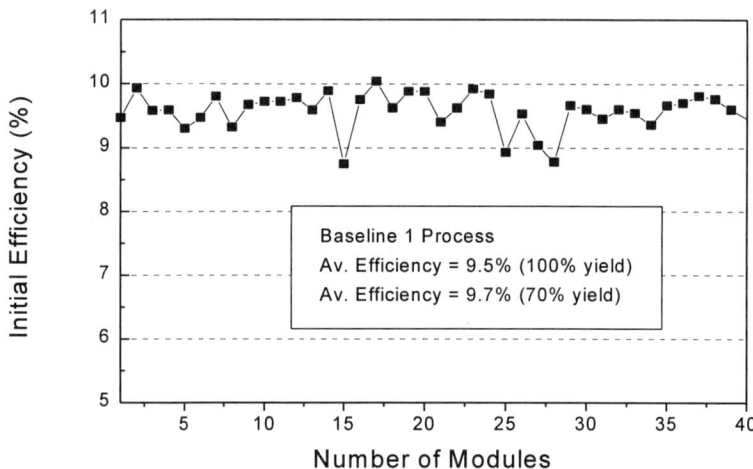

**FIGURE 2.** Initial conversion efficiency of baseline 1 process modules (4 ft$^2$).

**TABLE 1.** Indoor measurements made on 4 ft$^2$ modules at Solarex and at NREL

| Module # | Voc (V) | Isc (mA) | FF | Efficiency Aperture Area (3106 cm$^2$) (%) | Efficiency Total Area (3665 cm$^2$) (%) | P$_{max}$ (W) | Meas. At | Delta in Effi. |
|---|---|---|---|---|---|---|---|---|
| TT960 | 60.83 | 754.5 | 66.8 | **9.88** | 8.37 | 30.67 | NREL | |
| | 60.54 | 735.7 | 64.0 | 9.23 | **7.82** | 28.67 | SLX | 6.97% |
| TT961 | 60.54 | 753.3 | 66.8 | **9.81** | 8.32 | 30.48 | NREL | |
| | 60.26 | 732.7 | 65.0 | 9.23 | **7.82** | 28.66 | SLX | 6.35% |
| TT962 | 60.92 | 758.3 | 67.1 | **9.98** | 8.46 | 30.99 | NREL | |
| | 60.72 | 738.6 | 66.0 | 9.46 | **8.01** | 29.39 | SLX | 5.44% |
| TT963 | 60.53 | 761.4 | 67.2 | **9.98** | 8.46 | 30.99 | NREL | |
| | 60.41 | 741.9 | 65.0 | 9.37 | **7.94** | 29.1 | SLX | 6.49% |
| TT964 | 60.51 | 752.4 | 66.4 | **9.72** | 8.24 | 30.21 | NREL | |
| | 60.24 | 734.7 | 64.0 | 9.1 | **7.71** | 28.27 | SLX | 6.86% |
| TT965 | 60.88 | 758.6 | 67.6 | **10.05** | 8.52 | 31.23 | NREL | |
| | 60.60 | 742.1 | 66.0 | 9.55 | **8.09** | 29.67 | SLX | 5.25% |
| | | | | *bold calculated | | | | Avg. 6.22% |

# Amorphous Silicon Process Robustness

In the past two years we have produced more than 1000 tandem junction modules. This has allowed us to study the various aspects of different processes and develop processes that operate effectively with a wide window of tolerance.

## *Front Contact*

In the device structure shown in Figure 1, the first p-layer is deposited on textured tin oxide. The CTO/p interface is critical to the device performance and depends strongly on the electrical properties and the morphology of the tin oxide. In a manufacturing line it is important to have a robust p-layer which is insensitive to changes in the tin oxide surface. We have studied the influence of two types of tin oxide coated glass: (1) textured tin oxide deposited on low-iron glass at Solarex and (2) commercially available textured tin oxide deposited on sodalime glass.

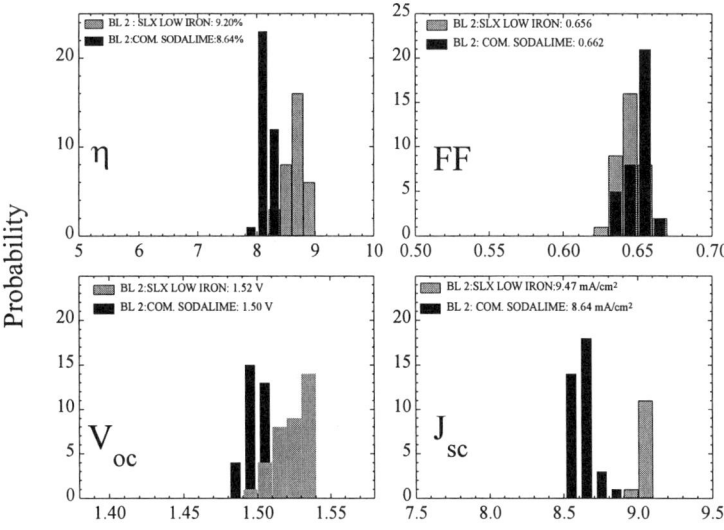

**FIGURE 3.** Amorphous silicon process baseline 2 on Solarex coated tin oxide on low iron glass and commercial tin oxide coated sodalime glass.

We have optimized the deposition of the first p-layer in such a way that essentially the same open-circuit voltage ($V_{oc}$) and fill factor (FF) are obtained on both types of CTO for tandem junction modules. Figure 3 shows the photovoltaic parameters of two batches (36 modules each), one batch deposited on Solarex CTO on low-iron glass and the other deposited on commercial CTO on sodalime glass. The $V_{oc}$ and the FF are essentially the same while the loss in $J_{sc}$ and consequently the loss in initial efficiency is entirely due to the optical properties of the glass (low-iron versus sodalime) and the optical properties of the tin oxide coatings (Solarex versus commercial).

## Amorphous Silicon Deposition Throughput and Material Utilization

The 10 MW/Year capacity of the first thin film plant is limited by the throughput of the PECVD amorphous silicon deposition machine. The other deposition processes, LPCVD, APCVD and sputtering as well as the laser scribing can accommodate higher throughputs than the amorphous silicon machine. Hence, reducing the deposition time for the tandem device structure is of paramount importance. The fraction of time consumed in the deposition of different parts of the tandem device structure are shown in Figure 4. The major thrust of this work has been to maintain the stabilized performance of the basic tandem junction while reducing the total deposition time.

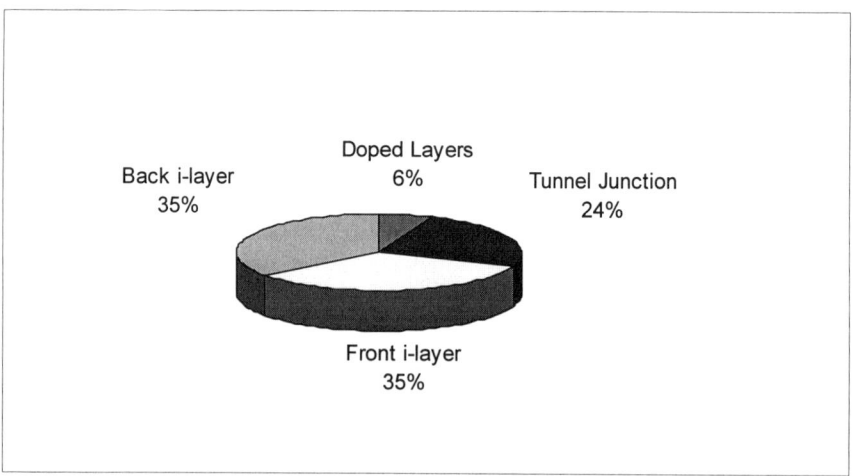

**FIGURE 4.** Fractional share of total deposition time for a-Si/a-SiGe tandem devices.

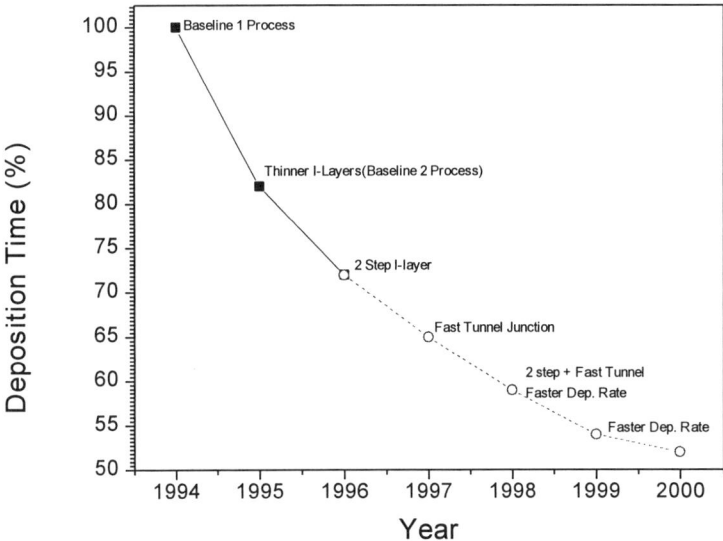

**FIGURE 5.** Chronological progress in the reduction of the total deposition time.

Since a large fraction of the time is spent on the deposition of the front and the back i-layers and the tunnel junction, we have concentrated our efforts in reducing these times. The chronological developments in the reduction of the total deposition time with a variety of process changes is shown in Figure 5.

While the baseline 1 process was a good starting foundation to gain manufacturing experience we have been seeking improvements in the manufacturing throughput and material costs. In 1995 we concentrated on reducing the deposition time by developing a tandem device with thinner i-layers. This decreased the deposition time by ~18% over that of the "baseline 1" process without a significant penalty in the stabilized efficiency. We optimized the tandem device by employing thinner i-layers and increasing the bandgap of the back junction by reducing the germanium content [4]. This "baseline 2" process has resulted in a reduction in the current generated by the tandem structure but has increased the fill-factor and the open-circuit voltage. Moreover, the stability of thin tandems have improved - the light-induced degradation has been reduced from ~17% to 14%-15%. The average stabilized efficiency (at 70% yield) is 7.8% which is close to that obtained with the baseline 1 process, but with a significantly improved cost-effectiveness. Figure 5 shows a comparison of the photovoltaic parameters of the thick (basic tandem junction) and the thin tandem junction.

Table 2 shows the relative reduction in the material usage and deposition time as well as the average performance of 4 ft$^2$ tandem modules using these two a-Si deposition processes.

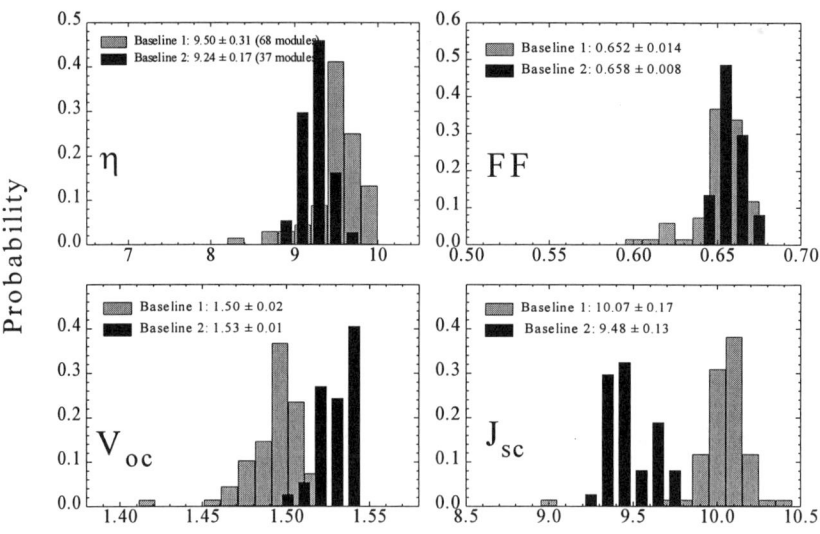

**FIGURE 6.** Comparison of the photovoltaic parameters of the baseline processes.

## *2-Step i-layer*

In the baseline 2 process the deposition of the entire intrinsic layer is carried out using silane heavily diluted in hydrogen. We have developed a process in which the high hydrogen dilution is limited to a narrow region near the p/i interface which preserves the advantage in device stability obtained with a high H-dilution throughout the i-layer but significantly decreases the total deposition time. This has allowed us to reduce the deposition time from baseline 2 process by ~ 12% without any penalty in the stabilized efficiency. This process has been successfully scaled-up to 4 ft$^2$ tandem modules. The results of 2 step i-layer are summarized in Table 2.

**TABLE 2.** Stabilized performance and throughput/ material utilization

| Process | Relative Material Usage (%) | Relative Deposition Time (%) | Avg. Stabilized Aperture Efficiency (%) | | Avg. Degradation (%) |
|---|---|---|---|---|---|
| | | | Low Iron | Commercial TCO | |
| Baseline 1 | 100 | 100 | 8.07 | - | ~17 |
| Baseline 2 | 62 | 82 | 7.80 | 7.3 | ~15 |
| 2-Step | 62 | 72 | -- | 7.5 | ~15 |

## Stabilized Performance

### ZnO Front Contact

While we have developed a baseline capability for the $SnO_2$ front contact deposition, we are also developing the next generation front contact, i.e. zinc oxide (ZnO). It has long been recognized that ZnO is potentially a superior TCO material to $SnO_2$ as a front contact. It is inherently more transparent than $SnO_2$ resulting in as much as a 10% increase in $J_{sc}$ for a-Si:H based multijunction solar cells. Depositing ZnO using an LPCVD process with diethylzinc, $H_2O$ and a dopant such as $B_2H_6$ at low substrate temperatures adds additional advantages over APCVD $SnO_2$ in controllability and ease of manufacturing. However, difficulties do exist in incorporating ZnO into modules: (1) there is a contact barrier at the ZnO/a-SiC:H p-layer interface which causes a lower FF and Voc if the same p-layer is used, (2) it is difficult to scribe ZnO cleanly and reliably by laser, (3) higher module shunt currents occur with ZnO which is probably due to surface debris generated from the LPCVD deposition process.

To address these issues, we have optimized ZnO and the p-layer deposition conditions to maximize the gain in $J_{sc}$ to 6-10% and to minimize the losses in $V_{oc}$ and FF due to a contact barrier to <2%. Figure 7 shows the current-voltage characteristics and the spectral response of a typical a-Si:H/a-SiGe:H tandem solar cell made on a textured ZnO substrate.

The conventional laser scribing process which produces clean $SnO_2$ scribes does not do an adequate job with ZnO front contacts due to its different optical, thermal and mechanical properties. The conventional laser scribing process leaves behind a severe molten edge along the scribe line. This molten edge has been identified as a significant source of shunts when incorporated into modules. To solve this problem, we have systematically investigated the laser scribing process for the ZnO front contact. The study revealed that the cleanliness of the scribe depends strongly on parameters such as laser intensity, beam shape, film/substrate position relative to the beam focal point, and adhesion of the ZnO film to the glass substrate. Module analysis has shown that shunt current due to the ZnO scribe is eliminated by a modified laser scribing process. Currently, we are working on minimizing the module shunt current resulting from surface debris by modifying the ZnO deposition process and enhancing the effectiveness of the substrate cleaning procedure.

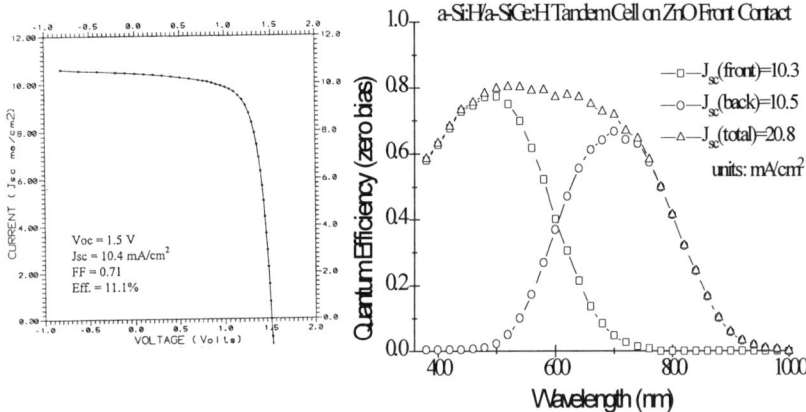

**FIGURE 7.** Illuminated current-voltage characteristics and spectral response of a typical tandem junction device made on a ZnO front contact.

## *Interface Treatment*

Experimental studies as well as device modeling have indicated that a high density of interface defects are present at the p/i interface which could significantly reduce the open-circuit voltage of a p-i-n solar cell. A bandgap graded buffer layer at the p/i interface can partially deflect the electron back diffusion current away from the interface and therefore increase $V_{oc}$. Another approach to solving this problem is to minimize the density of interface defects itself. We have developed a novel in-situ method which we believe passifies the interface defects as they are being formed. A significant increase in the open-circuit voltage was observed as a result of this interface treatment. Table 3 compares the small-area device parameters of tandem junction cells of which one has the baseline 2 type device structure and the other has an additional interface treatment step at the front junction p/i interface.

**TABLE 3.** Effect of interface treatment on tandem cells

| Interface Treatment | Voc (V) | Jsc (mA/cm$^2$) | FF | Effiency (%) |
|---|---|---|---|---|
| Yes | 1.63 | 9.5 | 0.72 | 11.2 |
| No | 1.58 | 9.4 | 0.73 | 10.8 |

## *Composite Rear Contact*

Our baseline a-Si:H/a-SiGe:H tandem structures employ ZnO/Al as the rear contact. Despite its poorer optical reflectivity, ZnO/Al rear contact has demonstrated much better robustness in terms of manufacturing yields as compared to a Zn/Ag rear contact.

In order to take advantage of the higher reflectivity of ZnO/Ag rear contact and hence gain $J_{sc}$ and efficiency without jeopardizing module yield due to shunting, we have explored ways to combine ZnO/Ag with other metal and/or metal-oxide layers to make composite back contacts, ZnO/Ag/X, where X is some "protective" layer or layers. In the ZnO/Ag/X configuration, a relatively thin layer of Ag (< 500 Å) is sandwiched between ZnO and X. The idea is to make the ZnO/Ag behave like an extension of the a-Si:H *n*-layer as far as laser scribing and electrical curing are concerned. The "protective" layer X must satisfy several conditions, such as being electrically conductive, shunt-resistant, and compatible with Ag in terms of adhesion and absence of inter-diffusion. Ideally, in the ZnO/Ag/X contact, the ZnO/Ag should dominate the optical but not the electrical behavior of the contact so as to minimize Ag-related shunting. The X layer should facilitate electrical contact and help minimize shunts, both intrinsic and scribing-related.

By optimizing the structure of this composite rear contact, we have obtained as much as 8% improvement in short circuit current for 1 ft$^2$ modules as compared to the standard baseline 2 tandem modules with ZnO/Al rear contact. The yield for this composite rear contact was also found to be as good as that for the standard ZnO/Al rear contact. Table 4 shows the comparison of typical 1 ft$^2$ modules made with the standard ZnO/Al rear contact and ZnO/Ag/X composite rear contact.

**TABLE 4.** Comparison of the photovoltaic parameters for a composite rear contact and the ZnO/Al rear contact for 1 ft$^2$ modules

| Rear contact | Voc | Jsc | FF | Effiency |
|---|---|---|---|---|
| Composite | 1.5 | 9.32 | 0.63 | 8.53 |
| ZnO/Al | 1.5 | 8.76 | 0.64 | 8.21 |

## FUNDAMENTAL STUDIES

While the current tandem devices only degrade about 15% upon prolonged exposure to 1 sun illumination, one could make significant improvements in the device performance by developing a stable material that would allow the fabrication of thicker devices. As part of our exploratory research to understand and minimize the light-induced degradation, we have made a surprising discovery - the light-induced degradation in a-Si:H solar cells can be reversed by the application of a strong reverse bias while the cells are exposed to intense illumination. The rate of this reversal of the degradation is strongly dependent on the strength of the electric field, the temperature and the light intensity.

Figure 9 shows some data for a single-junction a-Si:H p-i-n solar cell where the cell was subjected to 60 suns illumination for 20 minutes at 90$^0$C while at open circuit. While illuminated, the cell was then subjected to a reverse bias of -6 V, and as shown in the figure, the performance of the cell almost completely recovered in about 20 minutes. The recovery process is strongly field dependent and also exhibits an activation energy on the order of 1 eV. One possible interpretation is that the strong electric field induces the motion of hydrogen ions within microvoids thus allowing the metastable defects to return back to the ground state [5].

## SUMMARY

Solarex has focused on developing tandem junction modules that can be manufactured with high yields, good material utilization and high throughputs. As a result of an extensive research and development program, a robust manufacturing process has been developed that produces large-area tandem junction modules with stabilized conversion efficiencies of about 8%. In the beginning of 1997, Solarex will start producing 8 ft$^2$ tandem modules in James City County, Virginia for use in both remote and grid-tied applications.

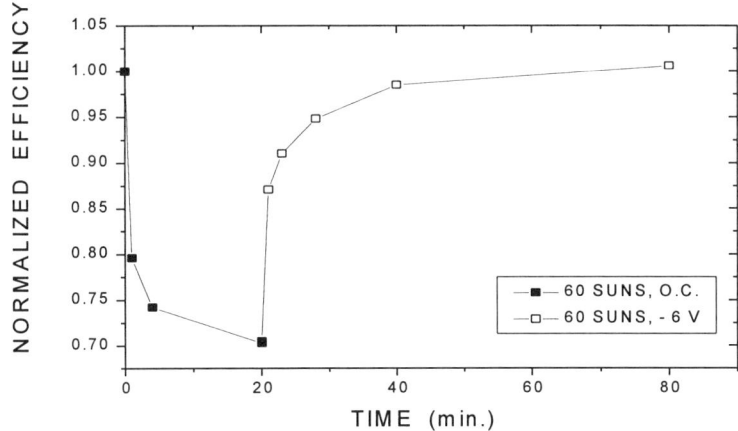

**FIGURE 9.** The normalized conversion efficiency as a function of time for an a-Si:H p-i-n cell that was exposed to 60 suns illumination at 90$^0$C under various bias conditions.

## ACKNOWLEDGMENTS

This work was partially supported by NREL under the subcontract No. Zan-4-11318-01 and No. ZM-2-11040-2.

## REFERENCES

1. D. E. Carlson, U. S. Patent No. 4,064,521(1977).
2. R. R. Arya, R. S. Oswald, Y. M. Li, N. Maley, K. Jansen, L. Yang, L.F. Chen, F. Willing, M. S. Bennett, J. Morris and D. E. Carlson, Proc. 24th IEEE Photovoltaic Specialists Conference (1994) p. 380.
3. D. E. Carlson, R. R. Arya, M. Bennett, L. F Chen, K. Jansen, Y.M. Li, J. Newton, K. Rajan, R. Romero, D. Talenti, E. Twesme, F. Willing and L. Yang, Proc. 25th IEEE Photovoltaic Specialists Conference (1996).
4. L. Yang, M. Bennett, L. Chen, K. Jansen, J. Kessler, Y.M. Li, J. Newton, K. Rajan, F. Willing, R. Arya, and D. Carlson., MRS Spring Meeting (1996).
5. D. E. Carlson and K. Rajan, to be published.

# Why Is CuInSe$_2$ Tolerant to Defects and What Is the Origin of "Ordered Defect Structures"

Alex Zunger, S. B. Zhang, and Su-Huai Wei

*National Renewable Energy Laboratory, Golden, CO 80401, U.S.A.*

**Abstract.** This paper explains both the (1) remarkable electronic passivity of CuInSe$_2$ to its many structural defects, and (2) the occurance of previously noted but unexplained series of structures CuIn$_5$Se$_8$, CuIn$_3$Se$_5$, Cu$_2$In$_4$Se$_7$, etc. in terms of the unusual stability of the charge-compensated defect pair $(2V_{Cu}^- + In_{Cu}^{2+})$.

## INTRODUCTION

The field of condensed matter physics of perfect crystalline lattices owes its relevance to experiment to the fact that the formation of native defects usually costs significant energy. Thus, perfect crystalline lattices should exist, at least in principle. One may contemplate, however, the possibility of *spontaneous formation* of native defects in crystalline lattices. If $\Delta H_f(\alpha, q)$ is the formation energy of a point defect of type $\alpha$ (vacancy, antisite, interstitial, ...) in charge state $q$, then even if $\Delta H_f(\alpha, q) > 0$ for a *single* defect it is possible that the formation energy of a pair/complex or array of interacting defects

$$\Delta H_f(\alpha + \beta) = [\Delta H_f(\alpha, q = 0) + \Delta H_f(\beta, q = 0)] + \delta H_{int} + \delta H_{ord} \qquad (1)$$

could be very small, or even negative. This could happen if the (positive) formation energy of the non-interacting defect pairs $[\Delta H_f(\alpha) + \Delta H_f(\beta)]$ is small, but the (attractive) defect interaction energy $\delta H_{int}$ (e.g., charge transfer and Coulomb) and/or the defect ordering energy $\delta H_{ord}$ are strongly stabilizing. It was expected [1] for example, that the energy of a non-interacting neutral donor-acceptor pair in insulators could be lowered by an amount $\delta H_{int} \sim qE_g + E_{Coul}$, where $E_g$ is the band gap energy and $E_{Coul}$ is the Coulomb attraction between charged defect pair, if $q$ electrons could drop from the near-conduction-band donor level to the near-valence-band acceptor level. First-principles calculations for GaAs [2] and ZnSe [3] have shown, however, that the lowest formation energy defect pair in GaAs — $[Ga_{As}^{2-} + As_{Ga}^{2+}]$ or in ZnSe — $[V_{Zn}^{2-} + Zn_i^{2+}]$ — still cost as much as 2-3 eV. Thus, defect pairs are unlikely to form spontaneously in ordinary semiconductors.

We have identified a semiconductor system – the *ternary* chalcopyrites of the $A^I B^{III} X_2^{VI}$ type [4] (e.g., CuInSe$_2$) where the formation of ordered arrays [5] of defect pairs can be made exothermic even at low temperatures. Unlike the case of *binary* zinc-blende semiconductors or their isovalent alloys, ternary chalcopyrites have two heterovalent cations so it is possible to form low energy, oppositely charged yet electrically compensated defects on the cation sublattice alone. Using local-density-approximation (LDA) [6] our first-principles total energy calculations show that (i) the formation of the *non-interacting* defect pair made of two Cu vacancies ($2V_{Cu}$) plus one In-on-Cu antisite ($In_{Cu}$) costs only 4.26 eV . (ii) The strong interaction $\delta H_{int} = -3.45$ eV between $2V_{Cu}^-$ and $In_{Cu}^{2+}$ reduces the formation energy to only 0.81 eV, significantly lower than for low-energy pairs in GaAs or ZnSe. And, (iii) repeating periodically m units of $(2V_{Cu}^- + In_{Cu}^{2+})$ for every n units of CuInSe$_2$, i.e.,

$$n(CuInSe_2) + m(In) \rightarrow Cu_{(n-3m)}In_{(n+m)}Se_{2n} + 3m(Cu) + \Delta H_f(n,m), \quad (2)$$

where m = 1, 2, 3, $\cdots$ and n = 3, 4, 5, $\cdots$, and where (In) and (Cu) denote In and Cu in their respective equilibrium chemical reservoirs, reduces the energy further by $\delta H_{ord} \sim -0.8$ eV/pair. Thus the formation energy $\Delta H_f(n,m)$ of the *"defect pair arrays"* [Eq. (2)] is close to zero and can even be made negative by a proper choice of the reservoir energies. Finally, (iv) the charge-compensated defect pair $(2V_{Cu}^- + In_{Cu}^{2+})$ is electrically inactive, i.e., it has no defect energy levels inside the fundamental band gap. Our findings can potentially explain two long-standing puzzles in the chalcopyrite material system [5,7]:

*First*, Cu$_2$Se + In$_2$Se$_3$ are known [5] to form a surprising series of compounds such as CuIn$_5$Se$_8$, CuIn$_3$Se$_5$, Cu$_2$In$_4$Se$_7$, $\cdots$ with hitherto unexplained Cu:In:Se ratios. We suggest that the extraordinarily low formation energy ($\Delta H_f = 0.81$ eV) of a single, interacting $(2V_{Cu}^- + In_{Cu}^{2+})$ pair and the significant ($\sim -0.8$ eV) pair-pair ordering energy leads to the formation of "ordered defect arrays" [Eq. (2)] such as CuIn$_5$Se$_8$ (m = 1, n = 4); CuIn$_3$Se$_5$ (m = 1, n = 5), Cu$_2$In$_4$Se$_7$ (m = 1, n = 7), etc.

*Second*, while in Si and in ordinary III-V semiconductors, polycrystallinity leads to a high concentration of electrically-active defects that have a detrimental effect on the performance of opto-electronic devices, polycrystalline CuInSe$_2$ is as good an electronic material as its single-crystal counterpart [7] (leading to a 17% efficient photovoltaic solar cell) [8], even though it has a huge amount of structural defects. This is explained by the attractive interaction between the (individually electrically active) constituents $V_{Cu}^-$ and $In_{Cu}^{2+}$ of the defect pair, leading to an effective electric annihilation of the recombination centers.

## METHODS OF CALCULATION

The formation energy $\Delta H_f(\alpha, q)$ of defect $\alpha$ in charge state $q$ depends on the Fermi energy $\epsilon_F^a$ (where $a$ denotes absolute values) as well as on the atomic chemical potentials $\mu^a$. In CuInSe$_2$,

$$\Delta H_f(\alpha, q) = E(\alpha, q) - E(CuInSe_2) + n_{Cu}\mu_{Cu}^a + n_{In}\mu_{In}^a + n_{Se}\mu_{Se}^a + q\epsilon_F^a, \quad (3)$$

where $E(\alpha, q)$ is the total energy of a supercell containing a defect of type $\alpha$ and charge $q$, $E(CuInSe_2)$ is the total energy for the same supercell in the absence of the defect, the n's are the numbers of Cu, In, Se atoms and q is the the number of electrons, transferred from the supercell to the reservoir in forming the defect cell. We will not consider Se-related defects in this study so we take $n_{Se} = 0$. Denoting

$$\Delta E(\alpha, q) = E(\alpha, q) - E(CuInSe_2) + n_{Cu}\,\mu_{Cu}^{solid} + n_{In}\,\mu_{In}^{solid} + q\,E_V, \quad (4)$$

Eq. (3) can be rewritten as

$$\Delta H_f(\alpha, q) = \Delta E(\alpha, q) + n_{Cu}\,\mu_{Cu} + n_{In}\,\mu_{In} + q\,\epsilon_F, \qquad (5)$$

where $\epsilon_F = \epsilon_F^a - E_V$, $\mu_{Cu} = \mu_{Cu}^a - \mu_{Cu}^{solid}$ and $\mu_{In} = \mu_{In}^a - \mu_{In}^{solid}$. Note that here the energy of the valence band maximum (VBM) $E_V$ is defined consistently as $E_V = E_0^N - E_+^{N+1} + \epsilon_{VBM}$ where $E_0^N = E(\text{CuInSe}_2)$ is the total energy of CuInSe$_2$ in its ground state, and $E_+^{N-1}$ is the total energy of CuInSe$_2$ with one hole in the VBM and an electron in the reservoir with an energy equal to the VBM eigenvalue ($\epsilon_{VBM}$). For ordinary semiconductors such as GaAs and ZnSe, we find that the correction $\delta = E_V - \epsilon_{VBM}$ is small (for example, $\delta = 0.03$ eV for ZnSe when the reservoir is represent by the "jellium" model). For semiconductors with localized $d$ character at the VBM, $\delta$ can be large, e.g., in CuInSe$_2$ $\delta = 0.24$ eV. This correction is applied to all calculated defect levels.

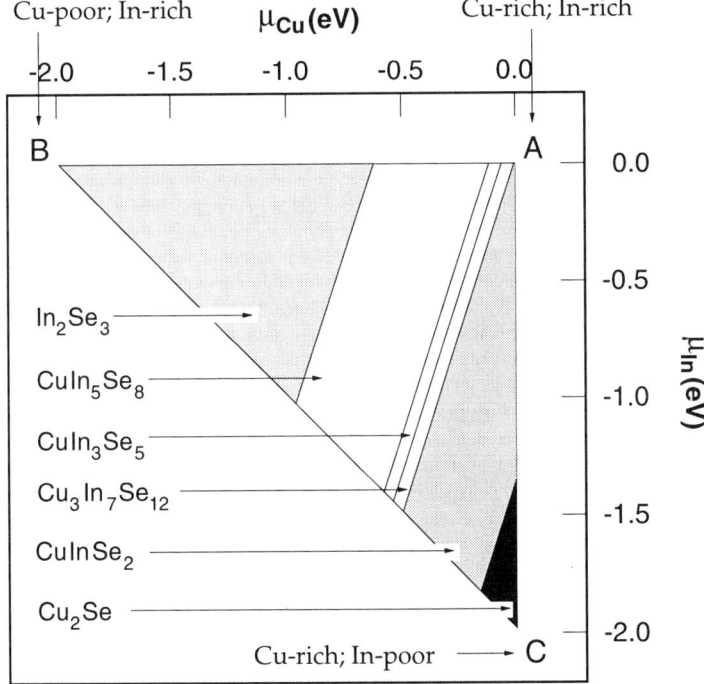

**Figure 1.** The calculated stability triangle of the Cu-In-Se systems [Eqs. (6) and (7)] in the ($\mu_{Cu}, \mu_{In}$)-plane. The vertices of the triangle are labeled as A, B and C corresponding to (A) Cu-rich, In-rich; (B) Cu-poor, In-rich and (C) Cu-rich, In-poor conditions, respectively.

There are some thermodynamic limits to $(\mu, \epsilon_F)$: $\epsilon_F$ is bound between the VBM and the conduction band minimum (CBM), and $\{\mu_{Cu}, \mu_{In}\}$ are bound by (i) the values that will cause precipitation of solid elemental Cu, In, and Se, so

$$\mu_{Cu} \leq 0 \,; \quad \mu_{In} \leq 0 \,; \quad \mu_{Se} \leq 0 \,, \tag{6}$$

(ii) by the values that maintain a stable CuInSe$_2$ compound, so

$$\mu_{Cu} + \mu_{In} + 2\mu_{Se} = \Delta H_f(CuInSe_2), \tag{7}$$

where $\Delta H_f(CuInSe_2) = -1.97$ eV is the calculated formation energy of solid CuInSe$_2$ from the elemental solids, and (iii) by the values that will cause formation of binaries (mainly $In_2Se_3$ and $Cu_2Se$), so

$$2\mu_{In} + 3\mu_{Se} \leq \Delta H_f(In_2Se_3)$$
$$2\mu_{Cu} + \mu_{Se} \leq \Delta H_f(Cu_2Se), \tag{8}$$

where $\Delta H_f(tetragonal, In_2Se_3) = -2.07$ eV [9] and $\Delta H_f(Cu_2Se) = -0.31$ eV, respectively is our calculated values. Figure 1 gives the calculated "stability triangle" in the two dimensional $(\mu_{Cu}, \mu_{In})$-plane as defined by Eqs. (6) and (7). The vertices are A (the Cu-rich and In-rich limit), B (the Cu-poor and In-rich limit), and C (the Cu-rich and In-poor limit). Vertex A corresponds to the special case where both the Cu and In chemical potentials are set at the values of the corresponding solid reservoirs. Equation (8) defines the regions where $In_2Se_3$ and $Cu_2Se$ are stable. The stability regions for ordered defect compounds shown in Fig. 1 will be discussed later.

We calculated $\Delta H_f(\alpha, q)$ for $\alpha = V_{Cu}, V_{In}, In_{Cu}, Cu_{In}$ and interstitial Cu ($Cu_i$). We place defect $\alpha$ at the center of a 32-atom tetragonal supercell with lattice vectors $(1,1,0)a$, $(-1,1,0)a$, and $(0,0,2\eta)a$, where $a = 5.784$ Å is the lattice parameter and $\eta = c/a = 1.004$ is the tetragonal ratio of CuInSe$_2$ [4]. The total energies are calculated using the local density functional formalism as implemented by the general potential linearized augmented plane wave (LAPW) method [10].

Using our approach the calculated formation energy of CuInSe$_2$ $\Delta H_f(CuInSe_2) = -1.97$ eV is close to the experimental value of -2.08 eV [11], as are the calculated formation energies for Cu$_2$Se and In$_2$Se$_3$ $\Delta H_f(Cu_2Se) = -0.31$ and $\Delta H_f(hexagonal, In_2Se_3) = -3.17$ eV, respectively, compared with experimental values of -0.68 and -3.38 eV [12]. We estimated that the uncertainty in our calculation is ±0.2 eV per point defect.

## RESULTS

### 1. Point defects in CuInSe$_2$

Figure 2 shows the defect formation energy $\Delta H_f(\alpha, q)$ for different types of single *isolated* defects, as a function of the electron Fermi energy $\epsilon_F$ at the chemical potential values A, B and C denoted in Fig. 1. The solid dots denote points where the slope of $\Delta H_f(\alpha, q)$ vs q changes; the corresponding value of $\epsilon_F$ is the defect transition energy $E_\alpha(q/q')$. Figure 2 shows that:

(i) The relative stability of various isolated single defects depends critically on the chemical potentials (regardless of the defects' charge states): $\Delta H_f(V_{Cu})$ can vary by as much as 2 eV from point A to B and $\Delta H_f(Cu_{In})$ can vary by as much as 4 eV from point B to C. This strong variation was ignored in previous calculations and discussions [13,14] of defect stability in CuInSe$_2$.

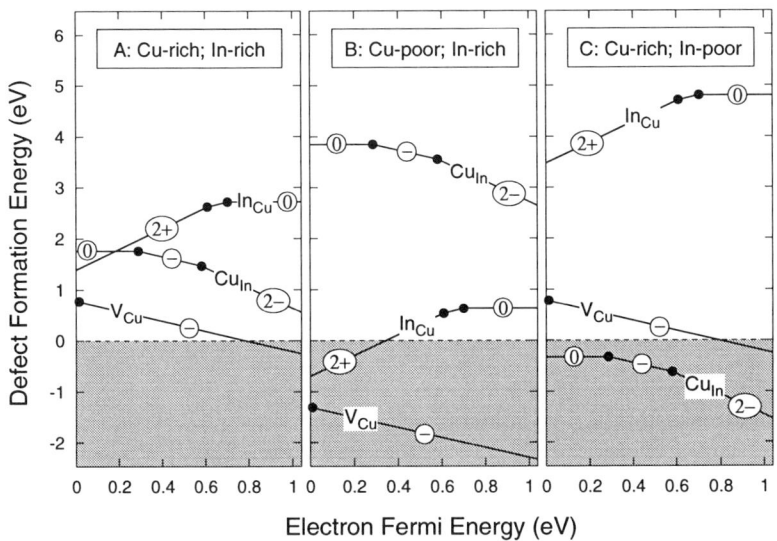

**Figure 2.** Formation energies [Eq. (5)] of $V_{Cu}$, $In_{Cu}$ and $Cu_{In}$ as a function of the electron Fermi energy, $E_F$ at the three characteristic chemical potentials A, B and C shown in Fig. 1. Charge state q which determines the slopes of each line segment is selectively shown. Solid dots denote the value of the Fermi energy where the charge state changes. The shaded area highlights negative formation energies.

(ii) Some of the formation energies of single neutral defects in CuInSe$_2$ are extraordinary low, e.g., $\Delta H_f(V_{Cu}^0) = -1.2$ eV (at B) and $\Delta H_f(Cu_{In}^0) = -0.3$ eV (at C). This reflects both the strong *anti-bonding* p-d character of the states near the VBM [15] in CuInSe$_2$, and the stability of solid Cu: $E_{cohesive}(Cu) = 3.49$ eV [16], compared with 1.35 eV for solid Zn [16], relevant to $V_{Zn}$ in ZnSe [17].

(iii) The formation energies have a significant dependence on the Fermi energy. This crucial dependence was ignored in almost all previous discussions of defects in CuInSe$_2$ [13,14]. In general, acceptor states such as $V_{Cu}^-$ form more easily in n-type material while donor states such as $In_{Cu}^{2+}$ form more easily in p-type material.

(iv) The predicted electrical defect levels are shown in Figure 3. The Cu vacancy is predicted to be a shallow acceptor $E(0/-) = E_V + 0.012$ eV, considerably shallower than the *isovalent* double-acceptor $E(-/2-) = E_V + 0.66$ eV of the Zn-vacancy in ZnSe [17]. The $In_{Cu}$ antisite is predicted to have two deep donor levels at $E(0/+) = E_C - 0.34$ eV and $E(+/2+) = E_C - 0.43$ eV, while $Cu_{In}$ have two deep acceptor levels at $E(0/-) = E_V + 0.29$ eV and $E(-/2-) = E_V + 0.58$ eV.

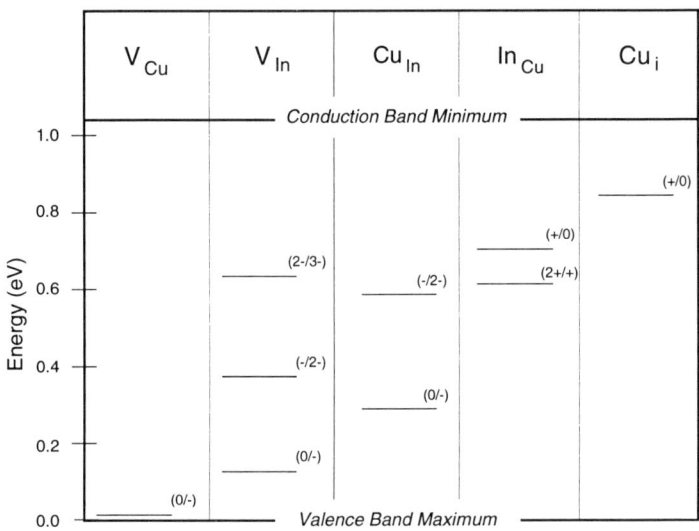

**Figure 3.** Fig. calculated defect transition electrical levels for single cation defects in CuInSe$_2$. Acceptors are (0/-), (-/2-), (2-/3-) and donors are (+/0), (2+/+).

## 2. Pairs of point defects in CuInSe$_2$.

Figure 2 further reveals the coexistence of several low energy point defects of opposite charges at the same $\epsilon_F$ and $\mu$. This allows the formation of charge-compensated defect pairs of low energies listed in Table I. Notable in Table I are the low formation energy of neutral pairs $(2V_{Cu}^0 + In_{Cu}^0)$ of -1.74 eV at B.

The formation energy of such non-interacting defect pairs can be lowered considerably through interaction and ordering:

(a) The interaction energy $\delta H_{int}$ between the components of a pair [Eq. (1)] is independent of the chemical potentials. It is calculated as the difference $\Delta H_f(\alpha+\beta) - \Delta H_f(\alpha) - \Delta H_f(\beta)$ (using 32-atom supercells) between a cell containing the pair and two cells containing one component each. The energy includes contribution from donor-acceptor compensation as well as the Coulomb attraction between charged defects. We find (Table I) for $(Cu_{In}^{2-} + 2Cu_i^+)$ that $\delta H_{int} = -2.61$ eV where the two Cu interstitials are collinear with the $Cu_{In}$ antisite. For $(2V_{Cu}^- + In_{Cu}^{2+})$, we find $\delta H_{int} = -3.45$ eV and that the two Cu vacancies are nearest fcc neighbors of the $In_{Cu}$ antisite. For $(Cu_{In}^{2-} + In_{Cu}^{2+})$, we find $\delta H_{int} = -3.67$ eV. The fact that $\delta H_{int} \ll 0$ is due largely to a strong Coulomb attraction between the charge components of the defect pairs, and in part due to the drop of electrons from higher energy donor levels to lower energy acceptor levels.

It is interesting to compare the formation energy of $(Cu_{In}^{2-} + In_{Cu}^{2+})$ in CuInSe$_2$ with that of $(Ga_{As}^{2-} + As_{Ga}^{2+})$ in GaAs: The formation energy of the interacting $(Cu_{In}^{2-} + In_{Cu}^{2+})$ pair is 0.8 eV/pair whereas that of $(Ga_{As}^{2-} + As_{Ga}^{2+})$ is much higher, being 1.8 eV [2]. It has been shown [18] that randomization of the GaAs costs only 0.72 eV/pair. For CuInSe$_2$ the randomization energy is only about 0.2 eV/pair [19]. Thus, we expect more antisite defects in CuInSe$_2$ than in GaAs and that in the former case those defects are easily randomized, forming "zinc-blende CuInSe$_2$" [19].

(b) The pair-pair ordering energy $\delta H_{ord} = -0.8$ eV (Table I) of the periodic $(2V_{Cu}^- + In_{Cu}^{2+})$ array was calculated by subtracting from the LAPW energy of the array the energy of the isolated $(2V_{Cu}^- + In_{Cu}^{2+})$ pairs [obtained in (a)]. We found that the ordering energies of the various arrangements of the defect pair arrays follow the trend set by the Madelung energy of the charged defects: the lower the Madelung energy is, the lower is the LAPW ordering energy. This observation allows us to search for the lowest-energy configuration from a large number of ordered $(2V_{Cu}^- + In_{Cu}^{2+})$ configurations using simply the point-ion model. Such a search shows that the lowest energy configuration for $CuIn_5Se_8$ (n=4, m=1) corresponds to a superlattice along the [110] direction in the Cu sublattice, i.e., $\cdots$Cu-$V_{Cu}$-$In_{Cu}$-$V_{Cu}$-Cu$\cdots$. This minimum energy structure is then examined and confirmed by our LAPW total energy calculation. We can see from Table I that the sum of the interaction and ordering energies $\delta H_{int} + \delta H_{ord}$ of Eq. (1) for the defect pair array $(In_{Cu}^{2+} + 2V_{Cu}^-)$ is about -4.25 eV, and that this cancels almost exactly the formation energy of the isolated neutral pair: $2\Delta H_f(V_{Cu}^0) + \Delta H_f(In_{Cu}^0) \simeq 4.26$ eV at A.

Table I. The calculated formation energies $\Delta H(\alpha) + \Delta H(\beta)$ of the neutral defects, the intra-pair interaction energies $\delta H_{int}$ (including charging energy and Coulomb attraction) and the pair-pair ordering energies $\delta H_{ord}$ (in eV) of a few low formation energy defect pairs at the three chemical potentials A, B and C shown in Fig. 1.

| | | $2V_{Cu}^0 + In_{Cu}^0$ | $Cu_{In}^0 + 2Cu_i^0$ | $V_{Cu}^0 + Cu_i^0$ | $In_{Cu}^0 + Cu_{In}^0$ |
|---|---|---|---|---|---|
| $\Delta H(\alpha) + \Delta H(\beta)$ | A: | 4.26 | 6.07 | 2.93 | 4.47 |
| | B: | -1.74 | 12.07 | 2.93 | 4.47 |
| | C: | 6.26 | 4.07 | 2.93 | 4.47 |
| $\delta H_{int}$ | | -3.45 | -2.61 | -1.13 | -3.67 |
| $\delta H_{ord}$ | | -0.80 | — | — | -0.8 |

## 3. Predicting "ordered defect arrays"

Table II show our calculated formation energies $\Delta H_f(n, m = 1)$ of a periodically repeated unit of $(2V_{Cu}^{-} + In_{Cu}^{2+})$ for every n units of CuInSe$_2$ for the chemical potential A, B and C, respectively, demonstrating that such defect pair arrays will form spontaneously for chemical potentials between A and B. The arrows in Fig. 1 point to the thermodynamic chemical potential domains where the "ordered defect arrays" are stable. We see that these compounds form a series connecting Cu$_2$Se at the (Cu-rich, In-poor) condition, to the In$_2$Se$_3$ at the (Cu-poor, In-rich) condition.

Table II. Calculated relative formation energies $\Delta H_f(n, m = 1)$ [Eq. (2)] (in eV) of the "ordered defect arrays" obtained by repeating one unit of $(2V_{Cu}^{-} + In_{Cu}^{2+})$ for every n units of CuInSe$_2$.

|  | n | $\mu = A$ | $\mu = B$ | $\mu = C$ |
|---|---|---|---|---|
| CuIn$_5$Se$_8$ | 4 | 0.05 | -5.95 | 2.05 |
| CuIn$_3$Se$_5$ | 5 | 0.03 | -5.97 | 2.03 |
| Cu$_3$In$_7$Se$_{12}$ | 6 | 0.02 | -5.98 | 2.02 |

Our predictions for the stable defect pair arrays can be used to understand the peculiar Cu-In-Se structures known to exist [5]. These can be divided into two classes (Table III): those that are on the Cu$_2$Se-In$_2$Se$_3$ tie-line [i.e., the compounds that can be written as (Cu$_2$Se)$_x$(In$_2$Se$_3$)$_{1-x}$ with $0 \leq x \leq 1$] and those that are not. We predict the stability of all In-rich tie-line compounds as resulting from repetition of m units of $(2V_{Cu}^{-} + In_{Cu}^{2+})$ in every n units of CuInSe$_2$. These are CuIn$_5$Se$_8$ (n=4,m=1), CuIn$_3$Se$_5$ (n=5,m=1), Cu$_2$In$_4$Se$_7$ (n=7,m=1), and Cu$_3$In$_5$Se$_9$ (n=9,m=1). There are also three observed off-the-tie-line compounds: CuIn$_7$Se$_{12}$, Cu$_4$In$_9$Se$_{16}$ and Cu$_3$In$_6$Se$_{11}$. In light of the low formation energy of neutral Cu vacancy (Fig. 2), we can rationalize the stabilities of these compounds as emerging from the creation of 2, 1 and 1 Cu vacancies per molecule in the tie-line compounds Cu$_3$In$_7$Se$_{12}$ (n=6,m=1), Cu$_5$In$_9$Se$_{16}$ (n=8,m=1) and Cu$_4$In$_6$Se$_{11}$ (n=11,m=1), respectively.

## 4. Why is CuInSe$_2$ electrically tolerant to structural defects.

To understand why "CuInSe$_2$" exhibits a surprising electric tolerance to > 1% structural point defects we have calculated electronic structure of the isolated $2V_{Cu} + In_{Cu}$ defect pair. We find that it has no defect gap levels akin to isolated defects in CuInSe$_2$. We have also calculated the electronic band structures of CuIn$_5$Se$_8$, CuIn$_3$Se$_5$ and Cu$_3$In$_7$Se$_{12}$, and find that the LDA-corrected band gaps are 1.38, 1.29 and 1.23 eV, respectively, all larger than the 1.04 eV gap of CuInSe$_2$.

Thus the attraction between the individually electrically active $V_{Cu}$ and $In_{Cu}$ defects (Fig. 3) give rises to a wider-gap, electrically inactive defect pairs, explaining the surprising electrical tolerance of non-stoichiometric CuInSe$_2$ to its structural defects [7]. The increase in the band gap is caused by a reduced Se p-Cu d interband repulsion due to the diminished d-character attendant upon forming Cu vacancies. This lowers, for example, the VBM of CuIn$_5$Se$_8$ by 0.42 eV. The calculated CBM of CuIn$_5$Se$_8$ is about 0.08 eV lower than that of CuInSe$_2$. This lowering is due to a combined effect of the Cu vacancies and $In_{Cu}$ antisites and is small because the effective electrostatic potentials of the vacancies and antisites have opposite signs, thus largely canceling each other.

Table III. Comparison between the predicted "ordered defect arrays" made by repeating one unit (m = 1) of $(2V_{Cu}^- + In_{Cu}^{2+})$ for every n units of CuInSe$_2$ [Eq. (2)] and the observed series of compounds [5].

| Predicted | | Observed [5] | |
|---|---|---|---|
| On tie-line | n | On tie-line | Off tie-line |
| $CuIn_5Se_8$ | 4 | $CuIn_5Se_8$ | — |
| $CuIn_3Se_5$ | 5 | $CuIn_3Se_5$ | — |
| $Cu_3In_7Se_{12}$ | 6 | — | $CuIn_7Se_{12}$ |
| $Cu_2In_4Se_7$ | 7 | $Cu_2In_4Se_7$ | — |
| $Cu_5In_9Se_{16}$ | 8 | — | $Cu_4In_9Se_{16}$ |
| $Cu_3In_5Se_9$ | 9 | $Cu_3In_5Se_9$ | — |
| $Cu_7In_{11}Se_{20}$ | 10 | — | — |
| $Cu_4In_6Se_{11}$ | 11 | — | $Cu_3In_6Se_{11}$ |

## SUMMARY

We showed that chalcopyrite semiconductor CuInSe$_2$ is qualitatively different from ordinary semiconductors such as GaAs and ZnSe in that charge compensated defect pairs $(2V_{Cu}^- + In_{Cu}^{2+})$ form in CuInSe$_2$ with an unusually low formation energy. This is due both to the low formation energy of Cu vacancies, resulting from the antibonding p-d character of the Cu-Se bonds, and due to the stabilizing charge transfer and associated Coulomb interactions between individual charged defects. We found that $(2V_{Cu}^- + In_{Cu}^{2+})$ is electrically inactive, thus explaining the surprising tolerance of CuInSe$_2$ to large concentration of defects. Attractive ordering energies were predicted for ordered $(2V_{Cu}^- + In_{Cu}^{2+})$ arrays, leading to net negative formation energies at certain chemical environments. This explains the occurrence of a series of hitherto surprising structures CuIn$_5$Se$_8$, CuIn$_3$Se$_5$, etc. For more references on this and related subjects, see the home page of the solid state theory group at **http://www.sst.nrel.gov**

## ACKNOWLEDGMENT

This work was supported by the U. S. Department of Energy, DOE-EE under contract No. DE-AC36-83CH10093.

## REFERENCES

1. J. A. Van Vechten, *Handbook of semiconductors*, vol. 3, Editor S. P. Keller, North-Holland (Amsterdam), p. 1 (1980).

2. S. B. Zhang and J. E. Northrup, Phys. Rev. Lett. **67**, 2339 (1991); G. A. Baraff and M. Schluter, Phys. Rev. Lett. **55**, 1327 (1985).

3. D. B. Laks, et al., Phys. Rev. B **45**, 10965 (1992) and C. Van De Walle, private communication for which we are grateful.

4. J. L. Shay and J. H. Wernick, *Ternary Chalcopyrite Semiconductors* (Pergamon, Oxford, 1975).

5. P. Villars and L. D. Calvert, *Pearson's Handbook of Crystallographic Data for Intermetallic Phases* (ASM International, Materials Park, Ohio, 1991), and references therein.

6. J. P. Perdew and A. Zunger, Phys. Rev. B **23**, 5048 (1981).

7. *Copper Indium Diselenide for Photovoltaic Applications*, edited by T. J. Coutts, L. L. Kazmerski and S. Wagner (Elsevier, Amsterdam, 1986).

8. A. M. Gabor, J. R. Tuttle, D. S. Albin, M. A. Contreras, R. Noufi and A. M. Hermann, Appl. Phys. Lett. **65**, 198 (1994).

9. The calculated formation energy of hexagonal $In_2Se_3$ is about 1.1 eV lower than that of the tetragonal phase, but we will not consider the hexagonal phase here since we expect a significant transformation energy barrier between the two phases.

10. S.-H. Wei and H. Krakauer, Phys. Rev. Lett. **55**, 1200 (1985).

11. D. Cahen and R. Noufi, J. Phys. Chem. Solids **53**, 991 (1992).

12. K. Mills, *Thermodynamic Date for Inorganic Salphides, Selenides and Tellurides* (Butterworth, London, 1974).

13. H. Neumann, E. Nowak and G. Kuhn, Res. Technol. **16**, 1369 (1981); H. Neumann, *ibid*, **18**, 483, 901 (1983).

14. C. Rincon and C. Bellabarba, Phys. Rev. B **33**, 7160 (1986).

15. S.-H. Wei and A. Zunger, Cryst. Res. Technol. **31**, 81 (1996).

16. C. Kittel, *Introduction to Solid State Physics* (6th ed., Wiley, Singapore, 1986), p. 55.

17. D. Y. Jeon, H. P. Gislason and G. D. Watkins, Phys. Rev. B **48**, 7872 (1993).

18. D. B. Laks, R. Magri and A. Zunger, Solid State Commun. **83**, 21 (1992).

19. S.-H. Wei, L. G. Ferreira, and A. Zunger, Phys. Rev. B **45**, 2533 (1992).

# THIN FILMS II

# Team-Based Thin-Film CIS Research Activities

## Harin S. Ullal

*National Renewable Energy Laboratory*
*1617 Cole Boulevard*
*Golden, CO 80401 USA*

**Abstact.** This paper describes the team-based thin-film copper indium diselenide (CIS) research activities. The CIS team was formed in December 1994 in Kona, Hawaii. Originally, the team had two working groups: the "Junction" and the "Absorber" groups. Currently, there are four working groups the Present Junction, New Junction, Substrate/Mo Impact, and Transient Effect groups. We have completed extensive data compilation of CIS-based films and solar cells using various techniques such as Auger, photoluminescence, scanning electron microscopy, secondary-ion mass spectrometry, X-ray photoelectron spectroscopy, X-ray diffraction, capacitance-voltage, light and dark current-voltage, and quantum efficiency. Studies are under way to understand the fundamental mechanisms that demonstrate a total-area, high efficiency of 17.7% in $CuInGaSe_2$ devices using chemical-bath deposition (CBD) CdS. Alternate buffer layers are also being investigated to replace the CBD CdS. The impact of various Mo substrates from the various industrial partners has been investigated, and the results are reported. A study is under way to investigate the transient effects in encapsulated/laminated thin-film CIS-based devices.

## INTRODUCTION

The copper indium diselenide (CIS) team was formed as part of the Thin-Film Photovoltaics Partnership Program (TFPPP) in Kona, Hawaii, in December 1994. The TFPPP has two objectives. The first is to support the successful introduction of U.S. thin-film products by addressing key near-term technical issues at U.S. businesses committed to thin-film PV production. The second is to support advanced (mid- and longer-term) thin-film research and development needed by industry for future competitiveness, including module performance, cost per kWh, and reliability of thin-film photovoltaic (PV) technologies. As part of the teaming effort, about 35 members from industry, universities, and NREL are actively participating in the various working groups. Originally, there were two working groups (WGs), the "Junction" group and the "Absorber" group. Currently, there are four WGs: the "Present Junction (PJ)," "New Junction (NJ)," "Substrate/Mo Impact (SI)," and "Transient Effects (TE)." The WGs were increased from two to four to make the groups smaller (so that they could focus on their respective problems) and to keep the groups more manageable. Each WG decides its group's objectives and tasks. The working group leader (WGL) is also selected by the WG. Also, ground rules have been established for all team members, and roles and responsibilities for the various personnel have also been established. Six team meeting have taken place at various locations in the United States.

Good working relationships have been established by the numerous members of the CIS team. The atmosphere has been one of collaboration with a spirit of cooperation, rather than a competitive atmosphere. Extensive data compilation for CIS-based films and devices has been completed and distributed among team members. Various characterization and measurement techniques such as Auger, photoluminescence (PL), scanning electron microscopy (SEM), secondary-ion mass spectrometry (SIMS), X-ray photoelectron spectroscopy (XPS), X-ray diffraction (XRD), light and dark current-voltage, and quantum efficiency have been used to understand the fundamental mechanisms for CIS films and devices.

## THE JUNCTION GROUP

The objectives of the CIS junction working group are to identify methods of forming junctions to CIS-based absorbers that are manufacturable, robust, and cost-effective, and to define diagnostic techniques and appropriate parameters to quantitatively compare junction quality. Toward this end, extensive experiments were planned and executed by the WG members. For example, a set of experiments decided by the WG members include the study of the surface and bulk compositions of absorbers, the chemical bath deposition (CBD) CdS/absorber interaction, the effect of CBD constituents and separate CBD reactants, physical vapor deposition, the CdS/absorber interaction, the CdS/I-ZnO interaction, the ZnO/absorber interaction (radiofrequency-sputtered ZnO), devices fabricated in all categories, and samples cut and distributed for analysis. Data were collected on the various films and devices and distributed to the all the CIS team members. Figure 1 shows a PL plot of a CIGS film subjected to various surface treatments. The PL signal is strongest for the sample that been CBD-CdS treated. The data indicate good electronic properties, presumably because of some surface passivation by the CBD process that results in a conformal coating on the CIGS film. Another possible explanation is that the CBD CdS acts as a diffusion barrier to the sputtered ZnO species.

## THE ABSORBER GROUP

The objectives of the CIS/absorber fabrication, modeling, and diagnostics group are to establish a useful correlation between absorber properties, device modeling, and

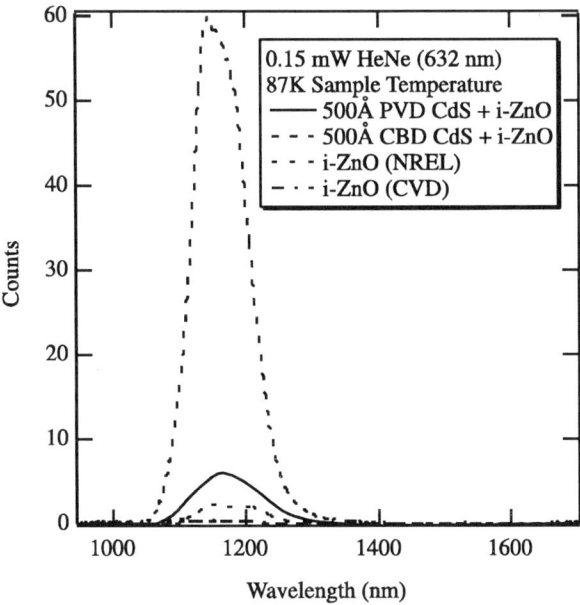

**Figure 1.** PL plots for four CIGS samples subjected to various surface treatments. The best PL signal is for the CIGS sample treated to a 500 Å CBD CdS and I-ZnO

actual device performance, and to develop "manufacturing friendly" material diagnostic tools that will lead to improved reproducibility and understanding of the physical and chemical nature of the absorber. The absorber group has two tasks. The first task is to identify the correlations between Ga and/or S profiles with the CIS-based absorber, the corresponding device performance, and performance predicted by modeling the equivalent device structure. This will be accomplished by participants fabricating CuInGaSSe$_2$ (CIGSS) absorbers, completing portions with CdS/ZnO/grids, and submitting it for film/device characterization. Characterization of the Ga and/or S profiles will subsequently be fed to device modeling, which will input measured Ga/S (bandgap) profiles into the code and compare the data to measured data. The second task is to correlate the chemical impurity content and structural nature of the absorber and absorber/back contact interface, and the device performance. This will be accomplished by participants fabricating CIGSS absorbers, completing portions with CdS/ZnO/grids, and submitting them for film/device characterization. Material characterization will target the identification of impurities that may originate external to the absorber, such as Na, K, and O, from the sodalime substrate, and the nature of the Mo/absorber interface after processing.

Figures 2a and 2b show the Auger depth profile for two Ga samples from International Solar Electric Technology (ISET) and Institutes of Energy Conversion (IEC) at the University of Delaware. The ISET sample shows a buildup of the Ga content at the CIGS/Mo interface, whereas the IEC sample has a more uniform distribution over the entire CIGS layer.

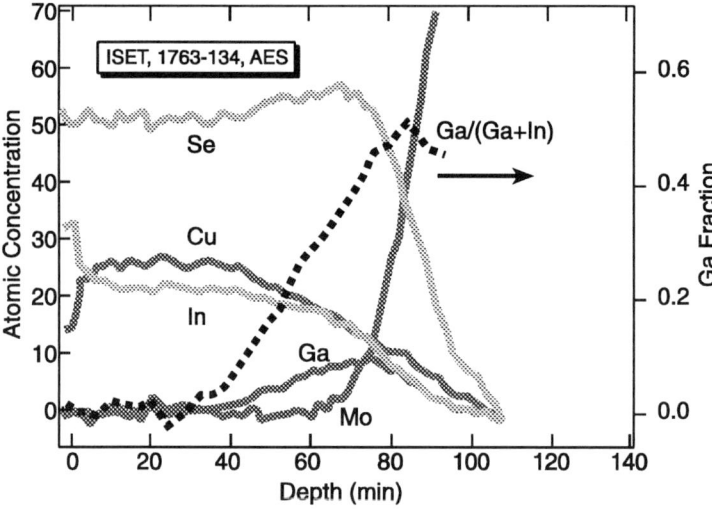

**Figure 2a.** Auger depth profile for an ISET CIGS absorber layer. The plot shows a build-up of the Ga at the CIGS/Mo interface.

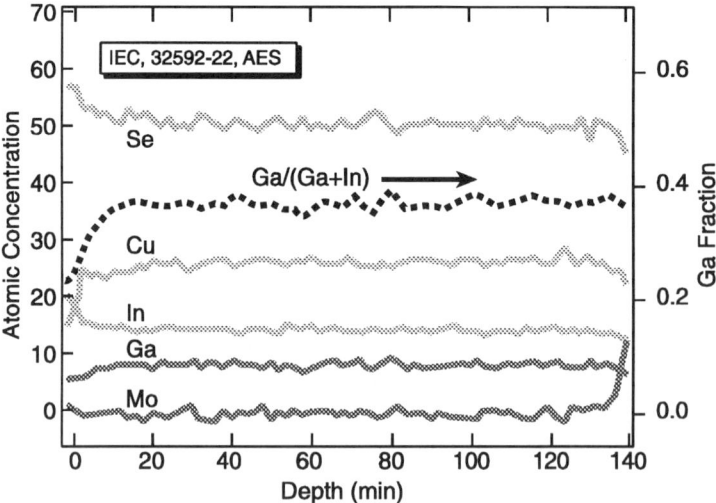

**Figure 2b.** Auger depth profile for a IEC CIGS absorber layer. The plot shows a more uniform distribution of the Ga in the CIGS absorber layer.

The effect of the Mo substrate was also studied. Mo substrates from six groups—Energy Photovoltaics (EPV), ISET, Lockheed Martin (LM), NREL, Siemens Solar Industries (SSI) and Solarex—were submitted to NREL. The CIGS absorber layer and device fabrication was done by NREL. Solar cell efficiencies varied from 4.3% to 17.3%. Extensive data have also been compiled by the Absorber group, and have been distributed to all team members.

## REENGINEERING THE GROUPS

After completing the first year of team work, the Guidance Team decided that the WGs were too large and lacked sufficient focus. The Guidance Team therefore recommended smaller working groups with focused objectives and tasks. The new WGs formed are the PJ, NJ, SI, and TE groups.

## Present Junction

The PJ is essentially a continuation of the earlier Junction group, but with fewer members. One common observation is that when a CBD CdS film is used as a buffer layer between the CIGS absorber layer and the I-ZnO layer, one gets the world-record, total-area solar cell efficiency of 17.7% verified by NREL. The exact role of the CBD CdS is not entirely clear. Some researchers have speculated that the CBD CdS acts as a passivating layer by forming a conformal coating on the CIGS absorber layer, which in turn reduces the surface recombination velocity and improves the device performance. Others have argued that the CBD CdS buffer layer acts as a diffusion barrier to the various ZnO species during the sputtering process.

The objective of the PJ WG is to understand the mechanisms by which CdS, especially CBD CdS, makes the most efficient CIS junction, and to compare the relative merits of different CdS deposition techniques.

## New Junction

There is an apparent consensus in the CIS community that the very thin CdS (500 Å) buffer layer needs to be replaced by a noncadmium containing layer to give the technology a "green" (environmentally friendly) appearance. Toward this end, there is a great deal of research under way in Japan, Germany, and the United States to replace the CBD CdS by other candidates, such as $In(OH)_2$, $In_2S_3$, $SnO_2$, ZnO, ZnSe, or ZnS. However, to date, the use of these alternate buffer layers has resulted in lower cell efficiency. NREL, in collaboration with Washington State University (WSU) has fabricated a CIGS solar cell without any CdS buffer layer, having a total-area efficiency of 12.7%. WSU researchers deposit the I-ZnO by metal organic chemical vapor deposition.

Another concern of the PV industry is the ZnO transparent conducting oxide (TCO). In a CIS module, the typical thickness of the ZnO layer is 2.0 to 2.5 µm to get the appropriate sheet resistance of about 8 ohms/sq. The lower sheet resistance is necessary to get lower series resistance and higher fill factor in a module. As the thickness of the ZnO increases, the amount of sunlight the cell absorbs decreases because of the optical losses in the ZnO. One of the challenges for this WG is to reduce the thickness of the ZnO layer while getting lower sheet resistance and higher transmission. In addition, it has been observed that the electronic and optical properties of ZnO deposited on sodalime glass are not the same as those when ZnO is deposited on the CIGS absorber layer. The NJ WG group is addressing these research issues.

The NJ WG has two objectives. The first objective is to develop noncadmium-containing buffer layers. The priority will be on vacuum processes that could potentially be incorporated into in-line manufacturing. CBD is not considered a primary option. The second objective is to develop improved TCO layers to minimize losses for module fabrication and quantify the effect of TCO layers on module performance. This task will focus only on the high-conductivity TCO layer.

## Substrate Mo/Impact

The previous Absorber WG conducted a set of experiments to evaluate the role of the Mo contact on the device performance. Six Mo substrates from the various industrial partners and NREL were part of this WG experiment. The CIGS was deposited by NREL, and device fabrication was also done by NREL. The solar cell efficiency varied from 4.5% to 17.3%, thus indicating a strong dependence on the Mo substrate and that not all Mo is created equal.

Another impact of the substrate is the out-diffusion of Na from the low-cost sodalime glass substrate through the polycrystalline Mo films and into the CIGS absorber layers. Thus far, Na has had a positive impact on the device performance. Scientists have argued that Na enhances the grain growth, acts as a dopant, improves the carrier concentration, and thus improves device performance. Although considerable empirical evidence exists, the exact role of Na is unclear. This becomes all the more important if alternate substrates such as alumina, polyamide, or stainless steel are used, because in these cases the extrinsic addition of Na is required to enhance device performance. Adding a diffusion barrier such as $SiO_2$ between the glass and Mo contact has been investigated thus far.

The objective of this WG is to study the impact of the glass substrate and Mo contact on the performance of CIS-based solar cells. The research includes correlation of Mo properties (chemical, mechanical, and structural) and the degree of Na diffusion through the layer; the study of substrates/Mo/CIS interaction through sample exchange between WG partners; and the possible diffusion barriers for Na.

## Transient Effects

Substantial development work done by Siemens Solar Industries to encapsulate/laminate CIS-based mini-modules has indicated an initial drop in the mini-module performance after encapsulation/lamination. However, when exposed to natural sunlight for a few weeks, the mini-modules regain or exceed initial performance. This effect is shown in Figure 3, and is the topic of investigation for the TE WG.

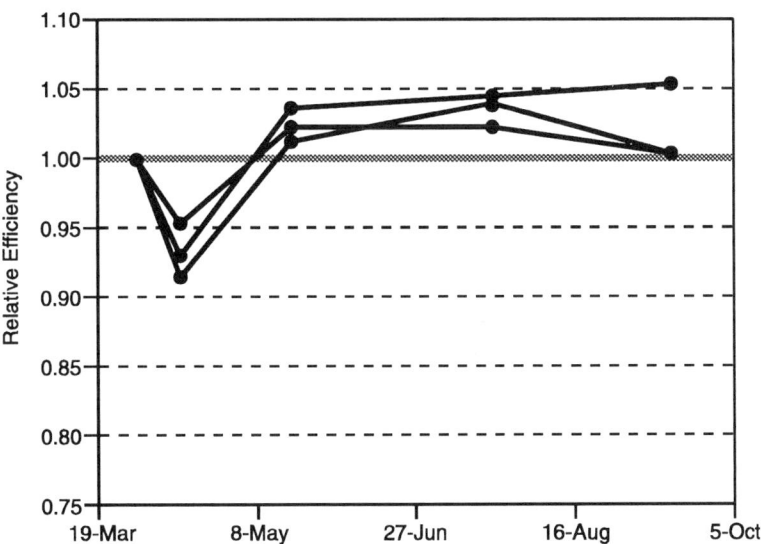

**Figure 3.** This plot shows the transient effects of a CIGSS mini-module fabricated by SSI after encapsulation/lamination and then exposed to natural light.

The objective of this WG is to study the transient effects in thin-film CIS-based devices. Studies will include solar cells, mini-modules, and modules before and after encapsulation/lamination. Major emphasis will be on understanding the fundamental mechanisms responsible for these effects in CIS-based devices.

## SUMMARY

The CIS team was established in December 1994. Since then, substantial technical progress has been made in better understanding the film and device properties of CIS. Extensive data using various techniques (Auger, PL, XPS, XRD, SEM, and SIMS, for films and; dark and light I-V, quantum efficiency, and capacitance-voltage for solar cells) have been compiled and distributed among the team members. There have been several exchanges of samples between team members. Discussions at team meetings have been candid, and there is a healthy spirit of cooperation among the team members. Working groups have been reengineered, so that smaller WGs focus on research problems that will help the PV industry solve relevant technical problems to commercialize thin-film CIS technology.

## ACKNOWLEDGMENTS

The author acknowledges the contribution of the National CIS team members as part of the TFPPP activity. The data included in this paper reflect the commitment, dedication, and hard work of the National CIS team members. This work was supported by the U.S. Department of Energy under contract No. DE-AC36-83CH10093.

# Materials and Processing Issues in Thin-Film Cu (In, Ga) Se$_2$-Based Solar Cells

J.R. Tuttle, M.A. Contreras, K.R. Ramanathan, S.E. Asher,
R. Bhattacharya, T.A. Berens, J. Keane, and R. Noufi

National Renewable Energy Laboratory, Golden, Colorado USA
(ph) 303-384-6510, (fax) 303-384-6430 (e-mail)
rommel_noufi@nrel.gov

## ABSTRACT

We have fabricated-high efficiency, Cu(In,Ga)Se$_2$ (CIGS)-based photovoltaic (PV) devices by four different processes. Each process may be characterized as either sequential or concurrent deposition of the metals with or without an activity of Se. A world-record, total-area efficiency of 17.7% has been achieved by the concurrent delivery of the metals in the presence of Se. Ga has been introduced into the device in a such a manner as to produce homogeneous, normal profiling, and double-profiling graded band gap structures. This has resulted in an open-circuit voltage ($V_{oc}$) parameter of 680 mV and a fill-factor over 78%. The quality of CIGS-based films and devices is becoming decoupled from the method of film delivery. This leads to novel, fast, and low-cost methods for abosrber fabrications. Two such deposition techniques, sputtering and electrodeposition, will also be discussed and results to date presented. Finally, a fabrication model has been developed allowing for simple translation of these processes to a manufacturing environment for the large-scale production of modules.

# INTRODUCTION

The National Renewable Energy Laboratory (NREL), under contract to the United States Department of Energy, has been involved in the research and development of thin-film PV since 1982. The primary charter of the Photovoltaics Program is to develop new and better PV technologies and to support industry in doing the same. The goal is to introduce PV as a cost-effective alternative to conventional utility power generation. This goal is accomplished by an approach that first considers basic materials research, followed by solar cell development, and concludes with technology transfer to industrial organizations and market development.

To establish cost effectiveness in PV technology, both performance and cost are considered. Solar cells and modules fabricated from polycrystalline CIGS-based thin films are strong candidates for high performance and low cost [1]. Laboratory-scale device efficiencies in excess of 15% have been reported by several groups [2-4]. The low-cost criterion is satisfied for most thin-film technologies through low materials usage, monolithic integration, and low manufacturing costs, to name a few. Several industrial groups have produced large-area (sub)modules with performance in excess of 7% [5,6]. One company has successfully produced a 10% module with an aperture area near 4000-$cm^2$ [4]. In this work, laboratory-scale device absorbers are fabricated by physical vapor deposition (PVD) processes that may be conducive to industrial scale up [7,8]. An additional advantage for the thin-film CIS technology developed at NREL is the potential for high yield through greater-than-average process tolerances and self-limiting process chemistry.

In this work, device performance and manufacturability issues will be discussed. This work is supported by extensive fundamental materials research that is reported elsewhere in the literature [9-16]. Four PVD absorber fabrication processes are presently being investigated in our laboratories. The associated champion device performance ranges from 12.6% to 17.7%. In each case, the process is described in a fashion that allows for transfer to industrial-scale deposition systems [17]. Growth models have also been developed to describe the formation chemistry [9]. The incorporation of Ga to raise the absorber band gap has been accomplished successfully and in such a manner that an $V_{oc}$ of 680 mV has been accomplished [8]. The higher $V_{oc}$'s and lower $J_{sc}$'s translate into lower interconnect losses at the module level.

In this paper, we also present a generic flowchart for the fabrication of CIGS absorbers, which takes into consideration the critical processing parameters. By breaking the process into stages, there are clear opportunities to use a variety of deposition techniques, separately and reduce the time of processing segments, and introduce intelligent process control.

**EXPERIMENTAL**

CIGS thin films are grown by physical vapor deposition (PVD) of the constituent elements under a vacuum of $10^{-8}$ Pa ($\approx 10^{-6}$ Torr) onto 5-cm x 5-cm (2-in x 2-in) Mo-coated soda-lime silica (SLS) glass. The PVD process may consist of co-evaporation of the four elements simultaneously, sequential evaporation of the metals followed by exposure to a Se species, or sequential evaporation of the metals in the presence of Se. Some details of the process specifics will be provided later in this paper. Control of the Cu, In, and Ga fluxes is accomplished by electron impact emission spectroscopy (EIES) of the vapor trail and of the Se flux by quartz crystal monitoring (QCM). Substrate temperatures of 300-600°C are achieved by heating from quartz-halogen lamps or resistive heating. Warping of the SLS substrate at temperatures above 500°C was minimized by the appropriate combination of SLS glass type and thickness and Mo film deposition parameters. In many cases, an intentional compositional gradient of about 2 at.%Cu was introduced across the 5-cm dimension in order to study material and device variations as a function of composition. Films were also deposited by sputtering and electrodeposition. The experimental details are given elsewhere[18,19].

PV devices are completed by chemical bath deposition (CBD) of about 500 Å of CdS followed by RF sputtering of 500 Å of intrinsic ZnO and 3000 Å of Al-doped ZnO. The CBD process has proven to be the only successful means of delivering a thin, conformal layer of CdS to the surface of the absorber. Several groups are looking at the role of CdS in the device [18] while others are looking to replace the CdS with a non-Cd-containing layer buffer layer by either chemical [19] or physical [20] deposition means. Ni/Al grid contacts are applied with approximately 4% coverage. I-V characterization is carried out at AM1.5 illumination, and device efficiencies are quoted as "total-area" to include grid losses. Quantum efficiency measurements are made in the dark and under voltage and light bias in the wavelength range 380-1500 nm.

## PVD PROCESS DESCRIPTION

The absorber fabrication process is defined by a variety of parameters. These include time-dependent profiles of the total Cu, In, Ga, and Se metal fluxes (deposition rate) in atoms/cm$^2$-sec (Å/s), Cu/(In+Ga) metal flux ratios, Se/(Cu+In+Ga) flux ratios, and substrate temperature. In Fig. 1, flux profiles for the fabrication of Cu(In,Ga)Se$_2$ are shown for three such processes [17]. The total metal flux determines the overall growth rate of the thin-film, which is about 15 Å/s. For processes where the metals are delivered sequentially in the presence of Se (Fig. 1a), "material delivery" may be accomplished much faster than "compound formation" resulting in "effective" growth rates. In future work, we wish to push the envelope on these delivery and formation rates in order to

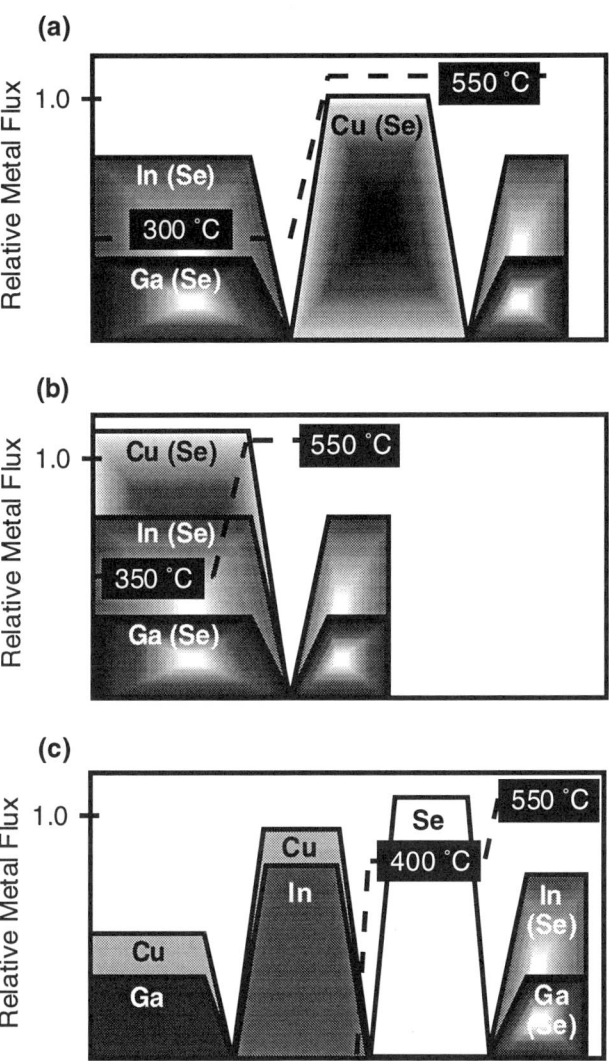

Fig. 1 Source flux profiles for the (a) 3-stage, (b) 2-stage, and (c) Se-vapor selenization processes

minimize total deposition time, a significant manufacturing issue. The Cu/(In+Ga) metal flux ratio controls the formation chemistry during growth and hence the resulting microstructure and electronic quality. The Se/metal flux ratio is held above the lower limit established for the formation of stable binary and ternary phases during growth. The ratio is typically 3:1 for the ternary, 5:1 for the (In,Ga):Se binaries, and 3:1 for the Cu:Se binaries. The Se overpressure is required to produce a filled anion sublattice, and, thus, a valent-neutral semiconductor. The substrate temperature governs the adatom mobility and the phase nature of the binary and ternary constituents during growth.

One way to visualize the formation of the compound from its constituent elements is to consider the Cu-In-Se ternary phase diagram (Fig. 2). At first glance, it simply appears as if stoichiometric amounts of Cu, In, and Se can be added together in any manner to produce the ternary compound. This is not the case. Instead, the "chemical reaction path" is described in terms of a formation chemistry, which is driven by thermodynamics [12,23] and kinetics. For example, in the selenization process (Fig. 2, #1, and Table 1[4]), there is a clear separation of the metal and selenide deposition steps. During the metal deposition, Cu and In interact to form Cu-rich and In-rich alloys [24]. Upon the introduction of the Se species, $In_ySe$ and $Cu_xSe$ binaries precede the $CuInSe_2$ (CIS) formation. The exact nature of the binaries is dependent upon time, temperature, and whether elemental Se or $H_2Se$ gas is used as the Se source [13]. In Table 1, the processes are described pictorially, with a suggestion as to the possible chemical reaction path leading to the compound. Here we also highlight the process issues that are attractive for manufacturing.

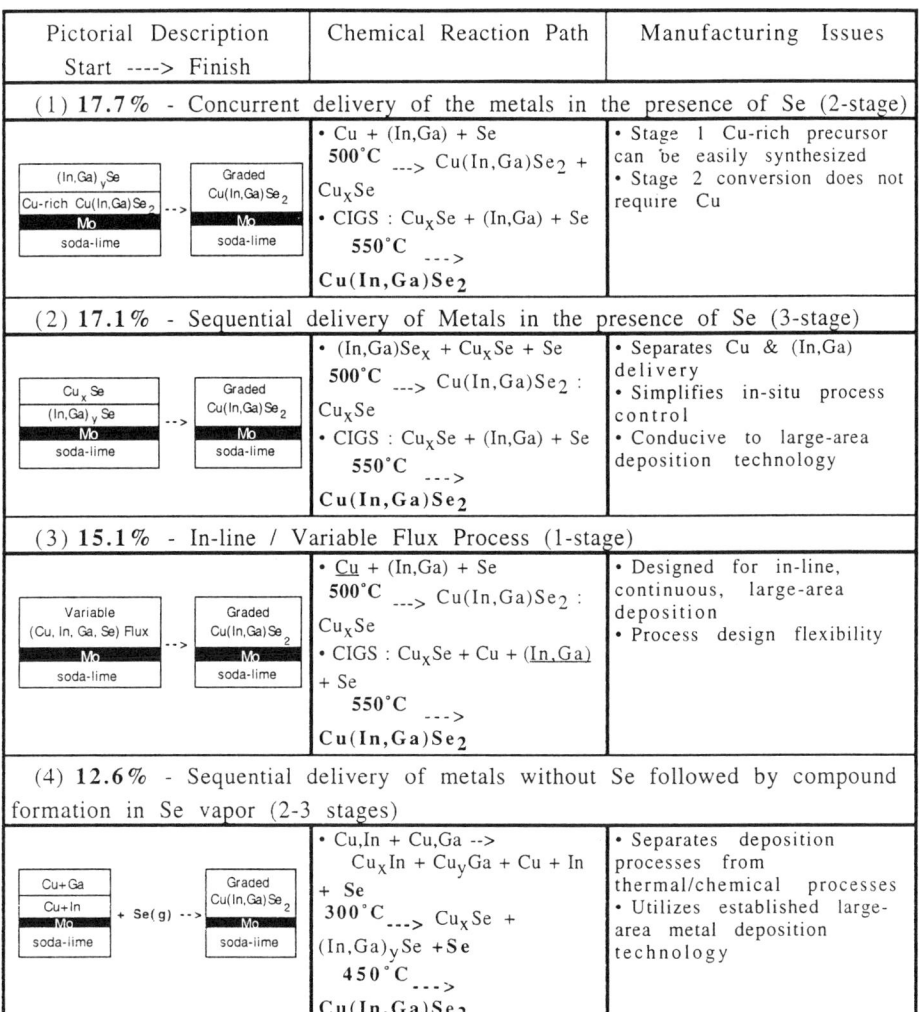

Table 1. Description of absorber processes utilized to fabricate CIGS absorbers

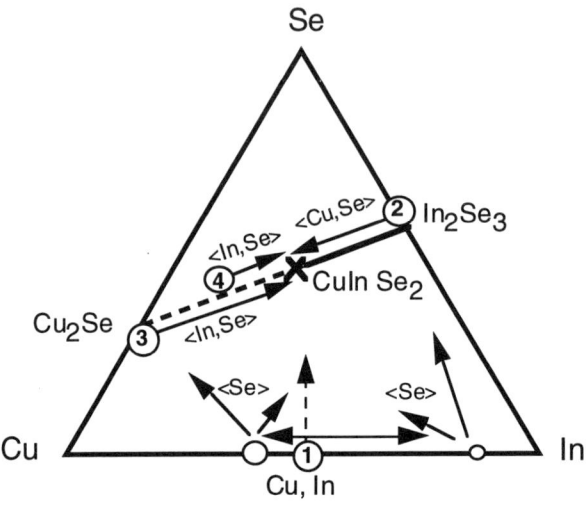

Fig. 2 Cu-In-Se ternary phase diagram with suggested chemical reaction paths to the CIS product

## RESULTS AND DISCUSSION

Our laboratory has investigated several absorber structures based upon the Cu(In,Ga)Se$_2$ material system: 1) homogeneous CIS ($E_g$=1.0 eV), 2) CIS on a CuGaSe$_2$ (CGS) buffer layer, 3) homogeneous CIGS ($E_g$=1.0-1.14 eV), 4) graded CIGS ($E_g$=1.0-1.7 eV) with a normal profile, and 5) graded CIGS with a double profile. These are represented pictorially in Fig. 3 as a depth profile of the semiconductor band gap. A discussion of the rationale behind these absorber designs is presented elsewhere [8,25]. In Fig. 4 and Table 2, the current-voltage (I-V) and quantum-efficiency (QE) results are presented for representative cells fabricated by the above-mentioned processes. The 17.7% total-area device performance level represents the NREL-confirmed world record for all polycrystalline and amorphous thin-film technologies [26]. We have, furthermore, reported a 1-cm$^2$ cell at 16.4% with very high $V_{oc}$, a 4.85-cm$^2$ cell at 15.3%, and a 0.074-cm$^2$ cell operated under 22-sun illumination at 17.7%, a 2.8% absolute improvement over the 1-sun control [2]. The exceptional parameter among theses cells is a 1-sun $V_{OC}$ value of 678 mV.

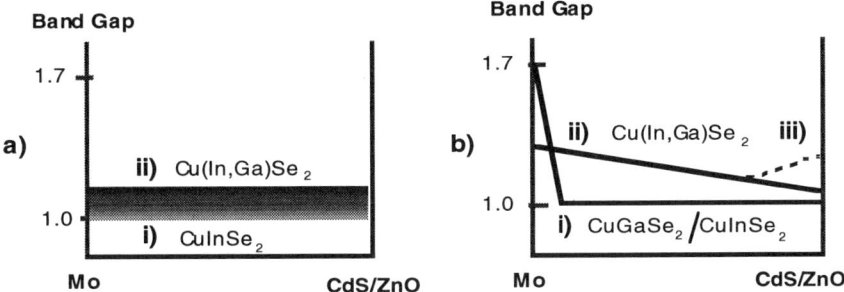

Fig. 3 Pictorial of band gap variation in CIGS absorber bulk. a) Homogeneous i) CIS and ii) CIGS, b) graded band gap i) CIS on CGS, ii) CIGS with a "normal" profile, and iii) CIGS with a "double" profile.

This has positive implications for module fabrication in that interconnect and series-resistance losses will decrease with larger cell widths and smaller operating currents, respectively.

Variations in performance are a result of both process and design variations. In Fig. 4 and Table 2, device (a) is a homogeneous CIS absorber, device (b) is a CIS absorber grown on 2000Å of CGS, and devices (c) and (d) are CIGS absorbers grown on a CGS buffer layer. Devices (b) and (d) were grown by process (1) in Table 1 (2-stage), while all others were grown by process (2) (3-stage). Note the improved spectral response and enhanced $V_{OC}$ in the CIS device incorporating the CGS buffer layer at the Mo back electrode (Fig. 4b). Only trace amounts of Ga are detected near the absorber surface in this device, suggesting a band gap effect unrelated to alloying. This phenomenon is presently under investigation.

| Sample | Area (cm$^2$) | $V_{oc}$ (mV) | $J_{sc}$ (mA/cm$^2$) | FF (%) | Total-Area Efficiency (%) | Comments |
|---|---|---|---|---|---|---|
| M1201 | 0.395 | 484 | 36.3 | 75.1 | 13.2 | 4 (a), CIS |
| S573 | 0.413 | 552 | 37.1 | 72.1 | 14.8 | 4 (b), CIS / CGS |
| C362 | 0.437 | 652 | 33.2 | 77.4 | 16.8 | 4 (c), CIGS / CGS, double |
| S773 | 0.414 | 674 | 34.0 | 77.2 | **17.7** | 4 (d), CIGS / CGS, normal |
| C371 | 1.025 | **678** | 32.0 | 75.8 | 16.4 | CIGS / CGS, double |
| C371 | **4.85** | 657 | 31.1 | 74.7 | 15.3 | CIGS / CGS, double |
| S773 | 0.103 | **714** | **628.4** | 78.6 | **17.7** | CIGS / CGS, double, 20-sun |

Table 2. Summary of device performance for champion cells made with different absorber structures by various processes. 4(a-d) refers to Fig. 4. "Double" and "normal" refer to the realized band gap profile. All I-V data are derived from official NREL measurements.

The details of the relationship between the intended device structure and the resulting Cu(In,Ga)Se$_2$ phase distribution in the absorber continue to be investigated 8,14,27. In general, the homogeneous structures are straightforward to fabricate and characterize. The spectral response and $V_{oc}$ are mutually consistent with the band gap of the absorber. This is not the case for the graded-band-gap absorbers. Growth of the absorber within the Cu(In,Ga)Se$_2$:Cu$_2$Se two-phase region for any period of time will enhance grain growth and In,Ga interdiffusion. Likewise, the more arduous the path from the metal constituents to the final compound, i.e. (In,Ga) -> (In,Ga)$_y$Se -> Cu(In,Ga)Se$_2$, the

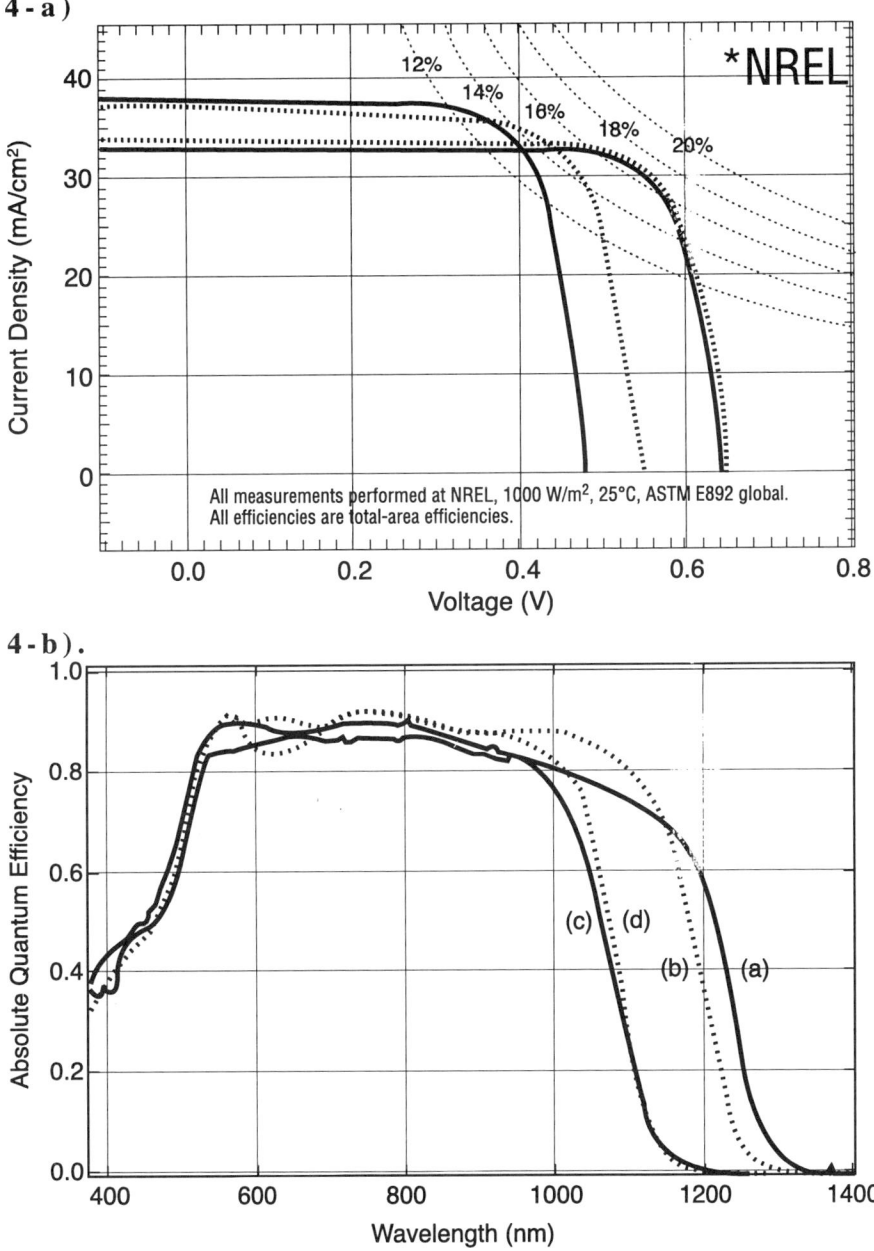

Fig. 4 I-V (a) and quantum-efficiency (b) measurements of CIGS-based device structures

more likely In and Ga will spatially polarize. Finally, the film roughness is influenced by both the degree of excess Cu and by the surface roughness of intermediate film layers that are present during growth. Smooth, specular films are desired in order to minimize junction area though reflection losses may be enhanced.

Absorber optimization will result from smooth surfaces and a controlled (In,Ga) profile throughout the absorber. Device optimization will result from improvements in the short-wavelength response (l 500nm). Comparison of the spectral response of our best cell with a champion Boeing cell [28] (Fig. 5) indicates room for improvement in the short-circuit current density ($J_{sc}$) with modifications to the window-layer processing. Combining the best parameters of our cells with the high $J_{sc}$ yields an 18.7% total-area performance level.

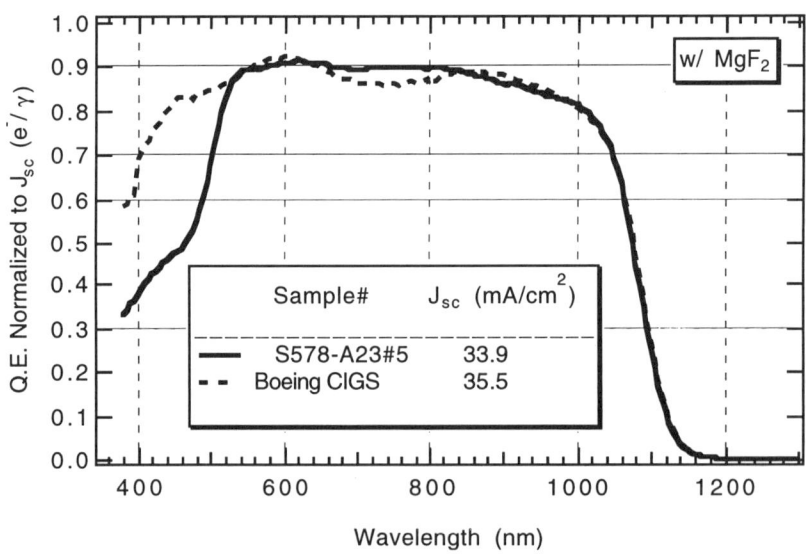

Fig. 5 External quantum efficiency comparison of two high-efficiency CIGS-based solar cells

## PROCESSING CHALLENGES

One difficult issue that is being addressed in our laboratory is irreproducibility associated with changes in the Mo/SLG substrate system. The irreproducibility may be characterized by a long time-constant that is related to process variations from one supplier, or a very short time-constant resulting from parallel processing of substrates from different suppliers. This issue has been investigated by a combined effort of material and device characterization. In Fig. 6, the results of a matrix experiment are presented whereby two sources of Mo/SLG substrates, A and B, from two time periods, past and present (A and A', B and B'), are processed into CIGS absorbers and devices. I-V and C-V measurements are performed on the devices to determine Voc and the carrier

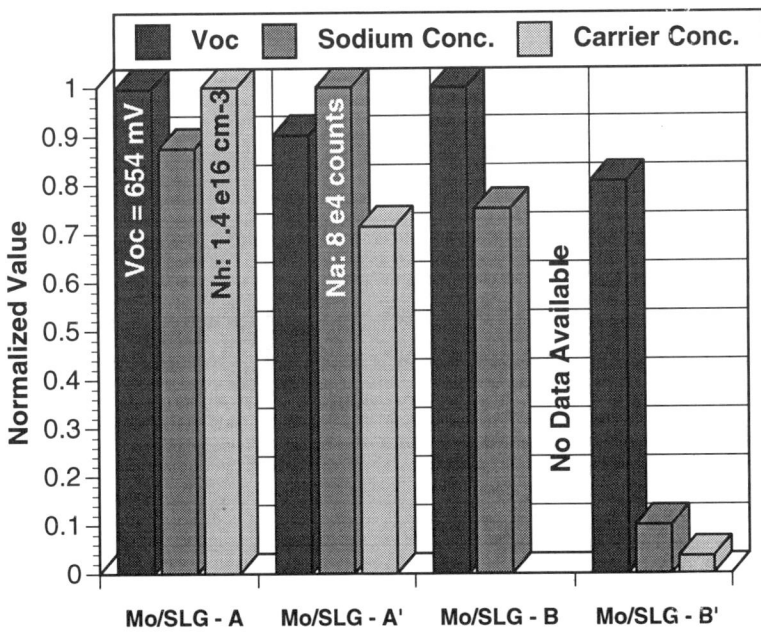

Fig. 6 Variation of Voc, carrier concentration ($N_h$), and Na content with substrate type and history.performed on the

concentration, $N_h$, at zero bias, respectively. SIMS analysis is absorber and Mo layers prior to device fabrication to quantify elemental (constituent and impurity) profiles. Devices fabricated from A and B were of very high quality at 17.7% and 16.8%. A' and B' were processed simultaneously.

Two conclusions can be drawn from these data. The first is that there is a relationship between the sodium (Na) concentration within the absorber and the resulting $N_h$ and Voc of the device. The second is that there is a level of consistency over time in the substrate A and inconsistency in B in terms of the Na migration from the substrate to the absorber film.

SIMS analysis (Fig. 7) of the CIGS/Mo/SLG stack suggests a possible cause of this phenomenon. Samples A, A', and B contain equivalent amounts of Na and a sharp transition between the CIGS and Mo layers (as measured by the simultaneous drop and rise in the Se and Mo signals, respectively, at the CIGS/Mo interface). Sample B', on the other hand, suggests the presence of a $Mo_xSe$ interlayer between the absorber and Mo. This sample contained an order of magnitude less Na in the absorber and, with it, an associated decrease in $N_h$ and device Voc. We conclude, therefore, that the nature of the Mo surface, and its reactivity with Se, can substantially influence the characteristics of the absorber and the performance of the device. Future work will focus on characterizing the Mo surface and identifying the characteristics that lead to this performance.

## NEW DIRECTIONS

The next-generation CIGS-based thin-film device will ideally have the following characteristics. The back contact and substrate combination will offer superior reproducibility to the present Mo/SLG system through controlled introduction of required impurities (e.g. Na, O). This will expand the list of potential substrates and back-contact metals to those that may

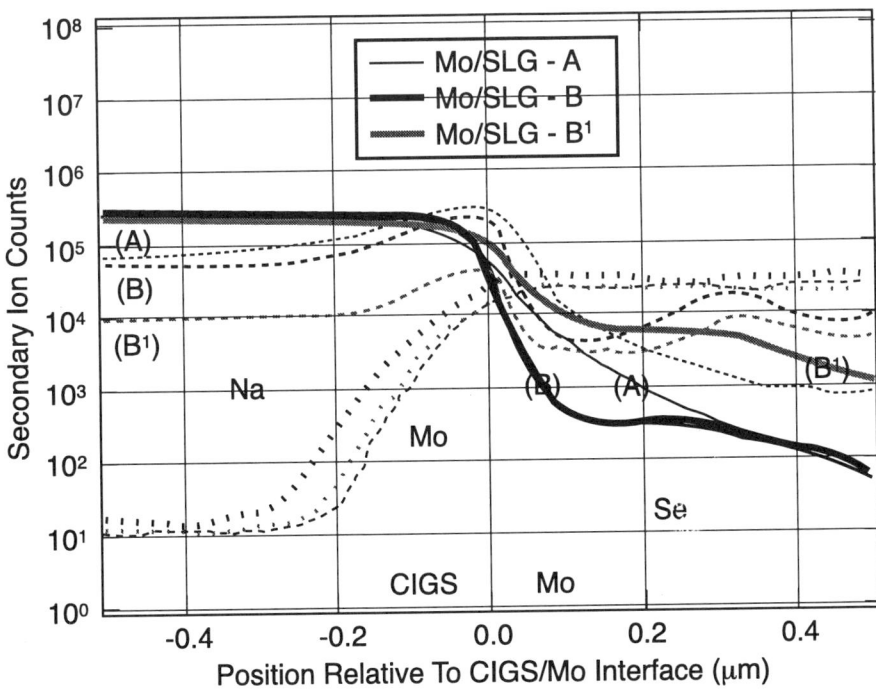

Fig. 7 SIMS analysis of CIGS/Mo interface. Sample B' has an greater than 10-times the Se content in the Mo back electrode.

be more optimally suited to CIGS thin-film processing. The absorber will be fabricated in a manner that minimizes in-situ process control and high-temperature processing. This will drastically reduce the cost of manufacturing equipment. Finally, the heterojunction partner will be formed in-situ with the absorber to relieve the necessity for a vacuum break and a CBD process. This will improve reliability and throughput and will reduce cost.

The critical absorber process parameters have been considered and a generic flowchart developed for the fabrication of CIGS absorbers (Fig. 8). By breaking the process into three or four independent stages, there are clear opportunities to use a variety of deposition techniques, separate and reduce the time for high-temperature (High-T) process segments, and introduce intelligent process control.

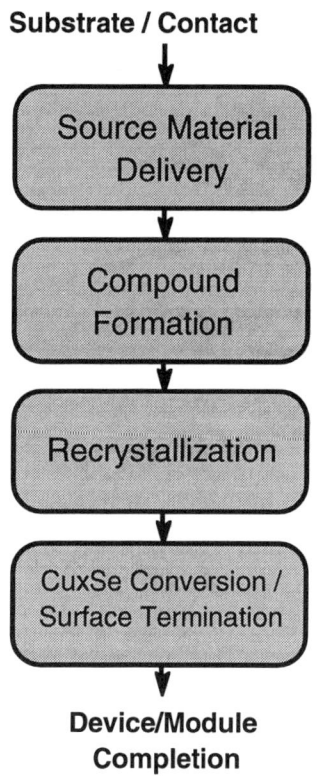

Fig. 8 CIGS absorber fabrication process flowchart.

In Fig. 9, we present two absorber processing scenarios that target manufacturability. In (a), CIGS source material is delivered in such a manner as to produce a low-quality, fine-grain precursor film. It is subsequently exposed to (Cu,Se) and (In,Ga,Se) at high-T to complete the absorber. In (b), the source material is a similar, low-quality CIGS:CS mixture. The CS is activated with Se activity at high T, followed by (In,Ga,Se). In these experiments, the process segment times are 3 min, resulting in a total high-T time of 6 min. This represents a 2-3x factor reduction compared to the two- and three-stage processes previously described.

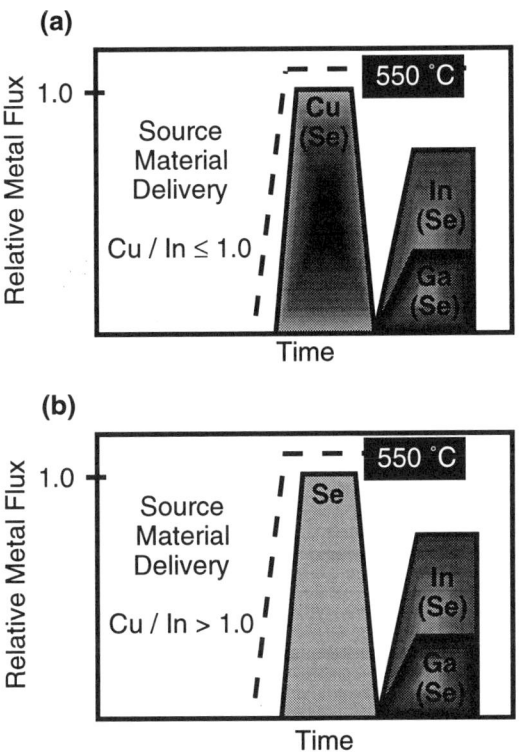

Fig. 9 Alternative "fast" CIGS absorber fabrication processes.

Three source-material delivery techniques are under investigation: rapid evaporation, sputtering, and electrodeposition. In Table III, we present the best results to date using these delivery techniques in conjunction with processing described in Fig. 9. The results reflect more of the relative time investments we have in each of the processes than their potential. We enthusiastically contend that each of these techniques, in conjunction with the appropriate recrystallization treatment, will produce material of equivalent quality.

| Process | $V_{oc}$ (mV) | $J_{sc}$ (mA/cm²) | FF (%) | η (%) | Comments |
|---|---|---|---|---|---|
| Evaporation | 623 | 32.9 | 75 | 15.3 | **CIGS** + (Cu,Se) + (In,Ga,Se) |
| Evaporation | 605 | 31.4 | 73 | 13.9 | **CIGS:CS** + (In,Ga,Se) |
| Sputtering | 508 | 24.3 | 57 | 7.0 | **CIGS** + Cu + Se + (In,Ga,Se) |
| Electrodeposition | 545 | 34.0 | 66 | 12.3 | **CIGS:CS** + (In,Ga,Se) |

Table 3. Device Parameters resulting from alternative processing of CIGS absorbers.

## GENERIC MANUFACTURING SCENARIO

A growth model describing the formation of $Cu(In,Ga)Se_2$ absorbers from a Cu-rich precursor has previously been presented [9]. A consequence of formulating this model is an understanding of how to translate these technologies from the laboratory to the factory. In Fig. 10, we present a manufacturing scenario that is generic to any of the above-mentioned processes. One of the exemplary aspects of this scenario is the inclusion of in-situ/in-line diagnostics and end-point detection. The diagnostics involve the non-destructive determination of Cu and (In,Ga) contents in the absorber film after intermediate deposition steps. This allows feedback into subsequent processing areas to assure an optimum composition in the final product. Likewise, the termination of the process can be detected by either a phase or compositional change in the absorber surface. In this way, the process is designed for maximum yield and optimum performance. Our laboratory is presently investigating industrial-scale processes at the 100-cm$^2$ sub-module level.

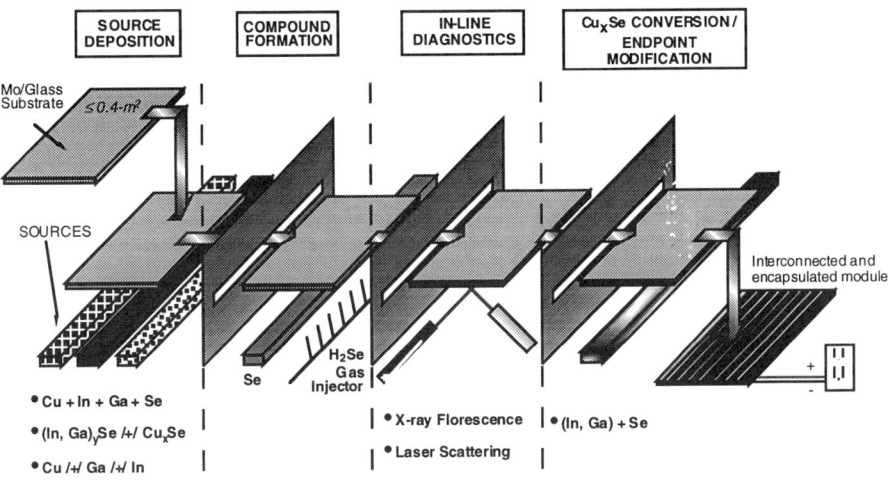

Fig. 10 Scenario for the manufacture of CIGS-based modules

## CONCLUDING REMARKS

As device efficiencies rise and processes are simplified, we can expect to see the introduction of high-performance, high-yield, reliable PV modules based upon the Cu(In,Ga)Se$_2$ absorber. Additional improvements will come with optimized device designs and a better understanding of the CuInSe$_2$:CuGaSe$_2$ alloying process. Our laboratory is dedicated to realizing this goal through continued research into the basic science of the materials and devices, as well as the development of process technologies that are transferable directly to industrial interests.

**ACKNOWLEDGMENTS**

The authors wish to thank J. Dolan and A. Mason for technical assistance. This work was supported by NREL under Contract No. DE-AC02-83CH10093 to the U.S. Department of Energy.

# REFERENCES

1. K. Zweibel, H.S. Ullal, B.G. von Roedern, R. Noufi, T.J. Coutts, and M.M. Al-Jassim, *Proceedings 23rd IEEE Photovoltaic Specialists Conference*, Louisville, KY, 1993, pp. 379-388.
2. J.R. Tuttle, M.A. Contreras, J.S. Ward, A.M Gabor, K.R. Ramanathan, A.L. Tennant, L. Wang, J. Keane, and R. Noufi, *Proceedings of the 1st World Conference on Photovoltaic Energy Conversion*, 5-9 December, 1994, Waikoloa, HI (in press).
3. W.H. Bloss, *Proceedings of the 12th European Photovoltaic Solar Energy Conference,* 11-15 April, 1994, (H.S. Stephens & Associates, U.K., 1994) pp. 37-43.
4. R. Gay, M. Dietrich, C. Fredric, C. Jensen, K. Knapp, D. Tarrant, and D. Willett, *Proceedings of the 12th European Photovoltaic Solar Energy Conference,* 11-15 April, 1994, (H.S. Stephens & Associates, U.K., 1994) pp. 935-938.
5. B.M. Basol, V.K. Kapur, C.R. Leidholm, A. Minnick, and A. Halani, AIP Conference Proceedings 306, *Proceedings of the 12th NREL Photovoltaic Program Review,* Denver, CO, October 13-15, 1993, pp.79-82.
6. A.E. Delahoy, J. Britt, F. Faras, F. Ziobro, A. Sizemore, G. Butler, and Z. Kiss, AIP Conference Proceedings 306, *Proceedings of the 12th NREL Photovoltaic Program Review,* Denver, CO, October 13-15, 1993, pp.370-381.
7. A.M. Gabor, J.R. Tuttle, D.S. Albin, M.A. Contreras, and R. Noufi, *Appl. Phys. Lett.* **65** (2) 1994, pp. 198-200.
8. M.A. Contreras, J.R. Tuttle, A.M. Gabor, A.L. Tennant, K.R. Ramanathan, S. Asher, A. Franz, J. Keane, L. Wang, and R. Noufi, *Proceedings of the 1st World Conference on Photovoltaic Energy Conversion*, 5-9 December, 1994, Waikoloa, HI (in press).
9. J.R. Tuttle, M. Contreras, M.H. Bode, D. Niles, D.S. Albin, R. Matson, A.M. Gabor, A. Tennant, and R. Noufi, *J. Appl. Phys.*, **77** (1), 1995, pp. 1-9.

10. J.R. Tuttle, M. Contreras, A. Tennant, D.S. Albin, and R. Noufi, *Proceedings 23rd IEEE Photovoltaic Specialists Conference*, Louisville, KY, 1993, pp. 415-421.
11. J.R. Tuttle, D.S. Albin, and R. Noufi, *Solar Cells*, **30** (1991), pp. 21-38.
12. D.S. Albin, J.R. Tuttle, and R. Noufi, *J.Elec.Materials*, **24**(4), 1995, p. 1.
13. R.W. Birkmire, J.E. Phillips, W.N. Shaferman, S.S. Hegedus, B.E. McCandless, and T.A. Yokimcus, "Polycrystalline Thin-Film Materials and Devices," Final Report to National Renewable Energy Laboratory under Subcontract No. XN-0-10023-1, 1/16/90-1/15/93.
14. A.M. Gabor, J.R. Tuttle, A. Schwartzlander, A.L. Tennant, M.A. Contreras, and R. Noufi, *Proceedings of the 1st World Conference on Photovoltaic Energy Conversion*, 5-9 December, 1994, Waikoloa, HI (in press).
15. R. Klenk, T. Walter, H.W. Schock and D. Cahen, *Adv. Mater.* **5**, (2), (1993), 114.
16. T. Walter and H.W. Schock, *Thin Solid Films*, **224**, (1993), 74.
17. M.A. Contreras, J.R. Tuttle, D.S. Albin, A. Tennant, and R. Noufi, *Proceedings 23rd IEEE Photovoltaic Specialists Conference*, Louisville, KY, 1993, pp. 486-490.
18. J.R. Tuttle, M.A. Contreras, K.R. Ramanathan, A. Tennant, S.A. Asher, D. Niles, R. Matson, J. Keane, and R. Noufi, *Proc. Of the 13th European PV Solar Energy Conf.*, Nice, France, 23-27 Oct., 1995, p. 2131.
19. R.N. Bhattacharya, A.M. Fernandez, M.A. Contreras, J. Keane, A.L. Tennant, R. Ramanathan, J.K. Tuttle, R. Noufi, and A.M. Hermann, *J. Electrochem Soc.,* **143(3),** March 1996, p. 854.
20. J. Kessler, K.O. Velthaus, M. Ruckh, R. Laichinger, H.W. Schock, D. Lincot, R. Ortega, and J. Vedel, *Proceedings of the 6th International Photovoltaic Science and Engineering Conference*, (PVSEC-6), New Delhi, 1992, 1005.

21. D. Hariskos, M. Ruckh, U. Rühle, and H.W Schock, *Proceedings of the 1st World Conference on Photovoltaic Energy Conversion*, 5-9 December, 1994, Waikoloa, HI (in press).

22. L.C. Olsen, F.W. Addis, D. Greer, H. Aguilar, W. Lei, and F. Abulfotuh, *Proceedings of the 1st World Conference on Photovoltaic Energy Conversion*, 5-9 December, 1994, Waikoloa, HI (in press).

23. D. Cahen and R. Noufi, *J. Phys. Chem. Solids*, **53** (8), 1992, pp. 991-1005.

24. D.S. Albin, G.D. Mooney, J.J. Carapella, A. Duda, J.R. Tuttle, R. Matson, and R. Noufi, *Solar Cells*, 30 (1991), pp. 41-46.

25. R.J. Schwartz and J.L. Gray, *Proceedings 21st IEEE Photovoltaic Specialists Conference*, Kissimmee, FL, 1990, pp. 570-574.

26. J.R. Tuttle, M.A. Contreras, T.J. Gillespie, K.R. Ramanathan, A.L. Tennant, J. Keane, A.M. Gabor, and R. Noufi, *Progress in Photovoltaics*, 1995 (in press).

27. J.R. Tuttle, D.S. Albin, A. Tennant, A.M. Gabor, M. Contreras, J.J. Carapella, Y. Qu, D. Du, and R. Noufi, *Proceedings of the 7th Photovoltaic Science and Engineering Conference*, Nagoya, Japan, Nov. 1993, *Sol. Energy Mats. and Sol. Cells*, **35** (1-4) 1994, pp. 193-202.

28. W.S. Chen, J.M. Stewart, R.A. Mickelson, W.E. Devaney, and B.J. Stanbery, *Final Technical Report for NREL Subcontract No. ZH-1-19019-6,*, September 1993, p. 16.

# Low Cost CIS Device Processing

B. M. Başol, V. K. Kapur, C. R. Leidholm, R. Roe, A. Halani and
G. Norsworthy

*International Solar Electric Technology (ISET)*
*8635 Aviation Blvd., Inglewood, CA 90301*

**Abstract.** CIS films were grown on soda-lime glass/Mo substrates using a low cost, non-vacuum technique. Morphology of the resulting layers was improved and solar cells with 12.4% total area efficiency were demonstrated on these films. A sub-module of about 25 cm$^2$ area was also fabricated with a conversion efficiency of 8.17%. Work is now in progress to grow films containing Ga and /or S and to take this technology to larger scale production.

## INTRODUCTION

CuInSe$_2$ (CIS) and related Group I-III-VI compound absorbers have been used in thin film solar cell structures and devices with over 17% conversion efficiency have been demonstrated (1). A review of the literature shows that high efficiency CIS cells have so far been fabricated on layers grown by the vacuum co-evaporation technique or by various versions of the two-stage selenization method employing evaporated or sputter deposited precursor films. Evaporation techniques are difficult to scale up because of film uniformity and material utilization considerations. Sputtering techniques are better suited for large area deposition, however, they require expensive vacuum equipment and sputtering targets. Therefore, a low cost, non-vacuum technique with a capability to process on large area substrates would be very attractive for the growth of CIS layers for photovoltaic applications. ISET has already reported on a two-stage technique that involved selenization of electrodeposited Cu and In layers, and demonstrated about 10% efficient devices using this non-vacuum approach (2). Compound electrodeposition was also used to obtain films which were further processed to yield over 9% efficient cells (3). In this paper, we present data on over 12% efficient devices fabricated on CIS layers grown by a non-vacuum deposition technique developed at ISET.

## EXPERIMENTAL

Soda-lime glass substrates were used in this work. Mo contact layers were deposited on the soda-lime glass substrates by D.C. magnetron sputtering. CIS films were grown on the Mo surface by a non-vacuum technique. Solar cell fabrication steps included deposition of a thin CdS layer by the commonly used chemical bath deposition approach. A ZnO layer was then deposited on the CdS surface using a MOCVD method. Solar cells of 0.09-1.00 cm$^2$ area were defined on the substrates using photolithographic techniques. Both films and devices were characterized to understand the nature of the material grown by the non-vacuum approach. In addition to solar cell fabrication, some work was also done on sub-module processing. The module integration approach was the same as the technique used on vacuum based devices. The Mo scribing was done by laser. The two scribes following CIS deposition were done mechanically. Details of the cell and sub-module processing steps can be found in our previous publications (4,5).

## RESULTS AND DISCUSSION

The solar cells fabricated on CIS layers obtained by ISET's non-vacuum technique (from now on referred to as the "non-vacuum films/cells") initially yielded efficiencies in the 8-10% range, clearly demonstrating the potential of this low cost process. The illuminated I-V characteristics and the relative spectral response of such an early device is shown in Fig. 1. This cell had a respectable $V_{oc}$ and a good fill factor value. The $J_{sc}$ value, however, was rather low at 32.78 mA/cm$^2$. An in-depth analysis of the device was then carried out to determine its parameters, the carrier density of the

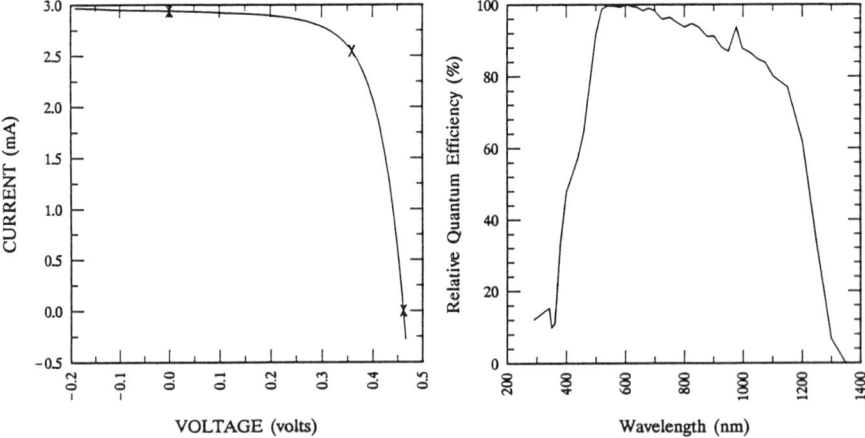

**FIGURE 1.** Light I-V characteristics and relative quantum efficiency of a 0.09 cm$^2$ device fabricated on non-vacuum CIS film. $V_{oc}$=0.4616 V, $J_{sc}$=32.78 mA/cm$^2$, FF=67.51%, η=10.2%.

CIS film and the photocurrent loss mechanisms causing the observed low $J_{sc}$ value. The results of this analysis is shown in Table 1 under the column labeled "10% cell". A review of the device parameters, the carrier density of the film, and the photocurrent losses due to the window layer absorption, reflection and deep penetrating photons, indicated that the early devices fabricated on films grown by the non-vacuum process were very much comparable with the CIS cells fabricated at ISET on evaporated and selenized absorber layers. The only notable difference, however, was the rather large unidentified photocurrent loss of about 7 mA/cm$^2$, which brought down the quantum efficiency curve uniformly across the whole spectrum of the measured response. SEM, EBIC and EDS studies were then initiated to understand this phenomenon.

**TABLE 1.** Device parameters, carrier densities and photocurrent loss mechanism analysis results for the two cells fabricated on non-vacuum CIS layers.

| Parameters | 10.2% cell | | 12.4% cell | |
|---|---|---|---|---|
| | Light | Dark | Light | Dark |
| Shunt r ($\Omega$-cm$^2$) | 700 | 1100 | 1300 | 3400 |
| Series R ($\Omega$-cm$^2$) | 0.2 | 0.2 | 0.3 | 0.5 |
| Diode Factor, A | 1.9 | 2.0 | 1.6 | 1.55 |
| p (x 10$^{16}$ cm$^{-3}$) | 2.7 | | 1 | |
| $J_{sc}$ Losses (mA/cm$^2$) | | | | |
| reflection | 3 | | 4 | |
| window loss | 2 | | 2 | |
| deep penetration | 3.5 | | 3.3 | |
| unidentified | 7 | | 1 | |

Figure 2 shows a surface EBIC image taken from the device of Fig. 1. Clearly, there was a high density of "low-response" regions in this cell, which appear as dark spots in the EBIC image. The dark regions constitute about 20% of the total device area shown in Fig. 2. If we assume that these areas do not contribute at all to photocurrent collection, this clearly explains the origin of the unidentified, wavelength independent current loss of 7 mA/cm$^2$ observed in the data of Table 1.

Most of the small-size low-response areas in Fig. 2 are round in shape with diameters ranging from 5$\mu$m to 30 $\mu$m. A closer look at the larger (up to about 150 $\mu$m in one dimension) low-response regions revealed that these areas were created by convergence of two or more small-size low-response areas, and that there was almost always a large protruding dome-like surface feature close to the center of each low-response area. A high magnification SEM of a 100 $\mu$m size low-response

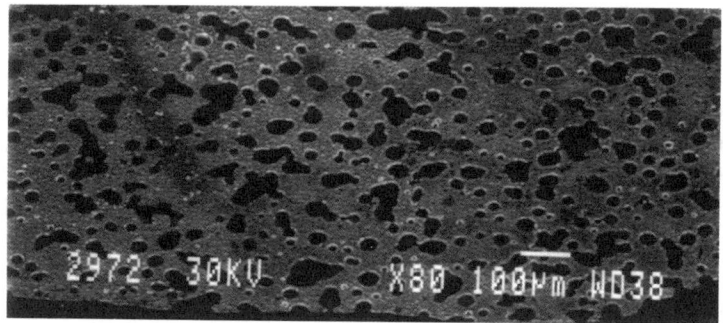

**FIGURE 2.** EBIC surface image of the cell of Figure 1.

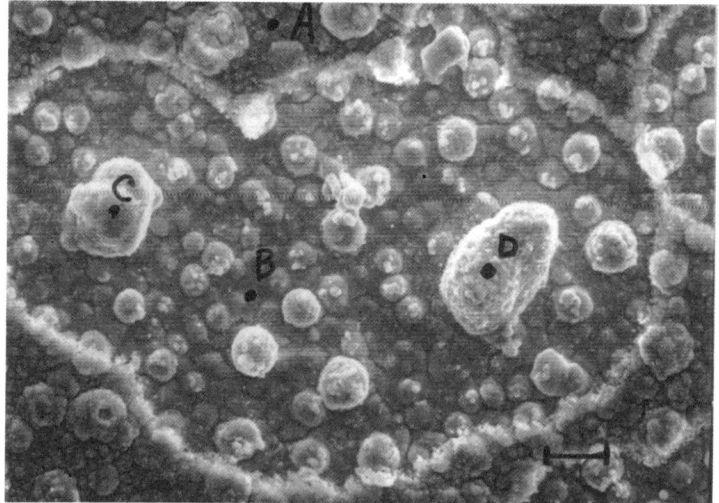

**FIGURE 3.** SEM of a low-response region showing the large surface features (C and D). EDS data was taken from locations A,B,C and D. The scale bar corresponds to 10 $\mu$m.

area is shown in Fig. 3. The large surface features labeled as C and D are clearly seen in this micrograph. EDS analysis made at the locations indicated on this figure showed that the active device region (A) contained both Zn and CIS, whereas, the spots within the low response region (B,C,D) contained only CIS. The chemical composition of the CIS material at points B,C and D were all the same. Therefore, it was concluded that the low response areas were caused by physical detachment of the CdS/ZnO window layer around the region where the dome-like surface features were present, and that the large surface features at C and D did not result from a compositional non-uniformity, such as excess Cu, in the CIS films obtained by the non-vacuum method. Cross sectional SEM was then used to investigate the nature of the dome-like features.

The micrograph of Fig. 4 shows the cross sectional SEM taken of one of the dome-like features of the sample of Fig. 1. This data suggested that the dome-like features resulted from the presence of near-spherical voids within the CIS layer. It was further observed that a 3 $\mu$m size void could effect an area of approximately 10 $\mu$m diameter in size, and that some of the domes (not shown in Fig. 4) lacked the ZnO layer on top of them. Having found the origin of the photogenerated current loss and the rough surface structure of the CIS layer giving rise to that loss mechanism, efforts were concentrated on the improvement of the morphology of CIS films grown by the non-vacuum technique. The stoichiometric uniformity and electronic quality of these films would clearly be adequate for the fabrication of higher efficiency devices once their morphology is improved.

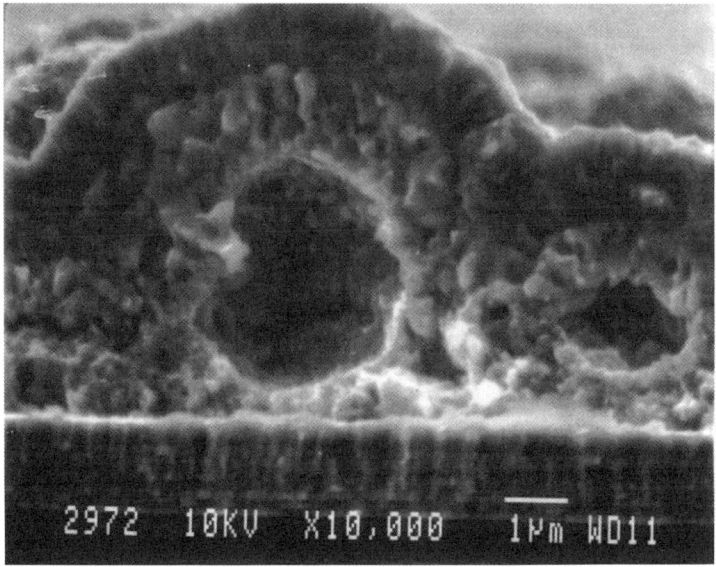

**FIGURE 4.** Cross sectional SEM of the sample of Fig. 1. The top 1.5$\mu$m layer is ZnO. The CIS layer is about 3 $\mu$m thick. The dome-like features have spherical voids under them as can be seen from this data.

The cross sectional SEM of Fig. 5 shows the morphology and grain structure of the new generation of films prepared by ISET's non-vacuum technique. Although there are still small voids present near the CIS/Mo interface, the surfaces of these layers are smoother than those of the early films discussed before. The grain structure observed in Fig. 5 is also improved. Well formed columnar grains of >1 $\mu$m size can be seen in this micrograph.

The illuminated I-V characteristics and the relative quantum efficiency of a device fabricated on the film of Fig. 5 are shown in Fig. 6. The short circuit current density of this 0.09 cm² area device is over 38 mA/cm² and the total area efficiency is 12.4%. The calculated active area efficiency is 13.3%. In-depth analysis of the cell of Fig. 6 was performed and the results are tabulated in Table 1, under the column

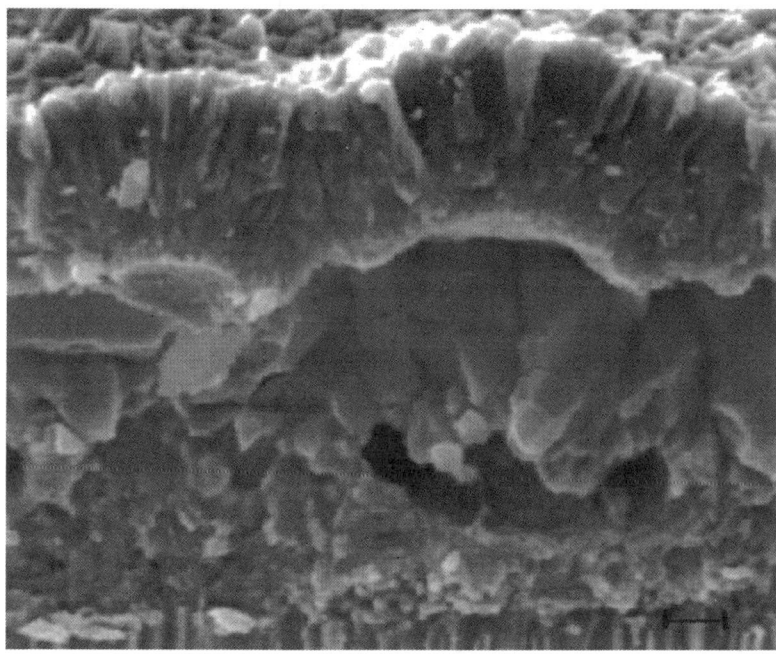

**FIGURE 5.** Cross sectional SEM taken from the improved device demonstrating lower void density and better film morphology. The scale bar corresponds to 1.0 μm.

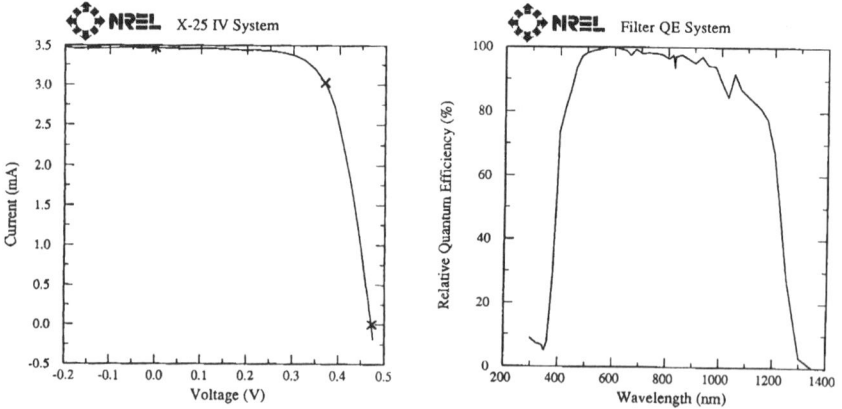

**FIGURE 6.** Light I-V characteristics and the relative Q.E. of a 0.09 cm² area cell fabricated on the sample of Fig. 5. $V_{oc}$=0.4728 V, $J_{sc}$=38.46 mA/cm², FF=67.97% and η=12.4%.

labeled "12.4% cell". This improved device had lower diode factor values and the unidentified current loss mechanism disappeared, as expected. When the quantum efficiency data of Fig. 1 and Fig. 6 are compared, it can be seen that the improved device has better long-wavelength response which is reflected in the smaller loss due to deep penetrating photons as indicated in Table 1.

In addition to discrete cell fabrication, work was also carried out to fabricate integrated sub-modules on CIS layers grown by the non-vacuum technique. The I-V characteristics of a 24.71 cm$^2$ area sub-module containing seven cells is shown in Fig. 7. The open circuit voltage per cell in this device was 0.438 V. The current density at the cell level was calculated to be 34.29 mA/cm$^2$ by taking into account 95% area utilization in the module structure. Fill factor was limited by the sheet resistance of the ZnO layer employed as well as the ZnO/Mo contact resistance.

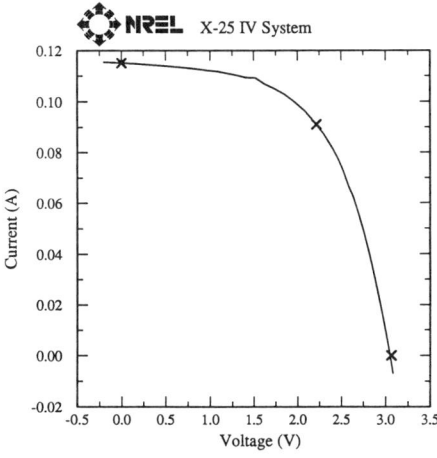

**FIGURE 7.** Illuminated I-V characteristics of a 24.71 cm$^2$ sub-module fabricated on a non-vacuum CIS layer. $V_{oc}$=3.063, $J_{sc}$=4.655 mA/cm$^2$, FF=57.27, η=8.17%.

## CONCLUSIONS

We fabricated over 10% efficient solar cells on CIS layers grown by a non-vacuum deposition technique. Improvement of the morphology and density of the CIS films resulted in devices with over 12% efficiency. To the best of our knowledge this result represents the highest efficiency cells ever produced on CIS films that have been processed entirely at atmospheric pressure. ISET's non-vacuum technique is low cost and it can effectively be applied to large area film deposition. A 25 cm$^2$ sub-module with over 8% conversion efficiency was fabricated in this work. Development is now in progress to establish a facility to produce up to 1 ft$^2$ size modules by the non-vacuum approach. Efforts are also under way to introduce Ga and S into the absorber layers for bandgap engineering and fabrication of higher efficiency cells.

## ACKNOWLEDGEMENTS

Authors gratefully acknowledge the device characterization work done at CSU by Prof. J. Sites' group, data supplied by R. Matson and K. Emery of NREL, and detailed technical discussions with H. Ullal, R. Noufi and K. Zweibel of NREL. This work was supported by NREL subcontract No. ZAF-5-14142-07.

## REFERENCES

1. Tuttle, J.R., et al., presented at the 1996 Spring MRS Meeting, San Francisco, CA, 8-12 April, 1996 (in press).
2. Kapur, V.K. et al., Final Technical Progress Report, DOE/SERI Contract No. XL-5-05036-1, September, 1988.
3. Bhattacharya, R.N., et al., J. Electrochem. Soc., **143**, 854 (1996).
4. Basol, B.M. et al., Proc. 11th European PVSEC, Montreux, Switzerland, 12-16 October, 1992, p. 803.
5. Basol, B.M., et al., Solar Energy Matl. and Solar Cells, **29**, 163 (1993).

# Technologies for Thin Film CIGS-Based Photovoltaics

J. S. Britt, A. E. Delahoy, and Z. J. Kiss

*Energy Photovoltaics, Inc.*
*276 Bakers Basin Road, Lawrenceville, NJ 08648*

**Abstract.** This paper reviews EPV's approach to CIGS formation by a hybrid of sputtering and evaporation processes, and describes a differential thermometry technique for performing time progressive reaction studies of the absorber formation process. Investigations of alternative window processes to replace CdS, the DC reactive sputtering of ZnO films, and recent module results are also discussed.

## INTRODUCTION

The commercialization of CIGS photovoltaic modules has been challenged by a number of technical and scientific issues. Perhaps the most difficult problem is the high rate deposition of CIGS films of uniformly good quality over large area substrates. EPV has squarely addressed this problem by defining and constructing large area vacuum deposition equipment capable of forming CIGS by a number of sequential routes (1). The equipment is even capable of depositing CIGS by coevaporation, but we have chosen not to pursue this route. Sequential deposition processes are advantageous because control of a single deposition rate is simpler than the simultaneous control of several rates required by coevaporation. The manufacturing equipment utilizes deposition by sputtering and by linear source evaporation in conjunction with substrate motion. This allows a number of hybrid recipes to be explored.

The ZnO transparent conductor presents a similar problem. Ceramic sputtering targets have unacceptably low deposition rates and are also expensive. Among other options, EPV has investigated the DC reactive sputtering from Zn:Al targets as a low-cost alternative. The presence of CdS in a module is considered to be undesirable, and its replacement or elimination would help alleviate concerns about consumer safety and module disposal. For this reason, alternative window layers and beneficial chemical treatments have been explored.

# ABSORBER DEVELOPMENT

EPV's manufacturing philosophy embraces all-vacuum deposition techniques such as evaporation and sputtering to provide high throughput and a high degree of film quality and process control. Cu is deposited in the large area deposition equipment by DC magnetron sputtering with the plate translated past the sputtering target. The selenides and Se are delivered to the substrate from linear evaporation sources while the plate is translated along its long axis perpendicular to the long axis of the linear sources. The deposition can occur simultaneously with substrate heating, even with the glass substrate heated beyond its softening point. These deposition techniques are best suited toward the implementation of recipes involving sequential deposition and reaction stages (2).

The linear evaporation source developed by EPV is capable of continuous, high rate deposition of films with uniform quality and can easily be scaled up to coat substrates of almost any practical width. We have recently succeeded in implementing a means for the continuous reloading of source material without breaking vacuum. The manufacture of the source itself is uncomplicated and the materials comprising it are inexpensive and easily procured. These advantages are expected to help keep manufacturing costs low.

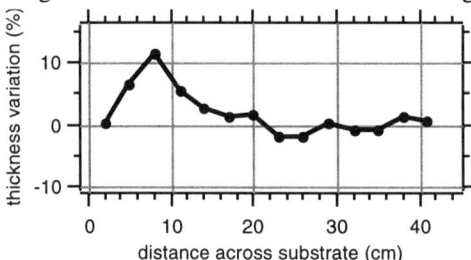

**Figure 1.** The variation in In-Se thickness across the width of a plate.

The design and use of the linear source is crucial for depositing films with uniform thickness along the length and across the width of a substrate. The linear source deposition pattern has been optimized empirically and with the aid of computer models. Fig. 1 shows the variation in thickness across the width of a plate for an $In_xSe_y$ film deposited from a linear source.

EPV has applied the general class of recipes within a small-area bell-jar deposition system. The highest total-area efficiency to date was 13.9% and was achieved with the structure soda lime glass / Mo / CIGS / thin CdS / ZnO / $MgF_2$. The NREL verified JV curve is shown in Fig. 2. Guided by these experiments, a new process has been defined for the large area system. CIGS films with excellent adhesion and uniformity are now made in the system from which small area devices with efficiencies over 9% have been extracted.

## Differential thermometry

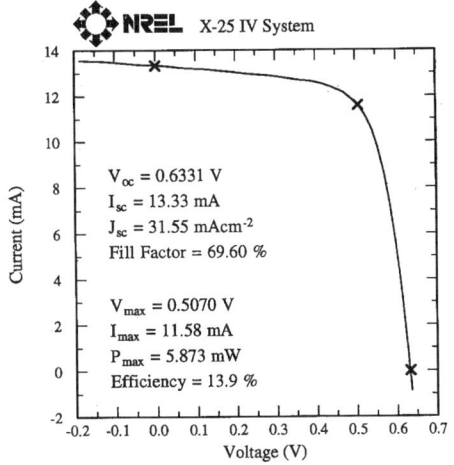

Figure 2. Small-area JV curve.

The formation of CIS by the selenization of metallic precursors has been demonstrated to proceed via a sequence of intermediate reactions (3). Understanding these intermediate reactions is critical to developing recipes that yield higher film quality, minimize processing time, and maximize Se utilization. The standard technique to determine what the intermediate products are and at what temperatures they form is stop the reaction at various points and assess the film crystallography by XRD. This technique provides valuable information but is limited by its punctuated nature. To aid in our recipe development, we have implemented a differential thermometry (DT) technique that allows time progressive reaction studies to be performed.

The technique consists of directly monitoring the film temperature of (Cu, In, Ga) films as they are being ramped in temperature and exposed to Se. To ensure accurate monitoring of the film temperature and to minimize thermal mass, the films are deposited directly onto a thin film RTD. The temperature is feedback controlled by an identical RTD (with no films) located adjacent to the first one. The resistances of the reference and sample RTDs are monitored in the four-wire configuration and the data is recorded on a PC. In one experiment, Cu and In were sequentially deposited by evaporation onto the thin film RTD. The relative thicknesses of the Cu and In films were adjusted to provide the slightly Cu-poor composition necessary for efficient devices. In another experiment, a film of In only was deposited with a thickness identical to the In film deposited in the first experiment. In both experiments, the films were annealed at 140°C for several minutes and then exposed to a Se flux of 3 nm/sec while the temperature was ramped at a constant rate to 530°C. After soaking at 530°C for several minutes, the RTDs were allowed to cool to 140°C and then heated in a time-temperature profile identical to the first one, still using the reference RTD for feedback control. The temperature of the sample RTD during this second temperature ramp was then subtracted from its temperature during the first ramp. The difference is plotted versus the reference RTD temperature. The results of these experiments are shown in Fig. 3.

A positive peak in the differential temperature indicates the occurrence of an exothermic reaction, a negative peak indicates an endothermic reaction. Changes in emissivity due to surface roughness, for example, may also be evident by this technique. At least three of the peaks appear to be common to both plots and may be attributed to the formation of indium selenides or oxides. Further analysis and a thorough literature review are necessary to definitively assign peaks to various reactions.

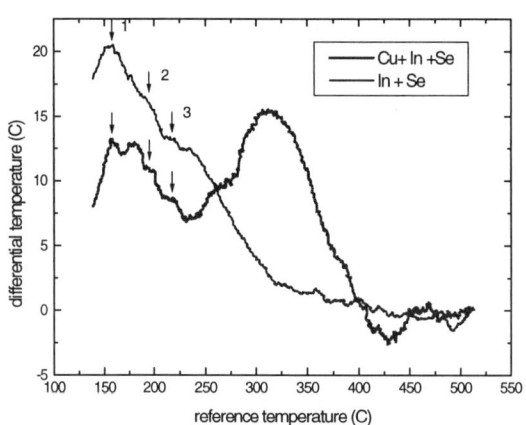

**Figure 3.** DT profiles of the selenization of (Cu,In) and In films.

## ALTERNATIVE WINDOW PROCESSES

Several groups have demonstrated efficient CIGS devices by replacing CdS with films of $In_xSe_y$, ZnSe, and other wide-bandgap materials (4,5). We have investigated $In_xS_y$ for this purpose because its bandgap is predicted to be larger than that of $In_xSe_y$ and the potentially beneficial presence of sulfur near the junction. Another topic of interest is the use of chemical treatments of the absorber surface to improve device photovoltaic characteristics. Some potentially beneficial effects of such chemical treatments are the removal of CuSe and native oxides, passivation of the surface and grain boundaries or other modification of electronic properties in the vicinity of the junction. Understanding the effect of such treatments could lead to their replication using vacuum techniques.

### $In_xS_y$ buffer layers

$In_2S_3$ source material was evaporated to completion from a graphite effusion cell to avoid drifting of the source composition. Witness glass slides were located next to the Mo/CIS substrate during the deposition for optical and thickness measurements. The composition of the films deposited in this manner are unlikely to be stoichiometric $In_2S_3$ because of the lack of a significant overpressure of S. Substrate temperatures were in the range 150°-250°C and film thicknesses ranged between 100-200nm. The transmission characteristics of $In_xS_y$ films deposited on glass were found to be poor at a substrate temperature of 150°C and improved with increasing deposition temperature.

The illuminated JV characteristics of devices with standard window layers and $In_xS_y$ buffer layers deposited on a common CIS film are listed in Table 1. The open-circuit voltages of the $In_xS_y$ devices were consistently lower than the CdS device, but the $J_{sc}$ and fill factor were similar in some cases. The quantum efficiencies of the best $In_xS_y$ devices are lower than that of the CdS devices at wavelengths greater than 520nm but are greater at shorter wavelengths such that the short-circuit current densities are comparable. Improving the transmission characteristics of the $In_xS_y$ films by optimizing the substrate temperature and film stoichiometry would allow thicker films to be deposited, with a potential improvement in device characteristics.

**TABLE 1.** JV characteristics of various device structures.

| Device structure | $V_{oc}$ (mV) | $J_{sc}$ (mA/cm$^2$) | FF (%) | Efficiency (%) |
|---|---|---|---|---|
| CIS/In$_x$S$_y$/ZnO | 382 | 34.7 | 62.5 | 8.3 |
| CIS/CBD CdS/ZnO | 431 | 34.3 | 65.5 | 9.7 |
| CIS/ZnO | 272 | 27.3 | 50.9 | 3.8 |

# Chemical Treatment

A novel surface treatment that consistently improves the photovoltaic characteristics of CIGS devices has been identified in the course of this investigation. The etchant is a cadmium-free compound, and the surface treatment consists of a brief dip in a heated solution of the compound, followed by a thorough water rinse. The improvement in efficiency was originally noted in direct CIS/ZnO devices where $V_{oc}$ and fill factor were enhanced. Further experiments were conducted comparing the effect of the surface treatment on devices with the standard CIS/(CBD) CdS/ZnO structure and again there was a significant improvement in efficiency (Table 2). In many cases, this surface treatment also allows the use of significantly thinner CdS films, approximately 1/5 the standard thickness, with only a small reduction in efficiency. Further work is required to understand the effect of this surface treatment and replicate it in a vacuum process.

**TABLE 2.** JV Characteristics of CIS/CBD CdS/ZnO Devices.

| Surface treated | $V_{oc}$ (mV) | $J_{sc}$ (mA/cm$^2$) | FF (%) | Efficiency (%) |
|---|---|---|---|---|
| no | 389 | 28.0 | 60.9 | 6.6 |
| yes | 442 | 34.9 | 70.7 | 10.9 |

## REACTIVE SPUTTERING OF ZnO

For deposition of a high quality ZnO:Al transparent conductor onto substrates up to 930 cm$^2$ in size we employ RF magnetron sputtering of a ZnO:Al$_2$O$_3$ target. This method is not suitable for a high throughput CIGS manufacturing plant because of inadequate deposition rate and high target cost. As an interim method of coating 0.43 m$^2$ modules we employ bipolar sputtering from an oxide target in an Ar/O$_2$ gas mixture. Film properties achievable by this method have so far been marginal (6).

For production, DC reactive sputtering from a Zn:Al target is preferred, and we have performed experiments to understand the nature of this process. The central result is that the type of film obtained (metallic, transparent/conducting, or insulating) is controlled by the oxidation state of the target. At low deposition rates, discharge voltage is a useful indication of the oxidation state, but at normal deposition rates strong hysteresis and runaway effects render the desired intermediate oxidation states inaccessible. This is illustrated in Fig. 4 in which stable operating conditions between points C and D (or between points E and F) cannot be achieved by any fixed value of oxygen flow. We found that the transition CD was heralded by a 25% increase in the optical emission of excited Zn. We conclude that control signals derived from

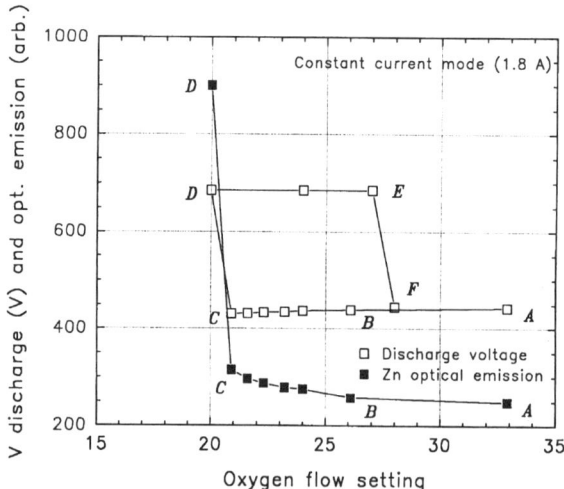

**Figure 4.** Discharge voltage and optical emission from excited Zn as a function of oxygen flow during reactive sputtering of Zn:Al in constant current mode.

direct measurement of the oxygen partial pressure, or of the Zn emission, could be used to maintain a stable process.

Film properties indistinguishable from those of RF sputtered ZnO:Al were achieved by reactive sputtering from a Zn:Al target, although only in a static mode. Substrate translation led to layering of favorable and unfavorable growth conditions, the latter apparently resulting from negative ion bombardment of the film. Additional work is required to completely implement the envisaged deposition scheme.

## MODULE ENCAPSULATION AND TERMINATION

After patterning, border isolation is accomplished by removing all thin films within 0.75" of the module perimeter by an abrasion technique (e.g. sandblasting). Electrical contacts are made by bonding two Al foil strips parallel to the scribes along opposite sides of the module. The foil strips are brought out behind the module through a hole drilled in the glass substrate. The CIS circuits are laminated to another glass plate with an EVA sheet (12 mil nominal thickness) in a vacuum laminator. A two conductor cable is attached to the foil strips and the entire lead-out area is electrically and environmentally isolated by forming a boot from a two-part thermally-cured silicone RTV.

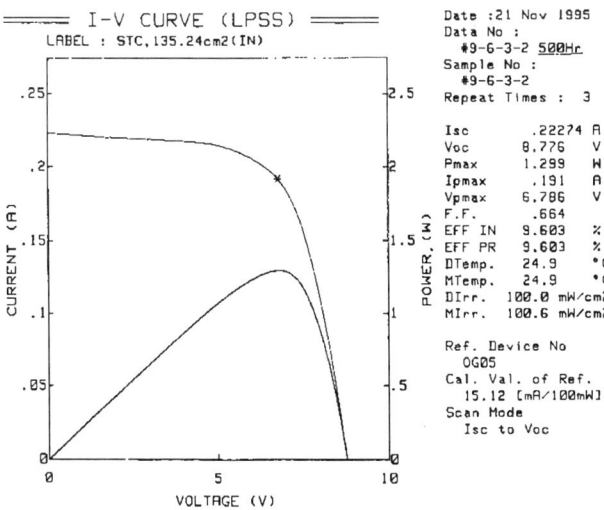

**Figure 5**. Laminated mini-module with an efficiency of 9.6% after 500 hours light-soaking.

A mini-module encapsulated and terminated using the described procedure has been tested after 500 hours of light-soaking. The aperture area efficiency of the mini-module was 9.6% (Fig. 5).

## STATUS AND FUTURE DIRECTIONS

All processing steps for large area modules (0.43 $m^2$) from film deposition to encapsulation and termination are conducted daily in our plant to improve each aspect of module fabrication. Improvements to our CIGS deposition process have led to better adhesion and uniformity while internal efficiencies have reached a level that will allow us to meet our goal of an 8% aperture area efficiency. We continue to wrestle with peripheral problems such as ZnO quality (transmission and conductivity) and Mo/ZnO contact resistance that currently limit module efficiencies, but are confident that these issues can be resolved in the near future.

## ACKNOWLEDGMENTS

The authors thank their colleagues at EPV (G. Butler, J. Esler, A. Kovacs, C. Seal, A. Sizemore, F. Ziobro, G. McComiskey, M. Doroba and others) for their contribution to this work. We thank Andrew Gabor for scientific contributions and the fabrication of high efficiency CIGS cells. We are indebted also to the scientific and program staff at NREL for their continued support and interest, especially R. Noufi, J. Tuttle, H. Ullal, and K. Zweibel. This work was partially funded by the U.S. DOE under subcontract ZAF-5-14142-04 from NREL.

## REFERENCES

1. Delahoy, A.E., Britt, J.S., Gabor, A. M., and Kiss, Z. J., "Large-Area, Thin-Film CIS Deposition and Module Fabrication," AIP Conference Proceedings Volume 353, 1993, pp. 3-11.
2. Gabor, A. M., Britt, J.S., Delahoy, A. E., Noufi, R., and Kiss, Z. J., "Manufacturing Compatible Methods for the Formation of Cu(In,Ga)Se$_2$ Thin Films," presented at the 25$^{th}$ IEEE Photovoltaic Specialists Conference, Washington DC, May 13-17, 1996.
3. Yamanaka, S., McCandless, B. E., Birkmire, R. W., "Reaction Chemistry of CuInSe$_2$ Formation by Selenization Using Elemental Se," in Proceedings of the 23$^{rd}$ IEEE Photovoltaic Specialists Conference, 1993, pp. 607-612.
4. Olsen, L.C., Addis, F.W., and Huber, D.A., "Investigation of polycrystalline thin film CuInSe2 solar cells based on ZnSe windows," in Proceedings of the 23$^{rd}$ IEEE Photovoltaic Specialists Conference, 1993, pp. 603-606.
5. Ohtake, Y., Ichikawa, M., Yamada, A., and Konagai, M., "Cadmium-free buffer layers for polycrystalline Cu(In,Ga)Se$_2$ thin-film solar cells," in Proceedings of the 13$^{th}$ European Photovoltaic Solar Energy Conference, 1995, pp. 2088-2091.
6. Delahoy, A.E., and Cherny, M., "Deposition Schemes for Low Cost Transparent Conductors for Photovoltaics," Mat. Res. Soc. Symp. Proc. **426**, 467 (1996).

# Fabrication and Characterization of Cu(InGa)Se$_2$ Solar Cells with Absorber Bandgap from 1.0 to 1.5 eV

W. N. Shafarman, R. W. Birkmire, M. Marudachalam,
B E. McCandless, and J. M Schultz[†]

*University Center of Excellence for Photovoltaic Research and Education*

*Institute of Energy Conversion, University of Delaware, Newark, Delaware 19716*
[†]*Dept. of Chemical Engineering, University of Delaware, Newark, Delaware 19716*

**Abstract.** Cu(InGa)Se$_2$ films were deposited by selenization of Cu/Ga/In precursor layers and by four source elemental evaporation. Characterization of films and devices is presented. The selenized films show that the process results in two phase films and the devices behave like CuInSe$_2$ devices. A high temperature anneal converts the films to single phase, resulting in an increased V$_{oc}$. The Ga and In are uniformly incorporated in the evaporated Cu(InGa)Se$_2$ films and consequently V$_{oc}$ increases as the Ga content increases. However, the performance of the evaporated devices with high Ga are limited by voltage dependent current collection. The orientation of Mo and Cu(InGa)Se$_2$ films are found to be related but in this case there is no correlation between the film characterization and device behavior. Cu(InGa)Se$_2$ films with the relative orientation differing by two orders of magnitude give nearly identical device results.

## INTRODUCTION

One of the challenges in the development of thin film photovoltaics is relating deposition and fabrication of films and devices with properties of the thin film materials and the behavior of completed devices. For film formation, critical control parameters include the delivery of species to the substrate and the reaction time-temperature profile. Film properties which are typically measured include the composition and structure of the films and their electronic and optical properties. Device characterization needs to be based on a model for device operation. Device measurements which could be related to the film properties and deposition include current-voltage (J-V) and quantum efficiency (QE) as well as more detailed device analyses.

In this work, some of these connections have been made with regard to the incorporation of Ga in Cu(InGa)Se$_2$ films containing a range of Ga content to give bandgaps of 1.0 to 1.5 eV. The increased bandgap (E$_g$) results in increased V$_{oc}$

and a decrease in $J_{sc}$ [1] which is advantageous for module design by reducing losses related to the transparent conducting oxides and interconnect spacing [2].

Cu(InGa)Se$_2$ films were deposited by selenization of Cu/Ga/In precursors and by four source elemental evaporation. Selenization is considered a promising process for large scale deposition of Cu(InGa)Se$_2$ in part because the metal precursors can be deposited using low temperature processes, eliminating the need for thermal evaporation of Cu at up to 1400°C. Sequential deposition of the metal layers onto an unheated substrate may also enable the use of deposition methods with demonstrated ability to achieve large area uniformity. The Se for reaction can be provided by either a chemical vapor delivery using H$_2$Se or physical vapor delivery from an elemental Se source.

Multisource evaporation utilizes simultaneous delivery of Cu, In, Ga, and Se onto a substrate heated to 400-600°C. This process has been used to fabricate the highest efficiency thin film solar cells. Evaporation allows good control of film composition and alloying with homogenous or compositionally graded absorber layers.

## DEPOSITION AND CHARACTERIZATION

For the selenized films, Cu/Ga/In precursor layers were sequentially deposited by dc sputtering on a glass/Mo substrate. These were reacted in a flowing H$_2$Se/Ar/O$_2$ atmosphere at 450°C for 90 min [3]. A Cu/In/Ga precursor stack resulted in poor morphology after selenization. The evaporated Cu(InGa)Se$_2$ deposition was described in [1]. The bilayer deposition incorporated a Cu-rich layer followed continuously by a layer containing no Cu so that the final film was slightly Cu-deficient. Ga, In and Se effusion rates were kept constant.

Device fabrication and characterization of films and devices were described previously [1]. Film composition and structure were characterized by energy dispersive x-ray spectroscopy (EDS), x-ray diffraction (XRD), and Auger depth profiles. Devices were completed with the structure glass / Mo / Cu(InGa)Se$_2$ / CdS / ZnO / top contact or grid. Device measurements include J-V characteristics under 100 mW/cm$^2$ AM1.5 illumination and QE as a function of voltage and light bias.

## SELENIZED Cu(InGa)Se$_2$

The selenization of Cu/Ga/In results in a two phase film with CuGaSe$_2$ near the Mo back contact and CuInSe$_2$ at the top of the film. This was shown by both XRD and Auger measurements [3]. In addition, EDS measurements which are sensitive to the top ~ 1 μm of the film, show little Ga. Devices made from these films have low $V_{oc}$, and behave similar to a $CuInSe_2$ device, consistent with the lack of Ga, and therefore low bandgap, in the front region of the absorber layer where the device behavior is controlled. Selenization at temperatures up to 650°C does not yield a single phase film. However, annealing the film at T ≥ 550°C in an inert atmosphere results in interdiffusion of the In and Ga, converting the film to single phase $Cu(InGa)Se_2$. This can be seen by the XRD results shown in Fig. 1 for a sample with $x \equiv Ga/(In+Ga) = 0.5$. The as-selenized film has peaks corresponding to $CuInSe_2$ and $CuGaSe_2$. After annealing at T = 600°C the peaks sharpen to show a single phase $Cu(InGa)Se_2$ with lattice spacing corresponding to the Ga content of the initial Cu/Ga/In precursor.

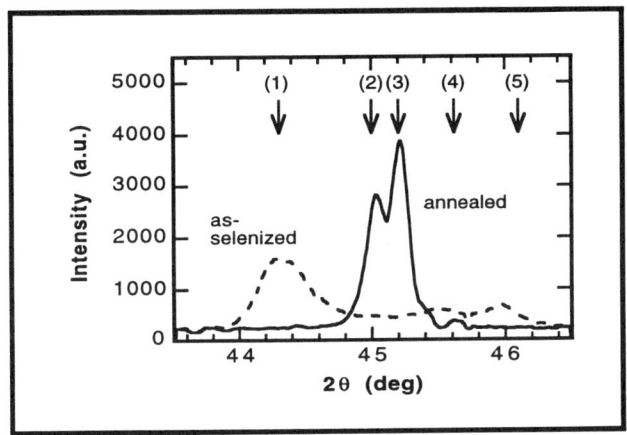

**FIGURE 1.** XRD scans of the (220)/(204) peaks of as-selenized and annealed films with x = 0.5. Arrows indicate the peak positions for (1) $CuInSe_2$ (220)/(204), (2) $Cu(InGa)Se_2$ (220), (3) $Cu(InGa)Se_2$ (204), (4) $CuGaSe_2$ (220), and (5) $CuGaSe_2$ (204).

As the film characterization shows interdiffusion with annealing, the device behavior shows a shift in the QE long wavelength fall-off and an increase in $V_{oc}$. The QE curves for devices with x = 0.5 as-selenized and with the high T anneal are compared in Fig. 2. The shift at long wavelength corresponds to an increase from ~1.0 eV to ~1.3 eV corresponding to a change in x from 0 to 0.5. The J-V parameters $V_{oc}$ and $J_{sc}$ without and with the annealing are listed in Table 1. $V_{oc}$ increases and $J_{sc}$ decreases as the bandgap at the front of the absorber layer increases.

A critical issue raised by the results of selenizing the Cu/Ga/In layers is the need to quantify and understand the interdiffusion of In and Ga in this system. The interdiffusion coefficients were determined using a $CuGaSe_2/CuInSe_2$ diffusion couple [4]. The samples were intentionally made Cu-deficient since interdiffusion of In and Ga was found to be enhanced by the presence of copper selenide in Cu-rich samples [5].

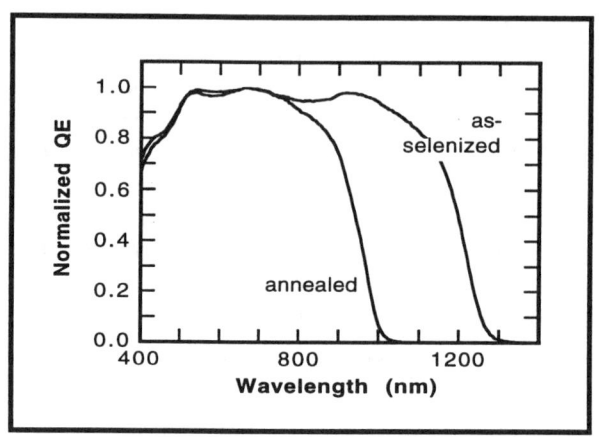

**FIGURE 2.** Quantum efficiency curves for devices with x = 0.5 as-selenized and after annealing.

**TABLE 1.** Device parameters with selenized Cu(InGa)Se$_2$ before and after annealing.

| x | Anneal (°C) | $V_{oc}$ (V) | $J_{sc}$ (mA/cm$^2$) |
|---|---|---|---|
| 0.25 | ... | 0.46 | 39 |
| 0.25 | 600 | 0.56 | 34 |
| 0.5 | ... | 0.53 | 38 |
| 0.5 | 600 | 0.59 | 30 |

The diffusion couple was annealed at temperatures up to 650°C for 30 min in Ar. Auger depth profiles of the In and Ga before and after annealing are shown in Fig. 3. The profiles of the total of In+Ga as well as Cu and Se remained constant and unchanged by the anneal. The interdiffusion coefficients were then determined from the profiles using Fick's second law of diffusion for the case of two semi-infinite layers. This gave $D_{In} = 1.5 \times 10^{-11}$ cm$^2$ s$^{-1}$ and $D_{Ga} = 4.0 \times 10^{-11}$ cm$^2$ s$^{-1}$. The similar interdiffusion coefficients of the In and Ga indicate that the gradient in the as-selenized films is not caused by different diffusion rates of In and Ga in the ternary phases. Instead, the gradient is probably related to differences in the surface energies of CuInSe$_2$ and CuGaSe$_2$ or different reaction rates leading to the formation of the ternary phases.

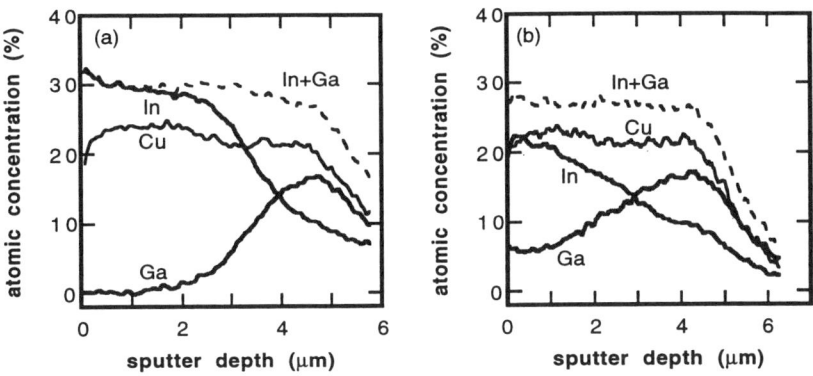

**FIGURE 3.** Auger depth profiles of the In and Ga in $CuGaSe_2/CuInSe_2$ diffusion couples, (a) as deposited and (b) after annealing at 650°C for 30 min.

## EVAPORATED Cu(InGa)Se$_2$

Cu(InGa)Se$_2$ films have been deposited with $x \equiv Ga/(In+Ga)$ ranging from 0.25 to 0.8, as determined by EDS. Unlike the as-selenized films, the evaporated films have the Ga and In uniformly distributed throughout the films. XRD peaks have no compositional broadening or asymmetry, as shown for the (220)/(204) peak doublet in Fig. 4. The shift in 2θ and doublet spacing change as the lattice parameter and c/a decrease with increasing x. Thus, the lattice parameter was used to determine x. This gave good agreement with the values determined by EDS and also with the optical bandgap measured on glass/Cu(InGa)Se$_2$ monitor pieces. Finally, the uniformity of the Ga and In was confirmed by Auger profiles.

Devices with the evaporated Cu(InGa)Se$_2$ had $V_{oc}$ increasing over the entire range of x as expected for the increase in bandgap with additional Ga incorporation [6]. The efficiency (η) was ~15 % for $E_g < 1.3$ eV but decreased for greater bandgap. The bandgap dependence of $V_{oc}$ and η is shown in Fig. 5.

A critical issue raised in this case is the mechanism for the decrease in efficiency with high bandgap. Detailed analysis of J-V data both in the dark and under illumination showed that all the results in the dark could be described by a standard diode equation. However, under illumination for $E_g > 1.3$ eV there was an additional loss attributed to a voltage dependent current collection which results in a loss in $J_{sc}$ and FF. This was also seen by an increasing voltage bias dependence to the QE. The voltage dependent collection is shown by the voltage dependence of the normalized current difference $(J_{illum} - J_{dark})/J_L$ where $J_{illum}$

and $J_{dark}$ are the currents measured under illumination and in the dark respectively. $J_L$ is taken as $J_{illum}$ at V = -1 V. This is shown in Fig. 6 for $E_g$ increasing from 1.21 to 1.54 eV. As the bandgap increases the voltage dependence of the current clearly increases. This behavior suggests that the minority carrier diffusion length is small in these devices so that the collection of light generated current is dependent on the space charge width which varies with the applied voltage.

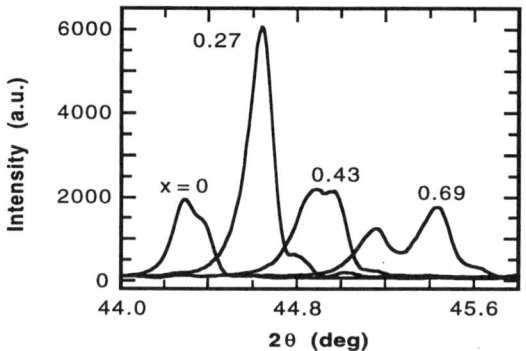

**FIGURE 4.** XRD scans of the (220)/(204) peaks of evaporated films with x increasing from 0 to 0.69.

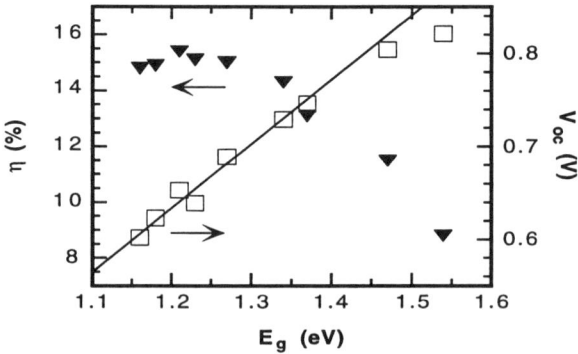

**FIGURE 5.** Bandgap dependence of η and $V_{oc}$ for devices with evaporated Cu(InGa)Se$_2$.

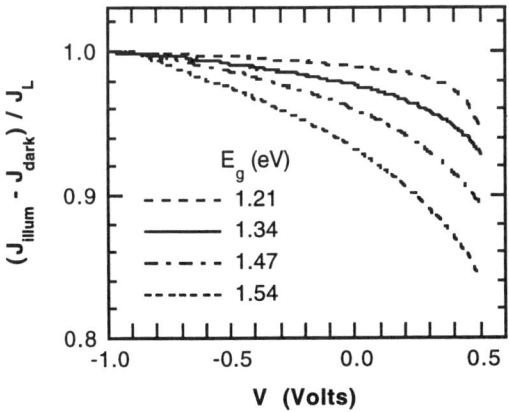

**FIGURE 6.** Voltage dependence of the normalized current difference ($J_{illum} - J_{dark}$)/$J_L$ with increasing bandgap.

## Mo/Cu(InGa)Se$_2$ ORIENTATION

The effect of the relative orientation of the Mo and Cu(InGa)Se$_2$ films, as measured by XRD, and its correlation to device performance has been evaluated. Cu(InGa)Se$_2$ films with $x = 0.3$ were deposited by evaporation on two different Mo films in a single deposition. The Mo films were deposited in different sputtering systems but both were 1 μm thick with sheet resistance 0.2 Ω/sq. The primary difference in the depositions was that one film was deposited in about 60 layers on a rotating substrate while the other was deposited in a single layer on a stationary substrate.

The orientations of the Mo and Cu(InGa)Se$_2$ films are characterized by the two strongest peaks for each in Table 2 and compared to the JCPDS powder diffraction standards. The multi-layer Mo film is more strongly oriented than the single layer film. The Cu(InGa)Se$_2$ film deposited on the more oriented Mo film is nearly randomly oriented with I(112)/I(220) comparable to the powder standard. However, the Cu(InGa)Se$_2$ film deposited on the less oriented Mo has a strong (112) orientation with I(112)/I(220) more than 2 orders of magnitude greater.

J-V parameters for devices made from these films are listed in Table 3. Despite the difference in the film orientation, the device results are nearly identical.

**TABLE 2.** XRD peak intensities, I, for Cu(InGa)Se$_2$ and Mo films from a single evaporation run. The JCPDS powder diffraction standards are listed for comparison.

|  | multi-layer Mo | single layer Mo | JCPDS |
|---|---|---|---|
| | Mo peak intensities | | |
| I(110) | 100 | 100 | 100 |
| I(211) | 1.3 | 6.2 | 39 |
| I(110)/I(211) | 79 | 16 | 2.6 |
| | Cu(InGa)Se$_2$ peak intensities | | |
| I(112) | 100 | 100 | 100 |
| I(220) | 78 | 0.46 | 40 |
| I(112)/I(220) | 1.3 | 217 | 2.5 |

**TABLE 3.** Device parameters with Cu(InGa)Se$_2$ deposited on different Mo films.

|  | $V_{oc}$ (mV) | $J_{sc}$ (mA/cm$^2$) | FF (%) | η (%) |
|---|---|---|---|---|
| multi-layer Mo | 646 | 31 | 72 | 14.5 |
| single layer Mo | 635 | 32 | 70 | 14.4 |

# CONCLUSIONS

The compositional distributions of In and Ga in Cu(InGa)Se$_2$ was characterized in films and devices. Films formed by selenization have a two phase CuGaSe$_2$/CuInSe$_2$ structure which results in devices which behave like CuInSe$_2$. The films are converted to single phase Cu(InGa)Se$_2$ by a high temperature anneal which increases $V_{oc}$ and shifts the QE edge as Ga increases the bandgap the near the front of the absorber. The In and Ga were found to interdiffuse with nearly equal rates in the ternary or quarternary films. Evaporated Cu(InGa)Se$_2$ films have uniformly distributed In and Ga and consequently $V_{oc}$ increases as the total Ga content increases. However, the devices with high Ga are limited by voltage dependent current collection.

A correlation was found between the relative orientation of the Mo and Cu(InGa)Se$_2$ films, as measured by XRD. However, unlike the compositional profile of the Cu(InGa)Se$_2$ films which shows a direct correlation to the device behavior, the orientation of the evaporated Cu(InGa)Se$_2$ shows no correlation to the device results.

## ACKNOWLEDGMENTS

The authors thank the entire staff at IEC including S. Gordon, K. Hart, H. Hichri, R. Klenk, and J. Phillips, and A. Swartzlander of NREL who provided Auger measurements. This work was supported by NREL subcontract XAV-3-13170-01 and EPRI subcontract 8063-03.

## REFERENCES

1. Shafarman, W.N., Klenk, R., and McCandless, B.E., "Device and material characterization of Cu(InGa)Se$_2$ solar cells with increasing band gap", *J. Appl. Phys.* **79**(9), 1996, p. 7324.
2. Kessler, J., Wiedeman, S., Russell, L., Fogleboch, J., Skibo, S., Arya, R. and Carlson, D., "Front Contact Optimization for Cu(In,Ga)Se$_2$ (Sub)Modules", in *Proc. of 25th IEEE PVSC*, 1996, p. 885.
3. Marudachalam, M., Hichri, H., Birkmire, R.W. and Schultz, J.M., "Preparation of homogeneous Cu(InGa)Se$_2$ films by selenization of metal precursors in H$_2$Se atmosphere", *Appl. Phys. Lett.* **67**, 1995, pp. 3978-3980.
4. Marudachalam, M., Hichri, H. Birkmire, R.W., Schultz, J.M., Swartzlander, A.B., Al-Jassim, M.M., "Diffusion of In and Ga in Selenized Cu-In and Cu-Ga Precursors", in *Proc. of 25th IEEE PVSC,* 1996, p. 805.
5. T. Walter and H.W. Schock, "Crystal growth and diffusion in Cu(In,Ga)Se$_2$ chalcopyrite thin films", *Thin Solid Films* **224**, 74-81 (1993).
6. Shafarman, W.N., Klenk, R., and McCandless, B.E., "Characterization of Cu(InGa)Se$_2$ Solar Cells with High Ga Content", in *Proc. of 25th IEEE PVSC,* 1996, p. 763.

# Progress in Thin Film CIGS Modules

S. Wiedeman, J. Kessler, L. Russell,
J. Fogleboch, S. Skibo, and R. Arya

*Solarex*
*(A business unit of Amoco/Enron Solar)*
*826 Newtown-Yardley Road*
*Newtown, PA 18940*
*(215) 860-0902*

**Abstract.** We have demonstrated 13% conversion efficiency in 40 cm$^2$ submodules based on CIGS. We review the developments that have made this possible, and identify several remaining areas where better understanding or further improvement would have a significant effect on the commercial viability of large area thin film modules based on CIGS. We report some recent results in these remaining areas, specifically on the achievement of larger area deposition for the absorber layer, the achievement of highly conductive ZnO, and some results on the function of the buffer and contacting layers in the CIGS based structure.

## Introduction

The demonstration of higher efficiency modules based on Cu(In,Ga)Se$_2$ (CIGS) hinges on the electronic quality and uniformity of the absorber layer, the reduction of electrical and optical losses due to the contacting layers, electrical and area losses in the scribe interconnect structure, and shunt losses in the module. After demonstration, or proof-of-concept, of high efficiency modules based on CIGS, commercialization further demands adequate reproducibility and control, low direct costs, and scale-up to large areas and high production rates. The issue of direct cost has been evaluated by others for CIS based PV to be in the range of, or less than, $1.00/watt [1,2]. The materials costs of the active thin film layers have been estimated [3] to be $5.60/m^2$, only a small fraction of the cited cost per watt above.

We have previously reported a method of electrical analysis for the module structure which gives the expected loss in a module design due to electrical and optical losses in the contacting layers and electrical and area losses in the scribe interconnect structure [4,5]. We have also reported the development of module interconnect scribes in monolithic structures [4,6] which minimize the resistive and area losses in the interconnects. Total interconnect widths as small as 150 microns [6,7] with an interconnect resistivity which is negligible in comparison to the resistance of other module elements have been achieved [6]. Using the parameters of these interconnects and of our standard RF sputtered ZnO from a ceramic target, the above cited analysis method indicates that an optimized design should produce module efficiencies which are only about one percentage point lower than those of small area devices produced from the same materials. Experimentally, as shown in Figure 1, a 40 cm$^2$ laminated submodule having an aperture area efficiency of 13% has been made from absorber layers on which the best small area cell was 15.5% efficient (total area basis), shown in Figure 2. Typically about 80 small area cells are produced on a 40 cm$^2$ area which average about one percentage point less in efficiency than the maximum value. It is therefore reasonable to consider that the overall "quality" of the absorber used for the 13% submodule would have resulted in small area cells typically between 14% and 14.5% efficiency, in agreement with the cited module loss analysis. As

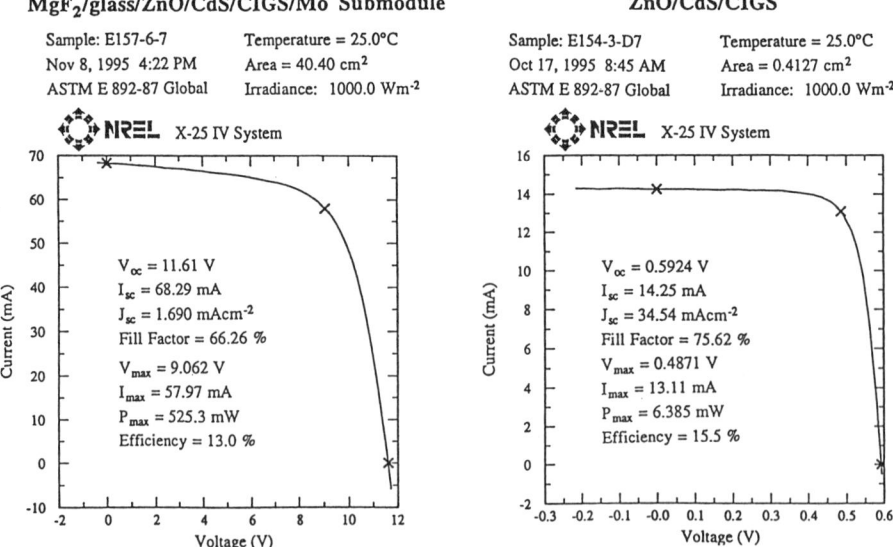

**Figure 1.** J-V characteristic of a 13% efficient laminated CIGS submodule.

**Figure 2.** J-V characteristic of a 15.5% efficient CIGS small area cell.

the control necessary to maintain the deposition conditions near the optimum, i.e. less total variation in the cell results is achieved, a corresponding improvement in module efficiency is also expected. Relatively large differences in fill factor between cells and modules indicate that gains in submodule efficiency can also be expected to result from better carrier collection through the front contact.

## Front Contact Development

The ZnO used to produce the 13% efficient submodule of Figure 1 was produced using RF sputtering from a ceramic target. Using this process, alteration of the thickness or doping density allows adjustment of the sheet resistivity, with attendant changes in the optical transparency which are shown in Figure 3. It is apparent that higher module efficiency is accessible through the use of better doped ZnO front contact layers, i.e. better ZnO conductivity for a given optical transparency. Toward this end, we are developing a reactive sputtering process for ZnO. Additional motivation for this approach is that the reactive process is much faster and less expensive [8]. ZnO films have been produced with the reactive process which are comparable to, or better than films from the standard (RF, ceramic target) process. As shown by the open symbols in Figure 3, these films have the same or lower optical loss for the same sheet resistivity.

The reactive process has yielded ZnO films with sheet resistivity as low as 3.4 ohms/square and excellent transparency in the visible. The optical behavior of a low resistivity, reactively sputtered film is shown Figure 4. Further development of this process is another factor which we expect will enable module efficiencies above 13%, as well as a method which is practical and cost effective for commercial production.

Although a ZnO front contact with sheet resistivity as low as 3.4 ohms/square would permit excellent current collection in a module structure, and also allow the use of wide module segments and thus fewer interconnects, the optical loss due to infrared absorption is significant. One approach which would circumvent this difficulty is the development of wider bandgap absorber layers. Some promising results have been announced in CIGS devices having an increased Ga content [9]. Assuming ZnO and interconnect parameters representative of those used at present, we have shown that, for equal cell efficiency on absorbers of different bandgaps, larger bandgap absorbers should result in improved module efficiency. The improvement in module efficiency is more significant for wider module segments, and also show a higher tolerance for variation in ZnO sheet resistivity [10]. For example, modules made using 1/4" segments would be expected to improve by more than one percentage point in efficiency if the absorber is 1.45 eV compared to 1.16 eV. The potential benefit offered by wider bandgap

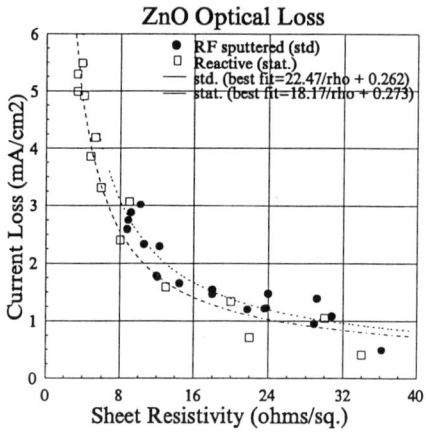
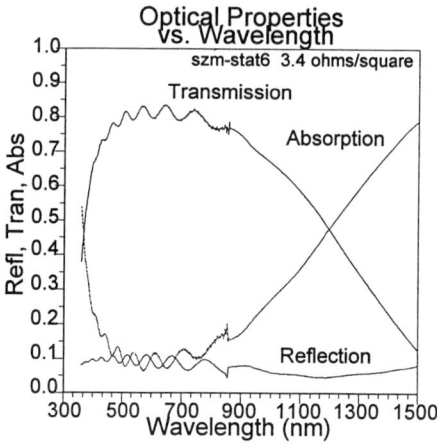

**Figure 3.** Typical behavior of Jsc loss, due to optical absorption, in the ZnO window layer versus its sheet resistivity, for two different processes.

**Figure 4.** Optical absorption, transmission and reflection of a highly conductive, reactively sputtered ZnO film (measured in air).

absorbers is contingent on the ability to produce them with sufficient quality (i.e. equally efficient devices at large bandgaps) using a manufacturable process.

In any case, the sheet resistivity and optical loss induced by the front contact is of significance for commercial development of CIGS based PV. It is common practice to measure the properties of the top contact layer on a glass substrate, often introduced as a witness piece to receive ZnO codeposited with that on a device or module. It is typically assumed that the electrical and optical properties of the codeposited ZnO on the glass witness piece are representative of what has been put on the device or module. Recent evidence [10] has shown that this assumption can be incorrect. Figure 5 shows the variation over an area including substrates of glass, and glass with either 'thin' or 'thick' CdS deposited by chemical bath deposition (CBD). The variation is significant, showing an average increase of 35% for substrates having thin CdS and 84% for those having thick CdS compared to glass with no CdS. The terms 'thin' and 'thick' refer to a single and double CBD treatment respectively under conditions typical of those used to complete the junction of a CIGS device or module. Our experience has been that the increase in sheet resistivity for ZnO sputtered onto actual device or module substrates is also greater than that on glass, similar to the CdS coated substrates.

Several critical issues are raised; from a pragmatic view, if the use of a glass witness is inappropriate to monitor ZnO properties, what method should be used? Also, could the gain in sheet resistivity be compensated for by simply increasing the front contact thickness? This latter approach would require that the more resistive ZnO films above also be correspondingly more transparent, both effects presumably due to a decrease in charged carrier density. From a scientific view, what is the origin of the change in ZnO resistivity? And is its occurrence general or specific to only certain methods or processes of ZnO deposition?

Optical measurement reveals that the absorption of ZnO films on glass, and on CdS coated glass differ little in the infrared, despite large differences in the sheet resistivity. If the optical absorption in the infrared was simply proportional to the number of free carriers in the film (i.e. mobility, refractive index are constant), then a plot of optical absorption vs. sheet conductivity would be linear with an intercept at zero. Variation in either carrier density or film thickness produced by variation of the underlying substrate material (CdS), would allow the data sets taken from different substrates to fit the same linear relationship. Figure 6 shows the optical absorption (at 1300 nm) versus sheet conductivity of the samples used for Figure 5. At 1300 nm the CdS and the glass substrates are virtually transparent. The data shown in this figure suggest that changes in the ZnO sheet resistivity are not due to changes in thickness or doping density alone. Hall effect measurements are planned to obtain further information about the carrier concentration and mobility in these films.

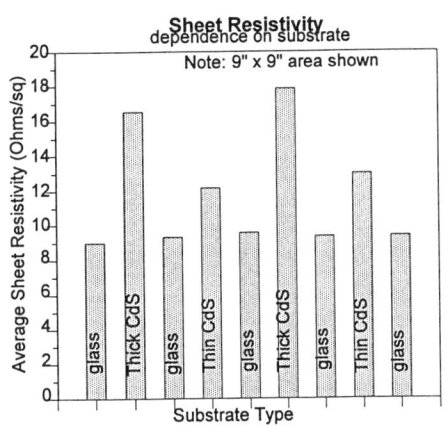
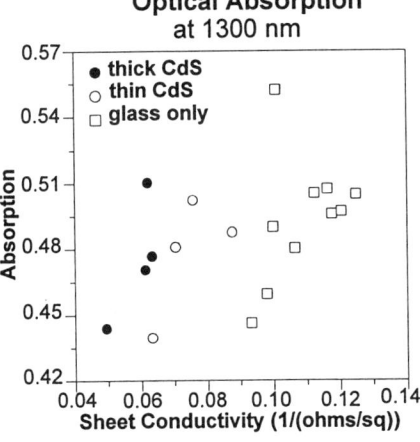

**Figure 5**. The variation in sheet resistivity, measured by 4 point probe, of ZnO films codeposited over a 9"x9" area on glass and glass coated with either thin or thick CdS. Each bar represents an average of three resistivity measurements.

**Figure 6**. The optical absorption of ZnO films deposited on glass, glass/'thin' CdS, and glass/'thick' CdS at 1300 nm, versus sheet conductivity.

# Buffer Layer/Heterojunction Development

Commercial prospects for CIGS based PV would be enhanced by the development of a Cd-free, dry process to replace the CdS CBD step presently used for heterojunction formation. This has been challenging, in no small part because the important function and requirements for junction formation are poorly understood. Some promising results have been reported [11,12] in the substitution of an intrinsic ZnO layer for the CdS. We have recently attempted variation of the CdS layer thickness, from 'none' to 'very thick' on cells with and without a buffer layer of intrinsic ZnO. As a further test, sputter deposited intrinsic ZnO of two types was used: ZnO from a pure ZnO target deposited in oxygen free conditions, and $ZnO:Al_2O_3$ target deposited with $O_2$ in the plasma. The resultant J-V parameters are shown in figure 7 as a function of CdS thickness. It seems apparent that, although the highest efficiency is achieved only by retaining some CdS, it may be possible to thin the CdS to a greater degree if the appropriate intrinsic ZnO is used at the junction [10,13]. Of the CdS-free cells, only those with ZnO deposited using $O_2$ free conditions had measurable performance and are plotted in Figure 7. The CdS-free cells made without any intrinsic ZnO or with resistive ZnO made with $O_2$ in the sputtering ambient had Voc < 100 mV and are not represented in Figure 7. This data suggests that not all

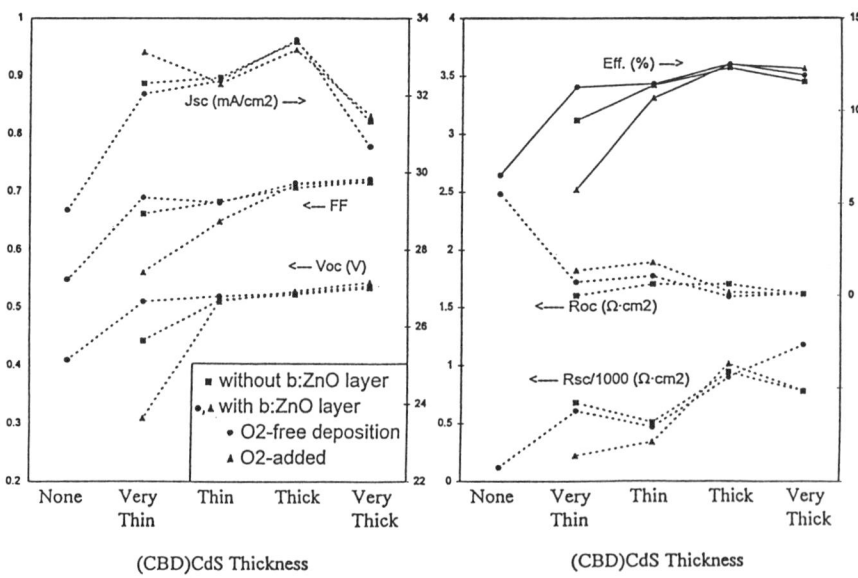

**Figure 7**. The photovoltaic parameters of cells made with two types of high resistivity ZnO and using several thicknesses of CBD CdS.

intrinsic zinc oxides are equivalent, and that, for supplantation of CdS, ZnO made in oxygen free conditions may be superior to that made with oxygen present.

## Progress in CIGS Deposition: Larger Area, Higher rates

Concurrent with the development of high efficiency, monolithic modules based on CIGS, progress in the capacity to deposit on large area substrates, with better control, adequate reproducibility and process speed are requirements for commercialization. Solarex has also made progress in these areas recently.

Previously, scale-up to 1000 cm$^2$ size has been demonstrated for all steps except the absorber layer [14]. In order to demonstrate scale-up of the absorber layers the first step has been the codeposition of nine 3"x3" substrates arranged in a 3x3 matrix corresponding to a deposition area of over 520cm$^2$. These nine substrates can be patterned to make either submodules or arrays of devices. Depositions have been accomplished in which the average efficiency of seven submodules and many devices on 3"x3" substrates are 12.3% and 12.9% respectively, with a variation of about one percentage point [6].

**Figure 8**. A photograph of a CIGS absorber deposited on a scribed, Mo/glass 9"x9" (522 cm$^2$) substrate.

Very recently the CIGS absorber layer has been successfully deposited on a single 9"x9" (520 cm$^2$) glass substrates with scribed and unscribed molybdenum. The depositions resulted in visually uniform, specular, well adherent films of about 2 microns thickness, shown in Figure 8. Compositional uniformity is shown in Figure 9. Previous work at Solarex has shown that high efficiency devices can be achieved over a broad range of Cu/(In + Ga) ratio, from 0.96 to 0.72, using the same process as for the large area depositions above. Although the composition of the large area sample shown in Figure 8 partially extends beyond this process window, the total variation in composition is no greater, and process adjustments can be expected to move the composition of the entire 520 cm$^2$ area into acceptable limits.

Since process speed is also a critical issue for commercial viability, this issue has been partially explored at Solarex. A modified process has been evaluated, which is similar in general nature to other processes used at Solarex, but significantly reduces the CIGS deposition time [6]. This process has successfully deposited the absorber layer in 10 to 30 minutes. A 20 minute absorber deposition has resulted in 15.2% device efficiencies (total area).

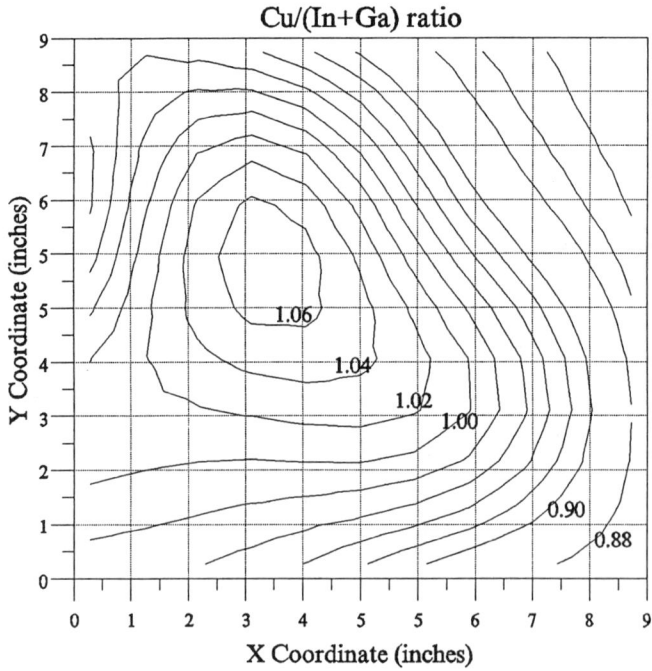

**Figure 9**. A contour plot showing the compositional variation in terms of the atomic percentage ratio of Cu/(In + Ga) over a 9"x9" (522cm$^2$) deposition area.

## Conclusion

Rapid progress in the last year has resulted in the demonstration of a 13% (aperture area) efficient submodule. Scale-up of this basic process to successfully deposit absorber layers on 9"x9" substrates has been accomplished. Process speed and costs have been shown to be attractive for this process or variants of it. Optimization of the absorber deposition to date has led to 15.5% efficiency in a small area device, and we believe further process improvement is possible. Progress in the development of transparent, front contacting layers has led to very conductive ZnO. Further progress in this area, evaluation of control and reproducibility of all steps, and development of a Cd-free, dry process are desirable for commercialization of PV product based on CIGS.

## Acknowledgments

The authors would like to acknowledge technical help and partial financial support from NREL under contract #ZAF-6-14142-12.

## References

[1] V.K. Kapur, B.M. Basol, A. Halani, C.R. Leidholm and A. Minnick, *Proc. 12th EU PVSEC*, Amsterdam (1994) p. 1608.

[2] K.E. Knapp, C. Eberspacher, D. Tarrant and G. Pollock, *Proc. 23rd IEEE PVSC*, Louisville KY (1993) p.1073.

[3] K. Zweibel and A.M. Barnett, *Renewable Energy*, Washington D.C., Island Press, 1993, edited by T.B. Johanasson, H. Kelly, A.K.N. Reddy, R.H. Williams, ch. 10 - "Polycrystalline Thin Film PV", p. 444.

[4] S. Wiedeman, J. Kessler, T. Lommasson, L. Russell, J. Fogleboch, S. Skibo and R. Arya, *Proc. 13th NREL PV Program Review,* AIP Conf. Proceedings **353**, Lakewood CO (1995) p.12.

[5] S. Wiedeman, J. Kessler, L. Russell, J. Fogleboch, S. Skibo and R. Arya, *Proc. 13th EU PVSEC, Nice* (1995) p.2095.

[6] J. Kessler, S. Wiedeman, L. Russell, J. Fogleboch, S. Skibo and R. Arya, *Proc. 25th IEEE PVSC,* Washington, D.C. (1996) p.813.

[7] J. Kessler, S. Wiedeman, L. Russell, T. Lommasson, S. Skibo, J. Fogleboch and R. Arya, *Proc. 1st WCPEC, Hawaii,* (1994) p.206.

[8] M. Ruckh, D. Hariskos, U. Ruhle, H.W. Schock, R. Menner and B. Dimmler, *Proc. 25th IEEE PVSC*, Washington, D.C., (1996) p.825.

[9] W.N. Shafarman, R. Klenk and B.E. McCandless, *25th IEEE PVSC*, Washington, D.C., (1996) p.763.
[10] J. Kessler, S. Wiedeman, L. Russell, J. Fogleboch, S. Skibo and R. Arya, *Proc. 25th IEEE PVSC,* Washington, D.C. (1996) p.885.
[11] J. Kessler, M. Ruckh, D. Hariskos, U. Ruhle, R. Menner and H.W. Schock, *Proc. 23rd IEEE PVSC*, Louisville KY (1993) p.447.
[12] L.C. Olsen, H. Aquilor, F.W. Addis, W. Lei and J. Li, *Proc. 25th IEEE PVSC*, Washington, D.C., (1996) p.997.
[13] T. Walter, D. Hariskos, R. Herberholz, V. Nadenau, R. Schaffer and H.W.Schock, *Proc. 13th EU PVSEC, Nice* (1995) p.1999.
[14] T.C. Lommasson, S. Wiedeman, L. Russell, J. Kessler and R.R. Arya, *Proc. 12th NREL PV Program Review,* AIP Conf. Proceedings **306**, Denver CO (1993) p.382.

# Progress in CIS-Based Module Development

Dale E. Tarrant, Jürgen Bauer, Ron Dearmore, Melinda E. Dietrich,
George T. Fernandez, Oswaldo D. Frausto, Christian V. Fredric, Cynthia L. Jensen,
Al R. Ramos, Jurge A. Schmitzberger, Robert D. Wieting, Dennis Willett,
Robert R. Gay

*Siemens Solar Industries, 4650 Adohr Lane, Camarillo, California 93012*

**Abstract.** Alloys of copper indium diselenide are the most promising candidates for reducing the cost of photovoltaics below the cost of crystalline silicon. Small area, fully integrated modules exceed 13% in efficiency and long-term outdoor stability has been demonstrated. The availability of natural resources and environmental impacts appear acceptable. Challenges remain to scale the process to larger area and to pass accelerated environmental testing. Process scale-up is systematically proceeding from the foundation of a reproducible small area module process with low variation; however, large-area circuit performance has not yet achieved the same level as the baseline process. Differences in performance between the baseline and large area processes have been isolated to differences in the equipment used to form absorbers and is the subject of current development. The impact of larger part size has been tested for each process step using the demonstrated baseline process. Results indicate that larger-area parts will achieve the performance level of the baseline process. This paper will outline the status and prospects for CIS-based photovoltaics and discuss efforts at Siemens Solar Industries to scale up to larger circuit area and to pass accelerated environmental tests.

## INTRODUCTION

Multinary $Cu(In,Ga)(Se,S)_2$ absorbers, which will be referred to in this paper as CIS-based absorbers, are promising candidates for reducing the cost of photovoltaics well below the cost of crystalline silicon. CIS champion efficiencies have continually improved recently culminating in 17+% efficient solar cells fabricated at NREL (1). Small area, fully integrated modules exceeding 13% in efficiency have been demonstrated by four groups (2). Long-term outdoor stability has been demonstrated at NREL by 1 ft. X 1ft. and 1 ft. X 4 ft. modules which have been in field service testing for as long as seven years (2,3). The availability of natural resources and the effect of manufacturing and deployment upon the environment both appear to be acceptable (2,3). Cost projections indicate about

$1.00/Wp based on projection of present processing to large area and high volumes and, based on the measured efficiency of small-area mini-modules (2,3).

However, a number of significant issues have been identified as challenges to the near-term commercialization of CIS. This paper will discuss the present status of CIS-based module development at SSI, the challenges to near term commercialization that are being addressed, and provide a perspective on the prospects of this material for near-term commercialization.

## CIS-BASED MODULE DEVELOPMENT STATUS

### Demonstrated Performance

SSI has abandoned development based on individual cells in favor of the "mini-module"; twelve cells monolithically interconnected on a 10 cm X 10 cm substrate (4). Four groups have now fabricated small area, fully integrated modules exceeding 13% in efficiency (Table 1). The high efficiency of these champion mini-modules provides an existence proof for a thin-film process capable of circuit efficiencies which approach commercial monocrystalline silicon. While these circuits demonstrate the potential of the technology, a rigorous measure of process reproducibility and yield is also necessary.

From an industrial perspective, the full process sequence anticipated for use in the final product must be mastered and rigorously demonstrated. To this end, SSI has repeatedly executed a baseline mini-module process. Statistical process control (SPC) techniques have been utilized as the measure of, and the tool for demonstrating, yield and process reproducibility (8). A reproducible low variation baseline is demonstrated for over two thousand mini-modules (Figure 1). Efficiency versus time is plotted with separations between groups of data indicating deliberate changes in the baseline process. Non-baseline experimental data is not

Table 1. Performance of mini-modules with efficiencies over 13%. Parameters have been converted to a per-cell, aperture-area basis.

| Manufacturer (Notes) | Area (cm2) | Cells | Voc/Cell (mV) | Jsc (mA/cm2) | FF | Eff |
|---|---|---|---|---|---|---|
| Showa Shell* (5) (Cd-free) | 51.9 | 12 | 560 | 36.9 | 68.1 | 14.1 |
| IPE/ZSW † (6) (AR coated) | 90.6 | 15 | 620 | 30.3 | 73.7 | 13.9 |
| SSI ‡ (2) | 50.3 | 12 | 567 | 33.8 | 71.0 | 13.6 |
| Solarex ‡ (7) (AR coated) | 40.4 | 20 | 581 | 33.8 | 66.3 | 13.0 |

*Measured at JQA, Japan † Measured at FhG ISE, Germany ‡ Measured at NREL

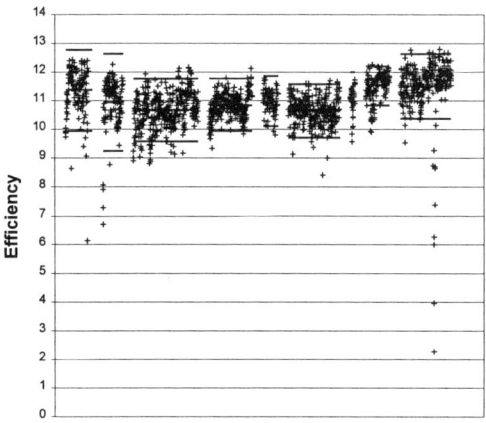

**Figure 1. Reproducible low variation baseline demonstrated for over two thousand unlaminated mini-modules.**

included and identified "special causes" have been removed.

After lamination and outdoor exposure, modules typically have even higher efficiencies. Laminated mini-module efficiencies after outdoor exposure typical of in-service conditions average 12.4% and laminated mini-modules over 13.5% efficient have been demonstrated (Figure 2).

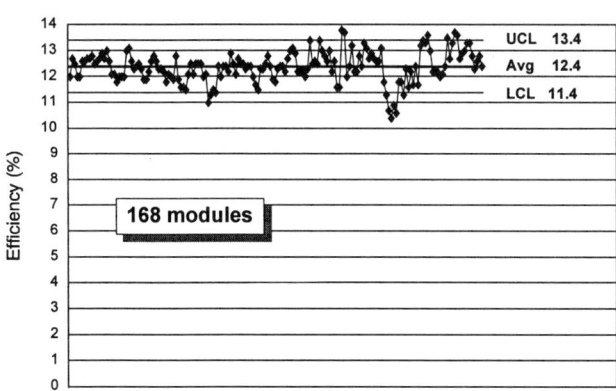

**Figure 2. Laminated mini-modules averaging 12.4% efficient after outdoor exposure typical of in-service conditions.**

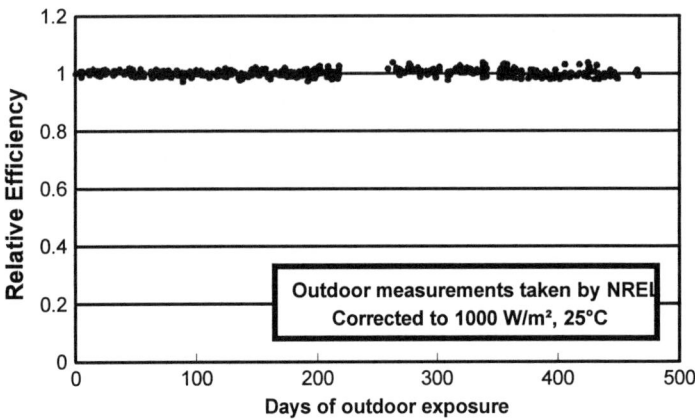

Figure 3. Field measurements of a CIS module at NREL, when corrected to standard test conditions, show good stability with no seasonal behavior

## Demonstrated Long-term Outdoor Stability

Long-term outdoor stability has been demonstrated at NREL where 1 ft. X 1ft. and 1 ft. X 4 ft. modules have been field tested for as long as seven years (1). For these measurements, the modules were brought indoors, the measurements were performed under standard test conditions, then modules were returned outdoors.

NREL is also monitoring an SSI provided 1-kW array of CIS modules with IV measurements in the field; the modules are not brought indoors for IV measurements (9). The modules are kept under load and measured every half-hour. With this in-service data corrected to standard conditions, both the modules and the array show good stability with essentially no daily or seasonal variation (Figure 3).

## CIS-BASED MODULE DEVELOPMENT CHALLENGES

### Process Scale-up

Process scale-up has proceeded from the foundation of the reproducible low variation baseline process. This has greatly aided characterization of large-area processing, and process development to isolate the sources of performance differences between baseline and large area processes. For each step in the process, the impact of the larger part size has been tested in the baseline. For example, by cutting 30 cm X 30 cm circuits into nine 10 cm X 10 cm circuits, the performance uniformity of the larger parts was measured in the baseline process. Such

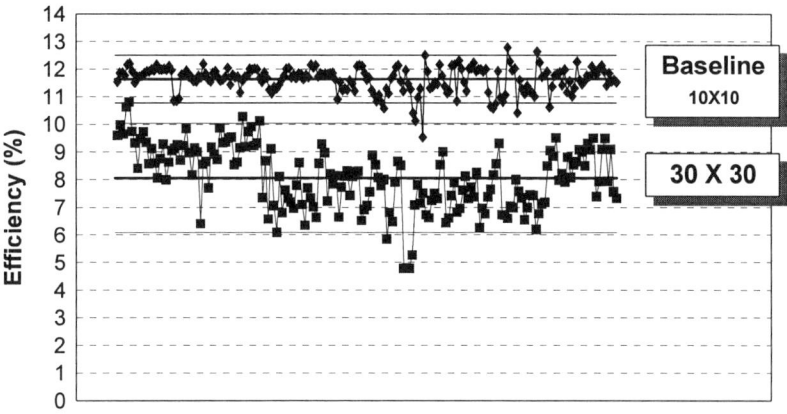

Figure 4. 30X30 circuits are inferior to baseline

experiments indicate that larger area parts should yield the same level of performance as small parts.

However, large-area circuit performance has not yet achieved the same level as the baseline. Baseline circuit plate efficiencies are about 11.5% whereas 30 cm X 30 cm performance is about at 8% (Figure 4). Absorber formation of both large and small circuit plates in the same reactor has proven informative. Identical performance is achieved for 10 cm X 10 cm and 30 cm X 30 cm circuits plates when fabricated in the large reactor (Figure 5). These results, combined with the results from evaluation of larger plates in the baseline process, indicate reactor

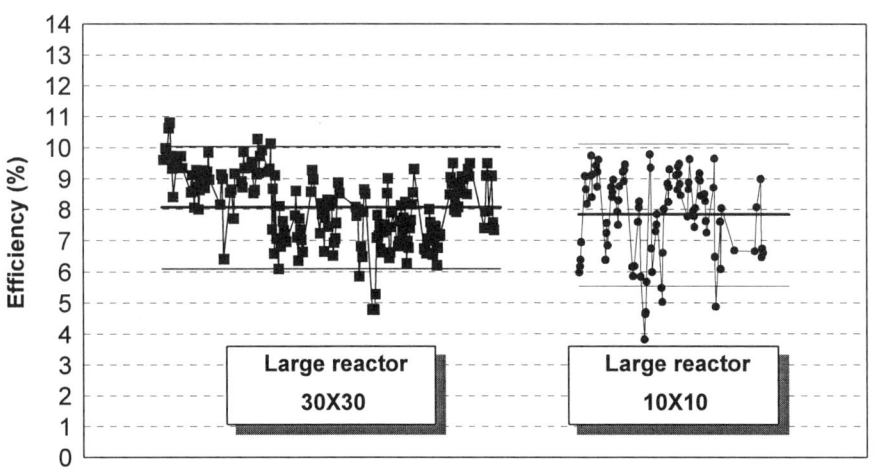

Figure 5. Part size itself is not the cause of the performance differences.

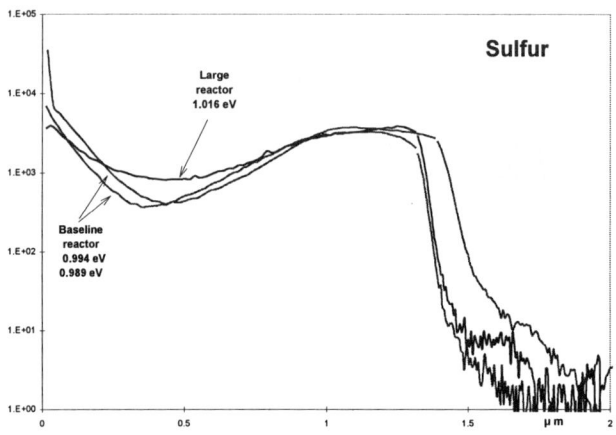

Figure 6. Sulfur profiles are different for absorbers from different reactors.

differences are responsible for performance differences; the part size itself is not the cause.

The SSI baseline process achieves an absorber structure with graded gallium and sulfur concentration profiles (4, 10). Absorbers obtained from the baseline process and from the large reactor have been characterized using bandgap measurements, SIMS, ICP, sheet resistance, contact resistance, etc. Bandgap measurements determined from spectral response indicate a difference in the structure of absorbers from the baseline and large area reactors. SIMS analysis (11) reveals that the profile of sulfur concentrations is the primary difference between absorbers from the two reactors (Figure 6). The concentration of sulfur is highest at the front of the absorber and decreases toward the center of the absorber for both reactors. The sulfur profile for baseline absorbers is "sharper." The sulfur profile has a higher gradient from the front to the center of the absorber, the concentration of sulfur at the front is higher, and the concentration near the middle of the absorber is lower for baseline absorbers.

These observed differences between absorbers from the two reactors have been related to the design of the large reactor. Hardware changes in the large reactor to mitigate these differences are underway. It should be emphasized that the impact of the larger part size has been tested in the baseline process and that the difference in performance is not related directly to part size. Such experiments indicate that larger-area parts should achieve the same level of performance as smaller parts.

# Accelerated Environmental Testing

Although stable for in-service conditions, circuits encapsulated in a standard, glass/EVA/circuit, laminate (4) have failed standard accelerated environmental tests. These tests highlight the sensitivity of these devices to heat and light. Efficiency typically increases with light exposure (12). Exposure to temperatures above those encountered during in-service conditions can cause the efficiency to fall. This loss then fully recovers even with low level or no light exposure. Because the standard environmental qualification tests involve exposure to 85°C (13), many modules show a loss of performance through the tests (4, 14) although post-test sun-soaking of the modules results in full recovery (4). Changes in the performance of CIS modules induced by light-soaking also represent a challenge to the proper testing and rating of these modules because the sun-soaking time required to stabilize the efficiency can be very long - from days to many weeks (4, 14).

Modules also fail the 1000-hour damp heat test (13) where moisture penetration into the package can cause irreversible loss of performance (14). Improved package designs are being pursued along with efforts to identify the origin of moisture sensitivity in the circuits by separating and examining the effects of the two major environmental stress factors: heat and humidity. Interconnect test structures were exposed to damp heat (85 °C, 85% relative humidity) (14). Interconnect test structures are essentially mini-module circuits but with double interconnects configured to allow independent measurement of ZnO to Mo contact resistance and ZnO sheet resistance (4). Both contact resistance and sheet resistance increase

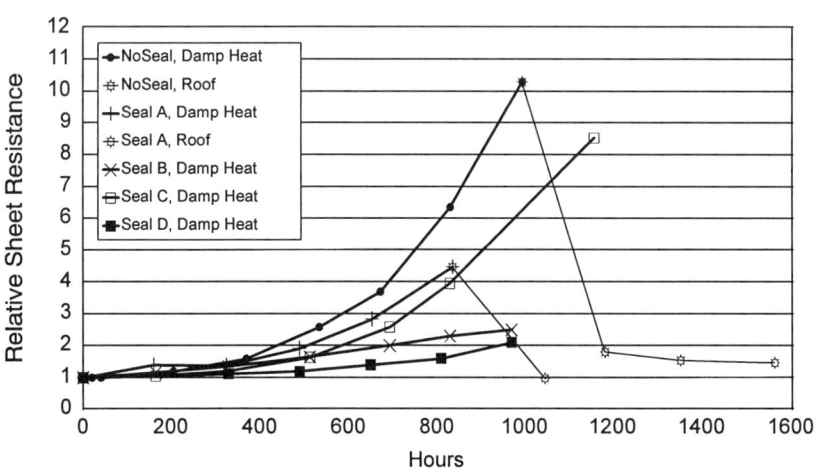

Figure 7. 30 X 30 cm ZnO Laminates

(degrade) with this exposure and the increase correlates with water penetration into the laminate observed as fogging of the EVA.

Based on these results, exploration of edge seal options to prevent water vapor ingress was performed using laminated 30 cm X 30 cm ZnO coated plates (Figure 7). The laminates included electrical contacts for ZnO sheet resistance measurement. Since the laminate is transparent, it is also possible to observe differences in water vapor ingress. Sheet resistance increases with damp heat exposure again correlating with EVA fogging. With subsequent outdoor exposure, the ZnO sheet resistance returned to pre damp-heat exposure values, though recovery was not accompanied by a visual decrease in EVA fogging. Edge seal options that significantly inhibit water vapor ingress and the resultant increase in sheet resistance have been demonstrated (Figure 7). Observations indicate that improving the adhesion between the edge seals and the laminates will further improve resistance to water vapor ingress.

In addition to testing at SSI, laminated mini-modules have been subjected to standard thermal cycling, humidity-freeze cycling, and damp heat exposure at NREL. The sequence of testing and measurements was arranged to gain information on the effects of light exposure before, after and during accelerated environmental testing. All modules were placed outdoors for at least two weeks prior to accelerated environmental testing and half of the modules went through the accelerated environmental exposure with simultaneous light exposure of about one sun. Results for thermal cycling are displayed in Figure 8 (15). First, efficiencies improve with two weeks of outdoor exposure. Then thermal cycling degrades

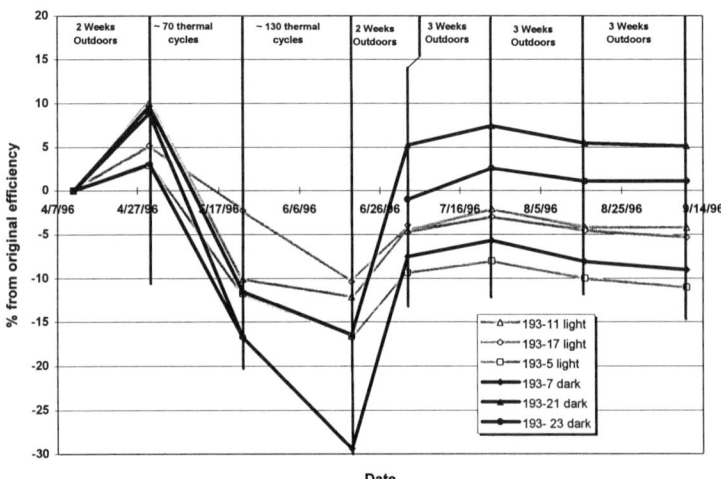

**Figure 8.**

performance by between 10 and 30 points. Less degradation is observed for the group with light exposure during thermal cycling. Performance recovers after subsequent outdoor exposure and is similar for both groups; with and without simultaneous light exposure during thermal cycling. Two of the three mini-modules in each group recover to within five points of initial performance.

Results of humidity-freeze cycling are similar. Performance degrades by between 10 and 30 points during accelerated testing, less degradation during cycling is observed for the group with light exposure, and two of the three mini-modules in each group recover to within five points of initial performance. Damp heat testing was conducted in the dark only and with only two mini-modules. Losses were about seventy points with less recovery than for either thermal cycling or humidity-freeze cycling. These mini-modules exhibited a fogging around the outside edge that did not disappear with outdoor exposure.

## SUMMARY

SSI has repeatedly executed a baseline 10 cm X 10 cm mini-module process rigorously demonstrating process reproducibility and yield for CIS-based devices. Laminated mini-module efficiencies after outdoor exposure typical of in-service conditions average 12.4% efficient, with champion mini-modules over 13.5% efficient. Excellent long-term outdoor stability has been demonstrated at NREL.

Process scale-up is systematically proceeding from the foundation of this reproducible low variation baseline process; however, large-area circuit performance has not yet achieved the same level as the baseline process. This performance difference has been characterized and related to design differences between baseline and larger area absorber formation reactors. The profile of sulfur concentration is the primary difference between absorbers from the two reactors. For all other process steps, the impact of the larger part size has been tested in the baseline process. This approach has demonstrated that performance is not directly related to part size; larger-area circuits should achieve the same level of performance as baseline circuits.

Although stable for in-service conditions, SSI CIS-based devices have failed standard accelerated environmental tests. SSI has separated and examined the effects of the two major stress factors during accelerated environmental tests: heat and humidity. Degradation induced by high temperatures during accelerated environmental testing recovers for normal in-service illumination and temperature conditions, while humidity induced degradation exhibits reversible and nonreversible components. Laminate edge seal options that significantly inhibit water vapor ingress have been demonstrated and, improved adhesion between the edge seals and the laminates is expected to further improve resistance to water vapor ingress.

# ACKNOWLEDGMENTS

The author gratefully acknowledges NREL for valuable measurement and analytical assistance. This work was supported by the National Renewable Energy Laboratory, Golden, CO, under subcontract No. ZAF-5-14142-03 of the Thin Film Photovoltaic Partnership Program through the U. S. Department of Energy.

# REFERENCES

1. Ken Zweibel, Harin S. Ullal, Bolko von Roedern, "Progress and Issues in Polycrystalline Thin-Film PV Technologies," 25th IEEE Photovoltaic Specialist Conference, 1996, pp. 745-750.

2. Robert R. Gay, "Status and Prospects for CIS-Based Photovoltaics," 9th International Photovoltaic Science and Engineering Conference, 11-15 November, 1996, Miyazaki, Japan

3. H. A. Aulich, "Advances in Thin Film PV-Technologies," 13th European PVSEC 1995, pp. 1441-1444.

4. D. E. Tarrant and R. R. Gay, "Research on High-Efficiency, Large-Area CuInSe2-Based Thin-Film Modules, Final Subcontract Report," NREL/TP-413-8121 (1995).

5. Katsumi Kushiya, "Fabrication of CIGS Thin-Film Mini-Modules With Zinc Compound Buffer," 9th International Photovoltaic Science and Engineering Conference, 11-15 November, 1996, Miyazaki, Japan

6. B. Dimmier, et al., "Thin Film Solar Modules Based on CIS Prepared by the Co-evaporation Method," 25th IEEE PVSC, 1996, pp. 757-762.

7. John Kessler, et al., "Cu(In,Ga)Se$_2$ Based Submodule Process Robustness," 25th IEEE PVSC, 1996, pp. 813-816.

8. R. Wieting, et al., "Progress in CIS-Based Photovoltaics through Statistical Process Control," 13th European PVSEC, 1995, pp. 1627-1630.

9. T. R. Strand, B. D. Kroposki, R. Hansen and D. Willett, "Siemens Solar CIS Photovoltaic Module and System Performance at the National Renewable Energy Laboratory," 25th IEEE PVSC, 1996, pp. 965-968.

10. D. Tarrant, J. Ermer, "I-III-VI2 Multinary Solar Cells Based on CuInSe2." 23rd IEEE PVSC, 1993, pp. 372-378.

11. Measurements by Sally Asher, NREL.

12. D. Willett and S. Kuriyagawa, "The Effects of Sweep Rate, Voltage Bias and Light Soaking on the Measurement of CIS-Based Solar Cell Characteristics," 23rd IEEE PVSC, 1993, pp. 495-500.

13. "Crystalline Silicon Terrestrial Photovoltaic (PV) Modules - Design Qualification and Type Approval," IEC-1215 (1993).

14. D. Willett, "Environmental Testing of CIS Based Modules," NREL PV Performance and Reliability Workshop (1995).

15. Measurements by Ben Kropowski, NREL.

# CdTe Team Activities

Peter V. Meyers

*ITN Energy Systems*
*12401 West 49th Ave.*
*Wheat Ridge, CO 80033 USA*

Abstract. In 1994 the CdTe Team was formed as part of DOE/NREL's Thin Film PV Partnership program. Team members are industrial, academic and governmental organizations with interests in thin film CdTe/CdS module technology. Ongoing activities of the CdTe Team's two working groups are directed toward 1) Development of a stability testing protocol for CdTe-based PV modules, and 2) Clarifying how CdS affects the trade-off between high Jsc and high Voc. Future activities will also include investigations of the CdTe to back electrode contact.

## BACKGROUND

The CdTe Team is part of DOE/NREL's Thin Film PV Partnership program. It was created in order to coordinate and utilize the abilities and resources of the academic, industrial and governmental CdTe PV technical community to promote the commercialization of CdTe photovoltaics through the solution of technological issues. The overall goal is achieved through the performance of tasks related to specific secondary goals which are identified and selected by all participants. Team activities are not all inclusive in that they are not intended to encompass all aspects of CdTe PV technology nor are they intended to replace research and development programs of the member organizations. Projects chosen address industry-wide issues (i.e., are not be process-specific) which can utilize the varied perspectives and expertise of Team members.

Membership is required of all NREL subcontractors who receive support for CdTe PV research or development subcontracts, but NREL support is not a requirement for membership and not all members receive NREL financial support. Members are required to play an active role and to provide active support for Team activities. Support may take the form of film or device fabrication or processing or testing or analysis; passive membership is not allowed.

CdTe team members are organized into working groups focused on specific tasks related to the technical goals selected by the members. Members must participate in at least one working group. A WGL (Working Group Leader) is elected by group members. Team activities are monitored by the CdTe Guidance Committee which represents NREL and DOE and which has authority over all

**TABLE 1.** CdTe Team Meetings.

| Date | Location |
|---|---|
| Dec 5-9, 1994 | Kona, HI |
| May 16, 1995 | Lakewood, CO |
| January 22, 1996 | Newark, DE |
| February 22, 1996* | Golden, CO |
| May 15, 1996 | Washington, DC |
| November 18, 1996 | Lakewood, CO |

* Guidance Committee Meeting

Team activities. The Guidance Committee Chairs also function as the chief executives of the CdTe team and provide guidance to WGLs, facilitate communication within the CdTe Team and communicate with the Guidance Committee. The WGLs assume responsibility for the coordination and monitoring of team activities and the establishment of individual member tasks. WGLs also keep the Guidance Committee Chairs informed of team activities and progress. Team members are responsible for the performance of agreed tasks, the reporting of results at team meetings, and the coordination of their activities with those of other team members. Progress on tasks is reviewed and reported at CdTe Team Meetings - held approximately every nine months - and at Guidance Committee Meetings held as required. (See Table 1.)

# STABILITY

## Statement of the Problem

In order for PV modules to obtain widespread commercial success it may be necessary to demonstrate stability for ten, twenty or even thirty years. Although long term field testing is the ultimate measure of stability, full lifetime testing of modules and cells is not economically practical. Nonetheless, the onus is on the CdTe community to provide technically sound procedures which will provide investors and potential customers with a reasonable estimation of module stability. In order to convincingly demonstrate this stability within a reasonable time, testing procedures must be developed which are based on an understanding of the mechanisms leading to device failure or degradation over time and the factors which affect them.

**TABLE 2.** Stability Working Group Overview

| Goal: Develop a stability testing protocol for CdTe-based PV modules | |
|---|---|
| Working Group Leader: Tom McMahon | |
| Organization | Members |
| Golden Photon, Inc. | Scot Albright, John Kester |
| ITN Energy Systems | Peter Meyers |
| NREL | Ben Kroposki, Tom McMahon |
| Solar Cells, Inc. | Gary Dorer, Rick Powell, Rick Sasala |

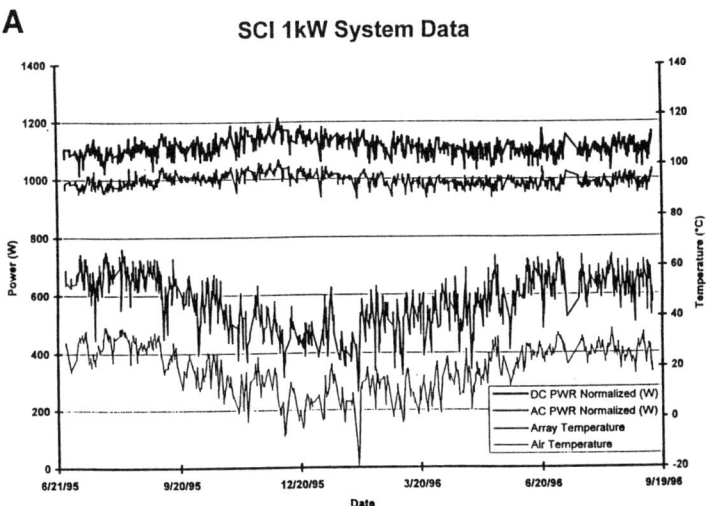

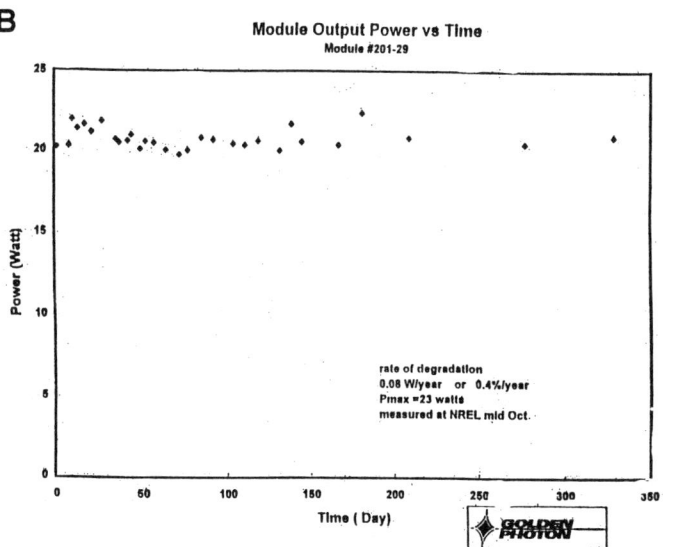

**FIGURE 1.** Typical results of field testing of CdTe PV arrays and modules. A) Solar Cells, Inc. (array) and B) Golden Photon, Inc. (module).

## Approach

Stability team activities include field testing of CdTe PV arrays and analysis of field performance, investigation of accelerated testing procedures, identification of potential degradation mechanisms, investigation of device and module properties and structures which affect stability, development of procedures for producing more stable modules, and ultimately the development of a stability testing protocol that will provide a reliable method for predicting the useful life of CdTe PV modules.

## Activities

Field testing of CdTe PV arrays supplied by SCI (Solar Cells, Inc.) and GPI (Golden Photon, Inc.) has been carried out by NREL for more than two years and more recently by FSEC. Factors affecting device performance including temperature, illumination level and time are being recorded and evaluated (1); representative data are displayed in Fig. 1. Results to date indicate that CdTe PV modules can be produced which are stable after more than a year and field testing is ongoing. Newer modules are being added to the field studies as suppliers modify and improve cell and module designs.

### *Stress testing of modules*

Submodules produced by SCI have been stressed at SCI and NREL by "light soaking" at various temperatures above those expected in the field and under electrical bias load lines including Jsc, Rmp (maximum power load) and Voc. Stress testing is intended to probe module designs for weaknesses - to uncover and accelerate any potential degradation mechanism. In the absence of an identification of the potential degradation mechanisms, the effect of these stresses is not known. The stresses chosen - heat, light and electrical bias - were selected as reasonable variables that might induce changes in the electrical or optical properties of the devices. Stress testing is therefore viewed as an iterative process in which the results of one round of testing will be used to determine stress conditions for the next round. At this point relatively few degradation mechanisms have been identified, and those that have been identified, e.g., degradation of the busbar contact, do not appear to be fundamental. Nonetheless, stress testing does result in changes to module performance and the goal is to relate these changes to specific mechanisms and to determine whether and how they relate to module lifetime (2).

### *Stress testing of cells*

Solar cells are simpler devices than modules (in that they do not have monolithic interconnects, busbars or encapsulants) and therefore changes in device performance are easier to interpret. Stress testing of CdTe cells performed at IEC suggested that specific devices exhibited at least two categories of change dependent upon temperature and electrical bias (3). These changes were classified as bulk effects - which were manifested as changes in the light-dark crossover of the I-V characteristics - and contact effects - manifested as a rollover of the light and dark I-V curves in the first quadrant (forward bias beyond Voc). See Fig. 2. Even these classifications are tentative, however, as a detailed model linking device characteristics to the physical and electronic nature of the device has not been developed. Furthermore, not all devices respond to stresses in the same way. At this point it appears that stability is a characteristic which can be optimized through appropriate cell design and processing.

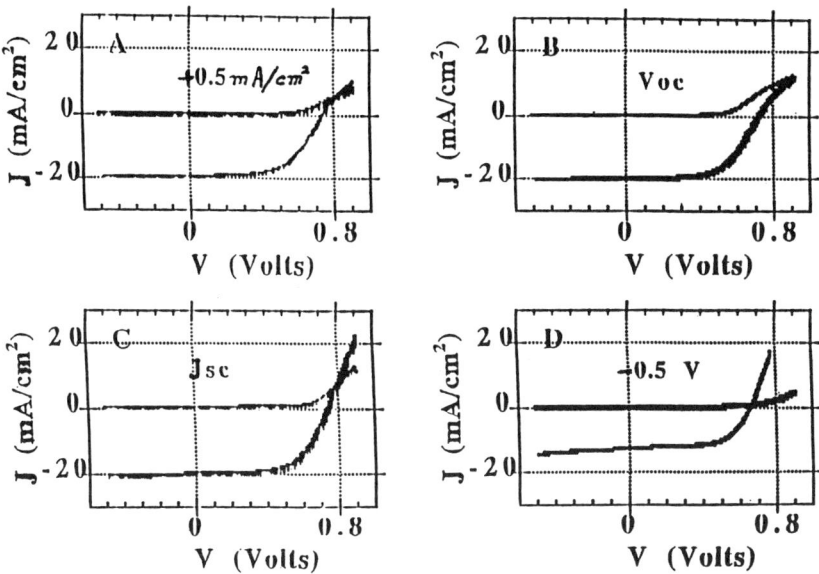

**FIGURE 2.** Light and dark J-V curves of devices stressed for 550 hrs at 92°C and 70 mw/cm$^2$ illumination with voltage bias of A) +5 mA/cm$^2$, B) Voc, C) Jsc and D) -0.5 V (3).

Teaming has provided a mechanism for collaboration among the various interested parties in that industrial groups (SCI, GPI) have provided modules and devices, a governmental group (NREL) provides independent testing, and university groups (CSU, IEC and Stanford) provide device analysis. In addition some work has been done in which back contacts have been fabricated by governmental and university groups on CdTe/CdS films prepared by the industrial members. This co-fabrication project, which was begun only recently, is an attempt to better understand the role of the back contact in device operation and ultimately its influence on device stability.

## THIN CADMIUM SULFIDE

### Statement of the problem

CdTe PV device efficiency is largely determined by the electrical and optical properties of the CdTe/CdS interface which are in turn greatly influenced by the thickness and other properties of the CdS layer. Thin CdS is desirable in order to minimize the loss of photocurrent due to light absorbed in the CdS window layer. Most, although not all, researchers discover that there is a lower limit to the

**TABLE 3.** Thin CdS Working Group Overview

| Goal: Clarify how CdS affects the trade-off between high Jsc and high Voc | |
|---|---|
| Working Group Leader: Chris Ferekides | |
| **Organization** | **Members** |
| Colorado School of Mines (CSM) | Duli Mao, John Trefney |
| Colorado State University (CSU) | Jim Sites |
| Florida Solar Energy Center (FSEC) | Neelkanth Dhere |
| Golden Photon, Inc. (GPI) | Scot Albright, John Kester |
| Institute of Energy Conversion (IEC) | Brian McCandless |
| National Renewable Energy Laboratory (NREL) | Dave Albin, Pete Sheldon, Ramesh Dhere |
| Solar Cells, Inc. (SCI) | Rick Sasala |
| Stanford University | Alan Fahrenbruch |
| University of South Florida (USF) | Chris Ferekides |
| University of Toledo (UT) | Al Compaan |

thickness of CdS. Below this limit devices display lower Voc and fill factor as compared to devices produced using identical procedures except for the CdS layer thickness. See Fig. 3. The fact that exceptions exist - i.e., that devices which simultaneously display Voc > 800 mV, FF > 0.70 and Jsc > 26 mA/cm$^2$, provides evidence that there is no fundamental mechanism controlling the trade-off between high Jsc and high Voc. Thus the team goal is to understand and quantify the role of CdS so that researchers and ultimately manufacturers can routinely produce high efficiency devices.

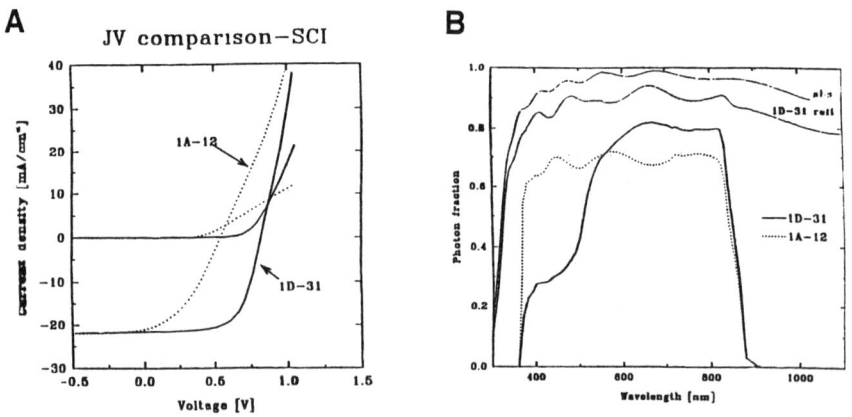

**FIGURE 3.** Typical A) J-V and B) QE results obtained from thick (800Å, solid line) and thin (200Å, dotted line) CdS films in CdTe/CdS solar cells.

## Approach

Thin CdS working group tasks are intended to identify the various factors which, singly or together, influence the trade-off between high Jsc and high Voc. Specific factors include 1) pinholes - which may provide regions of low Voc,

2) TCO and the existence of a high-resistivity layer between the TCO and CdS, 3) CdS/CdTe interdiffusion which may affect interface or grain boundary passivation as well as other junction parameters, and 4) CdS thickness - is there sufficient material to develop a junctions. Any of these factors can be influenced by the material properties of the CdS, CdTe or TCO.

## Activities

The general approach includes the fabrication of devices by various techniques by various team members using well characterized substrates and CdS thicknesses. These devices are then analyzed in a unified manner to establish the electrical, optical and chemical properties of the devices and to identify fundamental relationships related to CdS thickness and the mechanisms by which they affect device performance.

At this time the group has completed device preparation and analysis of two "sets" of thin CdS devices. In Set #1 CdS thickness were varied over the range of 500Å to 3000Å. Set #2 employed similar procedures except that CdS thicknesses were further reduced to nominal values of 200 Å and 800 Å. In both cases superstrates of $SnO_2$-coated borosilicate glass were prepared and distributed to team members for CdTe device fabrication using various thicknesses of CdS. Devices were produced by GPI (spray), SCI (CSS), USF (CSS), UT (sputtering), NREL (CSS), IEC (vacuum evaporation), and CSM (electrodeposition) and sent to CSU for measurements and analysis. Measurements at CSU included dark and light I-V, quantum efficiency, and C-V. Thus teaming has produced a standardized data set of device characteristics produced by a variety of means.

Team meetings have been a forum where team members bring present a wide variety of measurements and ideas relevant to the issue of thin CdS in CdTe devices. A few representative examples are listed below.

- IEC has characterized the quantity and size of pinholes in the CdS films deposited by chemical bath deposition, sputtering and vacuum evaporation. Device characteristics were then modeled as parallel CdTe/CdS and CdTe/TCO junctions with the latter having an area determined by the total pinhole area.
- IEC, UT and Stanford have performed studies of the interdiffusion of the CdTe and CdS layers and the formation and effects of Cd(Te,S) alloys.
- NREL has performed studies of the effect of an i-layer at the TCO - CdS interface and found that peak device performance is not increased but that process reproducibility is improved.
- NREL has also demonstrated that the $H_2$ anneal of $CdS/SnO_2$ layer results in the diffusion of Sn to the CdS surface and thereby affects the nucleation of the subsequently deposited CdTe layer.

## PLANS

At the Nov 96 meeting, the CdTe team discussed and adopted a statement of work (SOW) intended to serve as a blueprint for future team activities. Among the features of the SOW are the refocussing of the Stability working group onto the development of a stability testing protocol for CdTe modules with the ultimate aim of developing certification procedures that will be accepted industry-wide. In order to accomplish this goal, working group members plan to build upon results to date to develop and evaluate a stress and field testing program along a schedule as laid out in Table 4 A. In order to continue studies of the back contact, the Thin CdS working group has taken on additional tasks and been renamed the High Efficiency Device working group. Chris Ferekides will remain as WGL and team membership remains unchanged. An important element is the establishment of a working model to quantitatively explain device operation and relate it to the physical and electrical properties of the device. Key milestones are listed in Table 4 B.

# SUMMARY

The CdTe Team has served as an effective vehicle for addressing industry-wide technology issues related to CdTe-based thin film PV technology by bringing together the varied perspectives and capabilities of industrial, governmental and university organizations. The two working groups are actively pursuing tasks directed toward 1) Development of a stability testing protocol for CdTe-based PV modules, and 2) Clarifying how CdS affects the trade-off between high Jsc and high Voc. Future activities will also include investigations of the CdTe to back electrode contact.

**TABLE 4.** CdTe Statement of Work Milestones

**A   Stability Working Group**

| | |
|---|---|
| 11/96 | Select members to evaluate and recommend<br>• module parameters and methods of measurement to be used for analysis of module performance and indicators of module life.<br>• cell parameters and methods of measurements to be examined separately to sort out reliability problems associated with cell structures from those associated with module fabrication and cell interconnects<br>• testing procedures and stress parameters which may, alone or in combination, accelerate changes in some or all of the identified module performance parameters. |
| 1/30/97 | Adopt a preliminary stability testing protocol. |
| 2/28/97 | Module stress testing begins. |
| 11/30/97 | Status report on results of testing.<br>Stability team members evaluate progress and decide whether to continue present procedures or to adopt new ones. |
| 12/30/97 | Industrial members supply modules as required for modified stability testing protocol (STP). |
| 6/30/98 | Status report on results of testing<br>Recommendations for Standard STP presented for discussion and review. |
| 8/30/98 | Preliminary Proposed Standard STP agreed upon by Stability Team |
| 10/30/98 | Industrial members supply modules for evaluation of proposed STP. One year testing begins. Members are encouraged to perform in-house studies for comparison with Preliminary Proposed Standard STP. |
| 11/30/99 | Report on results of Preliminary Proposed Standard STP. |
| 12/30/99 | Stability Team adopts Standard Stability Testing Protocol. |

TABLE 4B   High Efficiency Device Working Group

**Junction**

| | |
|---|---|
| 5/97 | Based upon results to date, present a comprehensive model relating the physical structure, optical properties and electrical properties of the junction layers - TCO, i-layer, Cd(Te,S) alloys, and CdTe - to CdTe PV device operating parameters and suggest qualities that the junction layers should have in order to produce devices with high QE for $\lambda$ <525 nm |
| 5/97 | Members will report on their attempts to evaluate individual CdTe junctions and device fabrication procedures |
| 12/97 | A final report will be prepared defining the limitations of thin CdS and how these limits depend upon the properties of the CdS, CdTe, and device processing.<br>Members will each report on their attempt to minimize CdS absorption. Each report should address the issues of whether limiting optical absorption has been achieved and the factors used to reach that conclusion. |

**Back Contact**

| | |
|---|---|
| 5/97 | Members will submit for stability screening devices produced on SCI- and GPI-supplied CdTe/CdS which have been completed using the member's contacting procedure. |
| 5/97 | CSU will present the results of standardized measurements on Set #2 (200 and 800 Å CdS) and will identify and quantify apparent back contact effects |
| 9/97 | Analysis of results of stability screening procedures will be presented. |

# ACKNOWLEDGMENTS

The author gratefully acknowledges the contributions of his Co-Chairman, Harin Ullal, and those of all of the CdTe team members. This work was supported by the US Department of Energy under contract no. DE-AC36-83CH10093.

# REFERENCES

1. Kroposki, B., T. Strand, R. Hansen, R. Powell and R. Sasala, "Technical Evaluation of Solar Cells, Inc. CdTe Modules and Arrays at NREL", *Proc. 25th IEEE PVSC* (1996), pp 969-972.
2. Powell, R.C., R. Sasala, G. Rich, M. Steele, K. Bihn, N. Ritter, S. Cox, and G. Dorer, "Stability Testing of CdTe/CdS Thin Film Photovoltaic Modules", *Proc. 25th IEEE PVSC* (1996), pp 785-788.
3. Meyers, P.V., S. Asher and M. M. Al-Jassim, "A Search for Degradation Mechanisms of CdTe/CdS Solar Cells", *MRS Symposium Proc. Vol 426, "Thin Films for Photovoltaic and Related Device Applications"*, David Ginley, Anthony Catalano, Hans W. Schock, Chris Eberspacher, Terry M. Peterson, and Takahiro Wada, eds., (1996), pp 317-324.
4. Granata, J.E. and J.R. Sites, G. Contreras-Puente, A.D. Compaan, "Effect of CdS Thickness on CdS/CdTe Quantum Efficiency", *Proc. 25th IEEE PVSC* (1996), pp 853-856.

# CdTe Solar Cells: Electronic and Morphological Properties

John J. Kester, Scot Albright, Victor Kaydanov, Rosine Ribelin, Lawrence M. Woods, and Jeffrey A. Phillips

*Golden Photon, Inc., Golden, Colorado 80404*

**Abstract.** CdTe solar cells have been produced with record efficiency and excellent stability from a low cost manufacturing line. Efficiency of 14.7% has been attained by standard manufacturing film deposition processes on soda lime glass. Modules were produced that have 10.5% active area efficiency. Long term testing of modules have shown degradation rates of less than 0.5%/year. These improvements have been linked to changes in CdTe morphology produced by recrystallization time/temperature profiles.

## Introduction

A key issue in the manufacture of Cadmium Telluride (CdTe) thin films on Cadmium Sulfide (CdS) cells with high efficiency and long term stability is the morphology of the CdTe layer. Almost all manufacturing processes of CdTe cells include a heat treatment of the CdTe after its deposition. This heat treatment recrystallizes the CdTe producing an increase in grain size and homogeneity(1,2). The electronic properties of the resultant grain structure are dependent primarily on the recrystallization parameters of temperature and time at temperature.

The CdTe absorber layer must separate the electron hole pairs as uniformly as possible along its junction with CdS. In addition, the final device must have a low series resistance and stable contact with the back electrode. The recrystallization step seeks to optimize both processes simultaneously. Optimization includes not only the highest initial efficiency but also long term stability under use conditions. Often the initial efficiency of these materials is not indicative of their long term stability. Deployment of devices and panels in the field or accelerated stress testing is a necessary protocol for long term reliability of these materials.

This paper presents the measurements of CdTe efficiency and stability with a new and old recrystallization process. Uniformity of efficiency across a panel and within a batch is also presented. The effect of recrystallization on cell parameters and uniformity of the junction region is measured.

## Experimental

All devices described in this report were produced with films deposited by GPI's standard manufacturing technology. Current processing conditions include the use of a tin oxide on soda lime glass as a transparent conductive oxide with 10 ohms per square sheet resistance. The optimal CdS thickness that is deposited depends on the time/temperature profile that the CdTe receives during the recrystallization step. CdTe is deposited to achieve a final thickness of approximately 10 microns including the incorporation of $CdCl_2$ to aid in the recrystallization process. At this point the panels can be finished as small cells or a complete 2' x 2' panel. The panel completion process includes deposition of graphite and tin films and division into isolated strips. Small cells can be produced after the recrystallization step of the process by the application of a mask (typically 0.302 $cm^2$) for graphite and tin films. The primary difference in the production of small cells is that the division process is replaced with an alternate soldering technique for making electrical contact with the negative terminal of the cell.

Sheet resistance of CdTe was measured on a layer similar to the porous region (see below) of the CdTe device. Interdigitated electrodes were deposited 3 mm apart and 60 mm in length. Because the resistances were in the kilo-ohm to mega-ohm range sufficient accuracy can be achieved with a standard digital ohmmeter for DC measurements. A variable frequency LCR meter was used to separate contributions for intragrain and grain-to-grain contact. To assure that resistance of the CdTe layer was not affected by contact resistance, measurements were made across adjacent electrode pairs and then pairs with every other electrode connected yielding additive resistances.

## Cell and Panel Efficiency and Stability Results

Many improvements in the deposition and processing of device layers have led to significant improvements in cell performance. Figure 1 shows the current/voltage measurement of a record CdTe device efficiency of 14.7% on soda lime glass. As described above, this cell was produced with GPI's current panel manufacturing process through the recrystallization step and then finished as a small cell. No anti-reflection coating was applied. The cell diameter is approximately 0.66 cm and is centered on a 0.95 cm wide stripe of material. While the open circuit voltage (Voc) of 824 mV and fill factor of 70.2% have been attained previously, their combination with 25.4 mA/$cm^2$ for the short circuit current (Isc) pushed the efficiency to record for soda lime glass.

The high Isc was attained by having a thin CdS layer in the finished device. This is indicated by the constant level of 81% from 400 nm to 700 nm in the absolute

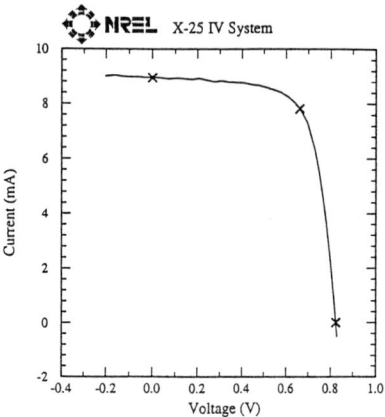

**Figure 1.** Current versus voltage of CdS/CdTe cell

quantum efficiency (QE) spectrum. The high quantum efficiency above the CdS bandgap near 500 nm suggests that the originally deposited CdS has been significantly interdiffused with the CdTe and no longer exists as a separate layer. Increased alloying is also indicated by the shift in the CdTe band edge to lower energy. The 50% level of the relative QE at the band edge occurs at approximately 860 nm. Similarly prepared devices have been examined by electron microprobe to determine the position of the sulfur within the structure. A line scan map of the elemental composition across the layers showed almost uniform distribution of sulfur in the CdTe layer. The lack of spatial resolution did not allow the determination of a residual CdS layer in the junction region. The combination of alloying and thin CdS is believed to be the primary cause of the increased photocurrent.

Completed modules with an area of 24 1/8" x 24 1/8" have been measured at NREL as having a maximum power ($P_{max}$) output of 29.3 watts. The practical efficiency is 7.8%. The aperture (3350 cm$^2$) efficiency is 8.7% and the active area (2793 cm$^2$) efficiency is 10.5%. The uniformity of efficiency across a panel can be measured by the production of 64 small cells from 13 areas across the panel. Recent measurements on a similar panel showed an average small cell efficiency of 10.7% with a standard deviation of 1.0%. Manufacturing viability depends on the uniformity of the efficiency output over all modules with a batch. A recent batch of 41 modules produced over 90% within $\pm$ 1.0 watt of the 24.5 watt batch average.

The long term stability of a module is shown in Figure 2. The module was encapsulated and within 2 days was placed on GPI's outdoor array under a 50 ohm load for almost one year. The rate of degradation of this panel is estimated from

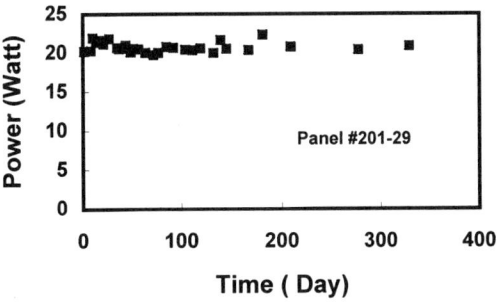

**Figure 2.** Module output power vs time

the slope of a least squares fit to be 0.08 watt/year or approximately 0.4%/year. This panel is now part of a long term test at NREL and was recently measured at $P_{max} = 23$ watts.

The panel above which measured $P_{max} = 29.3$ watts was done after over 75 days of exposure on GPI's outdoor array. The output power of this panel was only 1.7% lower than its initial value.

## CdTe Morphology

A significant improvement in cell parameters and stability has been produced by progress in the recrystallization of the CdTe film. The key parameters governing the recrystallization are the soak temperature and time at temperature. While changes in recrystallization require an optimization of other panel deposition processes, critical sensitivities can be determined from a test of a single parameter. Figure 3 shows the percentage change versus time of outdoor testing for the module with $P_{max} = 29.3$ watt in comparison with another from the same batch which had a different time/temperature profile during recrystallization. Except for the change in recrystallization, all other film depositions and processing steps for these panels were identical. The change in the time/temperature profile produced 4 panels with less than 2.7% degradation from their initial values The old recrystallization process produced 4 modules with greater than 25% degradation. The primary cell parameter responsible for this degradation is series resistance

The change in the time/temperature profile for CdTe recrystallization produces a dramatic change in morphology. The scanning electron micrograph (SEM) in Figure 4 compares the more stable CdTe material (Fig. 4b) with the less stable recrystallization (Fig. 4a). Two primary modes of sintering of films are volume diffusion and vapor phase transport (VPT). Volume diffusion is characterized by

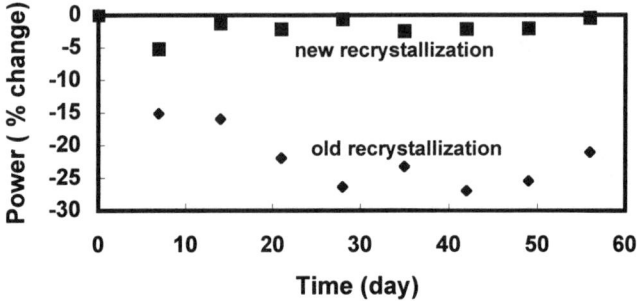

**Figure 3.** Change in output power versus time for different recrystallization processes

densification and shrinkage of the material causing coalescence into a dense homogeneous structure. The changes observed in GPI material are consistent with recrystallization that is dominated by vapor phase transport. VPT increases particle size but with no large scale densification or removal of pores as evidenced by the material microstructure.

Examination of the SEM in Figure 4b shows the improved time/temperature recrystallization profile has larger and more uniform grains than the old recrystallization profile in Figure 4a. Testing has been carried out that indicates that the growth of the material is indeed dominated by VPT during recrystallization. This growth of CdTe causes the formation of a dense region (3-4 µm) near the CdS interface and an adjacent porous region (6-7 µm). Extended recrystallization processes display an asymptotic growth in particle size of CdTe

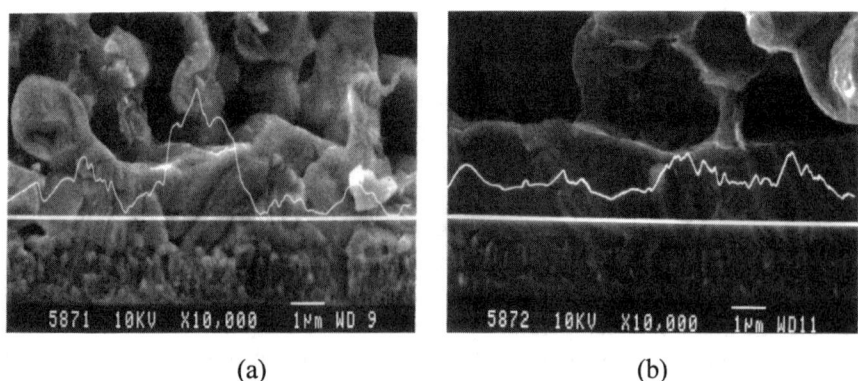

(a) (b)

**Figure 4.** Old recrystallization (a), and new recrystallization (b) including EBIC traces

without any reduction in the films total thickness or pore volume. Thermodynamic modeling is being pursued to optimize conditions which govern VPT and identify limits of densification.

Because of the device's stability dependence on its recrystallization procedure (Fig. 3) and because the stability performance over the first two months in the field is dominated by the series resistance, a correlation was sought between the CdTe sheet resistance and the series resistance of a device. A model for the electrical components of the porous region includes intragrain resistances, $R_1$, separated by the grain boundaries which can be represented by a parallel resistance and capacitance, $R_2$ and C, respectively. The measured resistance, $R_m$, is therefore a function of frequency, $\omega$, as given by

$$R_m = R_1 + \frac{R_2}{1 + (R_2\, C\, \omega)^2} \tag{1}$$

By measuring the frequency response the grain-to-grain and intragrain resistances can be separated. The intragrain resistivity was found to be insensitive to almost all processing conditions and is typically one to three orders of magnitude smaller than the grain-to-grain resistivity depending on processing conditions. This is consistent with previous studies of grain boundary effects(3). Resistance samples were treated in a manner similar to GPI's standard device processing. These samples were encapsulated with feedthoughs for measuring resistance and then placed on an outdoor array for a period of time sufficient to observe instabilities in series resistance in cells. The low frequency resistances before and after exposure for different recrystallizations are presented in Table 1. These results and others with different processing indicate that with the new recrystallization procedure the CdTe sheet resistance was more stable which correlates with the series resistance of similarly prepared cells.

Within the junction region the recrystallization profiles also produced different morphology and resultant electronic properties. Figure 4 shows the electron beam induced current (EBIC) line scans along with the scanning electron micrographs of the regions from which they were taken. The EBIC line scans were taken within the CdTe approximately one micron along a line parallel to the junction

**TABLE 1.** Sheet Resistance of CdTe Layer in M$\Omega$

| Recrystallization type | Illumination | Before | After | Change(%) |
|---|---|---|---|---|
| Old recrystallization | dark | 2.63 | 3.09 | 17 |
|  | light | 0.06 | 0.063 | 5 |
| New recrystallization | dark | 3.34 | 3.5 | 5 |
|  | light | 0.049 | 0.049 | 0 |

region. The line scans indicate the collection efficiency of electrons injected into the device. This efficiency for the collection of these electron induced currents is proportional to that produced by optical generation of photocurrents.

A comparison of the two types of recrystallization indicates that the new technique produces a more uniform collection efficiency than the older recrystallization profile(4). A digitization of the EBIC traces shows three times the standard deviation in photocurrent for the old crystallization above the new. While some of this variation is due to an interference by topographic features, a comparison of the secondary electron images and the electron beam induced currents show that, while related, topography does not dominate the EBIC results. In addition, the photoluminescent intensity at approximately 850 nm that is due to CdTe adjacent to the junction is greater for the new crystallization. This is consistent with a decreased dependence on surface or grain boundary states due to more uniform and increased grain sizes of the new recrystallization process.

## Conclusions

The electronic properties of devices and isolated layers of CdTe have shown the importance of morphology of CdTe to both the junction region and outside in the porous region of the material. In addition, improvement of initial device properties with new recrystallization time/temperature profiles resulted in a significant increase in short term stability of devices. The importance of process control was recognized by the great sensitivity of the sheet resistance of CdTe to standard processing conditions. The results seen on sheet resistance samples were similar to those found on devices. Improved recrystallization was also responsible for the increased efficiency that can be attributed to greater current collection efficiency from a more uniform junction region. Thus, critical changes in electronic properties have been tied directly to the CdTe morphology.

## Acknowledgements

This work was supported by the National Renewable Energy Laboratory under subcontract ZAF-5-14142-06. GPI acknowledges the technical support of by many NREL researchers, specifically, Rick Matson, Alice Mason, Dave Niles, Dean Levi, Brian Keyes, and Keith Emery.

## References

1. McCandless, B.E., Hichri, H., Hanket, G., and Birkmire, R.W., *Vapor Phase Treatment of CdTe/CdS Thin Films with $CdCl_2$:$O_2$*, in Proceedings of 1996 IEEE Photovoltaic Specialist Conference, pp. 781-784.

2. Park, J.W., Ahn, B.T., Im, H.B., and Kim, C.S., J. Electrochem. Soc. **139**, 3352-3356 (1992).
3. Thorpe, T.P., Fahrenbruch A.L., and Bube R.H., J. Appl. Phys. **60**, 3666 (1986).
4. Al-Jassim, M.M., Hasoon, F.S., Jones, K.M., Keyes, B.M., Matson, R.J., Moutinho, H.R., *The Morphology, Microstructure, and Luminescent Properties of CdS/CdTe Thin Film Solar Cells*, in Proceedings of 1993 IEEE Photovoltaic Specialist Conference, pp. 459-465.

# Recent Progress in CdTe Solar Cell Research at SCI

## R. A. Sasala, R. C. Powell, G. L. Dorer, N. Reiter

Solar Cells Inc., 1702 N. Westwood, Toledo, OH 43607

**Abstract.** Research at Solar Cells Inc. is focused on developing processes which will lead to high volume and low cost manufacturing of solar cells and to increase the performance of our present technology. The process research has focused on developing vapor transport deposition of the semiconductors, eliminating wet chemistry steps while minimizing the chloride treatment time, forming a low-loss back contact using only dry processing, and an improved interconnection technique. The performance improvement work has focused on the increase of the photocurrent by a combination of more transparent glass substrates and a thinner CdS window layer deposited on an i-$SnO_2$ buffer layer. SCI record 13.0% 1 $cm^2$ devices have been fabricated using these techniques. Stability monitoring continues and shows minimal degradation for over 20,000 hours of continuous light soak at 0.8 sun illumination.

## INTRODUCTION

The goal of Solar Cells Inc. is to develop the technology to produce photovoltaic modules in volumes sufficient to compete with the economics of conventional electric technologies. CdTe is the material of choice due to its demonstrated good performance both in the laboratory and in pilot production manufacturing. Furthermore, the deposition methods under development are consistent with eventual integration with a commercial scale glass float line. This approach shapes the direction of research and the priority that experiments are performed. SCI is investigating processes which shorten and/or simplify deposition and post-deposition procedures, flow in a logical manufacturing sequence and permit in-situ processing of all deposited layers. A specific example is the chloride treatment which is used by most CdTe workers to optimize device performance. The standard chloride treatment is typically 30 minutes long and requires several wet processing steps. Therefore, either a long in-line treatment chamber or a batched oven is required, both of which are undesirable for manufacturing. Development of a low-loss back contact formed with strictly dry processing and applied either under vacuum or at atmospheric pressure continues. Optimization of the performance and stability of devices made by processes altered for scale or speed remains an active area.

The focus of this paper will be to review the advances towards commercialization of CdTe photovoltaics that SCI has made over the past 18 months with the assistance of funds from NREL. The two areas to be addressed are device processing and device performance. The processing includes improved deposition, chloride treatment, back contact formation and interconnection. Performance improvements include increases in device efficiency and an update on continuing stability testing.

## CdTe DEPOSITION

SCI has been successful in implementing a large-scale modified close-spaced sublimation vacuum deposition process. Depositions are usually conducted at pressures near 1 Torr. The transport of material from the array of powder-filled source trays to the substrates is diffusion controlled. Likewise in laboratory close-spaced sublimation depositions, the transport of material is diffusion controlled but pressures of 10 to 30 Torr can be used due to closer source to substrate distances.

A commercial CdS/CdTe coater requires methods to supply raw material, to generate vapors, and to uniformly transport the vapors to the glass. Preferably the raw material supply is continuous and the system can operate for an extended duration. SCI has investigated several techniques for raw material introduction and for vapor generation.

The transport of vapors in most physical vapor deposition vacuum coaters (evaporation or sputtering) is primarily line-of-sight molecular flow. In the case of CdS and CdTe, however, the vapor pressures are sufficiently high at temperatures above 800 C, that significant quantities of material can be transported in heated conduits in the gas phase. Using inert carrier gas, Tuller et al. (1) and Chu et al. (2) have deposited CdTe at high pressure. Previously we have also deposited CdS and CdTe on a small scale at pressures up to 600 Torr using $N_2$ carrier gas (3).

SCI has tested a full-scale continuous-feed deposition apparatus. In this system, CdTe or CdS is sublimated continuously and the vapors are transported to the glass with the aid of an inert carrier gas. We will designate coatings deposited in this manner as vapor transport deposition coating (VTD).

Films have been deposited at pressures between 2 and 50 torr with growth rates exceeding 5 μm/min. Film thickness uniformity in the direction of glass travel has been excellent, typically ±5%. Cross web thickness uniformity has been found to depend on distributor geometry, gas flow and pressure.

The new deposition equipment provides a new level of control over film growth, particularly the local deposition rate, material utilization and film uniformity. Currently, the continuous sublimation system has been used to deposit onto a discontinuous stream of glass. Even under these conditions, films of both CdTe and CdS with microstructures similar to standard CSS material have been deposited. Electronic properties have been tested by combining CSS CdS with

VTD CdTe and vice versa. Small area devices have shown promising results. For example, 9% devices have been made using CSS CdS with VT CdTe. We believe that the continuous feed VTD system combined with a continuous stream of glass will produce steady state conditions and very reproducible films.

## CHLORIDE TREATMENT

Work on implementing the chloride treatment for manufacturing has focused on elimination of wet processes and reduction of treatment time. Successful use of HCl vapors as a chloride treatment agent has previously been reported using similar treatment temperatures and times as the conventional treatment (4). Recent work has examined the potential of treatment with $CdCl_2$ vapors and HCl for shorter duration.

The $CdCl_2$ vapor experiments were conducted in a quartz tube furnace which has separate heating zones for the $CdCl_2$ source material and the substrate The furnace also has provisions for the use of a carrier gas to assist the transport of the $CdCl_2$ vapors. The process parameters that were investigated include the temperature of the source, hence the partial pressure of the $CdCl_2$ vapor, the temperature of the substrate, total pressure, and time. Variations of the combinations of the process control variables were dictated by using design-of-experiment methods. This allowed a significant reduction in the number of experiments that needed to be conducted before trends can be evaluated and refinements made to the process.

The initial goal of the vapor $CdCl_2$ experiments was to determine the minimum treatment time interval which would produce the effect of a full standard wet $CdCl_2$ treatment. While there were some 9% efficient devices made with a treatment time as low as 60 s, consistent results required a minimum treatment time of 300 s. The 300 s treatment consistently produced devices in the 10 % range but did not produce open-circuit voltages as high as the standard treatment. The $V_{oc}$ was typically less than 800 mV. An advantage of the vapor $CdCl_2$ treatment is the potential of performing the treatment without leaving a residue which must be rinsed prior to contacting. Prevention of residue formation is accomplished by maintaining a sufficient substrate temperature during or after treatment.

Experiments have also been conducted to explore the reduction of process time for the HCl treatment in a similar manner as was found for the vapor $CdCl_2$ treatment. The HCl treatments were conducted at atmospheric pressure with a mostly inert ambient. The process control variables for the experiment include the substrate temperature, the HCl concentration and treatment time. HCl treatments tended to produce devices with higher $V_{oc}$'s than the vapor $CdCl_2$ method and 800 to 820 mV were typical. Device efficiencies over 10% were routinely produced. Similar to vapor $CdCl_2$ treatments we find 300 s to be the approximate minimum treatment time for reproducible high quality devices. We have found that optimum treatment substrate temperatures vary with the treatment medium. Vapor $CdCl_2$ treatments were found to work best with substrates between 420 and 440 °C. HCl, however, seems to require a somewhat lower substrate temperature - typically 380 -400 °C.

## INTERFACIAL LAYER

Formation of a low-loss contact to p-type CdTe requires special processing. It has been found that surface preparation of the CdTe is critical for fabrication of high quality devices and that, because of the polycrystallinity of these films, the contacting procedure can often affect the "bulk" properties of the films. These bulk effects have their most profound effect on the open-circuit voltage and the shunt resistance. Most of the traditional surface treatments have included wet etch steps which create a Te-rich layer on the CdTe surface. However, dry vapor processing is preferred for manufacturing. Thus, work has been performed on replacing the wet etch with direct deposition of Te. Initial efficiencies similar to standard wet treated devices have been observed in devices on which Te has been deposited with a variety of techniques including rf and magnetron sputtering, e-beam evaporation, and close-space sublimation. In fact, near record mini-module performance has been achieved with a deposited Te interfacial layer (IFL). Data for this device is shown in Figure 1.

Most of the work performed to date has been with rf-sputtered Te. The Te thickness is generally greater than 100Å, which is consistent with the required thickness reported by Niles et. al.(5). Greater thicknesses can be used without significant performance variations.

Recent work has focused on CSS deposition of Te. Because of the relatively high vapor pressure of Te, high deposition rates are readily achieved by simple sublimation. Deposition rates as high as 500Å/s have been observed and have been used to produce 10% efficient devices with characteristics similar to devices with sputtered Te films. Consequently, it appears that Te can be easily deposited in a manner similar to our CdS and CdTe deposition technology (VTD) and will not require more costly sputtering or e-beam evaporation.

Tests have been conducted to determine the ability of the Te IFL to withstand elevated temperatures which it may be subjected to during subsequent processing. The experiment was to simply heat, in air, substrates with rf-sputtered Te to temperatures from 200 - 325 °C for 30 minutes and then complete the devices in the standard manner. There was no observable difference in the performance of these devices which was outside the normal experimental variation. If the Te deposition is performed prior to the chloride treatment, the device performance will be extremely poor. Thus the preferred sequence is semiconductor deposition, chloride treatment and then Te deposition. We have found that the best depositions of Te onto CdTe using close-space sublimation are for substrate temperatures below 400 °C and therefore processing sequence and temperature requirements flow naturally. This is convenient for a in-line process as the Te can be deposited onto the chloride treated CdTe without requiring any additional heat to achieve the required substrate temperature.

In addition, use of a Te IFL leads to a notable improvement in the stability of devices tested under open circuit conditions in continuous light-soak. More will be discussed on this in a later section.

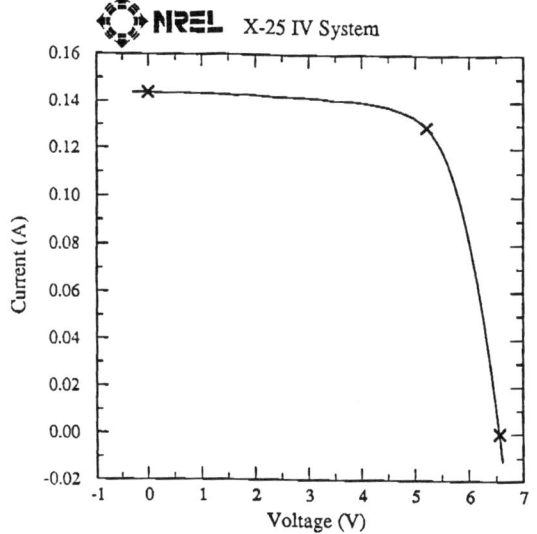

**Figure 1.** Near record mini-module (record = 10.6%) which has a Te IFL which was deposited using rf-sputtering.

## INTERCONNECTION

The present technique to isolate and interconnect adjacent cells in a module includes three separate laser scribes which must be precisely registered to each other. The pilot production laser system is a single beam Nd:YAG based laser system which requires more than one hour to complete the series of scribes. To transfer this technology onto a continuous manufacturing line will require multiple lasers each of which are split into multiple beams. This will undoubtedly be an expensive and challenging engineering project.

As an alternative interconnection technique, SCI is refining its patented "dot matrix"(6) interconnect structure. The basic principle behind dot matrix is to use a second metal in parallel with the TCO but isolated from the contact in intimate contact with the CdTe. The electrical connection between the TCO and the second back metal is made through a matrix of dots which are isolated from the semiconductors and first back metal. There are several advantages of the dot matrix technique when optimized. One is that the dots can occupy less space per unit area than the three laser scribes which results in a higher total area efficiency. Another is that the cell width is limited by the sheet resistance of the second back metal rather than the TCO. This permits the choice of cell width based upon the desired output voltage and not minimum $I^2R$ loss of the TCO. It also permits the use of a higher sheet resistance TCO which is both less expensive and has a higher optical transmission which leads to a higher efficiency. Another advantage of the dot matrix technique is that it can be made to be self registering.

To date, SCI has successfully made a 9.3% efficient (in-house measurement) 64 $cm^2$ interconnected mini-modules with 4 cm wide cells in the laboratory and is working on improved methodologies for the full size modules.

## EFFICIENCY IMPROVEMENTS

Analysis has shown that increased photocurrent could substantially increase device efficiency. SCI supplied substrates before and after each deposition step to researchers at Colorado State University to permit the separation of photon losses in each layer for a standard device with thick (3,000 Å) CdS. The result is shown in Figure 2 (7). The figure clearly shows that the major current losses are due to absorption in glass and low collection of carriers from light absorbed in the CdS (wavelengths below 520 nm). Since there has been little success in improving the collection efficiency of the CdS, the improvement strategy has been to minimize the window absorption by thinning the CdS. While this approach has resulted in increased current densities, reduced open-circuit voltage has lead to only marginal efficiency improvements. However, similar to results found at the University of

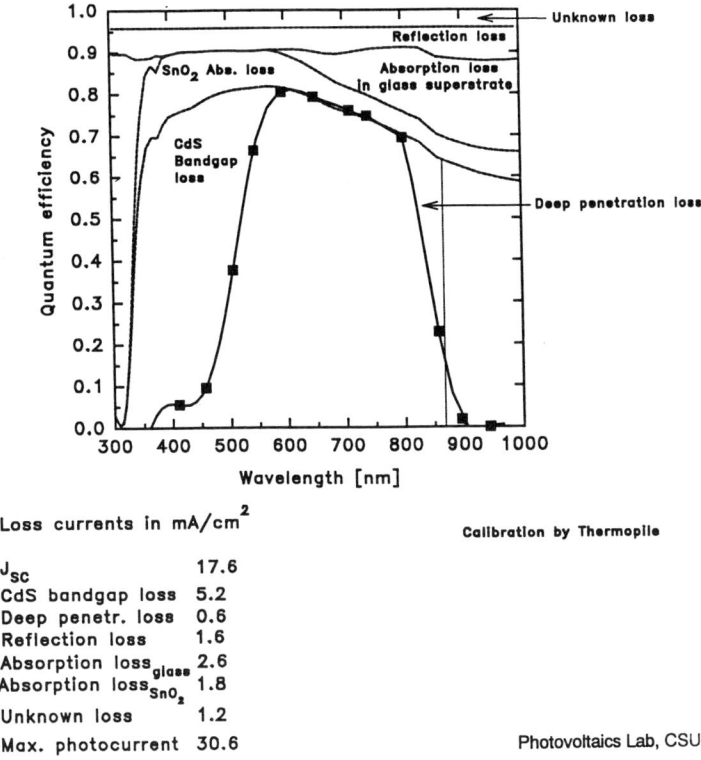

Figure 2. Separation of current loss mechanisms for a standard device with thick CdS. (From Gunther Stollwerck's MS thesis from Colorado State University.)

South Florida, the deposition of an intrinsic $SnO_2$ layer on top of the $SnO_2$:F permits the use of thin CdS without loss in $V_{oc}$. The intrinsic $SnO_2$ layer may reduce the effectiveness of high spots on the TCO and pinholes in the CdS as shunt paths. Work performed at SCI which supports this hypothesis includes comparisons of devices made with thin CdS on low and high haze TCO. Older LOF TEC8 glass has a haze up to 10% whereas newer TEC8 and TEC15 products have considerably less haze, as reported by LOF (8). Haze is a measure of the diffuse white light scattering according to ASTM D 1003. The haze is a direct result of surface roughness of the TCO and high haze corresponds to higher roughness. Devices made on the low haze TEC8 substrates have produced substantially higher open-circuit voltages than older substrates when using similar processing. An example of this is the SCI record 1 $cm^2$ cells which were recently made on 3mm thick low haze LOF TEC8 glass. The CdS thickness of this sample was approximately 400 Å (also with an i-$SnO_2$ layer) yet it still produced an open-circuit voltage near 840 mV, which is close to the SCI record with thick CdS. The I-V curve measured at NREL for this record device is shown in Figure 3. In addition to the high $V_{oc}$, the short-circuit current density of 22.3 mA/cm$^2$ is as high as has been made previously on soda lime glass at SCI. The $J_{sc}$ is higher than normal because of the reduced thickness of the CdS and glass. The quantum efficiency measured at NREL for an adjacent cell is shown in Figure 4. Comparison with Figure 2 shows that the response has increased in both the red and blue regions of the curve. The red response accounts for a 0.8 mA/cm$^2$ increase and is due to the thinner glass whereas the increased blue response accounts for a 2.7 mA/cm$^2$ increase and is due to the thinner CdS. The unknown loss of 1.2 mA/cm$^2$ reported in Figure 2 has been eliminated in this device. Using glass thinner than 3mm for full size modules (7200 cm$^2$) is not feasible because of mechanical strength requirements of installed modules. Therefore, reducing absorption loss in the glass can only be accomplished by lowering the Fe content in

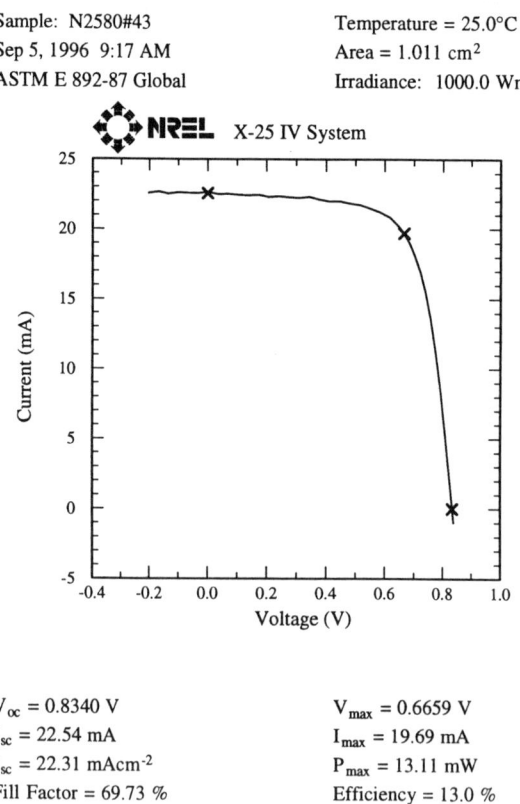

$V_{oc}$ = 0.8340 V
$I_{sc}$ = 22.54 mA
$J_{sc}$ = 22.31 mAcm$^{-2}$
Fill Factor = 69.73 %

$V_{max}$ = 0.6659 V
$I_{max}$ = 19.69 mA
$P_{max}$ = 13.11 mW
Efficiency = 13.0 %

**Figure 3.** I-V curve for record 1 cm$^2$ SCI device. The device has an i-SnO$_2$ buffer layer on the TCO and a 400 Å CdS thickness.

the glass. While uncoated low iron glass is commercially available, none is available with TCO coatings comparable to LOF TEC8 or TEC15 coatings. If a TCO coated water-white glass was available, one could expect an additional 1.8 mA/cm$^2$ of current.

SCI's highest efficiency devices were made of 1 mm thick soda lime glass with a TCO deposited by Asahi Glass. The I-V curve for this device is shown in Figure 5. The diode parameters are similar to those for the cell described above. The $V_{oc}$ is slightly higher and the current a little lower but the higher fill factor gives the cell the advantage and the slightly higher efficiency. Analysis of the quantum efficiency curve for this cell, shown in Figure 6., shows that the added current comes from the reduced glass thickness (the curve is flat out into the red) and an antireflection coating. The CdS thickness for this sample was approximately 1500Å which is approximately the minimum thickness, on high haze TCO, SCI has been successful using without an intrinsic $SnO_2$ layer and without a significant loss in $V_{oc}$. If the same CdS thickness was used on this cell as the cell in Figure 3, the Jsc would be 23.9 mA/cm$^2$. If all other parameters stayed constant, the efficiency would be approximately 14.5%.

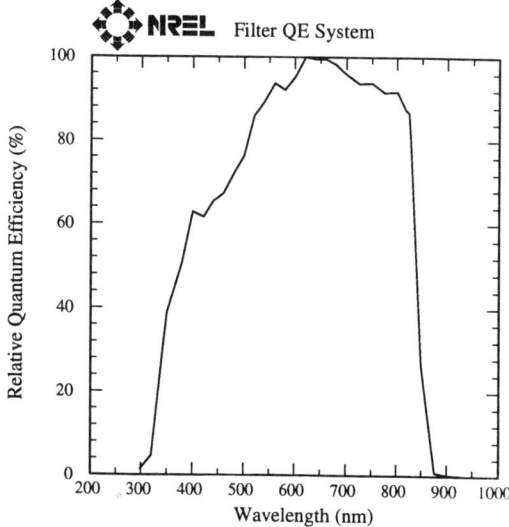

**Figure 4.** Quantum efficiency from SCI's record 1 cm$^2$ device deposited on 3mm soda lime glass. Note the large increase in blue response and the slightly higher red response due to thinner CdS and glass respectively.

# STABILITY

Photovoltaic module manufacturers must assure customers about the long-term power-delivery capability of their product. In the absence of extensive field history, methods of accelerated testing are needed. In addition, process development requires techniques to rapidly compare module stability differences resulting from process changes. Correlation of the results of accelerated tests and behavior in the field is complex but necessary.

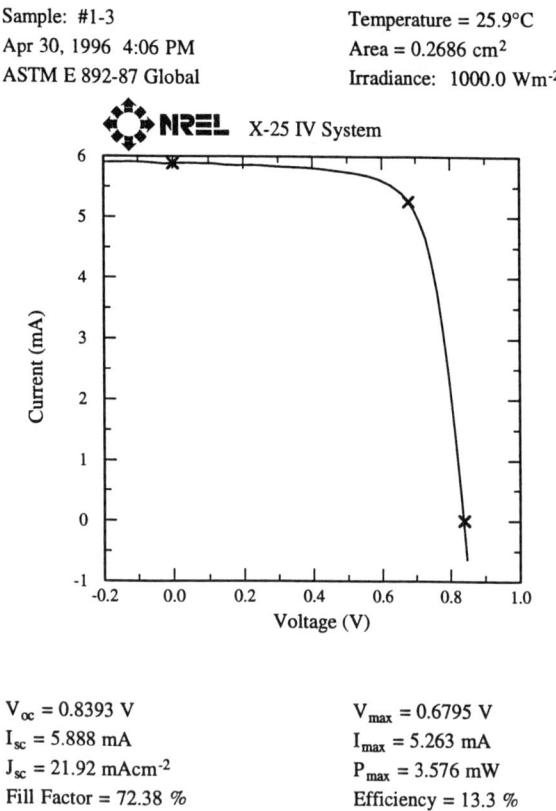

**Figure 5.** I-V curve for record small area SCI device. The device was deposited on 1mm soda lime glass with TCO deposited by Asahi Glass.

Many CdTe/CdS fabrication techniques are known. Some of these techniques result in devices with good stability. When observed, instability is usually attributed to the contact to p⁻-CdTe (9-11). While recent reports on CdTe module performance (12-15) have been encouraging, the data is limited and a more comprehensive testing methodology is needed. A complete description of the methods and results of this work has been previously reported (16). This section provides an update of the results from the first 2.5 years of these tests.

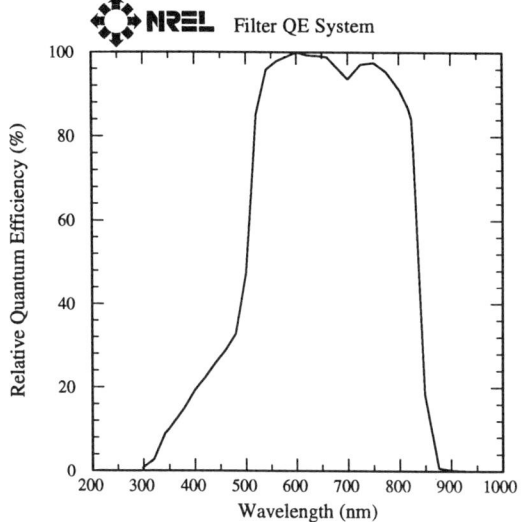

**Figure 6.** Quantum efficiency from SCI's record small area device deposited on 1 mm soda lime glass. Note the relatively flat red response due to the thin substrate.

## Light Soaking

The mainstay of the testing protocol, due to the similarity to field conditions, is continuous light soaking with resistive load. Furthermore, the majority of devices are biased near the maximum power point as they are under field conditions. Devices made with the standard process used in pilot production tested under resistive load are quite stable to greater than 20,000 hours of continuous light soaking at 0.8 sun illumination and 60 °C ambient, as shown by representative data in Figure 7. Devices held at short circuit conditions are also generally quite stable; however, devices held at open-circuit are less stable relative to the other bias conditions. We do not fully understand the mechanism responsible for the decreased stability at open circuit, but the effect appears to be rather general and consistently occurs for many, but not all, fabrication recipes. The exception is for devices with a Te IFL, as mentioned earlier. Figure 8 shows an example comparison of identical devices with and without a Te IFL and held at open-circuit under continuous illumination. The improvement with the Te is obvious. When held near the maximum power point, Te IFL devices perform as well as the standard device and tend to have a slightly better stability in the first several hundred hours of continuous light soak.

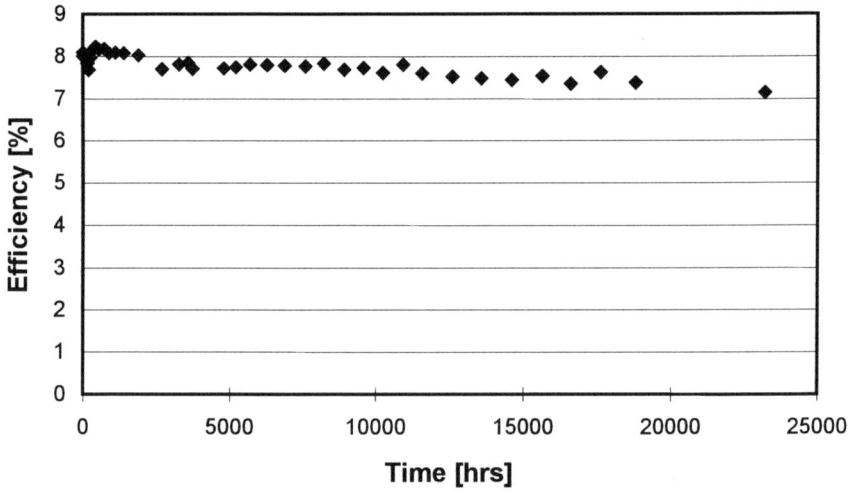

**Figure 7.** Efficiency of minimodules made with the standard process and subjected to continuous light soak while being held near the maximum power.

## Discussion

Stable CdTe-based photovoltaic modules have been made with our standard pilot production process. The stability of modules made using a particular process can only be ensured with long-term testing. Simple indoor continuous light soaking conducted for >5000 hours is the minimal testing recommended. Without such testing, erroneous conclusions about the superiority of a particular fabrication recipe can be made. The recommendation applies to individual cells as well as to modules. Elevating temperatures beyond normal operating conditions by 20 to 35 °C in light soaking may be effective in accelerating degradation but can only be used for recipe comparisons at this time.

In the first phase of protocol development we have found that long term device behavior depends on bias in light soaking stress tests. Prediction of outdoor field operation based on indoor continuous light soaking tests at constant temperature is complicated due to several factors. First, sufficient field data has not been accumulated. Second, the inherent variation of illumination and temperature in the field may be important. For example, we have observed partial performance recovery of some degraded devices after several days of interrupted indoor light exposure. Thus a simple illumination dose equivalence between continuous indoor light soaking and intermittent light exposure in the field may not be valid.

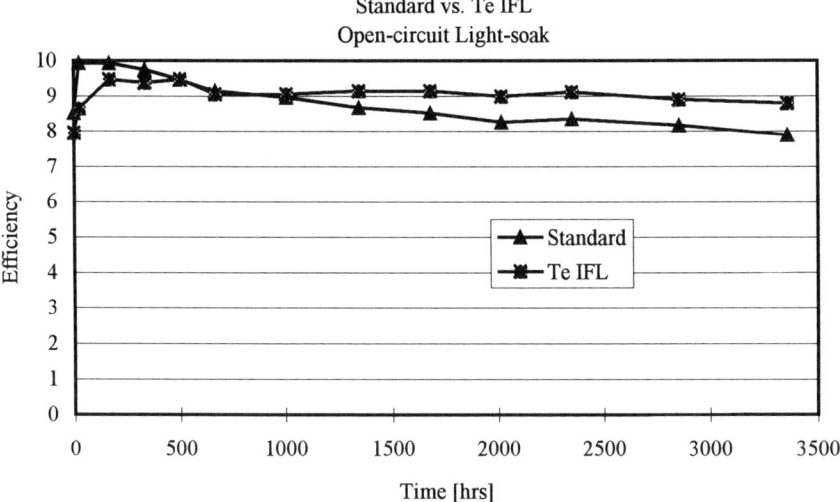

**Figure 8.** Comparison of device performance between standard wet process and dry Te IFL when held at open-circuit under continuous illumination.

The next phase of development will include some extension and revision of the initial approach. First, while we understand some of the operating physical mechanisms, more analysis of device operation and physical changes is needed. Second, more detailed conclusions will require a larger number of samples for improved statistics, and thus concentration to fewer stress conditions is likely. Third, since some evidence of dark relaxation has been observed, an intermittent light soaking test will be added.

## Outdoor Array Data

While all of the above described stability data for indoor tests are very important in device design and development, it alone is insufficient to predict module performance in the field. Additional information of actual field performance is required for two reasons. First, one must verify that the same phenomena is occurring in the indoor stress tests and outdoor operating conditions. For example, one must test that normal daily and seasonal temperature, humidity and illumination cycling do not induce additional degradation mechanisms not simulated in indoor tests. Second, one must use outdoor performance as a reference to determine any acceleration factor of indoor stress conditions.

SCI has detailed performance data on twenty-four months of data on our Westwood facility 1.2 kW array. Data includes current voltage data as well as illumination intensity and temperature data to permit the normalization of both intensity and temperature variations from standard reporting conditions. The data normalized for intensity from the SCI Westwood array is shown in Figure 9. The periodic nature of the curve is due to the seasonal variation of the ambient temperature. The rated output of the array is 1200 W at 25 °C which agrees well with the average between summer and winter output. The back-of-the-module temperature on days with low wind is typically 20 °C above the ambient which results in operating temperatures during winter months of 0 - 20 °C and summertime temperatures of 50 - 60 °C. A linear regression on temperature vs. maximum power (illumination normalized) yields a temperature coefficient of -0.32%/°C which is large enough to give a noticeable apparent seasonal performance dependence. This is consistent with the range of temperature coefficients measured for small area devices under carefully controlled laboratory conditions. Data taken over an eight month period was used to maximize the temperature range. The temperature coefficient calculated from data taken during any one month agree within 5%.

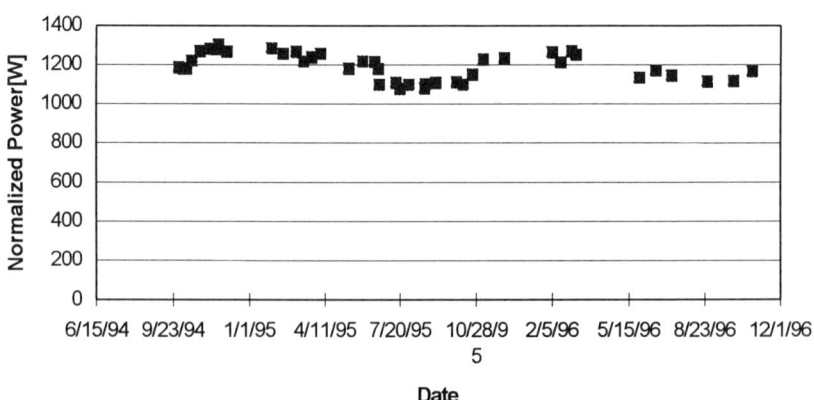

**Figure 9.** Power output of SCI Westwood 1.2 kW array normalized for intensity variations but not for temperature fluctuations.

Figure 10 shows the array data with the normalized power corrected for temperature away from 25 °C and this normalized output agrees with the initial array rating. With the temperature correction, the data shows good stability for its 2 years of testing.

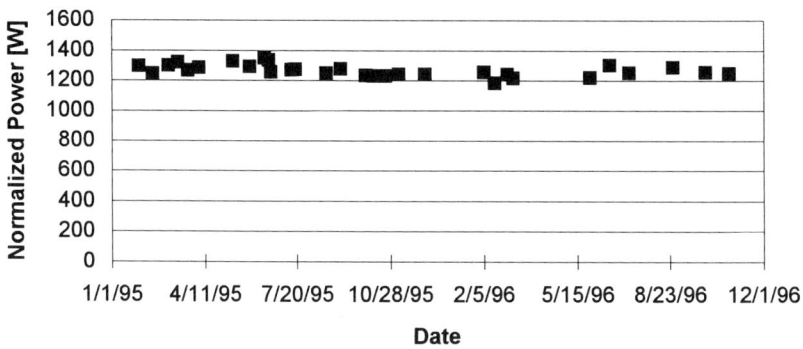

**Figure 10.** Power output of SCI Westwood 1.2 kW array normalized for intensity variations and corrected for temperature variations from 25 °C.

## SUMMARY

The research at SCI is focused on process and performance improvements that will result in high volume and low cost photovoltaic manufacturing. Significant progress on processing improvement has been made in the semiconductor deposition, chloride treatment, dry back contact processing, and interconnection. The deposition work has focused on developing vapor transport to allow for superior control of the semiconductor growth as well as continuous material feed. Devices made with material deposited with the new deposition apparatus appear to be similar to those deposited by the proven pilot production apparatus. The chloride treatment time has been reduced from 30 minutes to 5 minutes and shows some good device behavior with treatments as low as one minute. The chloride treatment has been carried out using HCl and $CdCl_2$ vapors as the source of chlorine. The back contact process has been improved with the use of a deposited Te interfacial layer to replace the standard wet etch. Device performance is similar to the SCI standard wet processed devices except they are more stable under continuous light soak under open-circuit conditions. Progress has also been made on an alternative interconnection technique which can potentially increase the efficiency while allowing lower cost, more transparent TCO glass. Additionally, the interconnection can be self registering. Minimodules with efficiency higher than 9% have been demonstrated utilizing this interconnect process.

Progress has also been made in performance improvements. The efficiency of small area devices (0.27 cm$^2$) has been raised to 13.3% by using both thinner glass and CdS. The efficiency of 1 cm$^2$ devices has been raised to 13% by using an i-SnO$_2$ layer coupled with a 400 Å thick CdS layer on commercially available TCO on soda-lime glass. The stability of standard process mini-modules has now been demonstrated for over 20,000 hours of continuous light soaking while held near the maximum power point. The two years of data on SCI's 1.2 kW array show good stability to corroborate the indoor test data.

## ACKNOWLEDGMENTS

This work has been partially supported by NREL subcontract ZAF-514142-05.

## REFERENCES

1) Tuller, H.L., Uematsu, K., and. Bowen, H.K, J. Crystal Growth **42**, 150 (1977).
2) Chu, T.L., Chu, S.S., Pauleau, Y., Murthy, K., Stokes, E.D., and Russell, P.E., J. Appl. Phys. **54**, 398 (1983).
3) Nolan, J. F., et al., Annual Technical Status Report, Subcontract ZR-1-11059-1, May 1993.
4) Zhou, T. X., Reiter, N., Powell, R. C., Sasala, R., Meyers, P. V., Proc. First WCPEC December 1994, pp 103-106.
5) Niles, D. W., Li, X., Sheldon, P., *J. Appl. Phys.* 77 (9), May 1995, pp 4489-4493.
6) McMaster, U.S. Patent 4,872,925.
7) Stollwerck, G., Colorado State University Master Thesis, unpublished.
8) Gerhardinger, P.F., and McCurdy, R.J., "Float Line Deposited Transparent Conductors - Implications for the PV Industry", Proc. Materials Research Society, **426**, Spring 1996.
9) Tyan, Y.-S., "Topics on Thin Film CdS/CdTe Solar Cells", *Solar Cells* **23**, 1988, pp. 19-29.
10) Fahrenbruch, A. F., "Ohmic Contacts and Doping of CdTe Solar Cells", *Solar Cells* **21**, 1987, pp. 399-412.
11) Szabo, L.F. and Biter, W. J., "Stable Ohmic Contacts to Thin Films of P-type Tellurium-Containing II-VI Semiconductors", U.S. Patent 4,735,662, 1988.
12) Zhou, T.X., Sasala, R.A., and Powell, R. C., "Fabrication of Stable Large-Area Thin-Film CdTe Photovoltaic Modules", in American Institute of Physics Conference Proceedings 353, 13th NREL Photovoltaics Program Review, H.S. Ullal and C. E. Witt Eds., Lakewood, CO, 1995, pp. 31-38.
13) Kroposki, B., Strand, T., Hansen, R., Powell, R., and Sasala, R., "Technical Evaluation of Solar Cells, Inc. CdTe Module and Array at NREL", 25th IEEE PVSC, 1996.
14) Ikegami, S., "CdS/CdTe Solar Cells by the Screen-Printing-Sintering Technique: Fabrication, Photovoltaic Properties and Applications", *Solar Cells* **23**, 1988, pp. 89-105.
15) Woodcock, J. M., et al., "Thin Film CdTe Photovoltaic Cells", 12th European Photovoltaic Solar Energy Conf., Amsterdam, April 1994.
16) Powell, R.C., et. Al., Proceedings of 25th IEEE PVSC, 1996, p. 785-788.

# CRYSTALLINE MATERIALS I

# The Crystalline-Silicon Photovoltaic R&D Project At NREL And SNL

James M. Gee and Ted F. Ciszek[*]

Sandia National Laboratories, Albuquerque, NM 87185-0752
[*]National Renewable Energy Laboratory, Golden, CO 80401

**Abstract.** This paper summarizes the U.S. Department of Energy R&D program in crystalline-silicon photovoltaic technology, which is jointly managed by Sandia National Laboratories and National Renewable Energy Laboratory. This program features a balance of basic and applied R&D, and of university, industry, and national laboratory R&D. The goal of the crystalline-silicon R&D program is to accelerate the commercial growth of crystalline-silicon photovoltaic technology, and four strategic objectives were identified to address this program goal. Technical progress towards meeting these objectives is reviewed.

## INTRODUCTION

While a wide variety of semiconductor materials have been examined and are still currently under development for photovoltaic (PV) modules, the dominant technology today still uses bulk crystalline-silicon (c-Si) substrates. Crystalline-silicon PV modules represented around 85% of the 81 $MW_p$ of PV modules sold in 1995 [1]. Despite the relative maturity of c-Si PV technology, industry continues to make improvements in their manufacturing processes and module design to reduce manufacturing cost and increase throughput. For example, Wohlgemuth *et al.* recently reported that Solarex is on target to reduce the manufacturing costs of their multicrystalline-silicon (mc-Si) module by a factor of 2 and increase their manufacturing capacity by a factor of 3, while Mitchell *et al.* described the relatively straightforward extensions of present technology necessary to reach a production level of 100 MWp per year using Czochralski (Cz) silicon [2,3]. Basore and Gee noted that c-Si PV is capable of meeting a residential PV system market of around 40 $GW_p$ in the United States (U.S.), while a recent European study found that conventional c-Si PV modules could reach direct manufacturing costs approaching $1 per $W_p$ at production levels of 500 $MW_p$ per year [4,5]. The net result is that c-Si PV technology has followed an aggressive learning curve, with various reports associating a reduction of 68% and

---

The work at Sandia National Laboratories was supported by the U.S. Department of Energy under contract DE-AC04-94AL85000.

83% in average selling price with each doubling of cumulative sales [6,7]. Extrapolation of these economic and technical trends with PV growth rates of between 20 and 40% achieved in recent years by U.S. PV manufacturers predict that c-Si PV modules will be able to meet U.S. Department of Energy goals for 2000 and 2010 as outlined in the recent five-year plan [8].

The past success and current improvements in c-Si PV technology has benefited greatly from the R&D investment in c-Si technology, and the U.S. Department of Energy supports a R&D project in c-Si PV technology to help continue this progress. This paper reviews the U.S. DOE crystalline-silicon PV R&D program. This program is jointly managed by researchers from the National Renewable Energy Laboratory (NREL) and Sandia National Laboratories (SNL), and features a balance of basic and applied R&D, and of university, industry, and national laboratory R&D. The paper first provides a description of the mission and strategic objectives, and then provides a description of the technical projects that address these strategic objectives. Included in the description of the technical projects are selected recent technical highlights from the national laboratories. For completeness, references are supplied for supporting work by university subcontractors.

## MISSION AND STRATEGIC OBJECTIVES

The goal of the Crystalline-Silicon R&D Project is to accelerate the development of c-Si photovoltaics and to enhance the United States' position in c-Si photovoltaics technology. This goal is approached through the following objectives:

- Improve the performance and/or reduce the cost of *present-generation commercial* c-Si solar cells and modules.
- Improve the *fundamental understanding of crystalline-silicon material*, with an emphasis on controlling the deleterious effects of impurities and defects in crystalline silicon.
- Develop *next-generation* c-Si PV technologies that significantly improve throughput, reduce energy consumption, and/or reduce manufacturing cost compared to the present technology.
- Coordinate national laboratory, university, and industry c-Si PV research, and non-PV c-Si R&D programs, through hosting of a workshop on c-Si processing and through operation of a c-Si R&D cooperative (Crystalline-silicon Research Cooperative).

## TECHNICAL PROJECTS AND HIGHLIGHTS

The mission goal of the project is approached through three technical tasks: silicon crystal growth, crystalline-silicon material science, and crystalline-silicon devices and processing. The goals and status of each of these tasks are reviewed.

# Silicon Crystal Growth

This task emphasizes research on innovative and novel approaches to Si crystal growth methods, which complements work in industry by examining next-generation technologies. These new crystal growth methods potentially have superior throughput, reduced energy and/or materials cost, and/or improved conversion efficiency compared to existing approaches.

There is currently considerable interest worldwide in thin-layer c-Si PV [9]. Thin-layer c-Si refers to crystalline-silicon layers with thicknesses less than 100 μm on a supporting substrate. The potential advantages of thin-layer c-Si PV include reduced material and energy usage compared to bulk c-Si PV, all planar processing for reduced manufacturing costs, and monolithic module integration. Our work (Ciszek *et al.*) in this task is examining growth of thin-layer c-Si films on low-cost metallurgical-grade c-Si substrates (MG-Si) by liquid phase epitaxy (LPE). LPE is an attractive technique because it has adequate growth rates (around 1 μm/min) at moderate temperatures (around 900°C), has high quality (near-equilibrium crystal growth), and uses relatively simple equipment. A conventional method to improve surface wetting in LPE is to melt back the surface in the solution. However, the melt-back step is not desirable with metallurgical-grade silicon substrates since impurities from the substrate would then contaminate the solution. We developed a Cu-Al-Si that is able to wet the surface well without a melt-back step. The high solubility of Si in Cu-Al (20-35%) at a growth temperature of ~900°C also creates an atomically rough solid/liquid interface, facilitating isotropic growth and a macroscopically smooth crystal surface (Figs.1 and 2). Details of this work is provided in another paper at this conference [10].

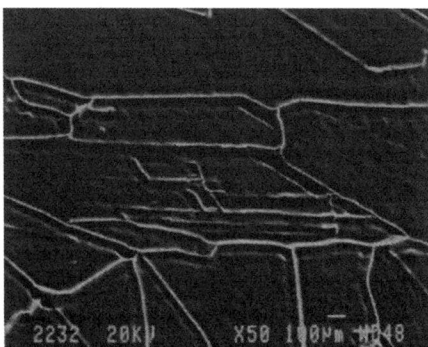 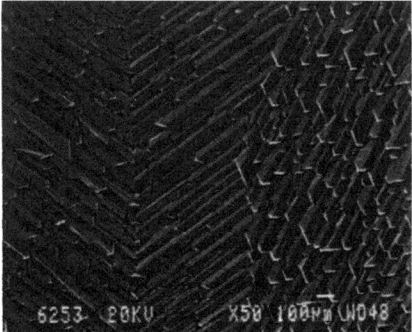

**Figure1**. Surface morphology of a LPE-Si/MG-Si layer grown from a 23%Si-28%Al-49%Cu solution.

**Figure 2.** Surface morphology of a LPE-Si/MG-Si layer grown from a 3%Si-97%In solution.

Other work in thin-layer c-Si PV supported by the SNL/NREL c-Si PV Project includes seeded crystal growth (California Institute of Technology) and optical confinement theory and experimentation (University of California/Los Angeles and NREL). Some of this work is summarized in other papers at this conference [11-13].

## Crystalline-Silicon Material Science

This task seeks to develop a fundamental understanding of the role of defects and impurities in c-Si materials and device processing. The goal of this research is to ameliorate the deleterious effects of defects and impurities in c-Si solar cells, which is an important objective with high industry relevance because the concentrations of defects and impurities are frequently increased by the use of lower cost Si feedstock and c-Si growth methods. Major activities in this task include generation and characterization of samples with controlled concentrations of dopants, impurities, and defects, research on the physics of and methods to getter impurities and passivate defects, and hosting of a workshop on crystalline-silicon PV technology. A major concern of this task is to improve the performance of large-area multicrystalline-silicon solar cells whose performance is reduced by the influence of bad grains that are difficult to getter using conventional methods. A summary of R&D requirements for improving large-area mc-Si cell performance is presented in another paper at this conference [14].

Iron is a very common metallic impurity in silicon that is also a recombination center. Developing a fundamental understanding of Fe in silicon, including interactions with crystal defects and with other impurities (dopants, oxygen, etc.), is of increasing importance due to the wider usage of less pure Si feedstock and interest in new thin-layer c-Si growth techniques using metallurgical-grade c-Si substrates. The float-zone (FZ) method for silicon crystal growth allows a high degree of control over background impurity and defect levels and is an excellent vehicle for controlled studies of deliberately introduced impurities and/or defects. At NREL, we (Ciszek *et al*) grew Fe-doped multicrystalline ingots by the FZ method to study Fe effects on minority charge carrier lifetime, grain structure, and electron-beam-induced current characteristics of multicrystalline silicon. Details of the growth and characterization is provided in Ciszek *et al.* [15].

Representative data is presented in Fig. 3. The minority-carrier lifetime decreased monotonically with increasing Fe content for similar grain sizes (from ~10 μs to 2 μs for < $10^{-3}$ cm$^2$ grains, from ~30 μs to 2 μs for ~5 x $10^{-3}$ cm$^2$ grains, and from ~300 μs to 2 μs for > $10^{-2}$ cm$^2$ grains) as the Fe content increased to $1 \cdot 10^{16}$ atoms/cm$^3$. We had previously observed that grain size alone has a strong effect on lifetime. We also saw evidence of constitutional supercooling in the heavily doped samples, with a dramatic accompanying effect on grain structure [15]. Such observations might aid in understanding the precipitation of Fe during ingot crystal growth and design of new thin-layer c-Si growth techniques.

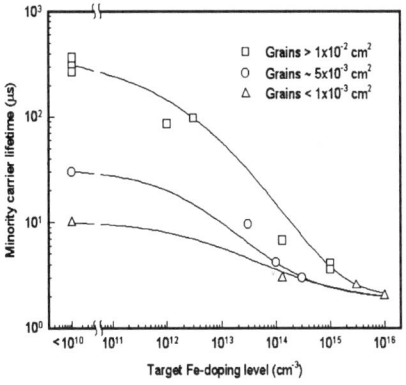

**Figure 3.** Measured bulk minority-carrier lifetime vs. target Fe-doping level for float-zoned multicrystalline-Si ingots with various grain sizes.

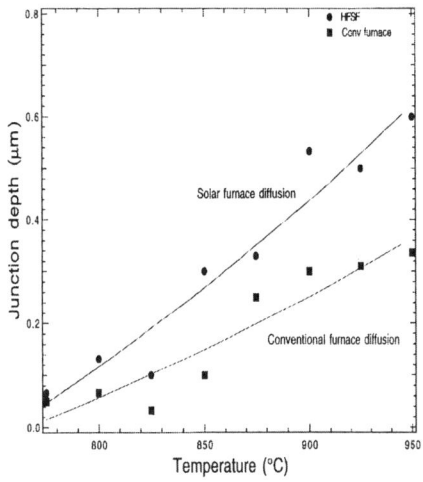

**Figure 4.** Junction depth of phosphorus diffusion performed with a solar furnace and with a conventional furnace.

Other work in this task includes the following: research on the physics of gettering and hydrogen passivation; development and characterization of gettering techniques on commercial c-Si materials; and development of new gettering techniques [16-23].

## Crystalline-Silicon Devices and Processing

This task develops new processes and device structures with a goal of improving commercial PV module performance (cost, efficiency, and reliability). This task emphasizes research activities with near-term impact that are identified collaboratively with major c-Si manufacturers. Recent work at the Georgia Institute of Technology and at SNL has found that the material quality of photovoltaic-grade c-Si substrates is capable of much higher performance than can be obtained with commercial fabrication processes, so that there is ample opportunity to improve the performance of commercial c-Si modules [24,25].

The goal of our process development work is to develop processes with significantly improved throughputs. Optical processing methods -- such as optical anneals, rapid thermal processing, or solar furnace processing – are of interest due to the reduced thermal budget and short process time. In particular, there may be possible kinetic advantages to optical processing. We (Tsuo *et al.*) recently demonstrated that dopant diffusions performed using high-intensity solar flux are deeper than obtained with the same time-temperature profile in an isothermal furnace (Fig. 4). This same work also demonstrated improved out-diffusion of impurities from a metallurgical-grade multicrystalline-silicon substrate [18].

Other work in this task includes development of plasma processes and novel doping processes for forming back-surface fields [26,24,29].

Major emphasis in cell development this year will be placed on two novel cell concepts (emitter wrap-through cell and self-aligned selective-emitter cell) that have the potential for significantly improved performance and reduced cost. The self-aligned selective-emitter cell uses plasma processing to achieve a low-recombination passivated emitter with commercial metallizations (Fig. 5). The cell is potentially low cost because the cell uses screen-printed grids, the emitter etch is self aligned, and the plasma etch and deposition can be performed in the same chamber. We (Ruby *et al.*) recently demonstrated an improvement in efficiency of 0.5% absolute with the new process compared to the baseline process on 103-cm$^2$ multicrystalline-silicon cells [26].

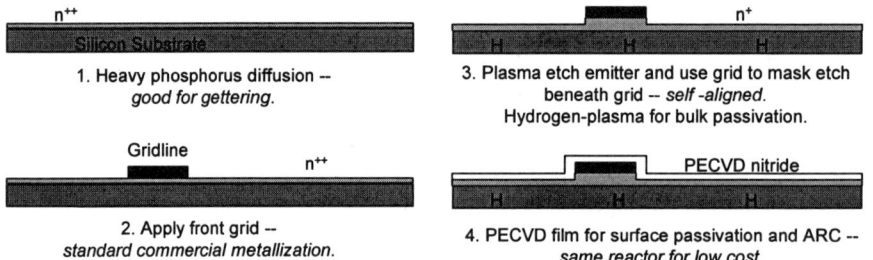

**Figure 5.** Process sequence for self-aligned selective-emitter cell. PECVD refers to plasm-enhanced chemical vapor deposition. A plasma hydrogenation step can also be included in the process.

The second cell concept under investigation is the emitter wrap-through (EWT) cell. The EWT cell has both contacts on the back surface, which is achieved by wrapping the emitter through laser-drilled holes from the front surface to the back surface (Fig. 6). The back-contact geometry has potentially higher performance due to no grid obscuration, and we (Gee *et al*) have projected efficiencies over 20% for 100-cm$^2$ EWT cell using photovoltaic-grade silicon [24]. The best result to date is 15.7% for a 42-cm$^2$ EWT cell with bifacial contacts and a photovoltaic-grade Cz silicon substrate [24].

The most significant advantage of the back-contact configuration is simplification of the module assembly. The present geometry with contacts on front and back surfaces is difficult to automate and module fabrication (including labor and materials) now accounts for nearly 50% of the finished module cost [27]. We (Gee *et al.*) are working on a new module assembly concept that encapsulates and electrically connects *all* the cells in the module *in a single step*. The key features of this new process include the following: (1) back-contact cells; (2) a module backplane that has both the electrical circuit and encapsulation material in a single piece; and (3) a single-step process for assembly of these components into a module (Fig. 7). This process reduces costs by reducing the number of steps, by eliminating low-throughput (e.g., individual cell tabbing, cell stringing, layout, etc.) steps, and by using completely planar processes that are easy to automate. We refer to this process as "monolithic module assembly" since it translates many of the advantages of monolithic module construction of thin-film PV to wafered c-Si PV. Simplifications in the module fabrication have been estimated to reduce the cost of module fabrication by up to 50%, which corresponds to a reduction of around 25% in the total manufacturing cost for the module [28]. To date, we have demonstrated a two-step assembly process where (1) the back-contact cells are soldered to the backsheet and (2) the cells/backsheet are then encapsulated in the module. The demonstration used Kapton™ for the backsheet, the circuit on the backsheet consisted of copper strips that was applied by a proprietary selective plating technique, and the backsheet did not yet include an encapsulation layer.

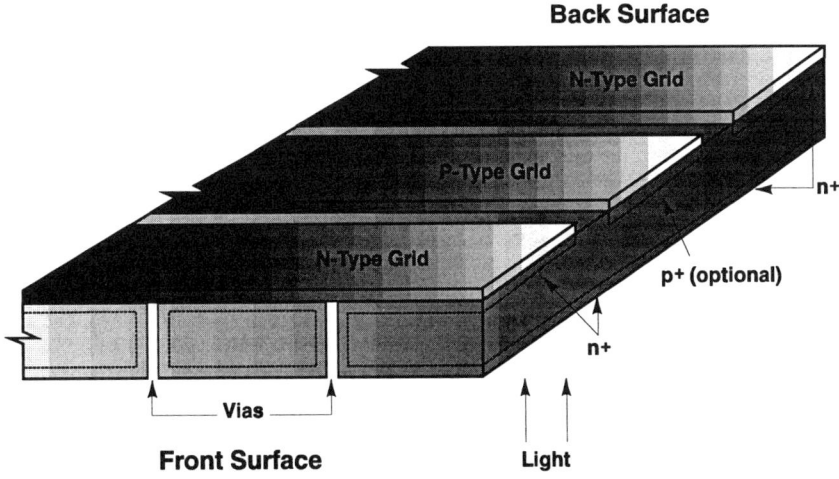

Figure 6. Illustration of an EWT cell.

Fundamental research to improve state-of-the-art device performance and investigate novel process and device concepts is addressed through support of a University Center of Excellence in Photovoltaics (UCEP) at the Georgia Institute of Technology. The work at UCEP is presently developing high-efficiency cells on commercial c-Si substrates, developing cell processes with potentially large throughput (e.g., screen-printed contacts and rapid-thermal processing), and examining advanced back-surface field processes. Some of this work is summarized in recent publications [25,29-31].

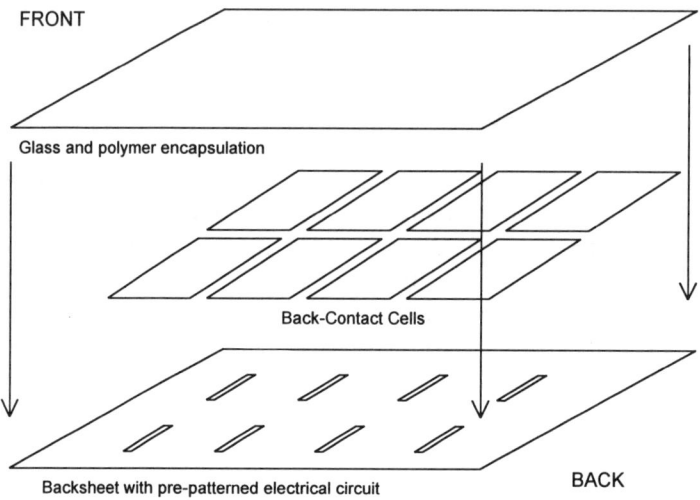

Figure 7. Illustration of monolithic module assembly.

## SUMMARY

Crystalline-silicon PV has the capability of meeting major Department of Energy goals for PV technology. This paper has summarized the structure and progress in the SNL/NREL c-Si PV R&D project. The project is divided into three tasks. Significant progress was achieved in all three tasks, including development of LPE growth on metallurgical-grade multicrystalline-silicon substrates, fundamental studies on the effect of Fe impurity in silicon, and development of novel cell and module concepts with the possibility of significant performance and cost improvements.

## ACKNOWLEDGEMENTS

There are many people who contribute to the c-Si R&D project. The authors would in particular like to acknowledge the many useful discussions with S. Tsuo, B.L. Sopori, and D.S. Ruby, and the many useful comments and suggestions from industrial and academic c-Si R&D colleagues that have helped develop the mission statement and the organization of this project. We would also like to acknowledge the support of J.P. Benner and M.L. Tatro for forming our interlaboratory project team.

## REFERENCES

1. **PV News**, February 1996, pp. 3-6.
2. J. Wohlgemuth, *et al.*, "Progress in the Solarex Crystalline Silicon PVMaT Program," 25th IEEE Photo. Spec. Conf. (PVSC), pp. 1181-1186 (1996).

3. K.W. Mitchell, *et al.*, 25$^{th}$ IEEE PVSC, pp. 541-543 (1996).
4. P.A. Basore and J.M. Gee, "Crystalline-Silicon Photovoltaics: Necessary and Sufficient," 1$^{st}$ World Conf. on PV Energy Conv. (WCPEC), pp. 2254-2257 (1994).
5. T.M. Bruton et al., "Multi-Megawatt Upscaling of Silicon and Thin-Film Solar Cell and Module Manufacturing - MUSIC FM," Eur. Conf. Renewable Energy Development, Venice, Italy, 22-25 November 1995.
6. E.J. Henderson and J.P. Kalejs, "The Road to Commercialization in the PV Industry: A Case Study of EFG Technology," 25$^{th}$ IEEE PVSC, pp. 1187-1190 (1996).
7. G. Cody and T. Tiedje, "A Learning Curve Approach to Projecting Cost and Performance in Thin-Film Photovoltaics," 25$^{th}$ IEEE PVSC, pp. 1521-1524 (1996).
8. **Photovoltaics – The Power of Choice: The National Photovoltaic Program Plan for 1996-2000**, DOE/GO-100096-017, January 1996.
9. See, for example, J.H Werner, R. Bergmann, and R. Brendel, "The Challenge of Crystalline Thin Film Silicon Solar Cells,' Festkörperprobleme/Advances in Solid State Physics, ed. By R. Helbig (Vieweg, Braunschweig,/Wiesbaden, 1994), vol. 34, pp. 115-146.
10. T. Wang and T.F. Ciszek, "Impurity Segregation in LPE Growth of Silicon from Cu-Al Solutions," this conference.
11. H. Atwater, this conference.
12. E. Yablonovitch, this conference.
13. B. Sopori, "Theory of Light Trapping in Thin Solar Cells," this conference.
14. B.L. Sopori, "18%-Efficient Commercial Silicon Solar Cells on Low-Cost Substrates: R&D Issues", this conference.
15. T. F. Ciszek, *et al.*, "Properties of Iron-Doped Multicrystalline Silicon Grown by the Float-Zone Technique," 25$^{th}$ IEEE PVSC, pp. 737-739 (1996).
16. T. Tan, "Physical and Numerical Modeling of Impurity Gettering in Silicon," this conference.
17. K. Kimerling, this conference.
18. S. Tsuo, "High-flux Solar Furnace Processing of Silicon Solar Cells," this conference.
19. W. Weber, "Precipitation Kinetics of Impurities in Solar Cell Silicon," this conference.
20. G. Rozgonyi, "Recombination at Decorated and Undecorated Dislocations," this conference.
21. S. Estreicher, "Molecular Dynamic Modeling of Hydrogen in Silicon," this conference.
22. L. Jastrzebski, "Gettering in Solar Cell Silicon," this conference.
23. W.K. Schubert and J.M. Gee, "Phosphorus and Aluminum Gettering – Investigation of Synergistic Effects in Single-Crystal and Multicrystalline Silicon," 25$^{th}$ IEEE PVSC, pp. 437-440 (1996).
24. J. M. Gee et al., "High-Efficiency Cell Structures and Processes Applied to Photovoltaic-Grade Czochralski Silicon," 25$^{th}$ IEEE PVSC, pp. 437-440 (1996).
25. A. Rohatgi et al., "Record High 18.6%-Efficient Solar Cell on HEM Multicrystalline Material," 25$^{th}$ IEEE PVSC, pp. 741-744 (1996).
26. D. S. Ruby et al., "Plasma Processing Applied to Crystalline-Silicon Solar Cells," this conference.
27. K.W. Mitchell et al., "The Reformation of Cz Si Photovoltaics," 1$^{st}$ World Conf. on PV Energy Conv., pp. 1266-1269 (1994).
28. F. Wald and J. Hanoka, private communication.
29. T. Krygowski et al., "A Novel Technology for the Simultaneous Diffusion of Boron, Aluminum, and Phosphorus in Silicon," 25$^{th}$ IEEE PVSC, pp. 393-396 (1996).
30. P. Doshi et al. "High-Efficiency Silicon Solar Cells by Low-Cost Rapid Thermal Processing, Screen Printing, and Plasma-Enhanced Chemical Vapor Deposition," 25$^{th}$ IEEE PVSC, pp. 421-424 (1996).
31. S. Narasimha et al., "The Optimization and Fabrication of High Efficiency HEM Multicrystalline Silicon Solar Cells," 25$^{th}$ IEEE PVSC pp. 449-452 (1996).

# Gettering and Passivation of High Efficiency Multicrystalline Silicon Solar Cells

A. Rohatgi, S. Narasimha, and L. Cai

*University Center for Excellence in Photovoltaic Research and Education,*
*Department of Electrical and Computer Engineering,*
*Georgia Institute of Technology, Atlanta, GA 30332-0250*

**Abstract.** A detailed study was conducted on aluminum and phosphorus gettering in HEM mc-Si and defect passivation by PECVD SiN in EFG mc-Si to achieve high efficiency solar cells on these promising photovoltaic materials. Solar cells with efficiencies as high as 18.6% (1 cm$^2$ area) were achieved on multicrystalline silicon (mc-Si) grown by the heat exchanger method (HEM) by a process which implements impurity gettering, an effective back surface field, front surface passivation, and forming gas annealing This represents the highest reported solar cell efficiency on mc-Si to date. PCD analysis revealed that the bulk lifetime in certain HEM samples after phosphorus gettering can be as high as 135 µs. By incorporating a deeper aluminum back surface field (Al-BSF), the back surface recombination velocity ($S_b$) for 0.65 Ω-cm HEM mc-Si solar cells was lowered from 10,000 cm/s to 2,000 cm/s resulting in the 18.6% efficient device. It was also observed that a screen-printed/RTP alloyed Al-BSF process could raise the efficiency of both float zone and relatively defect-free mc-Si solar cells by lowering $S_b$. However, this process was found to increase the electrical activity of extended defects so that mc-Si devices with a significant defect density showed an overall degradation in performance. In the case of EFG mc-Si, neural network modeling in conjunction with a study of post deposition annealing was used to provide guidelines for effective defect passivation by PECVD SiN films. Appropriate deposition and annealing conditions resulted in a 45% increase in cell efficiency due to AR coating and another 25-30% increase due to defect passivation by atomic hydrogen.

## INTRODUCTION

In order to achieve high efficiencies, conventional n$^+$-p-p$^+$ solar cells must exhibit high bulk lifetime ($\tau_b$) and low back surface recombination velocity ($S_b$). Traditionally, research on mc-Si material has concentrated on improving $\tau_b$ by implementing different crystal growth techniques or gettering and passivation procedures. Phosphorus and aluminum gettering are practical methods of improving $\tau_b$ in photovoltaic materials because they can be integrated cost effectively into the solar cell fabrication sequence. When gettering procedures successfully improve the average $\tau_b$, then $S_b$ becomes increasingly important to overall solar cell efficiency.

Recently, PECVD SiN passivation has drawn considerable attention because it is a low-temperature process which can give efficient AR coatings and effective bulk and surface passivation. However, no systematic study has been done to control deposition variables or post deposition anneal conditions in order to enhance the defect passivation by PECVD films.

Therefore, the objective of this paper is two-fold: 1) to understand and implement gettering treatments (phosphorus, aluminum, and co-gettering) and Al-BSF processes (conventional and screen-printed) in order to improve solar cell efficiencies on HEM mc-Si material, and 2) to use neural network modeling and post deposition annealing of PECVD SiN films to maximize the bulk defect passivation in EFG mc-Si ribbon.

## Gettering and Al-BSF Effects on HEM mc-Si Solar Cells

HEM mc-Si was used in this study because it is a directionally solidified material which has "cm-scale" grain size, low average dislocation density ($< 10^5$ cm$^{-2}$), and low oxygen content ($< 5$ ppm). All of these material attributes are important for attaining high cell efficiency. The particular HEM wafers used in this study were selected from the same starting ingot, with resistivities ranging between 0.65-0.80 Ω-cm.

Modeling curves for solar cells formed on 0.65 Ω-cm substrates are shown in Figure 1. The curves indicate that solar cell efficiencies should approach 18% for moderate $\tau_b$ levels ($\approx 35$ μs) even with a high recombination back surface ($S_b \approx$ 10,000 cm/s). However, in order to achieve cell efficiencies in excess of 19%, improvements must be made in both $\tau_b$ and $S_b$. Specifically, $S_b$ must be reduced to approximately 1,000 cm/s for 0.65 Ω-cm bulk silicon.

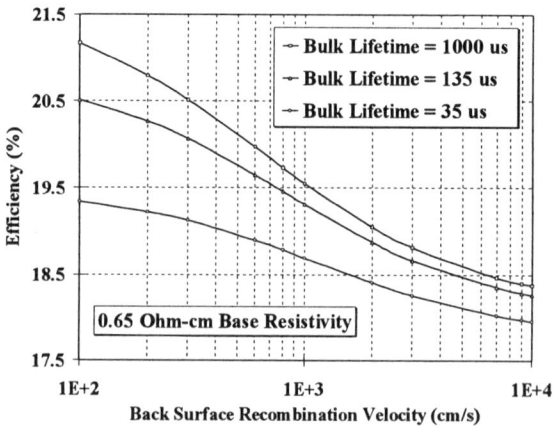

**FIGURE 1.** Model curves showing the effect of $\tau_b$ and $S_b$ on solar cell efficiency.

The as-grown lifetime of the HEM mc-Si used in this experiment was found to be 10-15 μs by photoconductance decay (PCD) measurements. This is well short of the $\tau_b$ required (> 35 μs) to achieve efficiencies in the range of 18-19%. Therefore, a controlled experiment was performed to determine the effects of four different gettering schemes on the $\tau_b$ in HEM mc-Si. The four gettering conditions were: 1) heat treatment alone in $N_2$, 2) phosphorus gettering alone, 3) aluminum gettering alone (in which 1 μm of Al was evaporated onto the material prior to the heat treatment), and 4) phosphorus and aluminum co-gettering (involving Al evaporation followed by a drive-in during the phosphorus diffusion). In all four cases, the thermal cycle was kept constant: 870°C/20 minutes. The results of this experiment are shown in Figure 2. Four important conclusions can be drawn from

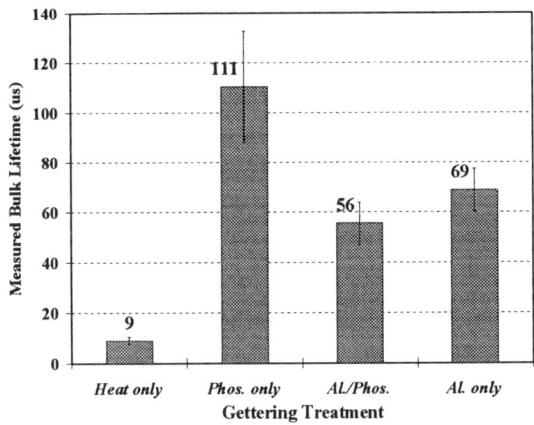

**FIGURE 2.** The effects of gettering treatments on the $\tau_b$ in HEM mc-Si.

this data. First, phosphorus gettering alone provided the most substantial improvement in minority carrier lifetime, raising $\tau_b$ to an average of 111 μs. Secondly, heat treatment alone did not provide any gettering action. Instead, the $\tau_b$ measured after this process fell below 10 μs, probably due to the dissolution of precipitates. Third, aluminum gettering alone raised the $\tau_b$ to 69 μs, which is less than the level achieved by phosphorus gettering alone. Finally, co-gettering raised the as-grown lifetime to only 56 μs, which is also considerably less than the lifetime level achieved by phosphorus gettering alone. This indicates that the presence of aluminum during co-gettering retards the gettering efficiency of phosphorus. A possible explanation for this finding is that silicon self-interstitials generated by phosphorus diffusion, which aid in the "kick-out" of substitutional impurities during gettering, are absorbed by vacancies generated by the aluminum layer during drive-in.

After determining the effectiveness of phosphorus gettering, its process parameters were optimized further. Figure 3 indicates an optimal gettering temperature of 900°C for the HEM mc-Si used here. (The results presented in Figure 3 are normalized; however, the maximum $\tau_b$ values reached 135 µs). The

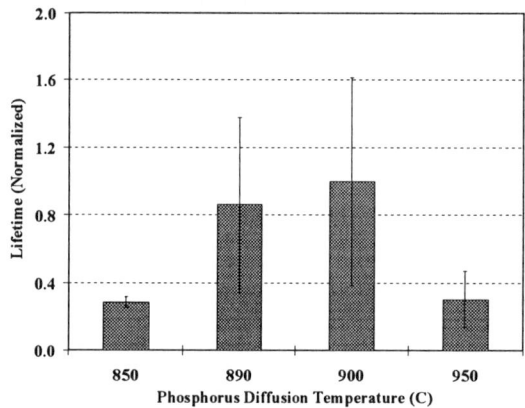

**FIGURE 3.** Effects of phosphorus diffusion temperature on the gettering efficiency of HEM mc-Si material.

existence of an optimal gettering temperature is evidence of the competition between gettering efficiency and the formation of lifetime reducing defects (i.e. dissolution of metallic precipitates or decoration of existing structural defects) as the gettering temperature is raised. Such behavior is typical for most mc-Si materials.

After achieving the desired $\tau_b$, HEM cells were fabricated using the following process sequence. First, the $n^+$ emitter was formed using the 900°C phosphorus diffusion cycle. The as-diffused emitter ($\approx$ 30 $\Omega$/sq.) was subsequently etched back to a reasonable sheet resistance ($\approx$ 80 $\Omega$/sq.). Next, 10 µm of aluminum (99.999%) was evaporated onto the backside of the device. The wafer was then annealed at 850°C for 30 minutes in order to form the Al-BSF. At the beginning of this cycle, a passivating thermal oxide was grown on the emitter by exposing the sample to oxygen for 10 minutes. The remaining 20 minutes of the anneal was done in nitrogen. Upon cooldown to 400°C, the wafer was annealed in forming gas in order to improve the oxide/silicon interface quality and provide atomic hydrogen for possible bulk defect passivation. Finally, contacts were formed by lift-off photolithography, and a $ZnS:MgF_2$ double-layer antireflection coating was applied.

## Characterization and Analysis of High Efficiency HEM Solar Cells

Figure 4 shows the IQE and output parameters of the best cell fabricated in this experiment with the 10 μm evaporated Al-BSF ($V_{oc}$ of 636.0 mV, $J_{sc}$ of 36.48 mA/cm$^2$, FF of 804, and efficiency of 18.6%, confirmed at the NREL). On the

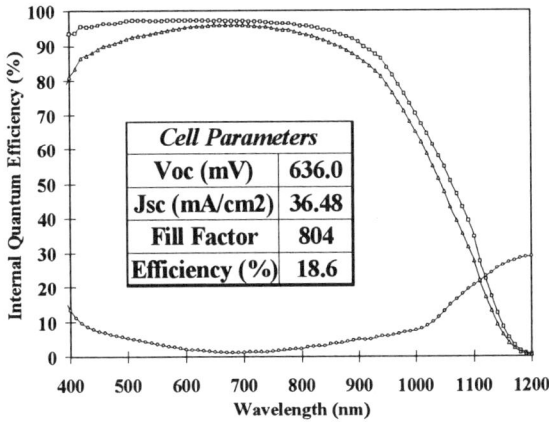

**FIGURE 4.** IQE curve and IV data for the record high HEM mc-Si solar cell.

same wafer, six adjacent 1 cm$^2$ devices had an efficiency range of 17.2% to 18.6% with an average efficiency of 18.0%. This suggested a nonuniform distribution of electrically active defects in the material. Light beam induced current (LBIC) scans were carried out on the high and low efficiency devices in order to understand the role of extended defects on cell performance. A comparison between the LBIC response of an 18.6% and a 17.2% cell is shown in Figure 5.

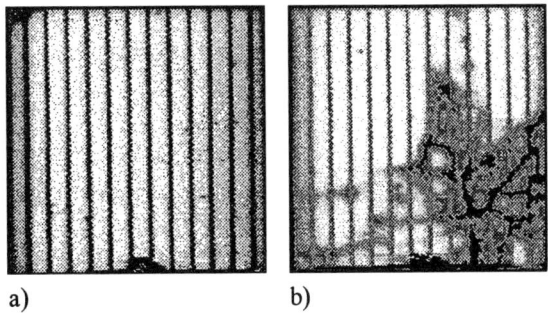

**FIGURE 5.** LBIC scans of the a) 18.6% HEM mc-Si solar cell and the b) 17.2% device.

The scan of the 18.6% device reveals that the active area is free from detectable electrically active defects. Furthermore, the high bulk lifetimes measured in the processed material indicate that phosphorus gettering was effective in removing impurities from the as-grown silicon. On the contrary, the LBIC response of the 17.2% cell shows the presence of a significant density of electrically active defects (primarily dislocation clusters decorated with impurities). The PDG, forming gas anneal, and aluminum treatments used in this study could not improve the current response in these defective regions. Forward biased current-voltage analysis of these devices showed that the defects lowered the cell performance by increasing both bulk and depletion region recombination. However, 70% of the efficiency degradation was due to the increased bulk recombination, or the $J_{o1}$ component of the total reverse saturation current, while the remaining 30% of the degradation was attributed to the increased junction recombination, or $J_{o2}$ [1].

The above mentioned cells were fabricated with 10 μm evaporated Al-BSFs. Analyzing the long wavelength IQE of the 18.6% device, and using the measured $\tau_b$ of 135 μs, the $S_b$ for this cell was found to be 2,000 cm/s. Figure 6 also shows that lowering $S_b$ to 800 cm/s would raise the efficiency of similar devices above 19.0%. Model calculation, performed elsewhere, have shown that the effectiveness of Al-BSFs improves with increasing amounts of aluminum deposited onto the wafer prior to the alloying step [2]. Screen-printing (SP) offers a convenient method of applying a significant amount of aluminum onto silicon, in order to achieve effective BSFs. Furthermore, the SP/RTP alloying process is inherently high-throughput, which allows for easy integration into an industrial production line. These qualities provided the motivation for investigating the effects of SP Al-BSFs on HEM mc-Si.

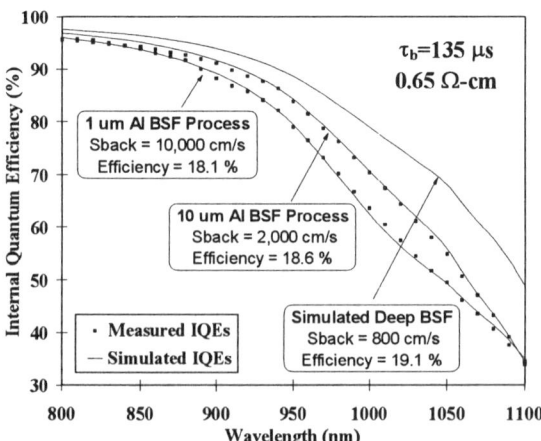

**FIGURE 6.** Long wavelength IQE data showing the effects of $S_b$ variation on cell performance.

Approximately 30 μm of aluminum conductor paste was screen printed onto silicon and fired for two minutes at 900°C in an RTP chamber. The inset of Figure 7 shows the corresponding Al-BSF profile with a junction depth of roughly 13 μm. Model calculations (also shown in Figure 7) performed using this BSF profile show that similar SP Al-BSFs should be able to reduce $S_b$ to ≈ 800 cm/s for 0.65 Ω-cm substrate material. The guidelines in Figure 1 indicate that this $S_b$ value coupled with $\tau_b$ values of 135 μs should yield solar cell efficiencies in excess of 19%.

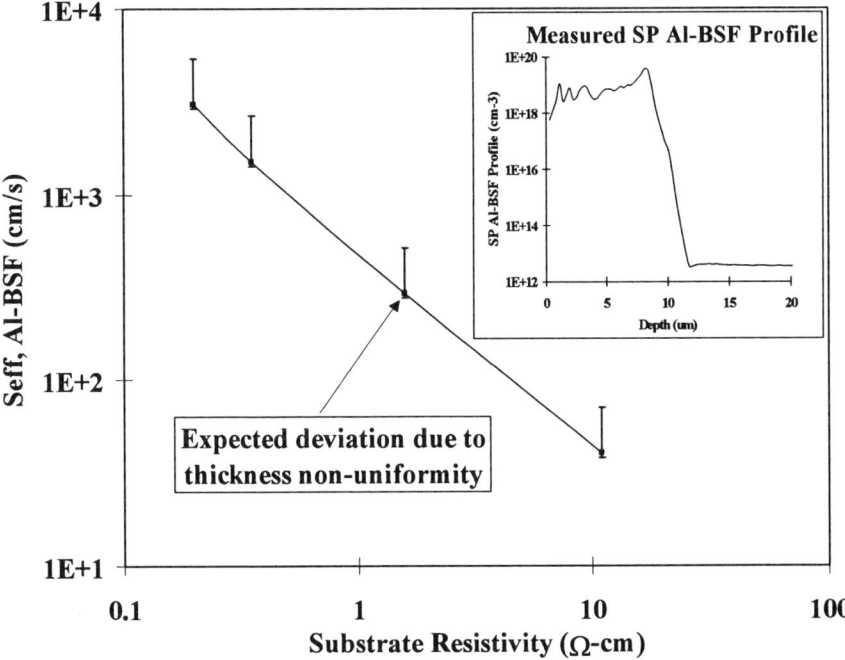

**FIGURE 7.** SP Al-BSF profile (inset) along with calculated $S_{eff}$ values as a function of substrate resistivity.

The SP/RTP Al-BSFs process was applied to HEM mc-Si wafers from the same ingot used in the first experiment. However, this group of wafers was found to yield a maximum baseline efficiency of 17.7% in comparison to the 18.2% determined in the first experiment. Figure 8 shows the efficiency comparison between cells made with the baseline Al-BSF and SP Al-BSFs fired at 800°C, 850°C, and 900°C for 11 sets of matched cells across four wafers consecutively

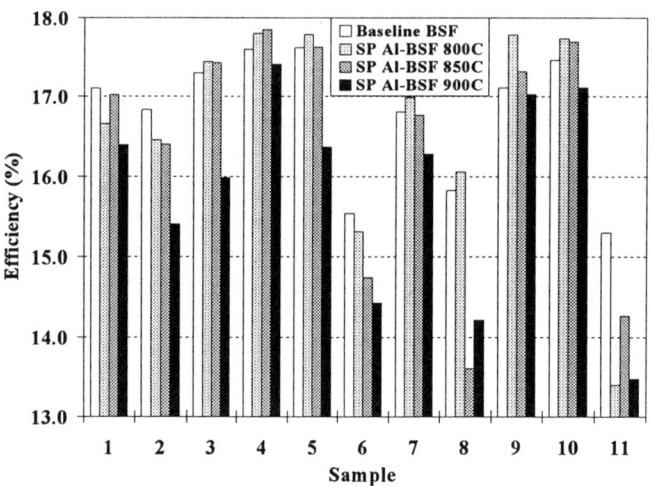

**FIGURE 8.** Efficiency variation of HEM mc-Si solar cells undergoing the baseline and SP/RTP alloyed Al-BSF treatments.

selected from the same HEM ingot. Three points are evident from the data. First of all, there is significant variation in cell performance across the wafer (efficiencies range from 17.7% to 15.3% for the baseline process). As the discussion above has determined, this variation is primarily due to the varying defect density found in the material. Second, in most cases, the efficiencies of cells with the SP/RTP Al-BSFs deteriorated with increased firing temperature. The 900°C firing temperature had a deleterious effect on the material quality. Third, in most regions of the wafer where the baseline process yielded efficiencies above 17.0% and $V_{oc}$ values above 615 mV (low defect density areas), the SP devices alloyed at 800°C and 850°C showed modest improvement. However, in regions where the baseline cells yielded efficiencies lower than 17.0% mV (higher defect density areas), the SP/RTP alloyed devices were worse than the baseline cells. This suggests a relationship between the amount of electrically active extended defects contained in the cell area and the effectiveness of the SP/RTP Al-BSF treatment. To further understand this relationship, LBIC analysis ($\lambda$=950 nm) was carried out on two solar cells formed at the same lateral location on adjacent wafers, one fabricated with the baseline process ($V_{oc}$ of 601 mV) and the other processed with the SP Al-BSF alloyed at 900°C ($V_{oc}$ of 580 mV). The LBIC scans (Figure 9) show that the electrical activity of the defective regions worsens with the SP/RTP process, leading to a loss in overall efficiency from 16.1% to 14.7%.

These results indicate that even though the SP/RTP Al-BSF process can substantially reduce $S_b$, it increases the electrical activity of extended defects in mc-Si. This degradation may be due to either the fast cooling rates used during RTP alloying or the injection of impurities from the aluminum conductor paste (formed from aluminum powder with purity of 99.7%) into the bulk silicon.

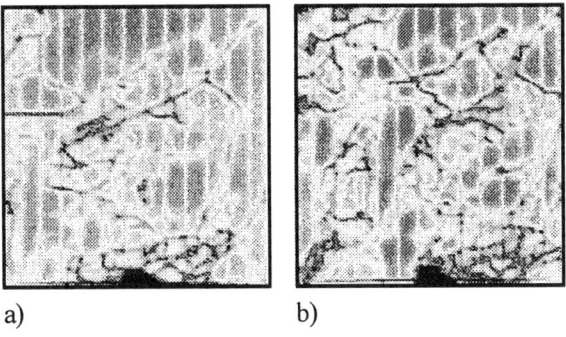

**FIGURE 9.** LBIC scans of solar cells undergoing the a) baseline (16.1%) and b) SP/RTP 900 (14.7%) Al-BSF process.

## Passivation of EFG mc-Si by PECVD SiN

PECVD SiN deposition process is quite complicated to fully understand from the first principles. Therefore, first neural network modeling was performed to establish the correlation between deposition parameters and film properties. Then, optimal parameters were used to maximize the influence on cell performance. In this study, six PECVD silicon nitride deposition parameters (substrate temperature, chamber pressure, RF power, and the flow rate of silane, ammonia, and nitrogen) were used as inputs to the neural network and positive charge, bonded hydrogen content, and effective lifetime were the desired outputs.

The silicon nitride films were deposited on 47 samples in a parallel plate Plasma-Therm 700 series batch reactor operating at 13.56 MHz. All the samples were annealed in nitrogen at 350°C for 20 minutes after the deposition. Feed-forward neural networks were trained using the error back-propagation (BP) algorithm. Details of neural network modeling are described elsewhere [3].

Positive charge in the SiN is important for surface passivation. Therefore, positive charge density in the SiN films was determined by C-V measurements of MOS capacitors in order to train and perform neural network modeling. The results of this modeling are shown in Figure 10a. It is clear that higher growth temperature and lower RF power are important for obtaining high positive charge density.

Hydrogen content in the SiN films can be beneficial for passivation of bulk defects in silicon. Therefore, FTIR measurements were performed to determine the hydrogen content of the film. Neural network modeling (Figure 10b) shows that, unlike positive charge density, the bonded hydrogen content (Si-H and N-H) in the

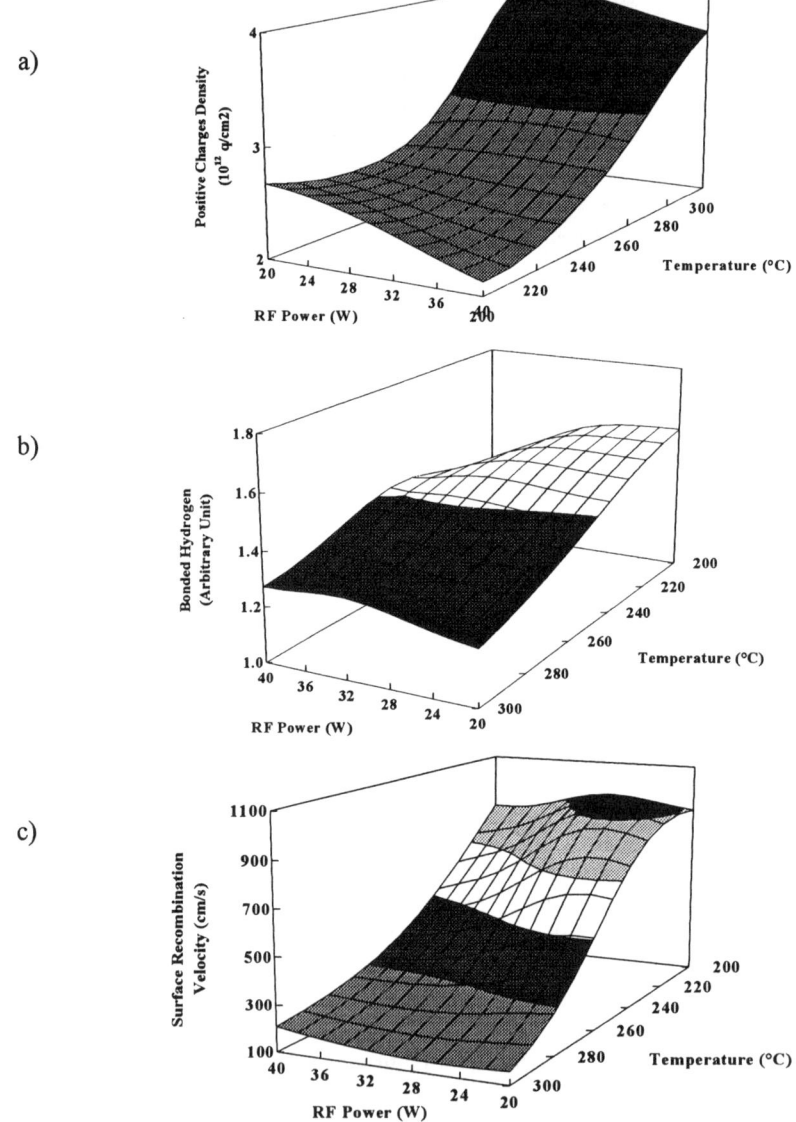

**FIGURE 10.** Neural network modeling relating PECVD deposition conditions to a) positive charge density, b) bonded hydrogen content, and c) surface recombination velocity.

SiN film can be enhanced by lowering the growth temperature. However, RF power seem to have very little effect on the hydrogen content.

In order to study bare silicon surface passivation, high quality 2 Ω-cm float zone silicon ($\tau_b > 1$ ms) was used. Figure 10c shows the effect of PECVD SiN deposition conditions on the surface passivation. It can been seen that increased substrate temperature decreases the surface recombination velocity (SRV), while pressure has relatively little effect on the surface passivation.

According to this analysis, a combination of higher deposition temperature and lower reactor pressure results in the lowest SRV value. In this experiment, SRV values as low as 100 cm/s were achieved on 2 Ω-cm silicon. It is important to realize that without any surface passivation, the SRV of a bare silicon surface exceeds $10^4$ cm/s. This shows that properly deposited PECVD SiN films can be very effective in passivating bare silicon surfaces.

Figure 11 shows a correlation between the refractive index and the bonded hydrogen content of the SiN films before and after a 650°C/1.5 minute photoanneal in a $N_2$ ambient. It is interesting to note that prior to photoannealing, the bonded hydrogen concentration in the as-deposited films increases rapidly up

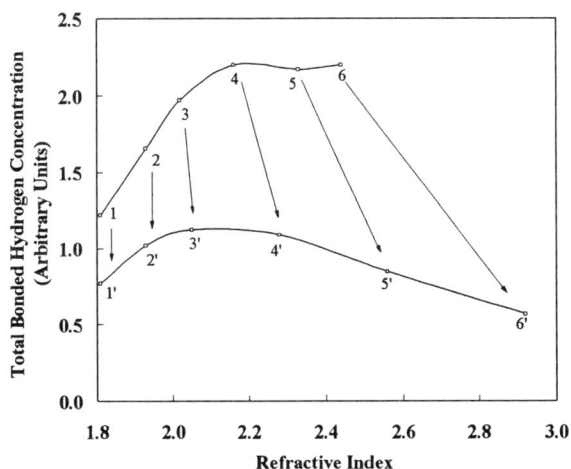

**FIGURE 11.** Change in bonded hydrogen content of PECVD SiN films after a 650°C/1.5 minute photoanneal.

to the as-deposited index of 2.2, and then tends to saturate. After the photoanneal, the films showed a decrease in bonded hydrogen content, especially in Si-H bond concentration. The lower curve in Figure 11 shows the correlation between the bonded hydrogen content and the refractive indices of the same films after the 650°C/1.5 minute anneal. Even though in all cases the bonded hydrogen concentration decreases after the anneal, films with higher as-deposited refractive

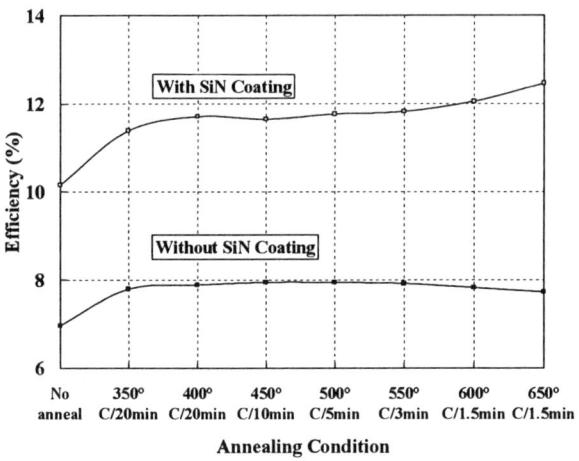

**FIGURE 12.** Change in efficiency of EFG solar cells with SiN films applied as a function of photoanneal temperature.

indices release more bonded hydrogen and show greater increase in the refractive index value. Thus, photoanneals result in an increase in refractive index, but decrease in hydrogen content of the silicon nitride film. In addition, the magnitude of this effect is proportional to the as-deposited refractive index value.

Figure 12 shows the change in average efficiency of eleven 1 cm$^2$ EFG cells (randomly selected from different wafers) as a function of annealing temperature. The average efficiency of the cells increased gradually with the increase in annealing temperature up to 650°C, and then the efficiency started to decrease at higher temperatures. It is important to note that a sharp increase in the average cell efficiency was observed after the first 350°C/20 minute anneal. Then, there was a plateau effect (in which the efficiencies did not change), followed by another increase in efficiency for photoanneal temperatures ranging from 500°C to 650°. For photoanneal temperatures above 650°C, device efficiencies degraded due to contact shunting effects. A process sequence involving contact formation after the PECVD deposition and anneal would result in even greater efficiency improvement due to hydrogenation. It should be recognized that due to the inhomogenous defect distribution in EFG material, the impact of the PECVD SiN deposition and RTP anneal could be different on different small area EFG cells. In another study, it was shown that the increase in the absolute cell efficiency after the 650°C post PECVD anneal can vary from 0.2% to 2%, depending upon on the location of a cell on the wafer [4].

In order to gain better understanding of whether the above beneficial effects are related to surface passivation, bulk defect passivation, or both, detailed IQE measurements were performed. IQE of a solar cell depends strongly on the surface recombination velocity and the bulk lifetime of the cell. For these EFG

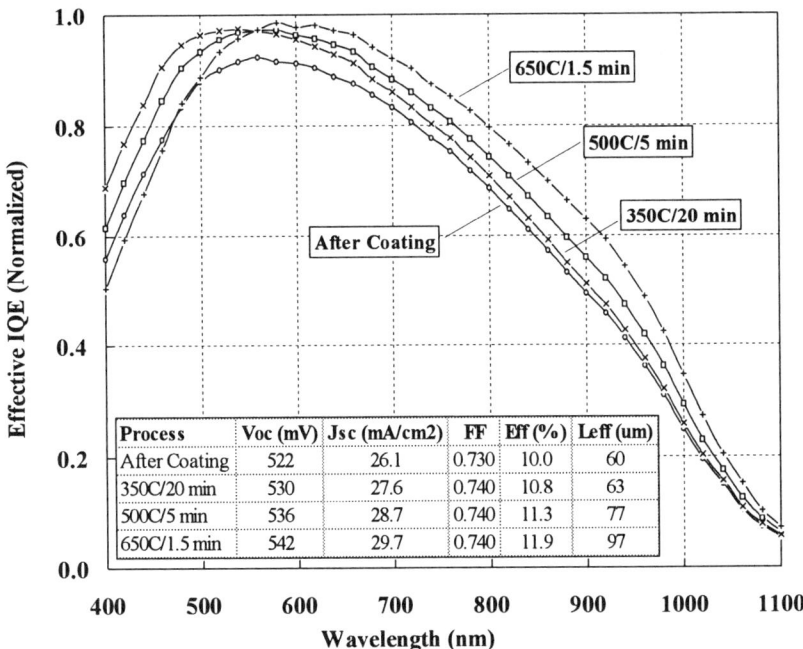

**FIGURE 13.** Effects of photoannealing SiN on the IQE response of EFG mc-Si solar cells.

cells, the short wavelength response is largely determined by front surface recombination velocity, while the long wavelength response is primarily affected by $\tau_b$. Therefore, by measuring and analyzing IQE, the effects of annealing on the surface and the bulk quality of the cells can be determined. It is important to note that high index $SiN_x$ will absorb short wavelength light (< 600 nm). Therefore, the IQE curves shown here are actually effective IQE measurements, with the effects of absorption embedded in the data.

Figure 13 shows the effect of higher temperature anneals on the IQE response of EFG cells. The anneal time was reduced to 5 minutes and 1.5 minutes for 500°C and 650°C anneals, respectively, to reduce junction leakage and shunting effects caused by contact metal diffusion. The IQE analysis showed that the 500°C anneal increased the diffusion length from 63 µm to 77 µm, and after the 650°C anneal, the diffusion length increased to 97 µm. Notice that the higher temperature anneals (> 350°C) reduce the short wavelength response, while the long wavelength response continues to increase.

The above trade off in long and short wavelength response tends to suggest the release of hydrogen (via breaking of Si-H bonds) from the nitride layer followed by hydrogen diffusion into the bulk. However, the decrease in short wavelength response cannot be explained on the basis of degradation in surface passivation because the $J_{oe}$ measurements show that annealing temperature has virtually no effect on emitter passivation after 350°C anneal [5]. However, the refractive index and the extinction coefficient continue to increase with temperature. This results in higher short wavelength absorption. Thus we conclude that the decrease in short wavelength response is largely due to increased absorption in the $SiN_x$, while the increase in long wavelength response is the result of bulk defect passivation.

## CONCLUSIONS

Solar cells with efficiencies as high as 18.6% (1 cm$^2$) have been achieved on mc-Si material grown by the HEM method. Bulk lifetime values in excess of 100 µs have been measured on the processed samples. For such high $\tau_b$ values, $S_b$ can have a large impact on the overall device performance. By implementing an improved (deeper) Al-BSF, the $S_b$ value in this study was reduced from 10,000 cm/s to approximately 2,000 cm/s for evaporated aluminum layers alloyed in a furnace. A further $S_b$ reduction to 200 cm/s was achieved on 2 Ω-cm float zone by using SP Al-BSFs alloyed in an RTP chamber. However, when applied to HEM mc-Si material, the SP/RTP process yielded mixed results. LBIC analysis shows that the electrical activity of extended defects increases after this process. For devices which are fairly defect free, the overall performance improves because of the $S_b$ reduction. However, for devices which contain a higher density of defects, the SP/RTP process leads to a degradation in efficiency. Since large area mc-Si devices naturally contain heavily defective regions, it is difficult to predict the effects of SP Al-BSFs in such cases. Work is in progress to investigate the response of large area mc-Si solar cells to SP Al-BSFs.

For the first time, neural network modeling is performed to correlate deposition parameters and the properties of SiN films that are relevant to solar cell performance. It is shown that higher growth temperature increases positive charge density which in turn improves surface passivation. Similarly, lower growth temperature and higher refractive index increase the bonded hydrogen content which can be released by annealing for bulk defect passivation. An appropriate SiN coating increased solar cell performance of EFG mc-Si solar cells by 45% (due to antireflection coating) and another 25-30% due to bulk defect passivation.

## ACKNOWLEDGMENTS

This work was supported by Sandia subcontract number AO-6162 and NREL subcontract number XD-2-11004-2.

# REFERENCES

1. Narasimha, S., Kamra, S., Rohatgi, A., Khattak, C.P., and Ruby, D., "The Optimization and Fabrication of High Efficiency HEM Multicrystalline Silicon Solar Cells," in *Proceedings of the 25th IEEE Photovoltaic Specialists Conference,* 1996, pp. 449-452.
2. Rohatgi, A. and Narasimha, S., "Design, Fabrication, and Analysis of Greater than 18% Efficiency Multicrystalline Silicon Solar Cells," *in the Technical Digest of the 9th International Photovoltaic Science and Engineering Conference*, 1996, pp. 85-89.
3. Han, S.S., Cai, L., May, G. and Rohatgi, A., "Modeling the Growth of PECVD Silicon Nitride Films for Polysilicon Solar Cell Applications Using Neural Networks," *IEEE Transactions on Semiconductor Manufacturing*, 9, 303-311 (1995).
4. Cai, L. and Rohatgi, A., "Effect of Post PECVD Photo-assisted Anneal on Multicrystalline Silicon Solar Cells," IEEE Transactions on Electron Devices, 44, 1-7 (1997).
5. Rohatgi, A., Doshi, P., Ropp, M., Cai, L., Doolittle, A., Narasimha, S., Krygowski, T., Rand, J., Ruby, D., and Meier, D.L., "Improved Understanding and Optimization of RTP and PECVD Processes for High Efficiency Silicon Solar Cells," *in Proceedings of the 25th IEEE Photovoltaic Specialists Conference,* 1996, pp. 449-452.

# PHYSICAL AND NUMERICAL MODELING OF IMPURITY GETTERING IN SILICON

T. Y. Tan, R. Gafiteanu, and U. M. Gösele

Department of Mechanical Engineering and Materials Science
Duke University
Durham, NC 27708-0300

## ABSTRACT

Physical and numerical modeling of gettering metallic impurities away from Si device active regions has been carried out. The modeled aspects include the dissolution of precipitated impurity (if any), diffusion of dissolved impurities from the gettered to the gettering region, and gettered impurity stabilization in the gettering region via segregation and/or precipitation processes. The modeled impurities include fast diffusing interstitial ($i$) species and substitutional-interstitial ($s$-$i$) species. A wafer backside Al layer, wafer frontside indiffusion of P, indiffusion of P together with an Al layer (P+Al), and intrinsic gettering (IG) are the modeled gettering techniques. The gettering of $i$ species may be rapidly accomplished using Al or IG, via segregation or precipitation processes. The gettering of $s$-$i$ species involves point defects and is best accomplished by the P+Al technique. If initially impurities are precipitated, they can only be efficiently gettered in a reasonable time at temperatures much higher than the precipitation temperature.

## INTRODUCTION

Gettering of metallic contaminants away from device active regions has already become an integral part of manufacturing integrated circuits (IC) using Czochralski (CZ) Si wafers [1,2], and is becoming increasingly important in Si solar cell processing for improving the cell efficiency [3-6]. Gettering consists of the creation of suitable gettering regions, and the subsequent gettering of contaminants. For intrinsic gettering (IG) scheme used for IC, the IG site/region properties has been extensively studied [2]. However, studies of the impurity gettering processes either experimentally or theoretically were scarce [7,8]. In the present paper, we report physical and numerical modeling results of the impurity gettering processes. We have chosen fast diffusing interstitial ($i$) species and substitutional-interstitial ($s$-$i$) species as the model impurities, and the modeled gettering techniques include the use of a wafer backside Al layer, wafer frontside indiffusion of P, indiffusion of P together with an Al layer (P+Al), and IG.

In IC cases using CZ wafers, gettering is used to guard against the minute amount of processing introduced contaminants. Because of the monolithic nature of IC devices, these contaminants can be gettered to the wafer bulk IG sites for which the metal atom diffusion distance is short, on the order of 10 μm. To manufacture commercial solar cells, low cost CZ wafers and large grain multicrystalline ribbons and wafers are used. The multicrystalline Si substrates contain grain boundaries, dislocations, precipitated and dissolved metals. Thus, gettering is needed to improve the Si substrate quality, which involves metal precipitate dissolution, and the dissolved metal atom diffusion distance is very long, on the order of 100-200 μm. Gettering is also needed for guarding against contamination during cell processing, because the manufacturing facility is usually not clean. In this study, we model the impurity gettering process according to the more stringent requirement for solar cell applications. Obviously, the results are applicable also to the IC cases.

## PHYSICAL PROCESSES AND MODELING

Gettering by P indiffusion from the Si wafer surface is particularly effective for substitutional-interstitial ($s$-$i$) metal impurities. e.g., Au [9], which dissolve in Si on both substitutional and interstitial sites, $A_s$ and $A_i$, respectively. The measured Au solubility is that of $A_s$, because it is much larger than that of $A_i$. But the mobility of $A_i$ is much higher so that the diffusion of $A_s$ is due to the migration of $A_i$ and the $A_s$-$A_i$ interchange. This interchange process involves Si self-interstitials $I$ via the *kick-out* mechanism [10]

$$A_i \Leftrightarrow A_s + I. \qquad (1)$$

The alternative mechanism involving the Si vacancy $V$ according to $A_i + V \Leftrightarrow A_s$, the Frank-Turnbull mechanism, is not consistent with experimental results [10]. Outdiffusion of $A_s$ is difficult because, via reaction (1), this will incur a large $I$ undersaturation which prevents $A_s$ to become $A_i$ to diffuse away. P indiffusion is highly effective in gettering $A_s$ because it injects $I$ to alleviate the $I$ undersaturation due to $A_s \rightarrow A_i$ changeover and the subsequent outdiffusion of $A_i$. Moreover, $A_s$ consists of acceptor species with tremendously large solubilities in the high concentration P diffused region where electrons are abundant. This provides a large segregation coefficient for stabilizing $A_s$ gettered to the high P concentration region. This is a diffusion-segregation process.

A layer of Al deposited on the Si wafer surface can provide a gettering effect because the solubility of other metals in Al is very high, reaching 1 at% at temperatures below the eutectic temperature ($T_e$) of 577°C, as has been first observed for Ni by Thompson and Tu [11]. Above $T_e$ a liquid Al-Si alloy forms, and in this liquid the solubility of a typical metal can exceed 10 at%, or exceed $5 \times 10^{21} \text{cm}^{-3}$. Since the solubility of metals in Si does not exceed $\sim 10^{17}$ cm$^{-3}$, the segregation coefficient m of the metal between the liquid and Si is larger than $10^4$, which provides a tremendously large driving force for the metal to segregate into the liquid. Thus, the elementary process involved in Al gettering is also impurity diffusion and segregation. Al gettering should be highly effective for interstitial species because of their large diffusivity values in Si, but less effective for the $s$-$i$ species because of the $I$ undersaturation cannot be effectively alleviated by Al which does not inject $I$ into Si.

The beneficial effects of using Al and/or P for improving the performance of solar cells by gettering have already been noticed [3-6]. In particular, there are experimental results [4,5] showing that P and Al gettering, when performed simultaneously, exhibit a synergistic effect, i.e., the gettering effectiveness exceeds the sum of performing the two gettering schemes in sequence.

To physically model and numerically simulate the gettering process of an atomically dissolved impurity species in Si, it is necessary to calculate the impurity diffusion and segregation processes simultaneously. For gettering a metal impurity interstitially dissolved in Si by Al, for which impurity segregation occurs abruptly at the Si-Al interface, the standard diffusion equation

$$\partial C/\partial t = D\, \partial^2 C/\partial x^2 \qquad (2)$$

can be used to simulate the diffusion processes in both Si and the Al layer. The empirical boundary condition of Antoniadis and Dutton [12], in the more accurate form of

$$F_s = (D^{\text{eff}}/\lambda)(C_1 - C_2/m_2), \qquad (3)$$

arrived at recently via a derivation [13], can be used as a computational criterion at the Al-Si interface to describe the segregation process. In Eq. (3) $F_s$ is the impurity flux at the interface, $C_1$ is the impurity concentration at the lattice plane in region 1 (e.g., Si) at the interface, $C_2$ is the same in region 2 (e.g., Al), $m_2$ is the equilibrium segregation coefficient of the species in region 2 relative to region 1 ($m_1=1$), $\lambda$ is the atomic plane spacing, $D^{\text{eff}}$ is an effective diffusivity given by $D^{\text{eff}} = \Gamma_{1\to 2}\lambda^2$ with $\Gamma_{1\to 2}$ being the atom jump frequency from region 1 to 2.

Equations (2) and (3) cannot be straightforwardly used to simulate the gettering process of the *s-i* impurities, since the impurity diffusion process is closely coupled to also the processes of Si native point defect diffusion and generation/annihilation. Additional diffusion equations and appropriate coupling terms in each of the equations are needed. For gettering by P, the gettered impurity solubilities and hence the corresponding segregation coefficients in the P diffused region are continuous functions of x due to the continuous changes in the P concentration profile as a function of x. This renders the use of Eq. (3) impractical. To treat such problems, a flux equation of diffusion-segregation (FEDS) and a diffusion-segregation equation (DSE) have been derived [14,15]:

$$J = -D[\partial C/\partial x - (C/m)(\partial m/\partial x)], \qquad (4)$$

$$\partial C/\partial t = \partial\{D[\partial C/\partial x - (C/m)(\partial m/\partial x)]\}/\partial x. \qquad (5)$$

In Eqs. (4) and (5), m may be regarded as either a continuously changing or fairly abruptly changing quantity. That is, insofar as the impurity or point defect segregation properties are concerned, the use of Eq. (5) is in principle sufficient for all gettering problems. This has been shown [16] by the fact that, as an approximation, Eq. (4) is reduced to adopt the form of Eq. (3) at an abrupt interface.

Solar grade multicrystalline Si substrates contain regions with high densities of dislocations and metal silicide precipitates. The minority carrier diffusion lengths in these regions are extremely low, and cannot be significantly improved by a normal gettering treatment. The reason may be that now the gettering process involves also precipitate dissolution which requires more extensive treatment. To model the impurity gettering process involving both precipitated and dissolved metal atoms in Si, a basic set of equations has been written by assuming that the precipitate dissolution process is limited by diffusion of the dissolved impurity atoms. These equations are

$$\partial C/\partial t = \partial\{D[\partial C/\partial x - (C/m)(\partial m/\partial x)]\}/\partial x + 4\pi r \rho D(C^* - C), \qquad (6)$$

$$C^* = C^{eq} \exp(2\Omega\sigma/rk_BT), \quad (7)$$

$$dr/dt = -\Omega D (C^*-C)/r. \quad (8)$$

In Eqs. (6)-(8) C is the impurity concentration in Si, r is the precipitate radius, $\rho$ is the precipitate density, $C^*$ is the impurity dynamic equilibrium concentration existing at the precipitate-Si interface, $C^{eq}$ is the impurity thermal equilibrium concentration in Si, $\Omega$ is the volume of one impurity atom in the precipitate, $\sigma$ is the precipitate-matrix interfacial energy density, and $k_B$ is Boltzmann's constant. It is noted that Eqs. (6)-(8) also apply to the precipitate growth process. The reason for regarding the precipitate dissolution process as diffusion limited is for the purpose of obtaining a best case estimate of the effectiveness of a gettering process, i.e., it is the fastest. This means that no kinetic barrier against the dissolution of the precipitate is assumed. Most silicide precipitation and precipitate dissolution processes are associated with a precipitate-matrix volume misfit, which leads to a kinetic barrier slowing down the process. However, the diffusion limited precipitate dissolution assumption should be fairly realistic for solar grade Si and for IG in CZ Si, because the high density of dislocations can serve to eliminate the kinetic barrier incurred by the misfit via mediations by point defects.

## NUMERICAL MODELING RESULTS

A simulation result of gettering a hypothetical interstitial impurity by an Al layer, obtained using Eqs. (1) and (2), is shown in Fig. 1. For this problem, the use of Eq. (5) yielded the same results.

In accordance with reaction (1) and Eq. (5), a complete set of differential equations have been written and solved [16] for the gettering of $Au_s$ away from the Si bulk by P indiffusion from the wafer front surface (Fig. 2a), by a layer of Al deposited on the wafer back surface (Fig. 2b), and by the combination of the two schemes (Fig. 3). In these equations we have accounted for the coupling among

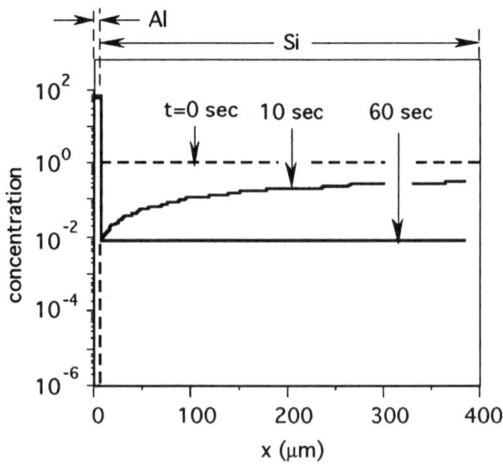

**Figure 1.** Simulation results of gettering a fast moving metallic impurity from the Si bulk using a 4 mm thick liquid Al layer on the left side of the Si wafer. The impurity atom diffusivity is assumed to be $10^{-4}$ cm$^2$s$^{-1}$ and the metal atom solubility in the liquid Al layer is assumed to be $10^4$ times of that in Si.

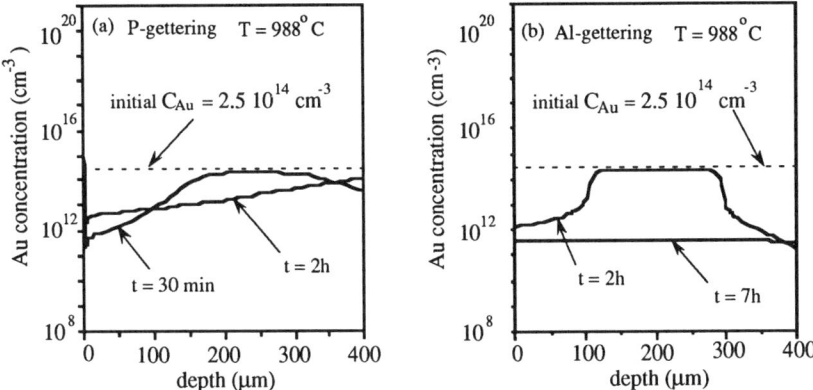

**Figure 2.** Simulation results of gettering Au by P and by Al. (a) Au concentration after 30 min and 2 h of P indiffusion gettering from the wafer left surface; (b) Au concentration after 2 and 7 h of gettering by an Al layer at the wafer right surface.

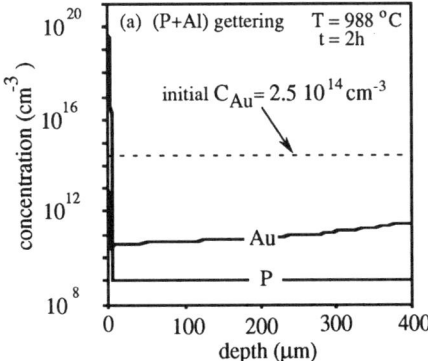

**Figure 3.** Au and P concentrations after 2 h of simultaneous P indiffusion gettering from the wafer left surface and Al layer gettering at the wafer right (back) surface.

the diffusing Au species and the point defect species $I$, and the diffusion-segregation properties of Au and of $I$. For P indiffusion and for Al gettering, the gettering sites are located at only one wafer surface, yet it is seen from Fig. 2 that the gettering of Au is more effective at both wafer surface regions than in the middle of the wafer. The results is not due to Au out-diffusion independent of gettering, because our simulation conditions are the same as in the experiments of Sveinbjörnsson et al. [17] of gettering Au by P indiffusion, for which the initial $Au_s$ concentration is the solubility value at the involved temperature. The accuracy of our model has been first checked by fitting their data. Now, gettering of Au in the Si wafer surface region joining the gettering region is more effective than in the wafer interior, because the gettered Au atoms need to only diffuse through a very short distance and because in the surface region $I$ is maintained near its thermal equilibrium value by the surface. Gettering of Au in the Si wafer surface region opposite to the gettering region location is also more effective, because $I$ is also maintained near its thermal equilibrium value by this surface. According to reaction (1), the gettering of $Au_s$ away from Si induces an $I$ undersaturation, which inhibits the further gettering of $Au_s$. In the wafer interior, this $I$ undersaturation is not effectively alleviated by $I$ indiffusion from the two wafer surfaces and therefore the gettering efficiency becomes lower than that at the two wafer surface regions. The *optimum* gettering times for the P indiffusion and Al gettering schemes are respectively 2 and 7 hours, indicating that P gettering is faster. This is because P indiffusion injects $I$ into Si which speeds up reaction (1) by alleviating the kick-out reaction generated $I$ undersaturation, while Al gettering does not inject $I$ into Si. On the other hand, the capacity of P gettering is smaller, and for longer gettering times P gettering becomes also less stable than Al gettering. This is because a finite P source has been used and hence the P profile has spread out for longer gettering times. Now, due to the decrease of the P concentrations, the Au solubility underneath the P profile is decreased and therefore the total capacity of P to stabilize Au has also decreased. The solubilities of Au in Si and in Al does not change with gettering time, and hence the Al gettering scheme is more stable. From Fig. 3, which shows the results of gettering Au by the combined method of P indiffusion and the Al layer, it is seen that it is fast and the gettering capacity is large. Furthermore, it is also as stable as Al gettering [16]. These points become particularly clear when the results due to all three gettering schemes are compared, as shown in Fig. 4. With these simulations, the synergistic effect of the combined P and Al gettering is explained, and it is clear that the combined P+Al scheme is the best by all measures.

Equations (6)-(8) have been used to numerically model the IG data of Gilles and Weber [7] obtained by measuring the concentrations of dissolved Fe atoms. In their experiments, Fe were diffused into CZ Si wafers with different densities of IG sites at 1100°C to the solubility value (monitored using FZ Si wafers). The wafers were then quenched to room temperature and subsequently annealed at 235°C for various times for Fe to precipitate out on IG sites. Figure 5 shows the data of Gilles and Weber [7] and our fitting curves. For simplicity, we used σ=0 in obtaining the fits. Their data showed that precipitation had already began during the quenching process, and this is accounted for by an effective previously existing precipitation time of ~35 min at 235°C.

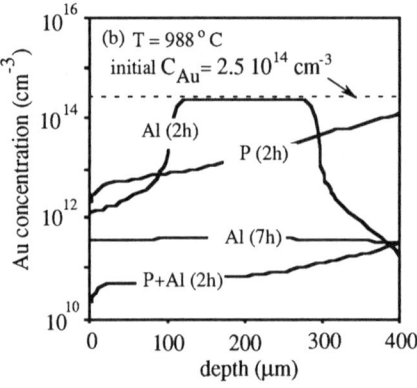

**Figure 4.** Comparison between the three gettering techniques (P, Al, P+Al) for 2 h gettering time, and Al gettering for 7 h.

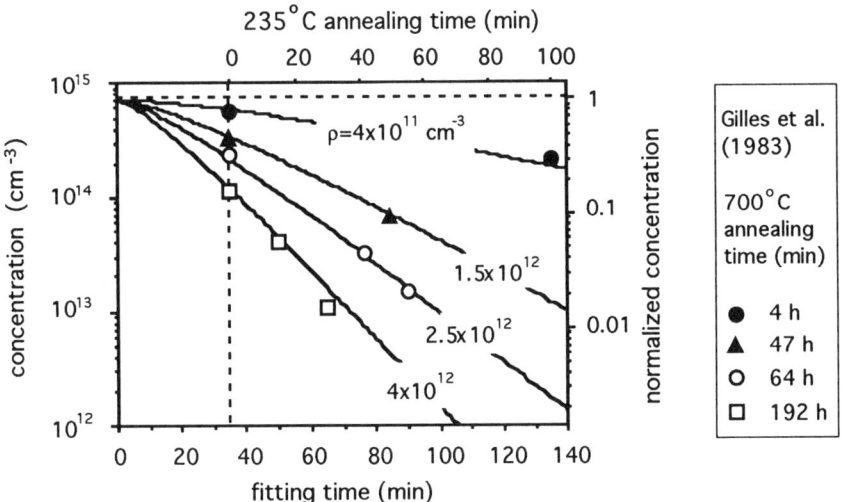

**Figure 5.** The Gilles and Weber [7] IG data of gettering Fe at 235°C (symbols), and fittings obtained using Eqs. (6)-(8). The CZ Si wafers were pre-annealed at 700°C for various times to create IG centers of different densities. The Fe was introduced at 1100°C to the solubility value (as monitored by FZ wafers), subsequently quenched to room temperature and annealed at 235°C for various times. The vertical axis shows Fe still in solution after the IG process. Parameters associated with each fitting curve designate the assumed precipitate density for obtaining the fitting.

Equations (6)-(8) have also been used to numerically model the gettering process of both dissolved and precipitated Fe. For this case a 200 μm thick Si wafer was assumed to have been saturated with Fe at 900°C and precipitated at 700°C to completion (no precipitation occurs anymore), and then gettered at different temperatures by a 2 μm thick Al layer situated at the backside of the wafer. We assumed σ=0, so that the precipitation completion condition is satisfied when Fe concentration has reached the 700°C thermal equilibrium value of ~$10^{11}$ cm$^{-3}$. Figure 6 shows the simulation results of gettering at 700°C for the ρ=$10^{11}$ cm$^{-3}$ case. Three outstanding features are noticed: (i) it takes ~60 h for the gettering process to reach the steady state for which the Fe concentration dropped to ~$10^7$ cm$^{-3}$; (ii) to reduce the Fe concentration everywhere in the Si wafer to below 99% of the 700°C Fe solubility value of ~$10^{11}$ cm$^{-3}$, it has taken already ~59 h of gettering time, with precipitates located at the wafer frontside still not totally dissolved; (iii) for shorter times, the Si wafer dissolved Fe concentration and precipitate size are not spatially uniform, they decrease monotonically from the frontside to the backside of the wafer. As the assumed precipitate density ρ decreases, the needed gettering time is increased. For example, the gettering time is increased to ~260 h for the ρ=$10^{10}$ cm$^{-3}$ case. For comparison, we have also simulated cases with the 900°C introduced Fe completely retained in solution in Si (as may be obtained by quenching experimentally) and precipitation did not occur during getter-

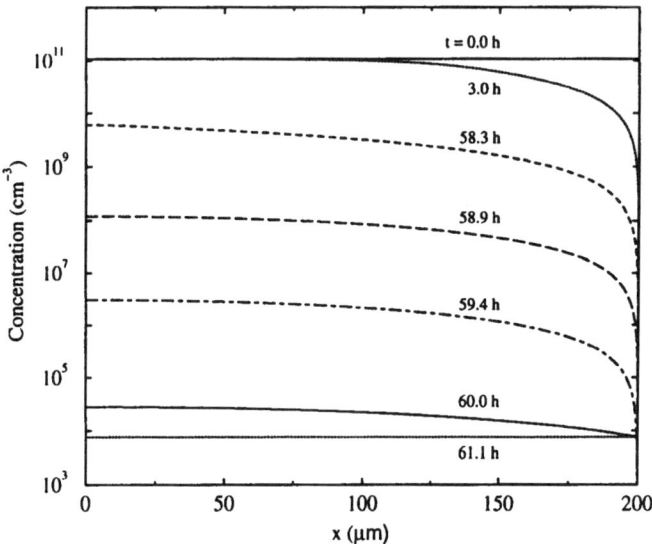

**Figure 6.** Calculated results of gettering by a Si wafer backside Al layer of Fe existing both in solution and in precipitates. The initial Fe concentration is assumed to be the thermal equilibrium value at 900°C, and was precipitated at 700°C to completion. The gettering annealing was assumed to be at also 700°C. The assumed Si wafer and Al layer thicknesses are 200 and 2 mm, respectively.

ing at 700°C. For this case it takes only ~2.7 h for the completion of gettering at 700°C. Moreover, to getter Fe to a concentration everywhere in the Si wafer to below 99% of the 700°C Fe solubility value of ~$10^{11}$ cm$^{-3}$, one needs only slightly more than 1 h. Clearly, it is a much more rapid process to getter the same total amount of Fe in dissolved form out of Si, because all Fe atoms are available to be gettered simultaneously. For the precipitated case shown in Fig. 6, long gettering times are needed, because of three factors, in descending order of importance: (i) the assumed gettering temperature is lower than the Fe introduction temperature, which limits the amount of dissolved Fe to be gettered at any time to a value corresponding to the gettering temperature $C^{eq}$ value, which is only a fraction of the total amount of Fe introduced; (ii) the total Fe precipitate dissolution surface is small, irrespective of the assumed high precipitate density; (iii) the dissolved Fe diffusivity is low. Thus, it will be quite ineffective to getter precipitated impurities at a temperature much lower than an effective temperature at which the impurity solubility corresponds to the total amount of the impurity present in the Si wafer. This suggests that a higher temperature may be more effective, since for a given precipitate density and size the needed gettering time is decreased. Figure 7 shows the summary of data obtained using the criterion that 99% of Fe introduced at 900°C is gettered into the Al layer, with gettering temperature and precipitate density used as modeling parameters. Clearly, the present simulation results have provided a plausible explanation to the fact that gettering by Al of multicrystalline

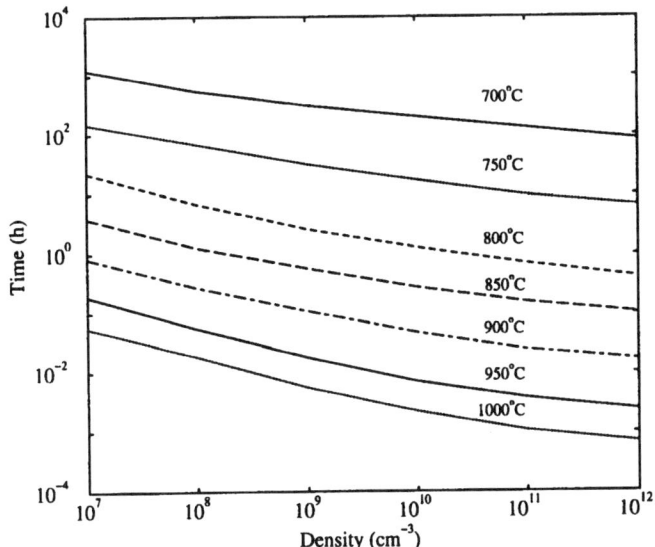

**Figure 7.** Calculated effective completion times of gettering dissolved and precipitated Fe by a Si wafer backside Al layer. Various precipitate density r and gettering temperature T values have been considered. Other conditions are the same as for the case shown in Fig. 5. The effective completion time is defined by the criterion that 99% of all Fe introduced at 900°C was gettered into the Al layer.

Si substrates may be ineffective. If carried out at a temperature somewhat higher than the precipitation completion temperature, the gettering process may even have an inverse effect. This is because on one hand the gettering time is too short and hence the process is not complete, and on the other hand the substrate dissolved impurity concentration may actually also have increased due to precipitate dissolution at the higher gettering temperature. In experiments, one should also be aware of the problem of further contamination in using an unclean furnace at a fairly high temperature, e.g., 1000°C.

## DISCUSSION

We believe that most of the basic physics and expressions for modeling the gettering process of a given impurity species, from the dissolution of the impurity precipitates to the diffusion and segregation (stabilization) of the impurity atoms, now exist. Accurate modeling parameters, such as the impurity charge states, level positions, solubilities, diffusivities, and interfacial energy densities between the silicide precipitates and Si are needed. Furthermore, we need also a realistic treatment of the precipitate nucleation problem.

## ACKNOWLEDGMENT

This work has been supported by National Renewable Energy Laboratory Subcontract No. XD-2-11004-1.

## REFERENCES

1. T. Y. Tan, E. E. Gardner, and W. K. Tice, Appl. Phys. Lett. 30, 175 (1977).
2. *Semiconductor Silicon 1994*, edited by H. R. Huff, W. Bergholz, and K. Sumino (The Electrochem. Soc., Pennington, NJ, 1994).
3. S. Narayanan, S. R. Wenham, and M. A. Green, IEEE Trans. Electron Dev. ED-37, 382 (1990).
4. B. Hartiti, A. Slaoui, J. C. Muller, and P. Siffert, Appl. Phys. Lett. 63, 1249 (1993).
5. P. Sana, A. Rohatgi, J. P. Kalejs, and R. O. Bell, Appl. Phys. Lett. 64, 97 (1994).
6. S. M. Joshi, U. M. Gösele, and T. Y. Tan, J. Appl. Phys. 77, 3858 (1995).
7. D. Gilles, E. R. Weber, and S. Hahn, Phys. Rev. Lett. 64, 196 (1990).
8. W. Schröter, M. Seibt, D. Gilles, Ch. 11 of "Electronic Structure and Properties of Semiconductors", Vol. 4 of "Materials Science and Technology: A Comprehensive Treatment" eds. R. W. Cahn, P. Haasen, and E. J. Kramer, Vol. 4 ed. W. Schröter (1991), p. 576.
9. D. Lecrosnier, J. Paugam, G. Pelous, F. Richou, amd M. Salvi, J. Appl. Phys. 52, 5090 (1981).
10. U. Gösele, W. Frank, A. Seeger, Appl. Phys., 23, 361 (1980).
11. R. D. Thompson and K. N. Tu, Appl. Phys. Lett. 41, 440 (1982).
12. D. A. Antoniadis and R. Dutton, IEEE Trans. Electron. Devices ED-26, 490 (1979).
13. T. Y. Tan and R. Gafiteanu, Second Annual Report, National Renewable Energy Laboratory Subcontract No. XD-2-11004-1 (1994), Chapter II.
14. H.-M. You, U. Gösele, and T. Y. Tan, J. Appl. Phys. 74, 2461 (1993).
15. T. Y. Tan, R. Gafiteanu, and U. M. Gösele, in *Semiconductor Silicon 1994*, edited by H. R. Huff, W. Bergholz, and K. Sumino (The Electrochem. Soc., Pennington NJ, 1994) p. 920.
16. R. Gafiteanu, U. Gösele, and T. Y. Tan, in *Defect and Impurity Engineered semiconductors and Devices*, eds. S. Ashok, I, Akasaki, J. Chevallier, N. M. Johnson, and B. L. Sopori, Mater. Res. Soc. Proc. 378 (Mater. Res. Soc., Pittsburgh, PA, 1995) p. 297.
17. E. Ö. Sveinbjörnsson, O. Engström and U. Södervall, J. Appl. Phys. 73, 7311 (1993).

# Carrier Recombination in Silicon Materials Used for Photovoltaic Devices

R. K. Ahrenkiel

*National Renewable Energy Laboratory*
*1617 Cole Boulevard, Golden, Colorado 80401*

**Abstract.** Recombination and transport parameters are critical to assessing photovoltaic parameters and predicting device performance. Here, I measured a group of silicon wafers of various origins, doping levels and mechanical structure, to demonstrate the range of phenomena that occur in photovoltaic materials. These various wafers had been the subject of a previous round-robin lifetime evaluation. These measurements demonstrate the difficulty and complexity of finding the true recombination rate when the material is incorporated into an operating device. The conclusion also supports the notion that characterizing a defect-dominated semiconductor with a single lifetime number is often inadequate.

## INTRODUCTION

The recombination rate of photo-generated charge carriers is a critical material parameter in photovoltaic (PV) devices. Both open-circuit voltage and short-circuit current are directly related to this important parameter. This paper will describe the experimental studies of carrier transport in a wide range of silicon materials that were used for a recent round-robin evaluation. The results of this round-robin exercise have recently been described [1]. The transport and recombination effects seen in this set of materials are representative of a wide range of silicon PV materials. This set of samples includes single-crystal silicon over a wide doping range. It also includes large-grain and small-grain polycrystalline silicon materials that are being developed for low-cost, flat-panel collectors. The transport in polycrystalline materials is complicated by trapping at grain boundaries. Defects types can be classified as either deep recombination levels or shallow traps. The presence and identification of these defects are determined by lifetime measurements that are available in our laboratory.

Two measurement techniques have been used at the National Renewable Energy Laboratory (NREL) to measure lifetime in a wide range of photovoltaic materials. The established technique for light-emitting, direct-bandgap materials is time-resolved photoluminescence (TRPL). This technique is only successful for measuring the lifetime of materials with bandgap ($E_g$) greater than about 1.0 eV. The range of materials measured by this technique include most of the III-V and II-VI compound semiconductors. The PV materials with indirect bandgaps or bandgaps less than 1.0 eV cannot currently be measured by TRPL. The latter includes both crystalline and polycrystalline silicon and $CuInSe_2$ (CIS) with $E_g < 1.0$ eV. To fill this gap in the measurement technology, I developed a radio-frequency photoconductive decay technique working in the ultrahigh-frequency range (UHFPCD). This contactless technique uses a small coil that is inductively coupled to the sample under test and was described previously [2,3]. The measurement frequency used in this UHFPCD technique is about 430 MHz. Recent work has shown that the system response is <u>linear</u> over three orders of magnitude [4]. The standard light source in these experimental studies is a pulsed YAG laser (1.064 μm, 7.0 ns FWHM) with doubling and tripling optics so that wavelengths of 0.532 μm and 0.355 μm are available. Solid-state sources such as diode lasers have also been used as sources when very low injection effects were desired. The infrared wavelength at 1.064 μm is very highly penetrating for silicon, for which the room-temperature absorption coefficient is 5.5 cm$^{-1}$. Thus, uniform volume excitation is easily achieved with this wavelength. If the surface recombination velocity effects need to be emphasized, the green 0.532-μm or ultraviolet 0.355-μm laser wavelengths are used because the absorption coefficients are 8.92 x 10$^3$ cm$^{-1}$ and 7.47 x 10$^5$ cm$^{-1}$, respectively.

Recently, an optical parameteric oscillator has been added to the laser system to make the output wavelengths tunable from 355 nm to about 2.0 μm. The transient photoconductance signal is proportional to the excess conductivity Δσ and decays with the recombination or minority-carrier lifetime. The technique has been used extensively for a wide range of materials, including heavily doped silicon, silicon spheres, InGaAs, and SiC. The lower limit of lifetime resolution for TRPL is about 100 ps, whereas that for RFPCD is about 10 ns.

## RECOMBINATION THEORY AND MINORITY-CARRIER TRANSPORT

The intrinsic recombination mechanisms in photovoltaic materials are Auger and radiative recombination which are a function of carrier concentration and injection level. In addition, Shockley-Read-Hall (SRH) recombination is a defect mechanism. The important defects are chemical impurities, mechanical defects, and grain boundaries. One may write the measured lifetime as:

$$\frac{1}{\tau} = \frac{1}{\tau_{SRH}} + \frac{1}{\tau_R} + \frac{1}{\tau_A} \qquad 1)$$

The low-injection radiative lifetime $\tau_R$ can be written as:

$$\frac{1}{\tau_R} = BN \qquad 2)$$

where B is a constant of a given material and N is the majority-carrier density. For GaAs and silicon, B values are 2 x 10$^{-10}$ cm$^3$ s$^{-1}$ and 1 x 10$^{-15}$ cm$^3$ s$^{-1}$, respectively. The low-injection Auger lifetime can be written as:

$$\frac{1}{\tau_A} = C_A N^2 \qquad 3)$$

Here, $C_A$ is the Auger coefficient specific to a semiconductor and carrier type, and N is the majority-carrier density. For p-type and n-type silicon, $C_A$ has been tabulated [5] as $1 \times 10^{-31}$ and $2.8 \times 10^{-31}$ cm$^6$/s, respectively, and is a dominant recombination mechanism for doping levels greater than $1 \times 10^{17}$ cm$^{-3}$.

One can decompose the SRH effect into surface and bulk recombination mechanisms. The surface component is usually described in terms of a surface recombination velocity, S. For a wafer of thickness d, the total SRH lifetime of Eqn. 1 is written as:

$$\frac{1}{\tau_{SRH}} = \frac{1}{\tau_B} + \frac{2S}{d} \qquad 4)$$

The total recombination lifetime, at low injection, can be written as:

$$\frac{1}{\tau} = \frac{1}{\tau_B} + \frac{2S}{d} + BN + C_A N^2 \qquad 5)$$

The lifetime as a function of injection level is much more complex. The intensity dependence of the SRH lifetime, assuming quasi-equilibrium, has been described by the author and coworkers in a previous work [6]. The recombination rate for a deep impurity in p-type material at energy $E_t$ and concentration $N_t$ is given by:

$$\frac{dn}{dt} = - \frac{\sigma_p \sigma_n v_{th} N_t (\rho N_A + \rho^2)}{\sigma_n \left(N_A + \rho + n_i e^{\frac{E_t - E_i}{kT}}\right) + \sigma_p \left(\rho + n_i e^{\frac{E_i - E_t}{kT}}\right)} \qquad 6)$$

where $\rho = \Delta n = \Delta p$, the density of light generated electron-hole pairs, respectively. Here $N_t$ is the density of a single defect type, $E_t$ is the energy level of the defect relative to the intrinsic level, and $\sigma_p$ and $\sigma_n$ are the hole and electron capture cross-sections, respectively. One sees from Eqn. 6 that the SRH recombination rate is a function of the electron and hole concentrations. <u>Thus, a single number for a lifetime does not exist for SRH-dominated materials</u>. To ascribe a single lifetime to a SRH material is therefore erroneous, unless the number is linked with injection conditions.

Our previous work gave the asymptotic SRH lifetimes as:

$$\tau_p = \frac{1}{\sigma_p N_t v_{th}} \qquad 7)$$

at low injection ($\rho < N_D$), when $E_t \sim E_i$.

At high injection ($\rho > N_D$), the <u>recombination lifetime</u> is:

$$\tau = \frac{1}{\sigma_p N_t v_{th}} + \frac{1}{\sigma_n N_t v_{th}} \qquad 8)$$

Thus, for a single deep level, one sees two SRH lifetimes depending on the injection level. At intermediate injection levels, one can detect a "lifetime" intermediate between these two asymptotes, depending on injection level.

Figure 1 shows a simulated result in p-type material as a function of injection level when $\sigma_n = 20\sigma_p$ and the background doping level is $1 \times 10^{16}$ cm$^{-3}$. Here, the model used $\tau_n$ of 5 µs and $\tau_p$ of 100 µs, values typical of moderately- doped silicon. Curve A through Curve D represent injection levels of $1 \times 10^{14}$, $1 \times 10^{15}$, $1 \times 10^{16}$, and $1 \times 10^{17}$ cm$^{-3}$, respectively. This behavior is often seen in SRH-dominated materials such as silicon and AlGaAs [7].

Rate equations may be written for the defect level when emission from the defect level is more probable than recombination. When $|E_t - E_i| \sim E_g$, a shallow defect level often produces carrier trapping but not significant recombination. These trapping effects have frequently been found in polycrystalline CdTe, polycrystalline silicon, and small (~ 1 mm diameter) silicon spheres [8]. Trapping behavior was recently reported in both polycrystalline CdTe [9,10] and small- grain silicon thin films [11,12]. These observations will be discussed in detail in this paper.

Shallow levels have been documented in the literature for many years [13] and have been described as "safe" traps. These defects trap mobile carriers, thereby affecting the excess carrier decay process, but they do not serve as recombination sites. As such, they do not affect the diffusion length. Misinterpreting trapping effects as recombination effects will seriously overestimate the diffusion length.

Assuming that the emission rate of captured minority carriers is $e_n$ (s$^{-1}$) per trapping site and that $n_t$ is the density of trapped carriers, then one can write a pair of coupled differential equations for the transient electron concentration:

$$\frac{dn}{dt} = -\frac{n}{\tau} - \sigma v_{th}(N_t - n_t)n + e_n n_t \qquad 9)$$

$$\frac{dn_t}{dt} = \sigma v_{th}(N_t - n_t)n - e_n n_t$$

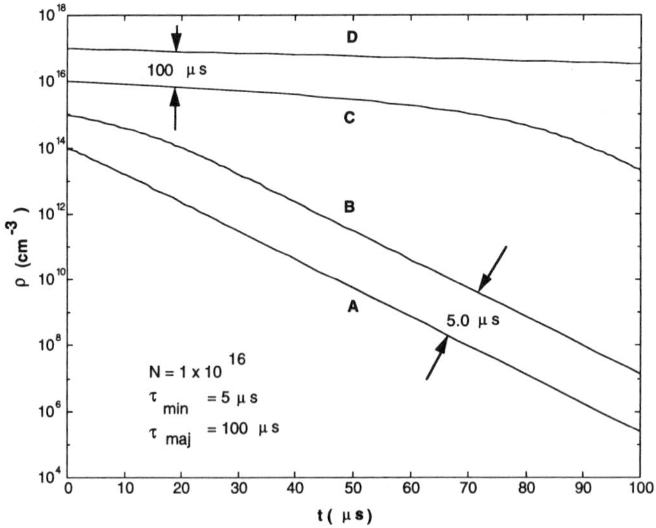

**FIGURE 1.** Solutions of Eqn. 1 with $E_t = E_i$, $N_A = 1 \times 10^{16}$ cm$^{-3}$, $\tau_n = 5 \propto \sigma$, and $\tau_p = 100 \propto \sigma$. The injection carrier density ρ is: A: $1 \times 10^{14}$ cm$^{-3}$, B: $1 \times 10^{15}$ cm$^{-3}$, C: $1 \times 10^{16}$ cm$^{-3}$, and D: $1 \times 10^{17}$ cm$^{-3}$.

Here, n(t) is the free minority-carrier density and $n_t(t)$ is the trapped minority-carrier density. Also, $\tau$ is the minority-carrier lifetime. One can define a trapping lifetime (the inverse of a trap capture rate) as:

$$\frac{1}{\tau_c} \equiv \sigma v_{th} N_t \qquad 10)$$

Figure 2 illustrates a solution of Eqn. 9 for a set of typical trapping and lifetime parameters that might be found in polycrystalline silicon. As the injection level increases, the traps may fill at which time the bulk recombination lifetime is observed.

Here, we set $\tau_R$ as 1 µs, $\tau_c$ as 1 µs and the long-term emission rate as $1 \times 10^5$ s$^{-1}$. The trap density is set to $1 \times 10^{14}$ cm$^{-3}$, and the injection levels are set to A: $5 \times 10^{13}$ cm$^{-3}$ and B: $5 \times 10^{14}$ cm$^{-3}$. Curve A is the calculated n(t) for low injection (the initial decay is approximately equal to $1/[1/\tau_c + 1/\tau_R]$. Curve B is the calculated trap occupation $n_t(t)$, and one sees that the trap filling occurs in about 1 µs. Curve C shows that the initial decay time increases to $\tau_R$ when the injection level is increased to $5 \times 10^{14}$ cm$^{-3}$, which is greater than $N_t$. Curve D shows $n_t(t)$ at the higher injection level, and the traps fill in about 0.2 µs. The traps stay filled for about 4 µs, and the signal is dominated by the recombination lifetime $\tau$. To extract the recombination lifetime from trap capture, trap filling is desirable.

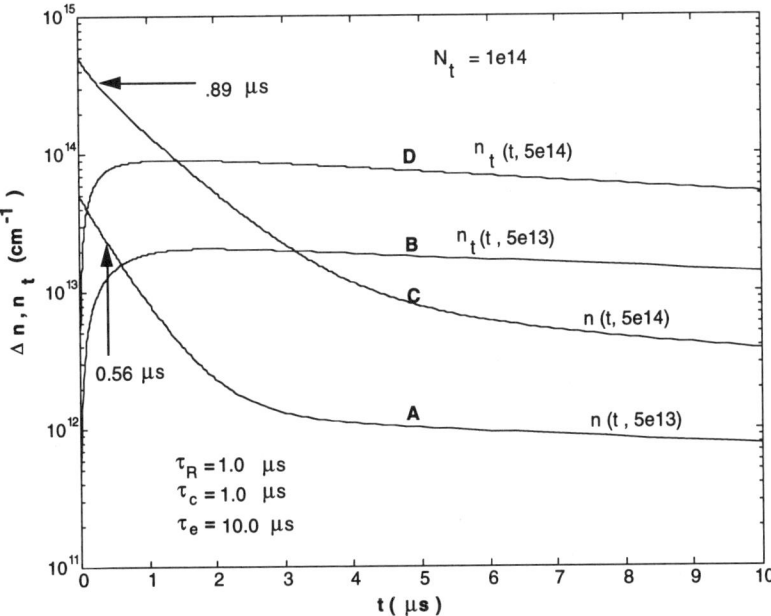

**FIGURE 2.** Solutions of Eqn. 9 with $N_t = 1 \times 10^{14}$ cm$^{-3}$, $\tau = 1 \, \infty\sigma$, $\tau_c = 1 \, \infty\sigma$, and $e_n = 1 \times 10^5$ s$^{-1}$. Curves A and B are n(t) and $n_t(t)$, respectively, with $\rho = 5 \times 10^{13}$ cm$^{-3}$. Curves C and D are n(t) and $n_t(t)$, respectively, with $\rho = 5 \times 10^{14}$ cm$^{-3}$.

As $e_n$ is temperature dependent, it exhibits a temperature dependence according to to the Arrhenius equation:

$$e_n = v_{th}\sigma N_c \exp\left(-\frac{\Delta E}{KT}\right) \qquad 11)$$

Here, $N_c$ is the effective density of conduction-band states (for trapped electrons), $v_{th}$ is the thermal velocity, and $\Delta E$ is the depth of the trap relative to the conduction band. Therefore, an indicator of defect trapping is that the temperature dependence of the slow decay follows Eqn. 11.

## EXPERIMENTAL TECHNIQUE

The silicon samples in this study were measured in both air and in iodine/methanol (I/M) solution [14]. The latter very effectively passivates the surface as will be shown in the lifetime data. The surface treatment prior to I/M treatments is described as follows. The sample was first cleaned in piranha solution (2:1, $H_2SO_4$ to $H_2O_2$) for 15 minutes. This was followed by a 5-minute rinse in deionized (DI) water. The samples were dipped in hydrofluoric acid (HF) for 1 minute, followed by a brief rinse in DI water. The sample was immediately placed in the M/I solution in a small teflon container. The latter had a measured molarity of $3.96 \times 10^{-3}$ M. For these measurements, the container was placed near the sensing coil, and the UHFPCD lifetime measurements were made with the sample in solution. Our samples were measured first, as received, and then cleaned and measured in M/I. The incident laser energy/pulse was measured with an energy meter. Using an absorption coefficient of 5.5 cm$^{-1}$ at 1064 nm, the injected carrier density could be calculated after measuring the sample thickness. The UHFPCD detection system was linear over at least three orders of injection level. After the lifetime measurement, the sample was placed on a mercury probe, and capacitance-voltage measurements (C-V) were made using a Keithley 590 Capacitance-Voltage Plotter.

## The Round-Robin Process

The first lifetime round-robin group met in December 1993, and a procedure was developed for comparing silicon lifetime measurements. As a result, three laboratories made lifetime measurements on the selected sample by various techniques. Both microwave reflection and RFPCD techniques were applied to these samples. The participants and their results are described in the NREL report [1]. They will be identified in this document as:
I. MIT Group: RFPCD measurements working with a probe frequency of about 50 MHz. Because of the lower probe frequency, decay times of less than about 1 μs were not resolvable [15].
II. Mobil Solar Group: Microwave reflection measurements with the samples immersed in hydrofluoric acid(HF).
III. North Carolina State University: Used a microwave photoconductivity technique, solid-state light sources (854 or 904 nm), and very low injection.

This author obtained the round-robin samples after the above measurements were completed but before obtaining the results of these measurements. These will be described below.

## Sample 1SB

This sample was a single-crystal wafer provided by MEMC Corporation. C-V measurements showed that the sample was p-type with a carrier concentration of $1.45 \times 10^{14}$ cm$^{-3}$. Assuming a hole mobility of 400 cm$^2$/Vs, the resistivity is calculated to be 108 ohm-cm. Figure 3 shows the sample response in air and in I/M solution at a single injection level. The single-pulse energy level is 1.72 mJ/cm$^2$. Using an absorption coefficient of 5.5 cm$^{-1}$ at 1064 µm, the injected carrier density is $2.8 \times 10^{16}$ cm$^{-3}$ for the data of Fig. 3. Curve A shows a higher injection lifetime of 30.5 µs, followed by a lowest injection lifetime of 7.2 µs. Curve B shows that the I/M surface passivation gives a constant lifetime of 69.3 µs. These data indicate that the low injection lifetime is dominated by surface recombination that is certainly reduced or eliminated by the I/M treatment. If one assumes that the very low-injection lifetime of Curve A is surface recombination dominated, then the surface recombination velocity of the sample in air is $4.7 \times 10^3$ cm/s using the sample thickness of 745 µm and the well-known relationship:

$$\frac{1}{\tau_s} = \frac{2S}{d} \qquad 12)$$

where d is the sample thickness and S is the recombination velocity. Here, the bulk lifetime contribution was ignored.

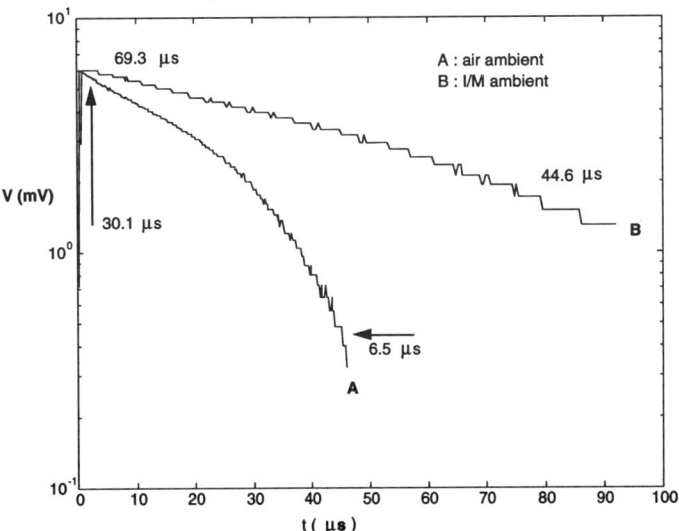

**FIGURE 3.** Lifetime data of round-robin sample 1SB. Curve A: sample lifetime in air-ambient with input pulse energy of 1.72 mJ/cm$^2$. Curve B: sample lifetime in I/M ambient with the sample pulse energy.

The effects of high injection are shown in Figure 4. Curve A is a measurement in air ambient with an incident pulse energy of 42 mJ/cm$^2$ corresponding to an injected carrier density of 6.9 x 10$^{17}$ cm$^{-3}$. At least two distinct regions with larger lifetime followed by smaller lifetime are observed with marked lifetime values. The intial lifetime of 50.4 µs is followed by an injection region with lifetime of 35.2 µs. A second region follows, with a 71-µs region followed by a 28-µs region. Such behavior is not predicted by the single-level SRH theory of Fig. 1. This photoconductive decay behavior may be the result of SRH saturation and shallow-trap emission or may be related to multiple deep centers with a range of capture cross-sections. The theory for the latter has not been developed. Curve B uses the same injection level as Curve A, but the wafer is immersed in the I/M passivating solution. A high-injection lifetime of several milliseconds is followed by 60.1 µs decay. The former is difficult to measure accurately because of the reponse exists for a very short time. The 60 µs-response is followed by a very long, 992-µs decay. It seems clear that the very long response of the latter is related to the reduction of the surface recombination velocity, which appears to dominate the lower injection levels. The overall behavior must be related to mulitiple, bulk SRH center effects.

For round-robin comparison, Group I found a lifetime of 33 µs and Group III found 1.6 µs, which likely compare with high and low-injection limits of Fig. 3, Curves A and B. Group II found a lifetime of 64 µs in HF that is certainly compatible with lifetime in I/M solution (Curve B) of Fig. 3.

## Sample 2SB

Sample 2SB is a polycrystalline silicon film product fabricated by AstroPower, Inc., using a proprietary process. The active layer in the material has mm size grains.

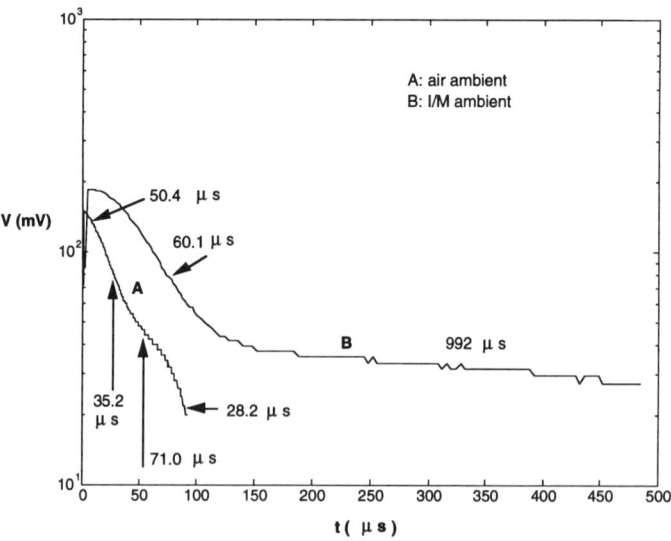

**FIGURE 4.** Lifetime data of round-robin sample 1SB with input pulse energy of of 42 mJ/cm$^2$.

The C-V measurement showed that the material is p-type, with a hole concentration of 2.3 x $10^{15}$ cm$^{-3}$. In Figure 5, Curve A shows the UHFPCD decay at the lowest injection levels that are obtainable with the system. The absolute injected concentration is not known because the active-layer thickness is unknown. However, the decay behavior is strongly indicative of the shallow-trapping model described by Fig. 2. The initial lifetime is about 0.5 µs, and the final lifetime is about 13 µs. The former is indicative of a combination of trap capture and recombination and the final lifetime is dominated by the trap-emission process. Curve B shows the response under increased injection and the initial decay time is 0.95 µs. Further increases in injected energy do not increase this lifetime. This high-injection response then is indicative of trap filling, and the lifetime is dominated by the recombination process in this regime, as shown by the model in Fig. 2.

These measurements were repeated with the sample immersed in I/M solution. The results were identical to those in Fig. 5. These data indicate that surface recombination is negligible in this fine-grain polycrystalline silicon and that grain boundaries are the dominant recombination sites.

The round-robin data on this sample basically agreed with Fig. 5. Group III did not report a result. Group II reported 1.2 µs for the initial component and 43 µs for the slow component. Group I only reported the slow component, with a time of 50.4 µs. As the emission signal is inherently nonexponential, measurement variation can easily be attributed to different injection levels.

## *Sample SB3*

Sample SB3 is a polished, polycrystalline wafer provided by Mobil Solar (now ASE Americas). This sample was grown by the edge-defined, film-fed growth (EFG) technique. This sample was not subjected to micro-examination by the author but

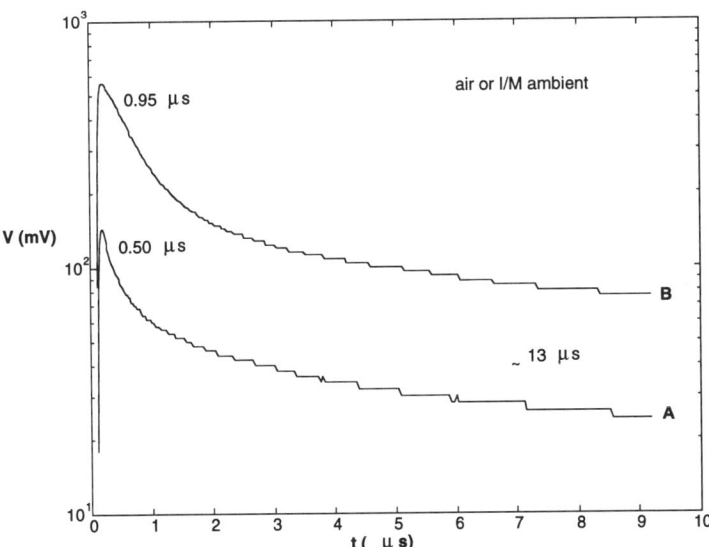

**FIGURE 5.** Lifetime data of round-robin sample 2SB with input pulse energies of: A: 1.72 mJ/cm$^2$; B: 42 mJ/cm$^2$.

the grain sizes in such samples are usually of mm dimension. C-V measurement indicated an electron concentration of $6.7 \times 10^{14}$ cm$^{-3}$. Figure 6 shows three UHFPCD measurements using a range of injection levels. Curve A is the decay curve of the sample in air ambient, with an incident pulse energy of 1.7 mJ/cm$^2$. The decay time here is 1.05 µs followed by a weak "tail" that could be related to grain boundary emission. Curve B is obtained by using the same incident pulse energy (1.7 mJ/cm$^2$) ), but the sample is immersed in I/M solution. The lifetime more than doubles, indicating that surface recombination is significant in this small-grain polycrystalline material. Finally, Curve C is obtained at very high injection with incident pulse energy of 42 mJ/cm$^2$ and the sample is immersed in I/M solution. The latter data indicate the SRH saturation of deep-level impurities and a subsequent large increase in lifetime to 14.5 µs. These bulk impurities could be related to grain boundary effects, but the decay follows the classic SRH behavior, a majority-carrier lifetime and a minority-carrier lifetime of 2.3 to 3.6 µs.

The round-robin results from Groups I, II, and III all indicated lifetimes of about 1 µs. This is in agreement with Curve A of the figure. The lifetime enhancement from surface passivation was apparently not seen by Group II.

## Sample SB4

This sample was provided for the round-robin study by the Osaka Titanic Corporation (OTC) and is a polycrystalline crystal grown with magnetic confinement of the melt. Visual inspection indicated this material to be large-grain polycrystalline silicon. C-V measurement showed a hole concentration varying from $2.8 \times 10^{15}$

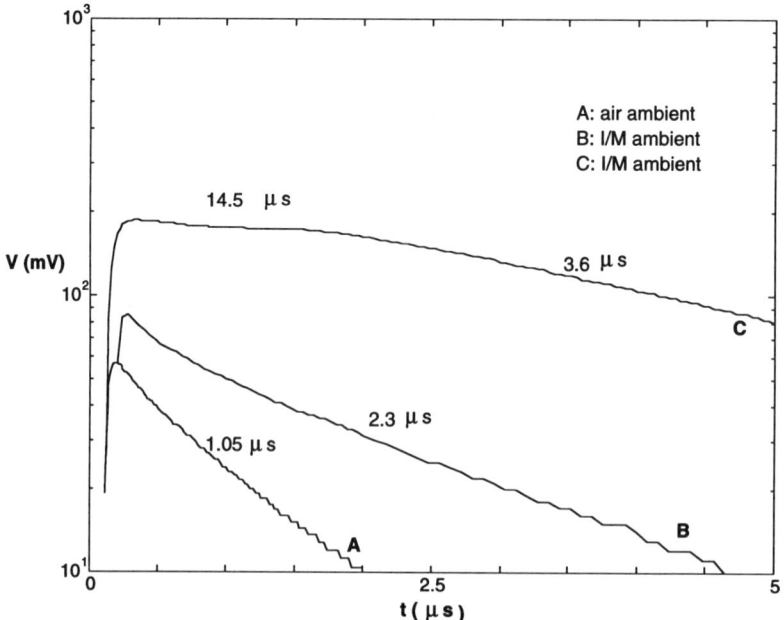

**FIGURE 6.** Lifetime data of round-robin sample 3SB with ambients as shown. The pulse energies are: A: 1.72 mJ/cm$^2$; B: 1.72 mJ/cm$^2$; C: 42 mJ/cm$^2$.

cm$^{-3}$ to 4.4 x 10$^{15}$ cm$^{-3}$ for this sample. The microcrystalline structure of this sample was not obtainable. In Fig. 7, Curve A is the UHFPCD data taken in air ambient with incident pulse energy of 1.7 mJ/cm$^2$. These data show the SRH saturation effect with a high-injection lifetime of 12.56 µs and a low-injection lifetime of 8.3 µs. The incident pulse energy was raised to 42 mJ/cm$^2$ for the data of Curve B, also measured in air ambient. The low-injection lifetime is equal to that of Curve A, and the high-injection lifetime rises to 19.8 µs, indicating further saturation of SRH defect. The sample was placed in I/M solution and the data of Curve C were obtained with incident pulse energy of 42 mJ/cm$^2$. The high-injection lifetime increases to 25.3 µs and the low-injection lifetime increases to 14.6 µs. These data indicate that surface recombination is the dominant mechanism at low-injection levels. These data also indicate some saturation of bulk defects, most likely at grain boundaries, for the higher injection levels.

The round-robin results from Group I indicated a lifetime of 20.3 µs, in agreement with Curve B (the I/M passivation measurement). Group I measured a lifetime of 11 µs, in agreeement with Curve A (air ambient). Group III measured 1.25 µs, which could be an asymptote of Curve A (8.3 µs).

## Sample SB5

Sample SB5 is a single-crystal silicon wafer supplied by Siemens Solar Inc. C-V measurements indicate hole concentrations ranging from 1.75 x 10$^{15}$ cm$^{-3}$ to 2.9 x 10$^{15}$ cm$^{-3}$. The UHFPCD data of Fig. 8 were obtained in air ambient, (Curve A), and in I/M ambient, (Curves B and C). For air ambient and lowest injection, the measurement produces a lifetime of 7.0 µs. Surface recombination is a factor, as

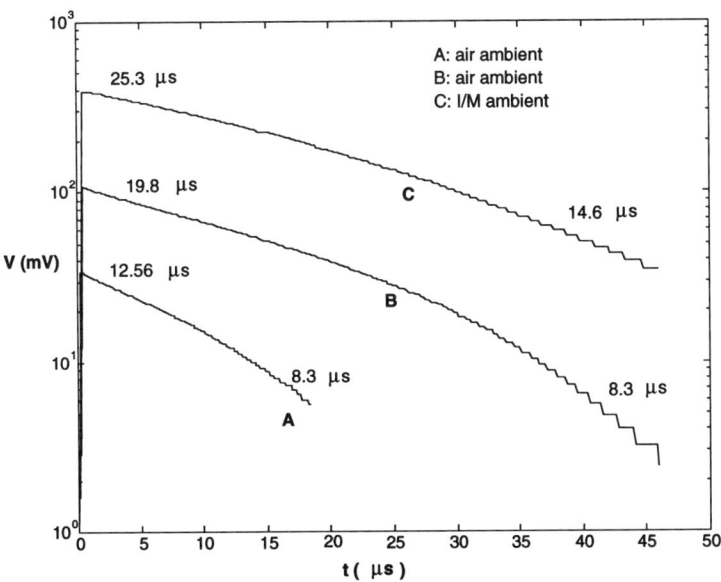

**FIGURE 7.** Lifetime data of round-robin sample 4SB with ambients as shown. The incident pulse energies are A: 1.72 mJ/cm$^2$; B: 42 mJ/cm$^2$; C: 42 mJ/cm$^2$.

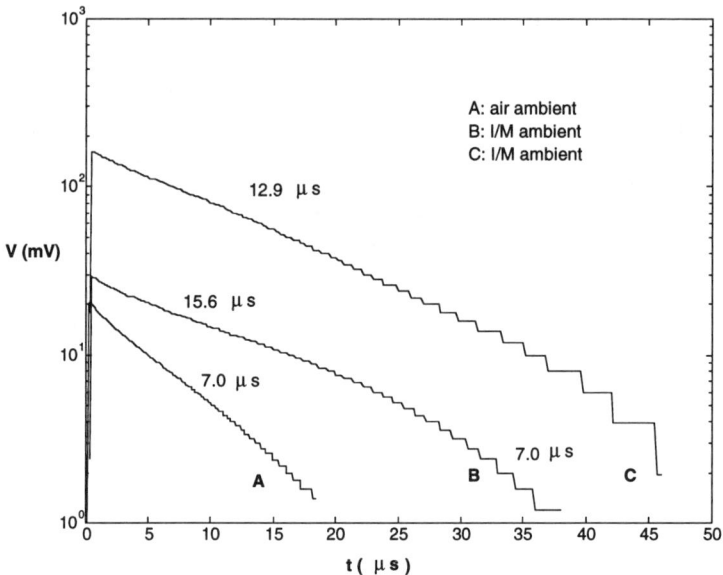

**FIGURE 8.** Lifetime data of round-robin sample 5SB with ambients as shown. The incident pulse energies are A: 1.72 mJ/cm$^2$; B: 1.7 mJ/cm$^2$; C: 42 mJ/cm$^2$.

seen in Curve B. Measurement in I/M solution increases the initial lifetime to 15.6 µs, which then falls to 7.0 µs near the end of the measured decay. Increasing the pulse energy to 42 mJ/cm$^2$ produces a nearly pure exponential decay with a lifetime of 12.9 µs. Because higher injection actually decreased the initial lifetime by a small amount, the dominant recombination mechanisms here may trigger the Auger effect.

The round-robin results were that Group II reported a lifetime of 17 µs in HF, which agrees with Curve B here. Group I reported a lifetime of less than 1 µs and Group II reported 2.1 µs. These do not agree with the lowest injection levels found here.

## Sample 6SB

Sample SB6 is an unpolished, polycrystalline wafer provided by Mobil Solar that was grown by the EFG technique. C-V measurements indicate hole concentrations varying from 8.9 x 10$^{14}$ cm$^{-3}$ to 1.4 x 10$^{15}$ cm$^{-3}$. There is a dramatic difference between this sample and the polished EFG sample SB3. The response in air ambient is not that different as shown in Curve A of Fig. 9. At an incident pulse energy of about 1 mJ/cm$^2$, the initial (high-injection) lifetime is 1.06 µs, as compared with 1.05 µs for Sample SB3. The low-injection portion of that response drops to 240 ns. However, on immersion in I/M solution, the lifetime does not increase dramatically as did SB3. Curve B shows the response with incident pulse energy of 42 mJ/cm$^2$, and the initial lifetime is 1.12 µs. A "tail" on the response, which may be indicative of grain-boundary trapping and emission, has a lifetime of 2.37 µs. However, the polished EFG sample SB3 showed a lifetime increase up to about 14 µs for the same

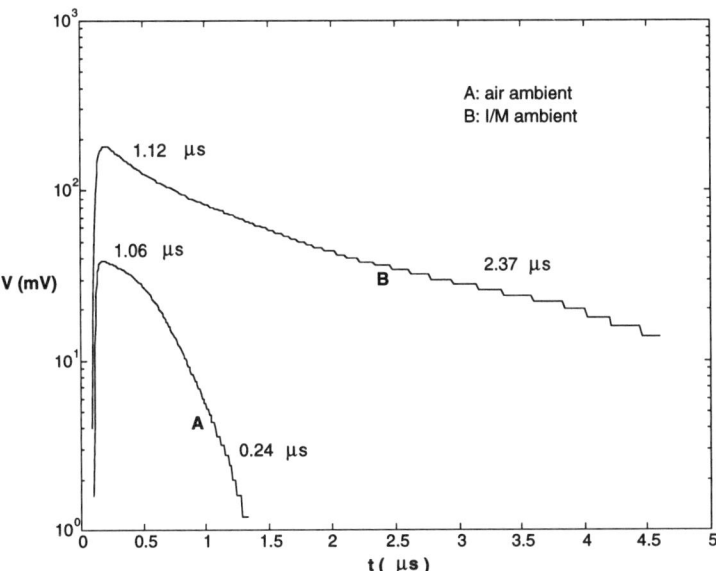

**FIGURE 9.** Lifetime data of round-robin sample 6SB with ambients as shown. The incident pulse energies are A: 1.0 mJ/cm$^2$; B: 42 mJ/cm$^2$.

pulse energy and surface treatment. The data may indicate that the effectiveness of the I/M passivation process is diminished by unpolished surfaces.

The round-robin results from the three groups are I: <1 µs, II: 0.6 µs, and III: <0.3 µs. These are in general agreement with the low-injection results of Curve A, or 240 ns. The I/M surface did not increase the lifetime appreciably, as shown.

## Sample 7SB

Sample 7SB is a large-grain polycrystalline sample supplied by Solarex Corporation. The material appears to be typical of the cast material used by Solarex in their photovoltaic product, with grain sizes varying from a few mm to a cm or more in cross-section. C-V measurements on the wafer showed hole concentrations ranging from 2 x 10$^{15}$ cm$^{-3}$ to 3 x 10$^{15}$ cm$^{-3}$. Some of the UHFPCD data for this sample is shown in Fig. 10. Curve A uses an input pulse energy of about 1.7 mJ/cm$^2$ with the sample in air ambient. One sees a SRH high-injection lifetime of 11.7 µs, followed by a low-injection lifetime of 1.62 µs. Curve B shows the strong effects of immersion in I/M solution. The high-injection lifetime increases to 15.1 µs, and the low-injection lifetime increases to 9.25 µs. These data indicate that the free surface of the material dominates the recombination process and that grain-boundary effects are small relative to the smaller-grain polycrytalliine materials described above.

The round-robin results are Group I: 15.4 µs and Group II: 11 µs. These are in good agreement with Curves A and B. Group III measured a low-injection lifetime of 1.87 µs that agrees quite well with the low-injection asymptote of Curve A.

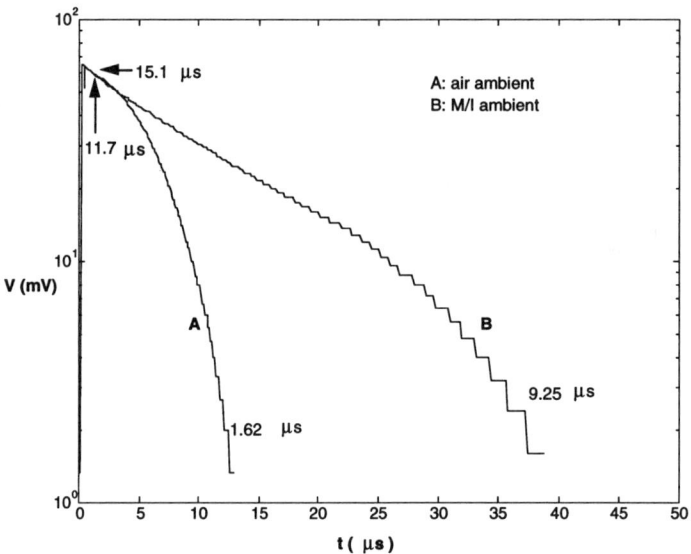

**FIGURE 10.** Lifetime data of round-robin sample 7SB with ambients as shown. The incident pulse energies are A: 1.72 mJ/cm$^2$; B: 1.72 mJ/cm$^2$.

## Sample 8SB

Sample 8SB was also provided by the Osaka Titanic Corporation (OTC) and is a large-grain, polycrystalline crystal grown without magnetic confinement. This sample should be compared with 4SB, which was grown with magnetic confinement (Fig. 7). C-V measurements indicated a very uniform hole concentration of $1.48 \times 10^{15}$ cm$^{-3}$ carriers. The UHFPCD decay data are shown in Fig. 11. Curve A was obtained in air ambient with an input pulse energy of about 1.7 mJ/cm$^2$. The initial, high-injection lifetime is 17.2 μs, followed by a low-injection lifetime of 9.0 μs. These data are very similar to Curve A of Fig. 7. Curve B was obtained from measurements in I/M solution at the same pulse energy and indicate an improved low- and high-injection lifetime. Therefore, one can assume that surface recombination is a significant contribution for the wafer in air ambient. Curve C shows a further improvement with incident pulse energy of 42 mJ/cm$^2$, which is indicative of volume saturation of SRH defects. Here, grain boundaries are the most likely candidates as recombination sites.

Group II measured a lifetime of 17 μs in HF, in good agreement with Curve B, if one looks at the low-injection asymptote. Group I measured a lifetime of 25.8 μs, which agrees qualitatively with Curve B measured in I/M solution. Group III measured 0.85 μs using very low injection.

## Sample 9SB

Sample 9SB is a very-large-grain polycrystalline wafer supplied by Crystal Systems, Inc. C-V measurements showed a hole concentration of $1.5 \times 10^{15}$ cm$^{-3}$, with a very uniform hole density. The UHFPCD data are shown in Fig. 12 under a range

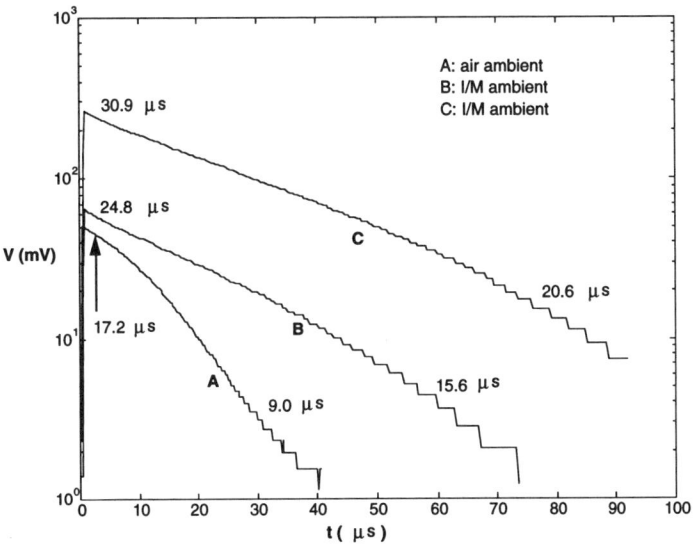

**FIGURE 11.** Lifetime data of round-robin sample 8SB with ambients as shown. The incident pulse energies are A: 1.72 mJ/cm$^2$; B: 1.72 mJ/cm$^2$; C: 42 mJ/cm$^2$.

of excitation and surface-treatment conditions. Curve A shows the data with the sample in air ambient and a pulse energy of 1.7 mJ/cm$^2$. The lifetime is constant and equal to about 1.2 μs. Curve B shows the response at the same input pulse energy, but with the sample in I/M solution. Here, the lifetime increases to about 2.6 μs, indicating a very significant decrease in the free-surface recombination velocity. Dramatic changes in the photoconductive response are shown when the pulse energy is increased to 42 mJ/cm$^2$, as shown in Curve C. An initial lifetime of 7.14 μs is followed by a very long electron lifetime at 126 μs. One could speculate that the initial response is Auger limited followed by a high-injection SRH response. Using the Auger coefficient described in the Introduction, one can calculate an injected density of 7 x 10$^{17}$ cm$^{-3}$ electrons. At 30 to 40 μs after the excitation pulse, the lifetime becomes 5.9 μs and is approaching the Curve B response. The very large intermediate lifetime might be related to saturation of grain-boundary defects in this very-large-grain material.

The round-robin report on this sample is Group I: 2.7 μs, Group II: 1.0 μs, and Group III: 1.63 μs. These are in qualitative agreement with Curves A and B here, the lowest-injection case.

## Sample 10SB

The sample 10SB is a single-crystal wafer supplied by Wacker Chemical Co. C-V measurements showed the hole concentration to be 1.9 x 10$^{16}$ cm$^{-3}$. In Fig. 13, Curve A, the UHFPCD signal decays, producing an initial lifetime of 9.9 μs, followed by a low-injection lifetime of 7.5 μs. The pulse energy for Curve A is 1.7 mJ/cm$^2$ and is the lowest injection condition measured here. Curve B shows the data

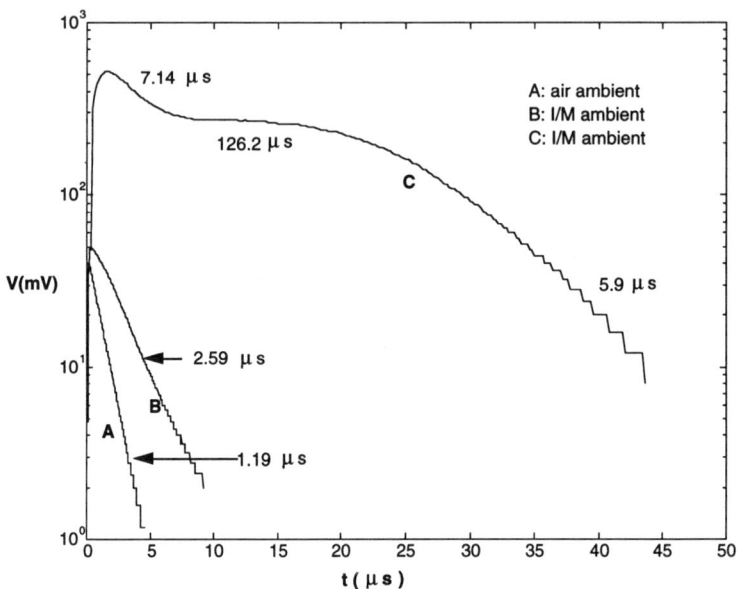

**FIGURE 12.** Lifetime data of round-robin sample 9SB with ambients as shown. The incident pulse energies are A: 1.72 mJ/cm$^2$; B: 1.72 mJ/cm$^2$; C: 42 mJ/cm$^2$.

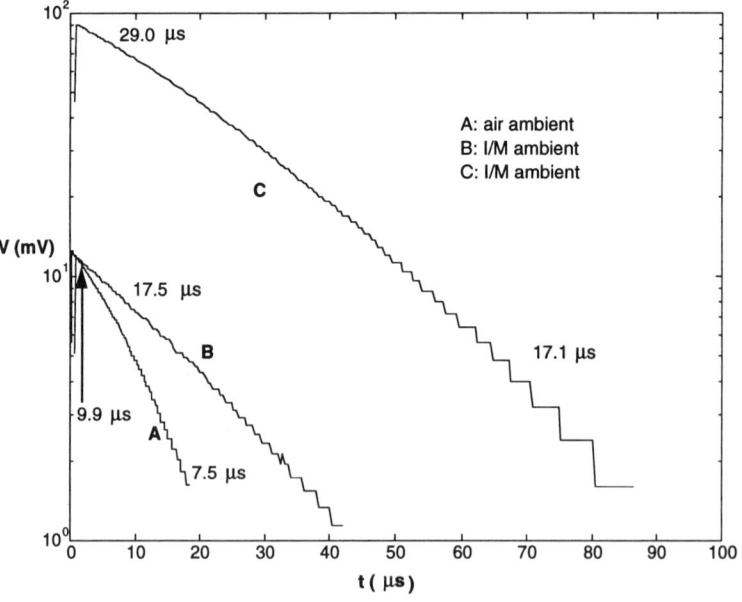

**FIGURE 13.** Lifetime data of round-robin sample 10SB with ambients as shown. The incident pulse energies are A: 1.72 mJ/cm$^2$; B: 1.72 mJ/cm$^2$; C: 42 mJ/cm$^2$.

for the same incident pulse energy and the sample immersed in I/M solution. The initial lifetime increases to 17.5 μs, and thus, surface recombination is a significant mechanism is this sample. Curve C shows the response with increased pulse energy to 42 mJ/cm², and the initial lifetime increases to 29.0 μs. These data are indicative of the saturation of bulk defects that are SRH recombination centers.

The round-robin report here is Group I: 20.3 μs, Group II: 45 μs, and Group III: <0.3 μs. I found no regions with lifetimes approaching 45 μs, even with high-injection and I/M passivation. The Group I result is consistent with Curve B if one assumes that higher injection "filled" some of the surface states.

## Sample 11SB

Sample 11SB is a single-crystal sample provided by the Polishing Corporation of America (PCA). C-V measurement showed a hole concentration of about 1.3 × $10^{15}$ cm$^{-3}$. In Fig. 14, Curve A shows UHFPCD data with an initial lifetime of 16.1 μs, followed by a SRH-type decay. The latter indicated a high-injection lifetime of 28.6 μs and a low-injection lifetime of 3.8 μs. The pulse energy for Curve A is 1.7 mJ/cm², and the measurements were made in air ambient. Curve B shows the response in air ambient with the pulse energy increased to 42 mJ/cm². Three distinct lifetime measurements can be seen from the data, with the low-injection lifetime increasing to 10.7 μs. The higher-injection portions of Curve B have approximately the same lifetimes as those of Curve A. Finally, Curve C shows the same injection

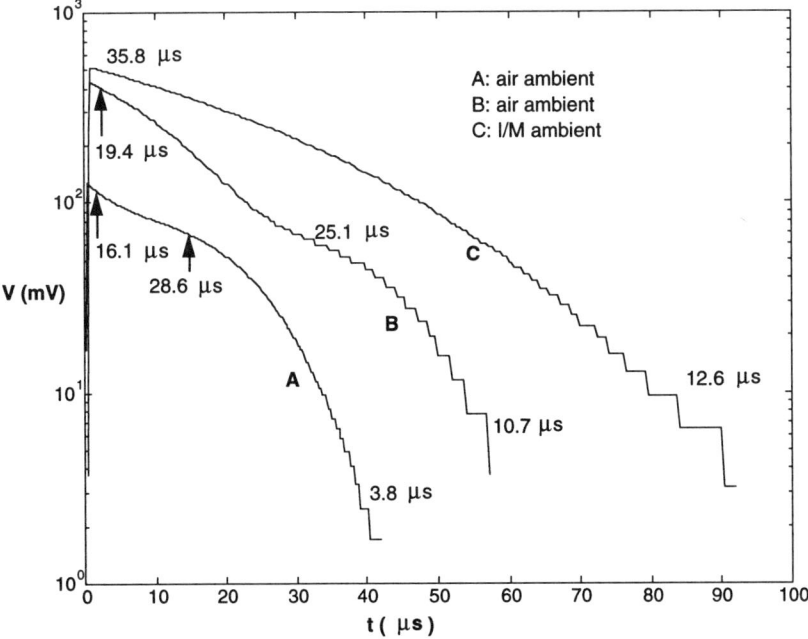

FIGURE 14. Lifetime data of round-robin sample 11SB with ambients as shown. The incident pulse energies are A: 1.72 mJ/cm²; B: 42 mJ/cm²; C: 42 mJ/cm².

level (42 mJ/cm$^2$), but with the sample immersed in I/M solution. A SRH-type response is seen with a high-injection lifetime of 35.8 µs and a low-injection lifetime of 12.6 µs. One concludes from these data that surface recombination is significant even at the highest injection levels when the sample surface is not passivated.

The round-robin results here are Group I: 16.9 µs, Group II: 55-83 µs, and Group III: 1.8 µs. The air-ambient result of Curve A is in fair agreement with the Group III result at the lowest-injection level. The Group I result agrees with Curve A (air ambient) at the initial injection level. No lifetime comparable to the Group II results are seen here at the high-injection level with I/M ambient.

## CONCLUSIONS

These round-robin data illustrate of the numerous effects of defects on carrier recombination. The round-robin report describes the variation between the three measurement groups on this sample set. However, I do not believe that there is any serious disagreement between the three data sets. One should expect such variability unless injection levels and surface passivation conditions are rigidly controlled. For example, Group III apparently operated at much lower injection levels than Groups I and II. Much of this work was done at injection levels near the doping level. If UHFPCD measurements were made at much lower levels, I might find much lower lifetimes than reported here. Again, quoting a single number for a minority-carrier lifetime in silicon is only meaningful when the above ancillary conditions are appended to the characterization data. In device applications, the standard operating injection levels will determine which number in the set of lifetime characterization numbers are relevant.

## ACKNOWLEDGEMENT

The author wishes to thank Dr. Michael Cudzinovic for providing the round-robin samples and for providing much useful information to the author. The author also wishes to thank Martha Symko for doing all of the etching, surface preparation, and iodine/methanol preparation for these studies. This work was supported by the U. S. Department of Energy under contract No. DE-AC36-83CH10093.

## REFERENCES

1. Cudzinovic, M., editor, "First Working Group Meeting on the Minority Carrier Diffusion Length/ Lifetime Measurement: Results of the Round Robin Lifetime/Diffusion Length Test," NREL/ TP-413-20470, November, 1995.

2. Ahrenkiel, R.K., Levi, D., and Arch, J., *Proceedings of the IEEE Photovoltaics Specialists Conference*, 1368–1371 (1994).

3. Ahrenkiel, R.K., *AIP conference Proceedings* **353**, 161 (1996).

4. Ahrenkiel, R.K., and Johnston, S. (to be published).

5. Dziewior, J., and Schmid, W., *Appl. Phys. Lett.* **31**, 346 (1977).

6. Ahrenkiel, R.K., Keyes, B.M., and Dunlavy, D.J., *J. Appl. Phys.* **70**, 225 (1991).

7. Ahrenkiel, R.K., Keyes, B.M., and Dunlavy, D.J., "Nonlinear Recombination Processes in Photovoltaic Semiconductors," *Solar Cells* **30**, 163 (1991).

8. Ahrenkiel, R.K., Levi, D., and Arch, J., "Recombination Lifetime Studies of Silicon Spheres," *Solar Energy Materials and Solar Cells* **41**, 171 (1996).

9. Ahrenkiel, R.K., Keyes, B.M., Wang, L., and Albright, S.P., "Minority-Carrier Lifetime of Polycrystalline CdTe in CdS/CdTe Solar Cells," *22nd IEEE Photovoltaics Specialists Conference*, 940 (1991).

10. Ahrenkiel, R.K., Keyes, B.M., Levi, D.L., Emery, K., Chu, T.L., and Chu, S.S., "The Spatial Uniformity of Minority-Carrier Lifetime in Polycrystalline CdTe Solar Cells," *Appl. Phys. Lett.* **64**, 2879 (1994).

11. Wallace, R.L., Anderson, W.A., Jones, K.M., and Ahrenkiel, R.K., *Conference Record of the Twenty Fifth IEEE Photovoltaics Specialists Conference*, 697 (1996).

12. Ahrenkiel, R.K., Keyes, B.M., and Levi, D.L., "Recombination Processes in Polycrystalline Photovoltaic Materials," *Proceedings of the Photovoltaic Solar Energy Conference*, 914 (1996).

13. Smith, R.A., *Semiconductors 2nd Edition*, England: Cambridge University Press, 1979.

14. M'saad, H., Norga, G.J., Michel, J., and Kimerling, L.C., *AIP Conference Proceedings* **306**, 471 (1994).

15. Kimerling, L. (private communication).

# CRYSTALLINE MATERIALS II

# Recent Developments in Terrestrial Concentrator Photovoltaics

Sarah R. Kurtz and Daniel Friedman

*National Renewable Energy Laboratory, Golden, CO 80401*

**Abstract.** This paper describes recent developments for the concentrator photovoltaic industry, including signs that the industry is progressing towards a significant share in traditional concentrator markets, as well as developing products for markets not traditionally thought to be accessible to concentrators. The question of why concentrators haven't developed significant market share is discussed and evidence that the time may be ripe for the growth of the concentrator industry is presented.

## INTRODUCTION

In 1989 Sandia National Laboratories (under authorization from DOE) launched the Photovoltaic Concentrator Initiative, a program designed to accelerate the achievement of a levelized-energy-cost of 10 to 15 cents/kWh (1). At that time, there were about a dozen companies pursuing various concentrator concepts, and the initiative was spurred on by technological advances including a 22% two-lens, two-cell minimodule and stacked cell efficiencies well over 30% (2). Also, a microprocessor-based tracker, glass secondaries and the use of copper heat spreaders added to the stimulus for the Initiative. The photovoltaics (PV) market was growing then, as now, faster than the supply, and it was projected that this demand would spill over into the concentrator industry. A number of companies were able to start developing products under the Initiative, but funding was terminated in 1991 because of a shortage of funds. At the same time, the Electric Power Research Institute (EPRI) funded its High-Concentration Photovoltaic (HCPV) Program with a similar goal. Funding for the HCPV Program was also cut because of a shortage of funds.

Funding for concentrator companies in the United States has remained available through technology-neutral programs like Photovoltaic Manufacturing Technology (PVMaT) and Technology Experience to Accelerate Markets in Utility Photovoltaics (TEAM-UP). However, concentrator installations have been limited primarily to demonstration projects, while sales for crystalline silicon PV have grown. In 1995, concentrators represented only 0.4 % of PV shipments (3), raising the question of when, if ever, concentrators will become an important player in the PV market.

Despite their slow growth, concentrators are attractive for a number of reasons. The primary advantage for concentrators has historically been their low cost when manufactured in very large volumes. Recent projections have shown that thin-film devices may eventually provide lower costs (4). However, the technological developments needed for low-cost, thin-film PV have not yet been realized, and concentrators can provide low costs in the near term, if a company is willing to commit to the needed commercialization program. In addition to low cost, concentrators offer high efficiency, ease of scale-up, and lower use of semiconductor materials (reducing the use of toxic compounds and other hazardous materials). In an accelerated market scenario, concentrators may be the only path to quickly reach a GW/year production level.

## WHY IS THE CONCENTRATOR MARKET SHARE SMALL?

Historically, concentrators were developed to give low prices in large volumes for utility-scale applications. However, this market has not yet materialized in the United States. This lack of a market for traditional concentrator PV has precluded the large production levels required to reduce costs, thus preventing small concentrator companies from competing for existing markets. For larger companies, the scale-up problem is much more easily solved.

Concentrators have several characteristics that restrict their suitability to certain markets: their large size (to justify the cost of the tracker), their poor response to diffuse light, the need for moving parts, and the need for maintenance of the moving parts.

Finally, many of the concentrator companies are still developing products. Only ENTECH has a product that has been tested in a significant number of installations; Midway and Amonix have both started testing reliable versions of their products within the last year. In general, the reliability of utility-scale (both concentrator and flat-plate) PV systems has not been adequately demonstrated (5). This has not been a major problem for flat-plate companies because the bulk of their sales are for small systems.

## CURRENT CONCENTRATOR PV INDUSTRY

There are today at least ten companies that are developing and/or manufacturing concentrator systems, a similar number that are interested in supplying cells to these companies, and a few research groups around the world that are developing new cell and concentrator system technologies.

### Concentrator System Companies

The most significant recent development in the concentrator industry is the entry of BP Solar (see discussion below for description of the hardware). BP Solar has projected that concentrators will reach 8-10 cents/kWh much sooner than flat-plate Si (6). For this reason, BP Solar has chosen concentrators as the appropriate technology for large central-power generation in sunny parts of the world. Also, three of the U.S. concentrator companies—ENTECH, Midway, and Amonix—are actively marketing systems. The concentrator company activities are summarized in Table 1.

Of the concentrator companies, **ENTECH**'s technology is the most mature (4th generation), using a linear fresnel lens with Si cells operating at about 20 X. ENTECH's 20 kW Photovoltaics for Utility-Scale Applications (PVUSA) array at Davis, CA, has had the highest performance (about 11% operational efficiency) and lowest degradation rate of all the PV technologies comprising this project over the past five years (5). The experiences from 600 kW of installed systems has allowed ENTECH to identify and fix most design problems. The tracker problems at both the PVUSA, Davis and Central & South West sites were remedied by adding motor brakes, but the Central & South West system is still

TABLE 1. Survey of Concentrator Company Activities.

| Company | Type | Concentration | Min. size (kW) | Recent failures/problems | New systems under test without serious failures | Comments |
|---|---|---|---|---|---|---|
| Alpha Solarco | point, fresnel | 250 X | 25 | multiple problems with original design | next generation to come online in early 1997 | developing new product with glass lenses; plant in China |
| Amonix | point, fresnel | 250 X | 20 | tracker and misc. problems with previous product (1995) (5) | 4 TEAM-UP installations, no serious problems reported so far (9) | are starting to market product |
| Australian National University | linear, trough | 25 X | 0.3 | | | currently building multi-kW system; plan to start commercialization in two years |
| BP Solar | linear, trough | 32 X | ~10 | | | planning 480 kW demonstration project in Tenerife (Canaries) 1997-8 |
| DayStar | linear, sheet | ~10 X | ~0.1 | | | Uses CIS-filament cells in flat-plate-like configuration; currently building and testing small prototypes. |
| EDTEK | point, dish | > 2000 X | 1 | | | innovative design; first prototype just completed |
| ENTECH | linear, fresnel | 20 X | 1 | a tracker problem at Central & South West PVUSA test (1995) (5) | 300 kW system without serious problems for 6 years (21); system in Minnesota without problems except on the coldest days (8) | just announced 1.5 MW for CSTRR |
| Midway Labs | point, fresnel | 300 X | 0.23 | previous generation product "didn't work" (1996) (9) | Commonwealth Edison (Illinois) reports no problems with the latest design (15) | competitive right now with 230 W and greater flat-plate market |
| Photovoltaics International | linear, fresnel | 15 X | 1.5 | | recent design under TEAM-UP test at APS is reported to be doing well (9,10) | new product ready to market in 1997 |
| Solar Res. Corp. | point, dish | 400 X | 1 | | | developing prototypes |
| TerraSun | novel | 10-200 X | 0.05 | | | building and testing small prototypes |

experiencing some problems (7). In the last year, a system was installed in Minnesota to test performance in the extreme cold. The system did experience some difficulties on the coldest days, but these have now been solved (8). Recently, ENTECH participated in a proposal to the Corporation for Solar Technology and Renewable Resources (CSTRR) to supply electricity for the Solar Enterprise Zone in Nevada. The prime bidder for this proposal was Nevada Power Corporation, with financing provided by Energy Unlimited (the wind developer on the team). CSTRR has announced plans to purchase power from Nevada Power, including a 1.5 MW PV power plant to be installed by ENTECH, with the understanding that the installation could grow to as much as 10 MW if CSTRR is able to find customers for the power. This increased production level should allow ENTECH to reduce manufacturing costs so as to be competitive with flat-plate systems and increase their market share.

**Photovoltaics International, (PVI) LLC** also uses a linear fresnel lens at a slightly lower concentration of 15 X. PVI's design collects some diffuse light, and allows the use of a simple tracker and Si cells coming off of a 1-sun production line. PVI reduces costs further by bringing lens production in-house and automating the assembly process. The company plans to start producing and marketing a 1.5-kW system in the first quarter of 1997, producing a few hundred kW in 1997, and scaling up to a 5-MW production level by 1999. The production levels will be limited by the market rather than the production capacity. PVI has installed a 3-kW system as part of a TEAM-UP project at Arizona Public Service (APS). APS reports that this system is "operating nominally" (9) and that they were pleasantly surprised at how few problems there were given that the design was new and hadn't previously been tested (10). PVI teamed with Kenetech for a CSTRR application and was chosen as one of four teams for the final consideration, but had to cancel when Kenetech declared bankruptcy. A 22-kW system was installed as part of a hydrogen production facility for the Clean Air Now/Xerox Solar Hydrogen Vehicle Project.

**Amonix** produces a 20-kW point-focus system using fresnel lenses, reflective secondaries and high efficiency silicon cells at about 250 X. Amonix has achieved very high system efficiencies, including 18% (PVUSA conditions) at its Arizona Public Service installation (11). Amonix' latest design is currently being tested at TEAM-UP installations at APS (two systems at APS's STAR facility), Central & South West Utilities, and Nevada Power. Its earlier design tested in Sacramento as part of the PVUSA project had (5) a few engineering problems with the trackers and water leakage, but the latest design has fixed these problems. There have been very few problems with the newer installations (10) and their selling price is projected to be < $2/W in high volume (12).

**Alpha Solarco** uses a similar point-focus design for a 25-kW unit operating at about 250 X. Alpha Solarco built a 4-MW/year production plant in China and is redesigning its system including replacement of the plastic lenses with cheaper, more transparent, and more stable glass lenses. Alpha Solarco expects to receive its first glass lenses in November and hopes to restart production in the near future. Alpha Solarco has already produced about 100 kW and advertises a price of $3000 per installed kW (for a minimum-size field of 500 kW).

**Midway Labs** also uses a point-focus design with fresnel lenses and Sunpower high-efficiency Si cells operating at about 300 X. Whereas all of the other companies have developed products in the 1-25 kW range, Midway currently markets a 230-W module that is cost competitive (13, 14) with all new, flat-plate mounted modules. By taking this approach, Midway can show a consistent and growing sales record. It has filled 40 orders in the last year, and is now producing modules at the rate of about 1/day, selling about 250 modules in a year. Although this still represents a miniscule fraction of the total worldwide PV sales (about 30 kW out of 100 MW), Midway has demonstrated a path to building up a sales volume. Midway's current product is being tested at Commonwealth Edison in Illinois and is working well (15), an important improvement over Midway's previous product (9). Because concentrators have traditionally been thought of only in the context of high power and grid support applications, with all other concentrator companies having developed products in the 1 to 25 kW range, Midway's low-power module, which does not depend on development of a utility PV market, is a promising new approach to concentrator market selection. At this conference Midway describes how it can use high voltage cells to make a competitive 140-W product for water pumping.

**EDTEK**, Inc. of Kent, WA and United Solar Technologies have just completed their first working 1-kW concentrator system. EDTEK takes an innovative approach allowing use of III-V cells at a concentration of 2400 X with tracking requirements similar to those of traditional designs. No other concentrator company is currently investigating a concentration level over 600 X. At 2400 X, the cost of III-V cells (predicted to be less than or about $10/cm$^2$) contributes less than $0.20/Watt (16). EDTEK's system uses cogeneration of electric (PV) and thermal power. Excessive heating of the cells is prevented by the secondary optic—a glass tube containing cooling water which absorbs infrared radiation while increasing the concentration ratio to 2400 X. The payback period (using Sacramento insolation and electric rates) is estimated as less than seven years (17).

**Solar Research Corporation (SRC)**. This Australian firm uses a parabolic dish concentrator at concentrations up to 500 suns, with the unused part of the spectrum being redirected for cogeneration (18). For proof of concept, they have demonstrated a 1.5-m-diameter dish with a total receiver efficiency of 31.8% at 280 suns, 18.4% of which is from a silicon cell module and the remainder is from cogeneration of steam at 1100°C. A 5-m-diameter (3.6 kW) dish has also been commissioned, and a 20-m-diameter dish is under development. The Australian government has very recently offered to invest in SRC's 40-kW grid-connected demonstration project (due for completion in 1997), and the Northern Territory Power and Water Authority has indicated strong interest in purchasing systems from SRC (19).

The **Australian National University** (ANU) is working on a reflecting parabolic trough concentrator for photovoltaic applications. The mirror size is about 2 m$^2$ and heat sinking is passive. Efficient, 25-sun solar cells are being fabricated at ANU. A 300-W single mirror system and a 3-kW system with 12 mirrors have been constructed. The 3-kW system is a module for a larger system that will be constructed over the next year. ANU's commercial partners are interested in commercialization in the next 2 years.

The European Union is currently funding a joint project of **BP Solar** and the **Institute of Solar Energy** of Madrid on PV concentration. The Institute of Solar Energy has developed in Madrid a parabolic reflective trough concentrator system. The geometric concentration ratio is 32X. The concentration cells are fabricated by BP Solar and properly encapsulated in concentration modules. The cost of this system could drop below $2/W (20). An agreement has been signed between the two partners giving to BP Solar exclusive rights for worldwide commercialization of this system. A demonstration plant of 480 kW will be installed in Tenerife (Canaries) during 1997-98. BP Solar plans to first complete demonstration projects, then to market concentrator systems for large, central-power generation around the world.

**TerraSun** is developing a novel concentrator PV product that is based on collecting and focusing spectrally selected light onto solar cells via unique optical elements. TerraSun's design is non-tracking and would be mounted like most flat-plate products, which should help to yield a low-cost module. PV modules can produce power in ranges as low as one watt to multiples of megawatts, depending on the specified size and configuration required. Concentration ratios are in the 10 X to 200 X range. TerraSun is fabricating and testing prototypes at this time and expects to have 50 W to 100 W PV modules in the marketplace within the next two years.

**DayStar** is a new company with a very new approach to concentrator technology. DayStar uses copper indium diselenide (CIS) cells deposited on stainless steel filaments that are integrated into an acrylic superstrate which generates a linear focus along the filaments. The thickness and cost of the acrylic is similar to those of the glass used in conventional flat-plate designs. The module cost (including inverters) will be significantly less than $2/W. Because of the reduced use of the semiconductor and the increased ease of coating filaments rather than flat sheets, the product cost is less than that for flat-plate CIS modules. The design is flexible and can be adapted for both tracking and non-tracking configurations by reducing the concentration ratio. DayStar is currently making 50-900 $cm^2$ prototypes and plans to produce 10-20 kW for a demonstration project by January of 1998 and to have modules and manufacturing systems for sale in July of 1998.

Although they don't assemble complete systems, **3M** is a very important player in the concentrator industry because it supplies the materials for the optics for almost all of the companies. 3M makes both linear- and point-focus plastic lenses as well as reflective films. 3M's web process has very high throughput. With its current equipment running one shift, 3M could make lenses for more than 100 MW/year.

## Reliability of Concentrator Systems

Reliability is a problem that has plagued all PV systems, including flat plate. The use of moving parts and the need for high accuracy tracking make reliability a more challenging problem for concentrators. However, it should be noted that many of the flat-plate systems at PVUSA are mounted on trackers, and also that ENRON is planning to use trackers on the 10-MW system for CSTRR, so tracker maintenance and reliability issues are not unique to concentrators.

Extensive prior market experience is not readily available for concentrator systems, and concluding that a technology is unreliable on a one-of-a-kind demonstration project is generally not logical. Often the failures are very easily fixed as in the case of ENTECH's 300 kW system in Austin, Texas. This system has operated for 6 years (21), but is being removed because of problems with peeling paint. (ENTECH's newer systems are galvanized, not painted.) An example of a concentrator system that has been operational for many years is the 350 kW system that began operation in 1981 in Saudi Arabia. This system experienced a 1% per year degradation rate because of failure of modules (breakage of bonds between the ceramic substrate and the cell) (22, 23). However, the system is still tracking and generating power after 15 years, despite its old design (24).

There are two key parts of the reliability question: (1) Are there any fundamental barriers to making concentrators reliable? (2) Are the products available today adequately reliable? The answer to question 1 is almost certainly that there are no fundamental barriers to reliability, given that the parts that distinguish concentrators from flat plate are extremely well understood. The wind industry has demonstrated that moving parts do not prevent production of a reliable technology.

Question (2) needs to be addressed for each of the concentrator companies. Table 1 includes a survey of recent concentrator reliability data. Most of the companies still have a long way to go to prove the reliability of their products, but all of the companies are making progress. PVUSA reports have consistently shown (5, 25) that the concentrator systems have higher maintenance costs than the flat-plate systems. The maintenance costs should decrease as the concentrator technologies mature, but the maintenance costs are always expected to be somewhat higher for systems with moving parts (including flat-plate systems mounted on trackers). PVUSA has reported maintenance costs as low as 0.36 cents/kWh (25) (at Kerman in 1994 for a system using one-axis passive tracking), but one of the utility-scale systems at Davis averaged 5.7 cents/kWh (5) in 1995. Clearly, these higher costs (note that the one-axis tracker was not the problem) are unacceptable and are a serious issue that both flat-plate and concentrator companies must take seriously.

## Concentrator Cell Companies

Table 2 summarizes the companies that are interested in supplying concentrator cells.

All of the one-sun crystalline Si companies can potentially make concentrator cells simply by changing the grid pattern. **BP Solar,** using their one-sun Saturn cells, can make concentrator cells with 19% efficiency at $2 W/cm^2$ irradiance. As described above, BP Solar is one of two companies with both cell and system capabilities and is planning a 480 kW installation in 1997-1998. They also supply cells to other concentrator companies. **Siemens Solar** has produced 6-sun Si concentrator cells with up to 16.6% aperture-area efficiency, and 19-sun cells up to 17.0% with the use of a prismatic cover. These cells use low-cost fabrication methods usually associated with one-sun cell technology, such as screen-printed metallization (26). **Solarex** developed buried-contact Si concentrator cells under the Sandia Concentrator Initiative. Although they haven't recently sold any concentrator cells, they would be interested in selling concentrator versions of their one-sun Si cells if a market materialized. **ASE** is an additional potential source for concentrator cells.

**SunPower** stands out as a company that has spent considerable effort developing and marketing concentrator cells. In 1987 Sinton and Swanson reported a 28% efficient concentrator (200X-500X) Si cell (27). Since then, problems with stability (28) have been overcome and the high efficiencies of the laboratory have been transferred to production cells. SunPower has sold over $300,000 worth of high-efficiency Si concentrator cells, and now has a production capability of 10 MW/year. Currently, the production cells have efficiencies > 25% and are sold for $0.75/W (assuming 25 $W/cm^2$ irradiance on the cell) (29, 30). With these cells SunPower built a lens-cell assembly with an efficiency of 21.6% (31).

TABLE 2. Survey of Companies Interested in Manufacturing Concentrator Cells.

| Company | Si | III-V | Low concentration | High concentration | Recent sales |
|---|---|---|---|---|---|
| Amonix | X | | | X | X |
| ASE | X | | X | | X |
| Astropower | X | X | X | X | |
| BP Solar | X | | X | | X |
| Photonic Power Systems | | X | | X | X |
| Siemens | X | | X | | X |
| Solarex | X | | X | | |
| Spectrolab | X | X | X | X | |
| SunPower | X | | | X | X |
| TECSTAR | X | X | | X | X |

**Amonix**, under contract with EPRI, also optimized the back-contact Si cell. While solving the same stability and production scale-up issues as SunPower, Amonix took a slightly different approach developing a process that could be used at commercial Si fabrication lines, providing Amonix with scale-up potential without the need of a large capital investment. Amonix is one of two companies that currently have both concentrator cell and system capability.

**TECSTAR** (Applied Solar Energy Corporation) has sold Si concentrator cells for a number of years. **Spectrolab** developed a concentrator cell capability under the Sandia Concentrator Initiative. In addition to the Si technology capabilities, both TECSTAR and Spectrolab are currently producing GaAs and GaInP/GaAs dual-junction cells. The III-V production capacity at Spectolab alone (32) is enough for 1GW/year @ 1000 X. Furthermore, the space PV industry is responding to its market demands for ever-higher efficiencies by performing the R&D required for future generations of cells. For example, Spectrolab and TECSTAR are developing a GaInP/GaAs/Ge 3-junction cell for the next generation of advanced high-efficiency devices (33, 34). Both companies view terrestrial concentrators as a potential market to service with their Si and III-V manufacturing lines.

**Photonic Power Systems** produces concentrator-type GaAs cells for conversion of power transmitted by optical fibers (35). They have been supplying GaAs cells to concentrator companies and are interested in expanding their GaAs cell capability to the GaInP/GaAs dual-junction technology.

**Astropower** is exploring monolithically connected, solar-cell arrays to alleviate resistive losses for concentrator cells. They presently have for sale a terrestrial monolithic concentrator solar cell product, SunVolts. This high voltage cell uses Si, but they are interested in developing a similar III-V capability in the future.

## GOVERNMENT-SPONSORED CONCENTRATOR R&D

A number of groups around the world are devloping high-efficiency cells for space applications. These may eventually be applicable to the terrestrial market, but we have not attempted to include them in this summary.

### Industry / NREL R&D Collaborations

In 1994, NREL demonstrated a GaInP/GaAs monolithic two-terminal tandem cell with an efficiency greater than 30% at 140-180 suns, and greater than 29% at 400 suns. More recently, NREL has sent GaInP/GaAs cells to Solar Research Corporation (SRC), Midway Labs, ENTECH, TerraSun, and soon to EDTEK. SRC has already returned measurement results: an efficiency for the GaInP/GaAs device of roughly 27% in the range of 80-400 suns on-sun; in other words, in a real-world testing environment that includes actual elevated operating temperatures and variations in spectrum (36). Furthermore, the devices shipped to SRC were unoptimized prototypes, and improvements in efficiency of a couple of percentage points should be achievable with future iterations of the device. ENTECH has measured an outdoor lens-cell efficiency of 26.5% using NREL dual-junction cells (37). Results from Midway and EDTEK should be forthcoming soon.

### R&D Efforts in Japan

The New Sunshine Project of Japan's Ministry of International Trade and Industry (MITI) (38) includes crystalline Si, amorphous Si, thin films, and single-crystal compound cells. This last category is intended for concentrator applications, and at least five different highly advanced designs are being developed in this category at a funding level of about $1M each. The MITI program is starting to realize successes: it has recently achieved (38) over 30% efficiency at 1 sun with a monolithic 2-terminal cell, breaking a record previously held by the NREL (39) with technology invented and developed at NREL. MITI's high-efficiency R&D program is the largest in the world, and is likely to result in Japan's taking a lead in the future concentrator industry.

### R&D Efforts in Germany

The Fraunhofer-Insitute for Solar Energy Systems in Germany is developing high-efficiency concentrator cells and has reported efficiencies as high as 26% (40, 41) The GaAs cells are grown by a low-cost, low-hazard, liquid-phase etch-back-regrowth method. Additionally, the Institute is studying the concentrating optics, including a 300X design that requires only one-axis tracking (34). A number of other groups are working on a variety of concentrator designs, including non-tracking designs (42-44).

## RECENT DEVELOPMENTS AFFECTING THE TERRESTRIAL CONCENTRATOR INDUSTRY

### Concentrator Companies are Progressing

The most significant recent development in the concentrator industry is the decision by a large, flat-plate company to enter the concentrator business. As described above, BP Solar has chosen the concentrator system developed at the Institute of Solar Energy in Madrid to meet the market it sees for large, central-power generation. Because of its size, BP Solar is likely to become a key player in the concentrator industry as soon as the demonstration phase of its commercialization plan is completed.

All of the concentrator companies are making progress in developing new and more reliable products. ENTECH, Midway Labs, and Amonix are actively marketing their products. ENTECH has just been promised its first > 1 MW order, allowing ENTECH to scale up production, reduce prices, and compete with other companies. Midway Labs has developed a 230-W module that is already cost-competitive with today's flat-plate systems. Thus, ENTECH and Midway Labs have improved their chances of entering the market in the near future.

### Growth of PV Market

Expansion of the PV market as a whole is beneficial for all the individual components of the market, including concentrators. The market has quadrupled over the past ten years (45). Expansion of the utility market, in particular, should be especially beneficial for concentrators given the emphasis on utility-scale applications by most of the concentrator companies. A partial list of indicators concerning the potential for such market expansion includes:

- A Utility Photovoltaic Group (UPVG) survey of the U.S. utilities showed their interest in purchasing 9 GW of PV if the price is $3/watt installed (46). Price predictions for large production levels are less than or equal to $3/watt for all of the concentrator companies. Outside of the U.S. there should be even more potential, especially wherever electricity is most expensive.

- The TEAM-UP program is giving the utilities hands-on experience with PV.

- In many areas, people are interested in using PV even if it increases what they pay for electricity. An example: Arizona has just passed a resolution requiring that 0.5% of electricity be generated by solar by the year 1999 and 1% by the year 2002, resulting in $400-800 million purchases of utility PV.

- PV on rooftops projects in Germany and Japan have provided a whole new market. A recent study by Wenger, et al. calculates that U.S. homeowners in 20 states could break even or come out ahead on purchases of 1-kW systems that are financed with a home purchase if the installed price is less than $3/W. Any company that makes a 1-kW system for $3/W should be able to develop a product for this market.

### Potential Shortage of Silicon

If a rapid expansion of the PV market does materialize, potential problems in the scalability of present PV manufacturing capacity must be considered. A very recent analysis by R. O. Johnson (45) concludes that *"the accelerated [market growth] scenario will be dependent upon phasing in a solution to the nearing shortage of silicon starting material. Relative growth of the crystalline wafer segment of the photovoltaic industry over the next five years will require an innovative solution to materials supply, to supplement existing channels of semiconductor scrap and off-grade materials."* One important candidate for a solution to this problem is thin-film PV. However, concentrators provide an additional option, which is important given that the promised price/performance/availability of thin films has not yet been realized in the marketplace.

## WILL THE CONCENTRATOR MARKET SHARE GROW?

The signs of growth in the utility market are encouraging, but sales to U.S. electricity generators are likely to stay low until after the utility restructuring is in place, the size of the near-term market being determined by public demand for "green" electricity. In the meantime, concentrator companies may need to look to other markets for building up sales. Utility-scale markets around the world can provide opportunities for companies with low prices and creative PV companies can always find ways to market products.

However, the existing flat-plate industry is strong and represents strong competition for concentrator companies, even in a growing market. Three small companies are ready to take on this competition, but the outcome for them is not yet clear. The diversification of large, flat-plate companies into the concentrator business, as BP Solar has chosen, could make it even more difficult for the existing small concentrator companies, but will bring a growing acceptance of concentrator PV.

## CONCLUSIONS

Concentrator companies are making progress toward increasing their market share by developing new products, increasing the reliability of their existing products, and finding ways to overcome barriers to market entry. They have a long way to go before they can claim a significant fraction of the PV market, but we have identified no insurmountable barriers. Japan's large R&D effort on high-efficiency cells is likely to result in Japan becoming an increasingly important player in what is now an industry dominated by the U.S. BP Solar's decision to enter the concentrator business radically changes the concentrator industry from one limited to very small companies to one that has the resources to evolve quite quickly.

## ACKNOWLEDGMENTS

The completion of this paper would not have been possible without the contributions of many people. We would like to thank all of those who took their time to educate us, but especially R. McConnell, A. Maish, M. O'Neill, M. Walpert, N. Kaminar, R. Hoffmann, V. Garboushian, D. Roubideaux, E. Schmidt, E. Horne, J. Lasich, R. Swanson, P. Iles, G. Ralph, T. Cavicchi, T. Bruton, R. King, P. Jaster, T. Hickman, B. Champion, J. Hoffner, A. Bett, E. Boes, E. DeMeo, C. Herig, and R. Taylor. This work was supported under DOE contract #DE-AC36-83CH10093.

## REFERENCES

1. Chamberlin, J. L. and D. L. King, "The Photovoltaic Concentrator Intiative," in *Proc. of the 21st IEEE PVSC*, 1990, 870-875.
2. Boes E. C., " Photovoltaic Concentrator Technology Development," in *Proc. of the 21st IEEE PVSC*, 1990, 944-951.
3. Maycock, P. D., Photovoltaic Technology, Performance, Cost and Market Forecast 1975-2010.
4. Stolte, W., R. Whisnant and C. McGowin, "Design, Performance and Cost of Energy from High Concentration and Flat-Plate Utility-Scale PV Systems," in *Proc. of the 23rd PVSC*, 1993, 1292-1297.
5. PVUSA Project Team, 1995 PVUSA Progress Report.
6. Bruton, T., private communication.
7. Champion, B., private communication.
8. "Minnesota Utility Tests Limits of Solar System in Extreme Cold," UPVG Record, Fall '96 p 2
9. Hickman T., P. Eckert and T. Lepley, "A Competition of Tracking Photovoltaic Systems in a Southwestern Electric Uility Transmission & Distribution Application," in *Proc. of the 25th IEEE PVSC*, 1996, 1381-1384.
10. Hickman, T., private communication.
11. Eckert, P., private communication.
12. Garboushian, V. et al, "An Evaluation of Integrated High-Concentration Photovoltaics for Large-Scale Grid Connected Applications," in *Proc. of the 25th IEEE PVSC*, 1996, 1373-76.

13. Perez, R., "The Fire Within," *Home Power* **40,** 28-31 (1994).
14. Jade Mountain advertises Midway's product as their highest efficiency, lowest cost, new PV module at $1998 for 230 W or $3169 for 460 W.
15. Radiewicz, M., private communication, reports that the new unit has been tracking for a couple of months without problems and that previous problems seem to have been fixed.
16. This calculation excludes the cost of the bus bars and assumes a system efficiency of 25%, consistent with measurements of dual-junction, III-V cells on other concentrator systems.
17. "PVT: A Solar Concentrator Total Energy System", for the California Energy Commission.
18. Lasich, J. B., et al, "Close-Packed Cell Arrays for Dish Concentrators," in *Proceedings of the First World Conference on Photovoltaic Energy Conversion*, 1994, 1938-1941.
19. Lasich, J., private communication.
20. Sala, G., et al, "The EUCLIDES Prototype: an Efficient Parabolic Trough for PV Concentration," in *Proc. of the 25th IEEE PVSC*, 1996, 1207-1210.
21. Hoffner, J. and P. Jaster, "3M Austin Concentrating Photovoltaic Plant Two-year Performance Report, 1992-1993"; lessons learned included: site selection for concentrator systems should include analysis of DNI, the roof mounted system was expensive, the sealant on some cables leaked, there were nuisance electronic problems, and the primary problem was peeling paint.
22. Alamoud, A. R. M., F. S. Huraib and A. A. Salim, "Reliability Prediction of the 350 kW Concentrating Photovoltaic Field," in *Proc. of the 8th European PVSEC*, 1988, 299-303.
23. Zakzouk, A. K. M. , A. R. M. Alamoud and B. H. Khoshaim, "Factors Affecting the Performance of a Photovoltaic Power System (PVPS)," *Int. J. Solar Energy* **5,** 67-81 (1987).
24. Alamoud, A. R. M., private communication.
25. PVUSA Project Team, 1994 PVUSA Progress Report.
26. King, R. R., et al, "Silicon Concentrator Solar Cells using Mass-produced, Flat-plate Cell Fabrication Technology," in *Proc. of the 23rd IEEE PVSC*, 1993, 167-171.
27. Sinton, R. A. and R. M. Swanson, "Design Criteria for Si Point-Contact Concentrator Solar Cells," *IEEE Trans. on Elec. Dev.* **ED-34,** 2116-2123 (1987).
28. Gruenbaum, P. E., R. A. Sinton and R. M. Swanson, "Stability Problems in Point Contact Solar Cells," in *Proc. of the 20th IEEE PVSC*, 1988, 423-428.
29. Crane, R. , P. Verlinden and R. Swanson, "Building a Cost-effective, Fabrication Facility for Silicon Solar Cell R & D and Production," in *Proc. of the 25th IEEE PVSC* , 1996, 529-532.
30. Swanson, R., private communication.
31. Verlinden, P. J., et al, "A 26.8% Efficiency Concentrator Point-Contact Solar Cell," in *Proc. of the 13th European PVSEC*, 1995, 1582-1585.
32. "Emcore Reactor #7 for Spectrolab," *Compound Semiconductor* **2,** 21 (1996).
33. Chiang, P. K., et al, "Experimental Results of GaInP/GaAs/Ge Triple Junction Cell Development for Space Power Systems," in *Proc. of the 25th IEEE PVSC* , 1996, 183-186.
34. Yeh, Y. C. M., et al, "Production Experience with Large Area, Dual Junction Space Cells," in *Proc. of the 25th PVSC*, 1996, 187-190.
35. Fahrenbruch, A., et al, "GaAs- and InAlGaAs-Based Concentrator-type Cells for Conversion of Power Transmitted by Optical Fibers," in *Proc. of the 25th IEEE PVSC*, 1996, 117-120.
36. Friedman, D. J., et al, "On-sun Concentrator Performance of GaInP/GaAs Tandem Cells," in *Proc. of the 25th IEEE PVSC*, 1996, 73-75.
37. O'Neill, M. private communication.
38. Yamaguchi, M. and S. Wakamatsu, "Super-high Efficiency Solar Cell R&D Program in Japan," in *Proc. of the 25th IEEE PVSC*, 1996, 9-11.
39. Bertness, K. A., et al, "29.5%-Efficient GaInP/GaAs Tandem Solar Cells," *Appl. Phys. Lett.* **65,** 989-991 (1994).
40. Blieske U., et al, "LPE-GaAs and LBSF-Si solar cells for tandem concentrator applications," in *Proc. of the First World Conference on PV Energy Conversion*, 1994, 1902-1905.
41. Baldus, A. et al, "GaAs One-Sun and Concentrator Solar Cells based on LPE-ER Grown structures," in *Proc. of the 1st World Conference on PV EC*, 1994, 1697-1700.
42. Sakuta, K., S. Sawata and M. Tanimoto, "Luminescent Concentrator Module of a Practical Size," in *Proc. of the 1st World Conference on PV Energy Conversion*, 1994, 1115-1118.
43. Yoshioka, K., et al, "An Optimum Design and Properties of a Static Concentrator with a Non-Imaging Lens," in *Proc. of the 1st World Conference on PV EC*, 1994, 1119-1122.
44. Bowden, S., et al, "High Efficiency Photovoltaic Roof Tile with Static Concentrator," in *Proc. of the 23rd IEEE Photovoltaic Specialists' Conference*, 1993, 1068-1072.
45. Johnson, R. O., "PV Industry Overview, Domestic & International," in *Proc. of the Annual Utility PV Experience Exhibition & Conference*, 1996, to be published.
46. D.O.E., "Photovoltaics, the Power of Choice, the National PV Program Plan for 1996-2000".

# High-Efficiency GaAs Solar Cells on mm and sub-mm Grain-Size Polycrystalline Ge Substrates

R. Venkatasubramanian, B. O'Quinn, and E. Siivola
Research Triangle Institute, Research Triangle Park, North Carolina 27709

## Abstract

GaAs material and device structure optimization studies on optical-grade, millimeter-and-less grain-size polycrystalline Ge substrates are presented. We discuss the growth of high-quality epitaxial layers across various crystalline orientations of a polycrystalline substrate; this is important for obtaining high-performance solar cells. The GaAs solar cell on n-type poly-Ge substrate is a p-on-n type, with an undoped spacer between the p-emitter and the n-base. An experimental study of dark currents in these junctions, with and without the spacer, as a function of temperature (77K to 288K) is presented; this study suggests that the spacer reduces the tunneling contribution to dark current. In addition, we describe device-structure optimization studies that have led us to achieve an open-circuit voltage ($V_{oc}$) exceeding 1 Volt and an AM1.5 efficiency of ~19% for a 4-cm$^2$-area GaAs cell on sub-mm grain-size poly-Ge. We have also observed an efficiency ~21% for a 0.25-cm$^2$-area cell on similar small-grain poly-Ge substrates.

## Introduction

High-efficiency single-junction GaAs (25.7% under AM1.5G) [1] and tandem GaInP$_2$/GaAs (29.5% under AM1.5G) [2] solar cells have been demonstrated on single-crystal GaAs substrates under 1-sun conditions. These cell efficiencies can be improved further under concentration (500-1000 suns) and are potentially useful in a variety of concentrator systems [3]. However, the cost of GaAs solar cells on single-crystal GaAs substrates appear to make it unsuitable for flat-plate terrestrial systems. Therefore, it is important to investigate and effect a transition of these high-efficiency cells on to low-cost, large-area substrates for flat-plate applications.

Several low-cost substrate alternatives have been considered recently for the demonstration of high-efficiency GaAs solar cells. One approach utilizes single-crystal Si substrates; this has remained a challenge due to the lattice and thermal mismatch between GaAs and Si substrates. The reported best 1-sun efficiency for a 0.25-cm$^2$-area GaAs solar cell on single-crystal Si substrate is 17.14% [3]. In contrast, single-crystal Ge substrates have emerged as an effective alternative to single-crystal GaAs substrates. For example, the highest AM1.5 efficiency for a GaAs solar cell (4-cm$^2$-area) on single-crystal Ge substrate is ~24% [3]. In addition, the use of Ge substrates lead to an enhanced yield in a manufacturing environment, when compared to using GaAs substrates. However, the cost associated with single-crystal Ge substrates are likely to be prohibitive for flat-plate applications.

Large-area poly-Ge substrates, with an average grain size of sub-mm to several mm and produced by a cast process for optical-window applications, appear attractive to meet the need of low-cost substrates for GaAs solar cells. An early attempt to develop GaAs solar cells on such cast optical-grade poly-Ge substrates was successful, with an efficiency of 15.8% for a 1-$cm^2$-area cell [4]. In addition, the availability of large-area (as much as 24" diameter) cast poly-Ge substrates, in combination with well-established large-scale MOCVD growth feasibility, would allow cell-processing costs to be drastically reduced; this is necessary to meet the long-term cost-goal for flat-plate applications. Further, the understanding developed in the amelioration of the detrimental effects of grain-boundaries in GaAs cells on mm and sub-mm grain-size poly-Ge substrates would be useful in the eventual transition of GaAs cells on to fine-grain polycrystalline templates. In this paper we describe the recent progress that we have made in the development of high-efficiency GaAs solar cells on mm and sub-mm grain-size poly-Ge substrates.

## Key Issues with Polycrystalline Substrates and Templates

There are two basic issues related to the development of GaAs solar cells, or for that matter solar cells in any material system, on polycrystalline substrates or templates as shown in Figure 1a. A polycrystalline template could be a regular array of polycrystalline grains on a non-crystalline substrate such as glass. One issue concerns the deposition of uniform, two-dimensional, layered materials onto each of the various grains; the second is related to the minimization of the deleterious effects of grain boundaries in photo-carrier recombination and dark-current generation in a device containing the grain-boundary structures.

We can consider two situations to the growth of materials on polycrystalline substrates - an *ideal*, two-dimensional, uniform, layered growth across the various orientations so that we essentially preserve the grain-structure of the starting substrate and do not create additional grain boundaries in the active regions of the solar cell as indicated in Figure. 1b. A second situation could be a *non-ideal*, three-dimensional, non-uniform growth across the various orientations leading to the formation of additional

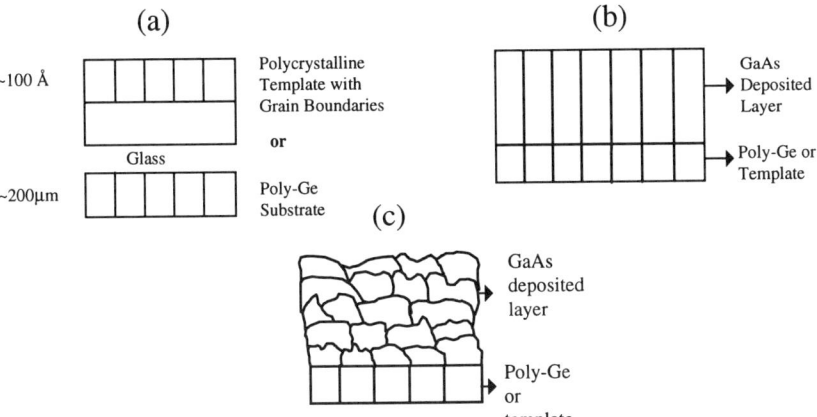

**FIGURE 1.** a) Schematic of a polycrystalline template or a substrate for cell growth; b) schematic of a layered, two-dimensional growth on poly-Ge or a polycrystalline template; c) schematic of a non-layered, three-dimensional, non-uniform growth on a polycrystalline template.

grain boundaries (Figure. 1c). Thus, the key to minimization of the grain-boundary effects in deposited polycrystalline materials appears to being able to develop uniform, two-dimensional, layered growth across the various crystalline orientations.

From a device performance point of view, in polycrystalline materials, the grain boundaries in the active regions of a solar cell can reduce photo-collection through minority-carrier recombination. However, if we develop ($p^+$-n) GaAs solar cells with a thin emitter, a majority of the photo-current would be generated in the n-base region. If a group-VI dopant like Se or S is chosen to dope the n- GaAs base region, it has been suggested [5] that a favorable $n/n^+$ minority-carrier mirror could be naturally formed at the grain-boundaries due to dopant segregation at these regions. These minority-carrier mirrors have been attributed to the relatively high short-circuit current density ($J_{sc}$) (> 20 mA/cm$^2$ under 1-sun) that can be readily obtained even in µm-grain-size poly-GaAs solar cells [5].

While high $J_{sc}$ values can be readily obtained in poly-GaAs solar cells, $V_{oc}$ values are typically low (~ 0.55 to 0.6 Volts) and the fill-factor has been low as well [5,6]. This is probably due to the generation of dark currents at grain boundaries. We have proposed the use of a "thin" undoped spacer at the depletion layer of the $p^+$-n junction and experimentally demonstrated that a significant reduction in dark current can be obtained leading to an increase in $V_{oc}$ and fill-factor of GaAs solar cells grown on large-grain poly-Ge substrates [4].

In this paper we discuss the mechanism behind dark-current reduction in $p^+$-n GaAs solar cells on poly-Ge substrates, with the use of the spacer, based on low-temperature dark I-V measurements. We also present results on specific approaches to enhance $J_{sc}$ and $V_{oc}$ in GaAs cells on mm and sub-mm grain-size poly-Ge substrates.

## GaAs Growth on Poly-Ge

High-quality growth of GaAs-AlGaAs layers across the various crystalline orientations of a polycrystalline Ge substrate is critical to obtaining high-performance GaAs solar cells. First of all, the finish of poly-Ge substrate prior to the MOCVD growth of the GaAs cell needed attention [7]. Further, polishing defects such as dents/ledges/steps have been found to be more detrimental, compared to grain-boundaries in the substrate, for GaAs solar cell performance. Although the $V_{oc}$ of the cells do improve with larger grain-sizes, the macro-polishing defects such as dents, ledges and steps in the cell active-area tend to reduce the $V_{oc}$ much more strongly than grain-boundaries. We believe this to be a result of presence of higher electric-fields (in the $p^+$n junction) at the vicinity of these steps and ledges, causing excessive leakage-currents.

In addition, we have optimized the MOCVD growth process on large-grain and small-grain poly-Ge substrates, recognizing the need for maintaining a higher concentration of As-growth species on the growth-surface to promote a two-dimensional layered growth [7]. Optimization studies of the minority-carrier properties of GaAs layers on poly-Ge substrates have revealed that lifetime-spread across various grains can be reduced through the use of lower doping for the $Al_{0.8}Ga_{0.2}As$ confinement layers [8,9].

## GaAs Solar Cell Device Optimization on Poly-Ge Substrates

Shown in Fig. 2, is a schematic of a GaAs solar cell device structure that we have developed [4] for use on polycrystalline Ge substrates. Note the undoped spacer between the base and the emitter, which has been shown to reduce the dark current and improve the cell $V_{oc}$ and fill-factor [4].

Electron-beam-induced-current (EBIC) scan on $p^+$-n GaAs junctions on poly-Ge substrates, with a spacer of 0.1 to 0.3 µm, have indicated hole diffusion lengths

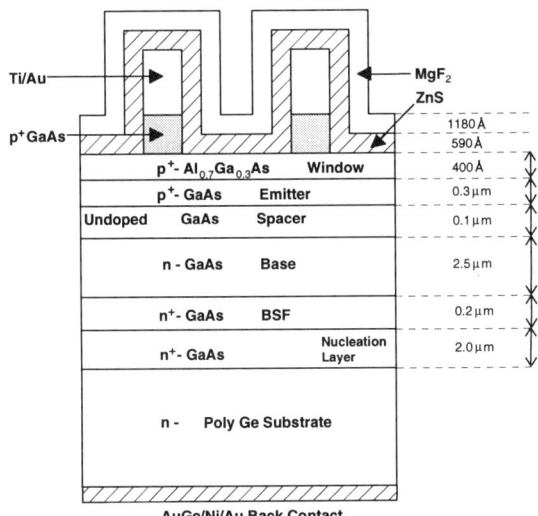

**FIGURE 2.** Schematic of a $p^+n$ GaAs solar cell device structure on poly-Ge substrate.

~1.2 µm in the vicinity of the depletion layer. We have also observed longer diffusion lengths away from the junction-region of the cell; this is consistent with the use of a spacer to reduce the doping near the depletion layer. We expect the regions away from the spacer to benefit from the formation of $n/n^+$ minority-carrier mirrors, leading to an enhancement in the minority-carrier diffusion lengths.

## Mechanism of Dark-Current Reduction With Spacer

As mentioned above, the use of a thin undoped spacer at the depletion-layer of the $p^+$-n junction tends to reduce the dark-current in GaAs cells on poly-Ge substrates. We have shown in Figure 3, the effect of an undoped spacer on dark-current reduction in GaAs cells on poly-Ge. For comparison, the I-V data of GaAs $p^+$-n junctions grown on single-crystal Ge, with and without the spacer, are also shown in this figure. We see no effect of the spacer on GaAs cells on single-crystal substrates, while there is a 40-fold reduction in the saturation dark current in GaAs cells with the spacer on poly-Ge substrates. The reduction in dark current with the use of spacer is apparently not unique to $p^+$-n GaAs junctions. We have also observed a similar effect with an undoped spacer on dark-current reduction in $p^+$-n $GaInP_2$ junctions on poly-Ge [9].

The reduction in dark current and the improvement in cell $V_{oc}$ are believed to be associated with the reduction of tunneling currents near the depletion-layer of the $p^+$-n junction in polycrystalline materials. We have conducted an experimental study of dark currents in $p^+$-n GaAs junctions as a function of temperature (77K to 288K) to support this hypothesis. The temperature dependence of saturation dark currents in $p^+$-n GaAs junctions, with and without the undoped spacer, are shown in Figure 4. We observe that the junction, without the spacer, shows a strong reduction with temperature. The activation energy for the saturation dark current with temperature is ~0.07 eV. We also note the apparent convergence of the two curves near zero Kelvin. We have seen a similar behavior of dark current variation with temperature and an apparent convergence of the two curves, with and without the spacer, for $p^+$-n $GaInP_2$ junctions as well. The activation energy for the saturation dark current with temperature for the $p^+$-n $GaInP_2$ junctions is ~ 0.04 eV.

The band diagram of a p-n junction near the vicinity of a grain boundary, due to n-type dopant accumulation at grain-boundaries as noted earlier, is expected to resemble

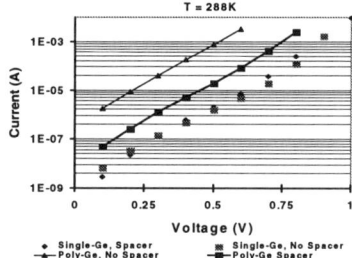

**Figure 3.**  Effect of an undoped spacer on dark current reduction in GaAs cells on single crystal Ge and poly-Ge substrates. Note all diodes have an area of 1cm$^2$.

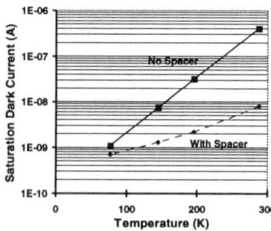

**Figure 4.**  Temperature dependence of saturation dark current in GaAs cells on poly-Ge, with and without the undoped spacer.

that of a p-n$^+$ junction. This would lead to electron tunneling (dark) current from the p-emitter to the n$^+$-base, depending on the empty states below the Fermi-energy ($E_f$) in the

$n^+$-region. Thus the dark-current activation energy will be related to the amount of degeneracy in the $n^+$-region. The measured activation energies, in the range of 0.07 eV to 0.04 eV for the GaAs and $GaInP_2$ junctions, respectively, are typical for the degeneracy.

Based on this model for tunneling dark-current, we can also explain the apparent convergence of the two curves near zero Kelvin in Fig. 4. We observe that the occupational probability of electrons below $E_f$ is given by the Fermi-Dirac function [10]. As we approach T~0, we expect no empty states below $E_f$ for the electrons to tunnel from the p to $n^+$-region. Thus the tunneling dark currents would go to zero and we expect no difference between the cells with and without the spacer. In essence, we observe that the key to reducing dark-currents in $p^+$-n GaAs and $GaInP_2$ solar cells on polycrystalline-Ge substrates may be to avoid carrier tunneling into the heavily-doped regions near the grain boundaries.

## GaAs Cell Structure Optimization

We have indicated in our recent effort [9] that the $p^+$-n GaAs cells on poly-Ge typically have a $J_{sc}$ of ~22 $mA/cm^2$ under AM1.5 illumination and that further improvement in the $J_{sc}$ (towards a near-ideal value such as ~28 $mA/cm^2$ reported in GaAs cells in Ref. 1) can be obtained mainly through improved blue-response. Typical spectral response data on the GaAs cells have indicated a roll-off in the blue-wavelength region. To improve the blue-response, we have investigated the effect of reducing the thickness of the $Al_xGa_{1-x}As$ window; this was considered necessary as a significant absorption of the high-energy photons can occur even in an indirect-gap window, not contributing to $J_{sc}$. We reduced the thickness of the window from ~450 Å to ~300 Å. The reduction in window thickness, in addition to keeping x in $Al_xGa_{1-x}As$ window around ~0.75, led to an improved $J_{sc}$ of ~24.5 $mA/cm^2$. Further reduction in window thickness from 300 Å to ~220 Å apparently does not improve the $J_{sc}$ any more. Thus, a 300 Å-thick window appears optimum for a window doping level of ~1E19 $cm^{-3}$.

In order to improve the $J_{sc}$, we considered the use of a thicker emitter as our other characterization studies had indicated ample electron diffusion lengths in the $p^+$-GaAs emitter. The cell $J_{sc}$ improved from ~24.5 $mA/cm^2$ to ~25.5 $mA/cm^2$, when the AR-coat was also optimized for the thicker emitter. In addition, we observed that the $J_{sc}$ can improve to as much as 26.3 $mA/cm^2$ by using a smaller emitter grid-metallization coverage.

Towards the improvement of $V_{oc}$ values, we indicated above that an undoped spacer is required to reduce the contribution of tunneling current to the dark current in GaAs cells on poly-Ge. However, it was necessary to consider the optimum spacer requirement. We investigated the effect of varying the spacer thickness on cell $V_{oc}$ values, for approximately similar grain-size in the cells. A typical set of data is shown in Table 1. We believe, as long as the spacer is thick enough to avoid tunneling of carriers between the heavily-doped emitter and the heavily-doped base regions formed near the grain-boundaries, a thinner spacer could be advantageous for lowering the dark current. In high-bandgap materials, like GaAs, the dark-current density ($J_{dark}$) is expected to be dominated by generation in the depletion-layer in comparison to the diffusion-limited dark-current generation from the quasi-neutral base region. The generation dark-current density from the depletion layer can be written as

$$J_{dark} = (q\, n_i\, W) / \tau_{sc}$$

where q is the electronic charge, $n_i$ is the intrinsic carrier concentration, W is the depletion-layer width and $\tau_{sc}$ is the effective space-charge lifetime. A thinner spacer would reduce the depletion layer width in the base region of the p$^+$-n junction and, therefore, will reduce $J_{dark}$; hence, the $V_{oc}$ is likely to improve with a smaller spacer. We also considered the effect of the emitter thickness and the amount of emitter-grid metallization coverage on the $V_{oc}$ of cells. Typical data shown in Table 2, suggests that a significant component to dark-current generation may be from the emitter-contact regions or from the $Al_xGa_{1-x}As$ window-GaAs cap heterointerface. We need to look into this aspect for further improvement in $V_{oc}$.

**Table 1.** Effect of spacer thickness on $V_{oc}$ of GaAs cells on mm-grain size poly-Ge substrates, with nearly all cell structure parameters held constant. All cell areas are ~4 cm$^2$.

| Cell # | Spacer Thickness | $V_{oc}$ (V) |
|---|---|---|
| 1-2766 | 0.22 | 0.90 |
| 1-2763 | 0.14 | 0.96 |
| 1-2782 | 0.14 | 0.96 |
| 1-2786 | 0.10 | 1.00 |

**Table 2.** Effect of emitter thickness on $V_{oc}$ of GaAs cells on mm-grain size poly-Ge substrates, with all other cell structure parameters held constant. All cell areas are ~4 cm$^2$.

| Cell # | Emitter Thickness | Grid Coverage (%) | $V_{oc}$ (Volt) |
|---|---|---|---|
| 1-2786 | 0.14 | 9 | 1.00 |
| 1-2804 | 0.22 | 9 | 1.012 |
| 1-2809 | 0.28 | 9 | 1.01 |
| 1-2840 | 0.28 | 5 | 1.03 |

### Efficiency of GaAs Solar cells on Poly-Ge Substrates

The material and device optimization studies have enabled us to achieve significant improvement in large-area GaAs cell efficiencies, under AM1.5 conditions, on mm and sub-mm grain-size poly-Ge substrates. The typical grain structure observed in the cell active-area is shown in Figure 5; this indicates that the size of various grains and other crystalline structures in the substrates are about 400 µm, about the spacing between the grid fingers in the cells.

400 µm

**Figure 5.** Typical grain-boundary structures on sub-mm grain poly-Ge substrate; note the spacing between adjacent grid fingers is ~ 400 microns.

The data shown in Table 3 indicates the progress that we have made at various stages in the last two years of the NREL-sponsored effort. We can note the recent progress in cell results are in spite of smaller, sub-mm grain-sizes in the starting poly-Ge been grown on higher-resistivity (~ 0.5 Ohm-cm) poly-Ge substrates which result in substrates. Also, we observe that the recent state-of-the-art GaAs cells have lower fill-factor for the cells. With the use of lower resistivity substrates, we hope to achieve higher cell efficiencies. The NREL-verified cell efficiency data on the 18.2% cell indicated in Table 3, is shown in Figure 6. The more recent cell efficiency measurements are in the process of being verified at NREL. The data in Table 3 also point to a direction that large-area GaAs cell efficiencies in excess of 20% are achievable on mm and sub-mm grain-size poly-Ge substrates. It is noteworthy that in a small-area, 21%-efficient cell, a $V_{oc}$ of as much as 1.04 Volt and a $J_{sc}$ ~ 27 mA/cm$^2$ have been observed, thereby approaching some of the best single-junction GaAs cell results on single-crystal GaAs substrates.

Table 3.    Progress of GaAs cell efficiency on poly-Ge substrates.

|  | Grain-Size (cm) | Cell Area (cm$^2$) | AM1.5 Eff. (%) | Ref. |
|---|---|---|---|---|
| First Demonstration | 0.5 | 1 | 15.8 | [4] |
| Phase 1 | 0.5 | 4 | 16.6 | [8] |
| Phase 2 Begin | 0.5 | 4 | 18.2 | [9] |
| Phase 2 End | 0.1 | 4 | 18.7 | |
| Phase 3 (Begin) | 0.05-0.1 | 4 | ~19 | |
| Current state-of-art | 0.05-0.1 | 0.25 | ~21 | |

**Summary**

Material and device-structure considerations that are important for the fabrication of high-efficiency GaAs solar cells on polycrystalline Ge substrates are discussed. An experimental study of dark currents in GaAs p$^+$-n junctions, with and without the spacer, as a function of temperature indicates that the spacer probably reduces the contribution of tunneling dark-currents associated with electron tunneling from the p$^+$-emitter to the heavily-doped n$^+$-regions (of the n-base) formed near the grain-boundaries. Device-structure optimization studies have led us to achieve a open-circuit voltage exceeding 1 Volt and an AM1.5 efficiency of 19% for a 4-cm$^2$-area GaAs cell on sub-mm grain-size poly-Ge. We have also observed efficiency near ~21% for a 0.25-cm$^2$-area cell on similar small-grain poly-Ge substrate.

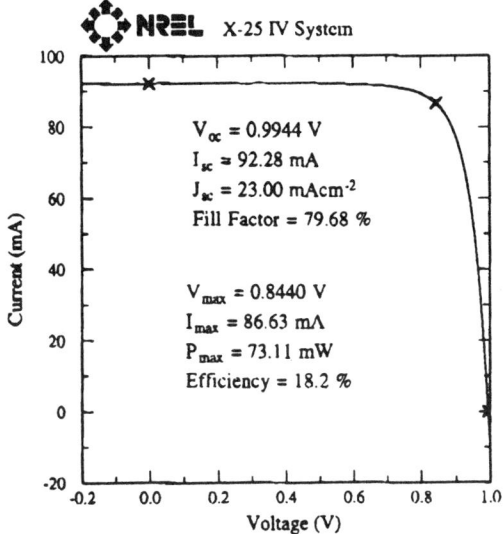

**Figure 6**   I-V characteristic of a NREL-verified, 18.2%-efficient, 4-cm$^2$-area GaAs cell on poly-Ge with a $V_{oc}$ of ~1Volt.

## Acknowledgments

This work has been performed under Subcontract No. YAL-3-1-3357-03 from the National Renewable Energy Laboratory. Dr. John Benner and Dr. Robert McConnell are the technical monitors for the program. We acknowledge valuable discussions with Dr. Benner.

## References

1) S.R. Kurtz, J.M. Olson, and A. Kibbler, Proc. of 21st IEEE Photovoltaic Specialists Conf. (IEEE Press, NY, 1990) p.138.

2) J.M. Olson, S.R. Kurtz, A.G. Kibbler, and P. Faine, Proc. of the 21st IEEE

Photovoltaic Specialists Conf. (IEEE Press, NY, 1990) p.24.

3) M. Green and K. Emery, Progress in Photovoltaics: Research and Applications, Vol 2, 231-234 (1994).

4) R. Venkatasubramanian, M.L. Timmons, P.R. Sharps, and J.A. Hutchby, Proc. of the 23rd IEEE Photovoltaic Specialists Conf. (IEEE Press, NY, 1993) p.691.

5) K.P. Pande, D. H. Reep, S.K. Shastry, A.S. Weiner, J.M. Borrego and S.K. Ghandhi, IEEE Trans. on Electron Devices, Vol. ED-27, 635 (1980).

6) S.S. Chu, T.L. Chu, and Y.T. Lee, IEEE Trans. on Electron Devices, Vol. ED-27, 640 (1980).

7) R. Venkatasubramanian, Unpublished Results.

8) R. Venkatasubramanian, B. O'Quinn, J. Hills, D. Malta, M.L. Timmons, J.A. Hutchby, R. Ahrenkiel and B.M. Keyes, in Proceedings of the 12th NREL Photovoltaic Program Review, Ed. by H.S. Ullal and C.E. Witt, AIP Conf. Proc. No. 353 (AIP, Denver, CO) 1995.

9) R. Venkatasubramanian, B. O'Quinn, J. Hills, P.R. Sharps, M.L. Timmons, J.A. Hutchby, H. Field, R. Ahrenkiel and B.M. Keyes, Proc. of 25th IEEE PVSC, 1996, IEEE, NY, pp. 31-36.

10) B.G. Steetman, *Solid-State Electronic Devices*, Prentice Hall, 71 (1980).

# The "Micromorph" cell: a New Way to High-Efficiency-Low-Temperature Crystalline Silicon Thin-Film Cell Manufacturing ?

H. Keppner, U. Kroll, P. Torres, J. Meier, R. Platz, D. Fischer,
N. Beck, S. Dubail, J. A. Anna Selvan, N. Pellaton Vaucher,
M. Goerlitzer, Y. Ziegler, R. Tscharner, Ch. Hof, M. Goetz, P. Pernet,
N. Wyrsch, J. Vuille, J. Cuperus, and A. Shah
J. Pohl*

Institut de Microtechnique, A.-L. Breguet 2, Université de Neuchâtel, CH-2000 Neuchâtel, Switzerland; *University of Konstanz, D-78434 Konstanz, Germany

**Abstract.** Hydrogenated microcrystalline Silicon (μc-Si:H) produced by the VHF-GD (Very High Frequency Glow Discharge) process can be considered to be a new base material for thin-film crystalline silicon solar cells. The most striking feature of such cells, in contrast to conventional amorphous silicon technology, is their stability under light-soaking. With respect to crystalline silicon technology, their most striking advantage is their low process temperature (220°C). The so called "micromorph" cell contains such a μc-Si:H based cell as bottom cell, whereas the top-cell consists of amorphous silicon. A stable efficiency of 10.7% (confirmed by ISE Freiburg) is reported in this paper.

At present, all solar cell concepts based on thin-film crystalline silicon have a common problem to overcome: namely, too long manufacturing times. In order to help in solving this problem for the particular case of plasma-deposited μc-Si:H, results on combined argon /hydrogen dilution of the feedgas (silane) are presented. It is shown that rates as high as 9.4 Å/s can be obtained; furthermore, a first solar cell deposited with 8.7 Å/s resulted in an efficiency of 3.1%.

## 1. INTRODUCTION

In recent work (1), hydrogenated microcrystalline silicon (μc-Si:H) has been successfully introduced as active semiconductor in entirely μc-Si:H p-i-n solar cells. The present paper will first review the development from the first "mid-gap" layers to entirely microcrystalline solar cells and the extension to the so-called "micromorph" tandem cell concept. Thereby these micromorph cells have already achieved, after 1000 hours light-soaking a stabilized efficiency of 10.7% (confirmed by ISE Freiburg, Germany).

The Very High Frequency Glow-Discharge (VHF-GD) technique has sofar played a key role in obtaining "solar grade" μc-Si:H material. With this technique μc-Si:H can be deposited at temperatures as low as 220°C, a fact that permits the use of low-cost substrates such as plastic, glass, aluminium etc., in the field of crystalline silicon. One may generally argue that due to that further cost reduction for future solar cell manufacturing can be achieved.

At the present state of development, μc-Si:H and micromorph solar cells have two main problems to overcome: first, their stable efficiency has to be further increased; second, the deposition rate of high-quality material has to be increased from presently 1-2 Å/s to more than 10 Å/s. It will be shown in the second part of this paper that by diluting silane in argon and hydrogen, deposition rates up to 9.4 Å/s can basically be obtained. The use of argon/hydrogen dilution increases the parameter space for optimization of the material quality. Thus, the full optimization of deposition will take more R&D time and is not yet completed. A first μc-Si:H solar cell deposited at 8.7 Å/s with an efficiency of 3.1% has already been achieved.

## 2. SOLAR CELLS BASED ON MICROCRYSTALLINE SILICON: REVIEW OF PAST WORK

Hydrogenated microcrystalline silicon is generally obtained in a plasma using silane strongly diluted with hydrogen (2). Apart from few isolated papers (3-5), μc-Si:H was in the past generally not seriously taken into account as a candidate for becoming an active semiconductor for solar cells. The novel inputs that have lead to the successful use of μc-Si:H in solar cells as reported in recent work (1, 6-9), can be reduced to two main characteristics:

1. The absorber material in μc-Si:H solar cell must be "mid-gap" material.
2. The material used must have a low grain-boundary trap density and a low defect density within the grains.

*"Mid-gap" material*

The deposition of "mid gap" μc-Si:H is not straightforward; as-deposited intrinsic μc-Si:H turns out to be slightly n-type; this is basically due to oxygen impurities, as reported by Veprek et al. (10). Two possibilities have in the past been used to overcome this problem:

**1. The compensation technique:** Small traces of diborane compensate the n-type behaviour (7). The compensation technique leads to surprisingly good results for layers and solar cells but it turns out to be technologically delicate.

**2. The purifying technique:** A gas purifier (oxygen getter) installed close to the reactor removes oxygen-containing impurities (8). Thereby, low dark conductivities and μc-Si:H single-junction solar cells with 7.7% efficiency could be obtained (6, 11). The purifying technique can completely replace the compensation technique.

## Low grain-boundary trap density

It can be assumed, that μc-Si:H deposited by VHF-GD has a very high fraction of crystalline silicon. The amorphous fraction is thereby essentially reduced to the grain boundaries. Using the grain boundary trapping model developed by Seto et al. (12), Meier et al. (13) estimated a grain-boundary trap density that is approximately 100 times lower than that what has been reported for polysilicon deposited by CVD, with roughly the same grain size (12). Infrared spectroscopy performed by Kroll et al. (14) showed that bonded hydrogen in VHF-GD μc-Si:H is preferentially located at crystallite surfaces. Thus, one may tentatively conclude that an efficient grain-boundary passivation due to "excess" atomic hydrogen takes place.

VHF-GD deposited μc-Si:H shows a pronounced absorption edge; thereby a gap-energy as low as 1 eV could be deduced (8). As a surprising fact, the overall absorption within the **whole** near-infrared range is higher than that obtained with monocystalline silicon. This particular feature makes μc-Si:H an attractive candidate for thin-film silicon solar cells.

## Selected properties of entirely microcrystalline cells

The most striking features of entirely microcrystalline solar cells as produced by VHF-GD are:
1. Complete stability under intense, long-time light-soaking (8).
2. As can be predicted by the absorption spectra, entirely microcrystalline solar cells show a spectral response that is strongly extended towards the infrared, when compared with amorphous silicon solar cells.
3. The manufacturing technique of microcrystalline silicon is compatible to that of amorphous silicon, i.e. the process temperatures do not exceed 250 °C.

## The "Micromorph" Concept

The combination of an entirely microcrystalline solar cell with an amorphous silicon solar cell results in a "true" tandem cell with **different** gap-energies for the top (1.76 eV) and the bottom cell (1.0 eV). This type of tandem cell has been called by the authors the "micromorph" cell. The stabilized efficiencies obtained sofar are summarized in Tab. 1:

**TABLE 1.** stabilized parameters of entirely μc-Si:H and of micromorph cells.

| cell type | η [%] | $I_{sc}$ | $V_{oc}$ | FF | area [cm$^2$] | light-soaking procedure |
|---|---|---|---|---|---|---|
| μc-Si:H p-i-n | 7.7 | 25.3 | 0.448 | 67.9 | 0.33 | 10 suns, 48°C, 400 h |
| Micromorph | 10.0* | 11.3 | 1.34 | 65.5 | 1.22 | AM 1.5; 50°C 1000 h |
| Micromorph | 10.7* | 11.9 | 1.34 | 66.7 | 0.13 | AM 1.5; 50°C 1000 h |

\* Confirmed by the Fraunhofer ISE PV- Kalibrationslabor

*Limits and potential for further improvements*

The $V_{OC}$-values obtained sofar for entirely microcrystalline cells (micromorph bottom cell) are substantially lower (448 mV) than the $V_{OC}$-values of typical a-Si:H (top) cells (890mV). Due to this fact, the top cell produces, within the micromorph tandem cell, two thirds of the total power of the cell (9). Because of this, the stability of the top cell will remain a key feature as long as the open circuit voltage of the bottom cell remains limited to such low values: One may conclude that a further increase of the $V_{OC}$-value of the bottom cell **alone** is the most important individual factor for further improvement in the stable efficiency of the micromorph cell.

The most striking limitation in the future industrial potential of any kind of thin-film crystalline silicon solar cell is, at present, the too long manufacturing times (recrystallization and/or deposition rates). Looking at the deposition rates of µc-Si:H, the VHF-GD process in itself brings sofar at least some improvement, as reported by Finger et al. (15). Attempts at obtaining yet higher deposition rates are described by Torres et al. (16). In this paper we present an alternative approach using argon plus hydrogen dilution of silane.

## 3. USE OF ARGON AS DILUTION GAS: GENERAL CONSIDERATIONS

Argon dilution and mixtures of Ar, $H_2$, and silane have often been reported in literature for a-Si:H deposition with the attempt to obtain more stable material (e.g. (17)). As a further beneficial effect, Kroll et al. (18, 19) report on significant silane savings using argon dilution. In another work, an increase of the deposition rate of a-Si:H is reported by Hautala et al (20). Sansonnens et al. (21) report on molecular quenching via metastables that leads to enhanced dissociation of silane. In all these cases, the authors used merely $Ar/SiH_4$ mixtures without hydrogen. Other authors report on properties of µc-S:H layers obtained using strongly argon-diluted silane plasmas (22). Further results on the increase of the deposition rate of µc-Si:H are reported by Imajyo et al. (23): they used a DC plasma gun for their experiments at high DC discharge currents. In this last case a mixture of Ar, $H_2$ and $SiH_4$ was applied.

In our work we have varied the composition of the diluting gases argon and hydrogen; thereby the VHF-GD technique was applied. In doing this, we emphasize on two aspects:

1. Reduction of the peak energy of ions that impinge on the growing layers; this constitutes, thus, an attempt to further improve the material quality,
2. increase of the deposition rate of VHF-GD µc-Si:H.

The full potential of µc-Si:H silicon to play a role in PV energy conversion gives still rise to speculations. One may assume that the plasma control parameters, that can lead to desirable growth mechanisms, are not fully optimized as yet. Looking at the "**ideal**" µc-Si:H layers with respect to an application in solar cells the following goals can be formulated:

a) The grains should be as large as possible (low grain boundary density).
b) The grains should have low defect densities (lattice defects and impurities).
c) Low grain-boundary trap density (highly passivated grain-boundaries).
d) No grain-boundaries perpendicular to the photocurrent.

The points a) and b) can be considered interdependent in the following way: High particle energies from the plasma create defects in the growing grains that lead, if the lattice damage is pronounced, to loss in the "information" as required for continuous "epitaxial-type" growth within the grains. A reduced grain-size with a higher grain-boundary density has then to be expected.

We should therefore look at the peak energies of particles to be found typically in a plasma used for µc-Si:H deposition. Due to the strong dilution of silane in hydrogen, the mean free path of silicon-based radicals (Si, SiH, $SiH_2$, $SiH_3$ etc.) is in the range of centimetres; for hydrogen atoms tenths of millimetres are typical values. Silicon-type radicals can therefore **not** contribute to thermalization of silicon type ions ($Si^+$, $SiH^+$, $SiH_2^+$, $SiH_3^+$, etc.) that are accelerated across the sheath. Now, for a sheath potential higher than the threshold energy $E_{th}$ = 16 eV, defect formation occurs in the growing layer, according to Veprek et al. (24). Values as high as 16 eV are easily obtained in a 13.56 MHz discharge, whereas the ions in a 70 MHz discharge have typically maximum ion energies that are just around this value (25, 26). Note, that the collisions between Si-type ions and hydrogen atoms randomizes only several ppm of the energy of the ion. In order to thermalize efficiently the high peak-energies of Si-type radicals one can therefore envisage diluting the plasma with argon as an inert gas; note, that argon has about the same mass as silicon-based radicals.

The points c) - d) deal with the aspect of increasing diffusion- (or drift) lengths for electronic transport; these lengths are affected either by defects in the grains or by trap-assisted recombination at grain-boundaries. Furthermore, charged traps in grain-boundaries have the tendency to screen locally the built-in field; this leads again to enhanced recombination and to a loss in the efficiency of the cell (27).

Looking at our demands (a-d) pronounced columnar grown i.e. crystalline silicon columns throughout the thickness of the cell would be ideal. Meier et al. (8) have shown, using SEM and TEM micrographs and X-ray measurements, that VHF-GD µc-Si:H layers possess, indeed, a pronounced columnar structure in the (220) direction. At present there are no experimental tools at hand that allows one to quantify the characteristics c)-d) in a microscopic way giving thereby the possibility to "tune" the deposition conditions for obtaining an "ideal" µc-Si:H as sketched above.

# 4. EXPERIMENTAL PROCEDURES USED FOR ARGON /HYDROGEN DILUTION

All the experiments were undertaken using the VHF-GD technique at 70 MHz. Three series were deposited, as shown in Table 2.

For the M-series (**M**ixing-series), the mixture of the diluting gases Ar and $H_2$ was varied at constant silane flow. (6 sccm). The sum of flows of the diluting gases (Ar + $H_2$) was kept at 150 sccm.

For the D-series (**D**ilution-series), the mixture of the two diluting gases was kept constant, at 90 sccm $H_2$ and 60 sccm Ar, whereas the silane flow was varied.

For the P-series the **P**ower of the plasma was varied for a fixed silane flow of 8.4 sccm in 90 sccm hydrogen and 60 sccm Argon.

**TABLE 2.** Strategy of $H_2$ / Ar dilution experiments for deposition of μc-Si:H.

| Serie | $\Phi(SiH_4)$ [sccm] | $\Phi(H_2)$ [sccm] | $\Phi(Ar)$ [sccm] | Power [W] | Pressure [mbar] |
|---|---|---|---|---|---|
| M | 6 | 140 - 0 | 10 - 150 | 30 | 0.8 mbar |
| D | 6-12 | 90 | 60 | 40 | 0.9 mbar |
| P | 8.4 | 90 | 60 | 10 - 70 | 0.9 mbar |

For our Ar / $H_2$ dilution experiments, looking at the above goals a)-d), we tried to quantify the columnar growth of the μc-Si:H layers using X-Ray diffraction.

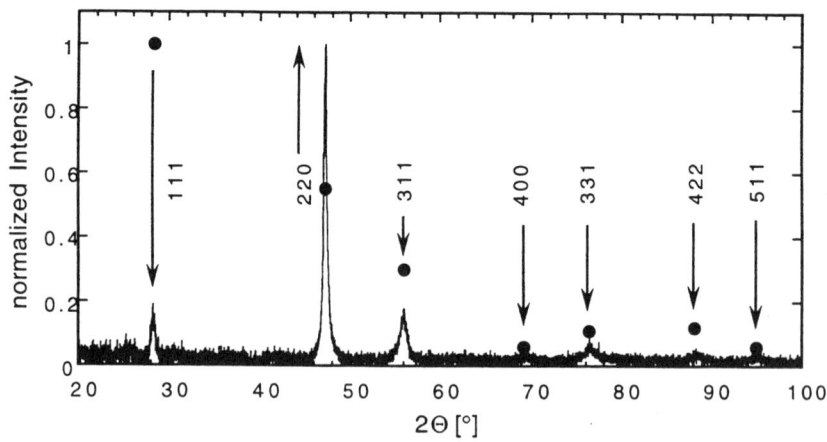

**FIGURE 1.** Typical X-Ray diffraction pattern of a μc-Si:H thin-film grown by VHF-GD deposition using Ar/$H_2$ dilution. The points refer to the powder sample according to the JCPDS (28).

To this end we suggest using a formula (1) that can serve to roughly indicate the preferential columnar growth of μc-Si:H layers; this formula is based on corresponding X-Ray diffraction parameters as found in the JCPDS (28) chart for isotropic powder samples.

$$F =: \frac{A(220)}{A(111)} \cdot \frac{A|_{powder}(111)}{A|_{powder}(220)} = \frac{A(220)}{A(111)} \cdot 1.81 \quad [1]$$

Here A(iii) represents the area under the respective diffraction; the values of the isotropic powder sample $A_{powder}$(iii) can be obtained from the JCPDS chart. A typical X-Ray diffraction pattern obtained for µc-Si:H is thereby given in Fig. 1.

From X-Ray measurements, the average grain-size was determined using Sherrer's formula. Doing this, one basically assumes that the crystallites have a spherical shape. As we know, this is definitively not the case, as the SEM, TEM and X-Ray diffraction patterns indeed confirm. One should consider the growth not to be spherical but in the form of a "cigar". Therefore, in addition to using the apparent grain-size as calculated by Sherrer's equation, we propose using the factor F (equation (1)), to roughly characterize structural properties.

## 5. RESULTS AND DISCUSSION.
### 5.1 Results on Layers.

The X-Ray diffraction pattern of the various samples of series M, D, P are similar to the one shown in Fig.1. As already observed in previous work (8) samples grown by $H_2$ dilution also show a strong anisotropical columnar growth in the (220) direction.

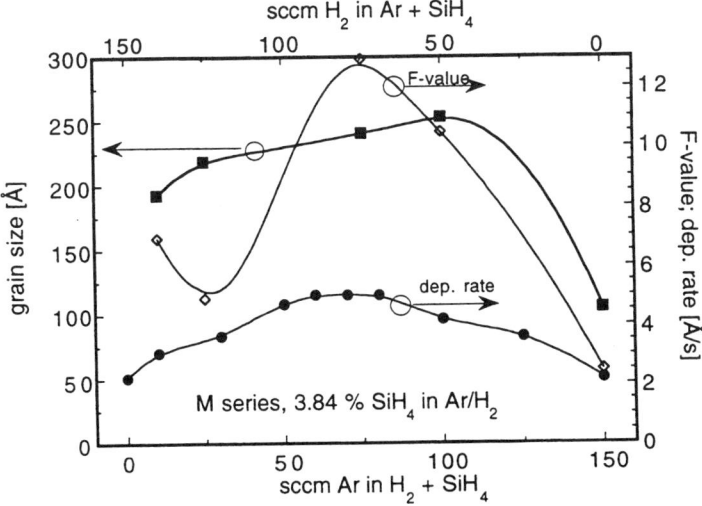

**FIGURE. 2.** Parameters obtained from X-ray diffraction patterns of the M-series and the deposition rate from the same series. The F-values were calculated from equation (1). The apparent "grain-size" was calculated from the (220) diffraction signal using Scherrer's equation.

All samples of the M-series were found to be microcristalline (Fig. 2). As the most striking result, a pronounced increase in deposition rate up to 5 Å/s was obtained for about equal parts of Ar and $H_2$ as dilution gases. The F-values show here a maximum: this leads us to assume that we have the most pronounced columnar growth under these deposition conditions. Looking at the "apparent grain-size" according to Scherrer's formula, a continuous increase can be found when increasing Ar dilution in the plasma. We assume that enhanced thermalization of high energy particles occur and that this effect allows crystallites to obtain the highest "apparent grain-size" values in this study (250 Å). For the D-series we applied the same dilution-values as those needed for obtaining the highest deposition rate within the M-series; however, we applied a slightly increased pressure (0.9 mbar). Fig 3. shows the results.

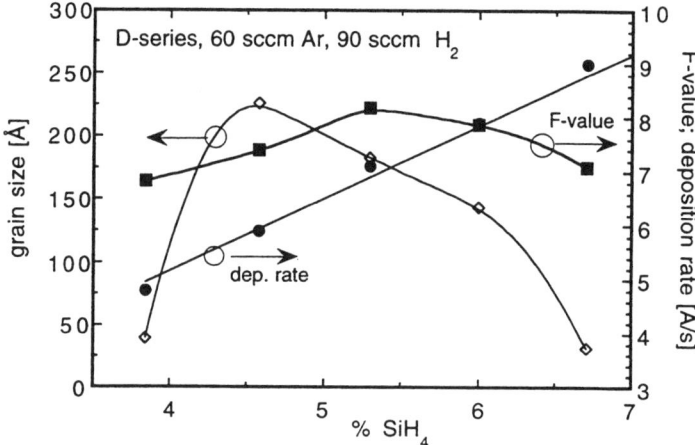

**FIGURE 3.** D-series: apparent "grain-size" as well as F-value and deposition rate in function of the dilution; thereby, the composition and the flow of the diluting gases was kept constant (90 sccm $H_2$, 60 sccm Ar).

All samples within the D-series were found to be microcrystalline. The F-values are roughly constant for all dilutions, whereas the "apparent average grain-size" (from Scherrer's equation) has a maximum at 4.5% silane. Adding more silane leads to a continuous decrease of grain-size. The deposition rate increases linearly up to the highest value of 9.4 Å/s obtained in this study.

In the P-series, the samples deposited at 10 and 20 W resulted in amorphous silicon. Surprisingly, the deposition rate of µc-Si:H obtained in this series has its highest value at 40 W and thereafter drops constantly for increasing power. The "apparent grain-size" shows a similar tendency. We can assume that a high value of ion peak-energy in the plasma creates defects which gives rise to an overall reduction in the growth of the crystallites. The anisotropy factor shows a sharp maximum around 50 W i.e. for larger power values than those studied in series M and D.

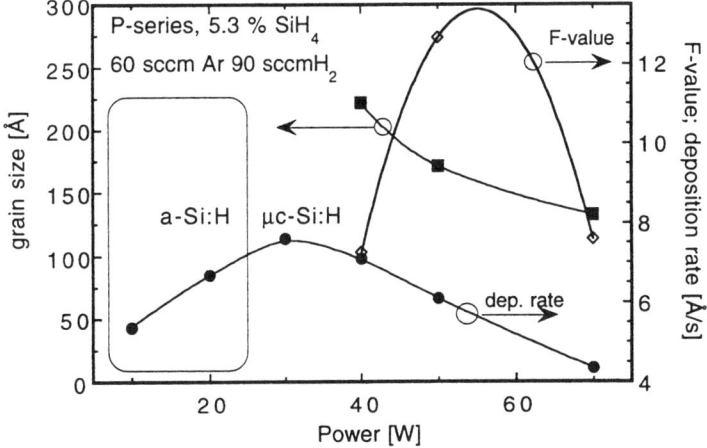

**FIGURE 4.** "Apparent grain-size", F-value and deposition rate in function of VHF- power. Note, that for reduced power, the layers become amorphous.

## 5.2. Results on Solar Cells.

In a next step the Ar / $H_2$ diluted layers that were deposited at high deposition rates, were incorporated in a solar cell with a i-layer thickness of 5.5 µm.

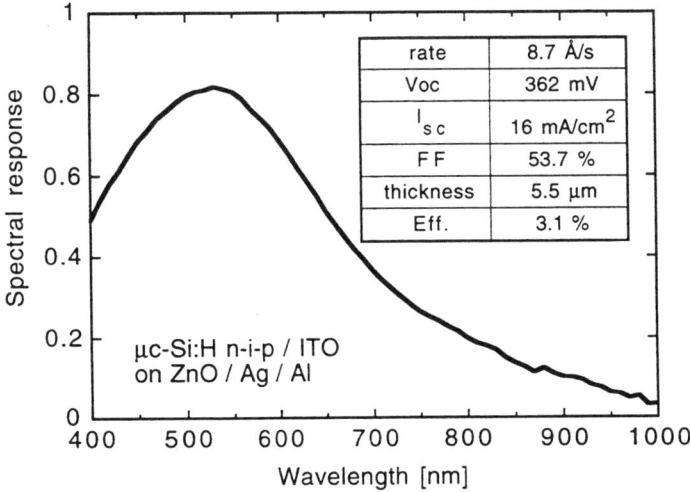

**FIGURE 5.** Spectral response and performance parameters of a first µc-Si:H n-i-p solar cell deposited at 8.7 Å/s.

Due to the strong tendency to peel off, an aluminium substrate and a n-i-p structure had to be used. Fig 5. shows the spectral response values and the solar cell parameters of this cell.

In contrast to the above individual layers of series M, D, and P, the entire cell was deposited using the purifying technique. It was found in a parallel study that this is crucial, because Ar-diluted µc-Si:H layers have the tendency to incorporate oxygen in a more pronounced manner than when using only $H_2$-dilution, according to Keppner et al. (29); a similar effect was already observed for a-Si:H layers by Kroll et al. (19). The cell presented in Fig. 5 shows a reduced spectral response in the near infrared; this is in agreement with PDS results (29) showing reduced infrared absorption around 1.2 eV. At this stage of the work it is not yet clear whether the reduced near infrared absorption is typical for layers produced at high rates with Ar-dilution or whether it is only due to the preparation techniques not being fully optimized.

## 6. CONCLUSIONS

A silicon-based tandem solar cell called the "micromorph" solar cell containing a microcrystalline bottom cell and an amorphous top cell has obtained a stable efficiency of more than 10 %. Because of the relatively low absorption coefficient in the near infrared, as obtaines in the µc-Si:H bottom cell, relatively high i-layer thicknesses are needed and the rather long deposition times remain a problem to be overcome. Argon / hydrogen dilution of silane, as the method was used here in combination with the VHF-GD process, basically provides the possibillity to achieve deposition rates up to 10 Å/s. A first trial check with such high-rate material already result e.g. in a cell with 3.15 % efficiency. Argon / hydrogen dilution experiments show, furthermore, that the columnar growth structure and the average grain-size of the crystallites can, indeed, be controlled within a certain range. Thanks to the addition of argon as dilution gas, the parameter space that allows one to "tune" µc-Si:H w.r.t solar cell requierements is strongly expanded.

## ACKNOWLEDGEMENTS

This work was supported by the Swiss Federal Energy Department under Research Grant EF-REN (93)032

## REFERENCES

1. J. Meier, R. Flückiger, H. Keppner, A. Shah, *Appl. Phys. Lett.*, Vol. 65 (7), pp. 860-862, 1994.
2. S. Veprek and V. Marecek *Solid State Electronics* Nr. 11, p. 683.
3. C. Wang and G. Lucowsky, *Proc. 21st IEEE Photovoltaic Specialists Conference, Orlando* 1990, Vol. 2, pp 1614-1618.

4. R. Flückiger, J. Meier, H. Keppner, U. Kroll, A. Shah, O. Greim, M. Morris, J. Pohl, P. Hapke, R. Carius, *Proceedings of the 11th EC Photovoltaic Solar Energy Conference, Montreux*, 1992, p. 617.
5. M. Faraj, S. Gokhale, S. M. Choudhari, and M. G. Takwale, *Appl. Phys. Lett.* 60, p. 3289,1992
6. J. Meier, S. Dubail, D. Fischer, J. A. Anna Selvan, N. Pellaton Vaucher, R. Platz, C. Hof, R. Flückiger, U. Kroll, N. Wyrsch, P. Torres, H. Keppner, A. Shah, K.-D. Ufert, *Proceedings of the 13th EC Photovoltaic Solar Energy Conference, Nice*, 1995, p. 1445.
7. J. Meier, S. Dubail, R. Flückiger, D. Fischer, H. Keppner, A. Shah, *Proceedings of the 1st World Conference on Photovoltaic Energy Conversion*, Hawaii, 1994, Vol. 1, pp. 409-412.
8. J. Meier, P. Torres, R. Platz, S. Dubail, U. Kroll, J.A. Anna Selvan, N. Pellaton-Vaucher, Ch. Hof, D. Fischer, H. Keppner, A. Shah, K.-D. Ufert, P. Giannoulès, J. Koehler, To be published in the *Proc. MRS 1996 Spring Meeting San Francisco*.
9. D. Fischer, S. Dubail, J.A. Anna Selvan, N. Pellaton-Vaucher, R. Platz, Ch. Hof, U. Kroll, J. Meier, P.Torres, H. Keppner, N. Wyrsch, M. Goetz, A. Shah, K.-D. Ufert, *Proc. 25st IEEE Photovoltaic Specialists Conference, Washington*, 1996, Vol. 2 pp. 1053-1056.
10. S. Veprek, Z. Iqbal, R. O. Kühne, P. Capezzuto, F-A Sarott and J. K. Gimzewski, *J. Phys. C: Solid State Physics,* 16, pp. 6241-6262, 1983.
11. P. Torres, J. Meier, R. Flückiger, U. Kroll, J.A. Anna Selvan, H. Keppner, and A. Shah, S. D Littlewood, I. E. Kelly, P. Giannoulès, *Appl. Phys. Lett.* 69, (10), pp. 1373-1375.
12. J. Y. W. Seto, *J. Appl. Phys.* 46, p. 5247, 1975.
13. J. Meier, S. Dubail, R. Platz, P. Torres, U. Kroll, J.A. Anna Selvan, N. Pellaton Vaucher, Ch. Hof, D. Fischer, H. Keppner, R. Flückiger, A.Shah, V. Sklover, K.-D. Ufert, to be published in the *Technical Digest of the 9. PVSEC Miyazaki, Japan*, 1996.
14. U. Kroll, J. Meier, and A. Shah, S. Mikhailow and J. Weber, *J. Appl. Phys.* 80(9), pp. 4971-4975, 1996.
15. F. Finger, .P.Hapke, M. Lysberg, R. Carius, H. Wagner, Appl. Phys. Lett. 65(20), p. 247,1994.
16. P. Torres, J. Meier, M. Goetz, N. Beck, U. Kroll, H. Keppner, and A. Shah to be published in the *Proceedings of the MRS 1996 Fall Meeting*, Boston.
17. A. Matsuda, S. Mashima, K. Hasezaki, A. Suzuki, S. Yamasaki and P.J. McElhenny, *Appl. Phys. Lett.,* 58, p. 2494, 1991.
18. U. Kroll, to be published in the *Technical Digest of the 9. PVSEC Miyazaki, Japan,* 1996.
19. U. Kroll, PhD thesis University of Neuchâtel, Hartung & Gorre Verlag, Konstanz, ISBN 3-89191-905-0, 1995.
20. J. Hautala, Z. Saleh, J.F.M. Westendorp, H. Meiling, S. Sherman, and S. Wagner, to be published in the Proceedings of the MRS Conference (420), 1996.
21. L. Sansonnens, A.A. Howling, Ch. Hollenstein, J-L. Dorier, and U. Kroll, *J. Phys. D: Appl. Phys.* 27, pp 1406-1411, 1994.
22. U. K. Das and P. Chaudhuri, S.T. Kshirsagar, J. Appl. Phys. 80(9) pp. 5389-5397, 1996
23. N. Imajyo, *J. of Non-Cryst. Solids,* 198-200, pp. 935-939, 1995
24. S.Veprek, F-A.Sarott, S.Rampert, E.Taglauer, *J. Vac. Sci Technol. A.* 7 (4), p. 2614, 1989,
25. H. Keppner, U. Kroll, P. Torres, J. Meier, D. Fischer, M. Goetz, R. Tscharner and A. Shah, *Proc. 25st IEEE Photovoltaic Specialists Conference, Washington,* 1996, p 669.
26. J. Dutta, U. Kroll, P. Chabloz, and A. Shah, J. Appl. Phys. 72 (7), p. 3220, 1992.
27. A. L. Fahrenbuch and R. H. Bube, "Fundamentals of solar cells", *Academic Press,* New York, London, p. 372, 1983.
28. JCPDS International Centre for Diffraction Data; ASTM chart Nr. 27-1402
29. H. Keppner, U. Kroll, P. Torres, J. Meier, R. Platz, D. Fischer, N. Beck, S. Dubail, J.A. Anna Selvan, N. Pellaton Vaucher, M. Goerlitzer, Y. Ziegler, R. Tscharner, Ch. Hof, M. Goetz, P. Pernet, N. Wyrsch, J. Vuille, and A. Shah, J. Pohl* and E. Bucher*, to be published in the *Proceedings of the MRS 1996 Fall Meeting*, Boston.

# COMPONENT AND SYSTEM EVALUATION AND RELIABILITY

# Photovoltaic Performance and Reliability Workshop Summary

Benjamin Kroposki

National Renewable Energy Laboratory
1617 Cole Blvd.
Golden, CO 80401

**Abstract.** The objective of the Photovoltaic Performance and Reliability Workshop was to provide a forum where the entire photovoltaic (PV) community (manufacturers, researchers, system designers, and customers) could get together and discuss technical issues relating to PV. The workshop included presentations from twenty-five speakers and had more than one hundred attendees. This workshop also included several open sessions in which the audience and speakers could discuss technical subjects in depth. Several major topics were discussed including: PV characterization and measurements, service lifetimes for PV devices, degradation and failure mechanisms for PV devices, standardization of testing procedures, AC module performance and reliability testing, inverter performance and reliability testing, standardization of utility interconnect requirements, experience from field deployed systems, and system certification.

## INTRODUCTION

The Photovoltaic Performance and Reliability Workshop was held in Lakewood, Colorado on September 4-6, 1996. The objective of this workshop was to provide a forum where the entire photovoltaic (PV) community (manufacturers, researchers, system designers, and customers) could get together and discuss technical issues relating to PV performance and reliability. The workshop included presentations from twenty-five speakers and had more than one hundred attendees. This workshop also included several open sessions in which the audience and speakers could discuss technical subjects in depth.

Several major topics were discussed, including: PV characterization and measurements, service lifetimes for PV devices, degradation and failure mechanisms for PV devices, standardization of testing procedures, AC module performance and reliability testing, inverter performance and reliability testing, standardization of utility interconnect requirements, experience from field deployed systems, and system certification.

This paper gives a summary of the major sessions of the workshop, titled Cell, Module, and System Performance; Cell and Module Reliability; System Reliability; and PV Applications and Field Experience.

# CELL, MODULE, AND SYSTEM PERFORMANCE

Presentations in this session focused on characterization and measurements of PV devices. One important aspect in determining PV cell and module efficiencies and service lifetimes are the uncertainties of the measurements. Currently, the key performance indicator for PV devices is peak power or efficiency. This is usually measured with an I-V curve taken at standard conditions or normalized to certain reference conditions. The uncertainties in these measurements are composed of random and systematic error sources. Because the changes in power or efficiency are often very small, the random errors must be minimized or "fossilized." Several other measurements may be useful when looking at PV device performance. These include: dark I-V curves, cell photo-response, shunt and series resistance measurements, fluorescence analysis, thermographs, and relative quantum efficiency measurements. Comprehensive baseline module testing is more time consuming, but it is essential to sorting out the contributions of various degradation mechanisms.[1]

Of the other measurements mentioned previously, one is of particular interest to thin-film technologies. This is the measurement of the internal series resistance of the cell. This measurement is valuable for spotting small changes in back contact resistance during accelerated testing and in identifying the source of any increased series resistance. A new way of measuring the internal series resistance of copper indium diselenide (CIS) and copper indium gallium diselendide (CIGS) samples was developed by A. Delahoy. Delahoy's method provides a better way to separate internal series resistance from sheet resistance effects[2].

Thin-film cells and modules often require different measurement methods. This is because thin-film cells are often gridless, are monolithically deposited, have long cell lengths (1-3 ft), and have different material properties. Thin-film module characterization requires the correct placement of current-voltage probes to avoid artificially high or low series resistance, consistent handling of voltage transients, low-circuit impedance and reasonable intensity for measuring quantum efficiency, uniform simulator intensity, and careful interpretation of selective illumination measurements[3].

In the area of system performance, the topic of system degradation was discussed. In studying the Photovoltaics for Utility-Scale Applications (PVUSA) arrays, a list of possible degradation mechanisms was given[4], which included: surface deterioration and/or soiling, cover material degradation, encapsulant degradation, sun-induced cell degradation, cell contact corrosion, balance-of-system (BOS) wiring corrosion, and inverter component aging. Two methods of estimating degradation were also examined: tracking monthly efficiency changes and periodically re-rating the system output with regression models. By using these methods, it was found that PVUSA systems experience a 2% per year rate of degradation. Also, the two methods for calculating degradation agree well.

Another system performance topic discussed was the comparison of energy production for different PV technologies. Work conducted at the National Renewable Energy Laboratory (NREL) examined the performance of dual-junction amorphous silicon (a-Si), CIS, cadmium telluride (CdTe), single crystalline silicon (x-Si), and multicrystalline silicon (m-Si). Conclusions demonstrated that narrower band-gap technologies (CIS, x-Si, and m-Si) showed a large seasonal swing (13-14%) in performance. Dual-junction a-Si demonstrated a 7% seasonal swing, while the CdTe array demonstrated only a 1% seasonal swing in performance[5].

Finally, in the module and system performance area, a lot of discussion took place on the topic of AC modules. AC modules are large-area PV modules with an integrated, individual inverter attached to the back. This allows for AC power to be produced directly at the module. The module's output can then be connected directly to a building's conventional AC distribution system without the need for any DC wiring, string combiners, DC ground-fault protection or additional power-conditioning equipment[6]. Because of their advantages over current DC PV modules, AC modules are well suited to be a major factor in building integrated systems. In such systems, the AC modules could displace the cost of architectural glass.

## CELL AND MODULE RELIABILITY

One of the most important tools, when examining cell and module reliability, is accelerated environmental testing (AET). AET can be used to obtain failure data, project failure rates, monitor reliability, and help predict service lifetimes. It is also important to understand the failure mechanisms and different methods to extrapolate results so that meaningful conclusions can be made[7].

Accelerated testing has been used to look for degradation mechanisms in all types of PV modules. Results from work conducted on CdTe modules was presented by P. Myers[8]. This work showed that no fundamental degradation mechanism for CdTe has been found when stressing devices by using combinations of illumination, temperature, and voltage bias. Nevertheless, not all CdTe devices display the same degree of stability nor do they all respond in the same way to the same stresses. This research has shown that the device stability of CdTe is not easily understood and may be process specific.

A new way of examining module reliability is to look at encapsulant adhesion in field-aged modules. This process developed at Sandia National Laboratories (SNL), allows for core samples of the module to be extracted. This is done to test solder bond integrity and adhesion strength. The solder bond integrity is critical to module performance and reliability. Solder bonds can age because of thermal cycling, which resultg in coarsening, cracking and failure. Adhesion strength is critical to keep modules from delaminating over time[9].

Module manufacturers are also looking at module reliability for nonstandard packaging designs. One of these is a nonglass construction from Photocomm. The manufacturer is currently conducting accelerated thermal cycling tests to provide a rapid means of examining different competing module designs. This work has shown that alternative interconnect materials provided substantial increases in fatigue strengths[10].

One of the most important steps in any PV module reliability program is the ability to pass a module qualification test. Currently, Arizona State University (ASU) has set up a laboratory to test modules to IEEE 1262 and UL 1703 standards. The laboratory quality system meets the suggested requirements of PV-1 and PV-1.1 for accredited laboratories in support of the Photovoltaic Module Certification Program[11]. ASU has also been participating in the Springborn Testing and Research Project. This project deals with testing new formulations of ethylene vinyl acetate (EVA). It is hoped that these new formulations will eliminate or reduce the browning/yellowing problem[12].

## SYSTEM RELIABILITY

Reliability issues in PV systems and components has been a topic of much discussion during the past year. While there have been great improvements in the reliability of modules because of documents like IEEE 1262 and UL 1703, BOS reliability is of great concern. Researchers at PVUSA presented data on the system and component failures they have experienced. These included diode failures, module junction box failures, combiner box failures, and metal oxide varistor (MOV) failures. The researchers believe that most of these potential reliability problems can be eliminated through good engineering practice, rigorous quality control of systems, and selecting standard, listed, field-proven components[13].

Inverter manufacturers have also started rigorous quality control programs to test their inverters before shipping. Trace Engineering is developing an advanced test and burn-in program that will be automated with a computer that will monitor critical variables, extensively test all operating modes, and collect data on each unit. This advanced testing procedure will allow for higher quality products, better reliability, lower cost, increased production capacity, and tracking of quality and performance variability[14].

Omnion Power Engineering Corporation is also examining inverter reliability. Omnion has been building a field failure database to help understand field failures and to suggest solutions to improve reliability. Omnion is also implementing extensive design testing, standardization of product, and minimizing parts counts in order to build a more reliable inverter[15].

Ascension Technology is also conducting work in inverter reliability by determining service lifetimes for the AC module by examining lifetimes for the electrolytic capacitors in the inverters. These studies focus on optimizing enclosure designs to reduce capacitor temperature and developing a model that will allow the use of solar insolation and ambient temperature data to predict capacitor temperature. These models will predict electrolytic capacitor lifetime, which is presumed to be a significant factor in inverter reliability[16].

Finally, in the area of system reliability are the issues of code and standard compliance. Article 690 of the National Electric Code (NEC) establishes requirements for the installation of field-installed and wired photovoltaic power systems. Underwriters Laboratories (UL) has a set of standards that tests the safety of PV modules and related components. There are cost increments inherent in installing PV systems that use UL-listed components and comply with the NEC. There may be small impacts on performance when installing systems to NEC, but the safety benefits should outweigh any performance losses. Also, PV systems that use listed components and are installed per NEC have the highest probability of success[17].

## PV APPLICATIONS AND FIELD EXPERIENCE

One of the issues facing consumers and utilities in the PV marketplace is the lack of standardized system packages in order to comparison shop. Another is educating people about photovoltaics. The Photovoltaic Service Network (PSN) is a nonprofit organization of utilities whose purpose is promoting PV technology. PSN educates and trains utilities about PV and creates a forum for information exchange between utilities and the PV industry. PSN has also developed a product list catalog that contains pre-qualified standardized PV systems. This document aids utilities in deciding which PV systems best fit their needs[18].

PV applications can be found in various regions around the United States. Several small stand-alone systems are used in the Upper Great Lakes region of Michigan. Most of these systems are used for providing potable water and ventilation of vault toilets. A presentation by C. Currin demonstrated how these PV systems meet the power needs for nine national and state parks and forests in the area[19]. Energy utilization in stand-alone systems is also a concern discussed by J. Stevens. Correctly sizing battery banks and using new set points can have a positive impact on energy usage[20].

Another important application for PV systems is in hybrid systems, in which PV is combined with wind, diesel, or another energy source. Some advantages of hybrid systems include reduced fuel requirements, lower maintenance costs, and improved system availability. The installations that were investigated in several presentations realize some of the advantages but have lower total system efficiencies than expected[21][22]. Hybrid modeling studies have shown that good PV array utilization can be maintained with proper operating strategy and BOS design[23].

## CONCLUSIONS

The PV Performance and Reliability Workshop was a very effective forum to discuss technical topics within the PV community. The workshop allowed manufacturers, installers, researchers, and users to concentrate on discussing important issues in the field of photovoltaics. The discussions have brought up several interesting topics that will remain focal points for the PV community. These include: creating better methods for PV device characterization; more detailed studies of field deployed modules and systems; accelerated environmental testing of PV modules, inverters, and BOS components; improved reliability for BOS components; AC module systems; and hybrid system performance.

## ACKNOWLEDGMENTS

The author would like to thank the Department of Energy, NREL, and SNL for their support in hosting this year's workshop. Thanks also to the workshop committee and chairmen. Finally, thanks to all of the contributing speakers who participated in this workshop.

This work is performed under Contract No. DE-AC36-83CH10093 to the U.S. Department of Energy.

## REFERENCES

1. Emery, K., "Uncertainties in PV Performance Indicators for Service Life Prediction" in *Proceedings of the Photovoltaic Performance and Reliability Workshop*, 1996, pp. 56-70.

2. Delahoy, A. and T. McMahon, "Determination of the Internal Series Resistance of Photovoltaic Cells" *in Proceedings of the Photovoltaic Performance and Reliability Workshop*, 1996, pp. 72-82.

3. Eisgruber, I., "Thin-Film Module Characterization" *in Proceedings of the Photovoltaic Performance and Reliability Workshop*, 1996, pp. 84-101.

4. Townsend, T., "An Evaluation of Long-Term System Degradation at PVUSA" *in Proceedings of the Photovoltaic Performance and Reliability Workshop*, 1996, pp. 102-113.

5. Strand, T. and R. Hansen, "Comparison of Array Performance at NREL" in *Proceedings of the Photovoltaic Performance and Reliability Workshop*, 1996, pp. 132-151.

6. Strong, S. and R. Wills, "The AC Photovoltaic Module" *in Proceedings of the Photovoltaic Performance and Reliability Workshop*, 1996, pp. 152-153.

7. Jorgensen, G., "Accelerated Environmental Testing for Screening and Lifetime Prediction" *in Proceedings of the Photovoltaic Performance and Reliability Workshop*, 1996, pp. 193-216.

8. Myers, P., "Stability Issues Relating to CdTe Modules" *in Proceedings of the Photovoltaic Performance and Reliability Workshop*, 1996, pp. 218-243.

9. Quintana, M., D. King, and N. Dhere, "Encapsulant Adhesion and Solder Bond Integrity in Field-Aged Modules" *in Proceedings of the Photovoltaic Performance and Reliability Workshop*, 1996, pp. 245-257.

10. Spotts, R., "Interconnect Reliability Tests for Alternative Module Construction" *in Proceedings of the Photovoltaic Performance and Reliability Workshop*, 1996, pp. 259-272.

11. Osterwald, C., et. al, "Photovoltaic Module Certification/Laboratory Accreditation Criteria Development: Implementation Handbook, NREL/TP-412-21291, August, 1996.

12. Hammond, R. K. Whitfield, and L. Ji, "PV Qualification Test Experiences and Results" in *Proceedings of the Photovoltaic Performance and Reliability Workshop*, 1996, pp. 273-280.

13. Whitaker, C. A. Reyes, P. Hutchinson, T. Townsend, J. Newmiller, and B. Farmer, "Reliability Issues in PV Systems and Components" *in Proceedings of the Photovoltaic Performance and Reliability Workshop*, 1996, pp. 282-306.

14. Freitas, C., "Advanced Test and Burn-in Program for Photovoltaic Inverters" in *Proceedings of the Photovoltaic Performance and Reliability Workshop*, 1996, pp. 318-329.

15. Dennis, K., "The PCS - An Integral Component of System Performance and Reliability" *in Proceedings of the Photovoltaic Performance and Reliability Workshop*, 1996, pp. 330-341.

16. Handleman, C. and D. Reinmuller, "Use of Irradiance and Temperature Data to Predict AC Module Lifetime" *in Proceedings of the Photovoltaic Performance and Reliability Workshop*, 1996, pp. 154-190.

17. Wiles, J., "Cost and Performance Consequences of NEC and UL Requirements" in *Proceedings of the Photovoltaic Performance and Reliability Workshop*, 1996, pp. 308-316.

18. Lane, C., "PSN's Recent Experience" *in Proceedings of the Photovoltaic Performance and Reliability Workshop*, 1996, pp. 357-372.

19. Currin, C., "Experiences with Upper Great Lakes Stand-Alone Systems" in Proceedings of the Photovoltaic Performance and Reliability Workshop, 1996, pp. 393-395.

20. Stevens, J., "Energy Utilization in PV Systems" *in Proceedings of the Photovoltaic Performance and Reliability Workshop*, 1996, pp. 373-391.

21. Durand, S., A. Rosenthal, and M. Thomas "Photovoltaic Hybrid System Performance Comparisons: Prediction Versus Field Results" *in Proceedings*

*of the Photovoltaic Performance and Reliability Workshop*, 1996, pp. 342-354.

22. Lambeth, R., "Photovoltaic-Diesel Hybrid Power System Carol Spring Mountain Telecommunications Site" *in Proceedings of the Photovoltaic Performance and Reliability Workshop*, 1996, pp. 397-405.

23. Champan, R., "Array Utilization in Hybrid Systems" *in Proceedings of the Photovoltaic Performance and Reliability Workshop*, 1996, pp. 114-129.

# Service Lifetime Prediction for Encapsulated Photovoltaic Cells/Minimodules

## A.W. Czanderna and G.J. Jorgensen

Center for Performance Engineering and Reliability
National Renewable Energy Laboratory, Golden, CO 80401-3393

**Abstract.** The overall purposes of this paper are to elucidate the crucial importance of predicting the service lifetime (SLP) for photovoltaics (PV) modules and to present an outline for developing a SLP methodology for encapsulated PV cells and minimodules. The specific objectives are (a) to illustrate the generic nature of SLP for several types of solar energy conversion or conservation devices, (b) to summarize the major durability issues concerned with these devices, (c) to justify using SLP in the triad of cost, performance, and durability instead of only durability, (d) to define and explain the seven major elements that comprise a generic SLP methodology, (e) to provide background about implementing the SLP methodology for PV cells and minimodules including the complexity of the encapsulation problems, (f) to summarize briefly the past focus of our task for improving and/or replacing ethylene vinyl acetate (EVA) as a PV pottant, and (g) to provide an outline of our present and future studies using encapsulated PV cells and minimodules for improving the encapsulation of PV cells and predicting a service lifetime for them using the SLP methodology outlined in objective (d). By using this methodology, our major conclusion is that predicting the service lifetime of PV cells and minimodules is possible.

## INTRODUCTION

The objectives of this paper are given in (a) through (g) in the Abstract, which are all driven by and related to achieving a goal of a 30-year service lifetime for PV systems (1). Our task goals are (i) to identify, understand, and then mitigate the causes of changes in module materials that alter crucial materials properties and reduce the performance and/or limit the service lifetime of cells/modules and (ii) to develop new or improved materials that offer greater promise for a module service life expectancy of over 30 years. These goals are generic for most multilayer, energy efficiency (e.g., conservation) or renewable energy (EERE) conversion devices and can be modified by simply changing "material" in (i) or (ii) to cell, array, or system for other PV specific goals or by changing "module" in (i) or (ii) to some other EERE device such as a solar mirror, electrochromic window, or flat-plate collector. For the service lifetime of other elements, the word materials may also be changed to be broader, e.g., component or subassembly. In keeping with the generality of the stated goals, we will discuss first the general principles of what is required to establish the service lifetime of EERE multilayer devices used for solar energy conversion or conservation and then show how these principles are being applied to PV cells and minimodules.

## The Solar Environment and Collecting Solar Energy

The major problem in solar energy technologies is not discovering how to collect the radiant flux, but how to collect it at a competitive cost (2). The latter is one of the reasons ethylene vinyl acetate (EVA) was chosen for use in PV modules rather than other more expensive <u>known</u> polymers with better properties (3). Solar energy reaching the Earth has a typical power density of 500 to 1000 W/m,$^2$ which means large collection areas are required for any solar technology (2). The cost of the materials utilized, device production processes, and the operation and maintenance of systems must be held to a minimum. This requires, for example, using multilayered stacks of superstrates, substrates, and the active thin (or thick) films or coatings for various collection schemes, e.g., mirrors, PV systems, electrochromic windows, and flat-plate collectors (as illustrated in Fig. 1), and that these be made from inexpensive, durable, and easily processed materials.

The materials chosen not only provide device-specific functions but also environmental protection, which is crucial for the long service lifetimes that will reduce life-cycle costs and increase the market value of the devices. When in use, man-made solar energy conversion systems are subjected to a unique set of "real-

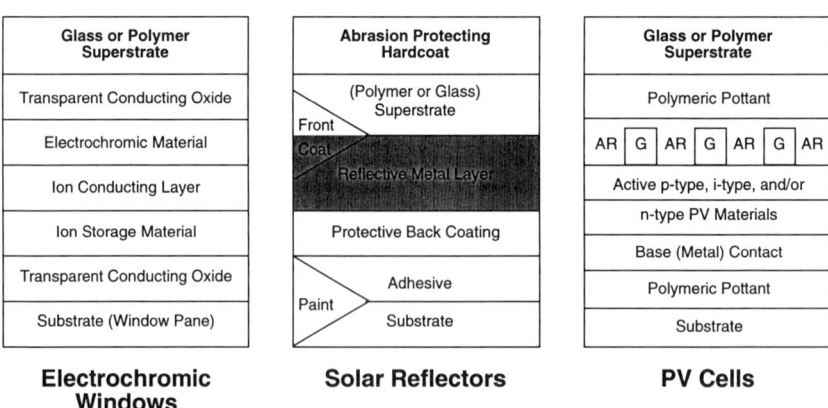

**FIGURE 1.** Cross sections of typical multilayer stacks used for solar reflectors, electrochromic (EC) windows, and PV cells. The front coat and paint layers are optional additions for solar reflectors. Different detailed mechanisms of failure are expected for the passive reflectors when compared with the active (ion or electron transport) PV or EC devices.

world" stresses that may alter their *stability* and, hence, their performance and life cycle costs in addition to the initial costs of the systems. These stresses include ultraviolet (UV) radiation, temperature, atmospheric gases and pollutants, diurnal and annual thermal cycles, and, in concentrating systems, a high-intensity solar irradiance. In addition, rain, hail, condensation and evaporation of water, dust, wind, thermal expansion mismatches, etc., may impose additional losses in the performance of a solar device. These stresses and factors must be considered not only individually, but also collectively for degradative effects that may result from their synergistic action on any part of the system. The first prerequisite is that the bulk properties of the superstrate, substrate, thin film, coating, and other materials be stable. After the requisite stability of the "bulk" materials is achieved, interface reactions are known to be thermodynamically driven because of the higher free energy state of atoms at interfaces (4). A further need may then be to choose the different materials carefully to permit achieving a 30-year "stability" (5a) or to modify the interfaces for attaining the same goal (5b). A service lifetime goal of over 20 or 30 years is targeted for all the devices in Fig. 1. For projecting a service lifetime to yield the desired time-dependent level of performance, much more SLP-directed work is needed. Furthermore, the detailed application of the SLP methodology will be more challenging for the active (PV and EC) devices than for the passive solar mirror constructions.

The goals cited in (i) and (ii) above are for the type of research needed to develop an understanding of the behavior of low-cost, high-performance, active and encapsulation materials that can be used to extend the service lifetime or to identify materials that offer new options for use in the device. For the conventional triad of requirements that includes low (initial) cost, high performance, and long-term durability (reliability), we substitute service lifetime to replace durability (reliability) as this is what is really desired. A service lifetime *prediction* (SLP) is the ability to project the future time dependence of the performance that defines the durability. Service lifetime must be known to determine the life-cycle cost for using a device of known initial cost and initial performance (i.e., efficiency in PV cells). The cost-effective deployment of any EERE device is partly limited by the durability and life-cycle cost of the materials used. Research on the active and encapsulating materials and studies that address the influence of the materials degradation on device performance are of critical importance, especially to understand soiling of surfaces, degradation of polymeric materials, the effects of oxygen and water vapor permeation, corrosion, the degradation of the active materials, and degradation at interfaces. The ultimate need is to identify materials that will not decrease the performance during exposure to actual use conditions for the desired/required service lifetime of the device. Establishing a service lifetime prediction requires a multidisciplinary team of experts plus supporting diagnostic expertise. These include people knowledgeable in the disciplines of materials science, materials engineering, surface science, corrosion science, polymer science, solid state physics, physics, physical and analytical chemistry, electrochemistry, statistical methods, theorists on lifetime prediction, etc., who have (or can access) sophisticated diagnostic and measuring equipment. Appropriate capabilities for accelerated and real time weathering of devices are also essential. If done properly, predicting a service lifetime of any device requires significant resources but is essential before major investment decisions will be made.

## SERVICE LIFETIME AND MAJOR ELEMENTS FOR PREDICTION

The service lifetime of materials, devices, or systems is the time at which its (time-averaged) performance degrades below a prescribed/required value, i.e., a failure or a failure to perform at the preassigned value. We deduce this definition from the American Society for Testing and Materials (6) definitions for durability, serviceability, and service life. Durability (6) is the capacity of maintaining the serviceability of a product, component, assembly, or construction over a specified period of time. Serviceability (6) is the capability of a product, component, assembly or construction to perform the function(s) for which it was designed and constructed. For EERE devices, the effective definition of durability is the capability of the device to perform its designed function, i.e., device performance vs time. (Reliability can be interchanged with this operative definition for durability.) Service life (6) is the period of time after installation during which all properties exceed the minimum acceptable values when routinely maintained. Thus, service life requires the selection of some minimum performance criteria, e.g., a PV module rated at 50 W at the normal operating temperature condition (NOTC) may be a "failure" when its power output falls below 40 W. *The minimum acceptable performance, i.e., "failure," needs to be defined for PV modules.* SLP is the estimated service life based on criteria and using the protocol outlined later in this section.

Desired lifetimes of typical EERE devices are as follows: polymeric or glass reflector constructions for mirror applications, > 20 years; PV modules, > 30 years; electrochromic windows, > 20 years; flat plate collectors, > 10 years; and Lo E coated windows, > 20 years. Because the desired lifetimes range from > 10 years to > 30 years, accelerated lifetime testing (ALT) in (simulated) weather environments and a predictive methodology must be used. The lifetimes of EERE devices are not unique in U.S. technology and several first-rate SLP groups have been developed at a few major U.S. corporations; as with EERE devices, U.S. industry (e.g., coatings, lighting, polymeric-based devices) cannot wait for the results from real-time testing (RTT) so must use ALT and SLP. Many U.S. companies are at a critical juncture for marketing products with a stated lifetime but need a SLP. Without a SLP, warranties will either be stated conservatively or have high risk.

A number of criteria are necessary for *accelerated testing* to be successful with a *goal* of making service lifetime predictions; these are discussed in some detail by Fischer et al. (7) and outlined in publications from various forums (6-8), as well as with the PV (9-12) and electrochromic windows (13) communities. These include, for example, that the accelerated test must not alter the degradation mechanism(s); the mechanisms and activation energies of the dominant reaction(s) at normal operating conditions and accelerated test conditions must be the same; both the specimens (including materials and components only) and accelerating parameters (UV, T, RH, product entrapment, etc.) must simulate reality; cells and/or modules that simulate reality must be used in the initial accelerated tests; and the time-dependent performance loss (e.g., power loss for PV modules) must be correlated with the degradative reactions. Ultimately and ideally, the accelerated tests must be made on commercial-scale modules that are the same size as those sold to the consumer, but this ideal *may* not be necessary if predictions from laboratory-scale specimens are reliable predictors of the commercial products. Obviously, a SLP requires a definition of "failure," i.e., what loss in efficiency is acceptable after how many years; failure needs to be defined for a PV module in keeping with the power losses of 1% to 2.5%/yr being observed in systems deployed in the terrestrial environment (1).

We now summarize the seven major elements of a service lifetime prediction methodology in which the first sentence states the element and subsequent comments clarify the element. The major advantage of the sequence given is that the first four elements can be used for improving multilayer devices until the optimum design and materials are obtained. Examples of how some of these elements have been used are available for mirrors (8, 12a, 12b, 14, 15), PV encapsulants (1, 16), and coatings (11).

SLP Element 1. The "final" design/materials selections are needed for the multilayer stack. For improving the durability of the device, each *prototype* design and the materials used can be considered as "final" for elements 1 through 4. When several prototype designs are studied, statistical methods are used to identify a test matrix of the best candidate combinations. Ultimately, a set of materials and a particular design will be identified that permits proceeding to element 5.

SLP Element 2. The "stresses" imposed on the device in real time use and the same types of stresses for ALT need to be identified and quantified. As discussed in the Introduction, the "stresses" have been identified for EERE devices used in a solar terrestrial environment. For accelerated environments and for simulating the reality of the solar UV and visible radiation, it is essential that any UV source match the wavelengths reaching the Earth's surface, which means having precise knowledge of the spectral irradiance incident on the EERE device, and that the UV source intensity be a reasonable multiple of the solar intensity. For these reasons, NREL scientists have used filtered Xe-arc lamp sources since 1978 (8, 17), and have rejected other sources such as fluorescent lamps because they do not simulate reality. Zussman indicates that the solar spectrum cut-off at sea level is

285 nm, and radiation between 290 nm and 300 nm is routinely incident at the Earth's surface (18). UV radiation can severely damage polymers if their activation spectra are at wavelengths from 290 to ca. 380 nm (19). With appropriate filters (20,21), the Xe-arc light source simulates the solar spectrum very well from 285 to 500 nm. The source intensities usually refer to the number of suns, which are simply multiples of the solar intensity in W/m$^2$ at the wavelengths of interest. The materials degradation from a Xe-arc light exposure may not match the in-service experience (18). This may result, in part, from the promotion of chemical effects of secondary processes in materials by the synergism of temperature, humidity, $O_2$, and other weathering factors (19). Similar detailed considerations are required for all imposed stresses unless it is shown that the degradation in performance is not related to a particular stress.

SLP Element 3. The complete devices are subjected to ALT and RTT to determine their durability and *the most sensitive measurement(s) of the performance loss* (or of a parameter that can be correlated to the performance) is measured. Typically, the device performance is evaluated periodically with time from measurements made by moving the samples to the instrument(s). Ideally, the measurement(s) should be made in situ either by using probes so the sample is never removed from its test location, i.e., an outdoor exposure rack or accelerated test chamber, or by using portable measuring equipment at the sample test location. The success in correlating ALT and RTT results depends crucially on the sensitivity, accuracy, and reproducibility of the measurement of the performance parameter, e.g., if a device performance is degrading at 1% per year, a measurement of the changes in the performance of 0.1% or even less is needed if the ALT data are to be correlated with the RTT data for "reasonable" RTT exposure times. For solar mirrors, specular reflectance is correlated to loss in performance and changes can be measured accurately and reproducibility (12a). *A measurement of PV performance with a sensitivity comparable to the specular reflectance of solar mirrors needs to be identified for PV cells or modules.*

SLP Element 4. The mechanisms of degradation of bulk materials and/or reactions at interfaces must be identified and understood. The degradation mechanism must result in a loss in performance of the device and/or compromise the materials function to be of concern. If the rate of performance loss from the degradation is fast relative to the expected service lifetime, the cause of degradation must be mitigated, and the sequence of elements 1 through 4 must be repeated for the new or modified materials or design used initially. If the rate of degradation is slow and the activation energy can be determined for the rate-controlling reaction, it is reasonable to proceed to element 5. At present, the design for silvered polymeric mirrors (Fig. 1) is the only EERE multilayer stack that is ready for proceeding to element 5. Substantial additional efforts are required with *PV cells* so we will be able to proceed to element 5 in 2002.

SLP Element 5. Models need to be developed for correlating ALT data and RTT data taken at several geographic sites with diverse stresses. The rate of degradation is site dependent because the stresses that cause degradation vary from site to site. For example, the total UV insolation in the sunny southwest deserts in the U.S. is a more aggressive stress than in the cloudy northeastern states. The models for correlating the ALT data and RTT data must be able to accommodate different magnitudes of the stresses including time-dependent variations and any synergism of the stresses. For a successful SLP, it is critical that correct mathematical interpretations be made of the experimental results that relate or correlate the key environmental stresses (e.g., UV, T, RH).

SLP Element 6. Stress and materials response data bases must be established that include data from different outdoor sites. This element follows directly from element 5. While some latitude may result from considering similarities in sites, enough data must be accumulated at sites with the climatic extremes and those in between to permit reasonable interpolation to any site for planned deployment of EERE devices.

SLP Element 7. Predictive service lifetime models are then developed from the data in 2 through 6 by using statistical approaches and life distribution models. A sufficient number of replicate samples must be part of the test matrix to deduce the life distribution model from the degradation (22). For example, an initial set of samples, which may range from a minimum of about 12 to 15 up to 50 and that all have "identical" performance, will degrade into a distribution of performances during use or aging. The Gaussian distribution, which is a special case of several types of distributions (7), can be used to illustrate this point. Initially, the Gaussian distribution is characterized by a full width at half maximum (FWHM) that is only limited by the uncertainty in measuring the initial performance parameter. As the sample set ages, the FWHM broadens because the performance of each individual device will degrade differently from others in the set (23). Thus, the distribution for aged samples will be the superposition of the distribution itself and that imposed by the uncertainty in measurement of the performance parameter(s). With the definition of "failure," the distribution of aged samples yields the time dependence of failures. Various types of models can be applied to describe the aged distribution (7, 22). For the best prediction results, large sample sets and ultrasensitive measurements of the performance parameter are required. Both these requirements increase the cost of making a SLP. The increased cost with increasing sample numbers is obvious. The performance parameter may require several measurements or developing a beyond-the-state-of-the-art measurement to achieve the desired result; in either case, the cost for making a SLP is increased. Therefore, it is critical to use efficient, statistical, experimental designs.

Obtaining a SLP for performance may be difficult for several reasons. These include the challenges of dealing with a large variability in failure times, determining the appropriate stresses causing performance degradation, extrapolating the results from ALT at elevated stress levels to the normal stress level, defining what is a failure of material(s) or system(s), having to use small lifetime data sets for economic reasons, and demonstrating that the degradation mechanism in ALT is the same as in RTT.

# SERVICE LIFETIME PREDICTION OF ENCAPSULATED PV CELLS AND MODULES

We illustrate the vision of being able to predict the service lifetime of an encapsulated PV module in Fig. 2. We have arbitrarily chosen a generic PV module with 100% of its rated output at NOTC. If no loss in performance occurs, the module will produce 100% forever. However, losses in PV *systems* range from a low of 1% per year to 2.5% per year (1), as shown by the solid lines in Fig. 2. The actual losses are shown for the Carissa Plains, CA, 5.2 MW system (24-26), which is the most extreme case of degradation reported. Because some of the modules were removed from the plant after 1991, the projection to seven years was made based on the efficiencies of the remaining modules. The losses in real systems are from *all* causes, and not just in the modules. Because it is not known how to project future output from a cell, module, or a system, several possible hypothetical projections are illustrated by the dashed lines for over nine years. These include projections with a simplistic linear extrapolation, with a decreasing rate of loss (perhaps from self-passivating reactions), and with an increasing rate of loss (perhaps from autocatalytic reactions). If the performance could be accurately predicted, the area under the projected curves would permit calculating the predicted output per year until failure is reached, and life cycle costs could then be calculated from the total power that would be produced and from the other life cycle costs, e.g., initial, maintenance, and operating costs. The major issue the PV community needs to resolve is what (time-averaged) loss in performance, i.e., power output, is permitted until the time of failure (in years) is reached.

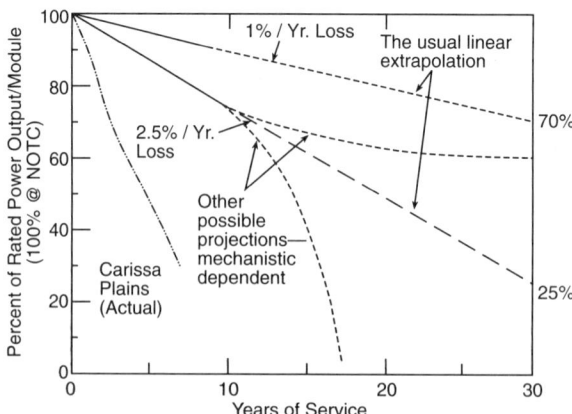

**FIGURE 2**. Actual and potential percentage losses in efficiency (performance) of PV *systems*.

Realizing the vision of being able to predict the power output for a system is clearly possible, but two significant problems must be resolved. First, the technical reasons for the power losses must be determined. The losses plotted in Fig. 2 are *system* losses; causes of performance losses need to be identified and then mitigated for cells, modules, or any other balance of systems components. No studies are known to be in progress that will establish relationships between the accelerated degradation of individual modules and RTT. In the last two years, we have gradually been able to direct the focus of our task, "Improving the Stability/Durability of Encapsulated PV Cells and Minimodules," to combine ALT and RTT of *individual* PV cells. In our prior work, we have clearly demonstrated some of the losses result from EVA browning (1, 27). Secondly, resources need to be increased substantially for a proactive technical approach that will result in improving *current* and *next-generation* PV products, e.g., by (1) monitoring the RTT performance of appropriate statistically-significant sets of individual PV cells, minimodules, and modules, (b) deducing causes of failure in *these* products and (c) by studying new/improved materials and designs at the cell and module level. A SLP then becomes possible by adopting such an approach for identifying and isolating failure modes or degradation mechanisms at the cell/minimodule, module, and other component levels as outlined (22). At present, RTT of *individual* module performance is being monitored at different sites for three cases (28-30), but without complete *initial* characterization before deployment. Eventually, ALT needs to be performed on sets of "identical" modules for accelerating the degradation of design/materials weaknesses and/or for comparing the rates with the RTT results (SLP Element 3); the RTT data needs to be taken at several environmentally diverse sites (SLP Element 5).

## Past Focus: Improving/Replacing EVA

In the last year, two key summary papers were published about EVA (1, 31). The first provides a critical review about using EVA as a pottant in encapsulated PV modules (1) and the second summarizes what can be done to retard the rate of discoloration (31). A summary of the qualitative reports of discoloration, quantitative reports of power losses in PV systems, EVA degradation mechanisms, the status of what we do and do not know about EVA discoloration, the inherent and process sources that result in accelerated discoloration, etc. are included (1). Critiques of why "lifetime projections" for EVA made in the 1980s are not valid and the reasons for the errors are also available (1,16). The most serious unrealistic projections made in the past can be avoided by determining and using actual activation energies from the Arrhenius equation or variants of it (7, 16) instead of the "rule of thumb" that reaction rates double for every 10 K increase, by considering the synergistic influences of UV, T, RH, etc. in laboratory test matrices (14-16), by simulating the reality of pottant confinement in ALT conditions (16), by operating the PV devices during ALT (16), and by using laboratory test samples that permit degradation products to accumulate when simulating hermetically sealed module designs (1,16). Our most recent progress for improving EVA or replacing it is given elsewhere in this volume by Pern (32) and Pern and Glick (33). As we have stated consistently since 1990, the stability of encapsulated PV modules is much more than a pottant degradation problem; our past and present work has addressed the EVA discoloration problem because it was identified as a major concern to the PV industry and is known to result in performance losses (27).

## Future Studies Using Encapsulated PV Cells and Minimodules

In earlier work, we established that the same type of EVA discoloration observed in field-degraded modules could be simulated in the laboratory by using *individually encapsulated PV cells* (27). The PV cell and module stability problem is not only more than a pottant degradation problem, but is also complex as shown in Fig. 3 (1, 9), which is a cross section of a contemporary multilayer PV cell. The glass cover plate may or may not contain a UV screen such as cerium dioxide, or a modified polymer may or may not be laminated between it and the pottant. A primer may or may not be used in the EVA formulation or be coated onto the glass substrate. The pottant in nearly all deployed monocrystalline (c-Si), or polycrystalline silicon (pc-Si) systems is EVA, and about 95% of the ca. 500 MW of installed PV capacity is pc-Si or c-Si. An antireflection (AR) coating (typically, 50 nm of $SiO_2$) may or may not be deposited onto the metalization or oxide surface of the Si solar cell(s). The active solar cell material(s) may be several multilayers and have a back or base contact. Another layer of EVA that

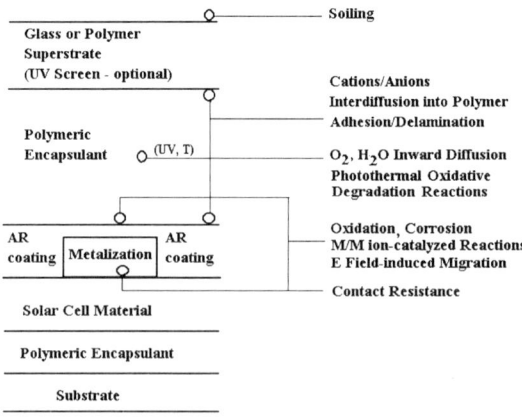

**FIGURE 3**. Schematic cross-section of an encapsulated PV cell and relevant reactions/processes that may reduce the cell performance and/or service lifetime (adapted from Refs. 1 and 9).

is shielded from exposure to UV, and the supporting substrate complete the module encapsulation. In a PV module, solar cells (e.g., 36 to 72 or more in a typical module) are joined by interconnects that are also embedded between the two EVA layers. Power output terminals are provided on each module.

As is also illustrated in Fig. 3, degradation can occur by weathering and/or soiling of the cover glass; photothermal, oxidative, or other degradation of the pottant (1, 16, 31); interdiffusion of ions into the pottant; metalization corrosion; electric field-induced ion migration or degradation; and polymer/metal oxide interface reactions or delaminations. Many of these processes may depend on initial impurity concentrations and trapped gases (vapors), and concentration changes during use. We emphasize this complexity of the entire module here because we have to establish which other degradative reactions must be mitigated (besides pottant discoloration) and which ones are too slow to impact the performance adversely over 30 years.

For individual cells, we reported on post-mortem results from a retrofited cell from Carrisa Plains (34). We have carried out one detailed study on single cell minimodules (27) as a precursor to future studies in which we plan to correlate performance changes with encapsulant and other degradation in the cells. Although degradation processes in cells are complex (Fig. 3), a number of complications from individual modules are eliminated, e.g., interconnect degradation, cell/module mismatch, and differences in degradation in each cell that are averaged for the entire module. Because of the presence of Cyasorb UV 531 in commercial EVA formulations, which absorbs UV light below 350 nm, and the 91% optical transmission of EVA that optically couples with the solar cells and soda-lime glass superstrate, a low percentage of the efficiency loss (ca.~1%-2%) that is measured for encapsulated modules results from the optical loss. Efficiency losses in c-Si and pc-Si solar cells resulting from direct contact with the EVA laminate have not been reported.

In a study comparing the effects of *accelerated,* simulated, thermal, and photothermal degradation on the EVA-encapsulated solar cells, Pern (27) measured losses of 13% in the short-circuit current ($I_{sc}$) and 19% in the efficiency that resulted directly from the reduction in light transmission through browned EVA, which was obtained by exposure to 85°C and UV from filtered RS-4 lamps for 198 days. Examples have been published showing the continuously decreasing spectral response (absolute quantum efficiency) as the EVA film discolored increasingly to a light yellow color in the solar cell heated in an 85°C oven for 198 days or to a brown color when exposed to an RS4 UV light source at 85°C for 198 days (1, 27). All solar cells showed little change in open-circuit voltage ($V_{oc}$) or fill factor (an important quantitative relationship for describing the performance of PV cells and modules), except for the noticeable decreases in $I_{sc}$ caused by EVA discoloration. Electrically, except for one solar cell, no significant change in the series resistance was measured (by dark I-V) for the solar cells studied over the 198-d period (27).

For future work our protocol will be based on preparing encapsulated PV cells and minimodules as active devices consisting of the multilayer stack as required for the seven elements of a SLP methodology. As a typical example of Element 1, the approach is illustrated in Fig. 4 for c-Si or pc-Si cells or minimodules. The multilayer stacks will consist of a glass or polymer superstrate with or without a UV screen/pottant polymer/active PV device, e.g., c-Si with a base contact and AR coating/polymer/substrate. The active devices will be of the same construction as those in contemporary modules and be a minimum of 3 cm x 3 cm and a maximum of 10 cm x 10 cm with output leads suitable for obtaining I-V and efficiency measurements. During this element, we are developing an ALT protocol and being challenged to prepare replicate test specimens. The stresses (SLP Element 2) have been identified and will be quantified for our ALT chambers (WeatherOmeters™ and Oriel solar simulators). After characterization with sensitive and other measures of performance behavior, sets of "identical" test specimens (Element 3) will be subjected to accelerated testing in controlled T and RH chambers, and with (a) a Xe-arc light source of 1 or 2 suns or (b) a condensed Xe-arc light source (solar simulator) of 5 to 17 suns from 290 nm to 400 nm in which all the test variables simulate reality. We would also like to be able to subject specimens to UV accelerated testing in an outdoor environment in which the minimodule T will be maintained at normal operating temperatures, but natural sunlight will be concentrated at 10 times e.g., by using *modified* DSET EMMA™ or EMMAQUA™ test capabilities that presently concentrate natural sunlight by about 5 times. When sufficient stability is demonstrated for the multilayer stacks made in SLP Element 2 and the degradation mechanisms have been mitigated or are sufficiently slow, we would then (SLP Element 5) deploy minimodules at six or more sites in the United States with representative and carefully recorded natural environmental exposure conditions. For SLP Element 6, we can benefit from NREL's present activities in establishing and using sites for testing candidate solar mirror materials and constructions (12a, 12b, 22, 35) and methodologies developed by them (15). Specimens at these "real-time testing" sites would be periodically monitored for their efficiency and other measurements that correlate with the cell/minimodule performance. Degradation mechanisms will be deduced from specimen "failures" from accelerated lifetime testing (ALT) and real-time testing (RTT). When they are the same, models will be developed to relate the complexity or simplicity of the multiplying factor from ALT to those for RTT, and the service lifetime will be estimated based on the interpretation of all the data acquired (SLP Element 7).

In the last three years, we have added to our task the necessary capabilities at NREL for Elements 1 through 3 with a miniextruder for extruding sheets of our own candidate pottants, a laminator for simulating industrial practice, additional characterization capabilities (I-V, yellowness index, quantum efficiency, etc.) to complement our UV-vis, color indices, and fluorescence analysis equipment for

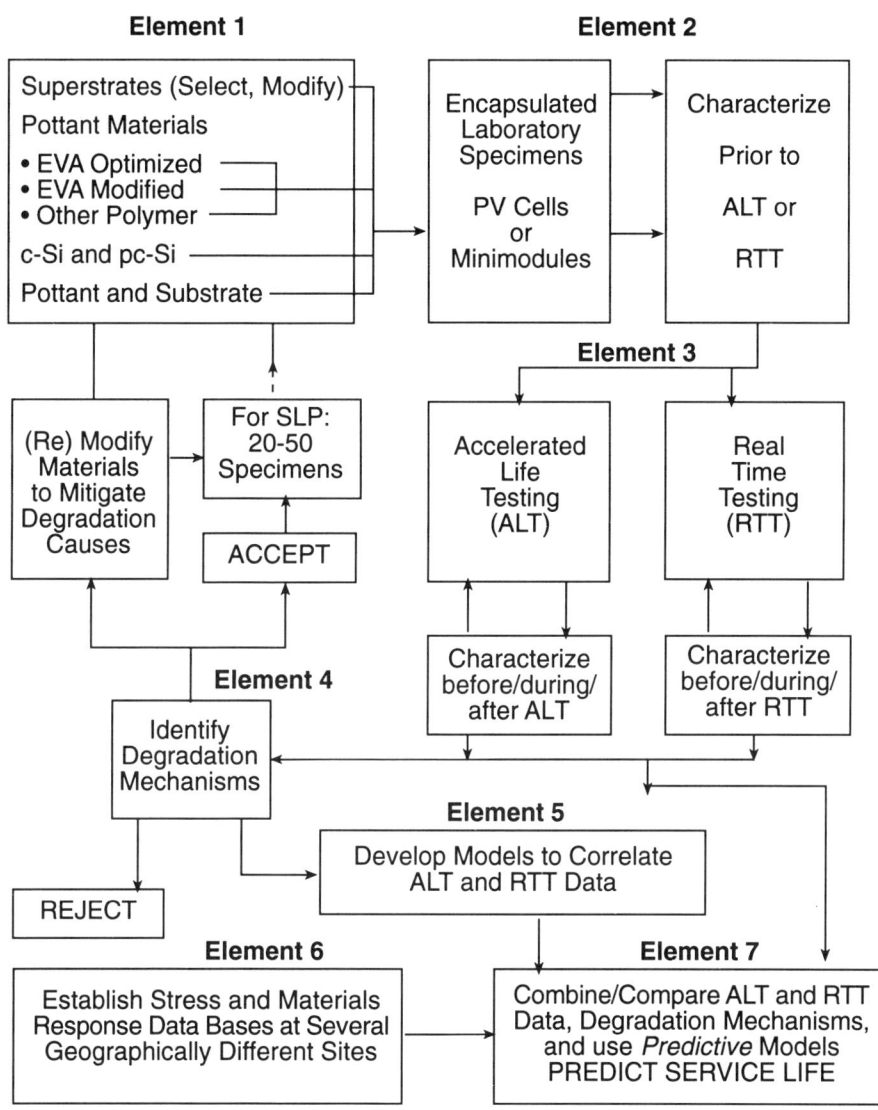

**FIGURE 4.** Technical approach for specifically applying SLP Elements 1 through 7 for PV cells and minimodules, but the scheme can be used for other PV components (e.g., modules) and EERE multilayered devices.

sensitive measures of PV cell performance losses, two WeatherOmeters to complement our Oriel condensed Xe-arc light sources, and DSET Suntest CPS table top units for conducting accelerated testing.

## CONCLUSIONS

A methodology for predicting the service lifetime of multilayered EERE devices has been outlined and related specifically to PV cells and minimodules. The SLP methodology is not limited to PV and EERE devices but also can be applied to U.S. industrial needs. Developing the technology base for predicting 30-year PV module lifetimes requires a multiyear research effort. A "failure" in the performance level (efficiency) needs to be defined for PV modules, and is necessary for making a SLP. Furthermore, an extremely sensitive measurement of a PV cell or module performance or one that is directly correlated to the performance also needs to be identified. The multiyear effort must also result in understanding degradative reaction mechanisms and their relative importance, establishing the expected levels of degradation, and utilizing the most appropriate experimental methods. Module service life prediction and material system concepts depend on correctly identified degradation mechanisms that reduce the performance or limit the service life of the module, and their appropriate applicability to reality. Long-term degradation mechanisms usually result from complex synergistic reactions between the environment and PV cell or module materials. The predominant degradative reactions may change during the module life, making analytical modeling extremely difficult. Degradation of polymeric materials can be catalyzed by their own reaction products, by solar cell metalization materials, or from ion transport into them that can eventually result in enhanced discoloration, cracking, moisture ingress, and failures in other module component materials. An increase of moisture in the encapsulant may facilitate electrochemical corrosion and progress to the point where dielectrical breakdown may occur between the cell circuit and the module ground. All of these and other degradative phenomena are critical to module durability. NREL now has the capabilities and equipment to proceed with SLP elements 1 through 4. NREL will make progress towards predicting service lifetime of PV cells and modules at a rate that depends on the resources available.

## ACKNOWLEDGMENTS

The authors are pleased to thank F.J. Pern and R. Hulstrom for their technical insight and careful review of the manuscript. We also thank Hag-Min Kim, Keith Emery, Steve Glick, and David King for their technical assistance. The authors are grateful to R. DeBlasio for his encouragement, C. Gay and B. Marshall for their support from the Directors Development Fund for developing SLP methodology, and the U.S. Department of Energy for their support of this work under Contract No. DE-AC36-83CH10093.

# REFERENCES

1.  Czanderna, A.W., and Pern, F.J., "Encapsulation of PV Modules Using Ethylene Vinyl Acetate Copolymer as a Pottant: A Critical Review," *Solar Energy Materials and Solar Cells*, **43**, pp. 101-183 (1996).

2.  Claasen, R.S., and Butler, B.L., "Introduction to Solar Materials Science," *Solar Materials Science*, L.E. Murr, ed., New York: Academic, 1980, pp. 3-51.

3.  Cuddihy, E., Coulbert, C., Gupta, A., and Liang, R., "Electricity from Photovoltaic Solar Cells," *Flat-Plate Solar Array Project, Final Report, Volume VII: Module Encapsulation*, JPL Publication 86-31, Pasadena, CA: Jet Propulsion Laboratory, (October 1986).

4.  Czanderna, A.W., "Surface and Interface Studies and the Stability of Solid Energy Materials," <u>Solar Materials Science</u>, L.E. Murr, ed., New York: Academic, 1980, pp. 93-143.

5.  a. Czanderna, A.W., and Gottschall, R., eds., "Basic Research Needs and Opportunities for Interfaces in Solar Energy Materials," *Materials Science and Engineering*, **53**, pp. 1-168, (1982); b. Czanderna, A.W., and Landgrebe, A.R., eds., *Current Status, Research Needs and Opportunities in Applications of Surface Processing to Transportation and Utilities Technologies*," NREL/CP-412-5007, Sept. 1992; 444 p., also *Critical Reviews in Surface Chemistry*, guest editors, **2**, Nos. 1-4, **3**, No. 1 (1993).

6.  "ASTM Book of Definitions," American Society for Testing and Materials, W. Conshohocken, PA, 1996.

7.  Fisher, R.M., Ketola, W.M., Martin, J., Jorgensen, G., Mertzel, E., Pernisz, U., and Zerlaut, G., "Accelerated Life Testing of Devices with S/S, S/L, and S/G Interfaces," *Critical Reviews Surface Chemistry*, **2**, pp. 317-330 (1993).

8.  Masterson, K., Czanderna, A.W., Blea, J., Goggin, R., Guiterrez, M., Jorgensen, G., and McFadden, J.D.O., *A Matrix Approach for Testing Mirrors-Part II*, Golden, CO: Solar Energy Research Institute, SERI/TP-255-1627, July 1983.

9.  Czanderna, A.W., "Overview of Possible Causes of EVA Degradation in PV Modules," *PV Module Reliability Workshop*, L. Mrig, ed., Golden, CO: Solar Energy Research Institute, SERI/CP-4079, 26 Oct. 1990, pp. 159-215.

10. Czanderna, A.W., "EVA Degradation Mechanisms: A Review of What is and is not Known," *Proceedings of a PV Module Reliability Workshop*, L. Mrig, ed., Golden, CO: National Renewable Energy Laboratory, NREL/CP-410-6033, 8-10 Sept. 1993, pp. 311-357.

11. Wineburg, J.P., "Can You Believe Lifetime Predictions from Accelerated Lifetime Testing," *Proceedings of the Photovoltaic Performance and Reliability Workshop*, 16-18 Sept. 1992, L. Mrig, ed., Golden, CO: National Renewable Energy Laboratory, NREL/CP-411-5184, September, 1992, pp. 365-375.

12. a. Jorgensen, G.J., "Durability Testing of Silvered Polymer Reflectors for Solar Concentrator Applications," *Photovoltaic Performance and Reliability Workshop*, L. Mrig, ed., Solar Energy Research Institute, SERI/CP-411-5184, Sept. 1992, pp. 345-364; b. Jorgensen, G.J., "An Overview of Service Lifetime Prediction (SLP):, *Photovoltaic Performance and Reliability Workshop*, L. Mrig, ed., NREL/CP-411-20379, Golden, CO: National Renewable Energy Laboratory, November, 1995, pp. 151-171; c. Putman, W., "Outdoor and Indoor UV Exposure Testing," op. cit. ref. 12a., pp. 279-310.

13. Czanderna, A.W., and Lampert, C., "Evaluation Criteria and Test Methods for Electrochromic Windows," SERI/PR-255-3537, July, 1990. Solar Energy Research Institute, Golden, CO.

14. Kim, H.-M., Jorgensen, G.J., King, D.E., and Czanderna, A.W., "Development of Methodology for Service Lifetime Prediction of Renewable Energy Devices," *Durability Testing of Nonmetallic Metals*, R. J. Herling, ed., Philadelphia, PA: American Society of Testing and Materials, ASTM STP 1294, 1996, pp. 171-189.

15. Jorgensen, G.J., Kim, H.-M., and Wendelin, T.J., "Durability Studies of Solar Reflector Materials Exposed to Environmental Stresses," *Durability Testing of Nonmetallic Materials*, R.J. Herling, ed., Philadelphia, PA: American Society for Testing and Materials, ASTM STP 1294, 1996, pp. 121-135.

16. Czanderna, A.W., Pern, F.J., "Estimating Lifetimes of a Polymer Encapsulant for Photovoltaic Modules from Accelerated Testing," *Durability Testing of Nonmetallic Materials*, R.J. Herling, ed., Philadelphia, PA: American Society for Testing and Materials, ASTM STP 1294, pp. 204-225.

17. Webb, J.D., and Czanderna, A.W., "Dependence of Predicted Outdoor Lifetime of Bisphenol-A Polycarbonates on the Terrestrial UV Irradiance Spectrum," *Solar Energy Materials*, **15**, 1-4, (1987).

18. Zussman, H.W., "Ultraviolet Absorbers for Stabilization of Materials and Screening Purposes," *Plastics Encyclopedia*, Sept., 1959, 1A, p. 372.

19. Searle, N.D., "Wavelength Sensitivity of Polymers," *Advances in the Stabilization and Controlled Degradation of Polymers*, Patsis, A.V., ed., Lancaster, PA: Technomic Pub. Co., 1986, **1**, pp. 62-74.

20. Webb, J.D., and Czanderna, A.W., "End Group Effects on the Wavelength Dependence of Laser Induced Photodegradation in Bisphenol-A Polycarbonate," *Macromolecules*, **19**, 2810-2825 (1986).

21. Webb, J.D., Czanderna, A.W., and Schissel, P., "Photodegradation of Polymer Films from Reflecting Substrates," *Polymer Stabilization and Degradation*, H.H.G. Jellinek, ed., Vol. 2, Amesterdam: Elsevier, 1989, pp. 373-431.

22. Jorgensen, G.J., "Accelerated Exposure Testing for Screening and Lifetime Prediction", *Photovoltaic Performance and Reliability Workshop*, B. Kroposki, ed., NREL/TP-411-21760, Golden, CO: National Renewable Energy Laboratory, October 1996, pp. 193-216.

23. Kim, H.M., and Jorgensen, G.J. "The Time-to-Failure Distribution of Renewable Energy Devices: Performance Based Approach", submitted to the <u>American Statistical Society</u>, presented at their meeting in Chicago, IL, August 4-8, 1996.

24. Gay, C.F., and Berman, E., "Performance of Large Photovoltaic Systems," *Chemtech*, pp. 182-186, (March 1990).

25. Rosenthal, A.L., and Lane, C.G., "Field Test Results for the 6 MW Carrizo Solar Photovoltaic Power Plant" *Solar Cells: Their Science, Technology, Applications and Economics*, Elsevier Sequoia, **30**, pp. 563-571, (1991).

26. Wenger, H.J., Schaefer, J., Rosenthal, A., Hammond, R., and Schlueter, L., "Decline of the Carrisa Plains PV Power Plant: The Impact of Concentrating Sunlight on Flat Plates," *Proc. 22nd IEEE Photovoltaic Specialists Conference (PVSC)*, New York, IEEE: 1991, pp. 586-591.

27. Pern, F.J., "A Comparative Study of Solar Cell Performance Under Thermal and Photothermal Tests," *Proc. PV Performance and Reliability Workshop*, L. Mrig, ed., NREL/CP-411-5148, Golden, CO: National Renewable Energy Laboratory, Sept. 1992, pp. 327-344.

28. Rosenthal, A., and Durand, S. "Long Term Effects on Roof-Mounted Photovoltaic Modules," in S. Smoller, coordinator, *NREL Photovoltaic Program FY 1995 Annual Report*, NREL/TP-410-21101, June 1996, pp. 377-380.

29. Berman, D., Biryakov, S., and Faiman, D., "Efficiency Loss Associated with EVA Laminate Browning Observed in the Negev Desert," *Solar Energy Materials and Solar Cells*, **36**, pp. 421-433 (1995).

30. Mrig, L., Strand, T., Kroposki, B., Hansen, R., and van Dyck, E., "Photovoltaic Module and System Technology Validation," S. Smoller, coordinator, *NREL Photovoltaic Program FY 1995 Annual Report*, NREL/TP-410-21101, June 1996, pp. 350-355.

31. Pern, F.J., "Factors that Affect the EVA Encapsulant Discoloration Rate upon Accelerated Exposure," *Solar Energy Materials and Solar Cells*, **41/42**, 587-615 (1996),

32. Pern, F.J., "Development of New EVA Formulations for Improved Performance at NREL," elsewhere in this volume.

33. Pern, F.J. and Glick, S.H., "A Study of Various Encapsulation Schemes for c-Si Solar Cells with EVA Encapsulants," elsewhere in this volume.

34. Pern, F.J., and Czanderna, A.W. "Characterization of Ethylene Vinyl Acetate (EVA) Encapsulant: Effects of Thermal Processing and Weathering Degradation on its Discoloration," *Solar Energy Materials and Solar Cells*, **25**, 3-25, (1992).

35. Jorgensen, G.J., Böhmer, M., Fend, T., and Sánchez, M., "International Collaborative Testing of Solar Reflectors", *Proceedings of the 8th International Symposium on Solar Thermal Concentrating Technologies*, October 6-11, 1996, Köln, Germany, to be published.

# Can the Staebler-Wronski Effect Account for the Long-Term Performance of a-Si PV Arrays?

Bolko von Roedern and Benjamin Kroposki

National Renewable Energy Laboratory (NREL), Golden, Co 80401-3393

**Abstract.** We suggest a model for the Staebler-Wronski degradation of a-Si-based solar cells that can account for long-term performance observations of deployed a-Si photovoltaic modules. The model suggests that the stabilization of the Staebler-Wronski degradation does not occur because an equilibrium between light-induced degradation and thermal or light-induced recovery is reached. Rather, stabilization occurs because the degradation phenomenon itself is self-limiting, i.e., once enough degradation has been introduced, the degradation process diminishes. This model shows that the module performance and the long-term power output of an a-Si PV array depend not only on the operating conditions, but also, on the temperature history of light exposure. This makes it difficult to accurately predict the exact amount of degradation in the field.

## INTRODUCTION

Light-induced degradation (Staebler-Wronski effect) of amorphous silicon (a-Si) solar cells and photovoltaic (PV) modules provides a serious challenge for this PV technology to achieve stabilized module efficiencies of 15%, the long-term Department of Energy (DOE) goal. Although light-induced performance changes may occur in many solar cells, only in a-Si is the effect of such magnitude that it has become a research topic of its own. It appears to be an intrinsic effect observable in all hydrogenated amorphous semiconductors, even though the details of the degradation phenomena depend significantly on the details of materials preparation. In most instances, the degradation is observed to stabilize after prolonged light exposure, which allows the fabrication of viable PV devices. However, the power rating of these devices must account for the expected degradation. This makes it important to know how much a given device will degrade.

The Staebler-Wronski effect is also observable in the degradation of fundamental material parameters such as the photoconductance (1), the midgap defect density (2), the luminescence (3), or the carrier transport mobilities (4). All these properties degrade with their own time constants (5), but the degradation often follows a stretched exponential relationship, i.e., after $10^2$ to $10^4$ hours of exposure to sunlight (100 mW/cm² intensity), the degradation diminishes. A very large portion or all of the observable degradation can be reversed upon annealing at temperatures above 150°C. This observation has led to a commonly accepted picture that the degradation stabilizes when an equilibrium with a counteracting annealing process is reached. Thermal anneal rates at typical module operating temperatures (<80°C) are often found to be too small to account for stabilization. Thus, ad hoc, a concept of light-induced annealing has been introduced (6). Our previous observations of controlled light-soaking experiments of a-Si PV modules have suggested, however, that the stabilized state of a photovoltaic module will be noticeably different depending on the temperature *history* of the light-soak procedure, rather than merely on the final operating conditions (7). This has led us to suggest that the stabilization occurs not because an equilibrium is reached, but rather, because the degradation process itself diminishes with time. This paper further discusses experimental details of the degradation of PV devices within the fairly well-established concept of "fast" and "slow" degradation and recovery components (8,9). We suggest that such a model can explain the degradation of a-Si PV devices, which can have a noticeable effect on the economics of a-Si PV arrays. The complex stabilization process determined by hierarchically constrained mechanisms (10) makes it difficult to quantitatively predict the long-term performance of a-Si PV arrays. However, no other degradation mechanisms have to be invoked to explain the performance of fielded a-Si PV arrays.

## STABILIZATION MODEL

Assessing and measuring the stabilized performance of a-Si photovoltaic devices is complicated by the fact that the device performance is affected not only by the measurement conditions but also by the short-term ("fast") and long-term ("slow") operating history of the device. For example, below are several examples where the a-Si array performance shows summer-to-winter efficiency fluctuations. The performance typically improves during the warmer summer months. Controlled light-soak tests have shown that although these changes in outdoor fielded arrays are observed on the timescale of months following the seasonal temperature changes, a change in the mean operating temperature leads to a *fast* performance change *that occurs within less than 100 hours* of operation (7). There is no reason to reject the hypothesis that a fast component of the Staebler-Wronski mechanism that can be annealed at temperatures <80°C would be responsible for these fluctuations.

In the following, we argue that the device performance stabilizes because the degradation process itself diminishes. We further suggest that the degradation rate will diminish after either the fast or the slow degradation component has been introduced into the device. The fast component of the degradation can be recovered by low-temperature annealing (T<80°C, i.e., actual operating temperatures). This will cause a quick partial recovery of the performance. Our previous controlled light-soak experiments (7) have shown that any recovery observable will saturate after less than 50 hours of exposure to a higher temperature. Subsequently, as the operating temperature is lowered again, new fast degradation can be reintroduced. We postulate, however, that the device, while it is in the recovered state, will also be susceptible for additional slow degradation. Presently, it is not clear whether such additional slow degradation will eventually alter the amount of fast degradation, which would result in a change of the summer-to-winter fluctuations; however, the long-term slope of the power output is well explained by this model. Thus, we suggest that the final power output from an a-Si array depends on the illumination history of the device throughout its deployment, rather than on the final operating conditions.

Research attempting to synthesize more stable a-Si:H devices and materials has shown that the details of the Staebler-Wronski degradation depend critically on the deposition parameters chosen. Some progress was made recently by using hydrogen dilution during the preparation of the intrinsic layers of such devices. Hydrogen dilution means that the intrinsic layer(s) of the devices are deposited from a >10:1 $H_2/SiH_4$ gas mixture (11), typically at somewhat lower substrate temperatures and lower deposition rates than when no hydrogen dilution is used. Controlled light-soaking experiments show that the degradation stabilizes faster in the hydrogen-diluted devices. This suggests that the details of the material deposition will affect the ratio of fast-to-slow degradation that is introduced upon light-soaking. A currently general observation is that devices that exhibit faster initial degradation also appear to stabilize more robustly.

Our suggested model allows us to explain, but not predict, the trends observed in the long-term performance of fielded a-Si PV arrays. Experimentally, we found that conducting indoor light-soaking experiments at 35° C, rather than at 50° C, leads to stabilized performance levels that correspond much better to those obtained outdoors. We will discuss several examples of fielded a-Si arrays, where we believe that the accuracy of data acquisition and the maintenance of the arrays allow us to deduce how the Staebler-Wronski degradation has affected long-term array performance.

## CASE STUDIES OF a-Si PV ARRAY PERFORMANCE

Of concern for the economics of a PV system is the question of how many percent per year the output, averaged over summer and winter, might degrade. Because of uncertainties in the performance measurements and the short-term performance fluctuations, this degradation can only be assessed after several years of observation (12). We now have more than 5 years of observations on several a-Si systems and can conclude that the average annual degradation amounts to 0.2%-4.8% (ignoring the substantial initial Staebler Wronski degradation during the first few months of exposure, which for many systems is not available because modules were exposed during installation, but data could only be taken after the entire system was commissioned). This compares to 1%-2% annual power loss in crystalline Si PV modules (13). The origin of these small power losses is presently not well understood.

In Table 1 we have compiled the average power losses per year for the arrays discussed in more detail below. Note that in the case of a-Si arrays, the best performance is typically obtained during the warmer summer months, whereas the performance of crystalline Si arrays typically improves during the winter month. Also note that these performance data relate to real operating efficiencies, i.e., they are normalized to the incident light intensity, but not corrected for the effects of temperature or the spectral content (14).

**TABLE 1.** Average Annual Degradation of Fielded PV Arrays. The data for the Siemens (ARCO) crystalline Si array are shown for comparison purposes.

| Array | module type | years of observation | present (dc) average efficiency | average annual degradation |
|---|---|---|---|---|
| PVUSA Siemens Solar (ARCO) | single crystal (CZ) silicon | 7.5 | 9.8% | 1.5($\pm$0.5)% |
| TISO | a-Si single jctn (ARCO G4000) | 7 | 3.3% | 2.3($\pm$0.5)% |
| PVUSA Sovonics (Davis) | a-Si dual jctn | 7.0 | 2.1% | 4.8($\pm$0.5)% |
| PVUSA UPG | a-Si dual jctn | 6.5 | 2.8% | 2.1($\pm$0.5)% |
| PVUSA Sovonics (Maui) | a-Si dual jctn | 6.5 | 3.3% | 1.8($\pm$0.5)% |
| NREL USSC array | a-Si dual jctn | 3.5 | 4.9 | 0.2($\pm$1.0)% |
| NREL USSC roofing system | a-Si dual jctn | 3 | 3.8 | 0.8($\pm$1.0)% |

Perhaps the best news is that in several instances, a-Si modules have shown long-term average degradation rates of less than 2% per year, thus providing evidence for long-term viability of this PV technology. However, the details of the degradation depend noticeably on the operating history of the modules and on the details of how the modules were manufactured.

## The TISO Array

This array was put up in conjunction with a long-term PV module testing program in the Swiss Canton of Ticino (15). Data are collected for string No. 1. The array is located in a southern alpine location with substantial summer-to-winter temperature changes. String performance and daily mean temperatures are shown in Fig. 1. The ARCO G4000 modules were commercially available for a period of time, but this product has been discontinued. The array provides a demonstration that performance of single-junction a-Si modules will stabilize and may not exhibit more long-term degradation than multijunction a-Si arrays. This observation is surprising many researchers who have argued that multijunctions would be inherently more stable. This may yet be true for the initial degradation during the first few months of exposure, but is has not always been substantiated by long-term outdoor observations. We have previously argued that the lower incident light-level in the bottom junction of multijunction devices may actually cause the stabilization of the bottom cell to be less robust (7).

## The PVUSA Arrays

The details of the PVUSA arrays can be found in their reports and publications (16). For the purpose of this paper, we focus on the older arrays.

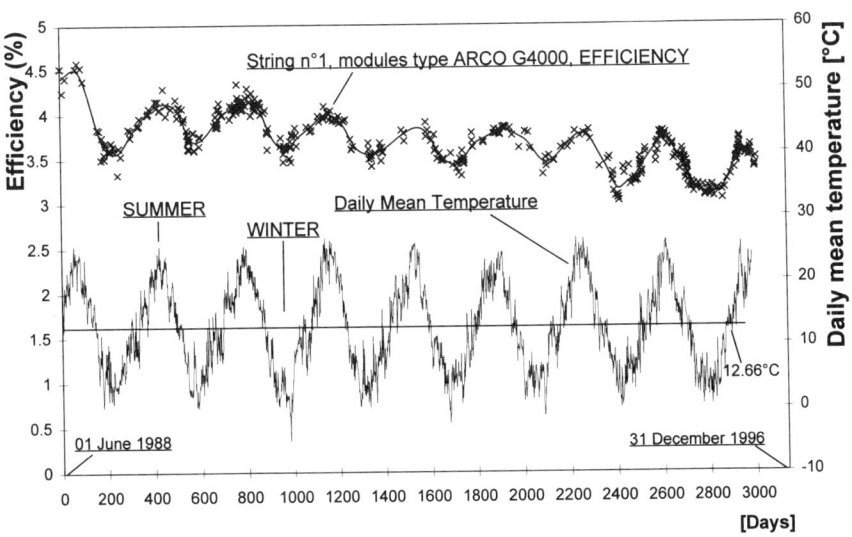

**FIGURE 1.** Historic Performance of string No. 1 of the TISO ARCO G4000 modules.

Specifically, we wish to focus on the differences in the degradation of the 2 similar (Sovonics) arrays at the Davis and Maui sites. Although the modules provided for the Maui array had a slightly improved Tefzel® front cover, leading to slightly better initial performance, we argue that this factor cannot explain the large difference in the degradation behavior. Rather, we suggest that most of the difference may be caused by the different climates at the two sites. The Davis site in California has a desert climate, with large swings in ambient temperature (minimum ambient T ~0°C, maximum ambient T 42°C; the Maui site has a year-round warm tropical climate (minimum ambient T 15°C, maximum 42°C). As seen in Fig. 2, the Sovonics array at Maui has no summer-to-winter fluctuations. It is noteworthy that the UPG array at Davis also does not show summer-to-winter fluctuations. This underpins our observation that aside from the exposure history,

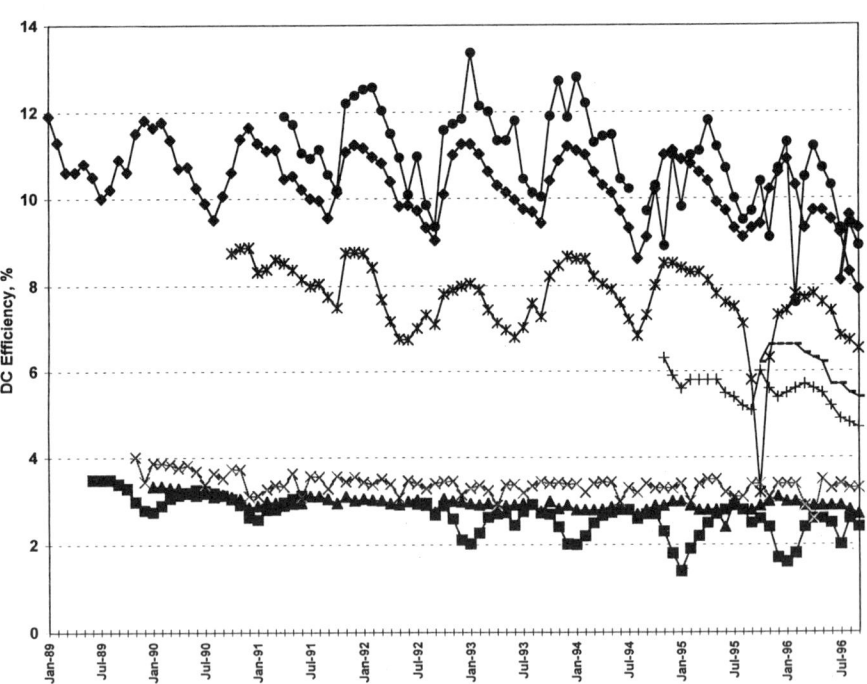

FIGURE 2. Historic Performance of the PVUSA "Emerging Technology" Arrays. Legend to the arrays: ♦ Siemens Solar (ARCO, single crystal Si), ■ Sovonics (Davis, dual-junction a-Si), ▲ Utility Power Group (UPG, dual junction a-Si), × Sovonics (Maui, dual-junction a-Si), * Solarex (bifacial poly-Si), ● ENTECH (concentrator), + AstroPower (polycrystalline Si deposited on low-cost substrates), - Solar Cells Inc. (SCI, CdTe thin film), ♦ (data since 1996) Amonix (concentrator).

the details of the module fabrication process also significantly affect the performance of a-Si:H modules.

It is of interest to note that between 5%-10% of the modules in the Sovonics Maui and Davis arrays have been replaced. Many replacements occurred because of junction-box failures (16). This factor cannot explain the differences in long-term performance. Of course, replacing modules in an a-Si array will have the effect of mitigating the overall Staebler-Wronski degradation. In comparing the Maui and Davis arrays, the number of replaced modules was actually larger in the Davis array (16), which should have helped this array degrade less.

### The NREL United Solar Arrays

The most recent performance of two NREL United Solar systems, Inc., (USSC) arrays are shown in Figures 3 and 4. The details of the 1.8-kW array and the roofing system are given elsewhere (17). We picked the arrays to demonstrate that low long-term degradation rates are possible for a-Si PV arrays, even in a Colorado climate. Because of the shorter exposure time, there is greater uncertainty in deriving a long-term degradation rate from the data. The efficiencies of the 1.8-kW and the roofing system cannot be compared because the roofing array contains a greater fraction of inactive area.

Presently, the degradation of the 1.8-kW system may be slightly higher than that of the roofing system. A possible explanation could be that the modules in the 1.8-kW array are thermally insulated on their back side to deliberately increase the operating temperature. Our model suggests that this could leave the array susceptible to having more long-term degradation introduced. However, the arrays have not been deployed for long enough times to assess which aspect of the increased operating temperature will dominate the power production over the life of the array, the increased performance due to enhanced recovery of fast defect, or the enhanced long-term degradation.

## A GEDANKEN EXPERIMENT

If our proposed models are applicable, we would like to make an interesting prediction about how the operating conditions might affect the performance. We suggest that if the PVUSA Sovonics arrays at Maui and Davis were swapped, the following effects of their power output would be observed: The Davis array moved to the Maui site would recover only insignificantly, whereas the Maui array moved to the Davis site would irreversibly degrade closer to the performance of the present Davis array. This would show how important the total exposure history is in comparison to the operating conditions of a-Si PV modules.

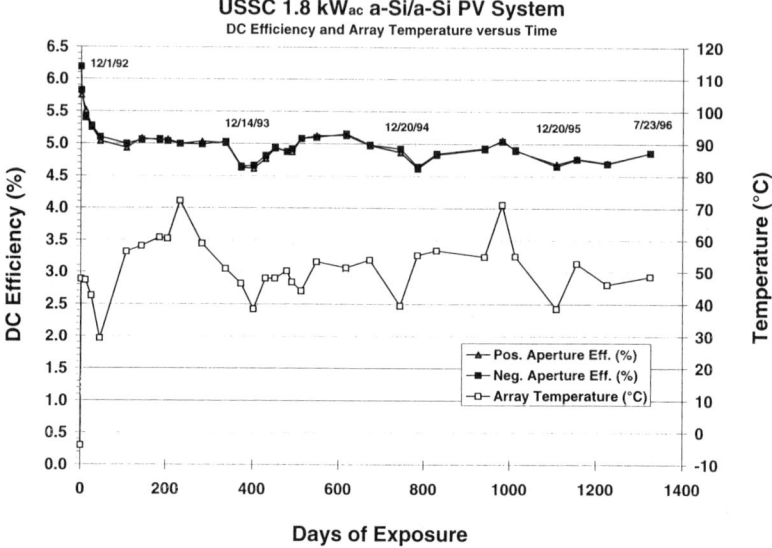

**FIGURE 3.** Historical Performance of NREL's 1.8-kW United Solar Systems Inc. Arrays

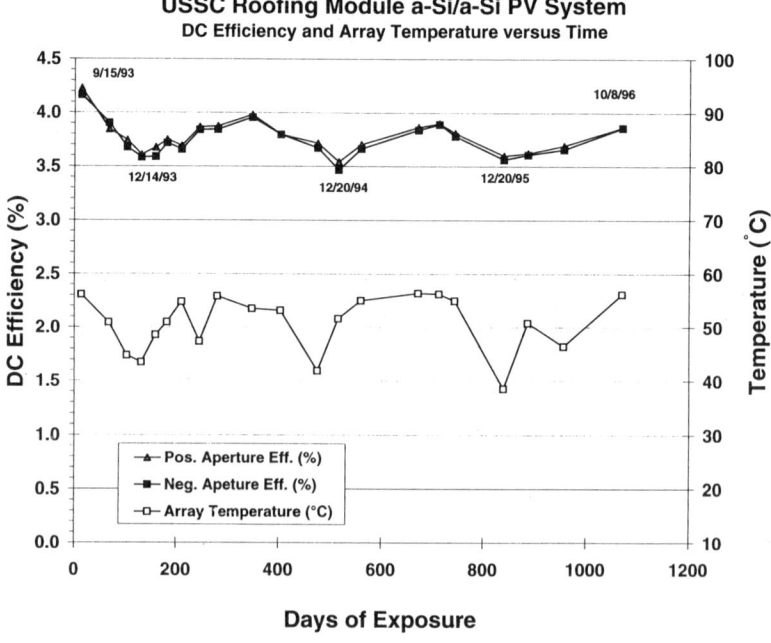

**FIGURE 4.** Historical Performance of NREL's United Solar Systems Inc. Roofing Arrays

# CONCLUSIONS

We propose to consider the stabilization of a-Si PV modules not as an equilibrium process between degradation and annealing, but rather, as a hierarchically constrained process where the degradation process itself diminishes. The presence of a fast and slow degradation process needs to be invoked. Some of the observations that could not be previously explained with the prevailing Staebler-Wronski models are:

(1) Exposure history, not the final operating conditions, determines the final power output of many a-Si modules.
(2) Some dual-junction arrays have as much long-term degradation as those made from single-junction devices.

Based on these observations, we conclude that it is presently impossible to quantitatively predict the long-term Staebler-Wronski degradation. The annual degradation rates reported in Table 1 may either increase or decrease with longer exposure time. However, except perhaps for the Sovonics PVUSA array at Davis, the observed long-term degradation rates do *not suggest* that long-term Staebler-Wronski degradation would prevent a-Si modules from being a viable PV product. In several instances, the degradation is, however, substantial enough to affect the lifetime economic viability of an array. Unfortunately, we feel that presently the degradation cannot be accurately predicted. It depends on manufacturing details that do not reveal themselves in either the initial device performance or in accelerated indoor light-soaking experiments. Thus, it is advisable to continue to monitor the long-term performance of established arrays, rather than abandoning these experiments when newer, improved modules become available. The same is true for crystalline Si arrays, for which presently no degradation model has been proposed.

It has long been suspected that a-Si:H modules will have a relative performance advantage when installed in relatively hot environments. We believe that observations justify this expectations. However, we suggest that higher operating temperatures will be most beneficial if the modules are *never* exposed to light at low temperatures (e.g., in a tropical climate as compared to a continental desert climate). Although our model also suggests the possibility of enhanced long-term degradation while the modules are annealed at higher operating temperatures, for practical purposes, this disadvantage often appears to be compensated for by the increased power output when the fast degradation has been annealed.

# ACKNOWLEDGEMENT

The authors appreciate the permission given by PVUSA and TISO to publish their up to date historic array performance figures. This work was supported by the U.S. Department of Energy under Contract No. DE-AC36-83CH10093.

## REFERENCES

1. Staebler, D.L., and Wronski, C.R., *Appl. Phys. Lett* **31**, 292-294 (1977).
2. Dersch, H., Stuke, J., and Beichler, J., *Appl. Phys. Lett.* **38**, 456-458 (1981).
3. Han, D., Wang, K., and von Roedern, B., *Phys. Rev. Lett.*, to be published
4. Dong, S., Liebe, J., Tang, Y., Braunstein, R., and von Roedern, B., this volume
5. von Roedern, B., *Appl. Phys. Lett.* **62**, 1368-1369 (1993).
6. H. Gleskova, P.A. Morin, and S. Wagner, *Materials Research Society Symposia Proceedings*, **297**, *Amorphous Silicon Technology - 1993*, pp. 589-594 (1993).
7. von Roedern, B., Kroposki, B., Strand, T., Mrig, L., "New Insights into the Staebler-Wronski Degradation Mechanism from Analyses of Solar Cell and Module Degradation Data," *Proc. 13th European Photovoltaic Solar Energy Conference*, Nice, France, October 23-27, pp. 1672-1676 (1995).
8. von Roedern, B., *Materials Research Society Symposia Proceedings* **219**, *Amorphous Silicon Technology - 1991*, pp. 493-498 (1991).
9. Yang, L., and Chen, L., *Appl. Phys. Lett.* **63**, 400-402 (1993).
10. von Roedern, B., *AIP Conference Proceedings* **234**, *Amorphous Silicon Materials and Solar Cells*, pp. 122-128 (1991)
11. Arya, R.R., Yang, L., Bennett, M., Newton, J., Li, Y.M., Fieselman, B., Chen, L.F., Rajan, K., Wood, G., Poplawski, C., Wilczynski, A., "Status, Progress and Challenges in High Performance, Stable Amorphous Silicon Alloy Based Triple Junction Modules," *Proceedings of the 23rd IEEE Photovoltaic Specialists Conference*, Louisville, KY, May 10-14, pp. 790-794 (1993).
12. Emery, K., "Uncertainties in PV Performance Indicators for Service Life Prediction," *Proceedings of the Photovoltaic Performance and Reliability Workshop*, Lakewood, CO, September 4-6, NREL /TP-411-21760, pp. 56-70 (1996).
13. Rosenthal, A.L., Thomas, M.G., and Durand, S.J., "A Ten-Year Review of Performance of Photovoltaic Systems," *Proceedings of the 23rd IEEE Photovoltaic Specialists Conference*, Louisville, KY, May 10-14, pp. 1289-1291 (1993).
14. Strand, T., Mrig, L., Hansen, R., and Emery, K., "Technical Evaluation of a Dual-Junction Same-Band-Gap Amorphous Silicon Photovoltaic System at NREL," *Proceedings of the IEEE 1st World Conference on Photovoltaic Energy Conversion*, Waikoloa, HI, December 5-9, pp. 850-853 (1994).
15. Chianese, D., Camani, M., Ceppi, P., and Iacobucci, D., "TISO: 4 KW Experimental Amorphous Silicon Power Plant," *Proceedings of the 10th European Photovoltaic Solar Energy Conference*, Lisbon, Portugal April 8-12, pp. 755-758 (1991).
16. Jennings, C., Farmer, B., Townsend, T., Hutchinson, P., Reyes, T., Whitaker, C., Gough, J., Shipman, D., Stolte, W., Wenger, H., and Hoff, T., "PVUSA - The First Decade of Experience," *Proceedings of the 25th IEEE Photovoltaic Specialists Conference*, Washington, DC, May 13-17, pp. 1513-1516 (1996).
17. Mrig, L., Hansen, R., Kroposki, B., and Strand, T., *AIP Conference Proceedings* **353**, *13th NREL Photovoltaics Program Review*, pp. 207-217 (1996).

# PV System Field Experience and Reliability

Steven Durand • Andrew Rosenthal
Southwest Technology Development Institute
PO Box 30001/Dept. 3 SOLAR
Las Cruces, New Mexico 88003-0001

Mike Thomas
Sandia National Laboratories
PO Box 5800, MS 0753
Albuquerque, New Mexico 87185

**Abstract**

Hybrid power systems consisting of battery inverters coupled with diesel, propane, or gasoline engine-driven electrical generators, and photovoltaic arrays are being used in many remote locations. The potential cost advantages of hybrid systems over simple engine-driven generator systems are causing hybrid systems to be considered for numerous applications including single-family residential, communications, and village power. This paper discusses the various design constraints of such systems and presents one technique for reducing hybrid system losses.

The Southwest Technology Development Institute under contract to the National Renewable Energy Laboratory and Sandia National Laboratories has been installing data acquisition systems (DAS) on a number of small and large hybrid PV systems. These systems range from small residential systems (1 kW PV - 7 kW generator), to medium sized systems (10 kW PV - 20 kW generator), to larger systems (100 kW PV - 200 kW generator). Even larger systems are being installed with hundreds of kilowatts of PV modules, multiple wind machines, and larger diesel generators.

## POTENTIAL COSTS SAVINGS

Hybrid systems are best when the demand on the power system matches the system production capabilities. Large demand during daylight hours, low demand at night, and large battery charging loads when the generator is running can all contribute to reducing system losses. Experience has shown that the hybrid systems are cost competitive when the system losses are less than 40% of total production. Of course, other constraints such as limited access or restricted emissions or reliability may drive the design instead of cost, but even in these cases, reducing system losses can be beneficial.

The task is to produce designs that make practical and efficient use of each subsystem while minimizing the disadvantages of having potentially redundant generation capacity. Compared to conventional, diesel-only generation, some of the potential advantages of hybrid systems include reduced fuel requirements, lower maintenance costs, and improved system availability.

Hybrid power systems reduce fuel consumption in two ways. First, the use of renewable energy generation offsets some of the fossil fuel requirements. Second, hybrid systems can be designed to operate the generator at higher, more fuel-efficient levels.

System availability may be improved by the inherent redundancy offered by hybrid power systems. If sized correctly, either the inverter or the generator can be used to meet the critical loads during either scheduled or unscheduled maintenance. Often, this requires automatic alarms that notify maintenance personnel when prompt repair is required.

Lower operation and maintenance costs are obtained by reducing the number of diesel generator runtime hours and by being able to predict when maintenance is required. Automatic or unmanned operation also has the potential for reducing costs.

The following sections review data from a number of installed hybrid systems to determine how well various designs have been able to realize these potential advantages. Evaluating hybrid systems by operational mode will also be introduced as a method of determining total system performance.

## EVALUATING HYBRID SYSTEMS BY OPERATIONAL MODE

A large part of the cost of a renewable-only system results from the need to size the array, batteries, and the inverter to support the load under worst-case weather and worst-case load conditions. In many applications, this maximum power may be less expensive if provided with a generator. Assessing system performance characteristics by operational mode is valuable for understanding hybrid system performance under widely variable loading. The first five modes of operation listed in Table 1 are normal hybrid operational states. The sixth mode represents operating periods not easily placed in any of the other operational states and includes non-hybrid, transitional, and fault operation. System operation is categorized into particular modes by determining the energy path. As an example, Mode 2 operation is selected when the PV array and the battery are both supplying power to the facility load through the inverter.

**Table 1.** Hybrid System Operational Modes

| Mode | Description |
|---|---|
| 1 | Source = Battery<br>Load = Facility |
| 2 | Source = PV + Battery<br>Load = Facility |
| 3 | Source = PV<br>Load = Facility + Battery |
| 4 | Source = PV + Generator<br>Load = Facility + Battery |
| 5 | Source = Generator<br>Load = Facility + Battery |
| 6 | Faults and Non-hybrid Operation |

System performance efficiency in any one the modes becomes more important as the time and energy transferred in that mode increases. An example of this type of analysis is given for the Carol Spring Mountain system.

## CAROL SPRING MOUNTAIN EXAMPLE

The Carol Spring Mountain (CSM) photovoltaic hybrid system is located in the Salt River Canyon Wilderness about 22 miles northeast of Globe, Arizona, along US Route 60. The site is located at latitude 33.67°, longitude 110.57°, and altitude 2012m. A variety of telecommunication loads are located on the top of Carol Spring Mountain including telephone and television repeaters. Arizona Public Service Company (APS) has historically provided power to the facility using a diesel generator for customers such as AT&T, Bell Atlantic, US West, Salt River Project, Department of Public Safety, and NewsWeek-Post cable television. A 25 kW photovoltaic system was installed on the south rim of the peak in July 1995.

The CSM hybrid system consists of the following equipment:

- An Abacus 639-4RA Bimode, three-phase, 30 kW (120/208 volts) inverter and battery charger
- A 53 kW diesel generator
- An Abacus maximum power tracker array controller
- 96 GNB Absolite IIP type 1-100A51 batteries
- 90 ASE Americas (ASE-300-DG/50) PV modules

Data collection from an on-site data acquisition system (DAS) started in January 1996. The DAS not only collects hourly performance data but also monitors system operation by mode. Figure 1 shows the percentage of time the system operated in each of the six modes for March 11-31, 1996.

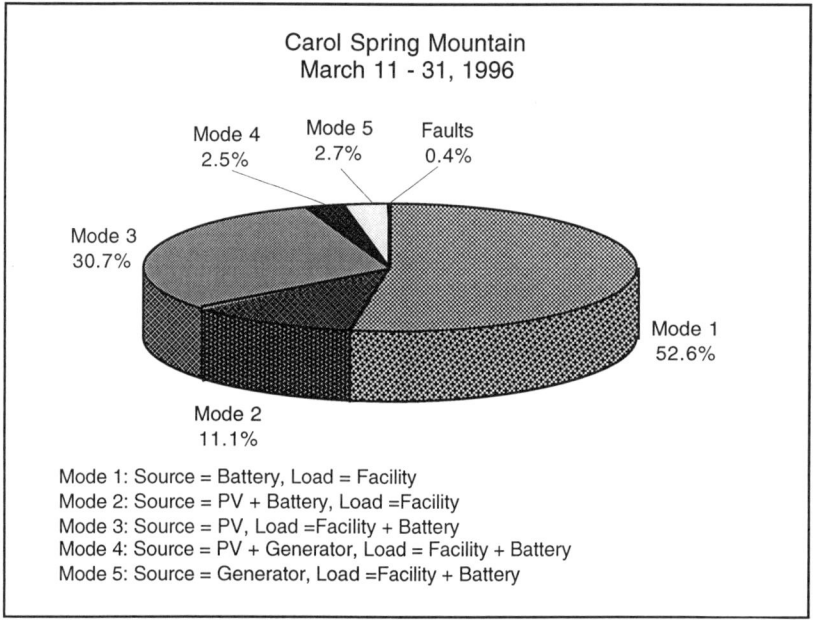

**Figure 1.** Percentage of Operating Time

Figure 2 shows the energy transferred in each of the five hybrid modes. Total system efficiency for the period monitored was 71%. Of the 4,420 kWh produced by the PV array and the diesel generator, 3,140 kWh were supplied to the facility load, while 1,280 kWh was dissipated in system losses, including battery roundtrip efficiency losses. To investigate the cause of these energy losses, the system performance by mode is examined.

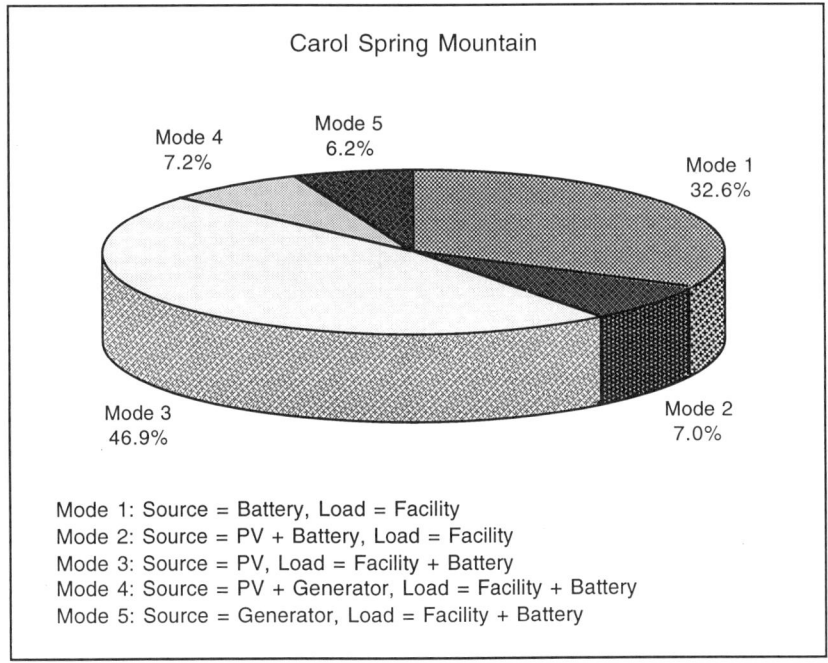

**Figure 2.** Energy Transferred by Mode

The system operated 52.5% of the time and transferred 33% of the energy in Mode 1. Mode 1 is primarily night-time operation at Carol Spring Mountain. This means the battery is being discharged and the ac load is being supplied solely by the inverter. The system efficiency in Mode 1 is 75% and is basically equal to the efficiency of the inverter. The inverter is being operated at only 22% loading (6.2kW out of 30kW) during the monitored period, which explains the relatively low inverter efficiency.

The percentage of operational time and energy transferred for all five modes of operation are given in Table 2. (Modes 2 and 4 are basically combinations of Modes 1, 3, and 5.) One issue illustrated by this example, is that the hybrid system was successful in reducing the engine runtime. The generator operated less than 14% of the time.

Table 2. Carol Spring Mountain Performance by Mode

| Mode | Description | % of Op Time | % of Energy Transferred |
|---|---|---|---|
| 1 | Source = Battery<br>Load = Facility | 52.5 | 32.6 |
| 2 | Source = PV + Battery<br>Load = Facility | 11.1 | 7.0 |
| 3 | Source = PV<br>Load = Facility + Battery | 30.7 | 46.9 |
| 4 | Source = PV + Generator<br>Load = Facility + Battery | 2.5 | 7.2 |
| 5 | Source = Generator<br>Load = Facility + Battery | 2.7 | 6.2 |

## DESERT STUDIES EXAMPLE

Southern California Edison installed a 10 kW photovoltaic/propane hybrid system at the Desert Studies Research site near Zzyyxx, California. The research facility was previously powered by a single diesel generator. The system consists of a 20 kW Onan propane generator, a 10 kW photovoltaic array composed of 160 Solarex MSX60 modules, a 48-volt, 5180 amp-hour battery bank composed of 32 GNB 3-85A31 assemblies, and two 8 kW Dimensions Unlimited inverters. The load consists of five residences and two laboratories. The system provides power 24 hours/day, and the system has about ten days of battery storage.

The total system efficiency for the period of January 1995 through December 1995 was 55%. This low efficiency is primarily the result of very low inverter loading. The average load is about 2 kW. This means that the two 8 kW inverters are only loaded to about 12 %.

## NYPA HYBRID SITE EXAMPLE

The New York State Research and Development Division, in conjunction with the New York Power Authority, installed a photovoltaic/propane hybrid system at a private residence near Ticonderoga, New York. The system is nominally comprised of a 5 kVA Onan propane generator, a 1 kW array consisting of 16 Siemens modules, a 48-volt, 450 amp-hour battery bank consisting of twenty four IBE 75N-13 cells, an Ananda Power Technologies system controller, and two 2.5 kW Trace Engineering Company, Model U2548SB inverter/battery chargers. The system was installed in late 1993 and has about one day of battery storage.

In this case, the total system efficiency for the two-year period, January 1994 through December 1995, was 64%. This efficiency value is primarily a function of the average load size. The average load for the period is about 0.3 kW. This loads the two 2.5 kW Trace inverters to about 6% of their rated output. The recorded inverter efficiency curve of the NYPA system for May 1994 is shown in Figure 3. This curve shows that the combined inverters operated below 600 watts 92% of the time.

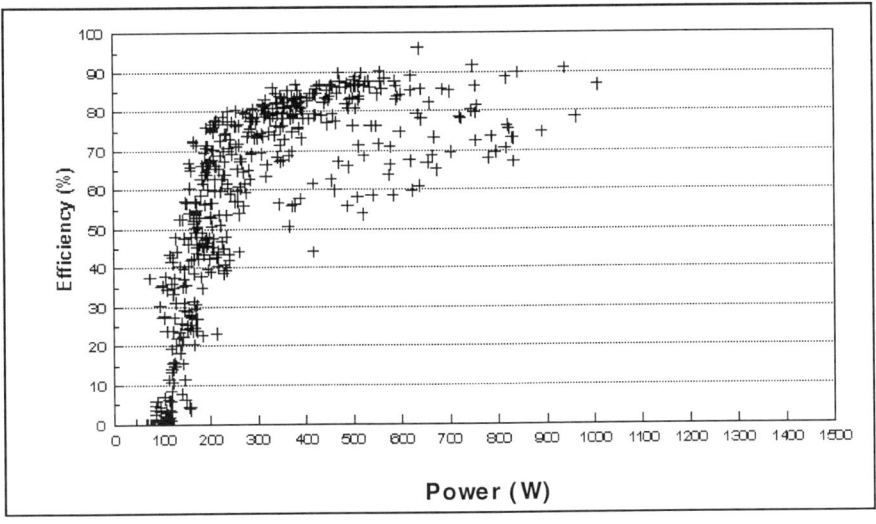

**Figure 3.** NYPA Site Trace Inverter Efficiency for May 1994

## NYSEG

The New York State Research and Development Division, in conjunction with New York State Energy and Gas (NYSEG), installed a photovoltaic/propane hybrid system at a private residence near Plattsburg, New York. The system is nominally comprised of a 4 kVA Kohler propane generator, a 900 W array consisting of 12 Siemens PC-4F75, 75-watt modules, a 48-volt, 450 amp-hour battery bank consisting of twenty four IBE 75N-13 cells, an Ananda Power Technologies system controller, and a 4 kW Trace Engineering Company, Model SW4048 inverter/battery charger. The system was installed in 1994 and has about one day of battery storage.

For this system, the overall operating efficiency for the period of January 1995 through December 1995 was 75%. This is higher than the NYPA system total system efficiency previously mentioned. The average load was about 0.3 kW.

The 4 kW Trace inverter was loaded on average to about 7%. This type of inverter has slightly higher efficiency at low loading, which greatly impacts the total system efficiency.

## DANGLING ROPE

The US National Park Service installed a Hybrid power system at the Dangling Rope Marina on Lake Powel near Page Arizona. The system was installed in August 1996. The system is nominally comprised of two Caterpillar 250 kVA propane generators, a 115 kW PV array consisting of 384 ASE DG/50 modules, a 396 -volt, 2.4 MWh battery bank consisting of 198 IC&D model 125-25 cells, a Kenetech Windpower HY-250 Power Processing Unit, which operates at 480 Vac, 3 phase. The system has about one day of battery storage.

For this system, the overall operating efficiency for the period of September 1996 was 87%. This is higher than the previous four systems.

The average load was about 1500 kWh per day. The system configuration is such that each of the individual components are operated efficiently. The daily load profile is centered around 1400 hours. In the morning, the load increases as the array production increases. The 100 kW array is well matched to the load, and very little PV energy is used to charge the batteries. In the early evening when array power can no longer supply the load, the battery bank sources the energy. The evening load averages about 78 kW. This loads the inverter to about 50%. The inverter efficiency is about 90% under this loading. Around midnight, the load decreases to below 50 kW. This would normally cause the inverter to operate in a less efficient region but, by midnight, the battery bank has been depleted; the propane generator takes over supplying the load and starts recharging the batteries. The battery-charging task loads the generator to about 80%; thus, the generator is also run in an efficient manner. Near sunrise, the generator has finished charging the battery and turns off. The load starts to increase and the cycle starts over.

The system sources are well matched to the load during the entire day. Very little PV power is used to charge the batteries and the inverter, and generators are loaded sufficiently to operate efficiently.

## GENERATION SYSTEM PERFORMANCE

The total system efficiencies for the five systems, including battery losses are shown in Table 3. All of the listed systems are well designed and built and operate as designed. The low efficiencies are recorded because in all cases, the installed inverters and generators were each sized to meet the peak load with no regard to power factor or power quality. The high Dangling Rope efficiency is cause by matching the loads to the sources.

**Table 3.** Measured System Efficiencies

| Site Name/ Location | Data Period | Total System % Efficiency |
|---|---|---|
| Carol Spring Mountain, AZ | Mar. 96 | 71 |
| Desert Studies, CA | Jan. 95 - Dec. 95 | 55 |
| NYPA, NY | Jan. 94 - Dec. 95 | 64 |
| NYSEG, NY | Jan 95 - Dec. 95 | 75 |
| Dangling Rope | Sept. 96 | 87 |

# EFFECTS OF SUBSYSTEM PERFORMANCE ON TOTAL SYSTEM PERFORMANCE

Typically, for diesel-only systems, the diesel generators operate on the low end of the efficiency curve most of the time (Figure 3). Fuel efficiencies of lower than 5 kWh/gallon are not uncommon. When these systems are converted to a hybrid system, the diesel generators are used primarily to charge the batteries and are operated at higher loads in a more fuel-efficient manner (Figure 4). Fuel efficiencies of greater than 10 kWh/gallon are attainable. This results in an approximate doubling of the useful energy produced for the same quantity of fuel consumed.

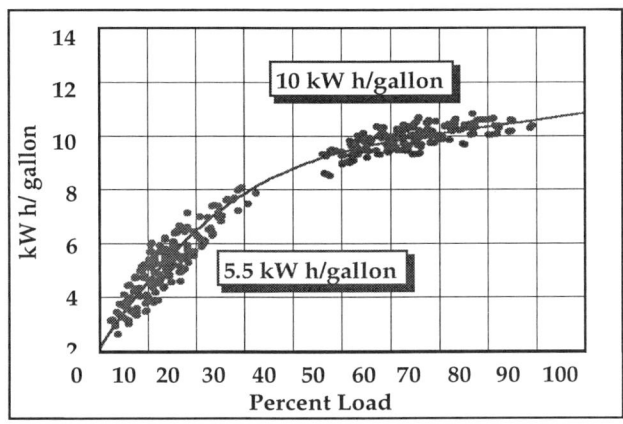

**Figure 4.** General Diesel Generator Fuel Efficiencies

## CONCLUSIONS

The goal of installing a hybrid system is to realize the potential advantages of reduced fuel use, lower operation and maintenance cost, and greater availability than diesel-only systems. The installed hybrid systems investigated in this study do realize most of these potential advantages but have lower total system efficiencies than expected. To investigate the determining factors of total system efficiency, the relationship of fuel consumption to renewable energy production was determined (Figure 5). To generate these data, the following assumptions were used. The diesel generator fuel efficiency as a diesel-only system averaged 5.5 kWh/gallon. The diesel generator fuel efficiency as a hybrid system averaged 10.0 kWh/gallon. These numbers were obtained from generalized Caterpillar production curves for a model C398 diesel generator.

Figure 5 indicates that, if the original goal was to reduce fuel consumption by 50% compared with the diesel-only system and the total system efficiency of the hybrid system is about 55% (similar to the NYPA system discussed earlier), then the renewable production required is about 90% of the pre-existing load. If the total system efficiency is improved to 70%, then the renewable production required can be reduced to about 50% of the pre-existing load.

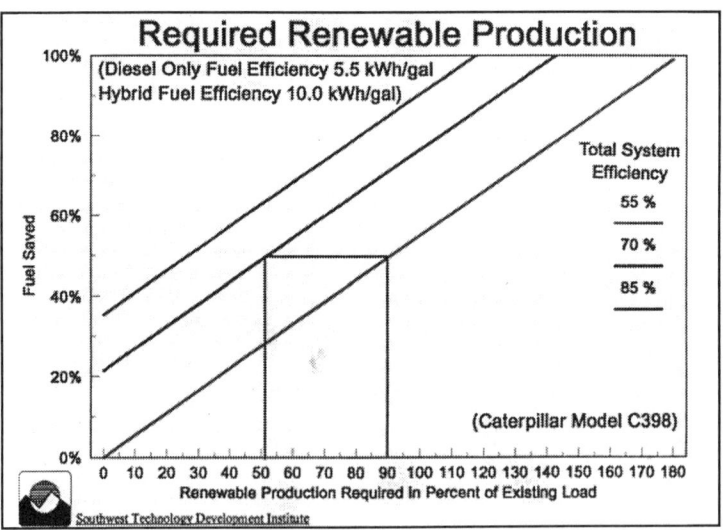

**Figure 5.** Fuel Saved vs. Required Renewable Production

# RECOMMENDATIONS

To improve hybrid system cost competitiveness, the overall system efficiency of the installed system must be increased. This can be accomplished by efficiently loading the power conditioning system, battery, and generator. A few methods are listed below.

Power conditioning systems could be designed to have higher overall operating efficiencies. One obvious solution would be to develop inverters with greater than 95% efficiency from 1% to 200% loading. At the present time, such hardware is not available. On the other hand, the present hardware could be operated in efficient regions a higher percentage of the time. The power conditioning system could also consist of two differently sized inverters—one that has high efficiency at light system loads and another that has high efficiency at high system loads. The PCS controller could automatically switch between the two inverters or even operate them in parallel, as needed.

PV energy should not be used to charge batteries. Every attempt should be made to match the resource to the load. The Dangling rope system has low system losses because most of the PV energy is supplied directly to the load.

The generators need to be loaded at more than 50% to obtain good fuel economy. This can be accomplished by using the generators to charge the battery systems and to supply large loads. If battery charging is performed during periods of low load, this will reduce the number of runtime hours the inverter will be operated at low loading and, thus, increase overall system efficiency.

# REFERENCES

1. Manwell, James F. et al, "Hybrid Systems Modeling: Development and Validation," pp 657—666. American Wind energy Association *WINDPOWER '94 Proceedings.* May 10-13, 1994.

2. Ramsey, Kay. *New York State Research & Development Authority System Monitoring, New York State Electric & Gas Corporation (NYSEG) Site, Monthly Performance Report.* Southwest Technology Development Institute, December 1995.

3. Durand, Steven and Hagee, Warren. *Sandia National Laboratories Photovoltaic Design Assistance Center Hybrid Monitoring Program Preliminary Results Xcalak, Mexico - SCE Desert Studies - NYR&D NYPA Site*. Southwest Technology Development Institute, July 1994.

4. Durand, Steven and Sansing, Bethany. *New York State Research & Development Authority System Monitoring New York Power Authority (NYPA) Site, Interim Report 1994, Residential Photovoltaic/Propane Hybrid Power Generation System*. Southwest Technology Development Institute, May 1995.

5. King, David L.; Ellis, Abraham; and Eckert, Peter E. *Carol Spring Mountain PV Array - Initial Performance Characterization*. Sandia National Laboratories, September 24, 1995.

6. Durand, Steven; Rosenthal, Andrew; and Thomas, Mike. "Photovoltaic Hybrid System Performance Comparisons: Prediction Versus Field Results," pp. 1353 - 1356, in *Twenty-Fifth IEEE Photovoltaic Specialists Conference Proceedings*. Institute of Electronic and Electrical Engineers, 1996.

# Inverter Testing at Sandia National Laboratories*

Jerry W. Ginn
Russell H. Bonn
Greg Sittler

Photovoltaic System Components Department
Sandia National Laboratories
PO Box 5800
Albuquerque, NM 87185-0752

**Abstract.** Inverters are key building blocks of photovoltaic (PV) systems that produce ac power. The balance of systems (BOS) portion of a PV system can account for up to 50% of the system cost, and its reliable operation is essential for a successful PV system. As part of its BOS program, Sandia National Laboratories (SNL) maintains a laboratory wherein accurate electrical measurements of power systems can be made under a variety of conditions. This paper outlines the work that is done in that laboratory.

## TESTING ACTIVITIES

Inverter testing at SNL thus far has fallen into one or more of the following three categories.

### Benchmark Testing

Tests have been designed and performed to benchmark the performance of inverters. The primary goal of benchmark testing is to provide information on

---

* This work was supported by the United States Department of Energy under Contract DE-AC04-94AL85000. Sandia is a multi-program laboratory operated by Sandia Corporation, a Lockheed Martin Company, for the United States Department of Energy.

inverter performance over a standardized set of tests. This is important because of the variations in the manner in which inverters have been rated and specified by the manufacturers. For example, inverter efficiency is often reported as a single number without regard for load characteristics which affect efficiency significantly. These tests are intended to provide information which is useful to PV system integrators in selecting an inverter for a specific application and in anticipating its performance under a variety of conditions.

## Development Testing

The purpose of development testing is to assist inverter manufacturers in their development of a technological innovation or refinement of their product. Consequently, the manufacturers are the primary customers for development test results. This service can be significant because creating, equipping, maintaining, and operating a test facility is, in many cases, prohibitively expensive. Key elements of the facility which can be useful include a variety of loads, dc and ac sources, and diagnostic equipment.

## Acceptance Testing

In a few cases, inverters have been tested to verify that their laboratory performance meets government contractual requirements. The customers in this case are the end users of the equipment. Since some of the capabilities of the inverter may be new, acceptance testing has typically followed a preliminary period of development testing. Two examples of acceptance testing are a 250-kVA Kenetech hybrid inverter for the Dangling Rope Marina at Lake Powell, Utah, and a 300-kVA Abacus Controls, Inc. hybrid inverter for the U.S. Navy's Superior Valley installation at China Lake, California.

# APPLICATION CATEGORIES

Inverter hardware can be grouped into the following four categories.

## Stand-Alone

Inverters operating in a stand-alone mode provide ac power from a dc battery. They range in power capability from hundreds of watts to a few kilowatts. Because their market is relatively substantial, including recreational vehicles, boats, off-grid homes, and any application requiring remote operation of ac equipment, many thousands of these units have been manufactured. Consequently, the technology involved in stand-alone inverters is relatively mature. SNL purchases these inverters for bench-mark testing. Issues of interest include load compatibility, power quality, and safety of operation.

## Small Grid-Tied

Grid-tied inverters take dc power either from a battery or directly from a PV array and provide ac power to the utility grid. The only grid-tied inverters that have been produced in quantity have had a capacity of a few kilowatts. Typically, they have been produced in response to projects that are least partially subsidized by a utility and/or by the government. The total number that have been manufactured is certainly less than the number of stand-alone inverters, but still in the hundreds. The technology associated with grid-tied inverters is relatively mature. SNL purchases these inverters for bench-mark testing. Issues of interest are those required to provide confidence on the part of potential utility customers, including power quality, safety of operation, islanding protection, and overall reliability.

## Small Hybrid

Small hybrids have evolved from small stand-alone inverters. They typically have single-phase ac outputs with power ratings of a few kilowatts. They can operate in a stand-alone mode but also have the capability to interact with a secondary ac source, such as the utility grid or a backup generator. The secondary source is requested by the inverter controls to recharge the battery and/or to power loads when necessary. Variations in control schemes are possible. The manufacturer typically provides the inverter to SNL for evaluation. Issues of interest include load compatibility, voltage regulation, power quality, site control, and safety.

## Large Hybrid

Large hybrid inverters can range in size from tens to hundreds of kilowatts and generally have three-phase outputs. Applications for these units are remote military installations and remote village power. Because only a few have been produced, the technology is still under development. The inverter is normally purchased by the end-user. Testing at SNL assists in the final development and may serve as a partial functional acceptance test prior to a final field acceptance test. Issues include load compatibility, voltage regulation, power quality, site control, and safety. Of these, the most challenging has been the control issue.

# SANDIA POWER-CONDITIONING SYSTEMS TEST FACILITY

## Data Acquisition System

Sandia maintains a laboratory capable of measuring performance of power-conditioning equipment ranging in size from a few hundred watts to hundreds of kW. Voltages and currents are acquired on both ac and dc sides of the equipment and are analyzed to evaluate key parameters including efficiency, distortion, output regulation, and load compatibility. Data is acquired in both averaging and high-speed waveform-acquisition modes. The data-acquisition system uses a 16-bit, 100 kHz digitizer controlled by a National Instruments Labview program. Data is plotted in Microsoft Excel format. Backup instruments for independent corroboration of data include: oscilloscopes, digital multimeters, dynamic signal analyzers, spectrum analyzers, and audio analyzers. Additional quantities which can be evaluated include conducted and radiated radio-frequency emissions, and audible noise.

## Hardware

### *Loads*

- Programmable resistive bank: 150 kW, 480 V, 3-phase
- Programmable inductive bank: 225 kVAR, 480 V, 3-phase
- Manual resistive bank: 360 kW, 480 V, 3-phase
- Nonlinear bank: 50 kVA, 277 V
- Motors: 3-phase to 10 hp, 1-phase to ¾ hp with dynamometer and computer control

## AC Sources

- Main power grid: 500 kVA, 480 V
- Separately-derived power grid: 50 kVA, 120/240 V
- Permanent onsite diesel generator: 92.5 kVA, 480 V with remote-start panel
- Temporary generator: wiring and switchgear provision for up to 500-kVA
- ac motor-generator: 150 kVA, 480 V

## DC Sources

- Photovoltaic arrays: 30 kW configurable (2 each)
- Photovoltaic simulators: 64 kW and 11 kW
- Power supplies: 350 V, 35 A (3 each) and 55 V, 180 A (2 each)
- dc motor-generator: 115 kW, 700 V

## Battery Storage

- 720 kWh bank of 288 cells with 1250 AH capacity, configurable to 576 Vdc in 24-V increments
- 52.8 kWh bank of 24 cells with 1100 AH capacity, configurable to 48 Vdc in 6-V increments

## Lightning Simulator

- Voltage and current surge generator: Velonix Model 587

# TESTING EXAMPLES

## Stand-Alone Example: Trace DR1524 Benchmark Test

The standardized set of tests for evaluation of stand-alone inverters results in a two-page test report. The report associated with the 1.5-kW quasi-sine wave Trace DR1524, is shown in the appendix. Efficiency is plotted for various loads. For all inverters, efficiency is not a single value, but is a function of load type and magnitude. Voltage and frequency regulation and voltage distortion are tabulated for a variety of loads. Regulation is very good, whereas distortion is

extremely high due to the quasi-sinusoidal waveform. If battery charging is a capability of the inverter, it is characterized in a table. Overloads are applied and the results are tabulated and compared to the manufacturer's specifications. Copies of the test reports for five different stand-alone inverters are available on Sandia's PV website:
http://www.sandia.gov/Renewable_Energy/photovoltaic/pv.html

## Large Hybrid Example: Kenetech Development/Acceptance Test

The standardized set of tests for evaluation of hybrid inverters results in a four-page test report. Tests of the inverter (dc-to-ac) mode of operation are analogous to those for stand-alone inverters. In addition, a series of tests summarizes interactions with the generator. Copies of the test reports for two different large hybrid inverters are available on the web site, one of which is that for the 250-kW Kenetech unit developed for Dangling Rope Marina.

Development testing was especially useful in this case, since this was a new product for Kenetech. Their company had extensive experience with wind-powered, grid-tied applications, but little with PV and none with batteries. As a result, their factory testing concentrated on verifying sub-system functionality. Compromises in their factory testing resulted from the lack of a generator, batteries, PV, or realistic nonlinear loads, all of which SNL was able to provide. Tracking of the maximum-power point was refined and tuned, the battery charge algorithm was updated, and a number of inverter/generator transfer issues were uncovered in a relatively short period of time.

This testing served as the laboratory portion of the contract acceptance tests. The inverter met all specifications; however, the voltage imbalance among phases was significant for severe load imbalances. This is shown in Figure 1. Kenetech was apprised of this result and was confident that by deriving feedback from the load side of the output transformer and changing the control algorithm, the voltage imbalance could be acceptably reduced in future products.

## Grid-Tied: Development of Benchmark Test

A test plan for grid-tied inverters has been developed and is available for review at Sandia's PV web site. Parameters of interest include efficiency, distortion, radio-frequency interference, maximum power tracking effectiveness, dc and ac operating ranges, acoustic noise, control features, anti-islanding effectiveness, and restart following utility outage.

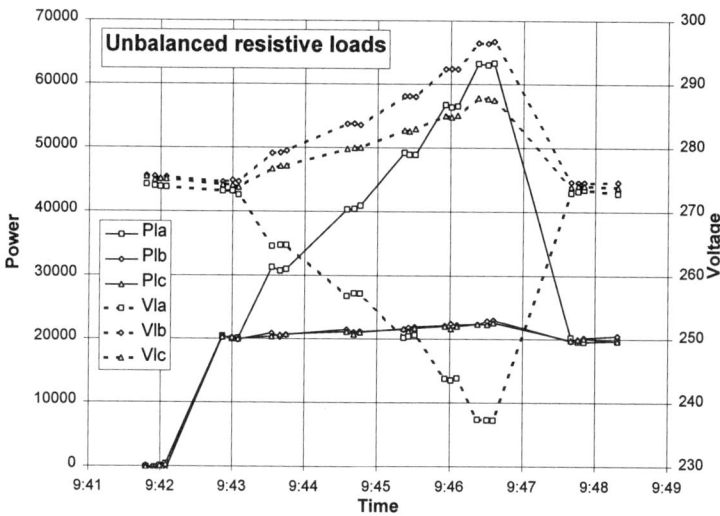

**FIGURE 1.** Unbalanced voltages resulting from unbalanced loads
(Phase A varied in 10 kW steps with a fixed resistive load on phases B and C)

temperature. Test results will be provided to manufacturers to assist their product development. Another goal of long-term testing is to obtain a statistically significant quantity of data for use in developing IEEE Standard 929, "IEEE Recommended Practice for Utility Interface of Residential and Intermediate Photovoltaic Systems." This data and IEEE 929 will help utilities assess reliability and any potential negative impact on their systems.

Grid-tied inverters to be characterized fall into two groups, both of which will be treated identically in the test. The first group includes inverters similar to those that have been installed by the Sacramento Municipal Utility District (SMUD), the Environmental Protection Agency (EPA) PV project and others. These include the 4-kW Omnion 2400, 3-kW Pacific Inverter PI-3000, 4-kW Trace 4024, 5-kW Abacus Sunverter, and a 2.2-kW inverter by Sanrex, a Japanese manufacturer. The second group includes module-scale inverters, which are mounted permanently as an integral part of a PV module. The two manufacturers that have developed hardware of this type for the PV Mat initiative are Ascension Technologies and Solar Design Associates, which uses an AES inverter. Module-scale inverters manufactured by Trace Engineering and Evergreen are also planned to be included.

## SUMMARY

The goal and future direction for the three types of SNL inverter testing are as follows.

Benchmark testing will:

1. result in a standardized method for evaluating inverters
2. influence government specifications
3. provide useful information to system integrators
4. reassure utilities that PV inverters will not interfere with their operation

Development testing will:
1. support PVMat, SNL R&D in reliability, and inverter manufacturers
2. be coordinated with system requirements to ensure best possible design

Laboratory acceptance testing will:
1. reduce field down time by early detection of problems
2. provide useful information on system performance
3. support users with potentially significant market impact

All SNL testing is under continuous development. Suggestions are actively solicited.

# Appendix: Stand-Alone Inverter Test Report of the Trace DR1524 1.5-kW Quasi-Sine-Wave Inverter

**manufacturer's specifications**

| Model Evaluated | DR1524 | Output voltage | 120 Vac |
|---|---|---|---|
| Rated Power | 1500 W | Distortion | N/A |
| Rated volt-amps | not specified | dc disconnect voltage | 22 Vdc |
| Surge power | 4500 W | dc float voltage | N/A |
| Efficiency | 94% (maximum) | Max charge rate | N/A |

**dc evaluation using 200 A-H battery (Trojan T-105)**

| Parameter | Quantity |
|---|---|
| Inverter Mode | |
|   battery disconnect voltage | 21.2 Vdc |
| Charge Mode | |
|   dc ripple voltage peak (float) | 1.1 V |
|   dc ripple current peak (float) | 28 A at 2.3A float |
|   battery float voltage | 26.6 Vdc |
|   charging efficiency | 91% average |

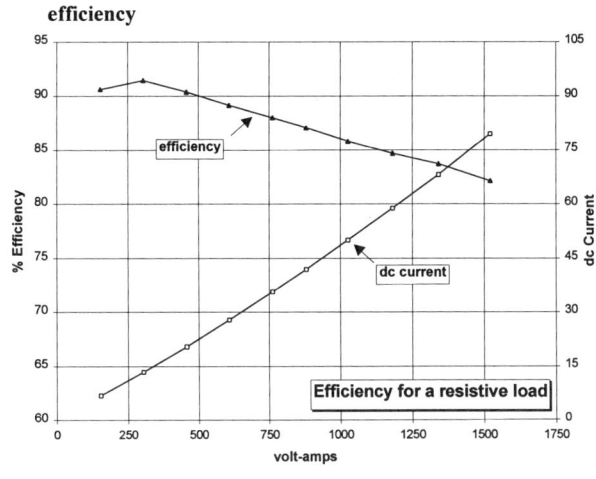

efficiency

Efficiency for a resistive load

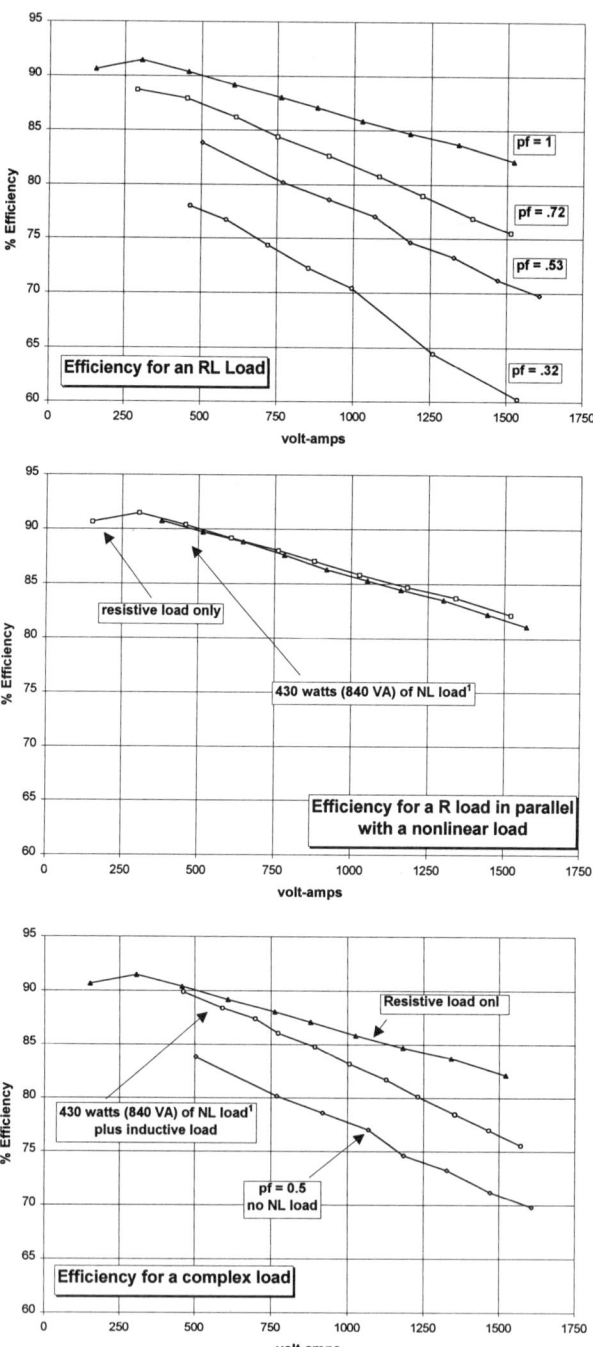

**voltage and frequency regulation, and distortion**

| Note: Regulation is referenced to no-load values of 118.7 Vac and of 60.00 Hz. | | | |
|---|---|---|---|
| Test Configuration | % voltage regulation | % frequency regulation | % voltage distortion |
| % full load (PF = 1.0) | | | |
| no load | 0 | 0 | 42 |
| 20% full load | 2.4 | 0.0 | 37 |
| 50% full load | 3.0 | 0.0 | 27 |
| 90% full load | 2.9 | 0.0 | 28 |
| 100% full load | 2.1 | 0.0 | 31 |
| full load reactive | | | |
| PF = .5 | 1.2 | 0.0 | 31 |
| PF = .72 | 1.1 | 0.0 | 32 |
| non-linear loads in parallel with R | | | |
| NL = 56% of rated VA (VA @ 75%) | 3.6 | 0.0 | 27 |
| NL = 56% of rated VA (VA @ 100%) | 3.5 | 0.0 | 33 |
| motor loads | | | |
| motor only | 2 | 2 | 2 |
| motor plus R = 100% full load | 2 | 2 | 2 |

**inverter overload**

| Test Configuration | Planned Load Duration | Measured Load Duration | Manufacturer's Specification | Measured Power Level |
|---|---|---|---|---|
| full load | 5 hours | 5 hours | 1500 W | 1550 W |
| full load + 20% | 15 minutes | 15 minutes | not specified | 1600 W[3] |
| full load + 50% | 1 minute | 1 minute | not specified | 1690 W[3] |
| full load + 100% | 30 seconds | 30 seconds | not specified | 1830 W[3] |

**motor starting**

| | SNL line motor only | inverter motor only | inverter fully loaded |
|---|---|---|---|
| initial ac voltage (rms) | 117 | 2 | 2 |
| voltage sag | 9% | 2 | 2 |
| voltage regulation | 1% | 2 | 2 |
| surge current (peak amps) | 85 | 2 | 2 |
| time to steady state (seconds) | 0.13 | 2 | 2 |
| steady-state voltage (rms) | 117 | 2 | 2 |
| steady-state current (rms) | 12 | 2 | 2 |

**Comments**

[1] Nonlinear load drew 430 W (840 VA) when connected to grid and 390 W (520 VA) when connected to inverter

[2] inverter would not start ¾-hp motor load

[3] ac output voltage was reduced; target power levels of 1800 W, 2250 W, and 3000 W could not be achieved

# Photovoltaic Module and Array Performance Characterization Methods for All System Operating Conditions

David L. King

*Sandia National Laboratories*
*Photovoltaic Systems Department, MS0752*
*Albuquerque, NM 87185*

**Abstract.** This paper provides new test methods and analytical procedures for characterizing the electrical performance of photovoltaic modules and arrays. The methods use outdoor measurements to provide performance parameters both at standard reporting conditions and for all operating conditions encountered by typical photovoltaic systems. Improvements over previously used test methods are identified, and examples of the successful application of the methodology are provided for crystalline- and amorphous-silicon modules and arrays. This work provides an improved understanding of module and array performance characteristics, and perhaps most importantly, a straight-forward yet rigorous model for predicting array performance at all operating conditions. For the first time, the influences of solar irradiance, operating temperature, solar spectrum, solar angle-of-incidence, and temperature coefficients are all addressed in a practical way that will benefit both designers and users of photovoltaics.

## INTRODUCTION

This work was motivated by a desire to improve the accuracy and versatility of methods currently used for characterizing the performance of photovoltaic arrays in their actual use environment. The resulting improvements will enhance industry's ability to design systems that meet performance specifications, to rate system performance after installation, and to continuously monitor performance over the system's life. In general, these improvements should help industry accelerate the commercialization of photovoltaic systems.

The current ASTM standard method for testing the electrical performance of modules and arrays (1) has served the industry well, but is best suited for determining module performance at only one operating condition, the "Standard Reporting Condition." Unfortunately, the standard reporting condition is at a temperature (25 °C) unrepresentative of actual operating conditions where

---

*This work was supported by the U.S. Department of Energy under contract DE-AC04-94AL8500.*

50 °C is more common. The ASTM method doesn't translate well to other operating conditions, doesn't address all factors involved in outdoor performance ratings, and is often considered no better than ±10% accurate when applied in the field to large photovoltaic arrays.

The limited versatility of the ASTM method led utilities to define what they considered a more realistic procedure for specifying system performance based on a month-long evaluation period with measurements translated to a specified solar irradiance, ambient temperature, and wind speed (2, 3). This procedure gave performance at a realistic operating temperature, but resulted in regression analyses that were less accurate than desired and limited in their ability to distinguish the interactive influences of solar irradiance, solar spectrum, solar angle-of-incidence, temperature coefficients, degree of thermal equilibrium, ambient temperature, and wind speed.

Recently, a "performance index" has been proposed as a means for continuously monitoring PV system performance. The index would provide the ratio of actual power to predicted power on a continuous basis. The value of this index, however, is dependent on the accuracy of the model used for predicting array performance (4). Today, data acquisition systems are often used to continuously monitor performance of large systems, but up to 15% loss in array output can go undetected due to the limitations of the predictive models used to estimate the expected array performance. In addition, almost ten years of system monitoring has been required before reliability analysts can confidently detect degradation in power output as large as 1 to 2% per year (5).

The testing methods, analytical procedures, and performance model described in this paper are the result of over 15 years of experience in outdoor testing of photovoltaic cells, modules, and arrays at Sandia National Laboratories. The resulting methodology for characterizing electrical performance is believed to be a significant improvement over previous methods, has been successfully applied to a wide variety of modules, and is now being applied by Sandia and others during the acceptance testing of large systems. Our goal is to validate and document the method, and submit it for consideration as a new test standard.

This paper will first describe the new array performance model, then illustrate the technical concepts that have led to improvements in outdoor testing methods, and finally illustrate the use of the new methodology in characterizing the performance of crystalline-silicon and amorphous-silicon arrays.

## ARRAY PERFORMANCE MODEL

Photovoltaic array (module) performance for an arbitrary operating condition can be described by Equations (1-5). The variables defining the operating condition are irradiance, cell temperature, absolute air mass, and solar angle-of-incidence on the array. The equations for short-circuit current ($I_{sc}$), maximum-power current ($I_{mp}$), open-circuit voltage ($V_{oc}$), and maximum-power voltage ($V_{mp}$) provide the four primary parameters from which others (fill factor, maximum power, efficiency) can be calculated. Equations (1,3, and 4) result in linear relationships closely related to the fundamental electrical characteristics of cells in the module. Equation (5) uses a second order relationship for $V_{mp}$ that implicitly contains the influences of factors such as module series resistance, wiring resistance, and non-ideal cell behavior at low light levels. Two additional empirical relationships, the "$AM_a$ Function" and the "AOI-Function" are used to compensate for the influences of the solar spectrum and solar angle-of-incidence (AOI) on the short-circuit current. The terminology used in the equations is consistent with that used in ASTM standard methods for testing cells, modules, and arrays (1, 6, 7).

A fundamental premise of this performance model is that the $I_{mp}$, $V_{mp}$, and $V_{oc}$ of a cell, module, or array are well behaved and predictable parameters when described as functions of $I_{sc}$ and cell temperature ($T_c$) only. In other words, for a given $I_{sc}$ and $T_c$, the shape of the current-voltage (I-V) curve will be the same for any solar spectrum and angle-of-incidence. When this premise is valid, the performance characterization of a module or array becomes simply a matter of determining the short-circuit current, $I_{sco}$, at a "reference operating condition," and then relating the other three performance parameters to this reference using the "effective irradiance" in Equation (2). One significant advantage of this approach is that compensating for the effects of solar spectrum and solar angle-of-incidence can be accomplished by adjusting only the $I_{sc}$ parameter, as in Equation (1).

$$I_{sc}(E, T_c, AM_a, AOI) = (E/E_o) \, f_1(AM_a) \, f_2(AOI) \, \{I_{sco} + \alpha_{Isc} (T_c - T_o)\} \quad (1)$$

$$E_e = I_{sc}(E, T_c = T_o, AM_a, AOI) / I_{sco} \quad (2)$$

$$I_{mp}(E_e, T_c,) = E_e \, \{I_{mpo} + \alpha_{Imp} (T_c - T_o)\} \quad (3)$$

$$V_{oc}(E_e, T_c,) = V_{oco} + C_1 \ln(E_e) + \beta_{Voc} (T_c - T_o) \quad (4)$$

$$V_{mp}(E_e, T_c,) = V_{mpo} + C_2 \ln(E_e) + C_3 \{\ln(E_e)\}^2 + \beta_{Vmp} (T_c - T_o) \quad (5)$$

Where:
E = Plane-of-array (POA) solar irradiance using broadband pyranometer measurement corrected for angle-of-incidence sensitivity, $W/m^2$
$E_e$ = "Effective" irradiance, dimensionless, or "suns"
$E_o$ = Reference "one sun" irradiance in plane-of-array, 1000 $W/m^2$
$f_1(AM_a)$ = Empirically determined "$AM_a$-Function" for solar spectral influence
$f_2(AOI)$ = Empirically determined "AOI-Function" for angle-of-incidence affects
$AM_a$ = Absolute air mass
AOI = Solar angle-of-incidence on module, degrees

$I_{sco} = I_{sc}(E = 1000 \text{ W/m}^2, T_c = T_o \text{ °C}, AM_a = 1.5, AOI = 0°)$
$I_{mpo} = I_{mp}(E_e = 1, T_c = T_o \text{ °C})$
$V_{oco} = V_{oc}(E_e = 1, T_c = T_o \text{ °C})$
$V_{mpo} = V_{mp}(E_e = 1, T_c = T_o \text{ °C})$
$T_c$ = Temperature of cells inside module, °C
$T_o$ = Reference temperature for cells in module, e.g., 25 or 50 °C
$\alpha_{Isc} = I_{sc}$ temperature coefficient, A/°C
$\alpha_{Imp} = I_{mp}$ temperature coefficient, A/°C
$\beta_{Voc} = V_{oc}$ temperature coefficient, V/°C
$\beta_{Vmp} = V_{mp}$ temperature coefficient, V/°C
$C_1$ = Empirically determined coefficient relating $V_{oc}$ to irradiance
$C_2, C_3$ = Empirically determined coefficients relating $V_{mp}$ to irradiance

The concept of "effective irradiance" is defined for photovoltaic devices in ASTM methods (6, 7) to account for the fact that the devices do not respond to all wavelengths of light contained in the solar spectrum. Thermopile-based pyranometers, like the Eppley PSP, measure the irradiance from the entire solar spectrum and are used for establishing the solar resource. Because power production from photovoltaic systems is based on this total solar resource, the concept of "effective irradiance" is used to describe the portion of the entire solar spectrum converted to electricity by the photovoltaic system. As used in this paper, the term is broadened to include not only the solar spectral influence, but also the optical effects related to solar angle-of-incidence. Thus, the effective irradiance, $E_e$, in Equation (2) depends on both the solar spectrum ($AM_a$) and the influence of AOI on the $I_{sc}$.

This new approach for modeling array or module performance has several important features when compared to other methods. Some of these features are summarized as follows:

1. The model provides a well defined approach for obtaining an array performance "rating" at any user specified operating condition, not just the ASTM standard reporting condition.
2. The model provides a predictive model for array performance at all operating conditions, including the effects of solar spectrum and angle-of-incidence.
3. The model is easily implemented in a common spreadsheet.
4. All parameters required in the model can be determined through straight forward outdoor measurement procedures.
5. The fundamental electrical behavior of solar cells is preserved.
6. Temperature coefficients are handled in a more rigorous way with separate coefficients for $I_{sc}$, $I_{mp}$, $V_{oc}$, and $V_{mp}$.
7. The accuracy of performance "ratings" is improved by emphasizing the determination of a reference short-circuit current ($I_{sco}$), and then relating the

other parameters ($I_{mp}$, $V_{oc}$, $V_{mp}$) to the ratio of the measured $I_{sc}$ and $I_{sco}$. This approach preserves the inherent self-consistent electrical behavior of the cells.
8. Performance is related to cell temperature inside the modules ($T_c$), rather than module temperature, thus compensating for the frequent situation where modules are not in thermal equilibrium.
9. When designing or predicting performance of arrays, the model gives $I_{sc}$ in terms of the variable most readily available from solar resource databases, the irradiance, E, as measured by a thermopile-based pyranometer.
10. This model, coupled with a solar resource database, could provide a practical method for calculating a daily, monthly, or annual "energy rating" (8).

Prior to illustrating the application of this methodology in the performance characterization of two photovoltaic arrays, the next six subsections of this paper will discuss the technical improvements and concepts that are required to take full advantage of the new performance model. The topics discussed will include: improvements in irradiance measurements using pyranometers, relating solar spectral influence to absolute air mass, quantifying the influence of AOI, a more rigorous approach to applying temperature coefficients, a method for calculating the temperature of cells inside modules, and an empirical relationship relating module temperature to irradiance, wind speed, and ambient temperature.

## Solar Irradiance Measurements

Historically, one of the largest errors in rating the performance of PV arrays has had nothing to do with the array itself. Rather, it has been due to a time-of-day dependent systematic error in measurements of the plane-of-array solar irradiance. The accuracy of an array performance rating is directly related to the accuracy of the solar irradiance measurement; therefore, systematic errors in irradiance measurements must be addressed as the first step in the rating process. The systematic error most often ignored in field performance measurements has been due to the influence of solar angle-of-incidence (AOI) on the response (calibration) of typical pyranometers. Academically this issue has been investigated (9, 10), but rarely are pyranometers calibrated as a function of AOI for general use.

Standard ASTM methods for calibrating pyranometers (11) typically result in a single "calibration number," often reported for an AOI = 45°. In addition, ASTM methods for reference cells (6, 7, 12) result in a calibration that is valid only for normal incidence. Figure 1 illustrates the magnitude of the correction required in irradiance measurements as a function of AOI for a typical Eppley PSP pyranometer. The curve shown in Figure 1 was generated by curve fitting the results of an AOI-dependent calibration performed at Sandia. Clearly, the first step in improving the performance rating of modules or arrays is to obtain an AOI-dependent calibration of the pyranometer used for field measurements.

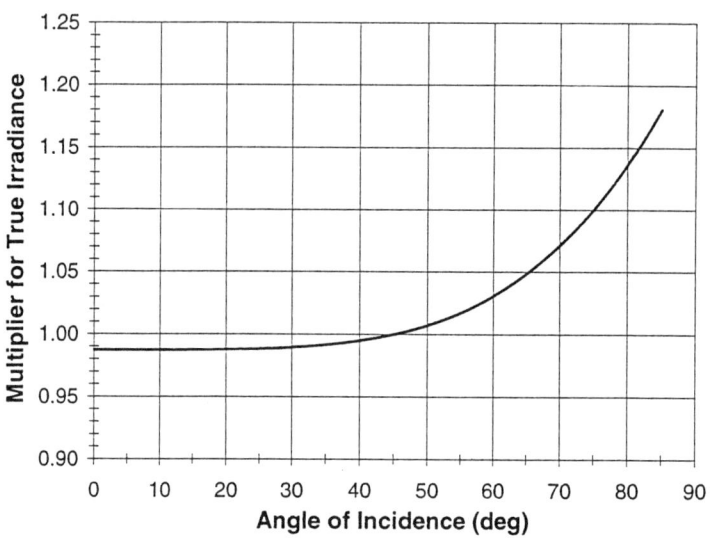

**FIGURE 1.** Influence of solar angle-of-incidence on the irradiance indicated by a typical Eppley PSP pyranometer.

## Influence of Solar Spectrum

As previously discussed, the current generated by a solar cell is influenced by the spectral distribution (spectrum) of sunshine. This is common knowledge to people familiar with photovoltaic technology. However, the magnitude of this effect and the real significance of the effect on the daily or annual energy production by a photovoltaic system is not well understood. Atmospheric scientists point out that the solar spectrum is influenced by a large number of variables including: absolute air mass, precipitable water, turbidity, clouds, dust, smoke, other aerosols, ground albedo, etc. (13). Nonetheless, testing experience at Sandia has indicated that, for the clear sky conditions typically present during performance rating, the majority of the solar spectral influence can be accounted for by considering only the effect of air mass on $I_{sc}$.

The solar spectral effect can be empirically related to absolute air mass, resulting in the "$AM_a$ Function", $f_1(AM_a)$. This function is technology specific, depending on the spectral response of the module, and also site specific, depending on the site's atmospheric characteristics. For clear sky conditions, however, experience has shown that an $AM_a$ Function determined for a crystalline silicon module in Albuquerque (NM) has little seasonal variation, and has been successfully applied to array measurements in Globe (AZ), Lake Powell (UT), Barstow (CA), and Sacramento (CA).

Air mass (AM) is the term used to describe the relative path length that the sun's rays have to traverse through the atmosphere before reaching the ground. An AM=1 condition occurs when the sun is directly overhead at a sea-level site; air mass values of 10 or greater occur near sunrise and sunset. Thus, air mass is a function of the position of the sun, which can be accurately calculated given site location, day of the year, and the time of day. To compensate for sites at altitudes other than sea level, the term "absolute air mass" ($AM_a$) is used. The absolute air mass is obtained by simply multiplying AM by the ratio of the site's atmospheric pressure (P) to that at sea level ($P_0$). If atmospheric pressure is not measured at the site, a simple exponential relationship used by the meteorological community can be used to calculate the pressure ratio using site altitude.

Equations (6-7) are commonly used for calculating the absolute air mass ($AM_a$) as a function of $Z_s$, the solar zenith angle (13, 14).

$$AM = \{\cos(Z_s) + 0.5057\,(96.080 - Z_s)^{-1.634}\}^{-1} \quad (6)$$

$$AM_a = (P/P_0)\,AM \qquad \text{where:}\quad P_0 = 760 \text{ mm Hg} \quad (7)$$

$$P/P_0 \approx e^{(-0.0001184\,h)} \qquad \text{where: } h = \text{altitude (m)} \quad (8)$$

Figure 2 shows the relative short-circuit current ($AM_a$ Function) for two different photovoltaic technologies, a Siemens (M55) crystalline-silicon module and a USSC (UPM-880) tandem amorphous-silicon module. Data in Figure 2 were measured with the modules on a solar tracker pointed normal to the sun from sunrise until sunset. The measured $I_{sc}$ was translated to 50 °C and normalized to 1000 W/m² using the irradiance as measured with an Eppley PSP pyranometer adjacent to the module. The thermopile detector in the Eppley pyranometer provided a spectrally-independent measurement of the total solar irradiance, E. Polynomial fits to the measured data provide the "AMa-Functions" that can be used in analyzing field performance measurements. These empirical functions are easily determined, requiring only a small solar tracker, a thermopile-based pyranometer, and a single solar cell with spectral response identical to those in the array being analyzed.

The $AM_a$ Function as used in the new model is straight forward to measure, widely applicable, and easily modeled knowing only the zenith angle of the sun and site altitude. Nonetheless, other analysts may desire to more rigorously model atmospheric effects and the resulting effect on $I_{sc}$. If this is done, the function $f_1(AM_a)$ in Equation (1) can simply be replaced, and the remainder of the performance model is still applicable. A specific example occurs when a reference cell is used to determine the reference $I_{sco}$ for a site where the solar spectrum present at the $AM_a=1.5$ condition differs significantly from the ASTM standard spectrum (15). In this case, the solar spectral irradiance occurring at $AM_a=1.5$ is measured with a spectral radiometer, and a spectral mismatch correction is calculated using a standard ASTM method (16). Including the spectral mismatch correction as part of the $f_1(AM_a)$ function will then relate measured performance back to the ASTM Standard Reporting Condition (1).

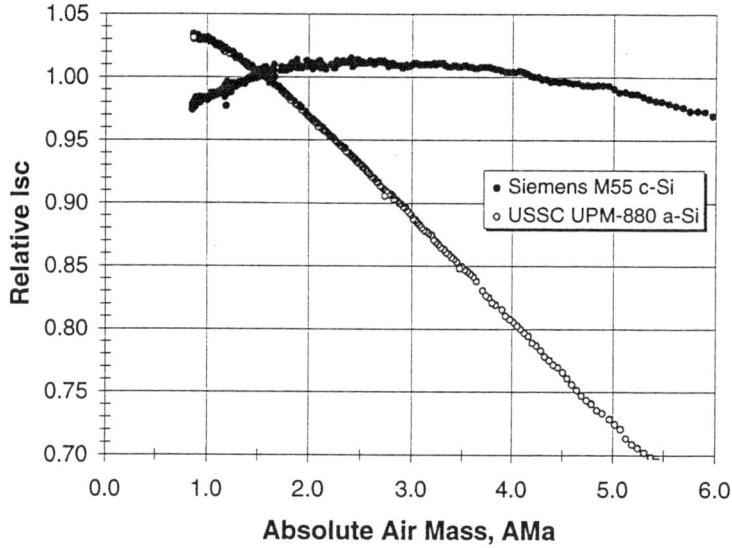

**FIGURE 2.** "$AM_a$ Functions" showing the influence of absolute air mass on $I_{sc}$ for a Siemens (M55) crystalline-silicon module and a USSC (UPM-880) tandem amorphous-silicon module.

## Influence of Solar Angle-of-Incidence

Photovoltaic modules have an AOI-dependent optical behavior that can be measured and used to improve the analysis of array performance. Like absolute air mass, solar angle-of-incidence is time-of-day dependent. Its affect on the short-circuit current ($I_{sc}$) of a photovoltaic module results from two causes. The first is familiar to solar enthusiasts as the "cosine effect." The "cosine effect" is independent of the module design, and is only geometry related. Due to the cosine effect, the $I_{sc}$ from a module varies directly with the cosine of the AOI. For example, at AOI = 60° the cosine effect reduces $I_{sc}$ by one half compared to the normal incidence condition. The second way $I_{sc}$ is affected by AOI is dependent on the module design. The optical characteristics of the module materials located between the sun and the solar cells cause the effect. For example, flat-plate modules typically have glass front surfaces; the dominant contributor to the "optical effect" in this case is reflectance from the front surface of the glass. This reflectance increases significantly for AOI greater than about 50 degrees.

Two different testing procedures can be used to measure these AOI-dependent influences on module performance. The first procedure uses POA irradiance measurements to remove the cosine effect and results in an empirical "AOI-Function" that quantifies only the "optical effect." The second procedure uses total (global) normal irradiance measurements and results in an alternative empirical function that contains both the cosine effect and the optical effect.

Figure 3 illustrates the "optical effect" measured for a typical ASE Americas (ASE-300-DG/50) crystalline-silicon module. A polynomial fit to the normalized $I_{sc}$ shown in the figure provides an "AOI Function" that can be applied in Equation (1). The measured data shown in Figure 3 were obtained at Sandia under clear sky conditions when the ratio of the direct normal irradiance (DNI) to total normal irradiance (TNI) was greater than 0.9. The measurements were made with the module on a computer-controlled solar tracker where AOI could be varied over a wide range in a short period of time, thus removing solar spectral influence from the data. The measured $I_{sc}$ was translated to 50 °C, normalized to 1000 W/m$^2$ using POA irradiance, and then divided by the $I_{sc}$ obtained at normal incidence. The pyranometer used to measure POA irradiance was calibrated as a function of AOI and provided the "true" irradiance on the module. The second relationship in Figure 3 describing the combined cosine and optical effects was measured during the same test, the difference being that the $I_{sc}$ was normalized using the total normal irradiance measured by a pyranometer on a separate tracker.

An additional clarification is perhaps needed here. The AOI Functions illustrated in Figure 3 are also dependent on the "clearness" of the sky, having the largest influence on $I_{sc}$ for test conditions where the ratio DNI/TNI is high. The clear sky situation is usually considered a prerequisite when conducting an outdoor rating of a module or array, so the functions are applicable. The opposite extreme occurs for a very overcast sky with perfectly diffuse illumination of the array; in this case, AOI has no influence whatsoever on the array's $I_{sc}$. Therefore, when the model is used for predicting array performance for a site with a diffuse solar resource, an additional relationship may be needed to diminish the optical effect measured under clear conditions.

## Calculating Solar Angle-of-Incidence

The solar angle-of-incidence on a module can be calculated using Equation (9). The values required in the equation are the azimuth and zenith angles defining the position of the sun ($AZ_s$, $Z_s$) and two angles that define the orientation of the module or array ($AZ_m$, $T_m$).

$$AOI = \cos^{-1}\{\cos(T_m)\cos(Z_s) + \sin(T_m)\sin(Z_s)\cos(AZ_s - AZ_m)\} \quad (9)$$

Where:
AOI = solar angle of incidence (deg)
$T_m$ = tilt angle of module (deg) (0° is horizontal)
$Z_s$ = zenith angle of sun (deg)
$AZ_m$ = azimuth angle of module (0° = North, 90° = East)
$AZ_s$ = azimuth angle of sun (0° = North, 90° = East)

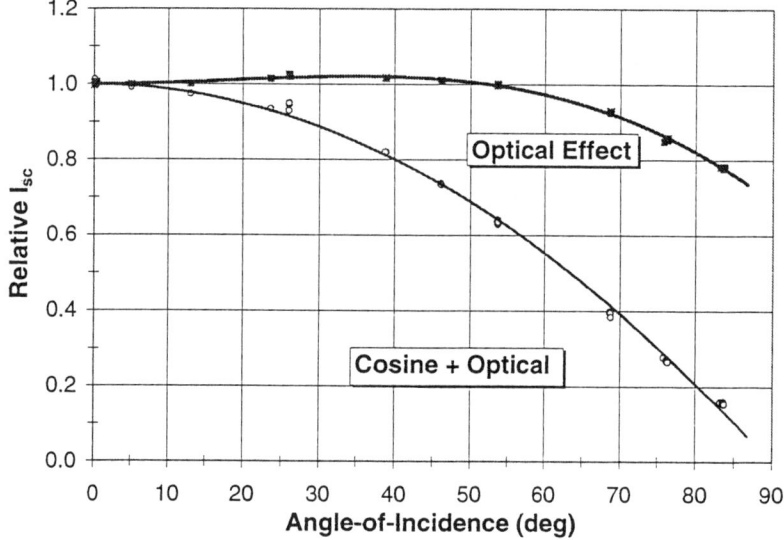

**FIGURE 3.** "AOI Functions" showing influence of solar angle-of-incidence on $I_{sc}$ for an ASE Americas (ASE-300-DG/50) module.

## Temperature Coefficients

Current ASTM standard methods make the assumption that all points on a measured current-voltage (I-V) curve can be translated to a different operating temperature by applying two temperature coefficients, one for current and one for voltage. Cell and module testing at Sandia has confirmed that this assumption is often invalid, and is one of the reasons that the I-V translation equations defined by ASTM (1) are less accurate than desired. For example, Table 1 gives the temperature coefficients measured at Sandia on four different commercial photovoltaic modules. The table illustrates that the coefficients at the maximum power condition can be significantly different from those at open-circuit and short-circuit. This observation led to the use of four separate temperature coefficients in the performance model defined by Equations (1-5).

**TABLE 1.** Measured temperature coefficients for typical commercial modules at ASTM Standard Reporting Conditions (1).

| Module | $dI_{sc}/dT$ (A/°C) | $dI_{mp}/dT$ (A/°C) | $dV_{oc}/dT$ (V/°C) | $dV_{mp}/dT$ (V/°C) | $dP_{mp}/dT$ (W/°C) |
|---|---|---|---|---|---|
| ASE-300-DG/50 | +.0059 | +.0021 | -0.23 | -0.24 | -1.3 |
| Siemens M55 | +.0013 | +.0001 | -.084 | -.085 | -.25 |
| Solarex MSX-64 | +.0022 | +.0002 | -.085 | -.086 | -.29 |
| USSC UPM-880 | +.0014 | +.0024 | -.085 | -.061 | -.044 |

The module temperature coefficients for $I_{sc}$, $I_{mp}$, $V_{oc}$, and $V_{mp}$ were measured at Sandia under outdoor conditions with high and stable solar irradiance (~1000 W/m$^2$), clear sky, and low wind speed (<2 m/s). Wind speeds above 2 m/s tend to increase the magnitude of the voltage coefficients measured. The module was first shaded until near ambient temperature was achieved, as indicated by multiple thermocouples attached to the back surface of the module. Then the module was quickly uncovered and I-V curves and temperature were measured every 20 seconds until the module reached its operating temperature (10 to 40 minutes depending on module design). Regression analysis was used to determine each of the four temperature coefficients.

The temperature coefficient for $P_{mp}$ in Table 1 was calculated rather than measured directly. $P_{mp}$ is the product of $I_{mp}$ and $V_{mp}$. Therefore, Equation (10) must be used to calculate this coefficient by using the temperature coefficients for $I_{mp}$ and $V_{mp}$ and the values for $V_{mp}$ and $I_{mp}$ at ASTM standard reporting conditions. The temperature coefficient for $P_{mp}$ varies with both irradiance level and module temperature. As a result, the common practice of assuming a constant $P_{mp}$ temperature coefficient should be used with caution. The -0.5 %/C value often used for crystalline silicon modules is only valid at 1000 W/m$^2$ irradiance and an operating temperature of 25 °C. Figure 4 dramatically illustrates this point for a tandem amorphous silicon module manufactured by United Solar Systems Corporation. At standard reporting conditions the module has a negative power coefficient of about -0.3 %/°C, but at cold low irradiance conditions the coefficient is positive, about +1.5 %/°C! A similar but less dramatic behavior is illustrated in Figure 5 for a crystalline silicon module.

To analyze array performance data, temperature coefficients appropriate for the entire array are required. Using predetermined temperature coefficients for modules, the array coefficients for voltage are determined by multiplying the module value by the number of modules connected in series in a module-string, and the array current coefficients are determined by multiplying the module value by the number of module-strings connected in parallel.

$$dP_{mp}/dT = V_{mp}\,(dI_{mp}/dT) + I_{mp}\,(dV_{mp}/dT) \tag{10}$$

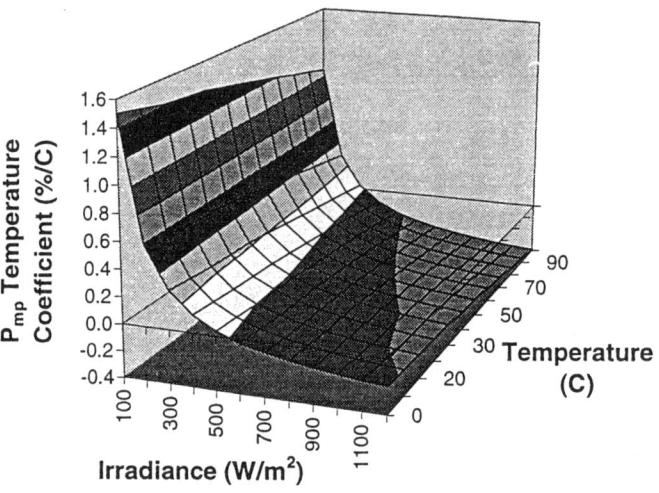

**FIGURE 4.** Temperature coefficient for $P_{mp}$ for a USSC UPM-880 tandem amorphous silicon module as a function of irradiance and cell temperature.

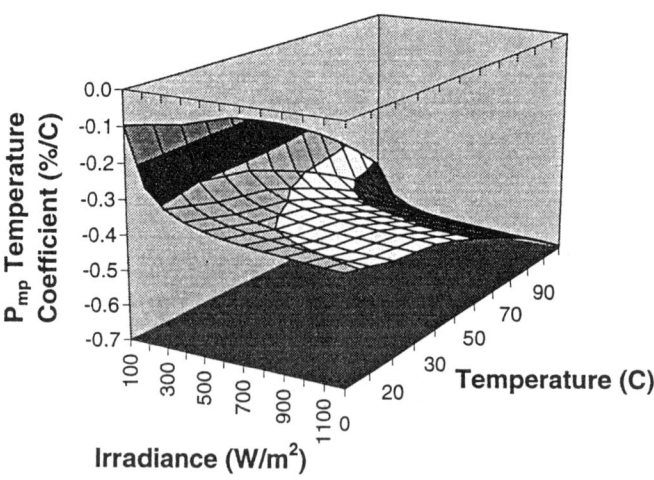

**FIGURE 5.** Temperature coefficient for $P_{mp}$ for an ASE Americas ASE-300-DG/50 module as a function of irradiance and cell temperature.

# Calculated Cell Temperature

Back-surface module temperatures are straight forward to measure, but they are not the best value to use in array performance characterization. The problem with using back-surface temperature is that the temperature difference between the back surface and the cell itself is neglected. This temperature difference arises from two factors: a temperature drop from the cell to back surface due to the thermal conductivity of materials between, and the frequent lack of thermal equilibrium in the outdoor environment. Lack of thermal equilibrium results from cloud passage, sunrise or sunset transitions, or sudden changes in wind speed or wind direction. One way to address these factors is to use a "calculated cell temperature" based on module $I_{sc}$ and $V_{oc}$, with the calculated value referenced to a known temperature when the module is in thermal equilibrium. A distinct advantage of this approach is that the calculated value provides an essentially instantaneous cell temperature, free of the time lag associated with the mass of the module. Thus, the bias errors in module performance measurements introduced by non-equilibrium conditions are avoided. The concept of calculated cell temperature is also used in commercial devices (ESTI sensors) used for measuring solar irradiance (17), and has been proposed as an alternative to actual temperature measurements during module testing (18).

Figure 6 illustrates an example of calculated cell temperature using data obtained during performance characterization of a 25-kW array of ASE Americas modules used in a telecommunications system located near Globe, AZ. By comparing the back-surface module temperature with the calculated cell temperature, it was determined that it took about 2.5 hours after sunrise for the array to reach quasi-thermal equilibrium. The equilibrium condition was maintained for about 2 hours until intermittent cloud cover occurred.

The relationship used to determine the "calculated cell temperature," $T_c$, is obtained by solving for $T_c$ in Equation (11). Basically, this equation gives an estimate of the average temperature for all the cells in the array. The reference values for $I_{scr}$, $V_{ocr}$, and $T_r$ are determined when the array is judged to be in a thermal equilibrium. In thermal equilibrium, the back surface temperature of the module can be used as a close approximation of the cell temperature, or an offset can be introduced to compensate for the small temperature gradient typically present between the back surface and the cell. Typical flat-plate crystalline silicon modules, in thermal equilibrium at 1000-W/m² irradiance and less than 3-m/s wind speed, usually have a cell temperature 2 to 3 °C higher than the back surface.

$$V_{oc}/N = V_{ocr}/N + (nkT_c/q)\ln(I_{sc}/I_{scr}) + \beta_{V_{oc}}(T_c - T_r) \qquad (11)$$

Where:
N = Total number of cells connected in series in the array
$I_{sc}$ = Measured array short-circuit current, (A)
$V_{oc}$ = Measured array open circuit voltage, (A)

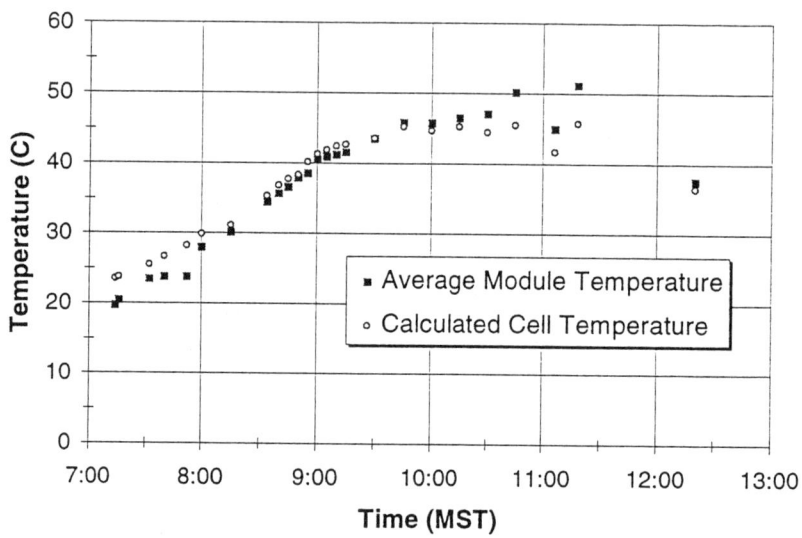

**FIGURE 6.** Calculated cell temperatures and module back-surface temperatures for a 25-kW array of ASE Americas (ASE-300-DG/50) modules tested on 9/25/95.

$I_{scr}$ = Array short-circuit current at the reference temperature, (A)
$V_{ocr}$ = Array open-circuit voltage at the reference temperature, (A)
$T_c$ = Temperature of cells inside module, (K)
$T_r$ = Arbitrary reference temperature for cells, (K)
$\beta_{Voc}$ = Temperature coefficient for $V_{oc}$ for individual cell, (V/°C)
n = Cell diode factor (n=1 can be assumed for typical silicon cells)
k = Boltzmann's constant, (1.38066E-23 J/K)
q = Elementary charge, (1.60218E-19 C)

## Module Operating Temperature

One final relationship is needed to make the performance model given by Equations (1-5) useful for system design and performance predictions prior to installation of an array. This relationship relates module temperature to the environmental variables typically available for system sizing calculations (solar irradiance, ambient temperature, and wind speed). Unfortunately, determining this relationship accurately is difficult because it depends not only on module design, but on mounting configuration (open rack or roof-integrated), wind direction, thermal radiation, and degree of thermal equilibrium. Figure 7 illustrates an empirically determined relationship for an ASE Americas module using measurements obtained over several months at Sandia. For example, at

800-W/m² irradiance and 1-m/s wind speed and ambient temperature of 20 °C, the module temperature is given as 44 °C. The data were screened to eliminate cloud transient effects, and thus represent near thermal equilibrium conditions. Although some work has been documented on the topic (19), module manufacturers and/or system designers will need to determine similar relationships for other module types and mounting configurations. On a positive note, if module operating temperature can be calculated within ±5 °C, then the resulting uncertainty in predicted power output should be less than 3%.

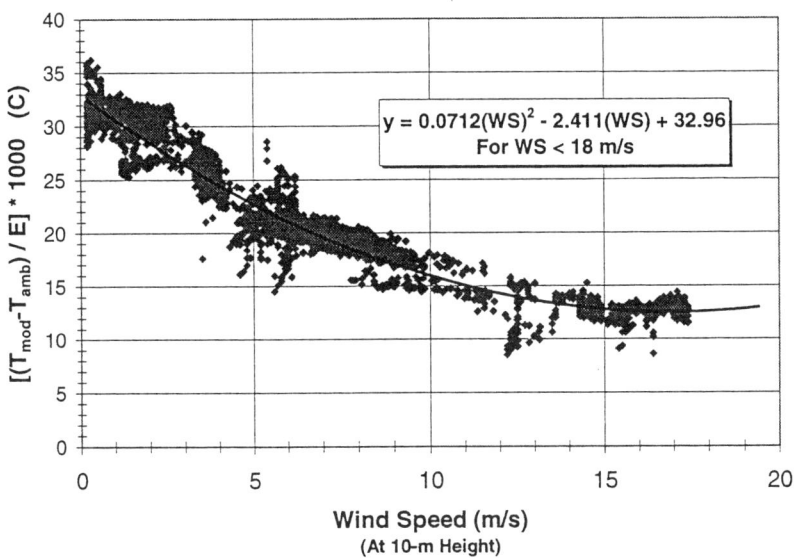

**FIGURE 7.** Empirical relationship providing module temperature, $T_{mod}$, as a function of ambient temperature, $T_{amb}$, POA irradiance (E), and wind speed for an ASE-300-DG/50 module mounted in an open rack.

## APPLICATION OF METHODOLOGY

The performance model and testing procedures previously discussed have been applied by Sandia to a wide variety of commercially available photovoltaic modules with good success. The methodology has also been applied during the performance rating of large arrays. To illustrate the accuracy and versatility of the methodology, the results of its application for two distinctly different photovoltaic arrays have been be provided. The first 25-kW array, owned by Arizona Public Service Company, was composed of 90 ASE-300-DG/50 modules configured in 15 parallel module-strings with 6 series connected modules in each. The second 1.3-kW roof-integrated array, owned by the Sacramento Municipal Utility

District, was composed of 96 USSC triple-junction amorphous-silicon modules (shingles) configured in 16 parallel module-strings with 6 series connected modules in each. In both cases, I-V curves were measured during a single day over a wide range of operating conditions using a DayStar (Model DS-100) curve tracer. Simultaneous measurements of module temperature and solar irradiance were recorded using a portable data acquisition system. A reference temperature of $T_o = 50°C$ was selected somewhat arbitrarily, but is representative of cell temperatures for typical flat-plate modules under PVUSA Test Conditions (2).

## Array Performance Characterization

The steps used in analyzing the array performance data were as follows:
1. Confirm calibration of all instruments and sensors used for field testing,
2. Correct POA irradiance measurements, E, for pyranometer's AOI dependent calibration constant,
3. Calculate the cell temperature, $T_c$, using Equation (11),
4. Translate the measured values for $I_{sc}$, $I_{mp}$, $V_{oc}$, and $V_{mp}$ to a reference cell temperature, $T_r = 50\ °C$, using Equations (12-15) and predetermined values for the temperature coefficients,

$$I_{sc}(50\ °C) = I_{sc} + \alpha_{Isc}\ E/E_o\ (50 - T_c) \tag{12}$$

$$I_{mp}(50\ °C) = I_{mp} + \alpha_{Imp}\ E/E_o\ (50 - T_c) \tag{13}$$

$$V_{oc}(50\ °C) = V_{oc} + \beta_{Voc}\ (50 - T_c) \tag{14}$$

$$V_{mp}(50\ °C) = V_{mp} + \beta_{Vmp}\ (50 - T_c) \tag{15}$$

5. Calculate the $AM_a$ and AOI using Equation (7) and Equation (9),
6. Adjust the $I_{sc}(50\ °C)$ values to $AM_a = 1.5$ and AOI = 0 degrees by dividing by the predetermined $f_1(AM_a)$ and $f_2(AOI)$ functions,
7. Plot $I_{sc}(50\ °C, AM_a=1.5, AOI=0°)$ versus POA irradiance and use a linear fit with zero intercept to obtain $I_{sco}$ needed in Equation (1),
8. Calculate the effective irradiance, $E_e$, by dividing measured $I_{sc}(50°C)$ by $I_{sco}$,
9. Plot $I_{mp}(50\ °C)$ versus $E_e$ and use a linear fit with zero intercept to obtain $I_{mpo}$ needed in Equation (3),
10. Plot $V_{oc}(50\ °C)$ versus $\ln(E_e)$ and use a linear fit to obtain $V_{oco}$ and $C_1$ needed in Equation (4),
11. Plot $V_{mp}(50\ °C)$ versus $\ln(E_e)$ and use a 2$^{nd}$ order polynomial fit to obtain $V_{mpo}$, $C_2$, and $C_3$ needed in Equation (5), and
12. Use performance model to calculate array performance for any desired POA irradiance, cell temperature, absolute air mass, and angle-of-incidence.

Figures 8-11 illustrate the results of applying the data analysis methodology to the ASE Americas crystalline silicon array. Figures 12-15 illustrate the results for the USSC triple-junction amorphous silicon array. In these figures, the measured values are shown along with the regression fits determined using the models described by Equations (1-5). The coefficients obtained from the regressions are the values needed in the performance model. In Figures 8 and 12, the irradiance indicated was corrected for the AOI-dependent behavior of the pyranometer.

The quality of the fits, the degree of linearity obtained, the zero intercepts for current versus irradiance, and the magnitude of the coefficient for $V_{oc}$ versus the logarithm of irradiance strongly indicate the validity of this performance characterization method. The model's versatility is illustrated by the success achieved for two distinctly different technologies under a wide range of operating conditions; irradiance from 100 to 1200 W/m$^2$, operating temperature from 10 to 65 °C, absolute air mass from 7 to 0.9, and angle-of-incidence from 0 to 75°. Figures 11 and 15 illustrate the measured $P_{mp}$ compared to $P_{mp}$ calculated using the performance model, over the duration of the test periods at both sites.

## CONCLUSIONS

The new performance characterization methodology described in this paper will enable the photovoltaic industry and its customers to accurately determine the performance of photovoltaic modules and arrays for all operating conditions. The methods can be used to design a new system, to rate the performance of an array after installation, to continuously monitor actual performance of an array relative to its anticipated performance, and to help evaluate the effectiveness of peak-power-trackers or the efficiency of inverters. For the first time, the method developed handles the influences of irradiance, temperature, solar spectrum, solar angle-of-incidence, and temperature coefficients in a practical yet rigorous way. The uncertainty in array performance rating has been reduced from perhaps ±10% to about ±3%, and the predictive model developed can also be used to calculate an "energy rating" for a module or array.

## ACKNOWLEDGEMENTS

A number of people have contributed significantly to the effort documented in this paper. The author is particularly grateful for the contributions of Jay Kratochvil and Bill Boyson (Sandia), Chuck Whitaker (Endecon Engineering), Pete Eckert (Arizona Public Service), and Steve Durand and Abraham Ellis (SW Technology Development Institute).

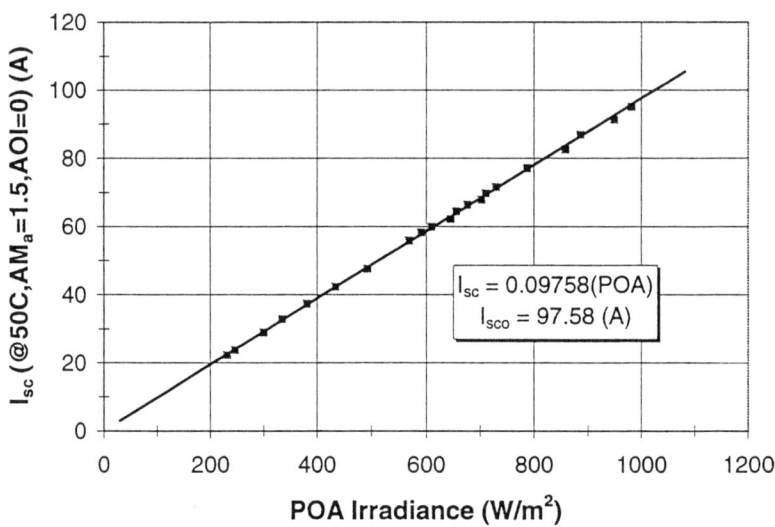

**FIGURE 8.** Measured $I_{sc}$ for 25-kW array of ASE Americas ASE-300-DG/50 modules. $I_{sc}$ is plotted versus the POA irradiance, E.

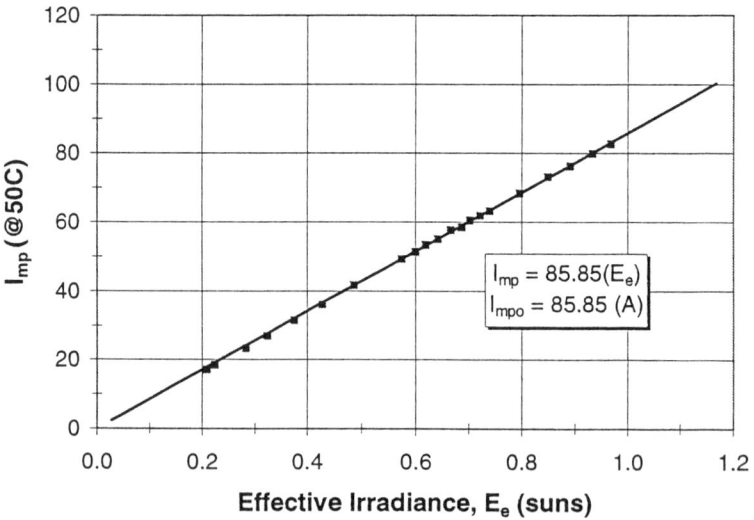

**FIGURE 9.** Measured $I_{mp}$ for 25-kW array of ASE Americas ASE-300-DG/50 modules. $I_{mp}$ is plotted versus the effective irradiance, $E_e$.

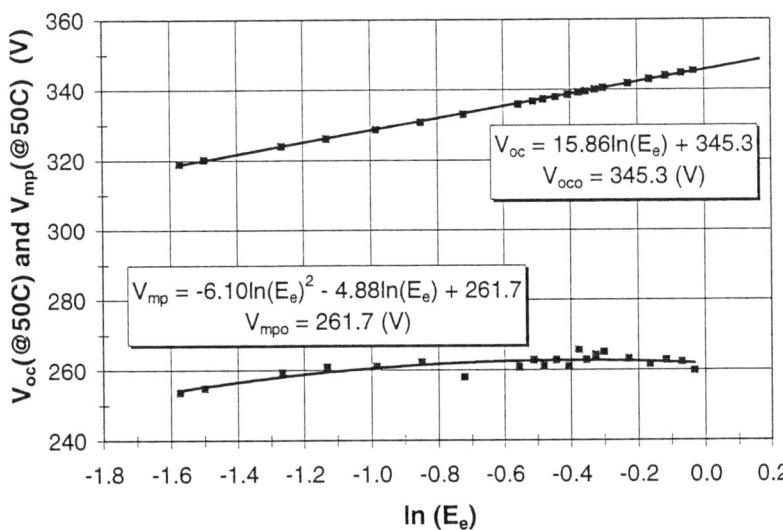

**FIGURE 10.** Measured $V_{oc}$ and $V_{mp}$ for 25-kW array of ASE Americas ASE-300-DG/50 modules including regression fits to measurements.

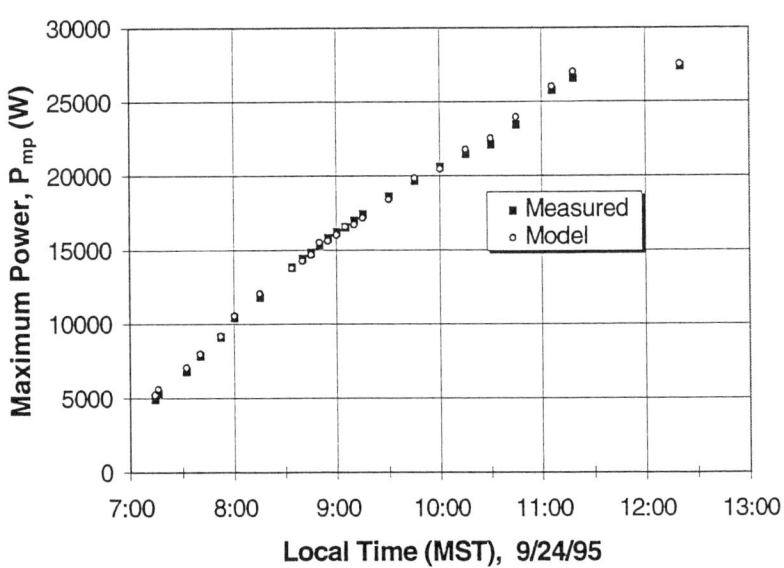

**FIGURE 11.** Measured $P_{mp}$ for 25-kW array of ASE Americas ASE-300-DG/50 modules compared to $P_{mp}$ given by the performance model.

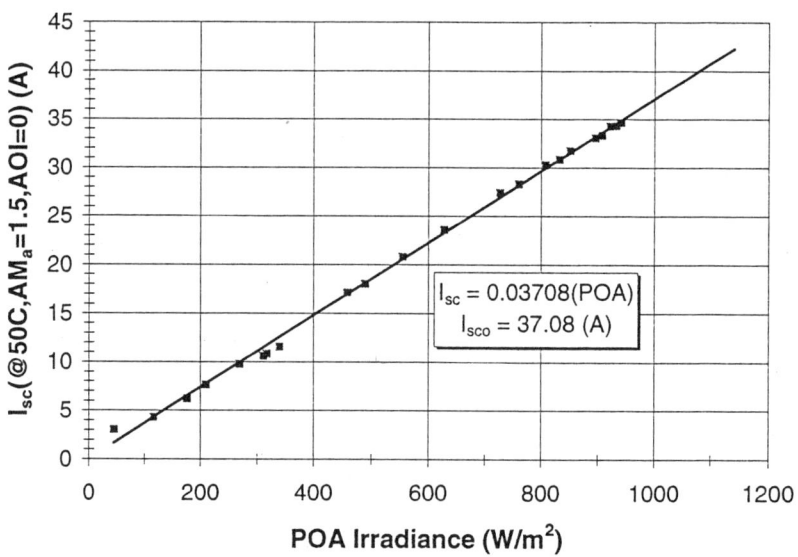

**FIGURE 12.** Measured $I_{sc}$ for 1.2-kW array of USSC triple-junction amorphous silicon modules. $I_{sc}$ is plotted versus the POA irradiance, E.

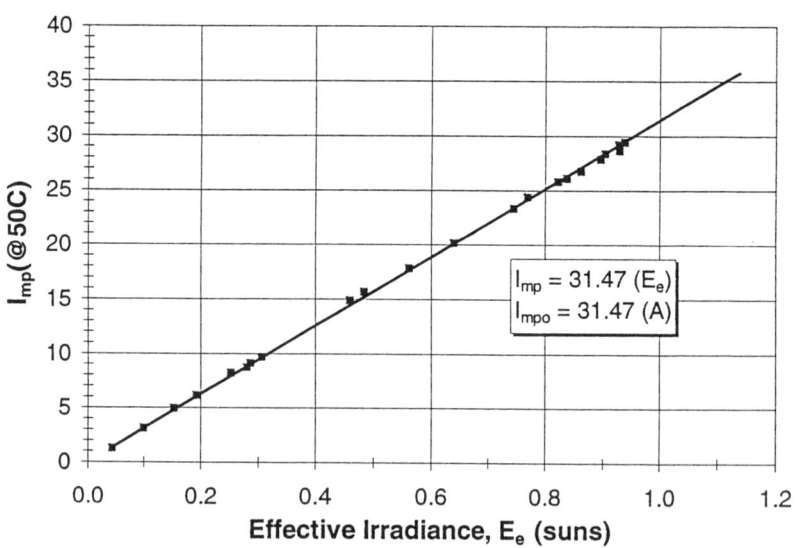

**FIGURE 13.** Measured $I_{mp}$ for 1.2-kW array of USSC triple-junction amorphous silicon modules. $I_{mp}$ is plotted versus the effective irradiance, $E_e$.

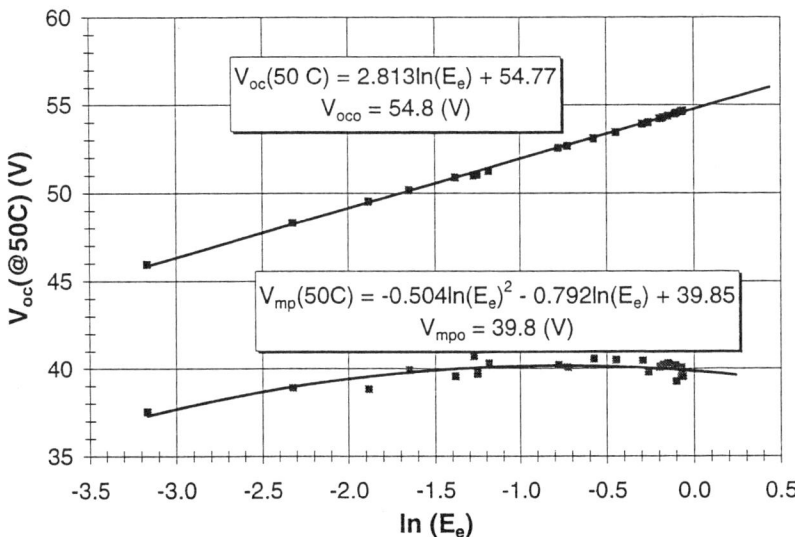

**FIGURE 14.** Measured $V_{oc}$ and $V_{mp}$ for 1.2-kW array of USSC triple-junction amorphous silicon modules, including regression fits to measurements.

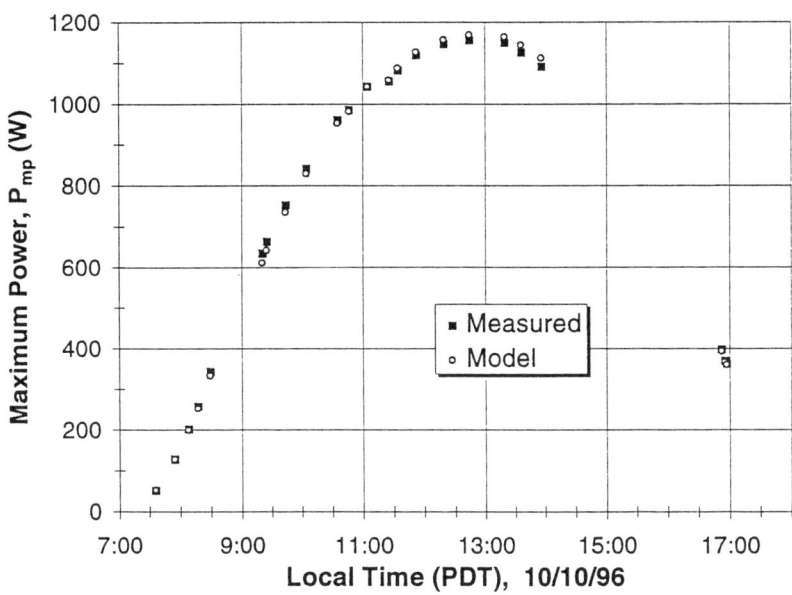

FIGURE 15. Measured $P_{mp}$ for 1.2-kW array of USSC triple-junction amorphous silicon modules compared to $P_{mp}$ given by the performance model.

# REFERENCES

1. ASTM E 1036, "Standard Methods for Testing Electrical Performance of Nonconcentrator Terrestrial PV Modules and Arrays Using Reference Cells."
2. T. Candelario, et al, "PVUSA - Performance, Experience, and Cost," *22nd IEEE PVSC*, Oct. 1991, pp. 493-500.
3. C. M. Whitaker, et.al., "Effects of Irradiance and Other Factors on PV Temperature Coefficients," $22^{nd}$ *IEEE PVSC*, Oct. 1991, pp. 608-613.
4. T. Townsend, et al, "A New Performance Index for PV System Analysis," *24th IEEE PVSC*, Dec. 1994, pp. 1036-1039.
5. A. Rosenthal, M. Thomas, and S. Durand, "A Ten Year Review of Performance of Photovoltaic Systems," *23rd IEEE PVSC*, May 1993, pp. 1289-1291.
6. ASTM E 1039, "Standard Method for Calibration and Characterization of Non-Concentrator Terrestrial Photovoltaic Reference Cells Under Global Irradiation."
7. ASTM E 1144, "Standard Method for Calibration of Non-Concentrator Terrestrial Photovoltaic Primary Reference Cells Under Direct Irradiance."
8. B. Kroposki, et.al., "Photovoltaic Module Energy Rating Methodology Development," $25^{th}$ *IEEE PVSC*, May 1996, pp. 1311-1314.
9. R. L. Hulstrom, editor, *Solar Resources*, MIT Press, Cambridge, 1989, ch. 5, pp. 174-308.
10. M. S. Imamura, et.al, "Assessment of Simplified Outdoor Calibration Methods for Solar Irradiance Sensors," *11th European PSEC Conference*, 1992, pp. 1638-1643.
11. ASTM E 941, "Standard Test Method for Calibration of Reference Pyranometers With Axis Tilted by Shading Method."
12. ASTM E 1125, " Standard Test Method for Calibration of Silicon Non-Concentrator Terrestrial Photovoltaic Reference Cells Using a Tabular Spectrum."
13. I. Zanesco and A. Krenzinger, "The Effects of Atmospheric Parameters on the Global Solar Irradiance and on the Current from a Silicon Solar Cell," *Progress in Photovoltaics*, Vol. 1, pp. 169-179, 1993.
14. F. Kasten and A. Young, *Applied Optics*, **28**, 4735-4738 (1989).
15. ASTM E 892, "Tables for Terrestrial Solar Spectral Irradiance at Air Mass 1.5 for 37° Tilted Surface."
16. ASTM E 973, "Test Method for Determination of a Spectral Mismatch Parameter Between a Photovoltaic Device and a Photovoltaic Reference Cell."
17. C. Helmke, W. Zaaiman, and H. Ossenbrink, "An Assessment of the Results of Calibrating 600 Silicon PV Reference Devices," $25^{th}$ *IEEE PVSC*, May 1996, pp. 1267-1270.
18. C. Leboeuf and H. Ossenbrink, "PV Module Power Output: Sensitivity and Uncertainty in Non-STC Measurements," $22^{nd}$ *IEEE PVSC*, Oct. 1991, pp. 614-619.
19. M. K. Fuentes, "A Simplified Thermal Model for Flat-Plate Photovoltaic Arrays," Sandia Report, SAND85-0330, May 1987.

# PV STANDARDS OVERVIEW

## Richard DeBlasio

National Renewable Energy Laboratory
1617 Cole Boulevard
Golden, Colorado 80401-3393

## ABSTRACT

A brief historical perspective and current status of the on going evolution of photovoltaic standards development and the use of these standards in promulgating accepted practices used in producing, measuring, and deploying Photovoltaic (PV) components and systems in the field. After nearly 20 years of experience in developing and writing domestic and international consensus PV standards the need and importance of standard methods and practices continues, as in the past, to be essential for a maturing PV industry. Part of this maturity has been in establishing and maintaining a common ground through the development of consensus standards and furthering the use of standards for PV commercialization in support of test facility accreditation, product certification, systems deployment, and safety code development to assure PV quality, performance, reliability, and safety.

## INTRODUCTION

In 1977 the U.S. government recognized the importance of establishing a common approach or uniform method of evaluating the performance of photovoltaic components and systems. The U.S. Department of Energy chartered the National Renewable Energy Laboratory (formerly the Solar Energy Research Laboratory) to develop an Interim Performance Criteria document for PV Energy Systems (1). This document, referred to by many as the IPC, laid the ground work for terrestrial PV component and system standards development and provided the basis for interim performance requirements and test methods used by the PV industry and government in the procurement process. It was also the forerunner to the PV standards writing bodies, and many of the criteria and test methods were adapted in principle from the IPC document printed in 1980 followed by the Performance Criteria document (2) printed in 1982. While not written under a formal consensus process, these documents represented the inputs of over 100 individuals from industry, government laboratories, and academia. The expertise provided from those active members and along with their recognized experience and knowledge gave the documents wide acceptance, and provided the catalyst for establishing a strong infrastructure of organized technical groups which ultimately became the voluntary consensus standards writing bodies (3).

Prior to 1978, consensus terrestrial PV standards activity was non existent, however considerable progress has been made since then, and a major reason for the success of these activities can be attributed to the stimulus provided by the DOE (Department of Energy) National PV Program and PV industry participation. Although the number of standards and codes organization active today is small, all have been working actively on photovoltaics. Major activities and groups include the Institute of Electrical and Electronic Engineers (IEEE), the American Society for Testing and Materials (ASTM), the International Electrotechnical Commission (IEC), the National Fire Protection Association (NFPA), and Underwriters Laboratories (UL).

## PV DOMESTIC STANDARDS

In 1978 the IEEE and ASTM, as well as UL in 1984, formally initiated consensus standards writing activities for photovoltaics as illustrated in Table 1. These initial activities were coordinated by the American National Standards Institute (ANSI). It was determined that the IEEE would produce the necessary standards related to PV module/array, power conditioning, storage, and systems, and the ASTM would address cell and cell/module standards development. While the ANSI no longer directly coordinates PV activities, the informal links and communications among these groups keep duplication of effort to a minimum. In fact in recent times, the IEEE and ASTM PV committees have held joint meetings in developing standards related to module qualification testing.

**TABLE 1.** Active Domestic PV Standards Organizations and Committees

| ANSI* | | |
|---|---|---|
| IEEE** | ASTM*** | UL**** |
| IEEE Standards Board Standards Coordinating Committee for PV Systems (SCC21) | Solar, Geothermal, and Other Alternative Energy Sources E44 Committee | Industrial Advisory Group on PV (PVIAG) |
| | E44.09 Subcommittee on PV Electric Power Conversion | Modules and Panels (UL 1703) |
| WG's: Systems, PCU's Storage, Module/Arrays | | Power Conditioning Units (UL Subject 1741) |

\* (American National Standards Institute)  \*\* (Institute of Electrical and Electronic Engineers)
\*\*\* (American Society for Testing and Materials)  \*\*\*\* (Underwriters Laboratories, Inc.)

The processes for initiating, developing, establishing a consensus, and obtaining approval are all unique within each organization. However, each group strives to include all parties representing the PV community (industry, end users, etc.). Also emphasized is liaison with other standards committees having the expertise needed to provide information representing the best practices and experience for the subject area being addressed by the sponsoring and writing committees.

The IEEE initially gave high priority to establishing a general overview document that defines systems and generic performance criteria and which would establish a basis for the development of all other IEEE PV standards. Procedures for an IEEE standard are rigorous. A specific standards development activity starts with a Project Authorization Request (PAR) from one of the working groups. After approval by SCC21 it must be approved by the New Standards Committee (NESCOM) and the Standards Board. This insures that proposed standards are not duplicated within the IEEE standards body. To further assist, liaison with the IEEE groups, including technical societies, are established. When the PAR is approved the working group writes the document and performs extensive reviews. Following formal review, balloting and final acceptance by the working group, the document is reviewed and balloted by SCC21. If approval is given it then is examined by the Review Committee (REVCOM) of IEEE. Final approval by the IEEE Standards Board is required before the document becomes an IEEE standard. Table 2 provides a list of IEEE Standards, status, and current activities.

The ASTM approach starts with identifying the need for a standard by the subcommittee (E44.09) whose members write and develop the standard with the approval of the executive committee of E44. After extensive reviews the document is balloted on by the subcommittee and any negative ballots received must be resolved before submittal to the full E44 committee, where again all negative ballots must be resolved. An accepted document is then balloted by the full ASTM membership, and if approval is obtained, is submitted to the committee on standards of ASTM for final approval and publication. Table 3 and 4 lists all of the ASTM PV standards and status.

The UL is a laboratory for examination and testing of devices, systems and materials to determine their relation to hazards to life and property, and to ascertain, define and publish standards, classifications and specifications for materials, devices, products, equipment, constructions, methods, and systems affecting such hazards. The standards developed for PV include UL Standard 1703 (Flat-Plate PV Modules and Panels) and UL Subject 1741 (Proposed draft of the Standard for Power Conditioner Units for use in Residential PV Power Systems) Table 5 provides a list of UL PV Standards and activities.

**TABLE 2.** IEEE SCC21 Standards and Status

| Number | Title | Status |
|---|---|---|
| 928 | IEEE Recommended Criteria for Terrestrial Photovoltaic Power Systems | Printed 1986<br>Reaffirmed 1991 |
| 929 | IEEE Recommended Practice for Utility Interface of Residential and Intermediate Photovoltaic Systems | Printed 1988<br>Reaffirmed 1991 |
| P929 | Recommended Practice for Utility Interface of Photovoltaic (PV) Systems | PAR* approval 9/96<br>work in progress |
| 937 | IEEE Recommended Practice for Installation and Maintenance of Lead Acid Batteries for Photovoltaic Systems | Printed 1987<br>Reaffirmed 1993 |
| 1013 | IEEE Recommended Practice for Sizing Lead Acid Batteries for Photovoltaic Systems | Printed 1990<br>Reaffirmation underway |
| 1144 | Sizing of Industrial Nickel Cadmium Batteries for Photovoltaic Systems | Printed 1987, Project Reaffirmation underway |
| 1145 | IEEE Recommended Practice for Installation and Maintenance of Nickel Cadmium Batteries for Photovoltaic Systems | Printed 1991<br>Reaffirmation underway |
| 1262 | Recommended Practice for Qualification of Photovoltaic Modules | Printed 1996 |
| P1361 | Recommended Practice for Determining Performance Characteristics and Suitability of Batteries in Photovoltaic Systems | PAR* approval 1993<br>work in progress |
| P1374 | Guide for Terrestrial Photovoltaic Power System Safety | PAR* approval 1993<br>work in progress |

* (Project Authorization Request to the IEEE Standards Board)

**TABLE 3.** ASTM E44.09 Standards and Status

| Number | Title | Status |
|---|---|---|
| E 927-91 | Specifications for Solar Simulation for Terrestrial Photovoltaic Testing | Printed/under revision |
| E 946-95 | Test Methods for Electrical Performance of Non-Concentrating Terrestrial Photovoltaic Cells using Reference Cell | Printed |
| E 973-91 | Test Method for Determination of the Spectral Mismatch Parameter between a PV Device and a PV Reference cell | Printed |
| E 1021-95 | Methods for Measuring the Spectral Response of Photovoltaic Cells | Printed/under revision |
| E 1036-85 | Methods of Testing Electrical Performance of Non-Concentrating Terrestrial PV Modules and Arrays using Reference Cells | Printed |
| E 1038-93 | Test Method for Determining Resistance of Photovoltaic Modules to Hail by Impact with Propelled Ice Balls | Printed |
| E 1039-94 | Method for Calibration and Characterization of Non-Concentrator Terrestrial PV Reference Cells under Global Irradiation | Printed |
| E 1040-93 | Specifications for Physical Characteristics of Non-Concentrator Terrestrial PV Reference Cells | Printed |
| E 1125-94 | Test Method for Calibration of Primary Non-Concentrator Terrestrial PV Reference Cells using Tabular Spectrum | Printed |
| E 1143-94 | Test Method for Determining the Linearity of a PV Device with respect to a Test Parameter | Printed |
| E 1171-93 | Test Method for PV Modules in Cyclic Temperature and Humidity Environments | Printed |
| E 1328-94 | Terminology Relating to PV Solar Energy Conversion | Printed |
| E 1362-95 | Test Method for the Calibration of Non-Concentrator Terrestrial PV Secondary Reference Cells | Printed |

TABLE 4. ASTM E44.09 Standards and Status (continued)

| Number | Title | Status |
|---|---|---|
| E 1462-95 | Test Method for Insulation Integrity and Ground Path Continuity of PV Modules | Printed |
| E 1524-93 | Test Method for Saltwater Immersion and Corrosion Testing of PV Modules for Marine Environment | Printed |
| E 1596-94 | Test Methods for Solar Radiation Weathering of PV Modules | Printed |
| E 1597-94 | Test Method for Saltwater Pressure Immersion and Temperature Testing of PV Modules for Marine Environments | Printed |
| E 1799-96 | Practice for Visual Inspection of PV Modules | Printed |
| E 1802-96 | Test Methods for Wet Insulation Integrity Testing of PV Modules | Printed |
| 131 | Test Method for Concentrator Devices | Work in progress |
| 200 | Test Method for Electrical Performance and Spectral Response of Multi-Junction PV cells and Modules | Work in progress |
| 201 | Test Method for Mechanical Integrity of PV Modules | Work in progress |

TABLE 5. UL (Underwriters Laboratories Inc). Standards and Status

| Number | Title | Status |
|---|---|---|
| UL-1703 | Flat-Plate Photovoltaic Modules and Panels | Printed 1986 first edition 1993, second addition |
| Subject -1741 | Proposed Draft of the Standard for Power Conditioner Units for Use in Residential PV Power Systems | Work in progress |

## INTERNATIONAL PV STANDARDS

Standards development activities in the international arena formally started in 1982 when the International Electrotechnical Commission formed Technical Committee 82 (TC-82) on Solar Photovoltaic Energy Systems. The committee is currently chaired by Japan and the United States holds the Secretariat position. Three working groups (WG) were established: WG-1 (Glossary), WG-2 (Modules), and WG-3 (Systems) and a fourth group on batteries works as part of a joint working group with TC-21 (Batteries). Current participating countries include: Australia, Austria, Canada, China, Czechoslovakia, Denmark, France, Germany, India, Italy, Japan, Romania, Russia, Spain, Switzerland, and the United States of America. As illustrated in Table 6 the ANSI and USNC provide the U.S. interface with the IEC and also the interface for U.S. coordination with respect to a technical advisor and advisory committee. The Solar Energy Industries Association (SEIA) coordinates the Secretariat and TAG (Technical Advisory Group) with funding support from the U.S. Department of Energy through the National Renewable Energy Laboratory (NREL).

**TABLE 6.** Active International PV Standards Organization and Committees

| ANSI[*] | USNC[**] | IEC[***] |
|---|---|---|
| | U.S. Technical Advisor | TC-82 Photovoltaics |
| | U.S. Technical Advisory Group (TAG) | Chair: Japan |
| | | Secretariat: U.S.A. |
| | | WG 1: Glossary |
| | | WG 2: Modules |
| | | WG 3: Systems |

[*] American National Standards Institute
[**] United States National Committee
[***] International Electrotechnical Commission

The process for international standards development is unique in that coordination must be maintained within each participating country through its TAG and requires strong cooperation and participation at TC82 meetings. The process starts when either the Secretariat or a National Committee prepares a draft document. The document is distributed to all National Committees for review and comment. The comments are forwarded to the Central Committee and the Secretariat. When sufficient agreement is reached, the document is released under the Accelerated Rule to all committees. If no substantive negative comments are received, the document is released by the Secretariat under the 6 month rule. If all issues are resolved by the participating countries, the Secretariat releases the document for final printing as an IEC standard.

International cooperation in this arena has been excellent over the years. Many of the IEEE and ASTM published standards have been used in developing a draft IEC standard with many of their specification adopted. Tables 7 and 8 below provides a list of published international standards and current activities.

**TABLE 7.** IEC TC-82 Standards and Status

| Number | Title | Status |
| --- | --- | --- |
| IEC-189 | Procedures for Temperature and Irradiance Corrections to Measured I/V Characteristics of Crystalline Silicon PV Devices | Printed |
| IEC-904-1 | Measurement of PV I/V Characteristics | Printed |
| IEC-904-2 | Requirements for Reference Solar Cells | Printed |
| IEC-904-3 | Measurement Principals for Terrestrial PV Solar Devices with Reference Spectral Irradiance Data | Printed |
| IEC-904-4 | On-Site Measurements of Crystalline Silicon PV Array I-V Characteristics | Being Printed |
| IEC-904-5 | Determination of the Equivalent Cell Temperature (ECT) of Photovoltaic (PV) Devices by the Open-Circuit Voltage Method | Printed |
| IEC-904-6 | Requirements for Reference Solar Modules | Printed |
| IEC-904-7 | Computation of Spectral Mismatch Error Introduced in the Testing of a PV Device | Approved for publication |
| IEC-904-8 | Guidance for Spectral Measurement of a PV Device | Approved for publication |
| IEC-904-9 | Solar Simulator Performance Requirements | Approved for Publication |
| IEC-1173 | Overvoltage Protection for PV Power Generating Systems | Printed |
| IEC-1194 | Characteristic Parameters of Stand-Alone PV Systems | Printed |
| IEC-1215 | Design and Type Approval of Crystalline Silicon Terrestrial PV Modules | Printed |
| IEC-1277 | Guide-General Description of PV Power Generating System | Being printed |

**TABLE 8.** IEC TC-82 Standards and Status (continued)

| Number   | Title                                                                          | Status        |
|----------|--------------------------------------------------------------------------------|---------------|
| IEC-1701 | Salt Mist Corrosion Testing of PV Modules                                      | Being Printed |
| IEC-1702 | Rating of Direct Coupled PV Pumping Systems                                    | Being printed |
| IEC-1721 | Susceptibility of a Module to Accidental Impact Damage (Resistance to Impact Test) | Being printed |
| IEC-1727 | PV Characteristics of the Utility Interface                                    | Being Printed |
| IEC-1829 | Crystalline Silicon PV Array-On Site Measurement of I-V Characteristics        | Being Printed |

**CODES AND CERTIFICATION**

While not within the scope of this paper, the following areas of activity are important to touch on. A major effort by the PV community over the past three years to revise electrical safety codes and establish a PV module certification and laboratory accreditation activity within the United States has been accomplished. The success of these activities can be attributed, in part, to the body of PV standards that have emerged in recent times. The efforts of IEEE, ASTM, IEC, UL, and NFPA (National Fire Protection Association) in providing the technical basis, knowledge, and body of expertise for revisions to the 1996 and 1999 National Electrical Code (Article 690) fully illustrates the importance of consensus standards in the commercialization of PV. In addition, the PV community participated in an NREL funded study conducted by Arizona State University (ASU) to develop and document the criteria (4) and implementation approach (5) for PV module certification and laboratory accreditation. The testing requirements recommended were based on ASTM and IEEE PV standards. With this came the incorporation of PowerMark Corporation in 1996, which has adopted and implementing the ASU results. Efforts on the part of PowerMark will further enhance and support the use and development of consensus standards through experience and improvements in current tests, testing procedures, and evaluation practices.

## CONCLUSION

The new issues facing the standards activities are of a global nature. A need to certify PV systems has been brought to the attention of the world PV community. Efforts have been underway to further explore what can be done and what needs to be done by the standards groups to help in evaluating system level performance and quality through test and test procedure standards. The future and success of PV commercialization will require continued participation by the PV community in standards and code development to assure product quality and safety. A common approach or uniform method, with sound technical basis found acceptable by all through consensus, as envisioned in 1978, still holds true today.

## ACKNOWLEDGMENTS

This work is supported under contract DE-AC36-83CH10093 with the U.S. Department of Energy. It is impossible to list all of the people who have contributed to the generation and implementation of standards and codes. Almost every organization even remotely related to photovoltaics has been represented and active at one time or another. Every major manufacturer has supported the present activities, and their continued support is the force which keeps the activities moving ahead.

## REFERENCES

1. DeBlasio, R., Forman, S., Hogan, S., Nuss, G., Post, H., Ross, R., Schafft, H., *"Interim Performance Criteria for Photovoltaic Systems"*, SERI/TR-742-654, 1980.

2. DeBlasio, R., Forman, S., Hogan, S., Hoffman, A., Hogan, S., Longrigg, P., Nuss, G., Post, H., Ross, R., Schafft, H., *" Performance Criteria for Photovoltaic Energy Systems"*, Volumes 1 and 2, SERI/TR-214-1567, 1982.

3. Hogan, S., DeBlasio, R.,*"Photovoltaic Standards, an Update"*, Photovoltaics International 1985.

4. Osterwald, C., Hammond, R., Wood, B., Backus, C., Sears, R., Zerlaut, G., D'Aiello, R., *"Photovoltaic Module Certification/Laboratory Accreditation Criteria Development"*, NREL/TP-412-7680, 1995.

5. Osterwald, C., Hammond, R., Wood, B., Backus, C., Sears, R., Zerlaut, G., D"Aiello, R., *"PhotovoltaicModuleCertification/LaboratoryAccreditationCriteriaDevelopment:Implementation Handbook"*, NREL/TP-412-21291, 1996.

# Battery Testing For Photovoltaic Applications

Tom Hund

Photovoltaic System Applications Department,
Sandia National Laboratories, Albuquerque, NM 87185-0753

**Abstract.** Battery testing for photovoltaic (PV) applications is funded at Sandia under the Department of Energy's (DOE) Photovoltaic Balance of Systems (BOS) Program. The goal of the PV BOS program is to improve PV system component design, operation, reliability, and to reduce overall life-cycle costs. The Sandia battery testing program consists of: 1) PV battery and charge controller market survey, 2) battery performance and life-cycle testing, 3) PV charge controller development, and 4) system field testing. Test results from this work have identified market size and trends, PV battery test procedures, application guidelines, and needed hardware improvements.

## INTRODUCTION

The Photovoltaic (PV) battery testing work at Sandia is funded by the Department of Energy's (DOE) Photovoltaic Balance of Systems (BOS) Program. Additional funding for the PV battery market survey has been provided by Sandia's Battery Analysis & Evaluation Department funded by the DOE Office of Utility Technology. In addition to work conducted at Sandia, the Florida Solar Energy Center (FSEC) is also funded by Sandia to conduct PV battery cycle-life testing and field evaluations. The PV battery testing at Sandia includes:

1) conducting a cost analysis and market survey of PV batteries and charge controllers to quantify the potential cost savings, market size, use patterns, and direction of growth in the PV industry (Sandia & Arizona State University),
2) conducting PV battery performance and cycle-life testing to identify specific battery charging characteristics that are important to improving battery cycle-life in PV systems (Sandia and FSEC),
3) working with PV charge controller manufacturers to improve component design, performance, and reliability (Sandia), and
4) conducting PV system field tests to verify and quantify potential performance improvements resulting from improved charge controllers and charging strategies (Sandia and FSEC).

The goal of this work is to improve PV system component design, operation, and reliability to reduce overall PV system life-cycle costs.

This work is supported by the Photovoltaic Energy Technology Division, US Department of Energy, contract DE-AC04-94AL85000.

# PV Battery Charging Overview

Batteries in PV systems are subject to performance losses that stem from five basic sources: 1) limited time and energy available to recharge the battery, 2) hardware failure, 3) inadequate battery maintenance, 3) improper system design, and 4) improper battery design. This report only focuses on PV battery charging issues.

Battery recharge in stand-alone PV systems is subject to solar resource that can be well below normal for weeks at a time resulting in a battery that is consistently under charged. The PV system battery recharge methodology is known in the battery industry as "opportunity charging". "Opportunity charging" differs greatly from traditional battery applications such as motive power deep-cycle, starting lighting and ignition (SLI), or uninterruptible power supply (UPS) float applications in that when the battery is under charge it may only receive a partial incomplete charge for long periods of time. This partial charge condition causes electrolyte stratification and/or irreversible sulfation in flooded deep-cycle lead-antimony batteries. After months in a partial charge condition, full recovery generally requires more charge than most PV/charge controller systems are capable of providing.[1] The degree to which premature PV battery capacity loss occurs in PV systems with flooded batteries is dependent on system design, regulation voltage, and battery design. Even well designed PV systems experience some premature battery capacity loss because of the lower PV regulation voltages and the lack of battery equalization.[2]

In addition to the problems associated with flooded batteries, the new gel or absorbed glass mat (AGM) valve regulated lead-acid (VRLA) batteries have additional PV battery charging difficulties. This is due primarily to the fact that VRLA batteries require a more precise charge control algorithm which varies depending on the manufacturer and type of VRLA battery. PV systems will typically use very simple on/off charge controllers with low reconnect voltages (~13.4 volts). The main problem is that these simple on/off charge controllers can be very inefficient because they tend to terminate charging prematurely. VRLA batteries are much more sensitive to regulation voltage (typical 14.1 to 14.4 volts at 25°C) and require much more time at the regulation voltage than flooded batteries. They are also much less tolerant of excess charge. VRLA battery manufacturers recommend constant voltage charging for their batteries, but most PV systems use on/off charge control.[3] The result is an under- or over-charged VRLA battery. If the VRLA battery is being charged at an excessively high voltage, then it will vent excess gas and lose capacity from loss of electrolyte or battery dry-out. Permanent capacity loss can also occur from under charging resulting in irreversible sulfation.[4] There is just a narrow window of charging parameters that a VRLA battery will operate under and still achieve rated cycle-life. Most PV charge controllers are not designed for the more rigorous requirements of a VRLA battery.

## POTENTIAL COST SAVINGS

Present PV systems using batteries for backup power add at least $0.10 to $0.50 per kWh to the system life-cycle cost (assuming 100% charge efficiency and full rated life of the battery).[5,6] These battery cost values vary due to the quality and type of battery used (flooded golf cart vs. industrial VRLA). If true battery charging efficiencies are factored (~70 to 90%) and battery life is cut in half, then battery costs can easily exceed $0.26 to $1.30 per kWh over the life of the system. This simple calculation indicates that battery costs can be equal to or greater than the life-cycle costs of the PV modules (assuming PV life-cycle costs of $0.25 per kWh or ~$6/Wp). Therefore the potential cost benefit to stand-alone PV systems is substantial if batteries meet their rated life expectancy.

## PV BATTERY AND CHARGE CONTROLLER SURVEY

This survey was funded by the DOE Office of Utility Technology through Sandia's Battery Analysis and Evaluation Department in support of the Photovoltaic System Applications Department. The survey was conducted by Bob Hammond of Arizona State University to better define the market size, hardware, operating environments, and future growth trends. Survey participants included:
1) 8 large PV system integrators,
2) 13 small PV system integrators,
3) 9 battery manufacturers, and
4) 8 charge controller manufacturers.

In addition to the survey, all participants were asked what they felt Sandia could do to expand the market for PV batteries. The completed results will be published by Sandia in December of 1996.[3]

Table 1 presents a preview from the battery survey listing total flooded and VRLA battery sales from the 21 PV system integrators for 1995. The data from the system integrators indicates that 64% of the PV batteries sold by them are VRLA with a dollar value of $3.4 million. A top-down market analysis, conducted by Bob Hammond using data from national PV sales and the survey, projects that the total US PV battery market should be about $34.7 million. Therefore the above PV systems houses share of this market is about 14%. The survey indicated that the majority of systems installed by the above integrators are using VRLA batteries, and with this in mind, it is very important to better understand the charging requirements of these batteries in PV systems.

**Table 1.** Total PV Battery Market For Surveyed System Integrators

| Battery Technology | 1995 | | | | | | |
|---|---|---|---|---|---|---|---|
| | # of Units | % of # | $ (Wholesale) | % of $ | kWhr | % of kWhr | $/kWhr |
| Valve Regulated | 16,846 | 64 | 3,390,782 | 71 | 26,524 | 57 | 128 |
| Flooded Vented | 9,462 | 36 | 1,370,060 | 29 | 20,012 | 43 | 68 |
| Total | 26,308 | 100 | 4,760,842 | 100 | 46,536 | 100 | 102 |

Included in the survey was a question directed to the PV charge controller manufacturers and system integrators requesting input on what Sandia could do to expand the PV battery market. The most predominant response was to provide assistance in the characterization of batteries for PV systems. Fifty-four percent of the small system integrators and 71% of the large system integrators indicated that this was their highest priority for support from Sandia.

## PV BATTERY LIFE-CYCLE TESTING

Battery manufacturers over the years have directed their design efforts toward 1) automotive starting, lighting, and ignition (SLI) batteries, 2) uninterruptible power supply (UPS) batteries, and 3) deep-cycle motive power batteries. All of these applications have very specific use requirements that may or may not be appropriate for PV systems. Much of the SLI and motive power battery performance data is obtained using Battery Council International (BCI) test procedures.[7] The BCI test procedures include specific tests for:
1) vehicular starting, lighting, and ignition (SLI) batteries,
2) electric vehicular and cycling batteries,
3) cycle-life testing of golf cart batteries,
4) cycle-life testing of deep cycle marine/RV batteries, and
5) cycle-life testing of floor maintenance batteries.

None of the above tests are specific to PV systems and the data that is generated is not directly applicable to PV systems. Controlled PV battery tests are not possible in fielded PV systems because of variations in solar resource, therefore, an automated battery tester was purchased from Firing Circuits to conduct PV battery testing at Sandia. The indoor test data can be used as a basis to modify system design, justify improved charge controllers, and evaluate new battery charging strategies. Improvements in PV battery performance will require a comprehensive approach that includes laboratory testing, improved system designs, improved battery application notes, improved charge controllers, and field verification. This is the approach taken at Sandia's PV System Applications Department.

## Test Procedure Development

Test results from previous indoor testing, using a 10% depth of discharge (DOD), identified several important points that need to be considered in any PV battery test procedure.[8,9] The test procedure considerations are:
1) Battery recharge is highly dependent on regulation voltage, charge/discharge rate, and time at regulation voltage/available energy.
2) Battery charge acceptance degrades in shallow cycling up to about 25 cycles.
3) Battery regulation voltage is dependent on battery temperature.

An example of the effect of charge rate on charge acceptance is in Figure 1. Figure 1 shows a GNB VRLA battery's overcharge (Ah charged - Ah load/Ah load x 100) as a function of the charge/discharge rate and the number of consecutive shallow cycles. The DOD and available Ah for recharge were the same for all test cases. As seen in Figure 1, the percentage overcharge with this test procedure is much higher (20 to 40%) for a 10-hour recharge and 8-hour discharge than for a 5-hour recharge and 2-hour discharge (5 to 12%). GNB recommends 5 to 10% overcharge for this 12-5000X VRLA battery. Also shown in Figure 1 is the effect of shallow cycling, which decreased overcharge or charge acceptance after each consecutive cycle. Figure 1 indicates that variations in battery charge acceptance will occur in PV systems because of changes in charge and discharge rates.

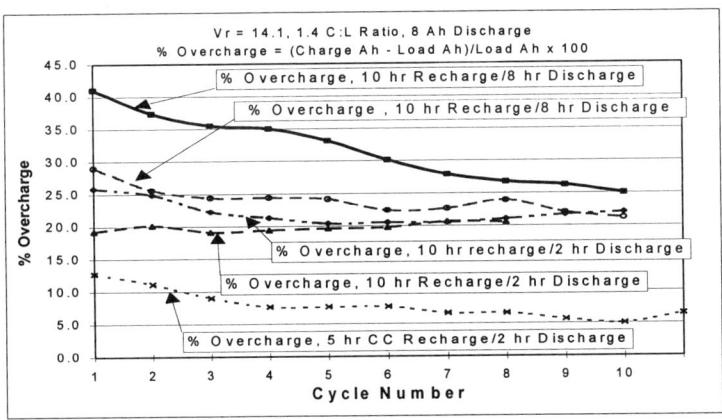

**Figure 1.** GNB 12-5000X Battery % Overcharge (Charge Acceptance) As Function of Charge/Discharge Rate And Cycle Number

In order to optimize battery performance in PV systems the effects of regulation voltage (Vr) set-points, charge/discharge rates, and array to load ratios must be well understood. It is also not known how the interaction of the above PV system parameters affect battery cycle-life. Higher battery cycle-life in PV systems is the end objective of the Sandia PV battery testing.

The most effective way to identify the effects and interactions of the Vr set-point, charge/discharge rate, and array to load ratio on battery cycle-life is the use of an experimental design and statistical analysis technique called Design of Experiments.[10] This technique will improve the quality of data, minimize testing, and identify the interactive relationships between the variables. In this case, the Design of Experiments procedure centers around a statistical analysis software package called MINITAB® that identifies the experimental design and processes the test results. The experimental model used for this evaluation is called the Central Composite Design and requires 20 test cases for three variables. MINITAB® is capable of designing the test plan and processing the cycle-life performance results to provide statistical diagnostics, main effects plots, and performance contour plots of all variables. This experimental technique is now being used at FSEC to provide a comprehensive cycle-life performance map of the GNB 12-5000X AGM VRLA battery which is very similar to the GNB Absolyte IIP battery. Both batteries are commonly used in PV systems.

## Test Procedure

1) **Initial Battery Charge** - Returns lost battery capacity from self discharge during storage after manufacturing.
2) **Initial Capacity Test** - Measures initial battery capacity at the test rate for one cycle. Abort voltages will be extracted from this test. The manufacturers recommended recharge procedure will be used. This test verifies battery capacity and health.
3) **Cycle-Life Tests** - Cycles batteries in a scenario that resembles the daily charge/discharge cycles of a PV system. (See Table 2 for test points and Figure 2 for example test sequence).
    a) 25-shallow cycles at specified variables. Cycle-life testing will be terminated when the battery is unable to deliver 80% of the Initial Capacity measured by end of load voltage (EODV).
    b) 4-deficit charge cycles discharging 60% of initial capacity. Abort test when the battery EODV indicates that 80% of its capacity has been discharged.
    c) 2-deficit charge cycles to 11.4 volts.
    d) Recovery charge cycles based on charge to load (C:L) ratio or regulation voltage. Recovery cycles will be calculated based on (Ah DOD/charge - discharge Ah) + 5 cycles.
    e) Sustaining shallow cycles using the specified variables in step (a). Cycle-life testing will be terminated when the battery is unable to deliver 80% of the Initial Capacity measured by EODV. The total number of cycles per

test sequence will be 91. (*Temperature is not varied in these tests, and all tests are to be performed at a nominal 25ºC +\-3ºC ambient temperature. Stabilized temperature baths or room temperature control are required.*)

4) **Data Processing** - Cycle-life results are input into MINITAB® from the 20 battery cycle-life test points for data processing.

The experimental design requires the identification of a low and a high value for regulation voltage, charge rate, and C:L ratio. The Design of Experiments software will identify the other points required for the Central Composite Design (see Table 2).

**Table 2.** Test Points For The 80 Ah GNB 12-5000X AGM VRLA Battery Test At FSEC

|  | Test Points | | | Axial Test Points | |
|---|---|---|---|---|---|
| Variable | High | Low | Center | High | Low |
| Voltage Set-Point | 14.4 | 13.8 | 14.1 | 14.6 | 13.6 |
| Charge/Discharge Rate (amps) | 4 | 1.6 | 2.3 | 8.2 | 1.3 |
| Charge to Load Ratio (Based On Time) | 1.5 | 1.1 | 1.3 | 1.64 | 0.96 |
| Temperature (Deg. C) | 25 | 25 | 25 | 25 | 25 |
| Depth of Discharge (%) | 20 | 20 | 20 | 20 | 20 |

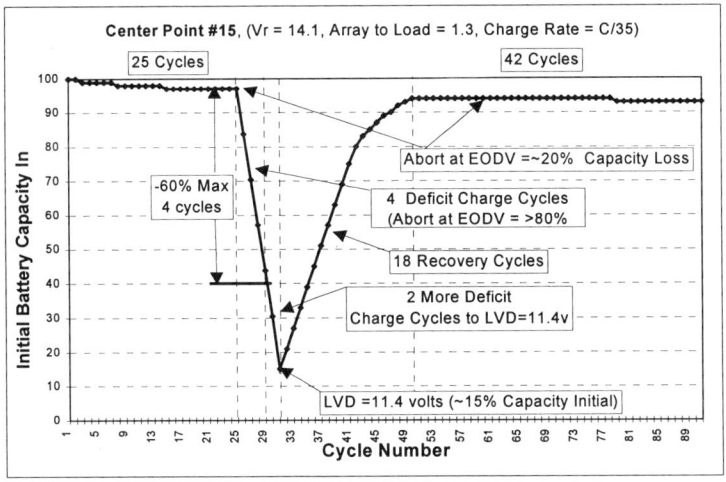

**Figure 2.** Simulated Example: PV Battery Cycle-Life Test Sequence
*Experimental Design/Data Reporting Format*

The Central Composite Design provides a higher quality prediction over the region of interest where main, interaction, and curvature effects are needed. All three of the above effects are used by MINITAB® to define the individual variable responses (Vr, rate, C:L), and identify the regression equations and coefficients necessary to map the response curvature which illustrates how factors influence the performance response (cycle-life). For this experimental design, MINITAB® has defined 20 tests for the given variables in Table 1. There are 8 tests to measure the response variable (Cycle-Life) at the corner (high and low) points, 6 tests to measure the response variable at the axial points (points on the axis outside the base cube), and 6 tests to measure the response variable at the center point. The high and low values are defined by the experimenter and MINITAB® calculates the other levels. The completed test sequence provides MINITAB® with all of the performance responses that are required to obtain the Residual Model Diagnostics (Figure 3 ), Main Effects Plot (Figure 4), and Performance Contour Plots (Figure 5). Three contour plots are required to fully map the interactions of the variables to the performance parameter (Cycle-Life).

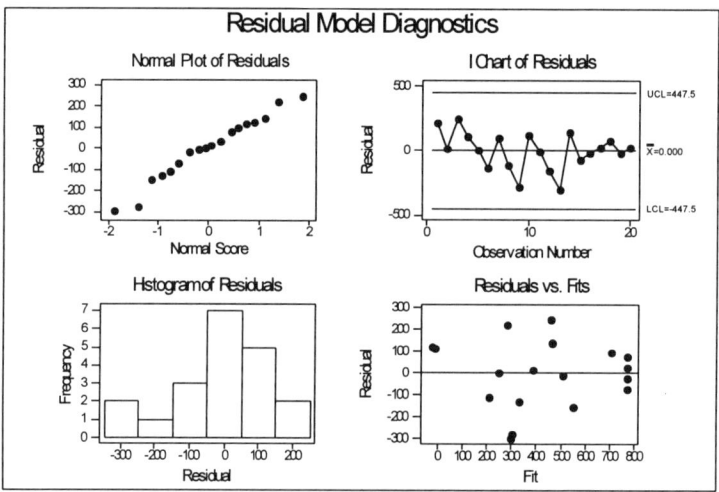

**Figure 3.** Simulated Residual Model Diagnostics

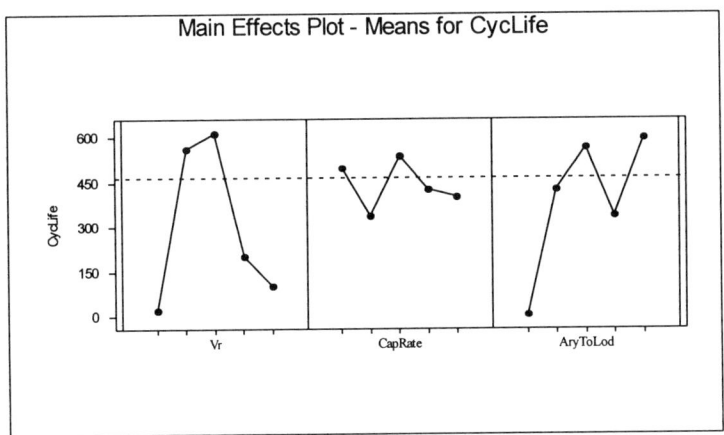

**Figure 4.** Simulated Main Effects Plot

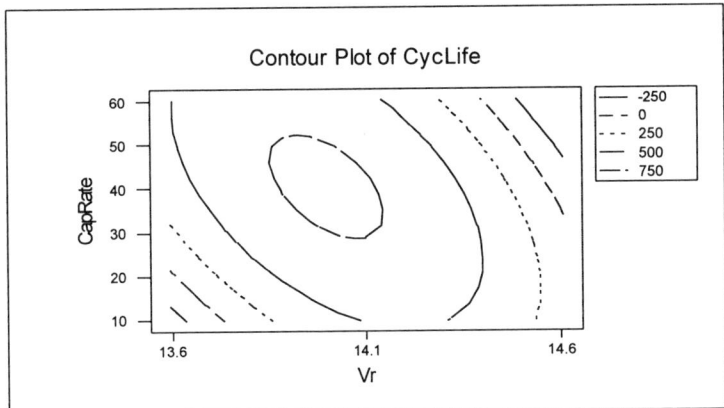

**Figure 5.** Simulated Contour Plot Of Cycle-Life (CycLife) Vs. Regulation Voltage (Vr) and Charge/Discharge Rate (CapRate)

## PV Battery Cycle-Life Test Results

Figure 6 and 7 show the preliminary battery overcharges and end of discharge voltages from test case #1 and #15. This test sequence is only being conducted using constant voltage charge control. At a later time the pulse width modulated charge control algorithm will be tested using a PV charge controller. In Figure 6, test case #1 (corner point), the battery is not recharging all of the deficit Ah. This is clearly indicated by the fact that in the first three cycles, the recharge Ah are between 97 to 100% and the EODV is dropping from 12.30 to 12.12 volts. It is interesting to note that after the first three cycles, the battery begins to receive a slight excess in recharge Ah (+2%). This is an indication of an initial capacity loss and improved charge acceptance at the lower battery capacity. The battery reached the abort criteria soon after this chart was prepared (-20% Capacity Loss).

In Figure 7, test case #15 (center point), the battery recharge Ah which are between 107 and 112% of the discharge Ah show that the battery is receiving about 8% overcharge after the initial 25 cycles. This is within the manufacturers specified 5 to 10% overcharge which indicates that the battery should see a much improved cycle-life compared to test case #1. Until this test is concluded it is difficult to say how long the battery will cycle. The test data here only indicates that this battery is being recharged within the recommended limits.

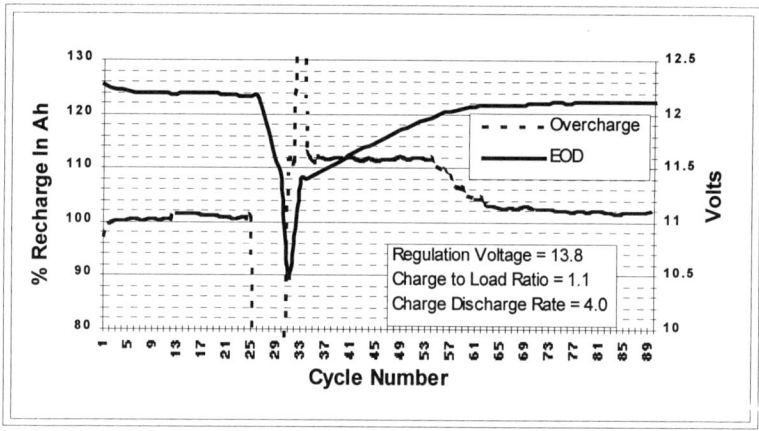

**Figure 6.** Test Case #1, Corner Point, PV Cycle-Life Test

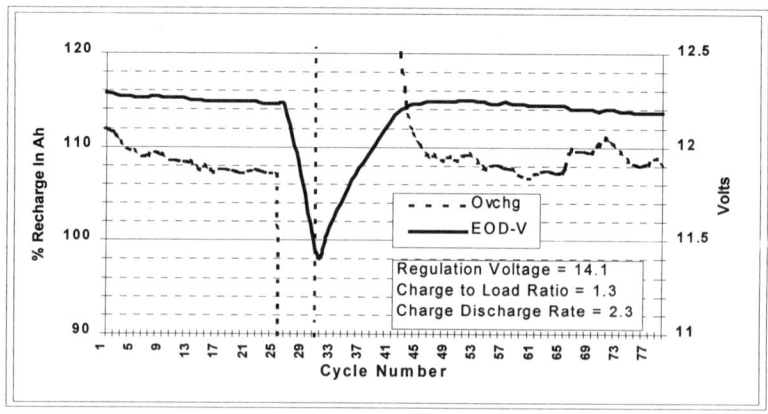

**Figure 7.** Test Case #15, Center Point, PV Cycle-Life Test

# PV CHARGE CONTROLLER AND FIELD TESTING

Sandia is now working with Morningstar and Digital Solar Technologies to test lower cost, improved performance, and higher reliability charge controllers such as the SunSaver PWM charge controller and the MPR-9400 microprocessor based PV charge controller. Some examples of this work are identified below.

## Morningstar SunSaver™

The development of this charge controller was partially funded by Sandia as part of DOE's Photovoltaic BOS Program. The SunSaver™ has been in field tests for over six months. The SunSaver™ has demonstrated exceptional performance and reliability for a small (<10 amps) low cost PV charge controller. It increases charge acceptance in VRLA batteries which should improve VRLA battery cycle-life. The increased charge acceptance is due to the PWM charge algorithm. Previous work[11] has indicated that improved charge acceptance is possible "...by discharging prior to charging or during the charging process". Figure 8 is an example of the PWM charging algorithm in the SunSaver™. The charge and discharge current pulses may be responsible for the improved charge acceptance seen when used with VRLA batteries. Field tests have shown that the SunSaver™ does provide VRLA batteries with a much higher overcharge or charge acceptance.

Figure 9 is an example of how the PV Battery Cycle-Life Test can be used to evaluate battery charge acceptance. In this test the Morningstar SunSaver™ was placed in series with the Firing Circuits automated battery tester to make a direct comparison between DC constant voltage charging and PWM charging using the same VRLA battery and test program for both charge algorithms. This battery survived over 200 cycles of the PV Battery Cycle-Life Test before it "permanently" lost over 20% of its capacity (99 to 73 Ah). The battery capacity could not be recovered with constant voltage charging using recommended charging procedures. SunSaver™ PV Battery Cycle-Life Test results now indicate that much of the battery capacity has been recovered. This is illustrated by the deficit charge cycle which removed 74 Ah at 11.46 volts. Previous capacity tests on this battery resulted in about 10 Ah of battery capacity between 11.5 and 10.5 volts. This indicates that an increase in battery capacity of about 11 Ah has occurred from the SunSaver™ PWM PV charge controller.

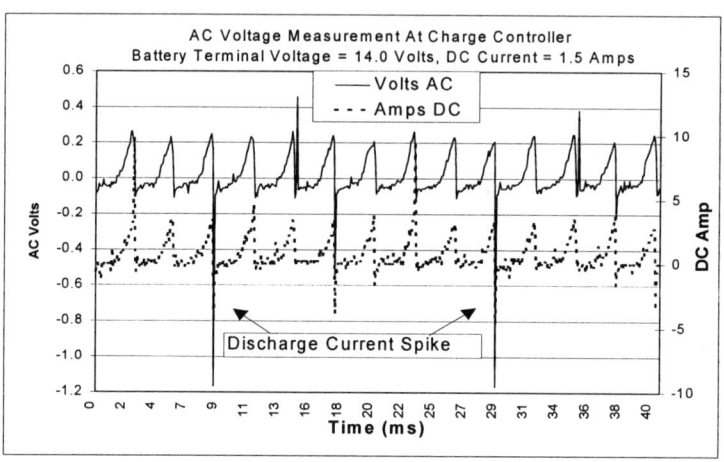

**Figure 8.** SunSaver™ Pulse Width Modulated Charging Showing Charge and Discharge Current Spikes

The test results in Figure 9 also indicate that the Morningstar SunSaver™ is providing the GNB 12-5000X AGM VRLA battery with 2 to 8% more overcharge compared to the constant voltage charge algorithm. In addition to the increased overcharge, the PWM charge controller is charging at 14.0 volts instead of 14.1 volts due to a voltage drop from the internal electrical resistance in the battery tester. The SunSaver™ is also charging the battery at about 25°C compared to the constant voltage data at about 27°C.

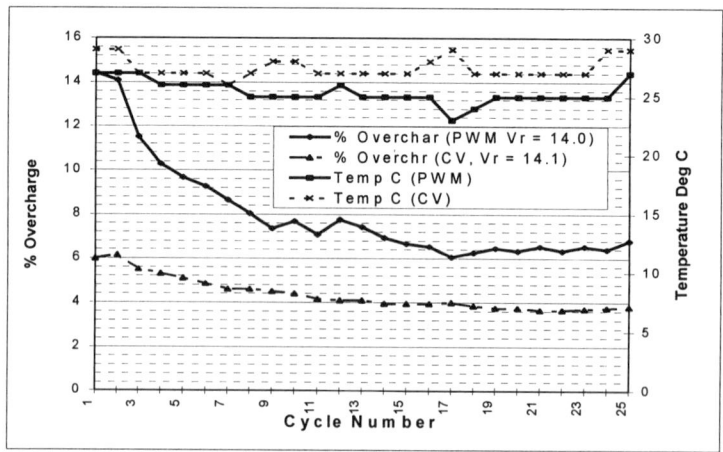

**Figure 9.** PV Battery Cycle-Life Test Comparing Constant Voltage Charging vs. The SunSaver™ Pulse Width Modulated Charging Using A GNB 12-5000X VRLA Battery

Both of these factors will reduce the overcharge for the PWM charger. In addition to the laboratory tests, field test results have shown that the SunSaver™ provides a much higher overcharge (>25%). The above test results certainly indicate that the PWM charger will charge the battery more efficiently with a lower voltage. Tests are continuing to evaluate the effect PWM and overcharge will have on battery cycle-life.

## Digital Solar Technologies MPR-9400

A new amp-hour (Ah) counting PV charge control algorithm developed in a cooperative effort with Digital Solar Technologies and implemented in their MPR-9400 microprocessor based PV charge controller is presently under test at Sandia. Previous Sandia analysis of PV hybrid power systems indicated that traditional voltage regulated charge control was not the most effectively way to charge batteries for hybrid power systems with multiple power sources and complex loads. Battery management in complex hybrid systems is difficult because of daily and seasonal load or DOD changes and fluctuations in solar resource. The new Digital Solar Technologies MPR-9400 Ah counting charge control algorithm and automatic battery equalization function have already demonstrated in preliminary testing that flooded deep-cycle batteries can be charged in a more effective way in hybrid power systems. Work is presently underway to demonstrate this new charge algorithm for the GNB VRLA battery technology.

The new Ah counting charge control algorithm requires the user to input several parameters into the MPR-9400. These parameters include:
1) **HVD-1 & 2** - Battery regulation voltage (Input by user)
2) **AHDOD** - Maximum depth of discharge per cycle (Counted by MPR-9400)
3) **%OVER** - Maximum overcharge above AHDOD (Input by user)
4) **%ADD** - Deficit or excess battery Ah at initial battery regulation voltage (Input by user)
5) **AHVRESET** - Battery voltage when new cycle or HVD's are reactivated (Input by user)
6) **BATAHINIT** - Battery capacity (Input by user)
7) **BatAH** - Maximum battery Ah recharged per cycle (Counted by MPR-9400)

Figure 10 shows the first test results for a GNB 12-5000X AGM VRLA battery bank. The Ah counting charge control disables the high voltage disconnect (HVD) 1 and 2 when the specified overcharge in Ah is reached (BatAH). The charge control algorithm requires the input of: 1) the battery capacity (BATAHINIT), 2) the preset maximum overcharge (%OVER), 3) a battery Ah adjustment for regulation voltage (%ADD), and 4) the cycle reset voltage (AHVRESET). The data in Figure 10 is calculated using the Campbell data logger and it indicates that the MPR-9400 is able to recharge the VRLA battery to within 1.5 Ah every day (253 to 254.5 Ah). This resulted in a battery overcharge based on depth of discharge of 7.9 to 12%, which is within the battery manufacturer's recommended values.

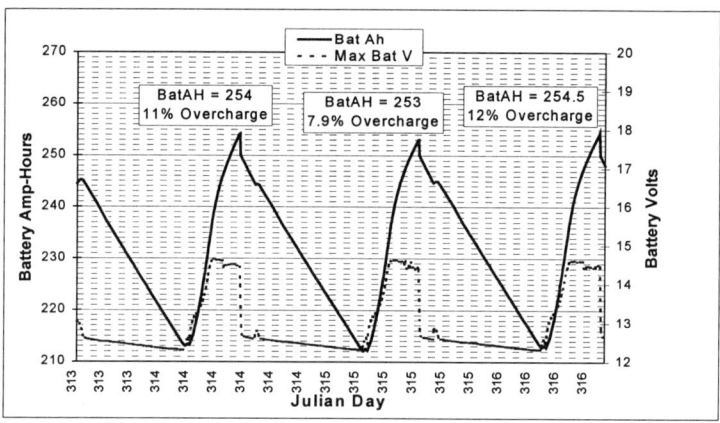

**Figure 10.** Ah Counting Charge Control on GNB 12-5000X VRLA Batteries. Battery Capacity Measured Using MPR-9400 Charge Control Algorithm. Battery Temperature is 13 to 15°C.

When the Ah counting charge control algorithm has completed its initial performance testing, it will be a significant advancement in PV hybrid charge control strategy. VRLA battery charge control will be greatly improved by providing the maximum available charge without under- or over-charging the batteries in variable resource and load periods. This will extend cycle-life by reducing dry-out and maximizing state of charge.

## SUMMARY

The Sandia PV battery testing program has already made significant steps toward its goal of improved PV battery cycle-life and lower life-cycle costs. This work has produced:
1) a detailed PV battery market survey,
2) PV Battery Test Procedure,
3) SunSaver charge controller
4) improved charge algorithm for the MPR-9400, and
5) improved PV system battery charging using the new charge controllers.

Work is now continuing to complete the cycle-life testing at Sandia and FSEC and to confirm field performance improvements using the new PV charge controllers.

The PV battery and charge controller survey, battery test results, and charge controller test results will be distributed to the PV industry to help facilitate improved system design, more reliable PV systems, and lower life-cycle costs.

## ACKNOWLEDGMENTS

Without the work and support of Garth Corey from Sandia's Battery Analysis & Evaluation Department, the battery survey would not have been done. This work was a significant and much appreciated addition to the PV battery testing program. Also the work of Ramu Swamy and Jim Dunlop at FSEC has been essential in conducting the PV Battery Life-Cycle Test.

## REFERENCES

1. Woodworth, J.R., Harrington, S.R., Dunlop, J.D., et al, "Evaluation of the Batteries and Charge Controllers in Small Stand-alone Photovoltaic Systems", First World Conference on Photovoltaic Energy Conversion, Hawaii, Dec. 1994.

2. Harrington S. R., Hund T. D., "Photovoltaic Lighting System Performance," 25th IEEE Photovoltaic Specialists Conference, Washington, May 1996. Pp. 1307-1400.

3. Hammond R. L., Turpin J. F., Corey G., Hund T., and Harrington S. R., "Photovoltaic Battery & Charge Controller Market & Applications Survey - An Evaluation Of The Photovoltaic System Market For 1995," Sandia Contractor Report # SAND96-xxxx, Dec. 1996.

4. Whitehead M. L., "Failure Mechanisms In VRLA Batteries," Proc. of 8th Battery Conference and Exhibition, Solihull, UK, May 1994, pp. 2.1.1 - 2.1.10.

5. Strong S., *The Solar Electric House*, "Voltage Regulators, Battery Chargers, Storage Batteries," Solar Design Associates, Inc., (508) 456-6855.

6. University of Cape Town Energy for Development Research Centre, *Remote Area Power Supply (RAPS) Design Manual*, ISBN:0 7992 1435 3, Ph. (021) 650-3230, Sept. 1992.

7. Battery Council International, *Battery Technical Manual 2nd Edition*, 401 North Michigan Av. Chicago, Illinois 60611, (312) 644-6610.

8. Dunlop J., and Swamy R., "Evaluation of Batteries and Charge Controllers In Small Stand-Alone PV Systems," Florida Solar Energy Center, Contractor Reports, Nov. 1996.

9. S.R. Harrington, and T. D. Hund, "Rating Batteries for Initial Capacity, Charging Parameters and Cycle Life in the Photovoltaic Application", PCIM , Sept., 1995

10. Box G.E.P., Hunter W.G., and Hunter J.S., *Statistics For Experimenters*, Wiley, NY, 1978.

11. Joseph A. Mas, "The Charging Process," Proc. 2nd Int. Electric Vehicle Symposium, Electric Vehicle Council, 1971, pp. 228-246.

# PV SOLAR RADIOMETRIC MEASUREMENTS

Daryl R. Myers, Theodore W. Cannon

*National Renewable Energy Laboratory, Golden, CO 80401*

**Abstract.** Radiometric measurements performed by the PV Solar Radiometric Measurements Task support NREL's centers for Measurements and Characterization, Performance Engineering and Reliability, and Renewable Energy Resources. The task provides characterization, measurements, testing, designs, and analysis of radiometric instrumentation and data for the performance of PV cells, modules, and systems. We describe recent characterization of the radiometric performance of pyranometers deployed for PV system testing at the NREL Outdoor Test Facility (OTF) and improvements undertaken in NREL broadband radiometer characterization. Typical measurement and calibration issues with diode array spectroradiometers used for absolute spectral measurements applied to PV performance and characterization are discussed.

## PV SYSTEM RADIOMETER EVALUATIONS

Direct-normal or beam radiation and global radiation on plane-of-array (POA) surfaces are typical broadband radiometric measurements for PV performance applications. These devices respond to the total radiation in the spectral region from 280 nm to 2800 nm. Pyrheliometers for measuring the direct beam have uncertainties of 1.5%-2.0% (1,2) and pyranometers for the global (total hemispherical) irradiance with uncertainties of 2%-5% (3).

PV system and module performance testing at NREL relies on well-calibrated individual radiometers assigned to each specific test activity. Each PV system is instrumented with a pyranometer integrated into the system data collection stream. An independent Reference Meteorological and Irradiance System (RMIS), described in previous review meetings(4,5), records 1-minute time-resolved data for quality assurance and special test applications.

In 1996, we evaluated the performance of three of the system radiometers versus the RMIS data under clear, partly cloudy, and overcast conditions for each of the four seasons. Diurnal profiles of RMIS data were examined to identify at least two specific clear, partly cloudy, and overcast days in the spring (March, April, May), summer (June, July, August), fall (September, October, November) and winter (December, January, February) of 1995. Figure 1 is the RMIS irradiance profile plot for Februrary 5th, 1995, the partly cloudy winter day selected. The plot displays (top to bottom) the 40° tilt, direct-normal, global-horizontal, and diffuse-horizontal irradiances. Table 1 lists the dates selected for this study. The radiometers evaluated were Kipp and Zonen Model CM-11 and CM-21 pyranometers. The pyranometers were installed on the United Solar

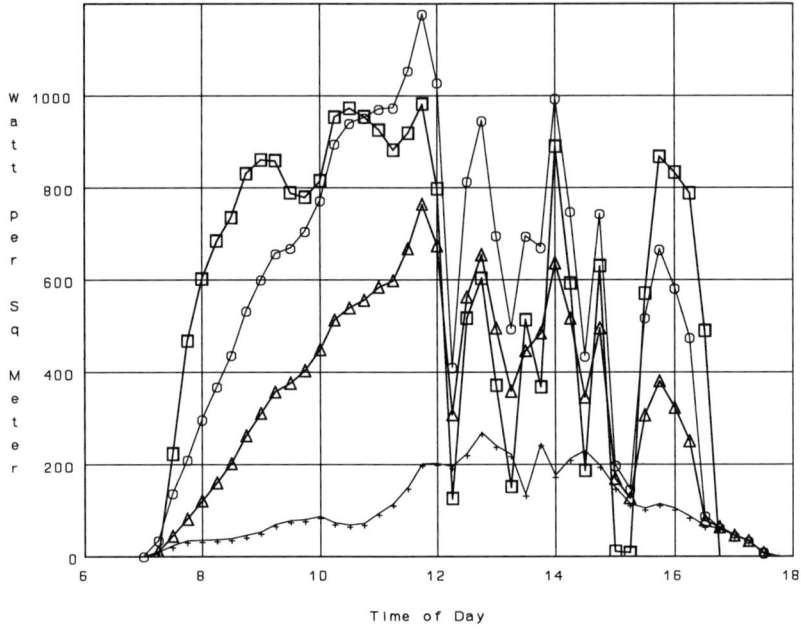

**Figure 1.** 40° Tilt, direct-beam, global-horizontal, and diffuse-horizontal (top to bottom) profiles for Feb 05, 1995, partly cloudy sky radiometer evaluation.

Table 1. Dates (1995) for Pyranometer Evaluation

| SEASON SKY> | Clear | Partly Cloudy | Cloudy |
|---|---|---|---|
| Spring | Mar 14, Apr 01 | Apr 30, May 12 | Mar 23, May 24 |
| Summer | Jun 18, Aug 01 | Jul 24, Aug 12 | Jul 30, Jul 31 |
| Fall | Oct 07, Oct 27 | Sep 23, Nov 20 | Oct 22, Nov 13 |
| Winter | Jan 22, Feb 25 | Jan 07, Feb 05 | Jan 16, Feb 10 |

Systems Corporation roofing modules, the Siemens Solar Industries cadmium indium diselenide (CIS) system, and an ASE Americas system(6,7). The radiometers were all installed at 40° (the latitude of the OTF test field is 39.74°N) facing south (azimuth 180° with respect to north=0°). The systems are located along the south edge of the OTF test field, at intervals of about 50 feet from west to east over a distance of about 200 feet. The RMIS station was located at the northwest corner of the test field, about 100 feet north of the PV system arrays.

RMIS one-minute data were integrated over the 15-minute period used for the three PV system data streams for each of the 24 days of the study. For each fifteen minute period, the average irradiance from the test pyranometers was divided by the RMIS 40°-tilt irradiance. Figures 2 to 4 show the ratios obtained.

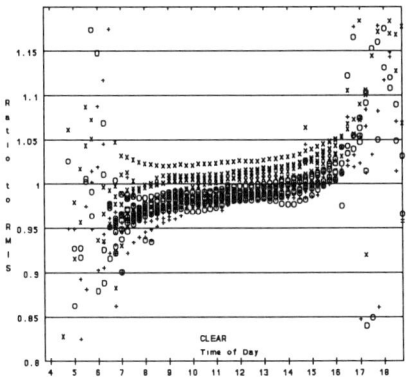
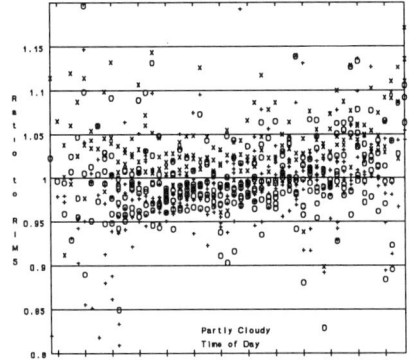

**Figure 2**. Clear sky condition PV system radiometer ratios to RMIS.

**Figure 3**. Partly cloudy sky condition PV system radiometer ratios to RMIS.

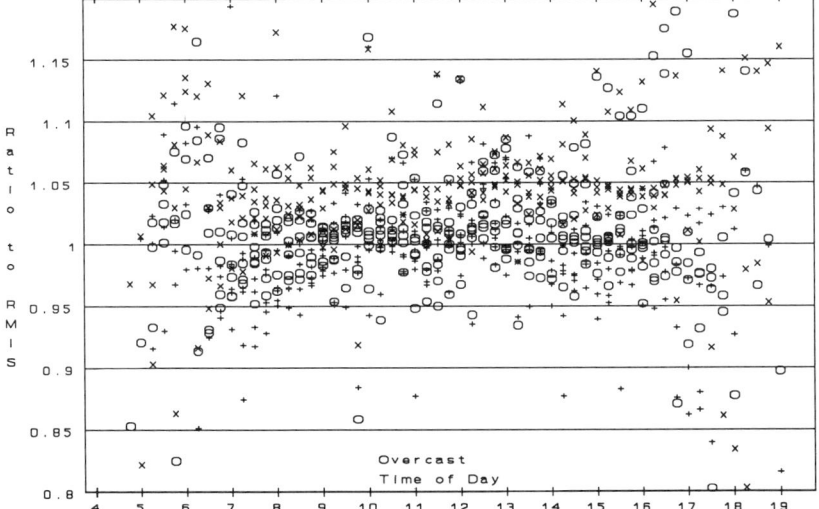

**Figure 4**. Overcast sky condition PV system radiometer (+=Roof, o=CIS, x=ASE) ratios to RMIS.

Each plot contains 2 days of ratios for each of the 3 systems, for all four seasons, or 24 curves. These plots indicate the following:

- For all clear-sky conditions, the four radiometers agreed to within 2.5% from 9 AM until 4 PM.

- Clear-sky range of ratios are within the ±3.0% uncertainties for the instruments when a single calibration responsivity is used.

- Increasing cloudiness results in ratios increasing to ±5% for partly cloudy and ±8% for totally overcast conditions

- The envelope of the ratios increases for all sky conditions in the early morning and late afternoon hours (before 9 AM and after 4 PM).

The increasing ratio envelope with increasing cloudiness is thought to be due to differences in the spatial and temporal distribution of the clouds from each radiometer's point of view (different locations of the instruments), and slight time response differences between the radiometers. Variations in radiometer temperatures and temperature response may also increase the ratio envelope under cloudier conditions. The wider envelope in the morning and afternoon ratios is attributed to north-south alignment and geometrical response differences. These are largest for large incidence angles. Figure 5 shows clear-sky data by season for the roofing system radiometer.

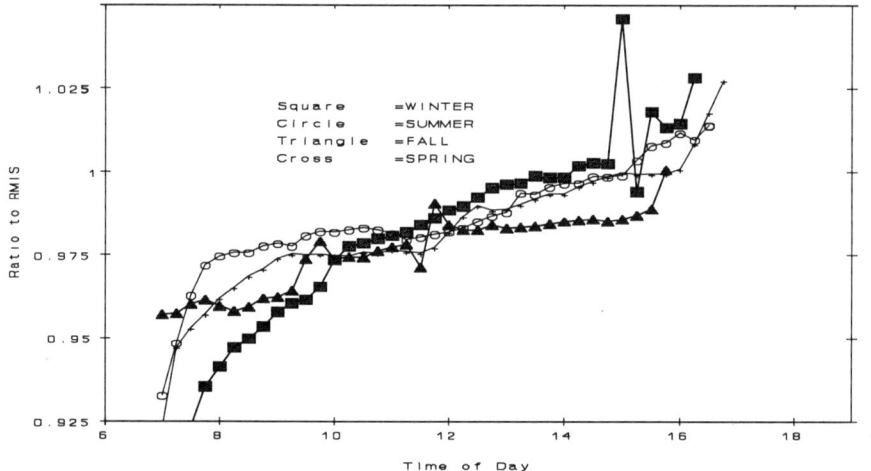

**Figure 5**. Seasonal variation in ratio of USSC roofing System radiometer to RMIS.

Noontime incidence angles are smallest in the spring and fall, the largest in the winter and summer.

To evaluate the radiometers use in OTF PV testing, we revised our broadband outdoor radiometer calibrations to map incidence angle response in the range from 85° to 0° incidence. We apply the component summation technique (compute the reference irradiance from direct and diffuse data) used for horizontal calibrations but tilt the radiometers to an angle equal to latitude minus the solar declination. The diffuse measurement is made with a pyranometer tilted at the same angle and a tracking shading disk to block the direct beam.

This technique achieves normal incident direct beam radiation at solar-noon and responsivity at off-normal incident angles for the rest of the day. We then compute average responsivity within incident angle bins (10 degrees wide) and produce a responsivity versus incident angle plot, with error bars spanning the range of responsivities within a bin.

Figure 6 illustrates the normalized (at 45°) incidence angle response of two different models of pyranometer arrived at using the technique in October of 1996. The unit represented by the open-circle symbols has a much flatter cosine response on average than that of the unit with the cross symbols. The spread of the data within any one range of incidence angles is consistently ±1.0%, representing the random components of uncertainty in the horizontal or this tilted calibration.

This technique maps the instrument response only along the east-west axis of the sensor. We will investigate the effects of varying the tilt angle to map more completely the cosine response of different regions of the sensor. We do not currently correct our PV system radiometer data for these effects, but hope to apply this calibration technique to all PV system radiometers at the OTF in the coming year.

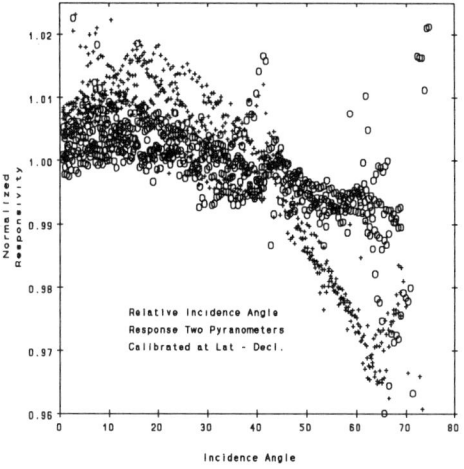

**Figure 6.** Incidence angle response for two pyranometers at latitude minus declination.

## ARRAY SPECTROGRAPH MEASUREMENT ISSUES

Diode array spectroradiometers are relatively inexpensive and readily available from many optical instrumentation manufacturers. These spectrographs acquire spectral information in milliseconds that would take minutes to acquire with traditional scanning grating spectro-radiometers as NREL uses and have described elsewhere (4,5,8,9). Using such an instrument, we became aware of the many difficulties associated in using such a device to measure spectra of pulse solar simulators.

The device used in our study covered the wavelength region from 300 nanometers (nm) to 1100 nm with an array of 1024 diode elements, or an equivalent spectral resolution of about 1 nm. The holographic grating was blazed at 1000 nm. The unit was calibrated for absolute spectral measurements by determining the responsivity (watts per square meter per nanometer)/(digital count) by measuring a 200-watt tungsten halogen lamp with a known spectral distribution. Spectral data were then collected under the NREL Spire 240A pulsed solar simulator. The 240A generates a 3-millisecond (ms) light pulse at 15 Hertz, or a period of 66.67 ms. Figure 7 compares the array radiometer and NREL scanning grating radiometer (Pulse Analysis Spectroradiometer System, PASS (8)) data, and shows the relative spectral distribution of the calibration lamp in the lower part of the figure.

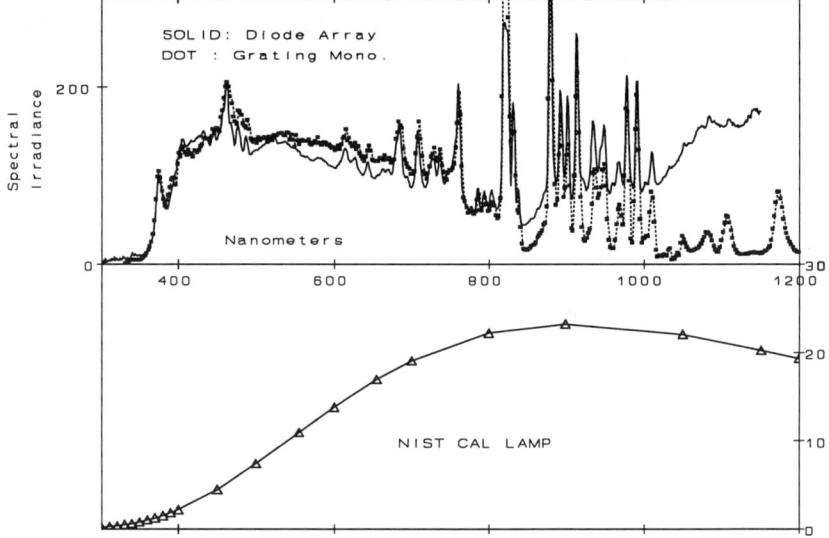

**Figure 7.** Relative spectral distribution of Spire 240A measured with diode array and scanning spectroradiometers and a calibration lamp spectrum.

Figure 7 shows that the array radiometer data greatly exceeds the PASS radiometer data in the region above 800 nanometers, and is lower than the PASS data in the region from 400 nm to 700 nm. The reasons for these discrepancies are: (1) lack of spectral order sorting filters in the diode array instrument, (2) great disparity in the spectral distributions of the calibration and simulator sources, (3) increased stray light due to the lack of an exit slit in front of the detector(s), and (4) time synchronization between the pulse source and the array data collection.

Gratings used in modern monochromators obey the grating equation, relating the grating spacing (d), the wavelength of impinging radiation ($\lambda$), the angle the radiation is incident at ($\phi$), and the angle the radiation is diffracted with respect to the normal ($\theta$), and the **order number** (m) where: $d (\sin \phi + \sin \theta) = m \lambda$. The equation indicates that **monochromatic** light will be diffracted in a number of different directions ($\theta_m$ for each order multiple m of the wavelength). When broadband light is diffracted from the grating higher orders of shorter wavelength light can overlap the first order of longer-wavelength light, and contribute to the signal in the longer-wavelength radiation. For example, second order 400-nm radiation will add to the first order 800-nm radiation seen by a detector selecting the 800-nm radiation. This is the primary cause of the excess radiation beyond 800 nm in the array data of Figure 7. The second-order radiation from the high irradiance levels at 400 nm to 800 nm overlap the first order radiation in the 800 nm to 1000 nm region, contributing a significant error to the signal beyond 800 nm.

Selection of an order-sorting (cut-on) filter to pass radiation of wavelength greater than about 600 nm and to reject radiation below that wavelength would allow only first-order 600-nm to 1100-nm light to reach the detector and produce a more accurate spectrum. This complicates the calibration and use of the array radiometer. At least two array spectral measurements are required, because the low wavelength data would have to be acquired without the filter in place and the longer wavlength data with the filter in place. This requires a calibration for the array radiometer with and without the filter in place and a combination of the two measured spectra to produce the final composite spectrum. Some array detectors are available with integral order-sorting filters deposited on the elements for detecting longer wavelengths.

The significantly different shapes of the calibration and measurement spectra, in conjunction with the problem of order sorting further complicates the interpretation of the measured data. There is relatively low energy in the 300 nm to 600 nm region of the standard lamp spectrum. The contribution of the higher orders of shortwave radiation to the first-order longer-wave radiation is much different than in the pulse simulator xenon source case. The calibration without the use of order sorting carries appropriate information only for sources with similar relative spectral distributions and not for any other unknown spectral distribution to be measured.

In addition to the above concerns, an understanding of the operation of this particular array radiometer is essential to obtaining meaningful data. A very simplified outline of the operation of the radiometer follows to illustrate the point.

The array used in the spectrograph is continually accumulating charge as long as the instrument is running. An operational cycle is carried out continuously by the instrument. The cycle consists of accumulating charge for a "Blanking time" of at least 6.84 ms, then "reading" the array by reading out (discharging) the array (so it does not eventually saturate) during the next 13.16 ms. The "blanking" time, or time of data acquisition can be extended by the user. When light reaches the detector array, charge builds up in each array element over the 6.84 blank time, and it is read out and processed during the 13.16-ms read time. Figure 8 is a diagram of this process compared to the duty cycle of a typical pulse simulator.

During calibration, the standard source is on continuously, so charge accumulates during each (minimum) 6.84-ms blanking interval. During the measurement of a 3-ms pulse from the Spire 240A, charge only builds up over the short 3-ms period, which may or may not occur totally within the blanking interval. The temporal duration of the charge build-up must be accounted for in the two different situations to get easily interpreted measured spectra. The array essentially produces a signal proportional to the *energy* seen during the blanking period. Without knowledge of the pulse shape and appropriate time

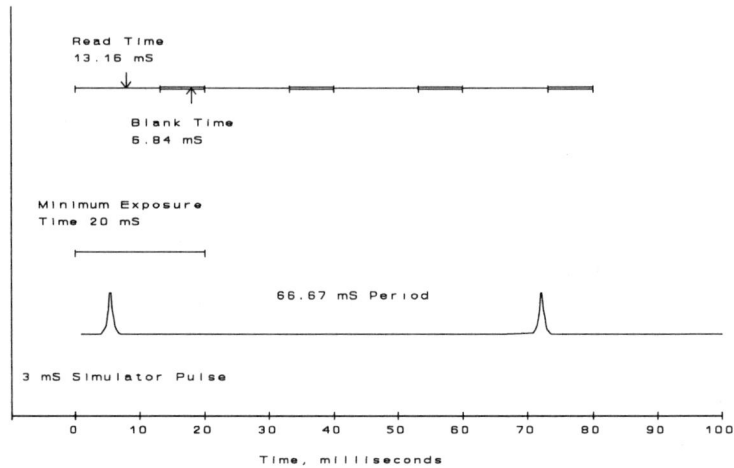

**Figure 8**. Timing diagram for array spectroradiometer data collection.

synchronization between the pulse and the opertional cycle of the array radiometer, only a portion, or none, of a pulse may be captured.

## CONCLUSION

The interpretation of PV module and system performance characterization at NREL requires accurate radiometric measurements. We intend to apply a modified component summation technique for outdoor broadband radiometer calibrations to map out and correct for geometrical response deviations in the pyranometers deployed at the NREL OTF test site. By calibrating radiometers (that are insensitive to tilt) tilted at latitude minus the solar declination, their response over a range of incidence angles from 0° to at least 85° can be obtained.

The advent of diode array spectrographic instruments provides the opportunity to acquire spectral data very quickly and in great quantity. However, even the makers of these instruments state that use of the instruments for absolute spectral radiometric measurements is "notoriously difficult". An understanding of the operational characteristics of such instrumentation, as well as the attention to the classical principles of diffraction grating dispersion is required to obtain data that is not confusing or, at best, difficult to interpret. The NREL PV Solar Radiometric Measurements task is working with instrument manufacturers and PV industry and research and development users to achieve useful diode array spectral data.

## ACKNOWLEDGEMENTS

Troy Strand, Kieth Emery, Benjamin Kroposki, Steven Rummel, and Robert Hansen helped greatly with data collection at the Outdoor Test Site. Steve Hogan of Spire Corporation loaned the array spectrogaph for testing. Richard DeBlasio, project leader, and Roland Hulstrom, Director for the Center for Performance and Reliability Engineering, provided helpful project guidance. This work was supported by U.S. Department of Energy contract No. DE-AC36-83CH10093.

## REFERENCES

1. Myers, D. R., Stoffel, T.L., "A Description of the Solar Radiometer Calibration (RADCAL) Process at SERI." *Proceedings of 1990 Annual Conference of the American Solar Energy Society. Austin, TX. March 19-22, 1990.* p. 171.

2. Myers, D. R., "Application of a Standard Method of Uncertainty Analysis to Solar Radiometer Calibrations." *Proceedings of 1989 Annual Conference of The American Solar Energy Society. Denver, CO. Jun 19-22, 1989.* p. 445.

3. Myers, D. R., Emery,K., Stoffel,T.L., "Uncertainty Estimates for Global Solar Irradiance Measurements Used to Evaluate PV Device Performance," *Solar Cells*, 27, 1989. p. 456.

4. Myers, D.R., Cannon, T., "Photovoltaic Radiometric Measurements and Evalutation", *13th NREL Photovoltaics Program Review, AIP Conference Proceedings 353.* American Institute of Physics, Woodbury, NY, 1995. p. 177.

5. Myers, D.R., and T. Cannon, "Technical Overview of Solar Radiation Research at NREL", *12th NREL Photovoltaics Program Review, AIP Conference Proceedings 306.* American Institute of Physics, Woodbury, NY, 1993. p. 137.

6. Mrig, L., Caiyem, Y., Rummel, S., Hansen, R., Kroposki, R., Strand, T., "Photovoltaic Module and System Performance Testing at NREL",*12th NREL Photovoltaics Program Review, AIP Conference Proceedings 306.* American Institute of Physics, Woodbury, NY, 1993. p. 164.

7. Mrig, L., Hansen, R.,Kroposki, B., Strand, T.,"Results of Module and System Testing at NREL", *13th NREL Photovoltaics Program Review, AIP Conference Proceedings 353.* American Institute of Physics, Woodbury, NY, 1995. p. 207.

8. Myers, D.R., Cannon, T., Trudell, D., "Radiometric Measurements for PV Characterization", *Proceedings of the PV Performance and Reliability Workshop, Sept. 8-10, 1993.* National Renewable Energy Laboratory, Golden Co. 80401.

9. Cannon, T. W., Hulstrom, R.L., Trudell, D.T., "New Instrumentation for Measuring Spectral Effects During Indoor and Outdoor PV Device Testing", *The Conference Record of the Twenty Third IEEE Photovoltaic Specialists Conference-1993, Louisville, KY. May 1993*, P. 1176.

# PHOTOVOLTAIC MANUFACTURING

# Photovoltaic Manufacturing Technology (PVMaT) Project - Latest Results

Richard L. Mitchell, C. Edwin Witt, and Holly P. Thomas
National Renewable Energy Laboratory, Golden, Colorado, 80401

## ABSTRACT

The Photovoltaic Manufacturing Technology (PVMaT) Project was initiated in 1990 to help the U.S. photovoltaic (PV) industry extend its world leadership role in manufacturing and the commercial development of PV modules and systems. It is being conducted in several staggered phases to support industry progress. The twelve subcontracts awarded under Phase 4A are now completing their first year of research. The subcontracts initiated in earlier phases are nearing completion, and their progress is summarized in this paper. An additional phase of PVMaT, Phase 4A, has been initiated and emphasizes product-driven manufacturing research and development. The intention of Phase 4A is to emphasize improvement and cost reduction in the manufacture of full-system PV products. The work areas include, but were not limited to, issues such as improving module manufacturing processes; system and system component packaging, integration, manufacturing, and assembly; product manufacturing flexibility; and balance-of-system development with the goal of product manufacturing improvements.

## INTRODUCTION

The Photovoltaic Manufacturing Technology (PVMaT) Project was initiated in 1990 to help the U.S. photovoltaic (PV) industry extend its world leadership role in PV manufacturing and the commercial development of PV modules and systems. As previously described [1,2,3,4], the PVMaT Project is a government/industry research and development (R&D) partnership between the U.S. federal government (through the U.S. Department of Energy [DOE]) and members of the U.S. PV industry. PVMaT is designed to accomplish its purpose by helping the U.S. PV industry improve manufacturing processes, accelerate manufacturing cost reductions for PV modules, improve commercial product performance, and lay the groundwork for a substantial scale-up in the capacity of U.S.-based PV manufacturing plants.

## Table 1. Current PVMaT Subcontractors

| Principal Subcontractor | Phase | Emphasis of Research | Investigator |
|---|---|---|---|
| Golden Photon | 2B | Commercial Scale-Up of Advanced Thin-Film Photovoltaic Technologies | Terry Brog |
| Solar Cells, Inc. | 2B | High-Throughput Manufacturing of Thin-Film CdTe Photovoltaic Modules | Dan Sandwisch |
| Solarex | 2B | Cast Polycrystalline Silicon PV Cell and Module Manufacturing Technology Improvements | John Wohlgemuth |
| Springborn | 3A | Development of Advanced PV Encapsulants | William Holley |
| Solar Design Associates, Inc. | 4A1 | The Development of Standardized, Low-Cost AC PV Systems | Steven Strong |
| Omnion Power Engineering | 4A1 | Three-Phase Power Conversion System for Utility Interconnected PV Applications | Hans Meyer |
| Utility Power Group | 4A1 | Development of a Low-Cost Integrated 15-kW AC Solar-Tracking Subarray for Grid-Connected PV Power System Applications | Michael Stern |
| Solar Electric Specialties | 4A1 | Design, Fabrication and Certification of Advanced Modular PV Power Systems | Glen Minyard |
| Trace Engin. | 4A1 | Modular, Bi-directional DC to AC Power Inverter Module for PV Applications | Christopher Freitas |
| Advanced Energy Systems | 4A1 | Next Generation Three-Phase Inverters | R.H. Wills |
| Ascension Technology | 4A1 | Manufacture of an AC Photovoltaic Module | Ed Kern |
| AstroPower. | 4A2 | Large Area Silicon-Film Panels and Solar Cells | James Rand |
| ASE Americas | 4A2 | Market Driven EFG Modules | Michael Kardauskas |
| Siemens Solar Industries | 4A2 | Photovoltaic Cz Silicon Module Improvements | Richard King |
| Iowa Thin Film Technologies | 4A2 | PVMaT Monolithic a-Si Modules on Continuous Polymer Substrates | Frank Jeffrey |
| Photovoltaic International | 4A2 | Manufacturing of the PVI Power Grid | Neil Kaminar |

The PVMaT Project is being carried out in five separate phases, designed to address separate R&D requirements. These phases are Phase 1, Phase 2A, Phase 2B, Phase 3A, and Phase 4A. A description of the focus and accomplishments for phases 1, 2A, 3A, and 2B have been detailed in previous papers [1,2,3]. The Phase 3A and 2B industrial participants that are currently active have been identified in Table 1 along with subcontracts awarded under Phase 4A. A further description of the R&D activities addressed by active PVMaT manufacturers is detailed in several other papers in these proceedings.

## PHASE 2B

Phase 2B consisted of four 3-year subcontracts, selected from 13 proposers and awarded in late 1993. These subcontracts addressed process-specific module manufacturing problems of individual manufacturers. They included work in CdTe module manufacturing, the manufacture of Spheral Solar™ cells and modules, as well as cast polysilicon wafers, cells, and module manufacturing. These subcontracts represented new technological additions to the PVMaT Project, and were cost-shared at 58% by the subcontractors. Three of these four subcontracts (Table 1) are currently active.

## PHASE 3A

Phase 3A consisted of two subcontracts, selected from 7 proposers and awarded in January of 1993. These subcontracts focused on module-related R&D problems that are common to several PV manufacturing groups. The remaining active subcontractor, Springborn, will complete field tests in early 1997.

## PHASE 4A

Phase 4A, Product-Driven Manufacturing, is the most recent phase of the PVMaT project. This is a broader approach, addressing the overall PVMaT goal of creating an improved U.S. market share by meeting the market challenges. The solicitation included responses from individual or teamed U.S. PV and related industries addressing the manufacture of PV modules and other end-products, as well as sub-elements of these products. Objectives included stimulating a broader interest in the production of PV products, encouraging and supporting risk-taking by industry to explore new manufacturing options and ideas for improved PV products or components, encouraging system and product integration, increasing module production capacity, reducing PV module production costs, and stimulating advances in balance-of-systems or developments in design leading to overall reduced system life-cycle costs of the PV end-product. Cost reduction, improved efficiency, and manufacturing flexibility and broader market applications for PV systems as a whole were emphasized. The subcontracts awarded in Phase 4A were divided into two parts, 4A1 and 4A2, and subcontractors in both parts are now completing their first year of the subcontract work.

## PHASE 4A1

Phase 4A1 addresses the product driven-system and component technology and includes manufacturing improvements directed toward innovative, low-cost, high-return, high-impact PV products. Proposals in this phase addressed manufacturing that was generally related to PV system components and aspects other than modules, system components such as inverters, and/or system integration efficiency and/or design improvements, with less focus on module manufacturing. In addition, proposals looked at issues in system/component integration to bring all elements together for a PV product offered on the market. Eight two-year subcontracts (listed in Table 3) were awarded from 31 Phase 4A proposals received. These subcontracts are cost-shared at 25% by the subcontractors.

## PHASE 4A2

Phase 4A2 is focused on product-driven PV module manufacturing technology. Subcontracts under this phase are directed at manufacturing flexibility and module manufacturing cost reduction for a wider range of PV products. Phase 4A2 consists of five three-year subcontracts (shown in Table 4), selected from 31 Phase 4A proposals. These subcontracted efforts include developing large-area Silicon-Film™ panel and cell manufacturing; edge-defined, film-fed growth module manufacturing; improvements in Czochralski crystalline silicon based modules; development of monolithic amorphous Silicon modules on continuous polymer substrates; and manufacturing of extruded concentrator modules. These subcontracts are cost-shared at 42% by the subcontractors.

## PROGRESS IN COST REDUCTION AND CAPACITY INCREASES

At the beginning of Phase 2A subcontracted research, information was collected to establish both the current and projected capacities and module costs for the eight participating manufacturers. The initial results represented both the diverse status of a still-maturing industry and the optimistic speculation of its members regarding the effects that the PVMaT project would have after their research efforts were allowed to take effect. Each year, the data have been updated based on information form the PV industrial participants.

This most recent cost and capacity data, shown in Fig. 1, represent an update of previous projections regarding these subcontracted efforts [4]. Data in this figure are based on each manufacturer's maximum production capacity during a given year. The "'average' module manufacturing costs" represent the average cost per watt of modules (weighted by each participant's capacity) produced by these twelve PVMaT industrial participants. Module cost estimates were then based on these manufacturing levels and included only those costs directly associated with the manufacturing of the modules (not marketing, administration, sales, etc.). It should be noted data associated with any particular point in time, represent a potential capability. Actual manufacturing production levels may be less (and concomitant costs higher) due to other considerations such as market conditions, available labor, etc. Figure 1 indicates that PV manufacturing capacity has increased by more than a factor of 4 in the last 4 years. Additionally, the "average" cost for manufacturing PV modules has been reduced by about 36%.

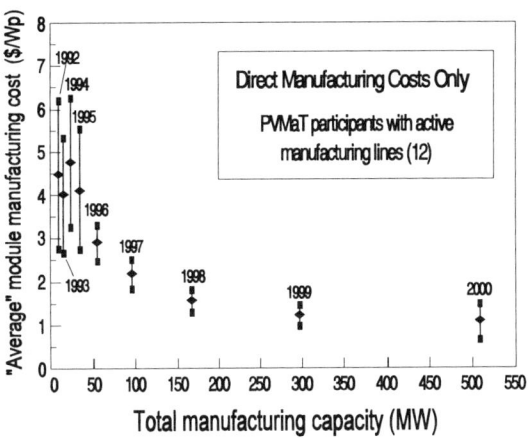

Figure 1 - PVMaT Manufacturing Cost/Capacity

Changes have been observed in annual projections for both "average" module manufacturing costs and manufacturing capacity from those of prior years. An evaluation of the 1996 modifications to the projections of 1995 and earlier, demonstrate several trends among the PVMaT manufacturing participants. The data collected in 1996 showed a 7% increase over the data collected in 1995 for the 1996 projected "average" module manufacturing costs and a 33% reduction in the projected 1996 manufacturing capacity. The increased cost adjustment resulted from; seven manufacturers increasing their module manufacturing costs from a range of 4% - 58%, and one decreasing its by 41%. The 1996 decreased capacity adjustment resulted from; one manufacturer increasing its manufacturing capacity by 3%, and five decreasing theirs in the range of 10% - 80%. The driving factors in these adjustments were: a few manufacturers pushing off their time tables for scale-up to their previously projected year-2000 levels; one manufacturer projecting a conversion of a product line to new product; and a significant reduction in the year 2000 capacity plan for another manufacturer.

Projections for 1998 also were revised, with "average" module manufacturing costs showing a 4% decrease over 1995 data and 1998 manufacturing capacity reflecting a 24% reduction. The decrease in 1998 projected costs resulted from three increasing their module manufacturing costs from a range of 14% - 42%, and three others decreasing theirs in the range of 2% - 40%. The decrease in 1998 projected capacity resulted from two increasing their manufacturing capacity in the range of 7% - 94%, and four decreasing theirs in the range of 23% - 50%. The driving factors in these adjustments were still modified time tables for scale-up to previously projected year-2000 levels; conversion of a product line to new product; and a reduction in planned year 2000 capacity.

Projections for 2000 "average" module manufacturing costs showed a 5% increase over 1995 data and projected 1998 manufacturing capacity reflected a 4% reduction. The increased 1998 projected costs reflect two manufacturers increasing their module manufacturing costs from a range of 11% - 46%, and four decreasing theirs in the range of 2% - 15%. The decrease in 1998 projected capacity resulted from two manufacturers increasing their capacity projections in the range of 30% - 300%, and four decreasing theirs in the range of 23% - 67%.

PVMaT participants reporting reductions in the "average" module manufacturing cost projections attributed the adjustments to better-than-expected improvements in module design, decreased materials and labor requirements, improved performance, and improved yields. The dominating factors in cost increases were identified as increased raw material costs, and more realistic estimates by the subcontractors in their actual manufacturing costs.

Manufacturers reporting reductions in their projected manufacturing capacity attributed these to several factors: 1) changing market conditions, 2) their having to devote unplanned time to solving product performance issues, 3) changes in business plans aimed at a more sustainable growth rate, and 4) problems with increasing materials costs. Increased projections in capacity were attributed to previously unforeseen market opportunities that have been larger and more promising than anticipated.

## CONCLUSIONS

The PVMaT Project is currently finishing the first year of subcontracts under Phase 4A1 and 4A2, with research in Phase 2B concluding their third year and Phase 3A now being completed. At this time, it is apparent from Fig. 1, that the U.S. PV industry involved in the PVMaT Project has made significant progress toward reducing manufacturing costs and increasing PV module manufacturing capacity. "Average" module manufacturing costs have been reduced 36% and total manufacturing capacity for 12 PVMaT industrial participants with active lines has increased by factor of more than 4. By 1998, projections are for a reduction in the costs of 65% at $1.64/Wp, and a factor of 14 increase in capacity by 1998 to 223MW. It has also indicated in both the industries future cost/capacity data and its technical projections that its optimism for continuing these improvements is high.

## ACKNOWLEDGEMENTS

This work is supported under DOE contract No. DE-AC36-83CH10093 with NREL. Many people have contributed to the development and implementation of the Photovoltaic Manufacturing Technology project and to the R&D efforts carried out in this program. The authors recognize this paper represents their work.

## REFERENCES

[1] Witt, C.E.; Herwig, L.O.; Mitchell, R.; and Money, G.D., "Status of the Photovoltaic Manufacturing Technology (PVMaT) Project" in *Proceedings of the 22nd IEEE Photovoltaics Specialists Conference*, Las Vegas, Nevada, October, 1991.

[2] Witt, C.E.; Mitchell, R.; Mooney, G.D.; Herwig, L.O.; Hasti, D.; and Sellers, R., "Progress in Phases 2 and 3 of the Photovoltaic Manufacturing Technology Project (PVMaT)." in *Proceedings of the 23rd IEEE Photovoltaics Specialists Conference*, Louisville, Kentucky, May, 1993,.

[3] Witt, C.E. Thomas, H.P., Herwig, L.O., Mitchell, R.L.; Ruby, D.S., and Aldrich, C.C., "Recent Progress in the Photovoltaic Manufacturing Technology Project (PVMaT)" in *Proceedings of the 1st World Conference on Photovoltaics*, Waikoloa, Hawaii, December, 1994.

[4] Mitchell, R.L.; Witt, C.E. Thomas, H.P., Herwig, L.O.,Ruby, D.S., and Aldrich, C.C., "Benefits from the U.S. Photovoltaic Manufacturing Technology Project" in *Proceedings of the 25th IEEE Photovoltaics Specialists Conference*, Washington, D.C., May, 1996.

# Improvements in Cast Polycrystalline Silicon PV Manufacturing Technology

John H. Wohlgemuth

*Solarex, A Business Unit of Amoco/Enron Solar, Frederick, Maryland*

**Abstract.** The objectives of this NREL sponsored Photovoltaic Manufacturing Technology (PVMaT) Program are to advance Solarex's cast polycrystalline silicon manufacturing technology, reduce module production cost in half, increase module performance and expand Solarex's commercial production capacity by a factor of three. To meet these objectives Solarex has: 1) Modified the casting process and stations and is now casting larger ingots in production; 2) Developed wire saw technology to cut wafers with less kerf loss and has transferred this technology to production; 3) Developed a laboratory process to increase cell efficiencies using back surface fields, mechanical texturing and gettering; 4) Modified the casting, wires saw and cell processes in order to fabricate larger (15.2 cm by 15.2 cm ) wafers and cells; 5) Improved the automated assembly of modules, reducing labor requirements and increasing throughput; and 6) Developed a frameless module with a lower cost backsheet and a simple, low cost electrical termination system. Solarex is now in the process of developing the equipment necessary for automated handling of thin 15.2 cm by 15.2 cm wafers and cells. This paper will discuss the efforts during the first two and a half years of the program.

## INTRODUCTION

The goal of Solarex's Crystalline PVMaT program is to improve the present Polycrystalline Silicon manufacturing facility to reduce cost, improve efficiency and increase production capacity. Key components of the program are:
- Casting of larger ingots.
- Use of wire saws to cut thinner, larger size wafers with less kerf loss.
- Transfer of higher efficiency cell processes to manufacturing.
- Increased automation in module assembly.
- High reliability mounting techniques for frameless modules.
- Automated handling of large, thin wafers.

The results of these efforts will be to reduce the module production cost in half, to increase the production capacity of Solarex's Frederick plant by a factor of 3 and to provide larger, higher efficiency modules that reduce the customer's balance of systems cost. All of this is to be achieved without sacrificing the high reliability already achieved with the crystalline modules in use today.

Solarex's Crystalline Silicon Technology is based on use of cast polycrystalline silicon wafers. At the beginning of the PVMaT program, the largest ingot that could be cast in the Solarex casting stations, produced four bricks 11.4 cm by 11.4

cm in cross-section.. Wafers were cut from the bricks using internal diameter (ID) saws. The cell process sequence was based on the use of screen printed thick film paste. Cells were tabbed and then matrixed using first generation automated equipment. The module construction consisted of a low iron glass superstrate, EVA encapsulant and a three part back sheet. The primary product was a module with 36 - 11.4 cm x 11.4 cm solar cells, producing 60 to 64 Watts under Standard Test Conditions (STC).

## CASTING

The original goal of the casting task was to develop the ability to cast ingots that yield four, 15 cm by 15 cm bricks with at least equivalent material quality as was achieved for the standard size. Modifications were necessary to the chamber so it would hold more silicon, and to the heaters and insulation. A laboratory process was developed, yielding equivalent quality ingots with a 73% increase in the useable silicon obtained from each casting.

Most of Solarex's products are still based on the use of 11.4 cm by 11.4 cm solar cells, so an effort was undertaken to develop casting for even larger ingots that would produce 9 - 11.4 cm by 11.4 cm bricks. Such an ingot requires approximately 20% more silicon than the PVMaT ingot. These "mongo" ingots required further changes in the insulation and receiver, but utilized the same casting stations as modified for the PVMaT ingots. The process for casting mongo ingots was successfully developed in the laboratory. It has been transferred to production with approximately half of the casting stations now modified to cast mongo ingots.

Analysis of the production data shows that mongo ingots have an equivalent yield of useable cast silicon to the smaller mega ingots. In addition there is no statistical difference in average cell efficiency between the mega and mongo processes. Casting of larger mongo ingots is being used to double Solarex's casting capacity at approximately 20% of the cost that would have been required, if the same amount of capacity was added by purchasing new casting stations.

## WIRE SAWS

The goal of this task is to develop the wire saw technology for cutting 15 cm by 15 cm polycrystalline wafers on 400 μm centers at lower cost per cut than achieved today on the ID saws. This represents a 50% increase in the useable silicon obtained from each cast and a 50% increase in the yield of wafers per purchased kilogram of Si feedstock.

Solarex is utilizing an HCT wire saw in this program. The saw has successfully demonstrated the ability to cut a variety of wafer sizes including 11.4 cm by 11.4 cm; 11.4 cm by 15.2 cm; and 15.2 cm by 15.2 cm on centers from 500 µm down to 400 µm. The saw has been operated in a production mode, producing more wafers than 16 ID saws, at better yields and lower per wafer cost than the ID saws. This saw has been so successful that Solarex has added additional HCT wire saws.

After wafers have been cut on the wire saw they must be removed from the hold down plate, placed in cassettes and cleaned. It is necessary to develop an automated process to reduce cost and increase yield especially as the volume of wire saw wafers increases and the thickness of the wafers decreases. The Automation and Robotics Research Institute (ARRI) at the University of Texas at Arlington is working as a subcontractor to Solarex on this project. ARRI has designed, built and tested prototype equipment to destack wet wafers into cassettes. The prototype worked very well for both dry and wet wafers. It has operated through thousands of cycles without breaking any wafers and with successful feeding of a single wafer into the slot more than 99.95% of the time. A full production machine will be built based on this design.

## CELL PROCESS

The goal of this task is to increase cell efficiencies to 15%, while decreasing the cost per watt at the module level. While a number of approaches to achieving high efficiency have been reported, many of these utilize processes and material that are not likely to be cost effective when applied to cast polycrystalline silicon in a manufacturing environment. The key to achieving the goal of this task is to select modifications to the present process that increase efficiency while lowering the cost per watt. Efforts have been in the areas of back surface fields (BSF), mechanical texturing, development of an integrated process and optimization of the design and process for larger (15.2 cm by 15.2 cm) cells.

### Back Surface Fields

Previous laboratory experiments demonstrated that an aluminum paste back surface field (BSF) can increase cell efficiency by approximately 5%. (1) Three BSF manufacturing trials were completed, with more than 40,000 BSF cells produced with all processes except BSF performed in production using production processes, equipment and personnel. The average cell efficiency for all of the BSF cells was 5 % higher than for the non-BSF cells produced during the same time period. Figure 1 shows the bin distribution for both the BSF cells and the standard production cells produced during these trials. The X bin has the highest efficiency with G bin the lowest. The BSF process has now been transferred to manufacturing for use on approximately 25% of all the cells produced. This will be expanded to 100% of the cells once the necessary equipment is in house.

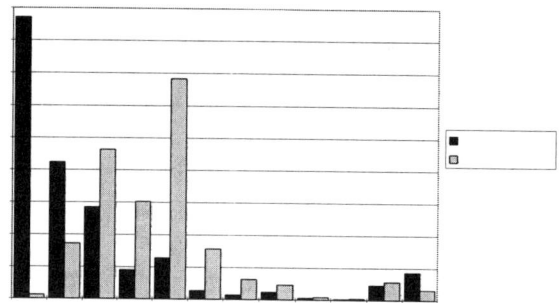

**FIGURE 1.** Cell Distribution Comparison, BSF versus Standard Production

One method of reducing the cost of PV modules is to cut thinner wafers, so cell process developments like BSF should be compatible with the use of thinner wafers. To evaluate the effect of the BSF process on wafer thickness, cells were fabricated on wafers with thickness between 200 and 300 μm. Figure 2 shows average cell efficiency as a function of thickness for non-BSF cells, while Figure 3 shows the same data for cells with BSF. Thinner non-BSF cells are less efficient, but thinner BSF cells are more efficient. The 5% efficiency gain observed for BSF cells with today's cell thickness increases to approximately 10% for the thinner cells now under development. (2)

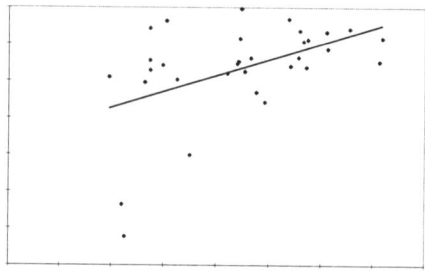

**FIGURE 2.** Efficiency versus Cell Thickness for Non-BSF Cells

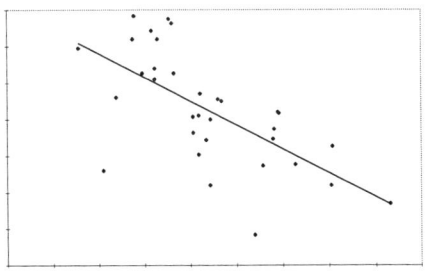

**FIGURE 3.** Efficiency versus Cell Thickness for BSF Cells

## Mechanical Texturing

A method to mechanically texture polycrystalline silicon has been developed during the program. Matched polycrystalline wafers were processed with and without mechanical texturing. The matched cells were measured, encapsulated behind low iron glass and EVA and then remeasured. The results are given in Table 1. Mechanical texturing results in a 3% gain in encapsulated cell efficiency. Efforts are now underway to develop a large area tool as part of a cost effective texturing process.

**TABLE 1.** Mechanical Texturing versus Planar Controls

| Cell Structure | Unencapsulated | | Encapsulated | |
|---|---|---|---|---|
| | Efficiency (%) | Jsc (mA/cm$^2$) | Efficiency (%) | Jsc (mA/cm$^2$) |
| Planar | 12.75 | 29.7 | 12.96 | 30.3 |
| Mechanical Texture | 13.19 | 30.7 | 13.35 | 31.1 |
| Difference | 3.4% | 3.4% | 3.0% | 2.6% |

## Integrated Cell Process Sequence

Back surface fields, mechanical texturing and phosphorus getting all increase cell efficiencies. An integrated cell process sequence has been developed to include these efficiency enhancing processes in a cost effective manner. The results for encapsulated single cell packages are shown in Table 2. This group of cells nearly met the program efficiency goal of 15%.

**TABLE 2.** Integrated Semicrystalline Silicon Cell Process Sequence

| Sample | Efficiency (%) | Jsc (mA/cm$^2$) | Voc (mV) | FF (%) |
|---|---|---|---|---|
| Planar | 14.00 | 32.28 | 599 | 72.5 |
| Mech. Tex. | 14.56 | 32.90 | 597 | 74.1 |
| Mech. Tex. + Gettered | 14.88 | 33.04 | 600 | 75.1 |

## MODULE ASSEMBLY

The initial goal of this task was to modify Solarex's module assembly system to increase throughput by 100% and decrease the labor requirement by 50%. The throughput goal has now been modified to increase module assembly by 200% to meet Solarex's manufacturing requirements.

Solarex subcontracted to the Automation and Robotics Research Institute (ARRI) at the University of Texas at Arlington to review and model the automated module assembly system in operation at the beginning of this PVMaT Program and to make recommendations for improving the equipment and/or process flow. The first step in this task was the development of a process flow chart detailing all of the module assembly steps. ARRI used this information to model and analyze the manufacturing process. As a short term goal the changes necessary to increase production capacity by 40% were identified and implemented during the first year of the contract.

In 1995 Solarex publicly announced plans to triple the capacity of the polycrystalline plant by 1999. This became the new goal for the module assembly task. ARRI used AT&T's discrete event simulation package called Witness to evaluate expansion scenarios. The Witness software is capable of modeling resource interactions in detail and providing an accurate representation of the factory using statistical analysis. The Witness software provided an analysis of the capacity and resource requirements for the different scenarios. A new factory concept was developed that allowed for incremental increases to meet the shorter term capacity requirements and that could ultimately result in the required tripling of module assembly capacity. The plan was based on replacing the back solder robots with XY positioners to increase the number of solder bonds made at one time from 2 to 4, thereby increasing the throughput by nearly a factor of two. This modification has now been implemented and the XY positioners are operating at the throughput level predicted.

The new factory configuration and Solarex's projected module build plan through 1999 were then used as inputs to the Witness program. As the volume of products ramps up, the model predicts when the various pieces of equipment reach maximum capacity. This analysis was utilized to develop a capital investment budget for the crystalline expansion. The implementation of this plan is now underway with capacity double what it was at the beginning of the PVMaT program, with a second doubling scheduled to be completed by 1999.

## FRAMELESS MODULE DEVELOPMENT

The goal of this task is to develop and qualify a frameless module design incorporating a lower cost back sheet material (less than $0.05/square foot) and user friendly, low cost electrical termination (less than $1.00/module). Since PVMaT is designed for large systems, the modules were designed to mount directly onto the support structure without integral frames. The first step has been the design of a compatible support structure and the identification of 3M Very High Bond Tape for mounting the modules to the structure.

This system was used on several large arrays, but major problems were encountered relative to the use of the tape. The problems appear to be related to either incorrect application by the installers or situations where the modules are mounted in such a way that there is a continual pressure on the beams. The failure mode is separation of the module from the beam. While correct assembly may alleviate this problem, it is likely that actual systems will not be assembled perfectly. We have therefore switched from using the tape to using RTV as the main adhesive, although we do use a small amount of tape to hold the beams in place while the RTV cures. The modules are panelized several days before installation, so the RTV can cure.

A key component in frameless module design is the backsheet, since the electrical termination and the support system itself must adhere to the backsheet. This also offered an opportunity to reduce cost from the backsheet being used at Solarex. A number of candidate materials were selected for evaluation in this program. Sample modules were made with these materials. These samples were subjected to a set of environmental qualification tests, in-house simulated UL fire tests, and exposure to 2 year equivalent UV in Phoenix, Arizona. The only material that met all of the technical requirements was Tedlar. Based on the PVMaT results, Solarex switched manufacturing to a single layer Tedlar backsheet. However, Tedlar will not meet the $0.05/square foot cost goal of this PVMaT program, although it does represents a significant reduction in back sheet cost over the three part material that was used before the PVMaT program.

Rather than use expensive junction boxes or connectors, the electrical connection from module to module will be made using pig tail wires with crimp connectors and shrinkable tubing insulation. This system has successfully passed IEC-1215 and IEEE-1262 without measurable change in internal resistance or leakage current during wet hi-pot testing at 2750 volts. This electrical termination system meets the technical requirements and the cost goals of the program.

## AUTOMATED CELL HANDLING

The goal of the automated cell handling task is to develop automated handling equipment for 200 μm thick 15 cm by 15 cm polycrystalline silicon wafers and cells with a high yield (less than 0.1% breakage per process handling step) at a throughput rate of at least 12 cells or wafers per minute. ARRI is also supporting Solarex in this task.

ARRI performed a set of wafer fracture experiments to measure the mechanical strength of typical cast polycrystalline wafers. (3) A finite element model was then developed to determine the distribution of stress and deflection in a wafer and to determine the probability of breakage of that wafer under the specified load. The model was used to simulate a typical wafer handling situation, to estimate the maximum load that can be applied during handling and the corresponding probability of breakage.

ARRI also performed a study of commercial and academic information sources related to the handling of silicon wafers. The purpose of this search was to gain an understanding of the wafer handling methods that are commercially available, or documented in the open research literature, in order to assist in the wafer handling design effort. Almost all of the wafer handling equipment has been developed for the semiconductor industry, where contamination prevention is one of the most important parameters in the design. This has lead to a preference for vacuum grippers that are allowed to come in full contact with the "back" face of the wafer, which is simply a featureless substrate. Non-vacuum methods involve arms that "push" the wafers in and out of their slot in the cassette, and are typically used in transfer and inspection operations; such arms come in contact with both sides of the wafer in a small ring-shaped section in the periphery of the wafer. The typical gripper design in semiconductor wafer sorting and transfer machines consists of flat prongs with embedded vacuum ports flush with the surface or slightly below it. This provides a greater surface-to-surface contact between the gripper and the wafer than using vacuum cups alone, which likely helps to restrict movement of the wafer during transport and minimizes breakage due to bending of the wafer. Vacuum sensors are used to detect that the wafer is latched onto the jaws. These concepts are now being utilized in discussions with handling equipment manufacturers.

## SUMMARY

This PVMaT Program has lead to the development of and/or improvements in processes, products and equipment. The following developments from this program have already been implemented in manufacturing:
- Casting of larger ingots;

- Use of wire saws;
- Addition of a back surface field on a significant fraction of all production cells;
- Introduction of a larger cell (11.4 cm by 15.2 cm);
- Doubling of production capacity in the module assembly area; and
- Use of a lower cost back sheet.

We expect that more PVMaT development efforts will be transferred to manufacturing in the near future.

At this time we believe that all of the major objectives of this PVMaT program (reduction of production costs, improved performance and tripling of capacity) will be met. Our preliminary analysis indicates that through the year 2000, Solarex will save $5.00 for every dollar it invested in PVMaT and our customers will save approximately $7.00 for every dollar that NREL invested in this PVMaT Program.

## ACKNOWLEDGMENTS

This work was supported in part by NREL contract ZAI-4-11294-01. The author wishes to thank the technical review committee of R. Mitchell, B. Sopori and J. Gee for their support and assistance. The author would also like to acknowledge the support and assistance of the other members of the technical staff at Solarex, as it is their work that is described in this paper.

## REFERENCES

1. J. H. Wohlgemuth, D. Whitehouse, T. Koval, J. Creager, F. Artigliere and M. Conway, "Solarex Crystalline PVMaT Program", *First World Conference on PV Energy Conversion*, 1994, p. 832.
2. T. Koval, J. Wohlgemuth and B. Kinsey, "Dependence of Cell Performance on Wafer Thickness for BSF and Non-BSF Cells", *Twenty-fifth IEEE Photovoltaic Specialist Conference*, 1996, p. 505.
3. J. Wohlgemuth, "Cast Polycrystalline Silicon Photovoltaic Module Manufacturing Technology Improvements", Semiannual Subcontract Report, 1 January, 1995 - 30 June, 1995, NREL/TP-411-20589.

# Progress in High-Throughput Manufacturing of Thin-Film CdTe Modules

## Dan W. Sandwisch

*Solar Cells, Inc., Toledo, Ohio 43607*

**Abstract.** Solar Cells, Inc. ("SCI") has manufactured and installed over 50kW of thin-film CdTe modules with an average total area efficiency of greater than 6.75%. These installations have demonstrated reliable performance at or above system ratings. In addition, modules routinely pass extensive stress testing. A champion 60cm x 120cm module has been measured at 61.0 watts or over 9.0% aperture area efficiency. SCI has initiated the scaling of its pilot line process to a multimegawatt level. This paper summarizes the status of these development activities including equipment specification, support programs, and product validation.

## INTRODUCTION

Cadmium telluride (CdTe) has become recognized as one of the leading materials for low-cost, thin-film photovoltaic (PV) modules. In 1991 SCI began a PV module manufacturing development program to demonstrate the technology's many advantages such as deposition rate, device stability, device performance, and process flexibility. Due to the success of these efforts, SCI began a manufacturing initiative in late 1993 in conjunction with support from The Department of Energy's PVMaT program. Current program activities include equipment specification, support programs development, and product validation.

## EQUIPMENT SPECIFICATION

The processes used to manufacture thin-film cadmium telluride modules are listed in Table 1. These processes have been demonstrated routinely on a 100 kW pilot production line. SCI has completed the specifications of all the major components of a 20MW manufacturing line. SCI has specifically concentrated on the advancement of a high-throughput semiconductor deposition system ("HTDS") and two industrial laser scribing machines. These steps present the most relevant issues related to scaling the pilot line to an industrial multimegawatt process such as cycle time, process control, and material utilization. These systems will be integrated with approximately 40 other pieces of equipment to produce modules at an annual nominal capacity of 20MW.

**TABLE 1.** Processes for Manufacturing Thin-Film CdTe Modules.

| | |
|---|---|
| TCO Inspection/Preparation | Scribe 3 |
| Scribe 1 | Busbar Application |
| Semiconductor Deposition | Encapsulation |
| Film Thickness Check | Potting |
| Post Deposition Treatment | Safety Checks |
| Scribe 2 | Pmax Test |
| Metal Deposition | Final Inspection |

As reported earlier, the HTDS (Fig. 1) incorporates many advantages over SCI's prototype deposition system including increased throughput and continuous operation (1). The production system has been built for continuous feed of 60cm x 120cm TCO-coated substrates at a throughput of one substrate/minute. The pressure control and conveyor systems were upgraded from the original construction to increase throughput and system reliability. Preliminary tests of these systems, along with the heating systems, resulted in a continuous 24-hour run in which over 1,400 60cm x 120cm substrates were processed with a 100% yield. The deposition components have also been upgraded to include design features which greatly enhance material utilization and deposition rate.

**FIGURE 1.** Photograph of high-throughput semiconductor deposition system.

The scribing systems are completely automated to scribe the 60cm x 120cm substrates (Fig. 2). The systems have four lasing beams that can scribe in either direction over the substrate so that various products (i.e. 17V and 65V) can be produced with low changeover time. The control features for these systems also allow for enhanced tracking of the process performance on a real-time basis. Due to developments on SCI's prototype laser system and the industrial laser systems, scribing cycle times have been reduced by over 90%.

**FIGURE 2.** Photograph of high-throughput laser system.

## SUPPORT PROGRAMS

An important part of the development effort is to establish programs which effectively handle safety, health, and environmental issues that accompany the production, deployment, and disposal of these modules. SCI has engaged outside agencies and consultants to conduct safety and health audits of the manufacturing facilities and to formulate appropriate programs and corrective actions. These programs include basic training programs as well as specific operational plans such as industrial hygiene and biological monitoring. Test results have confirmed that these programs are very successful in mitigating and eliminating plant risks.

Environmental efforts have focused on waste minimization and material reclamation and recycling (2). These process have been demonstrated at prototype levels. This work, begun under SCI's PVMaT subcontract, is now being commercialized with the support of a two-year, $925,000 Small Business Innovative Research subcontract. The objective of the work is to fully develop and commercialize a full-scale reclamation and recycling process primarily targeted at CdTe modules but flexible enough to be utilized in other recycling efforts such as florescent light tubes and video monitors.

## PRODUCT VALIDATION

The baseline SCI product is a 60cm x 120cm high-voltage ($65V_{max}$) module targeted at grid-connected applications. SCI projects that this product, along with a patented support structure, will reduce the cost of photovoltaic installations to below $3.00 per watt by the year 2000. This product has been installed at several sites (see Table 2).

TABLE 2. SCI Thin-Film CdTe Module Installations.

| Location | Size | Installation Date |
|---|---|---|
| Edwards AFB, CA | 25.0 kW | April 1996 |
| Toledo Edison, Ohio | 14.0 kW | October 1995 |
| PVUSA, Davis, CA | 10.8 kW | September 1995 |
| NREL, Golden, CO | 1.0 kW | June 1995 |
| STDI, Las Cruces, NM | 0.5 kW | February 1995 |
| Tunisia | 1.0 kW | January 1995 |
| Toledo, Ohio | 1.2 kW | August 1994 |

The installation at Edwards AFB incorporates 624 modules and has been measured at over 27.0kW at the DC junction box in mid-summer (Fig. 3). SCI supplied not only the modules but its proprietary panelization and support structure developed in part under its PVMaT subcontract. The project also confirmed that these designs result in greater than a 40% reduction in costs of encapsulation, potting, connections, and panelization when compared to its pre-PVMaT design. The majority of this reduction is the result of incorporating frameless panelization and replacing the junction box with a pigtail.

**FIGURE 3**. Photograph of 25.0kW array at Edwards AFB, California.

The Toledo Edison array consists of over 200 modules. The array is grid-connected and has been in operation for over twelve months. The array will be expanded to 100kW by the end of 1998.

**FIGURE 4**. Photograph of 14.0kW grid-connected array in Toledo, Ohio.

SCI also installed an array of 264 modules at the PVUSA demonstration site in Davis, California (Fig. 5). This array has been measured at 12.0kW at the DC junction box and has operated well since its installation in September 1995.

**FIGURE 5.** Photograph of 10.8kW grid-connected array at PVUSA in Davis, California.

## SUMMARY

Thin-film CdTe technology provides many advantages over other PV technologies and has demonstrated superior performance over other thin-film technologies. Key conclusions are:

- SCI has produced a 60.3W, 8.4% efficient module — the most powerful thin film polycrystalline module on record.
- SCI has defined and demonstrated cost effective thin-film CdTe modules capable of achieving installed costs of less than $3.00/W.
- Module durability has been demonstrated.
- The process for making these modules has been proven on a 100 kW pilot line.
- No technology barriers have been identified.

During the next eighteen months, SCI will focus on commercialization of this product. These efforts will include:

- Expansion of the current pilot line to 200 kW while installing a multimegawatt manufacturing facility.

- Continuation and expansion of module testing efforts including product certification.
- Testing of the recycling strategy in several established markets.
- Installation of a 10.8 kW array at PVUSA and expansion of the 0.5 kW array at Southwest Technology Development Institute to 1.0 kW.

## ACKNOWLEDGMENTS

The author is pleased to acknowledge the efforts of the SCI development team. The extraordinary capability and dedication of this team was essential in achieving the results described in this paper. The work reported here was supported in part by NREL under subcontracts No. ZAI-4-11294-02 and No. ZR-1-11059-1.

## REFERENCES

1. Sandwisch, D.W., "Development of CdTe Module Manufacturing", in *AIP Conference Proceedings 353, 13th NREL Photovoltaics Program Review,* Lakewood, Colorado, 1995, pp. 318 — 325.

2. Sasala, R. A., Zhou, T., and Kocher, W. M., "Environmentally Responsible Production, Use and Disposition of Cd-Bearing PV Modules", in *1994 IEEE First World Conference on Photovoltaic Energy Conversion, Conference Record of the Twenty-Fourth IEEE Photovoltaic Specialists Conference,* Waikoloa, Hawaii, 1994, Volume I, pp. 311 — 314.

# Improvements in Cz Silicon PV Module Manufacturing

Richard R. King, Kim W. Mitchell, Theresa L. Jester

*Siemens Solar Industries, 4650 Adohr Lane, Camarillo, CA 93012*

**Abstract.** Work focused on reducing the cost per watt of Cz Si photovoltaic modules under Phase I of Siemens Solar Industries' DOE/NREL PVMaT 4A subcontract is described. Module cost components are analyzed and solutions to high-cost items are discussed in terms of specific module designs. The approaches of using larger cells and modules to reduce per-part processing cost, and of minimizing yield loss are particularly leveraging. Yield components for various parts of the fabrication process and various types of defects are shown, and measurements of the force required to break wafers throughout the cell fabrication sequence are given. The most significant type of yield loss is mechanical breakage. The implementation of statistical process control on key manufacturing processes at Siemens Solar Industries is described. Module configurations prototyped during Phase I of this project and scheduled to begin production in Phase II have a projected cost per watt reduction of 19%.

## INTRODUCTION

The Photovoltaic Manufacturing Technology (PVMaT) project is sponsored by the U.S. Department of Energy (DOE) through the National Renewable Energy Laboratory (NREL) in order to assist the photovoltaics industry in improvement of module manufacturing, and reduction of module manufacturing cost. The objective of the DOE/NREL PVMaT subcontract with Siemens Solar Industries (SSI) is to continue the advancement of Siemens Solar Industries' photovoltaic manufacturing technology in order to achieve an 18% reduction in module cost per watt at the end of three phases of work, with each phase lasting a year as shown in Table 1. Phase I of this subcontract began in November 1995. The approaches for reaching this cost reduction goal are to analyze existing module cost structure and explore new module designs and materials, investigate the reduction of labor and improvement of yield, and to implement statistical process control (SPC) in module manufacturing.

TABLE 1. Goals of Siemens Solar Industries' PVMaT 4A subcontract from DOE/NREL.

|  | Phase I<br>1st Year | Phase II<br>2nd Year | Phase III<br>3rd Year |
|---|---|---|---|
| New module designs to reduce $/W | 6% reduction in module $/W | 12% reduction in module $/W | 18% reduction in module $/W |
| Improvement of yields and reduction of labor | 5% improvement in module mfg. yield<br><br>5% increase in module mfg. productivity | 10% improvement in module mfg. yield<br><br>10% increase in module mfg. productivity | 15% improvement in module mfg. yield<br><br>15% increase in module mfg. productivity |
| Improvement of module reliability | Implement SPC on 50% of appropriate mfg. processes | Implement SPC on 100% of appropriate mfg. processes | Assessment of SPC protocols, areas for improvement |

## MODULE COST ANALYSIS AND DESIGN

A first step toward reducing the cost per watt at the module level is to gain a thorough understanding of the present factors which dominate cost. This section gives examples of the cost structure of module manufacturing at SSI. Once the most significant costs are known, strategies for reducing them are formulated and discussed. Finally, specific module designs that address these cost issues are presented. These improvements in module configuration have been introduced into production in the SSI Camarillo plant during Phase I of the contract, or exist as prototypes with production scheduled to start during Phase II.

Figure 1 shows the division of cost per watt at the module level among the four major process segments of wafered silicon solar cell fabrication: ingot growth; wafering; cell fabrication; and module fabrication. Module fabrication has the largest cost of these, and so is a reasonable area on which to focus efforts to reduce the overall cost per watt. However, costs in the other areas are also very significant contributors to the cost per watt at the module level, and must be addressed as well.

In the module area, the costs can be broken down further, as shown in Figure 2. The cost of direct materials such as glass, frames, etc. is by far the largest component of the total costs in the module process area. Some labor, yield, maintenance costs may be more avoidable, however, and are still significant costs.

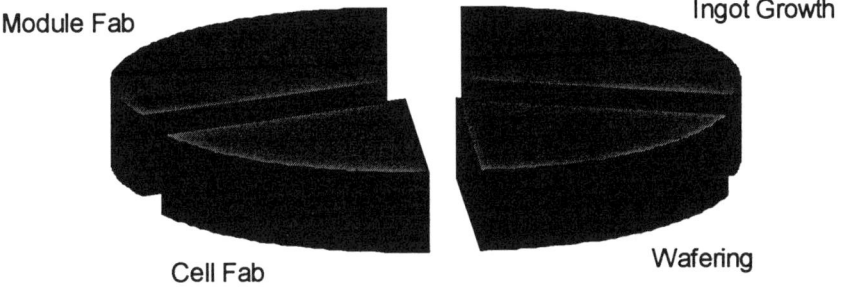

**FIGURE 1.** Approximate cost/watt by process area, normalized to module $/W.

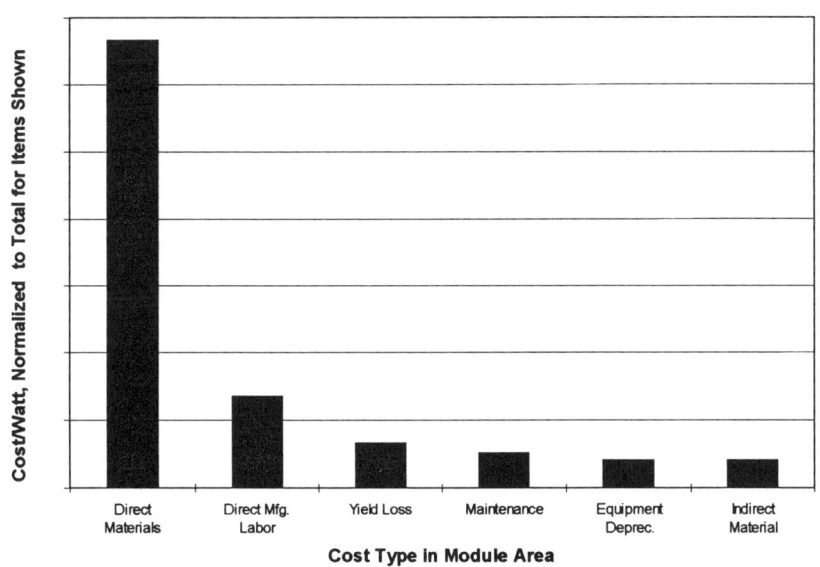

**FIGURE 2.** Costs in module fabrication area showing the large contribution of direct materials costs.

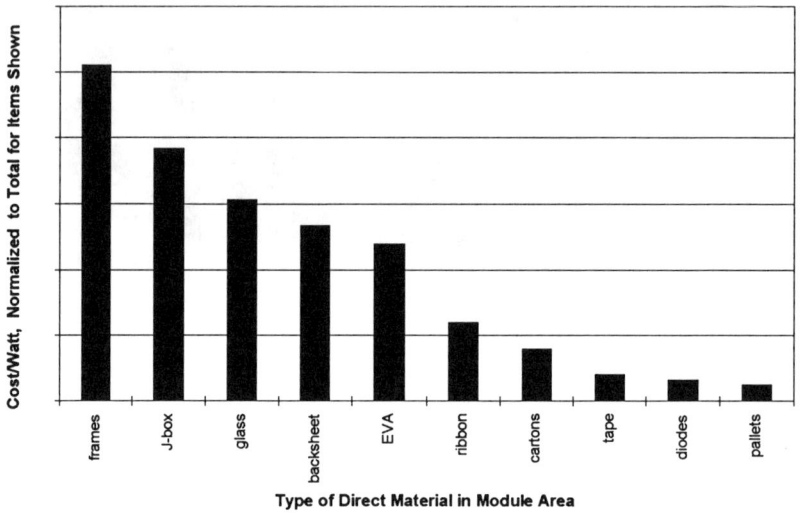

**FIGURE 3.** Cost/watt for direct material in the module fabrication area.

Zooming in once again to resolve the detail in the module direct costs, we see in Figure 3 that there are many contributors, but that the frames and junction boxes are the greatest module direct material costs. Similar analyses can be carried out for the other three process areas, *i.e.*, ingot, wafering, and cell fabrication.

## Solutions to Module Cost

Methods for reducing the cost per watt at the module level can be divided into at least three categories of reductions in: the cost of cells that go into the modules, the cost of module direct materials, and the cost of module manufacturing labor. Starting with the cost of cells, a principle used to great advantage in the integrated circuit industry is to increase the substrate size, thus producing a greater amount of the desired output (watts for solar cells) for each part processed. Another lesson from integrated circuit fabrication lines, as well as other types of manufacturing, is that yield is tremendously leveraging, particularly in processes with as many steps as there are between silicon feedstock and finished photovoltaic modules.

Several other issues are important for reducing cell cost. The cell shape should conform to the ingot in order to make the best use of the grown silicon. For Cz Si, this indicates round substrates. Decreasing cell material usage is a straightforward way to reduce cell costs, *e.g.*, through the use of thinner wafers. Increasing the cell power, *e.g.*, by reducing resistive losses, lowers the $/W by increasing the denominator of this ratio.

The themes of using larger parts and increasing yield are quite general and apply to modules as well. With regard to the cost of module direct materials, increasing the module size lowers the perimeter-to-area ratio, thus reducing the $/W contribution of the frames. Similarly, the cost of junction boxes can be spread out over a greater wattage by using larger modules. Simply finding less expensive materials and vendors for frames and junction boxes is always a valid method for reducing cost. The use of glass, backsheet, and EVA scale with the module area, but reduced sheet thicknesses can be used to conserve materials. As always, yield is highly leveraging for reducing direct materials costs.

For manufacturing direct labor, increased module size reduces the $/W of all processing steps whose costs scale with the number of parts processed rather than with the module area. Automated assembly, *e.g.*, the Spire automatic soldering equipment developed under another PVMaT subcontract, and semi-automated fixturing is important for managing the costs of a labor-intensive process like module fabrication. Once again, maintaining high yield is necessary for maximizing operator productivity and holding manufacturing labor costs low.

## Specific Module Designs

In May 1996, a new junction box design entered production, combining improved versatility for PV module users, and a cost that is 0.05 $/W lower for present module sizes. Figure 4 is a schematic of an SM110 module with 72 103-mm semi-square cells rather than the more usual 36. Production of this type of laminate began in the SSI Camarillo plant in August 1996. The larger module size and power (110 W) lower the $/W components from the frames, junction box, and module manufacturing labor, resulting in a 4% reduction in module $/W relative to 36-cell modules.

Figure 5 shows a new module configuration that uses 150-mm round cells in a 36-cell package. Prototypes of this module have been built during Phase I of this contract and shipped to NREL. The cost-per-watt is expected to be significantly lower for this type of module for a variety of reasons. The larger cell substrates used translate to an increase in surface area per wire saw run, and reduction of handling and equipment costs per watt for a given MW/year production level. The round shape of the 150-mm cell makes mode efficient use of the Cz ingot by eliminating slabbing of the ingot. The large module size also means that handling

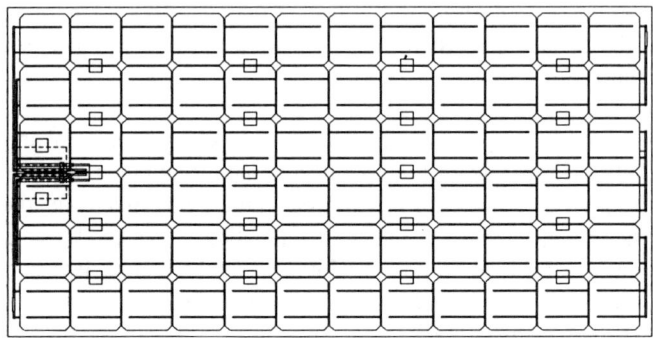

**FIGURE 4.** SM110 module with 72 103-mm semi-square cells reduces $/W contribution of the frames and the junction box.

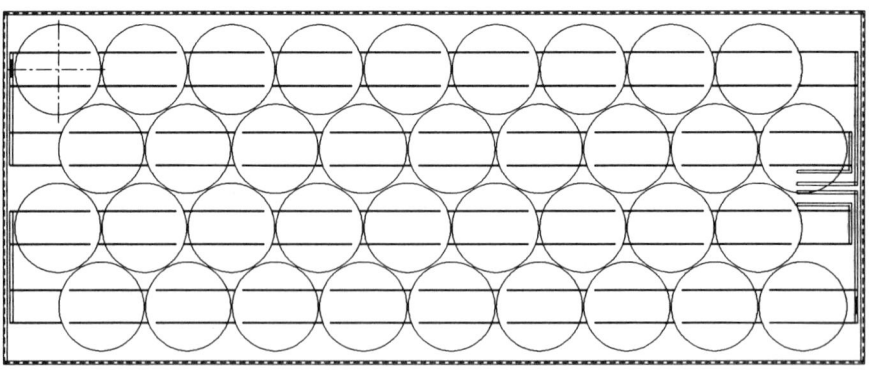

**FIGURE 5.** Large-area SM100 prototype module using 150-mm-diameter cells has a reduced cell contribution to $/W, as well as reduced frame and junction box $/W components.

and equipment costs per watt will be reduced for a given MW/year production level. All told, this module configuration with 150-mm round cells is projected to be capable of a 19% reduction in cost-per-watt.

## YIELD AND PRODUCTIVITY IMPROVEMENTS

In the discussion of cost in the last section, yield was found to impact nearly all types of potential cost reductions. Figure 6 plots the relative yield loss for the three process areas of wafering, cell fabrication, and module fabrication. Although the yield losses for the wafer and cell areas are the highest, a given yield loss in the module fabrication area is especially costly. The high yield losses in the wafer and cell areas are also important for the module cost per watt, since wafers and cells are constituents of the module. In Figure 7, the various types of yield loss in the line are plotted. Mechanical breakage is clearly the dominant type of defect causing yield loss. Most efforts at improving the yield have been aimed at reducing mechanical breakage.

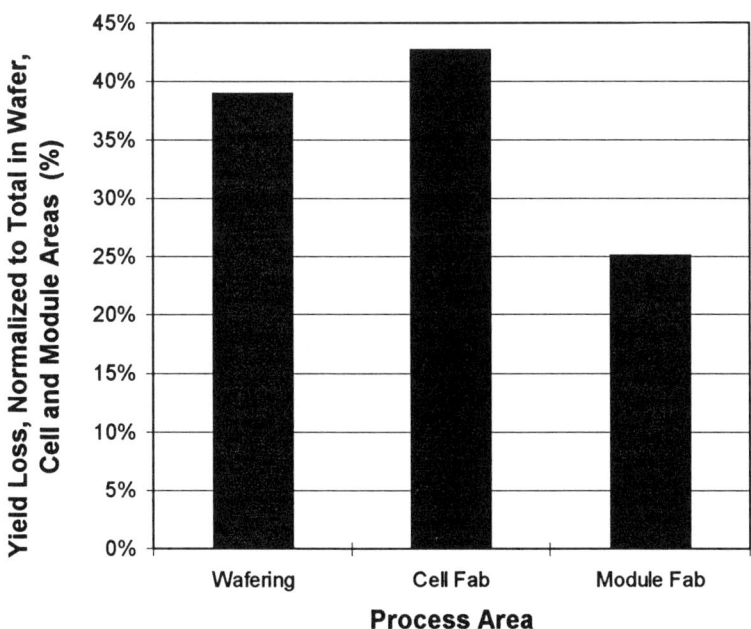

**FIGURE 6.** Yield loss by process area. Due to the value added at each step in the process, yield loss in the module area is especially costly.

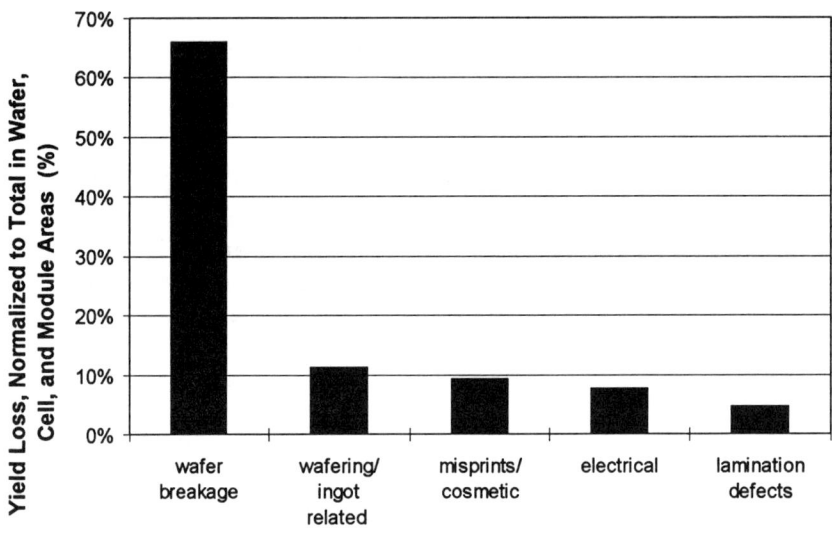

**FIGURE 7.** Yield loss by type of defect. Wafer breakage is by far the dominant loss mechanism for the wafering, cell fabrication, and module fabrication areas.

In order to provide more information about the root causes of wafer breakage, the average breakage force of wafers was measured by a stylus lowered at a controlled rate in the center of the wafer resting on two horizontal rails. The force required to break wafers in this configuration was measured for multicrystalline-Si (mc-Si) wafers and for single-crystal SSI Cz wafers, at 6 different points in the cell fabrication sequence, from as-sawn wafers to finished cells. The data is shown in Fig. 8. Each bar represents the average of breakage force measurements on 25 wafers, except for after Step 4, where the data set was only 10 wafers. The error bars indicate ± 1 standard deviation. These breakage force measurements may not highlight precisely the same mechanisms that cause breakage in the manufacturing line, but do provide a relative measure of wafer strength that can indicate major effects of wafer processing, and of different silicon materials.

Throughout the cell process, the breakage force of mc-Si is fairly similar to that of single-crystal SSI Cz. As sawn, the wafers are 50-60% weaker than after texture etching (after Step 1), and the mode of breakage is qualitatively different for the single-crystal Cz. Before etching, the Cz wafers tend to cleave into quarters, while after etching the wafer shatters into more small pieces. One explanation is that microfractures or other irregularities on the as-sawn wafer surface can serve as nucleation sites for larger fractures when the wafer is stressed during the breakage force measurement. The fracture nucleation sites on the as-sawn wafer are likely to be removed or their density decreased by the wet etching solutions.

For most of the cell process after etching, the breakage force stays fairly constant within the experimental uncertainty. On printed and fired cells (after Steps 5-8), however, the average breakage force drops by ~35% for both types of material. This could be due to the surface imperfections caused by the metal contacts. It is interesting to note that for all of the process steps after wet etching, the standard deviation of the breakage force measurements was higher for the mc-Si wafers than for single-crystal SSI Cz, perhaps due to the variation in crystal orientation in mc-Si.

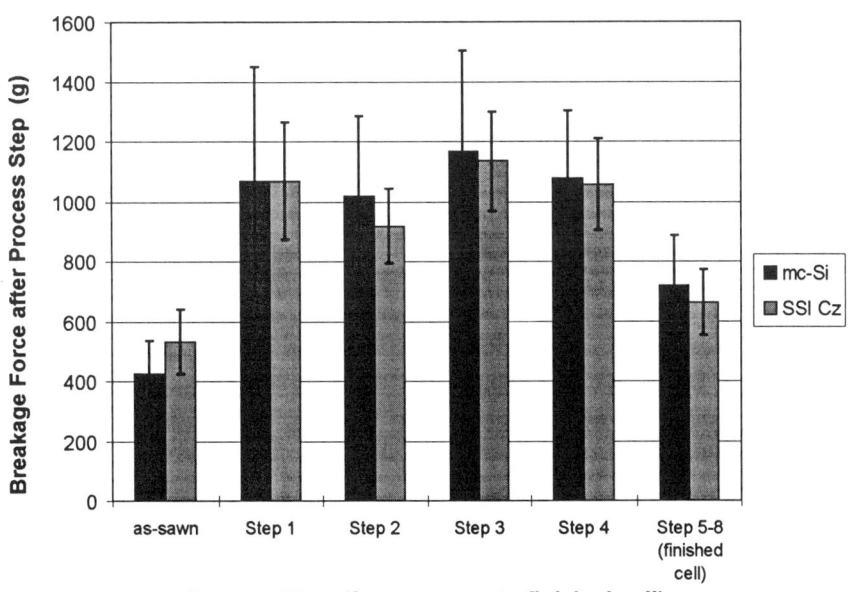

**FIGURE 8.** Wafer breakage force of multicrystalline-Si (mc-Si) and SSI CZ Si, as measured by a stylus for wafers resting on two horizontal rails. The error bars indicate ± 1 standard deviation in the data.

# MANUFACTURING SYSTEMS TO IMPROVE MODULE RELIABILITY

## ISO 9001 Certification

The SSI Camarillo plant received ISO 9001 certification in March 1996. This is a major milestone in the pursuit of quality manufacturing systems, and represents a very substantial effort by the company to establish and document procedures, operator work instructions, maintenance and calibration schedules, conduct operator training, etc. The benefit is a profoundly improved system to ensure manufacturing compliance and control of its processes. It is, however, only a beginning.

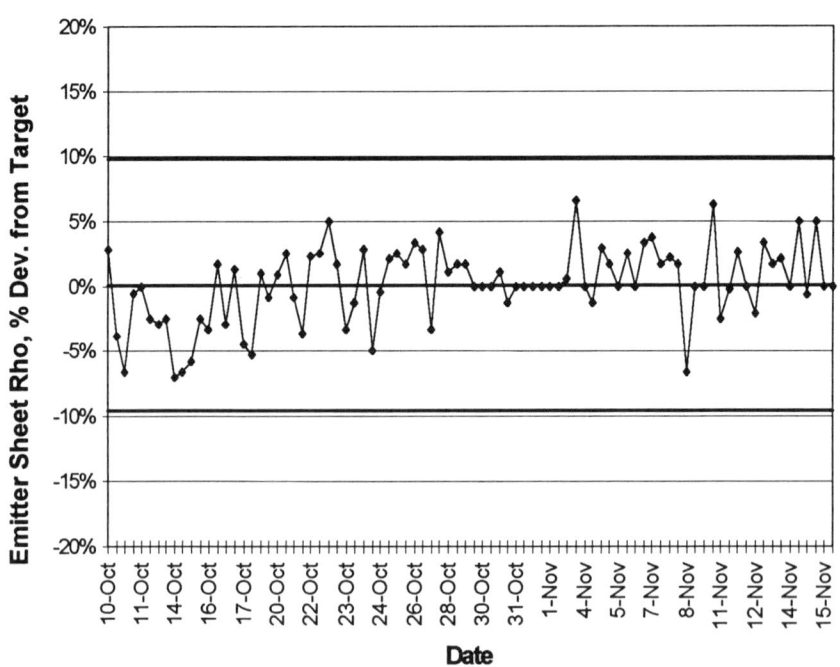

**FIGURE 9.** SPC charting of emitter sheet resistance after diffusion.

## Statistical Process Control Implementation

A goal of the PVMaT program at SSI is the identification of key processes in the fabrication sequence, e.g., wafer thickness variation, wafer cleaning, emitter sheet resistance, cell fill factor, incidence of wafer edge chips, solder bond pull strength, lamination defects, etc., and implementation of SPC charting at each key step at the end of Phase II. SPC methods are now in use at many sites in the plant, baseline data is being gathered, and a feedback mechanism for early detection of out-of-control process conditions exists. Figure 9 shows an SPC chart of emitter sheet resistance after diffusion as an example. The data is characteristic of an in-control process, well-centered on the target value. The regular charting of data by operators fosters greater participation, and provides a way to monitor progress in bringing a process parameter closer to a target value, tightening the distribution of values, or improving yield. Figure 10 gives an example of a shift in the average value of automatic soldering yield, and distribution tightening, coupled to events in the line.

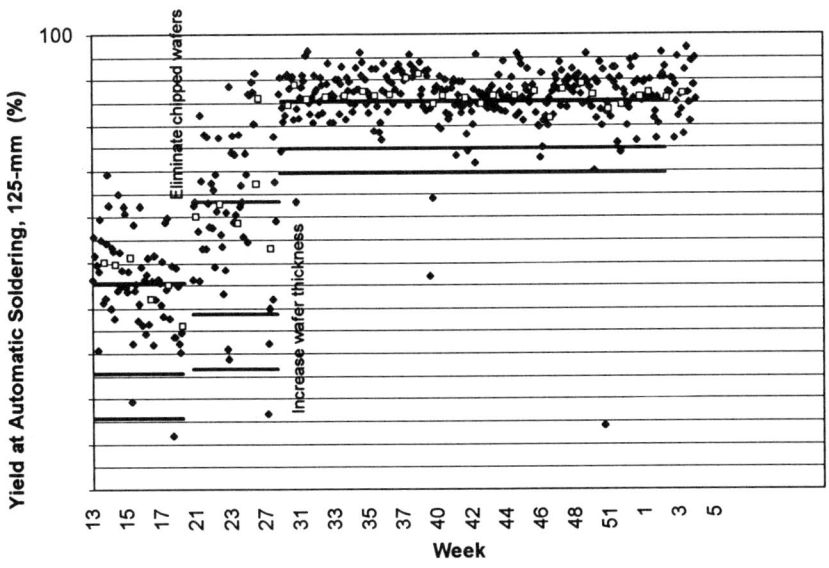

**FIGURE 10.** Charting of automatic soldering yield provides feedback mechanism to gauge the effect of process changes.

# SUMMARY

Cost drivers for Cz Si solar cell modules have been identified. Cell size and shape, material usage, module size, and cell and module yields have an especially strong influence on cost. Specific module designs to address cost issues have been implemented. Broader changes to the module configuration based on the 150-mm-diameter round cell are in progress during Phases II and III. Wafer breakage is the mechanism which dominates cell yields. Breakage is being reduced by understanding root causes, improved tracking of data, and by operator involvement. The development of systems to ensure compliance and control of module manufacturing includes ISO 9001 certification received by SSI in March 1996, and the ongoing implementation of SPC methods at key control points in manufacturing. As a result of yield gains and other improvements, a 5% reduction in cost-per-watt at the module level was achieved through September 1996. Based on large prototype modules built with 150-mm-diameter round cells, a 19% reduction in module cost per watt is projected, with production scheduled to begin in Phase II.

# ACKNOWLEDGMENTS

Many people have contributed to the work under this contract. Thanks are due especially to Rick Mitchell, NREL technical monitor, to Dave Bender, Ruben Balanga, Mark Crowder, Jean Hummel, Dave Jeffrey, Waltraut Klein, Greg Mihalik, Maria Tsimanis, Von Walters, Elena Woodard, Eugene Yamamoto, and others in the Engineering, Quality, and Finance groups at SSI.

This work was funded in part by DOE/NREL Subcontract # ZAF-5-14271-12.

# Large-Area Silicon-Film™ Manufacturing

D.H. Ford, A.M. Barnett, J.C. Checchi, J.S.Culik
R.B. Hall, E.L. Jackson, C.L. Kendall, J.A. Rand,

*AstroPower, Inc.
Solar Park
Newark, DE 19716-2000*

**Abstract.** The Silicon-Film™ process is on an accelerated path to large-scale manufacturing. A key element in that development is optimizing the specific geometry of both the Silicon-Film™ sheet and the resulting solar cell. That decision has been influenced by cost factors, engineering concerns, and marketing issues. The geometry investigation has focused first on sheet nominally 15 cm wide. This sheet generated solar cells with areas of 240 cm$^2$ and 675 cm$^2$. Most recently a new sheet fabrication machine was constructed that produces Silicon-Film™ with a width in excess of 30 cm. The results of testing have indicated that there is no limit to the width of sheet generated by this process. The new wide material has led to prototype solar cells with areas of 300, 400 and 1800 cm$^2$. Test results are available on some sizes, with all results expected later this year. Significant advances in solar cell processing have been developed in support of fabricating devices of this size including diffusion and application of the anti-reflection coating.

## INTRODUCTION

The Silicon-Film™ process is on an accelerated path to large-scale manufacturing. The existing process has been based on solar cells with an area of 240 cm$^2$ (the solar cell is referred to as the AP-225). Efficiencies in excess of 12% have been demonstrated on these large area cells. Smaller, laboratory devices have demonstrated efficiencies of 14.6% (see Figure 3).

---

This work is supported by the Photovoltaic Energy Technology Division, U.S. Department of Energy through its Photovoltaic Manufacturing Technology (PVMaT) Project administered by the National Renewable Energy Laboratory.

In support of the move to large-scale production, development work has been underway to determine the optimum width of the Silicon-Film™ sheet. The sheet presently in production is nominally 15 cm wide. Advantages to increased sheet width include reduced capital costs, improved material utilization, and the capability to increase solar cell size. Increased solar cell size has the advantage of higher power per piece with fixed cost for handling during solar cell and module fabrication.

The investigation into width has resulted in the construction of a new, wide, Silicon-Film™ machine. This equipment has demonstrated that Silicon-Film™ sheet material can be fabricated at a width greater than 30 cm. At this point, there is no known limit to the width capabilities of the Silicon-Film™ sheet growth process. This machine capability has permitted us to evaluate a number of different solar cell sizes. Three are pictured in Figure 1. Solar cells with areas of 240 cm$^2$, 300 cm$^2$, 400 cm$^2$, 675 cm$^2$, and 1800 cm$^2$ have been fabricated. A critical element in the optimization process is the capability of cutting finished solar cells into smaller sizes for different applications, as illustrated in Figure 2.

**FIGURE 1.** A photograph of the AP-225, the AP-400, and the AP-1800. The AP-1800 is believed to be the world's largest crystalline silicon solar cell, with dimensions of 30 cm by 60 cm. The ruler shown for reference is 12.5 cm long.

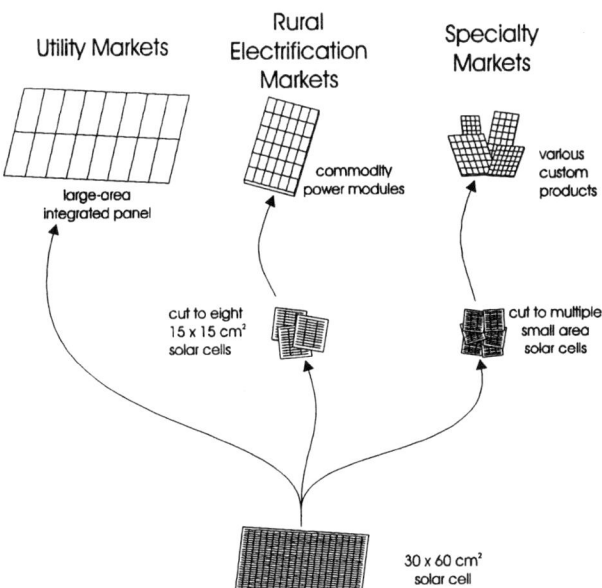

**FIGURE 2.** Flexibility of a 30 cm x 60 cm solar cell in meeting several different PV markets.

## PROCESSING DEVELOPMENTS

The flexibility in substrate size has led to the need for advances in fabricating emitters and anti-reflection coatings. In contrast to "traditional" solar cell manufacturing based on integrated-circuit technologies, these next-generation Silicon-Film™ processes are inherently free of substrate size boundary conditions. The use of low-cost, environmentally responsible precursors, and industrial coating application equipment with proven reliability, also make these advanced processes particularly attractive to large-scale adaptation.

Large area Silicon-Film™ emitter fabrication is accomplished by continuous-feed application of a phosphorous-based coating. This is followed by thermal cycling, using conditions approaching those encountered in rapid thermal processing (RTP). The resulting emitter sheet resistance is typically uniform to within 10% over the full device area. Wafer deglazing is straightforward, and requires no specialized equipment or chemicals.

For the anti-reflection coating, simple liquid precursors are applied to preheated solar cells, using commercial coating equipment, to form an oxide layer which is index-matched to module encapsulants. Advances made in this technology include the enhanced ability to chemically vary the index of refraction, and the improvement in coating characteristics using conductive and radiative heating methods.

Improvements in the characterization of Silicon-Film$^{TM}$ material have been made with the use of a RFPCD lifetime measurement technique developed by Dr. Richard Ahrenkiel of NREL. The measurement method allows a contactless measurement of the minority carrier lifetime in grown material with little or no sample preparation.

## DEVICE PERFORMANCE

A summary of the solar cells being evaluated is shown in Table 1. Some of the devices have been tested at NREL (see Note). Testing of these large devices has required the use of special testing apparatus. One such testing jig is being evaluated by NREL now with AP-225 and AP-300 solar cells. A second jig has been fabricated for testing the AP-1800 and will be supplied to NREL for performance testing.

**TABLE 1.** Summary of device data on different size solar cells.

| Product | Area | Power | Efficiency | Note |
|---|---|---|---|---|
| Laboratory Solar Cells | 1.0 cm$^2$ | 14.6 mW | 14.6% | (1) |
| AP-225 Solar Cell | 240 cm$^2$ | 2.93 W | 12.2% | (1) |
| AP-300 Solar Cell | 311 cm$^2$ | 3.48 W | 11.2% | (2) |
| AP-400 Solar Cell | 412 cm$^2$ | | | (3) |
| AP-6 75 Solar Cell | 676 cm$^2$ | 7.87 W | 11.6% | (1) |
| AP-1800 Solar Cell | 1800 cm$^2$ | | | (3) |

(1) NREL test , (2) AstroPower test, (3) Fabricated, but not yet tested

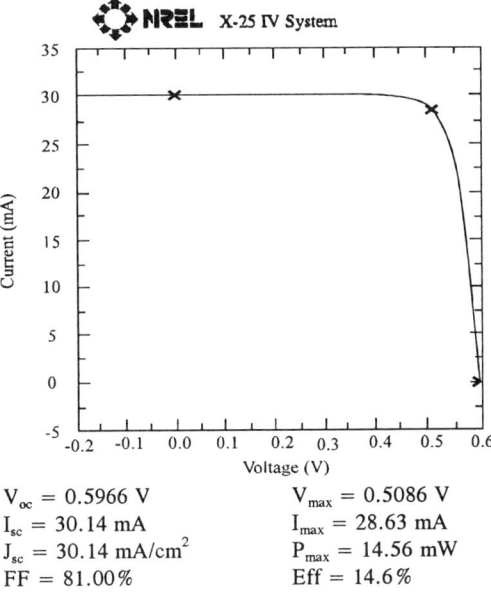

$V_{oc}$ = 0.5966 V  
$I_{sc}$ = 30.14 mA  
$J_{sc}$ = 30.14 mA/cm$^2$  
FF = 81.00%

$V_{max}$ = 0.5086 V  
$I_{max}$ = 28.63 mA  
$P_{max}$ = 14.56 mW  
Eff = 14.6%

**FIGURE 3.** Current-voltage curve for a 1-cm$^2$, Silicon-Film$^{TM}$ solar cell with an energy conversion efficiency of 14.6%.

## SUMMARY

The Silicon-Film$^{TM}$ process is presently in pilot production with a sheet nominally 15 cm wide. In planning for large-scale manufacturing, an effort has been undertaken to determine the limits to the sheet width. A new Silicon-Film$^{TM}$ sheet making facility was constructed that has successfully demonstrated sheets in excess of 30 cm wide. At this point there is known limit to sheet width. The enhanced machine capability has allowed us to evaluate a number of different solar cell sizes. The largest solar cell presently under evaluation has an area of 1800 cm$^2$.

## ACKNOWLEDGMENTS

The authors would like to acknowledge the contributions of Emanuel DelleDonne with device fabrication, Thomas Ellwood with Silicon-Film$^{TM}$ sheet manufacture, and Bill Bloothoofd and Chris Colgan with design and construction of the new Silicon-Film$^{TM}$ machine.

# PVMaT Improvements In Monolithic a-Si Modules on Continuous Polymer Substrates

Frank R. Jeffrey, Derrick P. Grimmer, Steven Brayman, Bradley Scandrett, Michael Thomas, Steven A.Martens, Wei Chen, and Max Noak*

*Iowa Thin Film Technologies Inc and *Iowa State University, Ames, Iowa 50010*

**Abstract** Iowa Thin Film Technologies, in conjunction with NREL, is engaged in a 3 year cost shared PVMaT program with goals to improve overall module performance, increase manufacturing throughput and to reduce the overall module manufacturing costs of the ITF production line by 68%. The first year of the contract has seen significant progress toward those goals with process improvements yielding a 30 % manufacturing cost reduction. These reductions come from improvements in the alignment processes in the patterning steps, development of a roll based lamination process, and improved area utilization.

## Introduction

The objective of ITFT's PVMaT program is to improve overall module performance, increase the throughput and to reduce the overall module manufacturing costs of the ITF production line by 68%. The Phase I efforts were aimed at a 30% to 40% overall reduction in the cost of module manufacturing for the ITF production line. Phase I project activities focused on efforts which included: increasing the throughput of the metalization, a-Si deposition, laser-scribing and welding processes; replacing the ITF TiN layer with a less absorbing ZnO layer; establishing new laser operating parameters to optimize the laser beam scan speed; and developing a new roll-based laminating process for the ITF production line. Phase I activities were divided into four tasks :

Task 1: Roll Based Deposition Throughput Improvement for Back Contact Layer
Task 2: Improvements in Laser-Scribing, Welding, and Printing Process Throughput
Task 3: Substrate and Materials Cost Reduction
Task 4: Development of a Roll Based Laminating Process

The overall development is comprised of many individual components. In the sections below, we discuss approaches and results for a number of these individual efforts.

## Texture Control

Total reflectance of light off the back contact and back contact texture are both critical to achieving the highest efficiency of collection of red light. High total reflectance minimizes loss of light by absorption at the back surface and texture provides scattering which increases the path length before reaching the front surface and creates the possibility of total internal reflection of light when it reaches that surface. In addition to the optical performance, the back contact must also be stable and have a good mechanical bond to the substrate.

To guarantee the best reflectance and optimum scattering, it is important to have a characterization method. It is one of the advantages of a roll based process that such a characterization method can be incorporated into the deposition system to monitor film quality and control deposition parameters.

To develop a monitor we initially evaluated the angular dependence of reflectance of good and poor substrate material using a He-Ne laser and a moveable silicon detector to establish the sensitivity of the method. Figure 1 shows the data for the "good" (top curve) and "poor" (bottom curve) samples in this experiment. It is evident that both the total specular reflectance and the ratio of diffuse to specular reflectance can be measured and correlated with substrate quality.

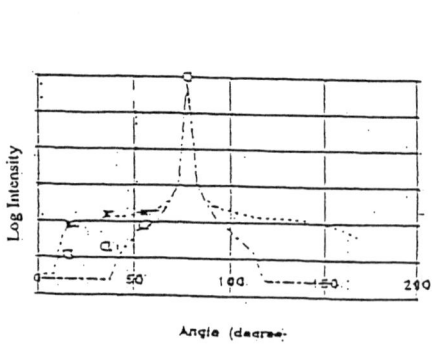 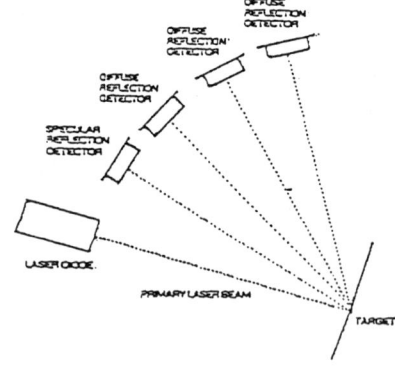

Figure 1. Angular distribution of 735 nm reflected light from properly textured (top curve) and poorly textured (bottom curve) substrate. Boxes are individual data points using in the situ detector for the respective samples.

Figure 2. A schematic diagram of the in-situ diffuse reflectance measurement system constructed using a laser diode and 4 detectors.

To establish a capability of monitoring texture during processing within the deposition chamber, a system was constructed which uses a laser diode along with 4 silicon detectors at fixed angular positions. A diagram of this system is shown in Figure 2. This system provides 4 data points which can be used to monitor the efficiency of scattering of red light. The individual data points on figure 1 show data from this setup overlaid on the prior data. From this it is clear that the 4 data points taken are adequate to indicate the full scattering response and be used as a quality control check on system operation. This system is one of a number or similar quality control monitors developed under the program.

## Back Surface Reflection Enhancement

Zinc oxide is being developed as an alternative to titanium nitride as the intermediate layer between the aluminum base metal/reflector/electrical conductor and the silicon layers for photovoltaic cells. ZnO has optical benefits that can enhance the reflectance of light from the base of the cell improving the overall efficiency of the cell

While laboratory results have shown that the ZnO layer incorporated between the back metal contact and the n+ layer can significantly enhance reflection and therefore the collection efficiency of red light. To date, ITF has been unable to incorporate this in our roll based pilot manufacturing line. Attempts to incorporate a sputtered ZnO layer have resulted in shunted cells or a very high leakage current while CVD ZnO has produced a barrier.

The prime candidate mechanisms for producing the high leakage currents were identified as follows: 1) The n layer deposition might reduce the ZnO and let Zn diffuse into the device destroying the barriers; 2) Spikes of ZnO may form which protrude through the cell to provide shunt paths; 3) Small pieces of the ZnO might separate from the underlying metal and flake of causing pinholes, shunting the device; 4) The ZnO might affect the nucleation of the micro-crystalline n+ layer, reducing the thickness below the needed level.

To help distinguish between these effects, SIMS was performed on samples of the devices fabricated on ZnO covered Al substrate. No Zn was detected before the back contact is reached and The back interface looked close to what is desired. If Zn had diffused far into the device, it should be visible in this scan.

If pinholes or ZnO spikes existed in the scanned area, one might expect a low level Zn signal to be evident early in the scan. As this was not apparent from the scan, if they do exist, they must be a very small percentage of the total area.

A wide range of experiments were carried out to determine the source of the shunts. Deposition temperature, power and ZnO film thickness were varied and devices examined for correlation's between these parameters and shunt level. No correlation was found. An amorphous n+ layer was substituted for the standard microcrystalline layer to investigate the possibility that the high deposition power and H2 etching associated with the microcrystalline layer were causing damage. A range of deposition powers and doping levels were used for the amorphous n+ layer. No reduction of shunt level was seen with these changes. A TiN layer was deposited over the ZnO layer as a barrier before fabrication of the silicon layers. This configuration was shown to have shunts also.

During the course of these experiments, one correlation was observed: the shunt level increased with the occurrence of physical damage that results from poor handling. While this is true for all substrate combinations, the sensitivity seemed much higher for the ZnO coated back material. It may be that the mechanical strength or interface bonding strength of the ZnO is enough lower than the TiN that it is susceptible to mechanical damage. This weakness would be more noticeable on our flexible substrate than on a more rigid material.

Our continued course of action will be to look at less conductive ZnO which would result in higher resistance leakage paths and to look for alternatives to ZnO which would have similar optical properties, but would have stronger mechanical properties. If the shunt paths are very small physically (indicated by the lack of Zn signal early on the SIMS), a higher resistance material should severely limit the amount of current that they could drain off. Prime candidates for replacing the ZnO are Titanium oxide and Titanium oxy-nitride. These were chosen because of the previously demonstrated ability of TiN.

## Improvements in Laser-Scribing, Welding, and Printing Process Throughput

A major part of our effort is aimed at improvements in laser-scribing, welding, and patterning system processes throughput. Several sensor technologies were evaluated for potential to improve alignment speed and reproducibility. Improved alignment reproducibility adequate to reduce line widths from 0.6 mm to 0.3 mm is desired. Improved alignment speed is expected to result in an increase in throughput of the laser and printer systems, and a 17% overall reduction in $/W cost of module manufacturing.

The scriber and printer have similar alignment technologies, both in hardware and software used. Common to both scriber and printer are stepper and servo motors for web motion over a platen and fiber optic sensors. The original alignment approach followed standard screen printing technology where the web is moved forward until the registration marks were found. The web was locked down at that position for processing. This process turned out to be very slow, as the web speed had to be slow when approaching the alignment position in order to get the required registration. We also determined that there was an accuracy limit of 100 microns using this system due to web "hopping" during slow movements (static friction). Under this PVMaT program, a new approach has been developed where the web is brought rapidly into approximate alignment and locked down. Sensors, now mounted on a processing head are used to determine the precise position of the web so that the head can be aligned to the workpiece before processing takes place. This process is much faster than the original because the head alignment motions are much faster than web motion alignment for the same level of accuracy. The final alignment reproducibility is improved to 10 microns from the earlier 100 microns.

Table 1 shows the original and current values for the alignment errors and processing speeds for the printer and scriber.

Table 1. Sources of alignment errors for the scriber and printer for original, current, and expected accuracy's.

## Scriber

|  | Web Hop | Head Location | Line Wobble | Line Bow | Line Spread |
|---|---|---|---|---|---|
| Original | +/- 0.1 mm | +/- 0.01 mm | +/- 0.02 mm | +/- .025 mm | +/- 0.05 mm |
| Current | - | +/- 0.01 mm | +/- 0.02 mm | +/- .025 mm | +/- 0.05 mm |
| Expected | - | +/- 0.01 mm | +/- 0.01 mm | +/- .01 mm | +/- 0.01 mm |

## Printer

|  | Web Hop | Screen Stretch | Screen Location |
|---|---|---|---|
| Original | +/- 0.1 mm | +/- 0.05 mm | - |
| Current | +/- 0.05 mm | +/- 0.05 mm | - |
| Expected | - | +/- 0.02 mm | +/- 0.01 mm |

# Development of a Roll Based Laminating Process

Efforts were aimed at evaluation of the potential of existing roll-based laminating technologies and equipment. Pressure sensitive adhesives (PSA) were evaluated taking into account the potential active species within the PSA that might lead to corrosion. Thermal adhesives were evaluated taking into account the limited film-compression time in a roll based lamination process. This effort is aimed at developing a roll based laminating process for the ITF production line and a significant reduction in module assembly costs.

The original lamination process (and lamination industry standard) involved: 1) cutting each one of the component pieces (including submodule, busbar, strain relief, encapsulant, and cover tape); 2) assembling the pieces; 3) vacuum laminating assembly; 4) trimming; and 5) attaching a wire connection. The new system consists of 1) cutting the sub module and busbar attachment; 2) roll laminating the assembly; 3) die cutting out the module; and 4) attaching a junction box.

The first part of our effort involved a search for a suitable roll laminator which could be incorporated into the present production line for ITF's module encapsulation. The most suitable for ITF lamination needs was a roll laminator, shown in Figure 3, that used shoe type preheaters to preheat the material before it was laminated. The shoe type preheater system minimizes the web tracking problem and the difficulty of applying heat to a rotating member. The one drawback to the shoe preheat roll laminator is that the plastic laminate is dragged across the preheaters under tension causing scratch marks on candidate plastic laminates including Dupont's Tefzel®.

The optimum benefit of roll laminating comes with use of pressure sensitive adhesives and thermal set adhesives. The process is as follows 1) place conductive metal foil tapes down on the modules buss areas 2) laminate in the roll laminator with shoe preheat 3) cut and test modules.

The testing and evaluation of alternative materials for roll lamination involved two adhesive types: thermal set adhesives (TSA) and pressure sensitive adhesives (PSA). Some of the candidate adhesives that were evaluated were: thin Ethylene Vinyl Acetate (EVA/TSA); Ultra High Adhesion rubber (UHA/PSA); Silicone (PSA); Acrylic (PSA); and Polyethylene (TSA).

The two TSA's Polyethylene and thin EVA (1.5 mil) were roll laminated on a thermal laminator. Both adhesives were used with a polyester film, the EVA was with 1.0 mil Poly, and the Polyethylene was used with UV stable polyester 1.0 mil and 2.0 mil. These laminates are commercially available and did not require that

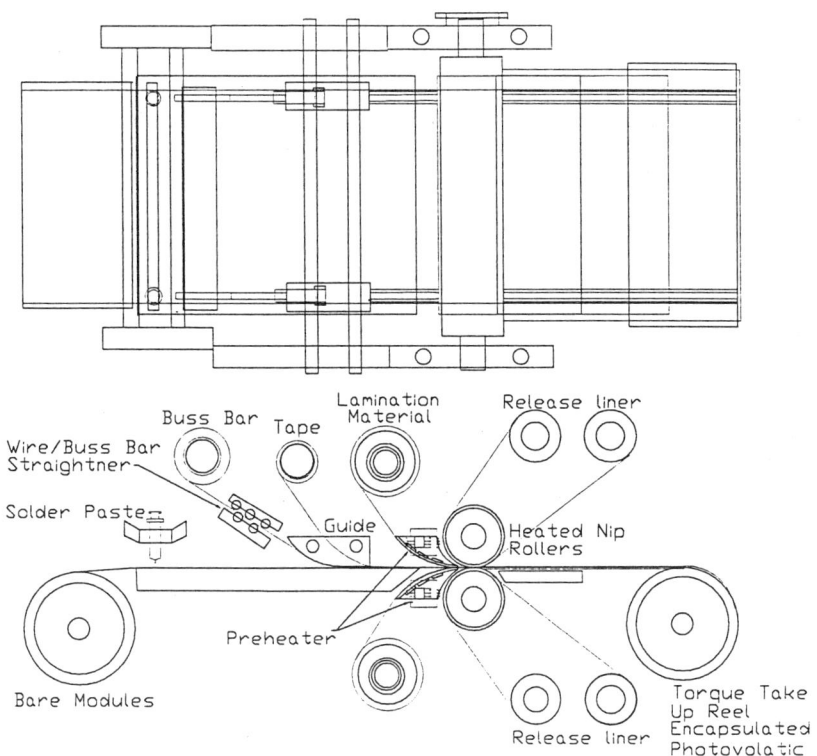

Figure 3. Schematic diagram of the roll based lamination system. The bottom encapsulant, top encapsulant, module material and bus bar matierial are all supplied from rolls and the final product is wound on a roll after lamination. Sheeting and terminal attachment are handled seperately.

the adhesive be laminated to the film. Both of these adhesives work excellently in the laminator and in the resultant solar modules in terms of clarity, uniformity, ease of lamination, adhesion, and overall appearance (wrinkle/bubbles). Samples were tested in the thermal cycler and 80°C oven. Modules were also optically checked with a Lambda 9 wavelength/intensity system, and pretested laminated and retested. Again both materials performed excellently.

To find a suitable pressure sensitive adhesive, ten different samples representing 3 classes of adhesive were selected for testing. The testing criteria included clarity, adhesion, appearance, and opposition to bubble formation. All of the ultra high adhesion (UHA) rubber samples failed all testing criteria The UHA rubbers were yellow in appearance and clarity and while adhesion was excellent during lamination, after one month of temperature cycling the adhesive separated from the module and the film laminate. Bubble formation in the UHA rubber adhesive appeared as the roll started out at the beginning of the process and after lamination the bubbles were still prevalent Thus, the UHA rubber adhesives failed all testing criteria.

The second adhesive type that was tested was the acrylic based PSA's. Testing produced varied results as shown in Table 2. There were two test variations for each laminate. Those listed under the first variation, "CYCLER," were temperature cycled for 4 weeks with a swing from -40°C to 80°C. The frequency for a complete cycle was 2 hours 30 minutes with 1 hour at -40°C followed by 10 minutes of ramp up to 80°C then 1 hour at 80°C followed by an additional 20 minutes of cool down to -40°C and repeat. This frequency provided 9.6 cycles/day, 67.2 cycles/week, and approximately 268.8 cycles/month. The second testing variation was an 80°C oven. This test is used for an extreme heat test which seems to have the most profound effect on the adhesive/film interaction. This test has run for 6 weeks and is continuing to date.

Two different laminate styles were also tried. The first, "trimmed," meant the laminates were trimmed to the edges of the module. This allowed examination of the adhesion of the adhesive to the module without the edge/adhesive to adhesive interaction to possibly mask results The second laminate style , "1/4" border," allowed for a 1/4" edge seal around the module more closely representing an actual laminated module.

Table 2. Acrylic PSA's: test results shown by category

| Testing conditions | | Adhesive trade name | Clarity | Appearance | Adhesion | Wrinkle/Bubble |
|---|---|---|---|---|---|---|
| Cycler | trimmed | MWC #190 | Clear | OK | FAIL | separated |
| Cycler | 1/4" border | MWC #190 | Clear | Good | EX | wrinkled |
| 80°C | trimmed | MWC #190 | Clear | OK | FAIL | separated |
| 80°C | 1/4" border | MWC #190 | Clear | Good | EX | wrinkled |
| Cycler | trimmed | Avery 1115 | Clear | Good | OK | few bubbles |
| Cycler | 1/4" border | Avery 1115 | Clear | Good | EX | wrinkled |
| 80°C | trimmed | Avery 1115 | Clear | Good | OK | few bubbles |
| 80°C | 1/4" border | Avery 1115 | Clear | Good | EX | wrinkled |
| Cycler | trimmed | Sun Poly 23 | Clear | EX | Good | Same as C.G. |
| Cycler | 1/4" border | Sun Poly 23 | Clear | EX | Good | Same as C.G. |
| 80°C | trimmed | Sun Poly 23 | Clear | EX | Good | Same as C.G. |
| 80°C | 1/4" border | Sun Poly 23 | Clear | EX | Good | Same as C.G. |
| Cycler | trimmed | 3M F9752PC | Clear | Good | EX | none |
| Cycler | 1/4" border | 3M F9752PC | VClear | Good | EX | wrinkled |
| 80°C | trimmed | 3M F9752PC | Clear | EX | EX | none |
| 80°C | 1/4" border | 3M F9752PC | Clear | Good | EX | wrinkled |
| Cycler | trimmed | 3M F9755PC | Clear | Good | Good | Separated ends |
| Cycler | 1/4" boarder | M F9755PC | Clear | Good | EX | wrinkled |
| 80°C | trimmed | 3M F9755PC | Clear | EX | Good | Separated ends |
| 80°C | 1/4" border | 3M F9755PC | Clear | Good | EX | wrinkled |
| Cycler | trimmed | 3M FC9469PC | Yellow | EX | Good | Separated ends |
| Cycler | 1/4" border | 3M FC9469PC | Yellow | Good | EX | wrinkled |
| 80°C | trimmed | 3M FC9469PC | Yellow | EX | Good | Separated ends |
| 80°C | 1/4" border | 3M FC9469PC | Yellow | Good | EX | wrinkled |

Test results shown above in Table 2 were given under each category. In order to pass to the next phase of testing, the adhesive's Clarity had to be "clear". For the Appearance, results above a "good" rating passed. All modules passed this criterion. As for Adhesion, nothing but "excellent" would be accepted. Under the next category, Wrinkle/Bubble, any module that had bubble formation during the testing phase was eliminated. As for wrinkles, it was determined that most of the wrinkle formation was due to the shrinkage of the Tefzel film and was not associated with the adhesive. This was the reason that a "good" rating passed in the appearance category. This testing criterion left the 3M 2 mil acrylic and 5 mil acrylic adhesives. The separated ends noted for the 3M samples were traced to the copper busbar and not due to the PSA under test.

Two silicone adhesives, a 4 mil silicone and a 2 mil silicone, also passed all criteria. Two additional silicone samples have begun the testing process.

Laminates which passed the above test were used in active module tests with the module performance checked before and after lamination. The performance change observed for all samples was an increase from unencapsulated to encapsulated-- which seemed to mimic the Tefzel®/EVA performance change. Fill factors increased 2-3 points and power point current increased form 2% to 10%.

The overall best performance PSA was the 4 mil silicone, but pricing makes the acrylic very attractive. The thin EVA's and Polyethylenes are far better than the best PSAs in terms of ease of assembly, but life expectancy is lower. Tefzel® film with 18 mil EVA still out performed all other materials in terms of optical clarity, UV resistance, and durability making this the best long term outdoor encapsulant. The UV stable polyesters with polyethylene TSAs appear the second best but further testing will be required as it could result that they might be a 5-10 year laminate with very low cost.

The roll laminators alone will not work for laminating modules in Tefzel® with the adhesive EVA (ethylene vinyl acetate) from Springborne Laboratories due to the adhesive's cure time. These adhesives require both pressure and heat for approximately 5 mins to reach an acceptable gel content, approximately 85%. This gel content is what is considered for "curing." Curing is impossible to accomplish in a nip roller where the contact time negligible. However Tefzel®/EVA encapsulation can be nip rolled to set the EVA adhesive and eliminate air pockets/bubbles. The remaining obstacle is curing the EVA.

Three methods were tested to cure the EVA after it had been roll sealed. The first involved sealing completed modules with Tefzel®/EVA in the heated roll laminator, stacking a large number of modules in a press, and placing them in a large oven for final curing. The second system involved placing the modules in a frame, and bladder apparatus which was pumped down and placed in an oven for the sealing and curing process. The third system used a process similar to the second but using a high temperature vacuum bag. Each system was rated based on appearance, clarity, and flatness as is described in Table 6 below. As can be seen in this table, the roll laminated module (#1) lost some clarity because of scratches

Table 3. Comparison of three methods tested to cure EVA.

| Procedure | Appearance | Clarity | Flatness |
|---|---|---|---|
| **Bubble Formation** | | | |
| #1 Roll laminated and pressed | | | |
|    Single | excellent | very clear | flat |
|    no bubbles | | | |
|    Multiple | good | OK | flat |
|    no bubbles | | | |
| #2 Vacuum laminated | | | |
|    frame and bladder | excellent | very clear | lumpy |
|    no bubbles | | | |
| #3 Vacuum laminated | | | |
|    vacuum bag | excellent | very clear | flat |
|    no bubbles | | | |

# CONCLUSIONS AND FUTURE WORK

Work done during Phase I, aimed at achieving the overall program goal of developing the most cost effective PV manufacturing process possible, resulted in significant improvements in throughput for the a number of processes including scribing, printing, and laminating. A number of quality control techniques were developed and implemented. an overall manufacturing cost reduction of 30% has been achieved through processing rate improvements and improved area utilization. A usable process for incorporating a back reflection enhancement layer still eludes us.

During the next year Iowa Thin Film Technologies, Inc. will continue efforts at implementing the back reflector; optimize performance and deposition rates in the new tandem machine; develop single pass capability in the TCO and metalization systems; and develop an automated cutting and bus bar attachment system to go with the roll laminator.

# The SunSine300 AC/PV Module

## Miles C. Russell, Gregory A. Kern and Clayton K. P. Handleman

*Ascension Technology, Inc. (http://www.ascensiontech.com)*
*P.O. Box 314, Lincoln Center, Massachusetts 01773*

**Abstract.** The SunSine300 is a fully-integrated AC/PV module developed by Ascension Technology, with support from the National Renewable Energy Laboratory, Sandia National Laboratories, New England Electric and ASE Americas. Rated output of the SunSine300 is 260 Watts ac, with the protective features and functions to comply with codes and achieve UL listing. Testing has been underway since autumn of 1996; pilot production is scheduled for March 1997, and commercial introduction anticipated in June 1997.

## WHY AN AC MODULE?

An AC module is a photovoltaic module with a mechanically and electrically integrated dc-to-ac inverter; the output of an AC module is ac power. AC modules simplify the design and reduce the cost of distributed residential and commercial PV systems. Among the many advantages offered by the AC module concept is its flexibility, allowing grid-tied PV systems to be configured in small power increments. Other advantages of an AC module are listed below:

- No dc wiring or dc components are present in an AC module system; ac wiring and ac-rated components are used. System wiring involves only ac-rated equipment, familiar to electrical contractors.
- Design time is reduced for an AC module PV system, since it is not necessary to configure a PV array to match the voltage, current and power input restrictions of a conventional power conditioner. The expense for custom PV system engineering is reduced or eliminated by the AC module. Array-to-inverter mismatch mistakes are not possible.
- Every AC module is individually maximum-power-point tracked. Sorting and matching of PV modules is not required, saving design time and

simplifying installation. Module mismatch losses that occur in conventional PV systems do not occur in an AC module system.
- Blocking diodes are not necessary in an AC module PV system. All AC modules are connected in parallel and reverse current flow prevented by the module inverter. The expense, energy losses and problems related to blocking diodes in conventional PV systems are eliminated with the AC module system.
- A grid-tied PV system can be configured using *any number* of AC modules, giving the PV industry many options for implementing distributed PV systems. This flexibility also means that grid-connected AC module PV systems can be installed in small sizes at lower absolute cost than is required for the smallest conventional grid-connected PV systems implemented today.

## BACKGROUND

In 1992 Ascension Technology began developing a 300 Watt inverter with support from New England Electric. The inverter was designed to operate with input from an ASE Americas' large-area 300 Watt photovoltaic module. Circuit topologies were investigated and preliminary circuit design was completed. In 1994, with support from Sandia National Laboratories, the inverter design was completed and a working "proof of concept" prototype was assembled and delivered to Sandia in April 1995.

Refinement and continued engineering of the early inverter began in August 1995, with support from the NREL PVMaT program. The focus of this work has been to complete the development and prepare for the manufacture of the SunSine300 AC/PV module. Nineteen electric utilities joined our PVMaT project as co-sponsors to provide technical guidance on the specifications of the SunSine300 and to purchase and evaluate beta-test units in the spring of 1997.

The development of the SunSine300 is also co-sponsored by ASE Americas, manufacturers of the 300 Watt PV laminate that is the platform for the SunSine300 AC/PV module. ASE's engineers have assisted in the mechanical and electrical integration of Ascension Technology's 300 Watt inverter with the PV laminate, and ASE has conducted environmental tests of SunSine300 prototypes.

## Technical Goals and Approach

Simplifying small, distributed grid-tied PV installations is, of course, the primary achievement of the AC module. However, throughout our engineering design process, of greatest importance to Ascension Technology have been the technical goals, and the primary technical goal for the SunSine300 is to attain high reliability. *The SunSine300 above all else, is extremely robust and of high engineering quality.*

Each SunSine300 contains all the protective features and functions required for code-compliant, safe grid-tied PV systems. The SunSine300 has been designed to comply with the following codes and standards:

- National Electrical Code, NFPA 70-1996 Article 690 (existing and proposed sections),
- IEEE 929-1988, Recommended Practice for Utility Interface of Residential and Intermediate PV Systems
- FCC Part 15B Conducted/Radiated Class B Electromagnetic Interference and Compatibility
- Underwriters Laboratory 1741, Proposed First Edition of the Standard for Power Conditioning Units for Use in Residential Photovoltaic Power Systems (note that the title of this standard will change in 1997)
- UL 1703, Standard for Flat-Plate Photovoltaic Modules and Panels
- Class A fire rating (achieved by the ASE module in UL1703 testing)

The SunSine300 is a true AC module; the inverter is integrated with the PV laminate in the factory. The PV laminate is not a module; the inverter replaces the standard diode housing found on the large-area PV module. The inverter is integral and not separable for use elsewhere. The dc source circuit is not user accessible, so the SunSine300 is therefore accurately described as a fully-integrated AC module.

## FEATURES

The SunSine300 AC module is based on a current-mode inverter. It has high bandwidth control of the sinusoidal output current and appears to the utility as an ac current source. The power electronics circuit in the SunSine300 is a simple and reliable topology. This topology was selected for its simplicity and efficiency after reviewing and modeling numerous candidate inverter circuits. The SunSine300 has only one fast active power switching device. All semiconductor devices were selected for industrial temperature range performance. The SunSine300 includes a 60 Hz transformer at its output, that helps with overall safety, performance and isolation from the utility grid. The SunSine300 has isolation between both the dc and ac circuits and ground of 2500 Volts, enhancing overall product safety.

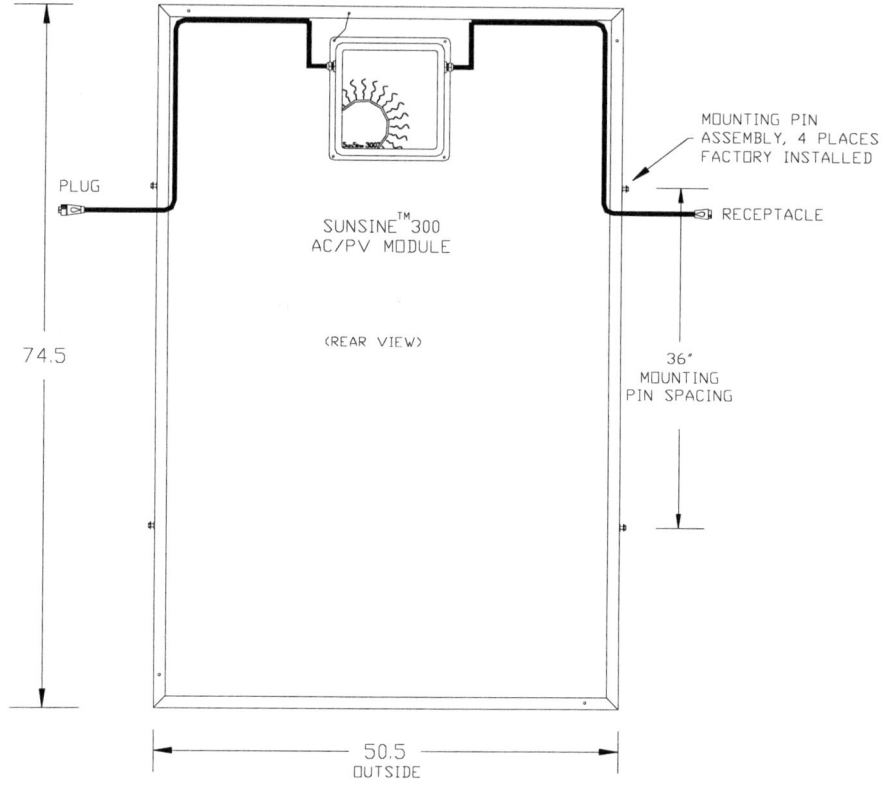

Figure 1. Dimensional drawing of the SunSine300 AC/PV Module.

## Controls

The SunSine300 control electronics track the maximum-power-point of the PV laminate to extract the maximum energy at all times. During steady-state conditions, the PV operating point is maintained within 1% of the maximum power point.

Anti-islanding is an area of great concern to electric utilities, who want assurance that when the grid goes down for any reason, a PV system will not continue to "run on". Such a condition is called "islanding" and would be a potential safety hazard to utility line personnel and also may have deleterious effects on connected equipment. The SunSine300 utilizes both active and passive measures to detect islanding and force shutdown of the unit. Passive methods are to monitor utility line voltage and frequency, and if either strays outside the normal operating limits, the unit will shut down. In addition the SunSine can rapidly detect a no-load condition (complete loss of utility voltage) and shut the unit down within 1 cycle.

Active methods of detecting an islanding condition that we considered include a frequency offset method and forcing a change in the output power. Although both may work under certain conditions, during the development of the SunSine300's inverter, we found that neither of these active methods was sufficient to prevent run-on in all load conditions for a single inverter. It has been our goal to insure that the SunSine300 does not run-on as a single unit or in systems with multiple units. Further, if the anti-islanding methods are not adequate for a single inverter, then they certainly will not be adequate for multiple inverter systems! To address islanding detection and prevention in a definitive way, we developed a proprietary method that we call the "Zebra" method.

## Integration

The 300 Watt inverter is assembled onto the specially prepared PV laminate in the factory. Electrically the PV circuit mates to the inverter by a blind-matable connector, so that when the inverter is mechanically attached, the electrical connection is also automatically made between the PV laminate and inverter. This simplifies the assembly of the SunSine300 and also facilitates field removal/swap-out of an inverter, should that ever be necessary.

No user-accessible dc wiring exists in the SunSine300. Ribbon cables emerge from the PV laminate and are factory soldered to a circuit board attached to the laminate. This circuit board also contains one gender of the connector that mates the PV circuit to the inverter when the two are joined.

The SunSine300 is equipped with quick connectors on the ac output, allowing rapid field wiring between units in an array. Simply plug SunSine300s together to parallel their outputs.

## BALANCE OF SYSTEM COMPONENTS

Ascension Technology engineered companion components for mounting and wiring to make complete SunSine300 systems. We evolved the design of our pitched-roof RoofJacks to reduce the torque on the attaching fasteners. We also expanded our family of UL-listed PV Source Circuit Protectors (PVSCPs) to include versions for SunSine300 PV systems. The PVSCP contains terminal blocks, surge suppression and can include fusing when necessary. The PVSCP comes pre-wired with quick connectors to mate to the connectors found on the SunSine300.

Ground fault protection is a requirement of the National Electrical Code for PV systems on dwellings, to prevent fire hazards. A standard off-the-shelf ground-fault-interrupting (GFI) circuit breaker should be used to interconnect the output of the SunSine300s to the electrical service in a dwelling. We have tested the performance of GFI breakers and found that they function normally when back-fed by the SunSine300. When a ground fault is imposed, the breaker opens and the SunSine300 shuts down within 1 cycle of the 60Hz waveform. With the ground fault removed, the breaker can be reset and normal operation can begin again.

Packaged systems will be available from Ascension Technology, that include the SunSine300 AC/PV module(s), RoofJack mounting hardware, a PV Source Circuit Protector and installation instructions/drawings. Field installation can be completed by an electrician and helper.

The SunSine300 is available with 120 Vac output (line - neutral) or 240 Vac output (line - line). Up to six SunSine300s can be paralleled for a 20 amp circuit breaker at 120 Vac, and 12 for a 20 amp circuit breaker at 240 Vac. Larger systems are easily configured also.

## SPECIFICATIONS AND PERFORMANCE

The table identifies the specifications of the SunSine300 AC/PV module.

Table 1. SunSine300 AC/PV Module Specifications

| AC Output | |
|---|---|
| Nominal Voltage and Range | 120 Vac or 240 Vac, 86-110% Vnom |
| Nominal Frequency and Range | 60 Hz, 59 - 61 Hz |
| Rated Power Output at STC | 260 Watts (± 4%) |
| Current Total Harmonic Distortion | < 5% (*typical <2%*) |
| Power Factor | >0.95 (*typical >0.995*) |
| Ambient Temperature Range | -40C (-40°F) to 50C (122 °F) |
| Dimensions, Weight | 74.5 x 50.5 in., 122 pounds |
| Acoustic Noise | Inaudible at full power |
| **Fully Automatic Operation, Protection and Reset** | |
| Operation within 1% of dc source circuit maximum power point | |
| Automatic morning start-up and evening shut down | |
| Passive and Proprietary Active anti-islanding protection | |
| No user controls | |

## Reliability

The electronics in the SunSine300 AC module are designed for exposure to a brutal environment. Everything from salt spray in seaside installations to desert heat to arctic temperatures will be encountered. Because of the inverter's close proximity to the cells in the module laminate, it will experience temperatures well above ambient during the day and below ambient at night due to radiative cooling of the laminate to the night sky.

We have utilized a variety of testing procedures and analyses for the SunSine300. These include thermal analysis of the enclosure to predict the lifetime of electrolytic capacitors, testing in environmental chambers and Highly Accelerated Lifetime Testing (HALT). HALT represents the state of the art in accelerated aging. The SunSine300 will be operated while exposed simultaneously to environmental extremes and vibration until it breaks, allowing identification of failure modes long before they would show up in the field or even in standard environmental testing.

Manufacturing quality control will include operational testing to full power, standard burn-in, and operational measurements to confirm critical component tolerances.

## SCHEDULE

In the third quarter of 1996 a number of SunSine300 prototypes were built for testing purposes. Over a six month period those units have been subjected to preliminary FCC testing, an Underwriters Laboratory preliminary investigation, and a battery of outdoor and indoor environmental tests (humidity/freeze, thermal cycling and damp heat). Based on these tests, minor modifications were made to the SunSine300.

Production of approximately 130 units is underway and will be completed by the end of the first quarter of 1997. These units will be delivered to our co-sponsoring electric utilities for beta testing. Units from this pilot production are also being tested by Underwriters Laboratories and confirmation of UL listing is anticipated by April 1997.

We will complete the system packaging and learn from the early experiences with the beta tests at the electric utility sites. Our manufacturing process will be streamlined as we head toward commercial introduction of the SunSine300 in the second quarter of 1997.

## ACKNOWLEDGMENTS

The SunSine300 represents the culmination of a tremendous engineering effort spanning a period of 5 years. Dr. John J. Bzura of New England Electric made it all possible at the start. Ascension Technology could not have accomplished such a feat without the unflagging support and encouragement of NREL and Sandia National Laboratories, and the special individuals who make our work such a pleasure. Thank you, Ward, Russ, Holly and Ben. Finally we are sincerely grateful to the pioneering electric utilities whose vision includes bringing solar electricity into the mainstream of their business in our lifetime.

# Design, Fabrication and Certification of Advanced Modular PV Power Systems

Glen E. Minyard and Timothy J. Lambarski

*Solar Electric Specialties Company*
*101 No. Main Street, Willits, California 95490*

Abstract. The Design, Fabrication and Certification of Advanced Modular PV Power Systems contract is a Photovoltaic Manufacturing Technology (PVMaT) cost-shared contract under Phase 4A1 for Product Driven System and Component Technologies. Phase 4A1 has the goals to improve the cost-effectiveness and manufacturing efficiency of PV end-products, optimize manufacturing and packaging methods, and generally improve balance-of-system performance, integration and manufacturing. This contract has the specific goal to reduce the installed PV system life cycle costs to the customer with the ultimate goal of increasing PV system marketability and customer acceptance.

The specific objectives of the project are to develop certified, standardized, modular, pre-engineered products lines of our main stand-alone systems, the Modular Autonomous PV Power Supply (MAPPS) and PV-Generator Hybrid System (Photogenset). To date, we have designed a 200 W MAPPS and a 1 kW Photogenset and are in the process of having the MAPPS certified by Underwriters Laboratories (UL Listed) and approved for hazardous locations by Factory Mutual (FM). We have also developed a manufacturing plan for product line expansion for the MAPPS. The Photogenset will be fabricated in February 1997 and will also be UL Listed. Functionality testing will be performed at NREL and Sandia with the intentions of providing verification of performance and reliability and of developing test-based performance specifications. In addition to an expansion on the goals, objectives and status of the project, specific accomplishments and benefits are also presented in this paper.

## BACKGROUND

Although most PV system integration companies advertise standard product lines, most PV systems sold today are still custom designed based on general product line designs. The fact that most systems houses have system product lines attests to the need to make the specifications simple and standardized so that they can be more easily understood by the potential customer and marketed by the sales engineer. This contract makes a concerted attempt to increase standardization and modularity as we see in other product categories such as computers whether for consumers or commercial applications. There will always be a need for custom designs, but true cost savings and customer acceptance require that systems be more technology transparent. "Turn it on and watch it work."

In addition, up until now no PV power systems have been certified by agencies outside the PV industry such as Underwriters Laboratories (UL) or Factory Mutual (FM). We are seeing a trend toward certification in components such as PV modules, charge controllers and inverters. More companies are getting their PV products UL or ETL Listed or FM Approved for hazardous locations. This project represents the first effort to have whole PV power systems certified.

## PROGRAM GOALS AND OBJECTIVES

The stated goal of the Design, Fabrication and Certification of Advanced Modular PV Power Systems contract is to significantly reduce the installed PV system life cycle costs to the customer. This is part of an overall goal of improving PV system marketability and customer acceptance. In line with the philosophy presented above, SES is attempting to make with happen by developing certified, standardized, modular, pre-engineered products lines of our main stand-alone systems, the Modular Autonomous PV Power Supply (MAPPS) and PV-Generator Hybrid System (Photogenset).

The technical and cost advantages of this approach are 1) shorter production lead times, 2) higher overall quality and system reliability, 3) decreased management and engineering time, 4) lower overhead and inventory costs, and 5) lower installation labor and material costs. The marketing advantages are that the systems can be specified and described more easily by the sales engineers and the customer can more easily assess the options and choose the system he needs. A corollary to this is greater customer acceptance of the PV solution and greater confidence among lending institutions.

## DEFINITION OF TERMS

The MAPPS is a small, stand-alone, low voltage DC power supply consisting of PV modules, a mount, a charge controller and batteries. These systems range from a single, small module unit to units with several, large modules. The rated power encompasses 20 W to about 300-400 W. The system voltages are 12, 24 and 48 V. They may also have an optional, small, DC-AC inverter for 120 Vac output.

The Photogenset is a mid-sized, stand-alone, AC power supply with engine generator for load and battery charging backup. It consists of PV modules, a mounting structure, a charge controller, batteries, a DC-AC inverter, an engine generator and a container. These systems range from under 1 kW to about 10 kW.

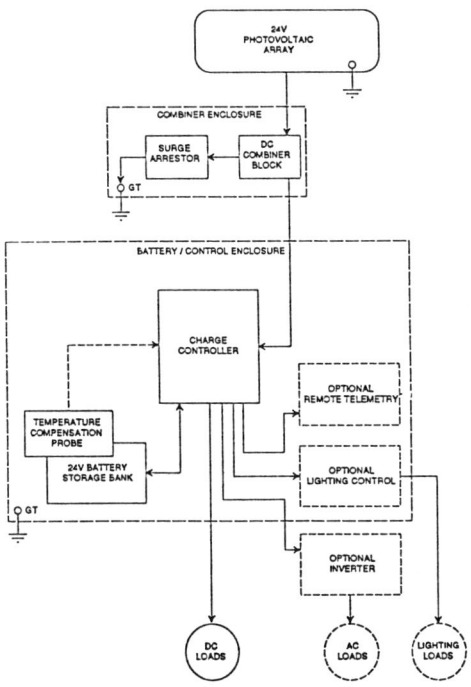

FIGURE 1. MAPPS Block Diagram

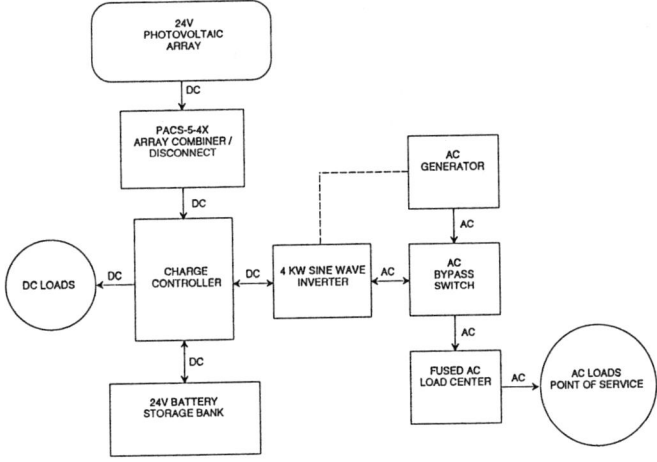

FIGURE 2. Photogenset Block Diagram

## STATUS OF ACTIVITIES

The task structure of the contract calls for development and fabrication of a 200 W MAPPS and a 1 kW Photogenset as the first prototypes for these respective product lines, certification of the prototypes and product lines, and functionality testing of the prototypes at NREL and Sandia. Within the charter was also room for development of new or improved components for the systems to improve performance and reliability while reducing costs or to facilitate certification.

The sizes of the prototypes were chosen for their high sales potential and manageable size and cost for fabrication and certification. The purpose of the functionality testing is to provide verification of performance and reliability and to develop test-based performance specifications.

## MAPPS Status

To date we have completed the design of the 200 W MAPPS and procured the components. We have developed a new enclosure for the batteries and control assembly. This enclosure costs less than the previous design and makes better use of internal space while providing a separate compartment for the control assembly. The design makes use of only UL Listed or UL Recognized components. We are also developing a new microprocessor-based controller for small systems. We expect to have this controller UL Listed when it is completed. It will be used as an option when greater control flexibility or remote monitoring is needed.

We have established the contracts for obtaining UL Listing the MAPPS and FM Approval Listing for use in hazardous locations. These locations, as defined in Article 500 of the National Electrical Code [1], have environments with incendiary gases present. The battery enclosure is currently undergoing a rain spray test at UL after which we will provide the electrical components for completing the evaluation. Following completion of the UL evaluation, the MAPPS will be sent to FM for its evaluation.

A second unit is being built for the functionality testing program. We are developing the test requirements and protocol in concert with NREL and expect the testing to begin in January 1997. We have also developed a manufacturing plan for development and certification of the MAPPS product line and expect to complete this effort by the end of the PVMaT contract.

## Photogenset Status

To date we have completed the design of the 1 kW Photogenset. The design incorporates a multi-platform layout. The Photogenset can be provided with either a skid-mounted building or in a trailer for mobile applications. We have attempted to make maximum use of UL Listed components; however, we are evaluating alternatives to the generator and inverter to facilitate Listing and reduce Listing evaluation costs. Because of the complexity of the Photogenset, the cost of Listing will be much higher than that for the MAPPS. We have obtained quotes from both UL and ETL (Electrical Testing Laboratories) to assess the costs and benefits of each. We expect to complete fabrication of the 1 kW prototype by the end of February by which time it will be scheduled for Listing evaluation and functionality testing.

## ACCOMPLISHMENTS AND BENEFITS

To recap current accomplishments, we have: 1) a UL-certifiable 200 W MAPPS with upgraded assembly process controls and installation instructions; 2) an understanding of certifications issues and requirements [2]; 3) a manufacturing plan for a standardized, modular MAPPS product line [3]; 4) a 1 kW Photogenset design with a multi-platform layout; and an improved, lower cost MAPPS battery enclosure.

Expected near term accomplishments include 1) UL Listing of the 200 W MAPPS; 2) a SES microprocessor-based small charge controller; 3) FM Approval of the 200 W MAPPS; and 4) an increased-reliability, UL certifiable 1 kW Photogenset. Prior to the end of the contract we expect to accomplish the following: 1) a standardized, UL Listed, MAPPS product line; 2) a certified 1 kW Photogenset; 3) a manufacturing plan for a standardized, modular Photogenset product; 4) functionality test results on both the MAPPS and Photogenset; and 5) test-based performance specifications.

We have realized a number of additional benefits from this contract including the prestige and marketing benefits of receiving a PVMaT contract, additions to the staff and consultants, increased staff experience and the knowledge gained from association with NREL, Sandia and Southwest Technology Development Institute. There are also benefits that are difficult to quantify such as the synergy between the PVMaT contract and our business efforts. We believe the experience gained through the PVMaT contract has been instrumental in obtaining some important new business, such as for 105 large charge controller systems for telecommunications in Peru and for three large MAPPS systems for a military application. At the same time the working of these other contracts has been instrumental in developing our new designs. We believe this PVMaT contract will provide important advances in PV power supply market penetration through lower costs, higher reliability, product standardization and product certification.

# REFERENCES

1. *National Electrical Code*, Quincy, MA, NFPA 1995
2. Lambarski, T. J., *Safety Requirements Report,* NREL Subcontract No. ZAF-5-14271-07, September 1996
3. Lambarski, T. J., *Manufacturing Plan for Standardized MAPPS Product Line,* NREL Subcontract No. ZAF -5-14271-07, awaiting publication

**MARKETS AND APPLICATIONS**

# Commercialization of Multijunction a-Si Modules

D. E. Carlson, R. R. Arya, L.-F. Chen, R. Oswald, J. Newton,
K. Rajan, R. Romero, F. Willing, and L. Yang

*Solarex, a Business Unit of Amoco/Enron Solar*
*826 Newtown-Yardley Road*
*Newtown, PA 18940*

**Abstract.** Solarex has just completed building a plant in James City County, Virginia that has the capacity to produce 10 MW per year of multijunction amorphous silicon PV modules. The plant will start commercial production of 8.6 ft$^2$ tandem modules in early 1997. The tandem device structure consists of two stacked p-i-n junctions, a front junction containing amorphous silicon and a back junction containing an amorphous silicon germanium alloy. All amorphous silicon alloys are deposited using plasma-enhanced chemical vapor deposition, and the large-area monolithic modules are interconnected using computerized laser scribing coupled with a machine vision system. The principle products will be monolithic modules (26" x 48") with nominal stabilized power ratings of 56, 50 and 43 peak watts All modules will be fabricated using a glass-EVA-glass encapsulation to ensure long-term reliability. These products are expected to be widely used in both remote and grid-tied applications.

## INTRODUCTION

The first amorphous silicon (a-Si) solar cell was fabricated at RCA Laboratories in 1974 (1). Sanyo was the first company to produce a commercial a-Si solar cell when they introduced a new solar-powered calculator in 1980. Solarex started the first commercial production of a-Si solar cells in the U.S. in 1984. These small-area single-junction devices were first made for low-light level, consumer applications, and then starting in 1986 Solarex started making 1 ft$^2$ modules for outdoor battery-charging applications. However, the performance of these single-junction devices has been limited to a stabilized conversion efficiency of about 5%.

In order to attain higher stabilized conversion efficiencies, Solarex as well as other companies started developing multijunction a-Si solar cell technology in the mid-1980's. Both tandem (two-junction) and triple-junction structures have been investigated, and while higher conversion efficiencies are possible with triple-junction devices, Solarex has decided to commercialize a monolithic tandem structure since it has the promise of better manufacturability and superior long-term reliability. Solarex has just completed building a thin-film PV manufacturing plant (TF1) in Virginia that will start commercial operation in early 1997 and will be capable of producing 10 MW/yr of 8.6 ft$^2$ tandem a-Si PV modules.

## DESCRIPTION OF THE TANDEM MODULE

Solarex has been investigating and developing both tandem and triple-junction a-Si device structures for over a decade. Since the manufacturing process is simpler for a tandem structure than for a triple-junction structure, the yields are expected to be higher, and moreover, less deposition time and feedstock materials are required. In addition, since the junctions are relatively thick in a tandem structure (the i-layers are ~ 200 nm thick) as compared to the top junction of a triple-junction structure (~ 80 - 100 nm thick), the tandem structure is likely to be more robust.

Once stabilized conversion efficiencies of about 9% were demonstrated in small-area tandem devices utilizing back junctions containing an amorphous silicon germanium alloy (a-SiGe), Solarex started producing 12" x 13" tandem modules in a pilot line in Newtown, Pennsylvania. After reasonable yields were demonstrated in pilot runs of the 12" x 13" modules, the pilot production process was scaled up to handle 4 ft$^2$ tandem modules (2). Once both high yields and stabilized conversion efficiencies of 8% were demonstrated for the 4 ft$^2$ tandem modules (3), Amoco/Enron Solar committed Solarex to build a large-scale manufacturing plant to produce 8.6 ft$^2$ tandem modules. Recently, Solarex has started producing limited quantities of 8.6 ft$^2$ tandem modules in a pilot line at the Newtown facility. (Figure 1 shows the various module configurations used by Solarex in developing the tandem module technology)

The tandem a-Si/a-SiGe structure is deposited on a tin oxide coated glass substrate that is 26" wide, 48" long and 3 mm thick.. Laser scribing is used to pattern the tin oxide, the a-Si/a-SiGe tandem structure and the back contact resulting in a series-connected monolithic module (see Fig. 2). The scribing is performed by computer-controlled, frequency-doubled Nd-Yag lasers, and a machine vision system is used to assure the alignment of the a-Si and back contact scribes with the tin oxide scribe in the interconnect region.

Currently Solarex is planning to make three different large-area tandem modules in the TF1 facility. They would be classified according to their nominal stabilized power ratings: MST-56MV (56 $W_p$), MST-50MV (50 $W_p$) and MST-43MV (43 $W_p$) and would be available either with or without frames. Unlike the early prototype modules shown in Figure 1, most of the production modules will be patterned with the laser scribes running parallel to the long edge of the module.

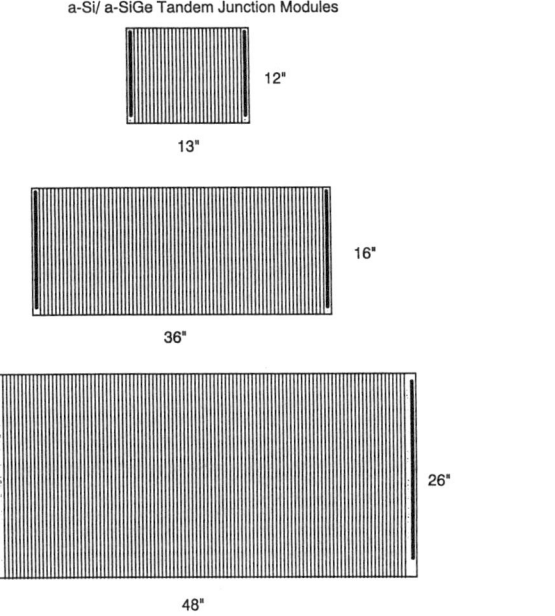

**FIGURE 1.** Solarex a-Si/a-SiGe tandem module configurations

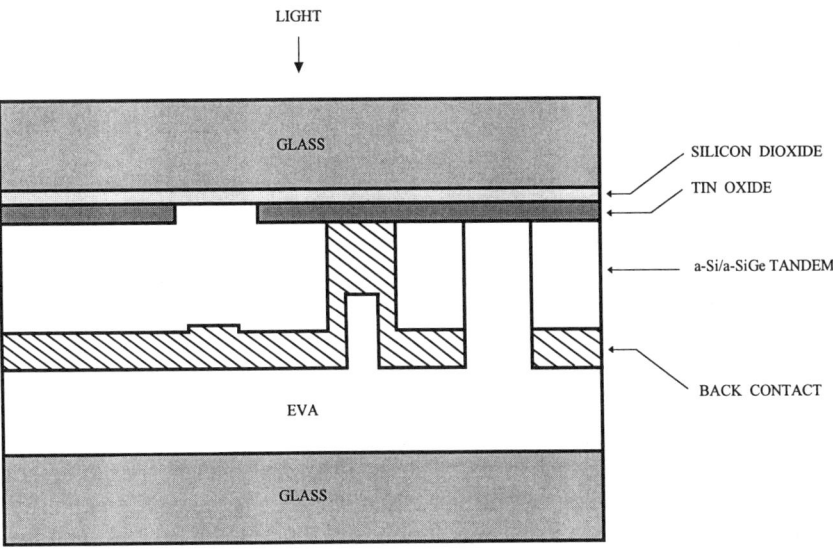

**FIGURE 2.** A schematic showing the interconnection for monolithic panels.

These modules would operate at relatively high voltages, ~80 volts at STC ($25^0$C) and ~ 70 volts at $60^0$C assuming 1 kW/m$^2$ of AM1.5 sunlight in both cases. The modules are designed to operate in arrays with system voltages up to 600 V.

## TANDEM MODULE PERFORMANCE

While thousands of 12" x 13" tandem modules and about three thousand 4 ft$^2$ tandem modules have been produced in the Solarex pilot lines at the Newtown facility, only a limited number of 8.6 ft$^2$ prototype modules have been made to date. The performance characteristics of four 8.6 ft$^2$ modules that were measured outdoors at NREL are listed in Table 1.

**TABLE 1.** Performance Characteristics for 8.6 ft$^2$ Prototype Tandem Modules Measured Outdoors at NREL

| Module # | $V_{OC}$ (V) | $I_{SC}$ (A) | FF | $P_{MAX}$ (W) |
|---|---|---|---|---|
| E1191 | 189.3 | 0.469 | 0.660 | 58.5 |
| E0935 | 185.4 | 0.464 | 0.656 | 56.4 |
| E0937 | 193.2 | 0.461 | 0.633 | 56.4 |
| E0938 | 190.4 | 0.459 | 0.629 | 54.9 |

The measurements were made outdoors in Golden, Colorado, on July 22, 1996 when the ambient temperature was $32^0$C. These early prototype modules were made in a high voltage configuration with the laser scribes running parallel to the short edge as shown in Figure 1.

The current-voltage characteristics of another 8.6 ft$^2$ prototype tandem module are shown in Figure 3. This module was patterned with the laser scribes running parallel to the long edge of the module so the open-circuit voltage ($V_{OC}$ = 103.8 V) is less than that of the modules measured at NREL (see Table 1). While the fill factor (FF) of this recent prototype module is only 0.572, further optimization of the process should lead to fill factors closer to 0.66 - 0.68 since these values were routinely obtained with 4 ft$^2$ tandem modules and with some of the high voltage 8.6 ft$^2$ modules.

The performance of the tandem a-Si modules does vary with the Air Mass, and they exhibit somewhat better relative performance than crystalline silicon modules at low Air Mass and are somewhat worse at high Air Mass, but for the range of AM0.5 to AM2.5, the variation in the short-circuit current ($I_{SC}$) is only on the order of 5% or less.

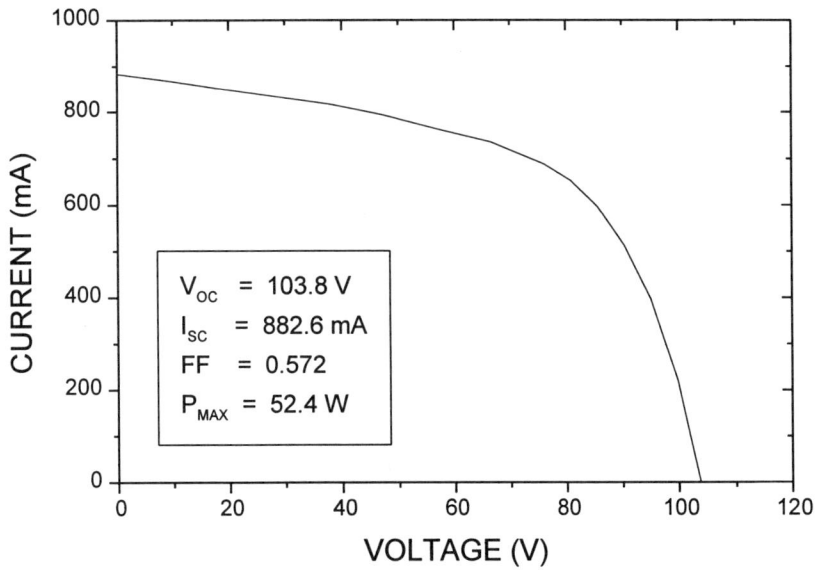

**FIGURE 3.** The current-voltage characteristics of a prototype tandem module under simulated AM1.5 illumination.

Based on measurements made on engineering prototypes, the temperature coefficients should be approximately: - 0.36%/$^0$C for the voltage, + 0.16%/$^0$C for the current and - 0.22%/$^0$C for the power. The coefficients do vary somewhat with irradiance and temperature. For an irradiance of 800 W/m$^2$, the temperature coefficient of the power was observed to vary from - 0.21%/$^0$C at 0$^0$C to - 0.31/$^0$C at 50$^0$C. At a temperature of 25$^0$C, the temperature coefficient of the power varied from - 0.20%/$^0$C at an irradiance of 600 W/m$^2$ to - 0.29%/$^0$C at an irradiance of 1000 W/m$^2$.

## LIGHT-INDUCED DEGRADATION

All a-Si solar cells exhibit light-induced degradation due to the Staebler-Wronski effect (4). While single-junction a-Si solar cells made in the early 1980's often degraded by ~ 30 - 40% before stabilizing, present commercial single-junction cells only degrade by ~ 20 - 25%. By using thin i-layers and hydrogen dilution during the growth of the i-layers, even more stable a-Si tandem devices have been demonstrated (2).

Test results on prototype 4 ft² tandem modules indicate that the initial conversion efficiency will decrease by about 15% or less due to light-induced degradation before stabilizing as shown in Figure 4. About half of the degradation is associated with a decrease in the fill factor of the modules.

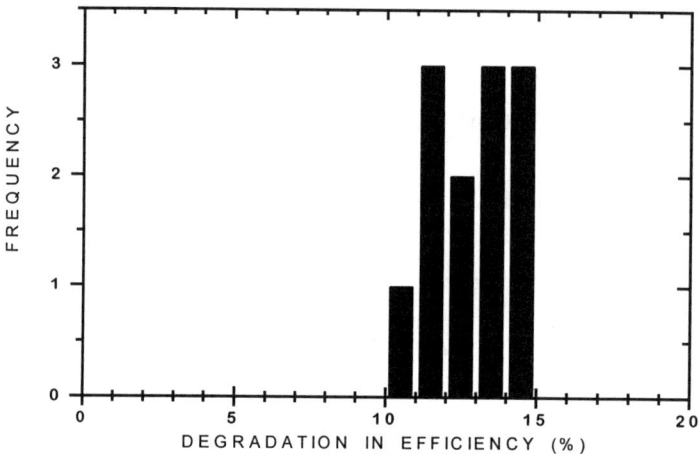

**FIGURE 4.** A histogram showing the degradation experienced by twelve 4 ft² tandem modules after exposure to 600 hours of simulated sunlight at 45°C.

The magnitude of the light-induced degradation depends not only on the rate at which defects are created but also on the rate at which defects are annealed out during operation. We have recently investigated the annealing of the light-induced degradation of a-Si solar cells at low temperatures (35 - 60°C). A number of p-i-n cells were made using different deposition conditions (the i-layers in each case were about 250 - 300 nm thick) and the cells were degraded by exposing them to 60 suns illumination for 30 minutes at 60°C. The cells were then annealed at either 35°C or at 60°C in the dark. (The tandem modules often reach temperatures of ~ 60°C in direct sunlight in hot climates.)

As shown in Figure 5, both the amount of degradation and the amount of recovery depend on the deposition conditions. The amount of degradation induced by the 60 suns exposure can be minimized by choosing the proper hydrogen dilution deposition conditions. While all the cells exhibit a significant amount of low-temperature recovery, the effect is much more significant in the hydrogen-diluted cells that did not degrade that much under the 60 suns exposure. In fact, the cells made with the "H-dilution (1)" process recovered about half of their performance loss after 820 hours at 35°C. When the same cells are annealed at 60°C, the recovery is even more rapid than that shown in Figure 5. After an exposure of 30 minutes to 60 suns at 60°C, the cells made with the "H-dilution (1)" process degraded to 78% of the average initial performance, and after 300 hours at 60°C in the dark, the cells had recovered to 95% of their initial performance.

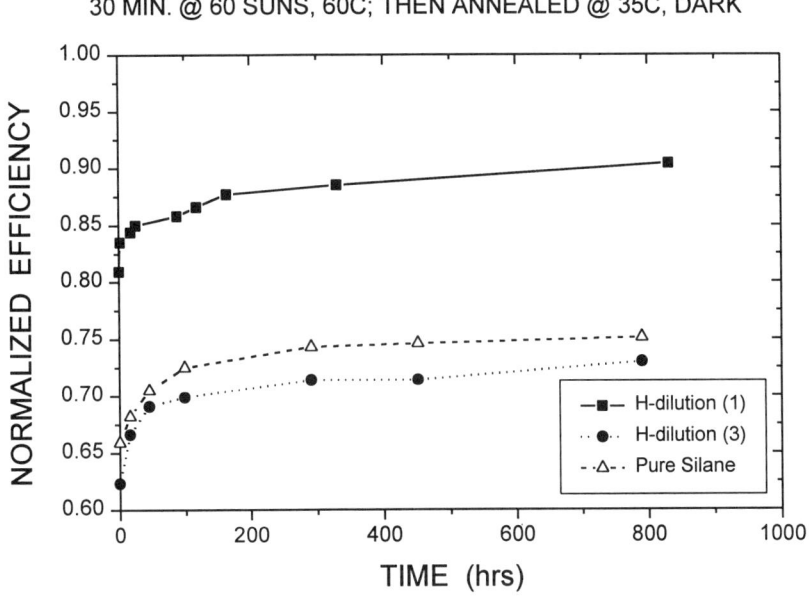

**FIGURE 5.** The normalized efficiency of p-i-n cells vs. time for annealing at 35°C.

All the a-Si cells studied exhibit a short period of relatively rapid recovery followed by a longer term slow recovery. The existence of "fast" and "slow" metastable defect states was proposed to explain the degradation behavior of a-Si solar cells several years ago (5). As is evident from the data in Figure 5, there is also clear evidence for both "fast" and "slow" states in the low-temperature annealing data. For example, in the case of the annealing of the "H-dilution (1)" cell at 60°C, the short term recovery has a characteristic recovery time of about 1.3 hours while the longer term recovery exhibits a recovery time of 199 hours. For a similar cell annealed at 35°C, the fast recovery occurs with a similar characteristic time while the longer term recovery exhibits a characteristic time of about 250 hours. Thus, there are apparently a large number of shallow defect levels that can readily anneal out at relatively low temperatures.

These low-temperature annealing effects can have a significant impact on the stabilized performance of amorphous silicon solar cells under normal operating conditions. In some situations such as flush-mounted roofing arrays, the operating temperature of an a-Si PV module can approach $70^0C$ or more in hot climates. To examine the extent of the low-temperature annealing effect on the stabilized conversion efficiency, we set up one light-soaking station where the 1 sun illumination would be on for 8 hours at a cell temperature of ~ $70^0C$ and then off for 18 hours at a temperature of ~ $40^0C$. Another station was set up where the temperature under 1 sun illumination was ~ $45^0C$ and the temperature in the dark was ~ $25^0C$. Four identical a-Si/a-SiGe tandem cells were placed in each light-soak station and were subjected to 110 cycles (110 days) of 1 sun illumination and dark annealing.

As shown in Fig. 6, the degradation of the cells in both stations leveled off after about 15 cycles. As predicted, the cells exposed to the cycling at higher temperatures exhibited significantly less degradation. These cells degraded only about 8 - 10% while the cells exposed to the lower temperature cycling degraded about 14 - 16%. Thus, a-Si tandem modules operating in very hot climates will exhibit significantly less degradation than those operating in more moderate climates due to the low-temperature annealing effects.

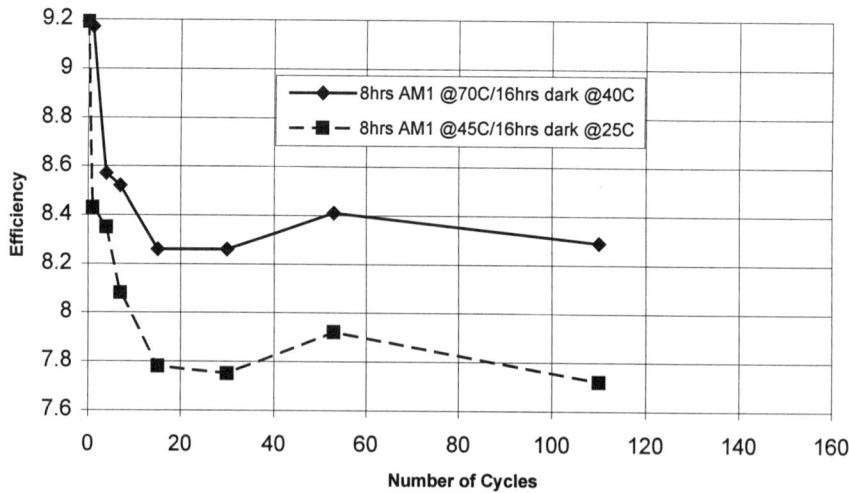

**FIGURE 6.** The normalized conversion efficiency of a-Si/a-SiGe tandem devices as a function of the number of light-soak cycles. (see the text)

## LONG-TERM RELIABILITY

Solarex's prototype tandem modules have passed an extensive battery of environmental and qualification tests (see Table 2). Each of these tests is based on a detailed procedure that has been established by either the Commission of European Communities or the NREL Interim Qualification Tests. For example, the Humidity Freeze Test involves cycling the modules between -$40^0$C and $85^0$C, 85% humidity for a total of 10 cycles; the module must exhibit less than 5% degradation in power to pass. The Wet Hi-Pot Test involves subjecting the modules to a voltage that is 1000 V plus twice the array voltage while the module is wet; the leakage current must be less than 50 microamps to pass.

**TABLE 2.** A List of the Solarex Qualification Tests

| Insulation | Damp Heat |
|---|---|
| Measure Temperature Coefficient | Robustness of Terminations |
| Measure NOTC | Twist |
| Performance at NOTC | Mechanical Load |
| Performance at Low Irradiation | Hail Impact |
| Outdoor Performance | Light Soaking |
| Hot Spot Endurance | Annealing |
| UV Exposure | Wet Leakage Current |
| Thermal Cycle | Wet Hi-Pot |
| Humidity Freeze | Dynamic Load |

The a-Si tandem structure is thermally quite stable compared to earlier single-junction devices. The single-junction cells will start to degrade at temperatures above ~ $150^0$C due to the interdiffusion of aluminum and silicon at the rear contact. The tandem structures utilize a ZnO/Al rear contact and are more thermally stable. When tandem cells were heated to $200^0$C for 1 hour, they only exhibited a 0.5% decrease in performance. At higher temperatures, the tandem devices start to degrade more rapidly due to hydrogen outdiffusion in the vicinity of the p/i interface (6), but this mechanism is negligible at normal operating temperatures and will not influence module life.

## MARKETS AND APPLICATIONS

Solarex is planning to use the new MST tandem modules in both remote and grid-tied applications. There are many remote applications where high conversion efficiencies are not essential and a low module price ($/W_p$) is the key driver. This is especially true for remote applications where the balance of systems costs are low such as water pumping.

Solarex is working with a number of U.S. utilities to develop grid-tied PV roofing arrays. The Solarex PV-VALUE Program involves a partnership between Solarex and several utilities to commercialize packaged, low-cost systems for the on-grid residential rooftop market. Under this program, homeowners would purchase a PV system in one of two sizes - either 1.9 kW(ac) or 3.2 kW(ac).

On another front, Amoco/Enron Solar Power Development is moving aggressively ahead to develop grid-tied solar farms. They are in final negotiations with Hawaiian Electric to construct a 4 $MW_p$ solar farm on the island of Hawaii and have executed a power purchase agreement with the state of Rajasthan in India to build a 50 $MW_p$ solar farm. More recently, Amoco/Enron Solar has been selected to develop a 10 $MW_p$ solar farm at the Nevada Test Site. This would be the largest PV power plant in the U.S. The company intends to use mostly MST tandem modules in these solar farms.

## SUMMARY

In the last two years, Solarex has focused on developing a tandem junction device structure that could be manufactured with high yields, good material utilization and high throughput. In the beginning of 1997, Solarex will start manufacturing 8.6 $ft^2$ tandem junction modules in James City County, Virginia with the goal of producing a high quality, cost-effective product for both remote and grid-connected applications.

## ACKNOWLEDGMENTS

This work was partially supported by NREL under subcontract Nos. ZAN-4-11318-01 and ZM-2-11040-2.

## REFERENCES

1. D. E. Carlson, U. S. Patent No. 4,064,521(1977).
2. R. R. Arya, R. S. Oswald, Y. M. Li, N. Maley, K. Jansen, L. Yang, L.F. Chen, F. Willing, M. S. Bennett, J. Morris and D. E. Carlson, Proc. 24th IEEE Photovoltaic Specialists Conference (1994) p. 380.

3. D. E. Carlson, R. R. Arya, M. Bennett, L. F Chen, K. Jansen, Y.M. Li, J. Newton, K. Rajan, R. Romero, D. Talenti, E. Twesme, F. Willing and L. Yang, Proc. 25th IEEE Photovoltaic Specialists Conference (1996).
4. D. L. Staebler and C. R. Wronski, Appl. Phys. Lett. **31**, 292 (1977).
5. L. Yang and L. Chen, Appl. Phys. Lett. **63**, 400 (1993).
6. D. E. Carlson and K. Rajan, Proceedings of the 13th European Photovoltaic Solar Energy Conference, (1995) p. 617.

# Photovoltaics in the Department of Defense[1]

Richard N. Chapman

Sandia National Laboratories
P.O. Box 5800, Albuquerque, New Mexico 87185-0753

**Abstract.** This paper documents the history of photovoltaic use within the Department of Defense leading up to the installation of 2.1 MW of photovoltaics underway today. This history describes the evolution of the Department of Defense's Tri-Service Photovoltaic Review Committee and the committee's strategic plan to realize photovoltaic's full potential through outreach, conditioning of the federal procurement system, and specific project development. The Photovoltaic Review Committee estimates photovoltaic's potential at nearly 4,000 MW, of which about 700 MW are considered to be cost-effective at today's prices. The paper describes photovoltaic's potential within the Department of Defense, the status and features of the 2.1-MW worth of photovoltaic systems under installation, and how these systems are selected and implemented. The paper also documents support provided to the Department of Defense by the Department of Energy dating back to the late 70s.

## INTRODUCTION

The Department of Defense (DoD) is the largest single user of energy in the United States. It spends more than $2 billion annually for about 35,000 GWh of utility-supplied electricity (1). The Department also generates substantial amounts of energy to power remote facilities isolated from commercial utility grids. Best current estimates indicate that the DoD spends about $1 billion annually for about 3,000 GWh of electricity for off-grid facilities (2). The DoD recognized over 20 years ago that photovoltaics could reduce the cost of electricity and reduce dependence on fossil fuels for many off-grid facilities. However, a lack of awareness of, and experience with, the technology inhibited the use of photovoltaics for these cost-effective applications. The DoD has

---

[1]This work was sponsored by the Strategic Environmental and Development Program project EN-046 and by the U. S. Department of Energy under contract DE-AC04-94AL85000 in partnership with the Department of Defense Photovoltaic Review Committee.

implemented and maintained programs and policies to minimize these obstacles and foster the use of photovoltaics, efforts which began in the late 70s with the participation of all the military branches in the Department of Energy's Federal Photovoltaic Utilization Program (FPUP).

# HISTORY OF PHOTOVOLTAICS IN THE DEPARTMENT OF DEFENSE

The Federal Photovoltaic Utilization Program was created to stimulate the use of photovoltaics in the federal sector and accelerate the growth of a commercially viable photovoltaic industry. Under FPUP, the DoD installed 218 systems at a cost of $5.8M (3). The photovoltaic arrays for these systems ranged from a single module up to 25 kW with the vast majority being less than 1 kW. Most were small dc systems for remote lighting, communications, instrumentation, battery charging, and range targets. There were several utility-tied systems as well as several stand-alone ac systems. The DoD's FPUP experience was the starting point of what has become a sustained and organized program within the DoD to use photovoltaics where appropriate. Working relationships formed during FPUP led to the creation of the DoD's Tri-Service Photovoltaic Review Committee in 1985.

The Photovoltaic Review Committee was chartered by the Office of the Secretary of Defense to encourage the use of photovoltaics as a standard technology where cost-effective and appropriate (4). The committee is chaired by Garyl Smith, of the Naval Air Weapons Station, China Lake, CA, and is composed of a representative from the Navy, Army, and Air Force. Current representatives are Chuck Combs, Naval Air Weapons Station, China Lake, CA, Roch Ducey, Army Construction Engineering Research Laboratory, Champaign, IL, and Larry Strother, Air Force Engineering and Services Center, Tyndall AFB, FL. Sandia National Laboratories has served as a technical consultant since the Photovoltaic Review Committee was chartered.

## Focus on Small Remote Applications from 1985-1992

The Photovoltaic Review Committee's activities have evolved as the integration of the technology into military practices has progressed. Initial activities focused on identifying applications. In 1986, the Energy Program Office at China Lake surveyed approximately 300 installations and identified an estimated 21,000 potential cost-effective systems in 33 application categories (5). Table 1 summarizes these systems and applications. The arrays for these systems ranged from 7 W to 21 kW. In 1987, the Army surveyed seven installations and identified cost-effective applications in 14 categories (6). The arrays for these applications, as shown in Table 2, ranged from 17 W to 9 kW. Overall, the Photovoltaic Review Committee estimated the potential for more than 50,000 small stand-alone systems with about 50 MW of photovoltaics (7). Having established the need for small remote photovoltaics systems, the Photovoltaic Review Committee shifted emphasis to outreach and education.

TABLE 1. Navy Survey on Potential Photovoltaic Applications

| Application | No. of Systems | Typical Capital Costs ($) | Average Life-Cycle Costs ($) | Simple Payback (Years) |
|---|---|---|---|---|
| Building Guard Station | 150 | 30,000 | 61,237 | 3.7 |
| Building Operations (Remote) | 20 | 248,000 | 60,636 | 9.9 |
| Cathodic Protection | 1,000 | 2,400 | 56,870 | 0.3 |
| Communications (Island) | 200 | 5,400 | 13,920 | 2.9 |
| Comm. Repeater (Remote-Large) | 300 | 25,600 | 70,080 | 2.9 |
| Comm. Repeater (Large) | 900 | 21,600 | 45,510 | 3.7 |
| Comm. Repeater (Small) | 900 | 10,800 | 48,144 | 2.0 |
| Comm. Repeater (Remote-Small) | 300 | 13,800 | 51,806 | 2.0 |
| Communication Emergency | 400 | 720 | 37,918 | 0.2 |
| Communication Telephone | 100 | 86,400 | 83,909 | 5.8 |
| Lights, Small Craft | 2,000 | 1,800 | 56,970 | 0.3 |
| Lights, Target | 100 | 14,400 | 66,643 | 1.9 |
| Lights, Warning | 1,500 | 1,440 | 116,766 | 0.1 |
| Lights, Wind-Sock | 200 | 7,200 | 2,176 | <0.1 |
| Meteorological Station (Large) | 200 | 21,600 | 141,074 | 1.4 |
| Meteorological Station (Small) | 1,500 | 140 | 11,564 | 0.1 |
| Aid to Navigation | 300 | 1,440 | 57,300 | 0.2 |
| Housing (Remote) | 100 | 15,000 | 66,164 | 2.0 |
| Instrumentation, Camera | 400 | 45,000 | 141,039 | 2.4 |
| Instrumentation, Offshore | 20 | 45,000 | 133,311 | 2.7 |
| Lights, Buoy | 1,500 | 1,000 | 111,079 | 0.1 |
| Lights, Heliport | 100 | 28,800 | 2,439 | <0.1 |
| Lights Magazine | 600 | 2,880 | 55,225 | 0.4 |
| Observation Tower | 200 | 72,000 | 100,344 | 4.6 |
| Radar, Beacon (Small) | 500 | 3,600 | 93,341 | 0.4 |
| Radar, Beacon (Large) | 300 | 14,400 | 68,974 | 1.9 |
| Radar, Target | 400 | 96,000 | 43,442 | 8.2 |
| Remote Bomb Scoring | 400 | 4,800 | 55,053 | 0.8 |
| Security, Sensor | 1,500 | 1,440 | 56,165 | 0.2 |
| Security, Visual | 600 | 14,400 | 69,974 | 1.7 |
| Security, Gate | 100 | 6,000 | 36,500 | 0.6 |
| Transportable Power | 3,000 | 48,000 | 138,600 | 2.6 |
| Water Pumping | 200 | 180,000 | 72,027 | 8.9 |

Personal networking and assessments of individual installations have been and will continue to be core Photovoltaic Review Committee functions. These functions are very effective in the development of photovoltaic advocates and in the development of applications that lead to actual systems. Obviously, this approach cannot reach all installations nor all personnel involved in energy supply. To address a larger audience, the Photovoltaic Review Committee held a series of user-directed workshops in Albuquerque, NM (November 1989), Atlanta, GA (November 1990), and in Yuma, AZ (May 1991). The Department of Energy, through Sandia National Laboratories, supported this outreach effort with the SOLTECH 88 conference and with military-specific publications. SOLTECH 88 included sessions dedicated to photovoltaic usage in the military (8). Sandia edited and published two Photovoltaic Review Committee reports specific to the DoD: *Photovoltaics for Military Applications* (9) in 1988 and *Maintenance and Operation Manual for Photovoltaics* (10) in 1990. Sandia also provided technical presentations at the three user-directed workshops. The Photovoltaic Review Committee has also published a number of articles that document its activities and progress (11-15).

**TABLE 2.** Typical Load and Power Characteristics of Potential Photovoltaic Applications in the Army

| Application | Approximate Array Size (W) |
|---|---|
| Test Range Equipment | 25-150 |
| Battery Chargers (for emergency power of water wells) | 17-20 |
| Clearance Lights (on water tanks) | 770-2500 |
| Global Positioning System Satellite Simulators | 2000-2500 |
| Mobile Firing Range | 80-90 |
| Mobile Generators | 500-9000 |
| Radio Repeaters | 1400-1600 |
| Firing Range Gun Position | 100-110 |
| Range Surveillance Video | 1000-3500 |
| Microwave Towers | 1500-1700 |
| Remote Data Acquisition | 300-500 |
| Meteorological Towers | 200-220 |
| Storage Facilities (Igloos) | 300-320 |
| Microwave Repeaters | 1200-1600 |

The identification of applications and the education efforts of the Photovoltaic Review Committee promoted the installation of about 2,000 systems with about 2 MW of photovoltaics by 1992 (7). At this point, the Photovoltaic Review Committee considered the small remote system as approaching a mature application and shifted emphasis to the development and implementation of larger applications. Two major sources of funding, the Energy Conservation Investment Program (ECIP) and the Strategic Environmental Research and Development Program (SERDP), allowed the Photovoltaic Review Committee to greatly accelerate this new emphasis.

# Shift to Larger Systems Beginning in 1992

The development and implementation of larger applications presented two major obstacles. One obstacle was the lack of suitable power processing hardware. The other was a ten-fold increase in the cost of larger systems compared with the small remote systems. The increased cost of larger systems includes both the engineering effort to develop larger applications and the cost of the systems themselves. SERDP provided funding to develop and test the required power processing hardware, and to develop the larger applications, while ECIP provided the funding for the larger systems.

It is important to understand the synergism of these two programs. ECIP is a military construction program. As a military construction program, ECIP funds can only be used for the design and construction of complete projects. They cannot be used for feasibility or engineering studies to develop projects and they cannot be used to develop or test new hardware. SERDP is a research and development program. As a research and development program, SERDP funds are intended to develop, test, and demonstrate new technologies. The DoD follows a standard progression to implement new technologies and applications. Figures 1a and 1b illustrate how SERDP and ECIP complemented each other to advance these larger photovoltaic systems toward implementation within this standard technology progression.

The first new application area addressed by the Photovoltaic Review Committee beyond small stand-alone was photovoltaic/diesel generator hybrids for remote facilities. (See the "Potential of Photovoltaics within the DoD" section for a description of application areas.) This intermediate remote application area required the development of 150-kW, three-phase power processing units with controls to integrate photovoltaics, batteries, and the diesel generator. Figure 1a shows that, prior to SERDP, this application area resided in what is referred to as the advanced development stage. This means that the required application functions had been developed (applied research) and that the switching and controls technologies required to perform those functions were available (basic research), but a functioning unit had not yet been fabricated and tested (advanced development).

There were no funds directed to this needed development and, as a result, the application area remained stagnant even though the Photovoltaic Review Committee had already identified 273 MW of cost-effective applications (16). SERDP provided $4M for the DoD photovoltaics program in FY92. Sandia used a portion of these funds to develop and test prototypes of this type of power processing hardware which, as shown in Figure 1b, advanced the intermediate remote application area from the advanced development area into the engineering development area, where the technology is evaluated in actual working systems. Several of these types of systems were then implemented under ECIP to verify their performance and economics. The combination of ECIP and SERDP enabled the intermediate application area far sooner than originally expected. In 1990, the Photovoltaic Review Committee estimated that this area would not be ready for implementation until FY98. Because of ECIP and SERDP, the intermediate application area was ready for implementation by FY95.

The photovoltaic SERDP project was awarded a total of $12.8M from FY92 through FY95. The FY93 through FY95 cycles addressed the power processing needs for the next DoD applications areas: large remote and isolated grid. (See the "Potential of Photovoltaics within the DoD" section for a description of application areas.) Of the $12.8M, $4.3M was invested in engineering and development and $8.5M was invested in systems to demonstrate the large remote and isolated grid technologies. Figures 1a and 1b show the status of the power processing technology for these applications areas before and after the photovoltaic SERDP project. The technology was advanced in increments, building on the advances from the prior application areas. The advance for the large remote area added the ability to parallel units to achieve power capacities beyond the intermediate remote applications. This technology will be demonstrated in the Superior Valley system. (See the section "Status of ECIP and SERDP Systems.") The advance for the isolated grid area added the ability to parallel with external sources (such as a generator or the utility). This technology will be demonstrated in the Yuma grid support system. These advances were evolutionary rather than revolutionary, in that existing technology was integrated to produce power processing hardware compatible with the DoD's needs and requirements.

The net result of this effort was an integrated system of building-block units that could serve a facility independent of any other source (stand-alone inverter mode), assist an external source to serve a facility (parallel inverter mode), and charge batteries from an external source (rectifier mode). The advances were to include gradual and seamless transfers between all modes. This was completed for transfers between the stand-alone inverter and rectifier modes but not for the transfers to and from the parallel inverter mode because of reduced funding and early termination. Details of these technology advances are documented elsewhere (16).

As stated earlier, SERDP and ECIP greatly accelerated the use of photovoltaics within the DoD. The Photovoltaic Review Committee estimates that SERDP and ECIP have reduced the time to implementation of the large remote application area by seven years (see Figure 2). A similar acceleration may have been achieved for the isolated grid application area even though the full development effort for this area was not completed under SERDP.

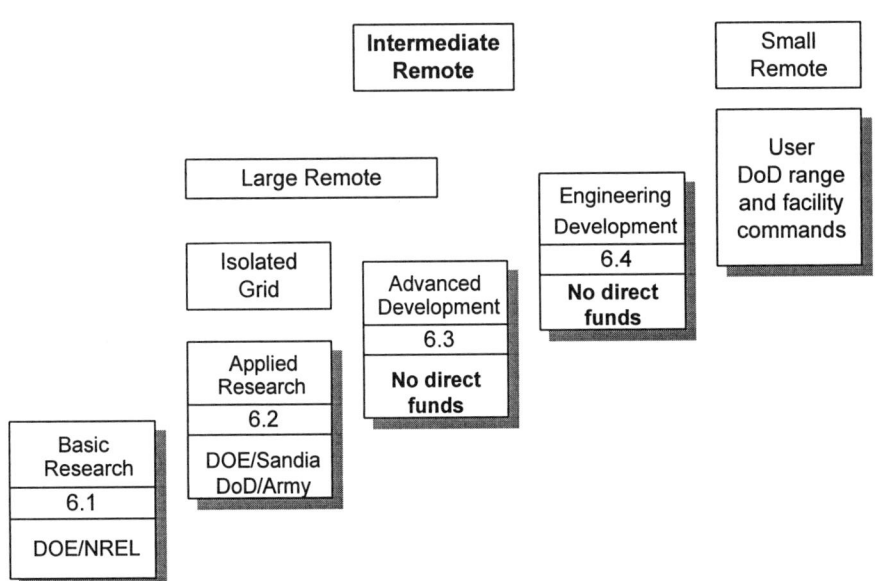

**FIGURE 1a.** Technology Status of Application Areas Before SERDP and ECIP.

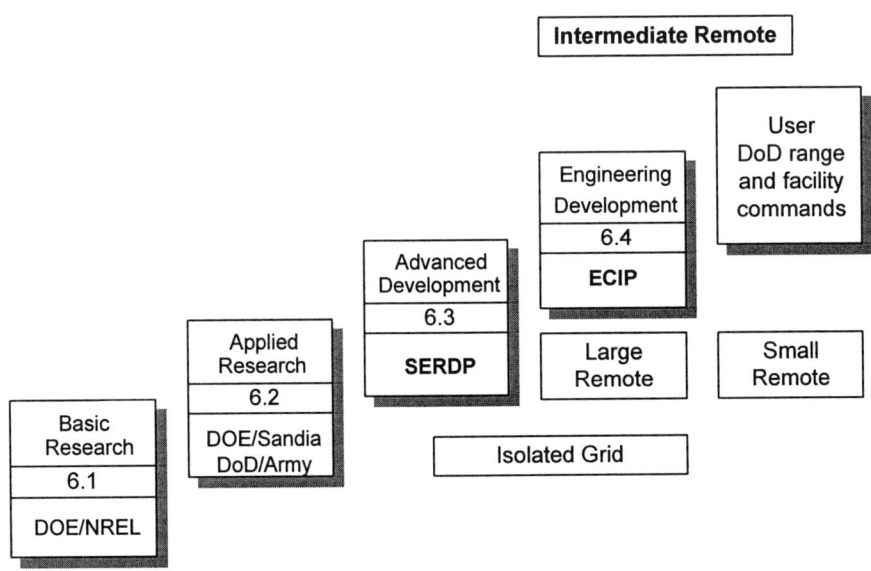

**FIGURE 1b.** Technology Status of Application Areas After SERDP and ECIP.

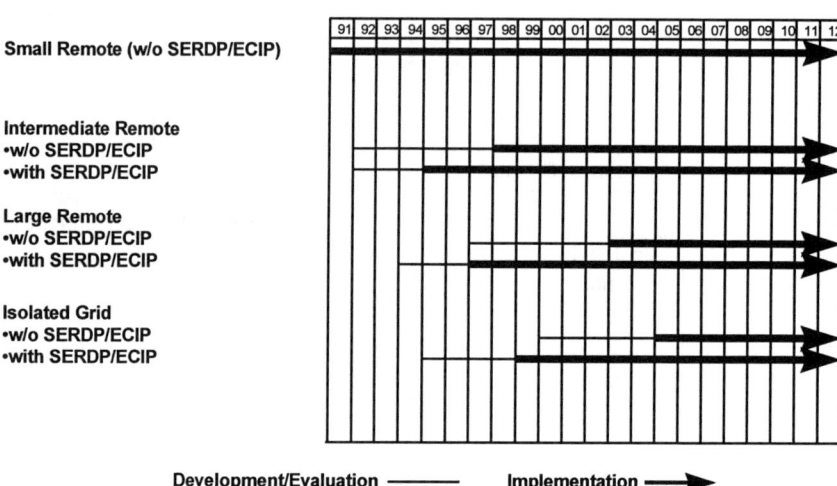

**FIGURE 2.** Effect of SERDP and ECIP on Implementation of Application Areas.

SERDP funds were also used for the creation of a living data base for DoD photovoltaic projects (17) and for the GloTech environmental assessment software (18). The data base was created to catalog potential photovoltaic projects and track photovoltaic systems. When fully developed, the data base will include a prioritized listing of approved projects. These projects can then be implemented as funding becomes available. The GloTech software, developed under the direction of the Environmental Protection Agency, quantifies environmental releases for both photovoltaic and conventional energy technologies. This is a cradle-to-grave analysis beginning with all materials and natural resources and ending with decommissioning and salvage. The software provides a means to compare the environmental consequences of one technology versus another.

## Current Activities of the Photovoltaic Review Committee

In 1993, the Photovoltaic Review Committee published its strategic plan for photovoltaics (4). This plan documents the objectives of the DoD photovoltaic program and the strategies for realizing these objectives. The plan describes three core activities for the Photovoltaic Review Committee: (1) Outreach, (2) Logistic System Conditioning, and (3) Project Development. The level of activity, and consequently the rate of progress toward achieving the objectives defined in the strategic plan, fluctuates with the resources provided. Prior to ECIP and SERDP, the Photovoltaic Review Committee depended on personnel resources provided by the three service branches with Department of Energy support through Sandia National Laboratories. The development and project funding from SERDP and ECIP accelerated the DoD photovoltaic program. However, funding cuts initiated by the 1994 Congress have again slowed the Photovoltaic Review Committee's progress. The photovoltaic SERDP project was dramatically cut in FY95 and then terminated in FY96 (the entire energy thrust area of SERDP was zeroed in FY97). ECIP funding of photovoltaic projects dropped from $5M in FY95 to $2.6M in FY96, and then to zero in FY97. The Photovoltaic Review Committee is pursuing financing mechanisms to offset these reductions in project funding (see the logistic system conditioning activity below). The Photovoltaic Review Committee's progress may fluctuate over time, but its direction and activities continue on the path defined in the strategic plan. These activities are described below.

Outreach, as described earlier, is designed to increase awareness of photovoltaic technology through networking, user workshops, and presentations and publications. The most recent outreach effort was the Renewable Energy Workshop at China Lake, CA (November 1995). Logistic System Conditioning is designed to integrate photovoltaics into the established federal procurement system to make the purchase of a photovoltaic system as routine as a conventional power source. The Federal Photovoltaic Utilization Program began this effort by encouraging federal agencies to integrate life-cycle costing into the economic evaluation process. Other efforts in this area include making photovoltaic modules and system components available through General Services Administration contracts and making photovoltaic systems available through streamlined procurement mechanisms like energy savings performance contracting (19) and utility off-grid tariff programs (20). Two of the Photovoltaic Review Committee's ECIP systems (Grasmere Point and Santa Cruz Island) were procured under utility off-grid tariff programs. Systems can also be partially or completely financed under these two mechanisms. Financing may become the primary mechanism to implement systems if the current environment of limited up-front funding continues.

The project development activity is designed to turn potential applications into actual systems. The activity focuses on helping installations identify potential applications, develop technical requirements, and satisfy institutional requirements. This activity has dominated the Photovoltaic Review Committee's activities since its emphasis shifted to implementation of larger systems in 1992. Nearly all the Photovoltaic Review Committee's personnel resources were required to identify and implement $28M worth of photovoltaic systems. At the same time, there has been a series of executive and congressional acts that mandate project development. The following paragraph summarizes these executive and congressional acts.

The DoD, like all federal agencies, is under continually increasing pressure to reduce energy consumption and control energy costs wherever possible. Executive Orders 12003 (1977), 12759 (1990) and 12902 (1994) set goals for increase in industrial energy efficiency for federal facilities. Order 12003 called for a 20% efficiency increase by 1985 (compared with 1975), Order 12759 called for an additional 20% by 2000 (compared with 1985), and Order 12902 extended this goal to 20% by 2005 (compared with 1990). (The goal from Order 12759 was reiterated in the Energy Policy Act passed by Congress in 1992.) In addition, the 1990 House Defense Authorization Act set a goal for DoD to implement 100 MW of renewables by 1996.

This treatise on the DoD's history with photovoltaics describes the evolution of the Photovoltaic Review Committee and the DoD photovoltaic program leading to the current emphasis on project development and implementation. The remainder of this paper focuses on the overall potential of photovoltaics within the DoD, the process used to select projects for implementation, and the status of 2.1 MW of systems resulting from ECIP and SERDP.

## POTENTIAL OF PHOTOVOLTAICS WITHIN THE DoD

Table 3 gives the Photovoltaic Review Committee estimates of the overall potential for photovoltaics within the DoD (2). The photovoltaic potential is listed according to application areas in order of cost-effectiveness (small remote is the most cost-effective and bulk power/peak shedding is the least cost-effective). The small remote application area includes stand-alone systems dedicated to a single load, such as lighting and communications. The intermediate and large remote area is primarily augmenting diesel generators (hybrid systems) for remote training, testing, and evaluation facilities. The isolated grid area is primarily augmenting diesel generator based power plants (again hybrid systems) that serve an entire installation (a micro-utility grid). Most of the small stand-alone systems and about one-half of the hybrid systems are cost effective today based on life-cycle cost comparisons including operation and maintenance savings and enhanced reliability. The remaining areas are utility interactive applications.

Grid support is strategically placed photovoltaic power stations that support weak utility feeders. Distributed load centers are large utility-tied uninterruptible power systems for ultra-high reliability at critical facilities. Bulk power/peak shedding is central photovoltaic power stations to displace bulk power and control peak demand from the utility. Most of these applications are not cost-effective today.

All except 4 of the 124 ECIP and SERDP systems are remote applications (small remote and intermediate/large hybrids). The predominance of the remote systems is the result of the ECIP regulation that photovoltaics must be the least-cost power source for the application. This trend will continue until costs drop significantly below the current $10 per watt of ac power for large utility-interactive systems (as was paid for the Yuma grid-tied photovoltaic power plant).

TABLE 3. Potential for Photovoltaics within the DoD

| Application Area | Power Capacity (kW) | No. of Systems | Photovoltaics Required (MW) |
|---|---|---|---|
| Small Remote | <25 | 51,625 | 51 |
| Intermediate/Large Remote | 25-1000 | 3,875 | 423 |
| Isolated Grid | >500 | 870 | 1,305 |
| Grid Support | >500 | 200 | 600 |
| Distributed Load Center | >500 | 800 | 250 |
| Bulk Power/Peak Shedding | >500 | 1,200 | 1,200 |
| *TOTALS* | --- | *58,570* | *3,829* |

# STATUS OF ECIP AND SERDP SYSTEMS

## Selection and Implementation Process

All Photovoltaic Review Committee members are tasked to identify and develop projects for their service branch. Projects are identified via surveys by the Photovoltaic Review Committee members and via requests for assistance from installations. The Photovoltaic Review Committee member first performs a preliminary assessment to establish that the project is technically viable and cost-effective. Subsequent activity depends on a set of factors including user support (support at the installation level and up through the chain of command), scope and cost, maturity of the technology, replicability within the DoD, and payback period. For small projects (in both scope and cost) with mature technology (small stand-alone systems), the Photovoltaic Review Committee provides information on how to develop system specifications and procure the system so the installation can pursue the project on its own. In general, the Photovoltaic Review Committee limits active involvement to projects that the installation cannot handle on its own.

In these cases, the Photovoltaic Review Committee, with support from Sandia when necessary, assists the installation to develop the system justification as required to obtain approval through the chain of command. This justification normally requires an economic comparison of the power alternatives, which in turn requires the development of a conceptual system and an estimate of the system performance. After the project is approved, it can be submitted to the appropriate funding or financing program.

Up to now, the primary funding program has been ECIP. Each fiscal year, the Photovoltaic Review Committee ranks the approved projects and submits the top projects for ECIP funding. The ranking is based on an evaluation of payback period, replicability, and user support. The number of projects submitted is limited by the available funding. The other project funding program was SERDP. These projects were selected to demonstrate specific technology advances.

After a project is selected and funded, it is assigned to the appropriate procurement agency and a technical working group is created to support the project through installation and start-up. The working group includes installation and Photovoltaic Review Committee personnel, with technical support from Sandia when necessary, and is tasked to develop system specifications, to help the procurement agency develop the solicitation package, to evaluate the technical content of proposals, to review and approve system design, and to define system acceptance and start-up criteria. The time required from developing the specifications to completion of construction varies widely depending on the complexity of the system and accessibility to the site. A reasonable time frame for a single system is about 6 months for contract award followed by 12 months to design and build it.

## System Status

Since 1992, the DoD has invested $28M for 124 photovoltaic systems through ECIP and SERDP. These systems represent a total of 2.1 MW of photovoltaics. $19.5M was provided by ECIP with the $8.5M balance from SERDP. Table 4 lists the main features and status of these projects. The majority of these are small stand-alone systems including 56 pop-up targets for the Pohakuloa Training Area (PTA ranges), 40 water pumps (Ft. Carson and Santa Cruz Island), and 12 range control towers (PTA ranges). Three are grid-interactive systems (Yuma grid-tied, Ascension Island, and PTA Bradshaw airfield). The Yuma grid-tied system is the most technically advanced and is representative of both the distributed load center and peak shedding applications areas. One is a regenerative fuel cell application (Edwards Air Force Base fuel cell). The remaining 12 are photovoltaic/diesel hybrid systems. A brief description of each system follows Table 4.

TABLE 4. Features and Status of ECIP and SERDP Systems

| Project | Array (kWp) | Battery (KWh) | PPU[a] (kW) | Status |
|---|---|---|---|---|
| Superior Valley, China Lake, CA, Navy | 344 | 3500 | 300 | operational |
| Grasmere, Mt Home AFB, ID, Air Force | 78 | 700 | 90 | operational |
| San Clemente Island, Navy | 94 | 2500 | 175 | construction |
| Range 500, 29 Palms, CA, Marines | 69 | 2000 | 180 | design |
| China Lake Hybrids, CA, Navy | | | | |
| • Junction Ranch | 130 | 2000 | 250 | construction |
| • Nato Site | 134 | 2000 | 250 | construction |
| • Kim Site | 235 | 2000 | 250 | construction |
| Pumps (39), Ft. Carson, CO, Army | <1 | --- | --- | operational |
| PTA ranges, Hawaii, Army | | | | |
| • Pop-up target (56) | 0.01 | 1 | --- | development |
| • Control Towers (12) | 0.4 | 6 | --- | development |
| • Track targets (2) | 5 | 600 | 30 | development |
| • Bradshaw airfield | 2 | 600 | 5 | development |
| Yuma grid-tied, Yuma, AZ, Army | 441 | 5600 | 900 | construction |
| Smart Munitions, Yuma, AZ, Army | 225 | 3500 | 225 | development |
| Mobile Power Center, Marines | 3.4 | 58 | 6 | testing |
| Ascension Island, Air Force | 86 | --- | 120 | construction |
| Fuel Cell, Edwards AFB, CA, Air Force | 50 | --- | --- | operational |
| Santa Cruz Island, Navy | | | | |
| • Communications site | 138 | 2700 | 150 | design |
| • Water Pump | 11 | --- | --- | design |

[a] - Capacity of power processing unit (PPU)

## *Superior Valley, China Lake, California*

This photovoltaic/diesel hybrid system powers the Naval Air Warfare Center's bombing complex at Superior Valley on the south range of China Lake. The prime contractor, Photocomm, Inc. was awarded $3.6M to design and install the system. The system includes 344 kW of ASE 300DG modules, a 300-kW Abacus power processing unit, and 3.5 MWh of C&D C170-19 batteries. This project is a SERDP-funded research and development project designed to advance power processing technology to allow "ganging" of multiple power processing units to achieve the power capacity required for larger DoD applications (see reference 21 for a detailed description of this system). The system also provides increased power capacity required by the expanding mission of the facility. The system was accepted by the Navy in September 1996.

## *Grasmere Range, Mountain Home Air Force Base, Idaho*

This project provides power to electronic equipment located on Grasmere Range, approximately 80 miles from Mountain Home Air Force Base and 40 miles from the nearest utility grid. Previously, power had been furnished by diesel engines. The expanding mission of the Grasmere facility now requires year-round power with 24-hour power during the winter. Because of the severe winter weather at the site, 24-hour power with diesels alone is impractical. This project allows the Air Force to reduce the diesel engine run time thereby providing year-round power at a reasonable cost. The prime contractor, Idaho Power, was awarded $1.3M to design and install the system. The system was procured through Idaho Power's off-grid tariff program and includes 78 kW of Solarex MSX-120 modules, a 90-kW Advanced Energy Systems power conditioning unit, and 700 kW of Hoppecke batteries (see reference 21 for a detailed description of this system). Idaho Power is also under contract to the Air Force to maintain and operate this system. The system became fully operational in February 1996.

## *Range Electronic Warfare Simulator, San Clemente Island, California*

Most of the facilities on San Clemente Island are served by a group of diesel generators that produce power for the island grid. However, the Range Electronic Warfare Simulator facility is isolated even from the island grid and depends on a series of generators to provide power to the equipment and housekeeping loads at the facility. The proposed system addresses the excessive costs associated with providing 24-hour power with diesel generators. The system will provide autonomous operation during weekends, which will substantially reduce operation costs. The system includes 94 kW of ASE 300DG modules, a 175-kW Kenetech power processing unit, and 2.5 MWh of GNB 85RC33 Resource Commander batteries. The procurement office is the Naval Facilities Engineering Command Southwest Division (SWDIV) in San Diego, and the Navy user is the Southern California Off-shore Range with headquarters at North Island in San Diego. Construction is 80% complete. The system is expected to be fully operational in January 1997. The prime contractor for this system is Integrated Power Corporation.

## *Range 500, Twenty-nine Palms, California*

Range 500 is a remotely located tank target practice range isolated from the grid. Presently, diesel generators power a mechanism that drives tank pop-up targets along a track. Photovoltaics will supplement those generators and provide improved power quality and reliability. The prime contractor is Utility Power Group. The system will include 69 kW of Siemens J4F modules, a 180-kW Kenetech power processing unit, and 2 MWh of C&D C125-33 batteries. The procuring agency is the Naval Facilities Engineering Command at San Diego (SWDIV). The project is in the final design stage.

## Three Hybrid Projects at China Lake, California

A contract for $6.3 million was recently awarded to Plateau Electric for the design and installation of three separate photovoltaic hybrid systems at China Lake, California. These systems are for existing, remotely located test sites and are currently under construction. They are:

Junction Ranch - 130 kW of photovoltaics
NATO Site - 134 kW of photovoltaics
Kim Site - 235 kW of photovoltaics

These systems are standardized as much as possible. Each system has an identical 250-kW Kenetech power processing unit and an identical 2-MWh battery bank (SEC C100-27 batteries). The systems have identical factory-built equipment enclosures to house the power processing units and associated controls and switchgear, and they use identical sub-array assemblies of ASE 300DG modules from Applied Power Corporation for the photovoltaic arrays. The procurement office is the Naval Air Warfare Center Weapons Division at China Lake.

## Fort Carson Water Pumping, Colorado

The Fort Carson photovoltaic project was approved as an ECIP project in 1992 to install some 39 different water pumping systems to replace aging windmills on the military reservation. In early 1995, Fort Carson personnel attended a photovoltaic water pumping workshop, hosted by members of the Photovoltaic Services Network (PSN). Working with the PSN, Fort Carson decided to turn its project into a demonstration of currently available photovoltaic water pumping technologies (22). Through summer-1995, eight different systems, using both ac and dc submersible pumps, were installed. The systems include: SolarJack (dc centrifugal), Golden Photon (ac centrifugal), A.Y. McDonald (dc centrifugal), Applied Power Corp. (ac centrifugal), EPV (ac centrifugal), Photocomm (ac centrifugal), and Robinson (dc centrifugal). Fort Carson has since selected the Photocomm system for the other sites. About 30 systems have been installed to date with the remainder to be installed by spring of 1997. This project is being implemented by Fort Carson's Directorate of Environmental Compliance and Management.

## Pohakuloa Training Area Ranges, Hawaii

A solicitation for multiple photovoltaic systems at the Army's Pohakuloa Training Area is in preparation and is scheduled for release by the end of 1996. The training area is a large range located between Mauna Kea and Mauna Loa in the interior of Hawaii. Approximately 70% of the range is not served by the utility grid. At this time, there are no plans to extend the grid into these remote areas and, as a result, the Army must provide its own power for training activities. This solicitation will provide photovoltaic power for control towers at 12 target ranges, it will convert 56 diesel powered pop-up targets to photovoltaic power, it will hybridize two diesel powered large track targets, and it will power runway lights for Bradshaw airfield. The control towers and pop-up targets systems will be powered by stand-alone photovoltaic systems. Each control tower will have about 400 W of photovoltaics and 6 kWh of battery to power warning beacons, interior lights, a public address system, portable computers, and radio communications. Each pop-up target will be fitted with a single 10-Watt photovoltaic module and 100-Ah 12-Volt battery. The track target hybrids will add about 5 kW of photovoltaics, 600 kWh of battery, and 30 kVA of power processing to the existing 60-kW generators. The runway light requirements have not yet been fully defined. The array will be between 1.5 - 2 kW and the battery will be about 600 kWh. The system will be backed up by the utility grid. The procuring agency is the Pacific Ocean Division of the US Army Corp of Engineers. The request for proposals is scheduled for release in December 1996.

## Yuma Proving Ground Grid-Tied System, Arizona

In early 1995, Utility Power Group won two prime contracts for a photovoltaic power station and a complementary battery storage/load leveling project for a total amount of $5.5M. The photovoltaic power station was funded at $3.9M by ECIP, and the battery storage/load leveling project was funded at $1.6M by SERDP. The two systems were integrated into a photovoltaic/battery-enhanced peaking station. The system includes 441 kW of Siemens M-55 modules, four 225-kW Kenetech power processing units, and 5.6 MWh of C&D C125-19 batteries. Under normal operating conditions during the summer peak demand season, the system will be capable of delivering up to 825 kW to the grid to help Yuma Proving Ground with high demand rates. The system will also be capable of operating in a stand-alone mode if an extended power outage is experienced. On a limited basis, the stand-alone system would continue to operate the Proving Ground's nearby water treatment plant and other emergency communications loads (see reference 21 for a detailed description of this system). Construction is 95% complete. The system is expected to be fully operational in January 1997.

## *Smart Munitions Complex, Yuma Proving Ground, Arizona*

Yuma Proving Ground is currently preparing specifications for a photovoltaic/diesel hybrid system using 225 kW of ASE 300DG modules and 3.5 MWh of Absolyte IIP 100A75 batteries that were purchased under the SERDP program. The hybrid system is the first element in the development of a new range area at the Proving Ground. It will initially power a new smart munitions test complex and will be expanded to serve as a micro-utility for multiple facilities in this developing range area. Yuma is contributing about $0.5M from its project funding accounts to create a functional system from the $1.5M worth of modules and batteries in its possession. This will be the first major DoD installation to use the valve-regulated battery technology. The system is expected to be completed in the summer of 1997.

## *Mobile Power Center, California*

A mobile power center has been developed for the 1st Marine Expeditionary Force at Camp Pendleton, California. It is designed to interface with either the commercial utility or generators and will have 3.2 kW of Solarex MSX-95ML modules, a 1-kW World Power Whisper 1000 wind turbine, a 6-kW Abacus inverter, a 3-kW IBE battery charger, and a 58-kWh C&D C125-21 battery. The mobile power center is a partially SERDP-funded research and development project designed to incorporate hybrid capabilities in a mobile package that is compatible with standard military equipment and suitably robust for military operations. The unit addresses the need to simplify the logistics of power requirements for tactical military exercises as well as enhance the reliability of the missions. The system was designed and constructed by Naval Research and Development in San Diego and is currently under test at Camp Pendleton.

## *Ascension Island Runway Lighting, Mid-Atlantic Ocean*

Integrated Power Corporation was awarded $1M to design and install a grid-interactive photovoltaic system for runway lighting at Ascension Island. The system includes 86 kW of ASE 300DG modules and a 120-kW Kenetech power processing unit. The system will supplement the island diesel generator power plant. It can also be upgraded with batteries to operate the runway lights independent of the island grid. The system has been designed, and hardware is being shipped to the island. Construction is scheduled to begin in February 1997, and the system is expected to be operational by the end of April 1997.

## *Solar Regenerative Fuel Cell Cooperative Project, California*

This is a cooperatively funded demonstration project with the Jet Propulsion Laboratory at Edwards Air Force Base. The Navy Energy Program Office based at China Lake procured and installed two different 25-kW Cadmium Telluride thin-film photovoltaic arrays - one manufactured by Solar Cells Inc. and the other by Golden Photon Inc. The system, which has been operational since August 1996, also includes two 25-kW Abacus maximum power trackers, an electrolyzer system, and a fuel cell. The direct current produced by the photovoltaic arrays powers an electrolyzer system that separates water into hydrogen and oxygen. The hydrogen and oxygen are recombined in the fuel cell to produce heat, electricity, and water. NASA regards a regenerative photovoltaic fuel cell power plant as a possible option for advanced human space exploration. The DoD will gain operational experience with an emerging, potentially lower-cost photovoltaic technology.

## *Santa Cruz Island, California*

This project was the only photovoltaic project approved in the FY 1996 ECIP budget cycle. The project was procured through Southern California Edison's off-grid tariff program and includes two systems: a photovoltaic/diesel hybrid for a mountain-top communications station, and a photovoltaic water pumping system. The hybrid system will include 138 kW of photovoltaic modules, 2700 kWh of batteries, and a 150-kW power processing unit. The water pumping system will include 11 kW of photovoltaic modules. Edison has solicited and evaluated bids from photovoltaic system suppliers for the design and installation of the two systems. The system suppliers will be announced in November 1996. The project is scheduled for completion by June 1997.

## SUMMARY

The first organized effort to integrate photovoltaics into military practices dates back to the late 70s when all branches of the DoD participated in the Department of Energy's Federal Photovoltaic Utilization Program. This led to the creation of the DoD's Tri-Service Photovoltaic Review Committee chartered in

1985 to foster the use of photovoltaics throughout the DoD. The Photovoltaic Review Committee has identified nearly a 4 GW potential for photovoltaics in stand-alone, photovoltaic/generator hybrids, and utility-interactive applications. The Photovoltaic Review Committee, with technical support from the Department of Energy through Sandia National Laboratories, works to integrate these applications into standard military practices so that they are used where cost-effective. The Photovoltaic Review Committee initially focused on the small remote applications area (stand-alone and hybrid systems less than 25 kW). By 1992, DoD had installed 2,000 small remote systems with 2 MW of photovoltaics.

In 1992, funding from SERDP and ECIP allowed the Photovoltaic Review Committee to shift emphasis to larger hybrid systems in the intermediate/large remote and isolated grid applications areas. Under Sandia direction, SERDP funds were used to develop, test, and demonstrate the power processing hardware required for larger hybrids and the ECIP funds were used to field 12 hybrid systems at installations covering a wide range of environmental and operating conditions. Today, intermediate sized hybrids are considered ready for implementation, with the larger hybrids only 2-3 years behind. A total of 124 systems with 2.1 MW of photovoltaics were procured with SERDP and ECIP funding. In addition to the 12 hybrids, these systems include 108 stand-alone systems, 3 utility-interactive systems, and a regenerative fuel cell application using the new Cadmium/Telluride thin-film photovoltaic module technology. The 120 systems funded by ECIP are cost-effective applications.

The Photovoltaic Review Committee will continue its work with larger hybrid systems and will move on to the utility-interactive applications as they become cost-effective. Sandia will continue to support the Photovoltaic Review Committee in both project and technology development. The level of activity, however, will vary with the available resources.

## REFERENCES

1. "DoD Energy at a Glance," Department of Defense, Office of the Deputy Assistant Secretary of Defense (Logistics), Energy Policy Directorate, July 1991.
2. "Photovoltaics for Military Applications - Strategic Research and Development Program (SERDP) Proposal for Fiscal Year 1994," prepared by Sandia National Laboratories for the Department of Energy, the Air and Energy Engineering Research Laboratory for the Environmental Protection Agency, and the Naval Air Weapons Station for the Department of Defense, SERDP Project EN-046, February 1994.
3. Pulscak, M., "Federal Photovoltaic Utilization Program," *Proceedings of the Federal Agencies and Industrial Review of the Federal Photovoltaic Utilization Program (5260-31)*, Albuquerque, New Mexico, pp. 2-1 to 2-13, October 1983.
4. "Department of Defense Photovoltaic Review Committee Strategic Plan," prepared by the Department of Defense Photovoltaic Review Committee, Chairman - Garyl Smith, Energy Program Office, China Lake, CA, October 1993.

5. Smith, G.D., "Navy Photovoltaic Utilization Survey," Department of Defense Photovoltaic Review Committee, Energy Program Office, China Lake, CA, June 1986 (Unpublished).
6. *Evaluation of International Photovoltaic Projects*, SAND85-7018, (Sandia National Laboratories, Albuquerque, New Mexico), September 1986.
7. "Photovoltaics for Military Applications - FY91 Briefing for Strategic Research and Development Program (SERDP) Scientific Advisory Board," prepared by Sandia National Laboratories for the Department of Energy, the Air and Energy Engineering Research Laboratory for the Environmental Protection Agency, and the Naval Air Weapons Station for the Department of Defense, SERDP Project EN-046, Washington, D.C., July 1992.
8. *1988 Photovoltaic Annual Systems Symposium: Agenda and Abstracts, February 16-18, 1988*, SAND88-0146, (Sandia National Laboratories, Albuquerque, New Mexico), February 1988.
9. *Photovoltaics for Military Applications - A Decision Makers Guide*, SAND87-7016, (Sandia National Laboratories, Albuquerque, New Mexico), January 1988.
10. *Maintenance and Operation of Stand-Alone Photovoltaic Systems*, prepared by Architectural Energy Corporation for Naval Facilities Engineering Command, Southern Division, and the Photovoltaic Review Committee, Department of Defense, Boulder, Colorado, December 1991.
11. Hoelscher, J., et al., "Photovoltaics in the United States Department of Defense," *25th Annual American Solar Energy Society Meeting*, Asheville, North Carolina, April 13-18, 1996.
12. Smith, G.D., et al., "The United States Department of Defense Terrestrial Photovoltaic Program - A 1994 Status Report," *First World Conference on Photovoltaic Energy Commission*, Waikoloa, Hawaii, December 5-9, 1994.
13. Smith, G.D., "Renewable Energy in the Department of Defense," *16th World Energy Engineering Congress*, Association of Energy Engineers Meeting in Atlanta, Georgia, October 28, 1993.
14. Hartzog, S., et al., "A Photovoltaics-Diesel Hybrid System for an Operational Navy Facility on San Clemente Island," *Proceedings of the Twenty-Third IEEE Photovoltaic Specialists Conference*, Louisville, Kentucky, May 10-14, 1993.
15. Ducey, R., et al., "Terrestrial Photovoltaic Systems for U.S. Military Applications," *Proceedings of the Twenty-First IEEE Photovoltaics Specialists Conference*, Kissimmee, Florida, May 21-25, 1990.
16. "Photovoltaics for Military Applications - FY95 Briefing for Strategic Research and Development Program (SERDP) Scientific Advisory Board," prepared by Sandia National Laboratories for the Department of Energy, the Air and Energy Engineering Research Laboratory for the Environmental Protection Agency, and the Naval Air Weapons Station for the Department of Defense, SERDP Project EN-046, Tyndall Air Force Base, Florida, February 7, 1995.
17. Hoelscher, J., "The Department of Defense Photovoltaic Review Committee Project Listing," prepared by Dyncorp for the Department of Defense Photovoltaic Review Committee, Arlington, VA, in preparation.
18. Beck, L.L., "GloED and GloTech: GLOBAL EMISSIONS AND TECHNOLOGY DATABASE SOFTWARE," *Proceedings: The 1995 Symposium on Greenhouse Gas Emissions and Mitigation Research,* Washington, D.C., June 27-29, 1995, pp. 2-74 to 2-84.
19. Dahle, D., Westby, R., and Ginsberg, M., ,"Alternative Financing for Federal Energy Projects: An Overview of the Federal Energy Management Program Role," National Renewable Energy Laboratory, (to be published in the *Proceedings of the 1996 ACEEE Summer Study*).
20. "Schedule No. 60 - Solar Photovoltaic Service Pilot Program," issued by Idaho Power Company, Idaho Public Utilities Commission, No. 25, Tariff No. 101, November 1992.

21. Chapman, R.N., "Hybrid Power Technology for Remote Military Facilities," *Proceedings of the Ninth International Powersystems World '96 Conference*, Las Vegas, Nevada, September 7-13, 1996, pp. 415-427.
22. "Assessment of Photovoltaic-Powered Pumping Systems at Fort Carson, Colorado," prepared by Architectural Energy Corporation for the U.S. Army Construction Engineering Research Laboratory, Contract No. DAC-94-D-0012, Boulder, Colorado, September 1996.

# The USAID/DOE Mexico Renewable Energy Program: Using Technology to Build New Markets

Charles J. Hanley

*Renewable Energy Office, Sandia National Laboratories, Albuquerque, NM 87185-0704*

**Abstract.** Under the Mexico Renewable Energy Program, managed by Sandia National Laboratories, sustainable markets for renewable energy technologies are developed through the implementation of pilot projects. Sandia provides technical assistance to several Mexican rural development organizations so they can gain the technical and institutional capability to appropriately utilize renewables within their ongoing programs. Activities in the area of water pumping have shown great replication potential, where the tremendous rural demand for water represents a potential renewable market of over $2 billion. Thirty-six photovoltaic water pumping projects have been installed thus far in the Mexican states of Chihuahua, Sonora, Baja California Sur, and Quintana Roo, and 60 more will be implemented this year. The majority of these projects are in partnership with the Mexican Trust for Shared Risk (FIRCO), which has asked Sandia for assistance in extending the program nationwide. This replication is beginning in five new states, and will continue to grow. Sandia is keeping the U.S. renewable energy industry involved in the program through facilitating partnerships between U.S. and Mexican vendors, and through commercialization assistance with new systems technologies. The program is sponsored by the Department of Energy and the U.S. Agency for International Development.

## BACKGROUND

Tremendous opportunities exist in Mexico for growth in the use of renewable energy technologies, especially photovoltaics. Over 5 million Mexican people do not have access to grid electricity. Mexico has over 100,000 rural communities in need of potable water, approximately 80,000 of which are not grid electrified. Over 600,000 rural ranches have existing water needs for livestock and irrigation. In terms of water pumping alone, this represents a potential demand for over two billion dollars of renewable energy equipment. This large and growing unelectrified rural population, coupled with abundant renewable resources, a strong government incentive to provide electricity to these populations, and a recovering economy, provides the basis for a renewable energy industry much larger than presently exists in Mexico.

Mexico's proximity to the U.S. makes it an especially attractive potential market for the U.S. renewables industry. The North American Free Trade Agreement facilitates the import of U.S.-produced renewable energy products into Mexico by effectively removing applicable tariffs. This proximity also facilitates the establishment of business relationships between U.S. and Mexican renewable equipment suppliers.

Although the Mexican federal government does not presently have a comprehensive policy geared toward the use of renewables, it does show a strong interest in the use of renewable technologies as part of its extensive rural development activities. Through the government-funded national solidarity program (PRONASOL), over 30,000 photovoltaic home lighting systems have been installed in the last five years. The National Commission on Energy Savings (CONAE) is geared toward reducing industrial consumption patterns, is actively considering the adoption of renewable energy technologies within its programs. Similarly, the Federal Electricity Commission (CFE) has recently started to consider renewables as a possible technology for rural electrification needs. In addition, renewables may play a significant role in a new agricultural development program called Alianza Para El Campo.

## PROGRAM OVERVIEW

In 1994, Sandia National Laboratories initiated its Mexico Renewable Energy Program, co-sponsored by the Department of Energy (DOE) and the U.S. Agency for International Development (USAID). Through a cooperative agreement, the program was structured in line with the missions of both organizations. The primary goals of the Sandia Mexico program are to increase the appropriate and sustainable use of renewable energy technologies, thereby expanding markets for the U.S. and Mexican industries and demonstrating the use of renewables in combating global climate change through offsetting emissions of greenhouse gases.

Sandia has developed a diverse, multi-organizational team to implement the program. The National Renewable Energy Laboratory (NREL) is a program partner and is the technical lead regarding solar and wind resource assessment and all wind-related projects, wind training, and other wind-related activities. The program team is also comprised of the Southwest Technology Development Institute of New Mexico State University, Enersol Associates, Dyncorp, and the National Rural Electric Cooperative Association. The Institute of International Education, a USAID-funded organization with worldwide experience in environmental and energy training, has partnered with Sandia in the sponsorship of program training activities.

### Emphasis on Sustainability

What may set the Mexico program apart from many other programs and approaches is that it does not seek to set up new organizations, programs, or projects centered around renewable energy per se. Instead, the program focuses on selected end-uses of energy, such as agricultural water pumping or remote communications, and incorporates the appropriate use of renewable energy into associated ongoing and funded development programs accordingly. This is an approach Sandia's Systems Assistance Center has used with considerable success for many years in U.S.-focused programs, such as with the National Park Service. Instead of having to build local capacity from ground-zero, the program augments existing local project implementation and technical capacity with the necessary training in how to assess, select, procure, and use renewable energy technologies successfully. This reduces the time required to implement viable, locally-championed renewable energy-based projects in the near-term and leads to earlier project replication.

Over $2.5 million in cost-shared pilot projects are being implemented as a mechanism to institutionalize the use of renewable energy technologies. Throughout the project implementation process, Sandia provides both formal and informal training and technical assistance regarding technologies, their applications, and the entire process of project implementation. This training is provided to both project implementers and local hardware suppliers to assure the long-term quality of installed systems. Each installed project is incorporated into the program monitoring plan, so that over time Sandia will continually update its "best practices" reports regarding the application of renewable energy technologies in international applications.

Project implementation activities under the program are focused on off-grid rural "productive use" applications, where the use of renewable energy technologies provides a measurable economic and/or social benefit to the end users. These applications are highly sustainable and replicable, because they provide a mechanism for paying for the renewable energy systems. Productive use examples include water pumping for livestock or crop irrigation, lighting for commercial or business activities, power for grain grinding or carpentry, and ecotourism. In addition, the promise of economic return facilitates the requirement that the end users contribute to overall project costs, and ultimately that projects be replicated without cost-share from the Sandia program. These end user contributions also help to assure long-term system performance, as owners actively play a critical role in long-term system maintenance and operation.

## In-Country Partnerships with Development Organizations

Technology application activities have been organized into several different partnerships, at both the state and federal levels in-country, and with representatives of both U.S. and Mexican industries. To date, the most significant partnership has been with the federal Trust for Shared Risk (FIRCO, by its Spanish acronym), which is an agricultural development organization under the Secretariat of Agriculture, and has offices in each of the 32 Mexican states. Sandia has established contracts with FIRCO in the states of Sonora, Baja California Sur, and Quintana Roo, through which hundreds of livestock watering projects are being identified and implemented. In Chihuahua, FIRCO is part of a renewable energy working group that was formed by 8 state agencies to manage the implementation of projects under the Sandia program. This working group has gained recognition within the state government as a model for cooperation between organizations with diverse, yet overlapping, interests. These partnerships, focused on a few specific types of applications in relatively concentrated geographical areas, enable the program to generate enough sales in a particular area to spark development (usually in partnership with the U.S. industry) of the local business infrastructure necessary to supply and maintain renewable energy products on an ongoing basis.

Sandia has also signed contracts with three international conservation organizations, The Nature Conservancy, Conservation International, and the World Wildlife Fund, to facilitate the use of renewables in the management of protected areas and as tools in the sustainable development of "buffer communities" - those that border ecologically sensitive regions. Sandia is working directly with these organizations and with their in-country partners to implement highly visible renewable energy projects. Since these organizations are less technically oriented than other partners, such as FIRCO, training and other project activities focus on identifying appropriate applications of renewable technologies, clearly defining energy demands, working with suppliers to complete projects, and long-term maintenance issues.

Sandia is also working to increase the involvement of the U.S. renewable energy industry in the Mexico program, through increased communications and the establishment of technology partnerships. The goal of these activities is to reduce barriers to commercialization, either of new products or in new or under-developed markets. Technical assistance is being provided to members of the U.S. utility industry who are exhibiting an active interest in extending their influence in Mexico in the identification and development of appropriate renewable energy projects. In addition, Sandia will use data that is presently being collected from project development and implementation activities to help guide members of the U.S. renewable energy industry in the development of new systems technologies, such as integrated refrigeration or ice making systems, and in the improvement of existing systems technologies, such as water pumping systems, for rural international applications.

## Cross-Cutting Activities

Several cross-cutting activities have been initiated to benefit the field components of the program. Formal training workshops have been presented to decision makers at both the federal and state levels, focused on the relative technical and economic benefits of renewables in many applications over other, more conventional, energy supply options, such as diesel generators. Technically-oriented workshops have been presented to project developers and implementers, as well as renewable energy equipment suppliers, throughout the country. These workshops generally culminate in the installation of one or more systems, allowing participants to gain hands-on experience and witness first-hand the benefits of renewable energy technologies.

Project analysis activities involve technical and economic feasibility, environmental impact, and follow-up monitoring and evaluation. These activities are conducted to assure that projects implemented meet pre-established criteria regarding sustainability and replicability. Follow-on activities will facilitate continued assessments of program success, while also providing technical feedback to the industry regarding the longevity of the various projects.

Resource assessment is being conducted at NREL to quantify the solar and wind resources throughout Mexico. Special attention is being given to present areas of programmatic activity, especially in the case of wind resource assessment in the states of Baja California Sur and Quintana Roo. Solar and wind resource maps with varying degrees of resolution are under development.

# PROJECT IMPLEMENTATION RESULTS

Through contracts established under this program, projects have been implemented in the Mexican states of Sonora, Chihuahua, Baja California Sur, Quintana Roo, Chiapas, Campeche, and Oaxaca. To date, 36 photovoltaic water pumping projects have been installed. Two wind systems have been installed under the program, one for a water pumping application in the state of Oaxaca and the other for facilities power at an ecotourism resort in Quintana Roo. In the Protected Areas part of the program, Sandia has helped partners realize PV projects for facilities power and communications in 5 locations, with projects in an additional 26 sites currently under development.

All water pumping projects have been coordinated with the ongoing activities of state rural development agencies, with the majority being through FIRCO. All of these projects, with the exception of those few that have accompanied formal training workshops, have been implemented through local competitive bid processes, purchased by the in-country partners, and have been supplied and installed by local vendors trained through the Sandia program. In many cases, Sandia has assisted these local suppliers in the formation of partnerships with members of U.S. industry to improve their system design, procurement, and installation capabilities. In addition, all projects have involved cost share contributions from Sandia, the local counterpart agencies, and the end users. The level of cost share from Sandia averages just above 50% overall.

A total of over 60 additional projects are underway in all of the above states. In those states where the Sandia program has had longer involvement, such as Chihuahua and Sonora, the Sandia engineers play only a minor supporting role as the in-country engineers work within the established framework to identify, develop, and implement new projects. In both of these states, engineers are beginning to implement other projects outside the Sandia program, due to the existing high levels of demand and their newly realized project implementation capacity.

Cooperative agreements have been signed with the state governments of Chihuahua and Sonora, and with the federal office of FIRCO, indicating mutual interest in pursuing further renewable energy applications in the future. Similar agreements are being negotiated in Baja California Sur, Quintana Roo, and Oaxaca. In addition, Sandia has also established agreements and business relationships with leading universities in Mexico that are involved in the exploration and advancement of renewable energy technologies, including the National Autonomous University of Mexico, and the University of Sonora. These agreements put Sandia in a key position to play a critical role in developments in both research and applications of renewables in Mexico in the future.

## Additional Technology Partnerships

Sandia has established a team to implement of a pilot hybrid (PV/diesel) ice making system in the state of Chihuahua as the start of a broad commercialization effort. The integrated hybrid energy supply component is a new product from SunWise Energy Systems Corp., and project costs are being shared by Sunwize, the New York State Research and Development Authority, the State of Chihuahua, and Sandia. A twin system will be installed at SWTDI in Las Cruces, New Mexico, and several others are planned for the state of New York. System operational and performance characteristics will be monitored extensively for at least one year, both in Las Cruces and Chihuahua, while commercialization plans are developed and finalized for the Mexican market.

Sandia has also been working with the Federal Electricity Commission (CFE) of Mexico, which is developing a partnership with Arizona Public Service in the development and implementation of a centralized hybrid community power system in the state of Baja California Sur. CFE is the steward of the federal government's recent mandate to electrify all remote communities with more than 100 inhabitants, and is therefore exhibiting a newly revitalized interest in centralized hybrid systems. NREL personnel, under the auspices of the Mexico program, have provided technical support regarding the relative effectiveness of different hybrid strategies for the town of San Juanico, the candidate site for this project. The involvement of the Sandia program thus far includes this technical support and an assessment of technical and institutional issues related to the implementation of a centralized power system in San Juanico. Sandia is presently conducting an investigation to determine if it will provide financial assistance toward equipment costs.

## NEW PROGRAM ACTIVITIES

Technical activities will continue through fiscal year 1997. Existing partnerships will be strengthened, and new partnerships initiated. Conditioning of both the supply and demand for photovoltaic systems, in terms of helping local project implementers and vendors gain the technical skills needed to implement high-quality projects, will continue as Sandia completes all existing project implementation commitments.

## Nationwide Replication of Projects

The strength of the technology partnership between Sandia and FIRCO has grown significantly from the project implementation activities that are presently underway in four states. Since the inception of the Sandia program in 1994, FIRCO's role has changed dramatically. Rather than implement its own rural development projects, FIRCO now plays a technical advisory role in the implementation of several federal programs. One of these is Alianza Para El Campo, in which end users apply for reimbursements from the federal government for private developments which meet certain criteria. Part of this program is a 10-state, $225 million, World Bank-funded rural development program, with livestock water pumping as a significant component. FIRCO has asked Sandia for technical assistance in the replication of its renewable energy activities to a nationwide level, starting with an expansion to these 10 states. Under this activity, Sandia will provide technical assistance and training to FIRCO engineers and technicians, who will then work with end users to identify and develop projects. Sandia and FIRCO are working together to include renewable energy equipment as one of the project components for which end users can receive government reimbursements. Sandia's contribution will be limited to this technical and programmatic assistance, and cost shares on a small number of projects that will be installed as part of formal training activities. All other projects will be implemented using end user, state, and federal contributions.

The early indications of the potential returns of this activity are very encouraging. After attending Sandia-sponsored workshops on renewables technologies and project implementation, two FIRCO state offices have each submitted proposals to the federal office for over 30 photovoltaic and wind-powered water pumping projects. Sandia plans to conduct additional training exercises in the states of Chiapas, Oaxaca, Veracruz, San Luis Potosí, and Yucatan, and expects to assist FIRCO in the identification and development of over 100 new projects in the next year.

Other promising replication possibilities exist with agencies other than FIRCO. For instance, CFE's electrification mandate may lead to the development of several large hybrid centralized systems in the next year, in the states of Baja California Sur and Sonora. Sandia has offered technical assistance in the development of all CFE projects involving renewable energy technologies. The Sandia team is also investigating the potential for renewables within the commercial fishing industry, which is experiencing a growing trend among the cooperatives toward incorporation and profit-orientation. These fishing cooperatives have significant energy needs for refrigeration, ice making, and water pumping. In cases such as village power and fisheries, the high maintenance and operational costs of large diesel generator systems make renewables a potentially economically attractive option. In addition to scouting potential partners and project sites, Sandia will also investigate the availability of in-country financing mechanisms to help improve the economic viability of these types of projects.

## Long Term System Monitoring

A comprehensive monitoring program has been established to facilitate the process of tracking system performance over time. Pertinent information regarding each installed project is entered into the program database, and the results of scheduled periodic technical assessments will be compiled. In addition, several select projects will be extensively monitored with on-site data acquisition hardware. These will include all applications of newly-developed systems technologies, such as the hybrid ice maker in Chihuahua, and some select water pumping systems to be used as baselines for long term study. The results of the monitoring program will be compiled into a "best practices" report, that will be available to U.S. industry for guidance on the implementation of photovoltaic projects in rural international applications worldwide.

## CONCLUSIONS

The Sandia Mexico Renewable Energy Program has made great progress in the establishment of a new, sustainable market for renewable energy technologies, especially photovoltaics, in Mexico. At this point, FIRCO directors, engineers, and technicians are playing central roles in determining the direction of the program and spreading technical capabilities throughout the country. As the Sandia team focuses its efforts on new geographic areas, project implementation continues in the states where program activities are well established, such as Chihuahua and Sonora.

The level of involvement of the U.S. industry in the Sandia program continues to grow as well. Through the implementation of water pumping projects, more local vendors are turning to U.S. suppliers for technical guidance. Also, the technology partnerships between Sandia and members of industry will result in integrated systems that will have applications not only throughout Mexico, which represents a large and accessible market, but across rest of the developing world as well.

# The Ramakrishna Mission PV Project — a Cooperation between India and the United States

Jack L. Stone and Harin S. Ullal

National Renewable Energy Laboratory
1617 Cole Boulevard
Golden, Colorado 80401 USA

**Abstract.** The Ministry of Non-Conventional Energy Sources and the National Renewable Energy Laboratory, a U. S. Department of Energy laboratory, agreed to cooperate in photovoltaic applications. A number of small-scale applications were identified for which both sides would cost share 50-50. This paper describes the sustainable village electrification project carried out in West Bengal in cooperation with the Ramakrishna Mission, the West Bengal Renewable Energy Development Agency, and Exide Industries Ltd. A number of cost-effective applications have been identified and are in the process of being installed in the Sundarbans region near Calcutta.

## BACKGROUND

Krishna Kumar, the former Minister of State for the Ministry of Non-Conventional Energy Sources (MNES), first visited the National Renewable Energy Laboratory (NREL) in 1993. Discussions there led to an agreement that MNES would collaborate with the U.S. Department of Energy (DOE), with NREL as the implementing agency for DOE in the general area of photovoltaics (PV). The details of the collaboration were finalized later in the year when Minister Kumar visited DOE's Secretary Hazel O'Leary in Washington. Three projects were chosen: (1) a small demonstration of PV village power applications; (2) integrated PV for low-cost housing; and (3) a hybrid system that employs PV-wind or PV-biomass. The two parties agreed to cost share the project equally, up to a maximum of $1.5M for each country's contribution. A Memorandum of Understanding (MOU) was also signed between NREL and MNES's Solar Energy Centre. The MOU provided for exchange of personnel, development of standard testing protocols for PV cells, modules, and systems, education and training activities, and other activities which would be mutually agreed to. About the time the project details were being developed, the DOE budget for international programs was severely curtailed. The agencies agreed to reduce the project to one activity at a $200K level for each party. The project chosen was a sustainable village power demonstration using non-grid connected PV. Discussions with MNES's financial arm, the Indian Renewable Energy Development

Agency (IREDA), led NREL to consider the Ramakrishna Mission as the nongovernment organization (NGO) partner. The Ramakrishna Mission (the "Mission") is a well-respected humanitarian organization in the Calcutta area that specializes in slum uplift. The attractive feature of working with the Mission was the established infrastructure, which includes training, education, and banking expertise. The Sundarbans area where a large number of isolated villages with no access to grid electricity, was chosen. The area is shown on the map in Figure 1.

**FIGURE 1.** The Sundarbans Region of West Bengal

Once the Mission was enlisted as the NGO partner, NREL released a public procurement to select the systems integrator for the project. Applied Power Corporation of Lacey, Washington, won the bid. Its subcontractor, Remote Power

International of Ft. Collins, Colorado, furnished the training component. MNES signed a contract with Exide Industries Ltd. in Calcutta, India, to furnish technical coordination on the ground. The West Bengal Renewable Energy Development Agency (WBREDA) of Calcutta was asked, as the state nodal agency, to participate in the project as well. NREL then contracted with the Tata Energy Research Institute (TERI) of New Delhi, India to perform a before-and-after study on the social and economic effects of the PV systems in the Sundarbans villages.

## PROJECT RESPONSIBILITIES

The project responsibilities of the various organizations involved are depicted in Figure 2. The agreement calls for 50-50 cost sharing: the United States provides the PV modules, charge controllers, a water pump, and the training; India provides the batteries, compact fluorescent lamps, lamp fixtures, a vaccine refrigerator, mounting structures, all balance of systems components, and solar lanterns, and pays all custom duties for the imported system components. The Ramakrishna Mission is responsible to identify the recipients of the various systems and participants in the NREL-furnished training sessions, provide follow-up training, maintenance, and replacement, and serve as the collector of revenues from the end users. NREL will also work with the Mission to identify potential private sector partners with whom proposals will be submitted to IREDA to move the project beyond the limited size possible from this initiative.

The following applications and involved villages were initially identified and agreed to. In the village of Gosaba (with 1000 families), the training center will be provided 10 lights for 4 hours of operation each night, two 30-watt wall sockets, a battery-charging station for 10 100 amp-hour batteries and 20 solar lanterns, and three stand-alone street lights with 11-watt compact fluorescent (CFL) lamps. The possibility of mounting the charger station on one of the Mission's boats will be investigated. This would allow the service to be transported throughout the island communities. The village of Katakhali, with 100 families, will be provided 100 domestic lighting units with one 11-watt CFL and one 30-watt socket per home. The youth club will have two 11-watt CFLs and one 30-watt wall socket. The village of Pakhirala will have its weaving center provided with three 11-watt CFLs, a community street light and eight 11-watt CFLs with two 30-watt wall sockets. These additions to the weaving center will extend the productivity hours by about four per night. The health clinic in Satyanaryanpur will receive a vaccine refrigerator and eight 11-watt CFLs with two 30-watt electrical sockets. A second battery-charging station for 10 100 amp-hour car batteries will be placed at the Chota Mollakhali youth center. The village of Kumirmari will have 100 home lighting systems installed. The village of Satjelia will be furnished 100 domestic home lighting systems with a 9-watt CFL and a 30-watt electrical socket. The location for the water pumping station will be

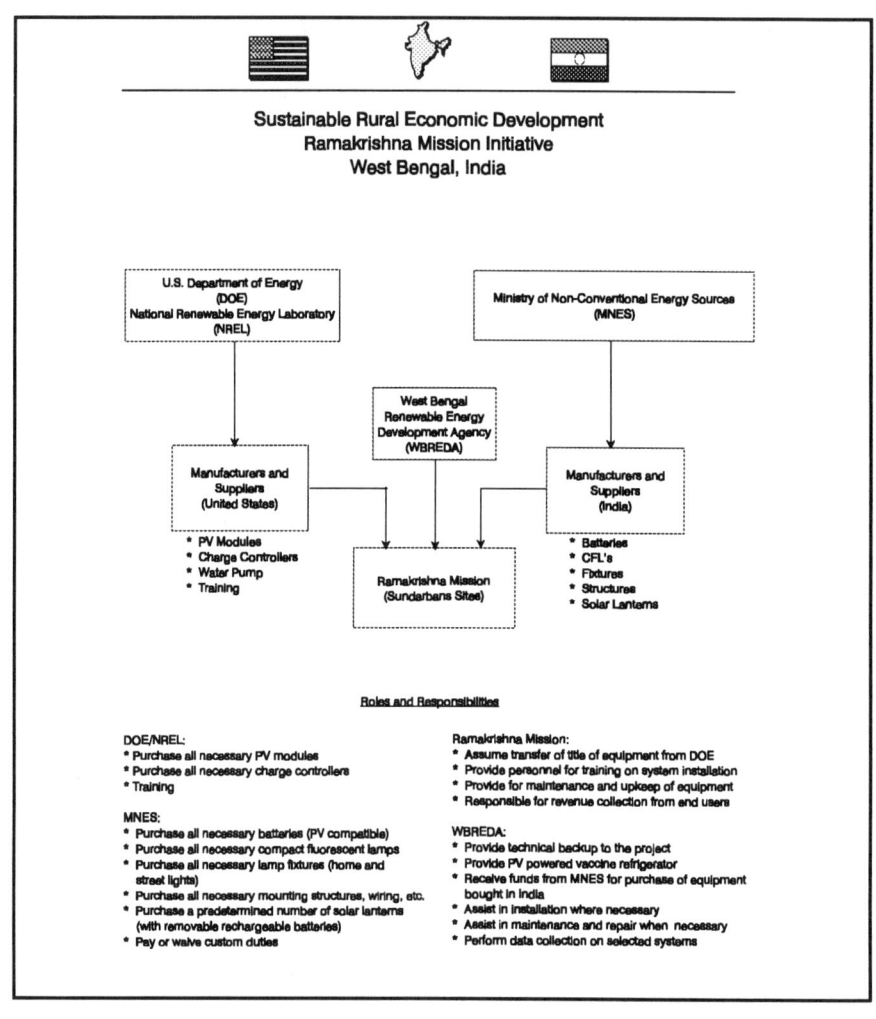

**FIGURE 2.** Roles and Responsibilities for Project Participants

determined as the installation time nears.

The two week training program will be available for 16 participants who have been chosen by the Mission for their background in basic electrical applications (including radio and television repair). The Mission has a very good reputation for providing high-quality training in a variety of areas. Remote Power International prepared a detailed training manual that will be left for the Mission to continue training sessions after the NREL-funded trainers leave the area. During the 2 weeks, the last week will be used to do hands-on installations in the island communities.

Remote Power International has prepared detailed schematics and installation procedures for all systems provided

## PROJECT FINANCING

Many renewable energy projects in India are heavily subsidized by the government. One concern of this project was that the beneficiaries of the PV systems are responsible to properly use and care for their systems. To guarantee this, NREL insists that the users invest their own money in the systems, according to their abilities to pay. This has the added advantage of not distorting the markets. Clearly subsidies in India will not sustain themselves far into the future. Therefore, the PV systems will have to be economical on their own merits.

In India the domestic unit of two lights plus one wall socket along with the necessary PV panel, battery, and accessories cost approximately Rs. 14,000 ( 1 $ ≈ Rs. 35 ). Out of this amount Rs. 6,000 is available as a government subsidy. Hence, the amount to be borne by the user is Rs. 8,000 per unit. The end user will be asked to provide a down payment of Rs. 3,500 at the time of installation. The rest of the amount (Rs. 4,500) will be treated as a low-interest loan to be repaid in monthly installments of Rs. 40 per month over 10 years. In this way Rs. 40 x 12 months x 10 years or Rs. 4,800 will be realized—Rs. 4,500 against the loan and Rs. 300 as interest. In addition to this an amount of Rs. 20 per month will be charged for each unit as maintenance charges for which the users will receive free service at their doorsteps. However, the costs for spares will be at the owners' expense. Thus the users will pay a total of Rs. 60 per month for 10 years. They may also opt to pay Rs. 100 per month (80+20) for 5 years. For a few beneficiaries of special category who are not in a position to make the Rs. 3,500 down payment, provision will be made to pay Rs. 500 only during installation and the rest of the amount will be treated as a loan to be repaid in 5 to 10 years. After the loan is liquidated, ownership will be transferred to the users. The amount recovered from the end users in the form of the down payments and loan interest will form the "Revolving Fund Capital" for the project, which in turn will be used to replicate the program to other villages and other beneficiaries of the same village.

## SOME SYSTEMS DETAILS

The minimal home lighting systems consists of one PV module (Solarex VLX-53, 17.2 V, 3.08 A, 53-W, polycrystalline silicon), one Morningstar SunSaver-6LVD charge controller, one 11-watt CFL, and one 75-amp-hour battery. The system is sized such that 4 hours of operation of the lights and auxiliary socket result in a 20% daily depth of discharge. The charge controller is set to cut out when a total depth of discharge of 80% is reached. This should allow a system autonomy of a little more than 5 days. The battery discharge characteristics should lead to battery life in excess of 5 years.

The street lighting systems are powered by two 50-watt Solarex PV modules that use a Trace C-12 controller. A 100 amp-hour battery is used to allow the 11-watt CFL to be used from sundown to sunup. The system is sized to provide 4 days of system autonomy. The same depth of discharge as the home lighting systems is maintained for these systems as well.

The battery charging stations are sized to provide charging for 10 completely discharged 100 amp-hour car batteries. Each battery to be charged has its own charge controller, Prostar-30 MLCD, with eight 50-watt PV modules. The total charging station requires 80, 50-watt Solarex PV modules. Given the solar resource in the region of operation, the charging is expected to take place during a 1-day period.

The water pump is a Grundfos, SP5A-7. Sixteen, 50-watt PV modules are required. A flotation collar is provided for the pump which allows the pump to be floated directly in the river. The pump is ac powered and requires a Grundfos SA-1500 inverter.

## SUMMARY

The Sustainable Rural Economic Development Ramakrishna Mission PV Initiative was conceived as a small-scale demonstration project that would show the economic viability of PV systems in the Sundarbans region of West Bengal. The viability was to be predicated on the systems being economical without substantial subsidy, and eventually without any subsidy at all. The operation and maintenance of the systems were to be the responsibility of the chosen NGO, the Ramakrishna Mission. Mission personnel were to identify beneficiaries of the PV systems, define a financing arrangement that would be acceptable and sustainable to the villagers of the region, and serve as a banker to collect revenues from the end users. The potential for expanding the project beyond the limited demonstration was also a prime consideration. With U.S. systems used for the installations, our industry would also have the advantage of satisfying any future PV system purchases. The cooperative nature of the project would also be expected to lead to improved relationships between our two countries and lead to further trade expansion. The project was also designed so as not to distort market forces, i.e., true costs. Further, without excessive subsidies, and with end-user money required for participation, the systems were expected to have the best of care. And most importantly, the benefits of electricity would be made available to those who in the past had little or no access. Improvements in educational opportunities, health care, productivity, and entrepreneurship would be standards for success of the project. Finally, the project should be self-sustaining. An infrastructure should remain that would support further applications including financing, education, training, repair, and maintenance. Successful PV deployment under these most difficult circumstances would pave the way to acceptance of the technologies as a way to fulfill the tremendous need for energy in the developing world.

## ACKNOWLEDGMENTS

The authors would like to acknowledge the assistance of the many who made the success of this project possible. Because you are too numerous to mention as well as to avoid running the risk of overlooking someone, we simply say thank you to everyone who had a part. You know who you are.

This work was supported under contract DE-AC36-83CH10093 with the U. S. Department of Energy.

# Sino/American Cooperation for Rural Electrification in China

## William L. Wallace and Y. Simon Tsuo
National Renewable Energy Laboratory, 1617 Cole Blvd., Golden, Colorado 80401, USA

**Abstract:** Rapid growth in economic development, coupled with the absence of an electric grid in large areas of the rural countryside, have created a need for new energy sources both in urban centers and rural areas in China. There is a very large need for new sources of energy for rural electrification in China as represented by 120 million people in remote regions who do not have access to an electric grid and by over 300 coastal islands in China that are unelectrified. In heavily populated regions in China where there is an electric grid, there are still severe shortages of electric power and limited access to the grid by village populations. In order to meet energy demands in rural China, renewable energy in the form of solar, wind, and biomass resources are being utilized as a cost effective alternative to grid extension and use of diesel and gasoline generators. An Energy Efficiency and Renewable Energy Protocol Agreement was signed by the U.S. Department of Energy with the Chinese State Science and Technology Commission in Beijing in February, 1995. Under this agreement, projects using photovoltaics for rural electrification are being conducted in Gansu Province in western China and Inner Mongolia in northern China, providing the basis for much wider deployment and use of photovoltaics for meeting the growing rural energy demands of China.

## BACKGROUND

China has an abundance of renewable-energy resources in the form of solar, wind, biomass, hydro, geothermal, and ocean tidal resources. China is also already one of the world's largest users of renewables, primarily in the form of hydropower and biomass energy, and the development of large wind farms for grid power and the use of solar and wind energy for rural energy development has also been given a high priority (1). The solar and wind resources of China are enormous. The potential of wind energy alone has been estimated at about 240 GW, which is approximately 10% of the estimated total wind resources in China.

Solar and wind resources are also strategically located in areas of greatest need in terms of rural energy development. More than 120 million rural people, mainly in northern and western China, and more than 300 coastal islands currently have no access to the electric power grid and no near-term prospects for grid connection. There is an excellent match of solar and wind resources to meet these rural electrification needs. For example, the richest solar energy resources in China are located in Inner Mongolia, the Qinghai-Tibet Plateau region, Ningxia, and Gansu. These are regions where the population density is low, and it is often too costly or impractical for grid extension to reach many of the potential users. There are also good solar resources in the coastal region of China. The regions of China which have good solar resources also tend to overlap with regions of high wind availability.

In February, 1995, the U.S. Department of Energy (DOE) signed the Energy Efficiency and Renewable Energy Protocol Agreement with the State Science and Technology Commission in Beijing. This Protocol established a broad umbrella for Sino/American cooperation to develop renewable energy technologies and markets in China. Under this Protocol, an annex agreement was signed with the Chinese Ministry of Agriculture (MOA) in June, 1995 establishing joint U.S./China cooperation for rural energy development. The cooperation with the MOA focuses on the use of photovoltaic and wind technologies for remote rural household and village electrification, and the development of village-scale biogasification systems for electric generation and thermal applications.

## STATUS OF PV DEVELOPMENT IN CHINA

The development of terrestrial applications of PV in China was initiated more than 20 years ago, in 1974, with the introduction of small systems for remote applications. Commercial production of terrestrial solar cells began in 1976. The current installed capacity of PV systems in China is small, but is growing rapidly. In 1993, the installed capacity of PV systems was about 3.8 $MW_p$; in 1994, the installed capacity was about 5.1 $MW_p$; and in 1995, the installed capacity was about 6.6 $MW_p$. Some 65% of this capacity is power for telecommunications applications (2). About 1.1 $MW_p$ of the installed PV generating capacity, about 16%, is installed in remote household and village power applications, for which opportunities exist throughout China. The remainder is installed in remote agricultural and industrial applications.

As of 1995, there were more than 32,000 rural household systems installed in China. There are 10,000 household PV systems installed in Qinghai alone, with the remainder distributed in Inner Mongolia, Tibet, Xinjiang, and Gansu. Household systems generally range from 20 to 80 W and are used for lighting and small consumer electronics. Larger hybrid systems are being developed in Inner Mongolia in the 400 Watt to 500 Watt range, that consist of a 300 Watt wind energy generator combined with PV capacity. Such systems will support additional loads, such as a small freezer and washing machine. The potential market is very large for increased use of PV solar home systems in China's northwest and western provinces and autonomous regions, where a minimum of 2 million unelectrified households have been identified as a near term remote market by local agencies (3).

There are six stand-alone PV power stations in China in the range of 7 kW to 25 kW, five of which are in Tibet and one in Gansu. A 30 kW, stand-alone power station is under construction in Tibet. China also has experience with wind/PV hybrid systems in the range of 200 W to 35 kW. In Inner Mongolia there are also at least twelve village power hybrid systems based on wind and/or PV systems containing battery storage, and some containing back-up diesel generators. There is no grid-connected experience with PV in China to date. However, the quality of grid-connected electricity is a pervasive problem, and the use of PV for grid-support, uninterruptible power supplies, and peak-shaving applications in the potential urban market is of great interest.

The total manufacturing capacity for PV modules in China is about 5.5 $MW_p$ in six imported production lines. Most PV module production in China is based on single-crystal silicon technology, with some amorphous silicon production. However, most manufacturing facilities are not operating at full capacity because of a combination of the following: i) outdated equipment since all of China's PV cell and module production lines were imported before 1991, ii) high manufacturing costs due to lack of automation and small-scale production, and iii) a shortage of silicon wafers. The current PV market is also limited, but is growing at about 30% per year. Presently, only six organizations in China have an annual production level of PV modules of more than 200 $kW_p$, including, the Qinhuangdao Huamei Photovoltaics Electronics Corporation Ltd., Yunnan Semiconductor Device Factory, Kaifeng Solar Cell Factory, Ningbo Solar Power Supply Factory, Shenzhen-Y.K. Solar Energy Company Ltd., and Harbin-Chronar Solar Energy Electricity Corporation.

Chinese-made modules have a significant price advantage over American-made modules when the module size is less than 50 $W_p$ because of lower labor costs, use of small diameter 3 inch ingot silicon crystal growth, and a combined 30% import tariff and value-added tax for imported modules. When the module size is 50 $W_p$ or greater, U.S. manufactured modules have a price advantage. Presently, Chinese module production cannot keep up with demand, and the average sales price has steadily increased during the last 2 years.

## RURAL ELECTRIFICATION IN WESTERN CHINA

High cost, lack of a marketing and distribution infrastructure, and variable quality of modules and balance-of-system components are barriers to the widespread deployment of photovoltaics in China. Several cooperative projects are being conducted in China to address these problems. Under the Energy Efficiency and Renewable Energy Protocol agreement, NREL is working with the MOA and the State Council Office for Poverty Alleviation and Rural Development in Beijing to develop a cost-shared program to provide household PV electricity systems to rural families in western China. This solar home system project is being conducted with the Solar Electric Light Fund (SELF), a non profit organization in Washington D.C., and the Gansu Solar Electric Light Fund (GSELF) a non profit organization in Lanzhou in the province of Gansu in western China.

The project in Gansu is designed to expand and strengthen the distribution and post-sales support infrastructure previously established in Gansu Province to promote the commercialization of PV for remote solar home systems. This infrastructure involves a partnership with several organizations, including: i) the rural energy office network supported by the MOA throughout China at the county and township level, ii) provincial government agencies associated with the poverty alleviation program in China, and iii) local PV system integrators operating in the province of Gansu. Rural energy offices exist at the township, district, and county levels and offices are found in 1,800 of the 2,300 total counties in China. The rural energy office network can help facilitate rural electrification projects throughout China, providing a widespread infrastructure for technology deployment. The use of revolving funds for credit and cash sales for financing the purchase of household systems to expand the market for PV is a critical component of the project.

The total value of Gansu solar home system project is $440,000, cost shared 50/50 by the U.S. DOE and Chinese partners. The Chinese partners in the project include the Gansu Provincial Poverty Alleviation and Rural Development Office (54%), the Gansu Planning Commission (18%), the Gansu State and Economic Trade Commission (18%), and GSELF (10%). The project is managed by the Solar Electric Light Fund in the United States and by the Gansu Solar Electric Light Fund in China. GSELF was established in 1993 for the specific purpose of promoting solar home systems in western China. With previous funding from the United Nations Development Program and the Rockefeller Foundation in the United States, SELF and GSELF had installed over 400 solar home systems in Gansu prior to the DOE project.

The current DOE project is based primarily on cash sales, with a 80% overall cost recovery for the project based on an average selling price for local vendors of solar home systems. A revolving-fund account has been set up at the Lanzhou Branch of the China Construction Bank for purchasing additional systems. During the project cycle of 18 months (April 30, 1996 to October 31, 1997), at least 600 solar home systems will be installed. These systems are based on nominal 20 Watt solar home lighting systems that include a 20 Watt PV module, a 12 volt/38 amp-hour battery, a charge controller, and two 8 Watt fluorescent lights. Several 50 Watt school systems are also being installed as part of a renewable energy education program in Gansu.

PV panels and sealed lead-acid batteries are being purchased from the United States and other balance-of-system components (including charge controllers, compact fluorescent lights, and wiring) are being provided by three local system integrators: the Gansu PV Company, the Gansu Zi Neng Automation Engineering Company, and the Zhong Xing Electronic Instruments Factory. Some complete systems are also being supplied by the United States. For the project, a set of test procedures have been established for certification of system components. The training of rural technicians and management staff of local service networks is included in the project.

## PV CASE STUDIES IN INNER MONGOLIA

In collaboration with the Chinese Academy of Sciences (CAS) in Beijing and the Center for Energy and Environmental Policy (CEEP) at the University of Delaware, NREL is also working with several agencies of the Inner Mongolian government to develop PV/wind hybrid projects in Inner Mongolia. The government of Inner Mongolia is committed to a village electrification program over the next five years that will, in the near term, electrify 38 villages and township centers by the end of 1997 using renewable energy hybrid village power systems. Inner Mongolia also has a well developed distribution infrastructure at the district, county, and township level consisting of new energy service stations that deploy renewable energy systems for remote households. To date, over 118,000 small wind generators and 3,800 PV systems have been installed for remote household applications using this network. By the end of 2000, Inner Mongolia has a goal of installing approximately another 80,000 household systems using a combination of wind, PV, and PV/wind hybrid systems.

In 1995 and 1996, the CAS, CEEP, and NREL performed a rural electrification options analysis for household rural electrification examining renewable energy and conventional fossil energy (based on diesel and gas gen-sets) case studies (4). Detailed case study data was collected from four counties in central and northern Inner Mongolia, including wind and solar resource data and performance/load data from 10 PV systems, 22 wind systems, 6 PV/wind hybrid systems, and 3 wind/gasoline gen-set systems which were in the 22 Watt to 600 Watt size range. Four sizes of gasoline gen-sets in the size range of 450 Watts to 1 kW were examined for comparison. Data was also collected for a wind/diesel village power system. All systems include battery storage.

Levelized cost analyses have been performed using the data collected for the household systems. Analyses indicated that for stand-alone electrical generation, wind generators are the least cost option for household electricity at $0.21 to $0.38 per kWh for the four counties. Small wind generators in the 100 Watt, 200 Watt, and 300 Watt size range are manufactured locally in Inner Mongolia for the household market. Small PV/wind hybrid systems were in the range of $0.30 to $0.55 per kWh. Small PV systems alone were in the range of $0.68 to $0.71 per kWh. Levelized costs of the four gasoline gen-sets for household use were in the range of $0.70 to $1.10 per kWh as a function of how the gen-sets were utilized.

While wind energy tends to be the lowest cost option for rural applications in the case studies examined, the seasonal variability of wind and solar resources makes hybrid PV/wind systems with battery storage for household and village power systems an attractive option in areas where there is a seasonal complementarity between solar and wind resources. This complementarity of resources exists in Inner Mongolia and may be important in other provinces of China as well. Current cooperation with the Inner Mongolia government involves consultation on the design of hybrid systems optimized for local wind and solar resource compositions at the county level.

## ACKNOWLEDGEMENT

This work was supported by the U.S. Department of Energy under contract number DE-AC36-83CH10093 to the National Renewable Energy

Laboratory.

## REFERENCES

(1) Q. Zheng, *Solar Energy in China*, L. Yan and L. Kong, eds., Beijing, China, April, 1995, pp. 38-52.

(2) Wang Sicheng, *Solar Energy in China*, L. Yan and L. Kong, eds., Beijing, China, April, 1995, pp. 105-111, updated with private communications.

(3) A. Cabraal, et al., "China Renewable Energy for Electric Power," September, 1996, World Bank Report No. 15592-CHA.

(4) J. Byrne, et. al., "Levelized Cost Analyses of Small-Scale, Off-Grid Photovoltaic, Wind, and PV-Wind Hybrid Systems for Inner Mongolia, China," March 1996, Center for Energy and Environmental Policy, University of Delaware, Newark, Delaware.

# POSTERS
# THIN FILMS (P)

# Potential Fluctuations in Intrinsic Hydrogenated Amorphous Silicon

S. Dong[*], J. Liebe[*], Y. Tang[*], R. Braunstein[*] and B. von Roedern[#]

[*]Department of Physics, University of California, Los Angeles, CA 90024, USA
[#]National Renewable Energy Laboratory, Golden, CO 80401-3393, USA

**Abstract.** The dangling bonds in hydrogenated amorphous silicon (a-Si:H) that can be thermally generated or light-induced result in charged defects. Inhomogeneities of a random distribution of charged defects may lead to the formation of long-range potential fluctuations which would influence the charge transport in amorphous semiconductors and limit solar cell performance. By employing the technique of photoconductive frequency mixing, the drift mobility ($\mu_d$) and photomixing lifetime ($\tau$) were determined separately. Strong evidence for the existence of long-range potential fluctuations in intrinsic a-Si:H has been found from the measurements of drift mobility in light induced degradation, as a function of applied electric field and as a function of illumination light intensity.

## INTRODUCTION

In order to further our understanding of the Staebler-Wronski degradation in a-Si:H, charge transport measurements were performed using the photomixing technique[1-4] to determine the drift mobility, $\mu_d$, and photomixing lifetime, $\tau$; the technique is based on the idea of heterodyne detection for photoconductors. When two similarly polarized monochromatic optical beams of slightly different frequencies are incident upon a photoconductor, the generation rate of electron-hole pairs and therefore the photocurrent produced, when a dc bias is applied, will contain components resulting from the square of the sum of the incident electrical fields. Consequently, a photocurrent composed of a dc and a microwave current due to the beat frequency of the incident fields will be produced; these two photocurrents allow a separate determination of the drift mobility and photomixing lifetime of the photo-generated carriers. The drift mobility, $\mu_d$, is an experimentally measurable quantity with no need for any detailed knowledge of trapping and thermal emission, i.e.:

$$\mu_d = \omega\sqrt{\langle\sigma_{ac}^2\rangle}/(\sqrt{2}eG_0\lambda), \qquad (1)$$

where $\omega$ is the photomixing angular frequency. $G_0$ is the dc electron-hole pair generation rate and $\sqrt{\langle\sigma_{ac}^2\rangle}$ is the root-mean-square ac photoconductivity which is determined from the power of the microwave photomixing signal. In the present work the longitudinal modes of a He-Ne laser were employed to yield a frequency difference, the beat frequency signal at 252 MHz was used ($\omega \sim 1.58$ GHz, corresponding to a time scale of ~ 630 ps), $\lambda$ (~ 5.05%) is an effective modulation index. The photomixing lifetime $\tau$ corresponding to $\mu_d$ can be written as

$$\tau = \frac{\mu_0\tau_r}{\mu_d} = \frac{\sigma_{dc}}{eG_0\mu_d}, \qquad (2)$$

where $\mu_0$ is the extended state mobility, $\tau_r$ is the recombination lifetime and $\sigma_{dc}$ is the dc photoconductivity. The photomixing lifetime and the drift mobility determined in the present work are both for electrons, as they are the dominant photoconductive charge carriers in intrinsic a-Si:H films. The photomixing technique has been successfully applied to single crystalline,[1] polycrystalline[2] and amorphous semiconductors[3-4].

The dangling bonds in a-Si:H result in charged defects that can be thermally generated or light-induced. Those charged defects may lead to a formation of long-rang potential fluctuations[5-8] which then would influence the charge transport in amorphous semiconductors. The present paper is focused on the behavior of the drift mobility during the light-soaked process and its dependence on the electric field and the illumination light intensity. These measurements yield evidence for the existence of long-range potential fluctuations (LRPF) in intrinsic a-Si:H materials. It was found that the decay of photoconductivity is not only due to a continuous decay of photomixing lifetime under light soaking but also due to a continuous decay of drift mobility. This decay of the drift mobility can be explained by the existence of LRPF. An increase of the applied electric field and/or the illumination light intensity reduces the effect of LRPF and subsequently increases the drift mobility, this results from lowering of the potential barrier by the external electric field and enhanced screening of the potential by the light generated free carriers.

## EXPERIMENTAL SETUP

The dc photo-signal was measured by a Keithley 617 Programmable Digital Electrometer, and the photomixing signal, i.e., the ac photo-signal, was measured by a Tektronix 492P Spectrum Analyzer. A Spectra-Physics 125A He-Ne laser was used as a light source for photomixing as well as for light soaking. A stepped neutral density filter was applied for intensity changes. All equipment was controlled by an IBM PC through a National Instruments Lab-PC card.

## RESULTS AND DISCUSSION

The samples used in this study were produced by the hot-wire assisted chemical vapor deposition technique at NREL. The films were about 2 μm thick deposited on 7059 Corning glass and co-planar electrodes of chromium were evaporated on top of these films with a separation of 0.4 mm. The results of one of these samples is reported as an example in this paper. This sample (THD58) was produced at a substrate temperature of 350° C yielding hydrogen content of 2-3%. All measurements were performed at room temperature. Prior to the measurements, the samples were annealed for 1 hours at 150° C.

The light soaking were performed *in situ* employing a He-Ne laser with an intensity of 7 suns for 4 hours. Figure 1 shows the decay of $\mu_d$, $\tau$ and $\sigma_{dc}$ due to light soaking.

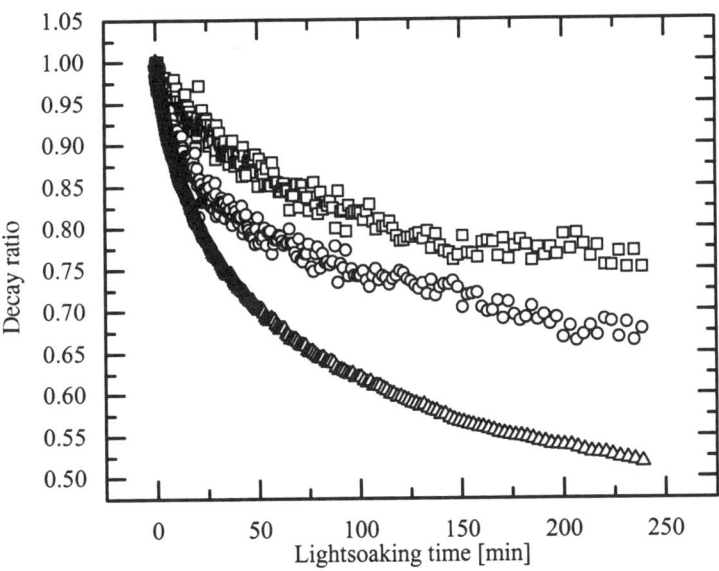

**FIGURE 1**: Decay ratio of $\tau$ (squares), $\mu_d$ (circles), $\sigma_{dc}$ (triangles) vs. light-soaking time.

For intrinsic a-Si:H, the drift mobility is determined by trapping of electrons into the conduction band tail and by scattering of electrons due to the intrinsic disorder. Both enhanced trapping and scattering can result in a decay of the drift mobility. Although we believe the enhanced scattering has to be the dominant mechanism for the light induced decay of the drift mobility,[3] our studies so far do not reject any of the existing microscopic models[9-12] for the Staebler-Wronski effect, such as weak bond breaking [9-10] and charge trapping [11] models. Rather our studies indicate that combinations of different models may be necessary to explain the generation of defects with different characteristics. Upon light soaking, in addition to the generation of defects, the defects that serve as deep trapping or recombination centers can be charged, since electrons and holes are trapped to them. This results in enhanced scattering and thus the decay of the drift mobility for electrons. The charged defects may become quasi-stable through certain relaxation processes and can also form long-range potential fluctuations, if they are not spatially correlated.

The electric field dependence of $\mu_d$ and $\tau$ were measured in the annealed state and the light-soaked state. Figure 2 shows the electric field dependence of the drift mobility.

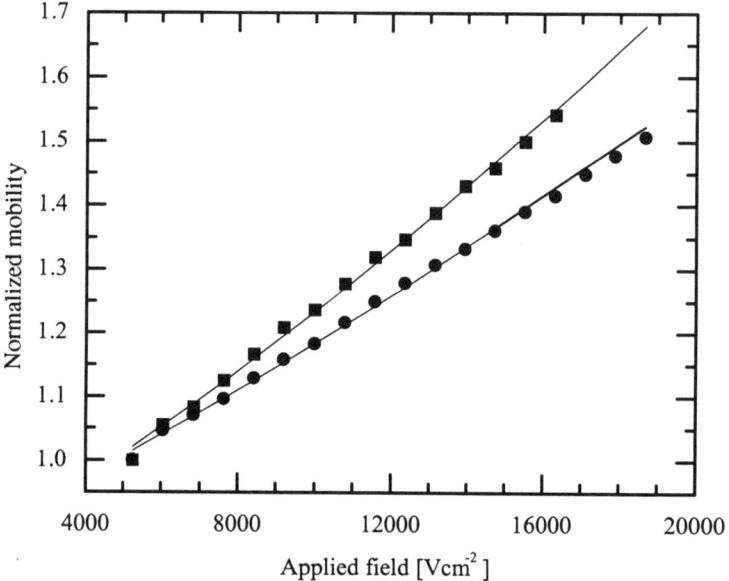

**FIGURE 2:** Drift mobility vs. applied electric field; annealed state (squares), light-soaked state (circles).

The squares and circles are the experimental points for the annealed and light-soaked states respectively. The solid curves were obtained by a curve fitting procedure to a model of transport through potential barriers[4]. For continuity of presentation, we present the salient features of the model in the following:

The light-induced defects as well as the native defects, which serve as recombination centers and trapping centers, can be charged and can form potential barriers or fluctuations. In the transport process, the mobile carriers can either go over the potential barrier through thermal activation, go around the potential barrier through scattering or tunnel through the potential barrier. If the former dominates the latter, the drift mobility should be proportional to the probability of a carrier going over the potential barrier ($V_p$) through thermal activation, that means:

$$\mu_d/\mu_d^0 = \int_{V_p}^{\infty} N(E)\,dE \Big/ \int_0^{\infty} N(E)\,dE \quad , \tag{3}$$

Where $\mu_d^0$ is the drift mobility without a potential barrier. The role of the external field (F) is to change the density of states N(E) inside the well, thus to change the mobility. N(E) is assumed to be proportional to the spacial range (L) at energy E inside the well (determined by the geometry of the well)

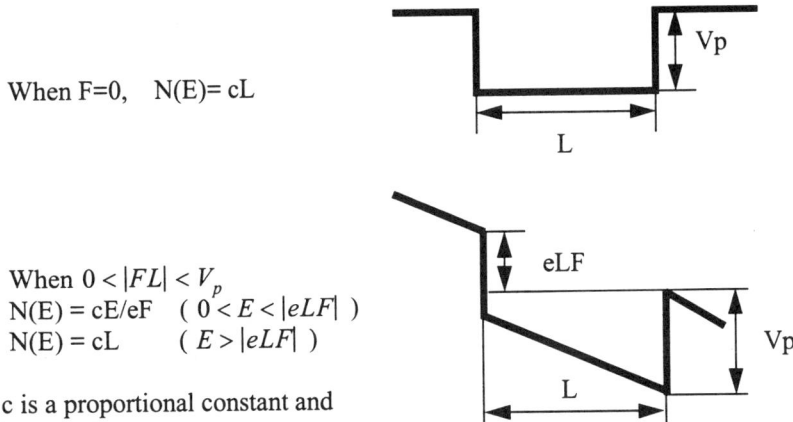

When F=0,   N(E)= cL

When $0 < |FL| < V_p$
N(E) = cE/eF   ( $0 < E < |eLF|$ )
N(E) = cL      ( $E > |eLF|$ )

where c is a proportional constant and e the electron charge.

Through simple statistical calculations one can obtain the following electric field (F) dependence of the drift mobility $\mu_d(F)$[4]:

$$\mu_d(F) = \mu_d^0 \exp\left(-\frac{eV_p}{kT}\right) \frac{eLF}{kT\left[1-\exp\left(-\frac{eLF}{kT}\right)\right]} \qquad (if\ |LF| \leq V_p) , \tag{4}$$

In our field dependence measurements, the product of eLF was always smaller than 0.04 eV. According to the curve fitting routine (see Figure 2) the range of the potential fluctuations in the annealed state and in the light-soaked state are $L_{ann}$ =23.9 nm and $L_{ls}$ =18.7 nm respectively. These results are expected assuming that the density of charged defects is determined by: $n_D \propto V_p^2/L$ [13]. Therefore due to the increase of $n_D$ during the light soaking process the magnitude ($V_p$) of the potential fluctuations has a tendency to increase, whereas the range (L) of the potential fluctuations has a tendency to decrease.

It was found that the $\mu_d$ increases while the $\tau$ decreases with an increasing electric field, where the $\sigma_{dc}$ is essentially independent of the electric field in the range from 1000 Vcm$^{-1}$ to 15,000 Vcm$^{-1}$. The fact that the lifetime decreases while the drift mobility increases indicates the existence of a diffusion limited transport and recombination[14]. The increase in $\mu_d$ is compensated by the corresponding decrease in $\tau$, which can result in a field independent $\mu_d\tau$ product or the commonly observed ohmic behavior of the photocurrent. Due to the presence of long-range potential fluctuations, an obvious consequence is that the drift mobility should increase with an applied field, because the external field offsets the internal field and reduces the magnitude of the potential barrier. Such increase in $\mu_d$ does not necessarily lead to an increase in $\sigma_{dc}$, since one commonly observes a corresponding decrease in $\tau$.

On the other hand, a variation of the light intensity (I) changes the generation rate and therefore the concentration of free carriers created during the photomixing measurement. The varied free carrier concentration can contribute to the screening of the potential, which leads to a change in the magnitude and the range of the potential fluctuations. The output of He-Ne laser was varied using a stepped neutral density filter resulting in light intensities ranging from 1000 mWcm$^{-2}$ to 10,000 mWcm$^{-2}$.

The photoconductivity both in the annealed state and in the light-soaked state increases with an increasing intensity, as can be seen in Figure 3.

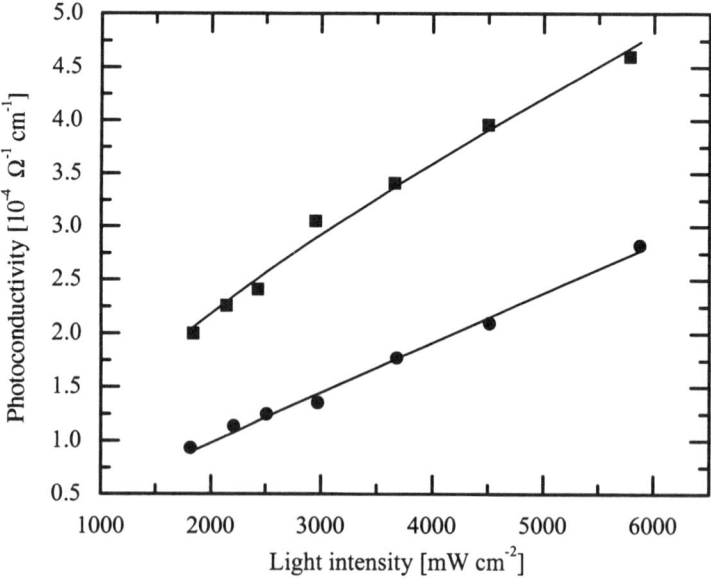

**FIGURE 3:** Photoconductivity vs. light intensity; annealed state (squares), light-soaked state (circles).

The solid lines are the results of a curve fitting procedure according to the power law dependency ($\sigma_{dc} \propto I^\gamma$) which corresponds to the results of other publications[15,16]. The exponent $\gamma$ in the annealed state and the light soaked state are $\gamma_{ann}$= 0.78 and $\gamma_{ls}$= 0.94 respectively.

Figure 4 shows an approximately linear relationship between log($\mu_d\tau$) and log(I).

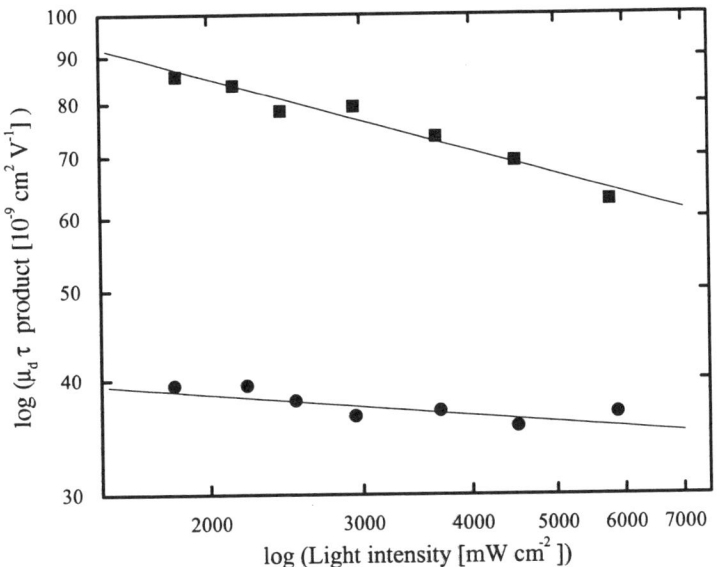

**FIGURE 4:** The $\mu_d\tau$ product vs. light intensity; annealed state (squares), light-soaked state (circles).

The individual intensity dependence of $\mu_d$ and $\tau$ are shown in Figure 5. $\mu_d$ increases and $\tau$ decreases with the increase of light intensity. In order to understand the phenomena, $\mu_d$ was measured as a function of electric field for different light intensities.

**TABLE 1.** Ranges of potential fluctuations for different light intensities

| intensity [mWcm$^{-2}$] | 1820 | 2514 | 2974 | 3686 |
|---|---|---|---|---|
| range [nm] | 11.7 | 13.3 | 15.9 | 18.6 |

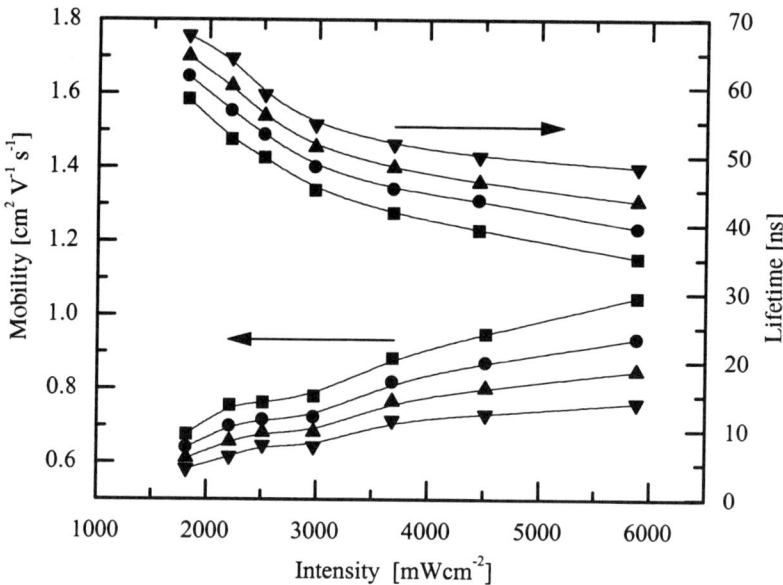

**FIGURE 5:** Drift mobility and photomixing lifetime vs. light intensity in the light-soaked state for various applied electic fields; 1820 mWcm$^{-2}$ (down triangle), 2514 mWcm$^{-2}$ (up triangles), 2974 mWcm$^{-2}$ (circles), 3686 mWcm$^{-2}$ (squares).

The experimental results in the light soaked state is shown in Figure 6. The ranges of the potential fluctuations for the different intensities obtained from the curve fitting routine (solid line in Figure 6) are listed in table 1.

As expected, when a higher intensity laser beam was used as photomixing light source, more carriers were generated and contributed to the screening. Therefore the magnitude of the potential fluctuations should decrease and the range should increase. Consequently, the drift mobility increases and photomixing lifetime decreases, which is shown in Figure 5. In the annealed state, we obtained a similar results. Further investigations on the intensity and electric field dependence of $\mu_d$ and $\tau$ will be reported at a later date.

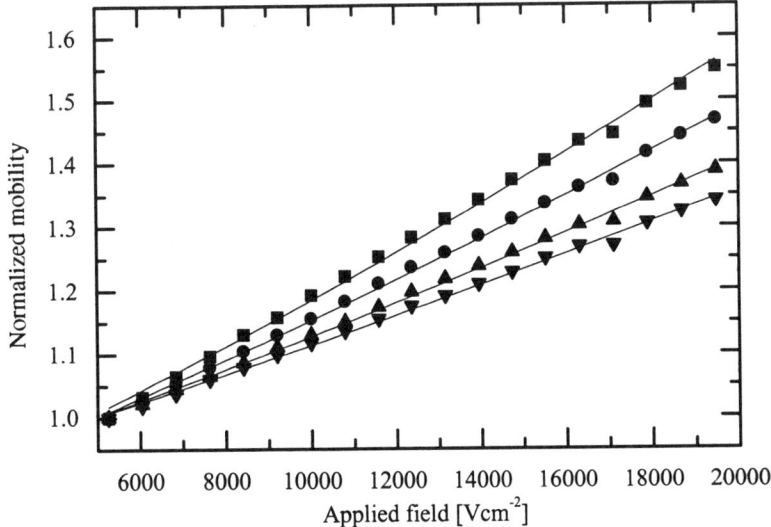

**FIGURE 6:** Drift mobility vs. applied electric field for different light intensities; 1820 mWcm$^{-2}$ (down triangle), 2514 mWcm$^{-2}$ (up triangles), 2974 mWcm$^{-2}$ (circles), 3686 mWcm$^{-2}$ (squares).

## CONCLUSIONS

The observed decay of the drift mobility indicates that in the light soaking process charged defects are induced. Photomixing measurements of the drift mobility and the photomixing lifetime as a function of electric field and light intensity provide evidence for the existence of long-range potential fluctuations due to charged defects. The results are consistent with a statistical model of transport over potential barriers which enable a determination of range of these potential fluctuations and the effects of screening by light generated mobile charge carriers. The present work should lead to a further understanding of the effect of the Staebler-Wronski instability on the charge transport parameters in a-Si:H.

## ACKNOWLEDGEMENT

We thank A.H. Mahan, R.S. Crandall and B.P. Nelson of the National Renwable Energy Laboratory for supplying the samples. J. Liebe gratefully acknowldges the support by the Deutscher Akademischer Austauschdienst. This work was supported by the National Renwable Energy Laboratory subcontract XAN-4-13318-10.

## REFERENCES

1. E. R. Giessinger, R. Braunstein, S. Dong, and B. G. Martin, *J. Appl. Phys.* **69**, 1469 (1991).
2. Yi Tang, R. Braunstein, and B. von Roedern, *Appl. Phys. Lett.* **63** (17), 2393 (1992).
3. Yi Tang, R. Braunstein, *J. Appl. Phys.* **79**, 850 (1996).
4. Yi Tang, S. Dong, R. Braunstein and B. von Roedern, *Appl. Phys. Lett.* **68**, 640 (1996).
5. H. Fritzsche, *J. Non-Cryst. Solids* **6**, 49 (1971).
6. H. M. Branz and M. Silver, *Phys. Rev. B* **42**, 7420 (1990).
7. B. von Roedern and A. Madan, *Phil. Mag. B* **63**, 293 (1991)
8. T. Overhof, *Mat. Res. Soc. Symp. Proc.* **258**, 681 (1992).
9. M. Stutzmann, *Phil. Mag. B* **60**, 531 (1989).
10. W. B. Jackson, *Phys. Rev. B* **41**, 10257 (1990).
11. D. Adler, *Solar Cells* **9**, 133 (1982).
12. R. H. Bube, L. Benatar, and D. Redfield, *J. Appl. Phys.* **75**, 1571 (1994)
13. S. D. Baranovskii and M. Silver, *Phil. Mag. Lett.* **61**, 77 (1990).
14. M. Silver and R. C. Jarnagin, *Mol. Cryst.* **3**, 461 (1968).
15. Y. Almerioh, J. Bullot, P. Cordier, M Gauther and G. Mawawa, *Phil. Mag. B* **63**, 1015 (1991)
16. G.J. Andriassens, S.D. Baranovskii, W. Fuhs, J. Jansen, O. Oktu, *J. Non-Cryst. Solids* **198**, 271 (1996)

# Transparent conducting oxide contacts for n-i-p and p-i-n amorphous silicon solar cells

Steven S. Hegedus, Wayne A. Buchanan, Erten Eser,
James E. Phillips, and William N. Shafarman

*University Center of Excellence
For Photovoltaic Research and Education*

*Institute of Energy Conversion
University of Delaware
Newark, DE 19716 USA*

**Abstract**: We investigate the effect of sputtered transparent conducting oxide (TCO) contacts on the device performance of ss/n-i-p/TCO and glass/SnO$_2$/p-i-n/TCO/Ag solar cells. TCO materials ITO and ZnO are compared, and found to have very similar transparency at the same sheet resistance. Sputtering ZnO with O$_2$ in the Ar reduces FF for ss/n-i-p/ZnO devices, compared to sputtering without O$_2$. This is attributed to an interface not bulk effect. Sputtering ITO with O$_2$ on the same devices increases J$_{sc}$ due to higher ITO transparency, compared to sputtering without O$_2$, but has no effect on FF. Based on curvature in the J(V) curve around V$_{oc}$, the ZnO/p layer contact appears to be non-ohmic. For p-i-n/TCO/Ag devices, µc-Si n-layers have much higher V$_{oc}$, J$_{sc}$, and FF for all variations of TCO/Ag back reflectors compared to an a-Si n-layer. Devices with ITO/Ag have lower V$_{oc}$ and J$_{sc}$ compared to devices with ZnO/Ag. Sputtering ZnO with O$_2$ has no detrimental effect on devices with µc-Si n-layers but severely reduces FF in devices with a-Si n-layers.

## INTRODUCTION

Transparent conducting oxide (TCO) materials are critical in fabricating thin film solar cells. They provide low sheet resistance contacts for lateral current flow while maintaining high transparency. TCO materials are used to provide a window layer for the incident illumination at the front of a device. In this configuration, they may also serve as an antireflection coating since $n_{air} < n_{TCO} < n_{sc}$ where the index of refraction of the TCO is between that of air (or glass) and that of the semiconductor (1). They are also used in a-Si solar cells at the back contact as a dielectric buffer layer between the a-Si or µc-Si doped contact and the metal contact (2-5). Additionally, TCO's can be textured to provide scattering which increases the optical absorption of weakly absorbed light.

The two most commonly used TCO materials are indium tin oxide (ITO), and zinc oxide (ZnO). Both are degenerate n-type semiconductors which can be optimized to have bandgaps greater than 3.3 eV, with absorption less than 10% over the visible spectrum and resistivities less than $10^{-3}$ $\Omega$-cm (6,7). The most common deposition techniques for ITO and ZnO for solar cell applications are chemical vapor deposition, evaporation and sputtering. In this paper, the solar cell performance of single junction a-Si devices having sputtered ITO and ZnO contacts are compared. The effect of $O_2$ in the sputter discharge is investigated. Both ITO and ZnO are applied as contacts to substrate type n-i-p devices and superstrate p-i-n devices. The impact of the TCO on the optical and electrical performance of both types of devices is presented.

## TCO DEPOSITION AND DEVICE FABRICATION

TCO materials were r.f. sputtered in Ar or $Ar/O_2$ onto unheated substrates of 7059 glass, for optical and electrical characterization, or onto a-Si devices. The ITO target was $In_2O_3$ mixed with 9% $SnO_2$. The ZnO target was ZnO mixed with 1% $Al_2O_3$. Nearly optimum film properties were found by sputtering the ITO at 700W and the ZnO at 900W. Deposition rates were approximately 0.1 µm/min. Sputtering in an $Ar/O_2$ mixture improved the transparency of the TCO films with a slight loss in resistivity. ITO and ZnO films were sputtered with and without $O_2$ in the Ar. $O_2$ had a negative effect on devices in some cases as discussed below. TCO and Ag contacts were deposited through masks onto devices.

Two different a-Si device structures were investigated. Substrate type devices with ss/n-i-p/TCO configuration (ss=stainless steel) were deposited by ECD. The p-layer was µc-Si. These devices were intended to duplicate the top cell of a triple junction device. Therefore, they lacked any optical enhancement such as texture or a back reflector, and the i-layer was very thin (<0.1 µm). The $J_{sc}$ from these cells was around 8 mA/cm² as required for a top cell (8). Superstrate type devices with glass/$SnO_2$/p-i-n/TCO/Ag structure were deposited at IEC. The $SnO_2$ (Asahi type U) was textured. The i-layers were 0.5 µm thick. The p-i-n cells had a-SiC p-layers and C graded buffers between the p and i-layers. Otherwise identical devices with a-Si or µc-Si n-layers were deposited for this study. The µc-Si n-layer was deposited at high hydrogen dilution ($H_2$:$SiH_4$ of 100:1) and higher power and pressure than the a-Si layer. The high conductivity (1 S/cm) and low activation energy (0.05 eV) confirm the microcrystalline nature of the n-layer. After sputtering TCO on the p-i-n cells, a 0.6 µm Ag layer was evaporated through the same mask for a low resistance highly reflective TCO/Ag contact. Cells were 0.4 cm² in area. Current voltage measurements were made with AM1.5 global illumination from an Oriel simulator at 28° C.

## TCO MATERIAL PROPERTIES

TCO layers were sputtered over a range of rf power and oxygen partial pressure. Optimum ITO and ZnO film properties were obtained at 700 W and 900 W with 0.9 and 0.1% $O_2$ in Ar, respectively. Table I shows the thickness D, sheet resistance $R_{sq}$, resistivity, and normalized transmission for ITO and ZnO films deposited with and without $O_2$. The normalized transmission $T_n$ =T/(1-R) was averaged over 400-900 nm.

The presence of $O_2$ during sputtering has a major impact on the ITO film properties (6). The ITO film without $O_2$ was brownish and had high absorption, unsuitable for solar cell applications. $O_2$ has a much smaller effect on the ZnO films. Note that the ITO and ZnO films deposited with $O_2$ have the same optical transmission at the same $R_{sh}$, indicating equivalent thickness trade-off between the need for low lateral current carrying resistance losses (thicker TCO) and high window layer transparency (thinner TCO). The resistivity of optimized ZnO is 2.5 times higher than that of ITO.

**TABLE I.** Sputtered TCO properties

| TCO | $O_2$ in Ar (%) | D (μm) | Rsq (Ω/sq) | resistivity (Ω-cm) | avg. $T_n$ 400-900nm |
|---|---|---|---|---|---|
| ITO | 0 | 0.2 | 35 | 7E-4 | 0.72 |
| ITO | 0.9 | 0.2 | 19 | 4E-4 | 0.96 |
| ZnO | 0 | 0.4 | 15 | 6E-4 | 0.94 |
| ZnO | 0.1 | 0.5 | 20 | 1E-3 | 0.96 |

## TCO CONTACTS ON ss/n-i-p/TCO DEVICES

ITO and ZnO were sputtered with and without $O_2$ in the Ar discharge onto the μc-Si p-layers of ss/n-i-p substrates. The a-Si devices were essentially identical since they had been cut out of a large area sheet. The TCO films were typically of the thickness range shown in Table II. This relatively low sheet resistance allowed the cells to be fabricated without grids since the area was small. However, thinner TCO films (~70 nm) are typically used in this application primarily for their AR effect.

Results are shown in Table II. The most important observation is that devices with ITO have higher FF than devices with ZnO. The decrease in FF is accompanied by an increase in resistance $R_{oc}$ which is dV/dJ at $V_{oc}$. Figure 1 shows the reduced FF for devices with ZnO is due to curvature near $V_{oc}$ (i.e. increasing $R_{oc}$), suggesting a second junction opposing the n-i-p cell. Table II shows that the effect of sputtering in an $O_2$ atmosphere is quite different between ITO and ZnO. Adding $O_2$ during ITO sputtering increases $J_{sc}$, as expected from the difference in $T_n$ in Table I, but has no effect on $V_{oc}$ or FF. Adding $O_2$ during ZnO sputtering reduces $V_{oc}$ and FF but has no effect on $J_{sc}$.

**TABLE II.** Performance of ss/n-i-p/TCO devices with ITO or ZnO contacts.

| TCO | TCO process | Rsq (Ω/sq) | $V_{oc}$ (V) | $J_{sc}$ (mA/cm$^2$) | FF (%) | $R_{oc}$ (Ω-cm$^2$) | Eff. (%) |
|---|---|---|---|---|---|---|---|
| ITO | no $O_2$ | 35 | 0.844 | 7.3 | 66.0 | 14 | 4.1 |
| ITO | $O_2$ | 19 | 0.860 | 8.3 | 65.1 | 13 | 4.6 |
| ZnO | no $O_2$ | 15 | 0.840 | 8.4 | 60.0 | 26 | 4.4 |
| ZnO | $O_2$ | 20 | 0.750 | 8.5 | 52.2 | 39 | 3.3 |

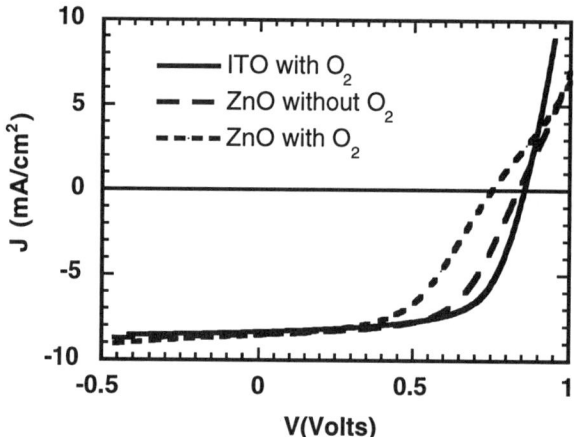

**FIGURE 1.** J(V) curves for ss/n-i-p/TCO devices with ITO and ZnO. Cell performance is shown in Table II.

## TCO CONTACTS ON glass/SnO$_2$/p-i-n/TCO/Ag DEVICES

Superstrate devices were deposited on glass/textured SnO$_2$ substrates. Four devices were deposited with µc-Si n-layers in one run and four with a-Si n-layers the following run. ZnO was sputtered with or without O$_2$ and ITO was sputtered with O$_2$ on both types of devices through a mask. Ag was evaporated through a mask on all TCO layers. For comparison, Ti/Ag contacts were also evaporated on both types of devices.

Results with µc-Si and a-Si n-layers with all four types of contacts are shown in Table III. Devices with the Ti/Ag metal contacts had FF~70% for both a-Si and µc-Si n-layers. However, for all three TCO/Ag contacts, the µc-Si n-layer had higher FF than the a-Si n-layer. The µc-Si n-layer also had higher $V_{oc}$ and $J_{sc}$. The higher $V_{oc}$ with the µc-Si n-layer suggests better pinning of the Fermi level which increases the built-in voltage. The µc-Si n-layer increases $J_{sc}$ by ~ 0.5 mA/cm$^2$ for Ti/Ag contacts and by ~1 mA/cm$^2$ with TCO/Ag contacts. Both ITO and ZnO give comparable FF on µc-Si n-layers (FF~68-69%) with or without $O_2$, but ZnO gives higher $V_{oc}$ and $J_{sc}$.

Figure 2 shows that sputtering ITO or ZnO with $O_2$ on an a-Si n-layer gives curvature at $V_{oc}$, suggesting a second junction. This curvature is not present on the µc-Si n-layers as shown in Figure 2. Comparing Figure 2 with Table III, the large values of $R_{oc}$ on these devices are indicative of curvature around $V_{oc}$ and the possible formation of a second junction.

The effect of sputtering ZnO with or without $O_2$ is shown in Figure 3. Sputtering ZnO with $O_2$ onto an a-Si n-layer gives strong curvature around $V_{oc}$ and low FF (52.0%), where as sputtering without $O_2$ gives less curvature, and a higher FF (62.6%). Sputtering with $O_2$ onto a µc-Si n-layer gives no evidence of a second junction, with FF=67.8%.

**TABLE III.** Cell performance of glass/SnO$_2$/p-i-n/TCO/Ag devices with either a-Si or µc-Si n-layers.

| back contact | TCO process | n-layer | $V_{oc}$ (V) | $J_{sc}$ (mA/cm$^2$) | FF (%) | $R_{oc}$ ($\Omega$/cm$^2$) | Eff. (%) |
|---|---|---|---|---|---|---|---|
| Ti/Ag | none | a-Si | 0.863 | 13.5 | 70.2 | 6.3 | 8.2 |
| Ti/Ag | none | µc-Si | 0.875 | 14.0 | 69.8 | 6.0 | 8.5 |
| ITO/Ag | w/ $O_2$ | a-Si | 0.853 | 13.8 | 60.0 | 23.1 | 7.0 |
| ITO/Ag | w/ $O_2$ | µc-Si | 0.865 | 14.9 | 69.1 | 5.9 | 8.9 |
| ZnO/Ag | w/ $O_2$ | a-Si | 0.864 | 14.9 | 52.0 | 55.7 | 6.7 |
| ZnO/Ag | w/ $O_2$ | µc-Si | 0.885 | 16.1 | 67.8 | 6.0 | 9.7 |
| ZnO/Ag | w/o $O_2$ | a-Si | 0.872 | 15.0 | 62.6 | 14.5 | 8.2 |
| ZnO/Ag | w/o $O_2$ | µc-Si | 0.881 | 16.2 | 67.7 | 5.8 | 9.7 |

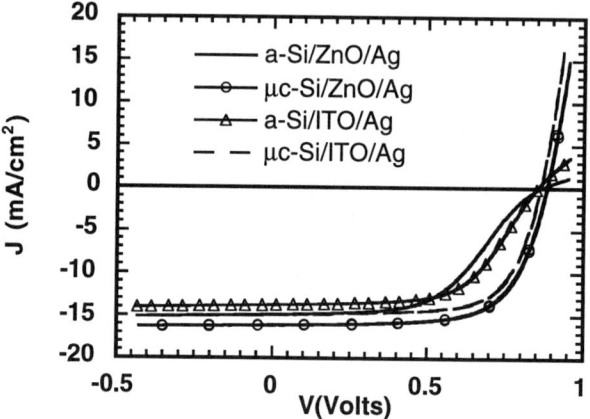

**FIGURE 2.** J(V) curves for glass/SnO$_2$/p-i-n/TCO/Ag devices with a-Si or µc-Si n-layers. ITO and ZnO were sputtered with O$_2$. Cell performance is shown in Table III.

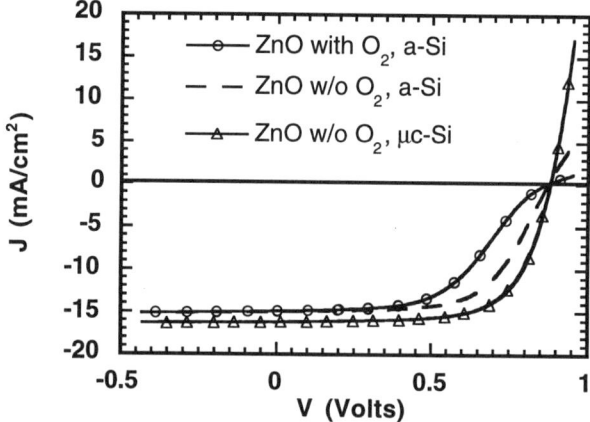

**FIGURE 3.** J(V) curves for glass/SnO$_2$/p-i-n/ZnO/Ag devices showing the effect of sputtering with or without O$_2$ on a-Si or µc-Si n-layers. Solar cell performance is shown in Table III.

# DISCUSSION AND CONCLUSIONS

We found poor FF and evidence for a second junction with ZnO/p contacts on n-i-p devices. Although this is the first time such behavior has been reported for the n-i-p device structure, poor electrical contacts leading to low FF have been observed in glass/ZnO/p-i-n devices deposited on a glass/ZnO substrate (9-11). In the p-i-n structure, the ZnO was typically deposited by APCVD or LPCVD, and the p-layer, typically a-SiC, is grown on the ZnO at 150-250°C. In our work, the ZnO is sputtered onto an unheated μc-Si p-layer, which has much higher conductivity than the a-SiC layer. Yet, the two device structures exhibit the same problem, despite the fact that the ZnO, the p-layer, and ZnO/p contact are formed from very different conditions. This strongly suggests that the ZnO/p contact is fundamentally a poor ohmic contact. We speculate that an oxide layer may contribute to this problem since sputtering ZnO in $O_2$ gave much poorer FF than with no $O_2$ yet the bulk resistivity of ZnO with $O_2$ is only a factor of 2 higher. There is no technical reason to switch from the standard ITO contact to ZnO for n-i-p devices in view of the poorer contact. However, re-optimization of the μc-Si p-layer with ZnO instead of ITO may improve the contact.

Regarding the p-i-n devices, we found significant differences between a-Si or μc-Si n-layers. The a-Si n-layer is more absorbing than the μc-Si layer. The larger improvement in $J_{sc}$ for TCO/Ag with the less absorbing μc-Si n-layer is consistent with increased multiple reflections and higher back reflection for the TCO/Ag contact compared to Ti/Ag. As optical enhancement and light trapping in the device improves, it becomes more crucial to reduce parasitic absorption at contacts and interfaces. Table III and Figure 2 also clearly show the improved electrical contact properties with a μc-Si n-layer compared to an a-Si n-layer when contacted with either ITO or ZnO. These results are consistent with those of Hayashi et al (5) who investigated ZnO/Al or Al contacts on a-Si or μc-Si n-layers and found poorest FF with ZnO/Al on an a-Si n-layer.

The effect of sputtering ZnO with or without $O_2$ is quite different between n-i-p and p-i-n cells. Sputtering ZnO onto μc-Si n-layers in p-i-n cells gave equivalent device results with or without $O_2$, whereas sputtering ZnO with $O_2$ was clearly detrimental in contacting μc-Si p-layers in n-i-p devices.

Barriers between TCO and doped (12, 13) or undoped (12, 14) a-Si layers have been previously reported. However, those interfaces were formed by depositing the a-Si onto the TCO in order to study the interaction between the plasma and the TCO, and subsequent metallic Schottky barrier formation. In our study, the TCO was sputtered onto the a-Si, which eliminates the question of TCO reduction and Schottky barrier formation. It is unclear that previous results for the TCO/a-Si interface (12-14) apply to our a-Si/TCO contacts. Computer modeling of p-i-n/ZnO devices (15) suggests that very thin n-layers (a few nanometers) will improve FF and $V_{oc}$. This is about 10 times thinner than typical n-layers, as we used in our devices. Among ZnO, ITO, and $SnO_2$, ZnO has the lowest work function difference with n-type a-Si (15). Thus, ZnO should have a negligible barrier with the a-Si n-layer and effectively pin the Fermi energy.

However, the model results find a very thin n-layer remains necessary to compensate for the negative interface space charge. In contrast to the modeling results (15), experimental results (4) showed a better FF with ZnO by increasing the n-layer thickness 3 times greater than their standard thickness for metal contacts. Clearly, the a-Si/sputtered TCO contact needs further study to understand the fundamental junction and contact properties.

Regarding sputtered TCO contacts for a-Si devices, we conclude: 1) ITO is a better top contact for µc-Si p-layers in ss/n-i-p/TCO devices than ZnO; 2) ZnO sputtered with or without $O_2$ is a better back contact compared to ITO for µc-Si n-layers for p-i-n/TCO/Ag devices; 3) ZnO sputtered with $O_2$ is detrimental to obtaining a low resistance contact to µc-Si p-layers and a-Si n-layers but has little effect on contacting µc-Si n-layers.

## ACKNOWLEDGMENTS

We wish to thank Kevin Hart and Ron Dozier of IEC for technical assistance and contributions to the a-Si program at IEC, and Xunming Deng at Energy Conversion Devices for supplying the ss/n-i-p devices. This work was supported by NREL under subcontract #XAV-3-13170-01.

## REFERENCES

1. Banerjee, A. and Guha, S., "Improved Blue Response of Amorphous Silicon Alloy Solar Cells" in *Mat. Res. Soc. Symp. Proc.* **192**, 1990, pp. 57-62.
2. Morris, J., Arya, R.R., O'Dowd, J.G. and Wiedeman, S., "Absorption enhancement in hydrogenated amorphous silicon-based solar cells", *J. Appl. Phys.* **67**(2),1990, pp.1079-1087.
3. Tao, G., Girwar, B.S., Landweer, G.E.N., Zeman, M. and Metselaar, J.W., "Enhanced Light Absorption in a-Si:H Layers of Solar Cells by Applying TCO/Metal Back Contacts", *Mat. Res. Soc. Symp. Proc.* **297**, 1993, pp. 845 849.
4. Terzini, E., Rubino, A., DeRosa, R., and Addonizio, M.L., "The Effect of ZnO Sputtering Deposition Parameters on the Performances of Back Reflector Enhanced Amorphous Silicon Solar Cells", *Mat. Res. Soc. Symp. Proc.* **377**, 1995, pp. 681-686.
5. Hayashi, K., Masataka, K., Ishikawa, A., and Yamagishi, H., "ZnO/Ag Sputtering Deposition on a-Si Solar Cells", in *IEEE First World Conf. on Photovoltaic Energy Conversion Proc."*, 1994, p. 674.
6. Buchanan, M., Webb, J.B., and Williams, D.F., "Preparation of conducting and transparent thin films of tin-doped indium oxide by magnetron sputtering", *Appl. Phys. Lett.* **37**(2), 1980, pp. 213-215.

7. Schropp, R.E.I. and Madan, A., "Properties of conductive zinc oxide films for transparent electrode applications prepared by rf magnetron sputtering", *J. Appl. Phys.* **66**(5), 1989, pp. 2027-2031.
8. Luft, W., Branz, H.M., Dalal, V.L., Hegedus, S.S., and Schiff, E.A., "Recent Progress in Amorphous Silicon PV Technology", *AIP*, 1994, pp. 31-45.
9. Kubon, M., Boehmer, E., Gastel, M., Siebke, F., Beyer, W., Beneking, C. and Wagner, H., "Solution of the ZnO/p Contact Problem in a-Si:H Solar Cells", *IEEE First World Conf. on Photovoltaic Energy Conversion Proc.*, 1994, p. 500.
10. Hegedus, S. and Buchanan, Liu, X. and Gordon, R., "Effect of Textured Tin Oxide and Zinc Oxide Substrates on the Current Generation in Amorphous Silicon Solar Cells", in *Proc. 25th IEEE PVSC*, 1996, p. 1129.
11. Hu, Jinahua and Gordon, R., "Textured fluorine-doped ZnO films by atmospheric pressure chemical vapor deposition and their use in amorphous silicon solar cells", *Solar Cells*, **30**, 1991, pp. 437-450.
12. Sanchez, R. Sinencio and Williams, Richard, "Barrier at the interface between amorphous silicon and transparent conducting oxides and its influence on solar cell performance", *J. Appl. Phys.* **54**(5), 1983, pp. 2757-2760.
13. Itoh, K., Matsumoto, H., Kobata, T., and Fijishima, A., "Determining height of a leaky Schottky barrier existing in the junction between $SnO_2$ and a highly doped p-type amorphous SiC by using the pulsed laser-induced transient photopotential technique", *Appl. Phys. Lett.* **51**(21), 1987, p. 1695.
14. Drevillon, B., Kumar, S., i Cabarrocas, P.R., and Siefert, J.M., "*In situ* investigation of the optoelectronic properties of transparent conducting oxide/amorphous silicon interfaces", *Appl. Phys. Lett.* **54**(21), 1989, pp.2088- 2090.
15. Smole, F., Topic, M. and Furlan, J., "Correlation between TCO/p and p/i Heterojunction and Effect of n/TCO Heterojunction on a-Si:H Solar Cell Performance", in *IEEE First World Conf. on Photovoltaic Energy Conversion Proc.*, 1994, p. 496.

# The Potential of Hydrogenated Amorphous Silicon-Chalcogen Alloys for Photovoltaic Applications: The Role of Persistent Photoconductivity

S. L Wang[*], J. M. Viner[*], P. C. Taylor[*], T. Itoh[†], and S. Nitta[†]

[*]Department of Physics, University of Utah, Salt Lake City, UT 84112, USA
[†]Department of Electronic and Computer Engineering, Gifu University,
1-1 Yanagido, Gifu, 501-11, Japan

The potential improvement in stability of hydrogenated silicon-sulfur alloys (a-SiS$_x$:H) with respect to ordinary hydrogenated amorphous silicon (a-Si:H) has been attributed to the introduction of an additional metastability known as persistent photoconductivity (PPC). In order to examine the PPC process in more detail we examine a series of alloys with large sulfur concentrations (x > 0.01). Although these alloys are not useful in photovoltaic devices, the high sulfur concentrations accentuate the PPC effect and allow one to study this effect with little competition from the ordinary Staebler-Wronski effect that dominates the metastable processes that occur in a-Si:H.

## INTRODUCTION

Recent results on the effects of alloying hydrogenated amorphous silicon (a-Si:H) with sulfur have shown that the addition of sulfur increases the n-type conductivity and lowers the conductivity activation energy.[1,2] Above about 1 at. % of sulfur concentration, the doping process is no longer effective. It is at approximately this sulfur concentration that an effect known as persistent photoconductivity (PPC) becomes the dominant metastable mechanism. The PPC effect is a metastable increase in the dark conductivity after exposure to band-gap light. The n-type doping mechanism in a-SiS$_x$:H for x < 0.01 has been attributed to an inefficient process where sulfur acts as a double donor much as it does in crystalline silicon. Perhaps the most compelling evidence for this doping mechanism is the compensation that sulfur provides for p-type dopants, such as boron.[3,4] At larger sulfur concentrations (x > 0.01) it is clear that the metastabilities introduced by alloying with sulfur are different than those that commonly occur in a-Si:H. In particular, the Staebler-Wronski effect no longer occurs, but instead there exists the metastable effect described above as the PPC effect.[5,6] A recent review describes the current understanding of the n-type doping mechanism for sulfur and the transition from the doping to the alloying regime where the PPC effect dominates.[7]

In the present paper we describe the effects that occur in samples of a-SiS$_x$:H where x ≥ 0.01 and the PPC effect dominates the metastable processes. This effect is similar to that which occurs in doping-modulated multilayers and in compensated samples of a-Si:H. It is interesting to not that the PPC effect has, in general, the opposite effect on both the dark and photoconductivities from that of the well-known Staebler Wronski effect. Also, the PPC effect tends to anneal away at temperatures that are considerably lower that those at which the Staebler-Wronski effect anneals. The exact annealing temperature, however, depends on the sulfur concentration.

It has been suggested that the PPC effect is simply the optical activation of sulfur-related donors that have been passivated by hydrogen during growth.[2,7] The idea, which is derived from the behavior os sulfur double-donors in crystalline silicon, is simply that a fraction of the sulfur donors is passivated by the presence of nearby hydrogen atoms during growth. Optical excitation can then activate these sites by dislodging the hydrogen to other sites in the material.[8] The process is reversible since annealing at elevated temperatures can restore the dark and photoconductivities to their original values. According to this ;hypothesis the PPC effect in a-SiS$_x$:H is just the optical activation of inefficient sulfur donors. At low sulfur concentrations the PPC effect is at least partially masked by the presence of the ordinary Staebler-Wronski effect, which has the opposite effect on both the dark and photoconductivities.

## EXPERIMENTAL DETAILS

The samples were made using the standard plasma-assisted chemical vapor deposition (PECVD) technique with SiH$_4$ and H$_2$S as source gases. Substrate temperatures of the films discussed in this paper were 220 C. Sulfur ;concentrations have been measured using secondary ion mass spectroscopy (SIMS) or electron microprobe analysis. Sample preparation details are available elsewhere.[7] Samples were annealed in vacuum at approximately 200 C for up to an hour before measurement. Illumination was accomplished with white light from a tungsten-halogen lamp whose intensity was 100 mW/cm$^2$. A filter was employed to eliminate infrared light from reaching the samples. The excess conductivity Δσ (or PPC effect) is defined as the difference in the conductivity measured 4 minutes after the end of illumination and the conductivity in the annealed state. Previous results have shown that the hydrogen concentration is mainly dependent on the deposition temperature and is essentially independent of the sulfur concentration.[5,9]

Photoluminescence (PL) measurements were made using an Ar$^+$-pumped dye or Ti:sapphire laser. Conductivities were measured using Mg electrodes covered with Au. Annealing experiments were performed using an evacuated furnace or a furnace filled with inert gas. Optical absorption experiments were performed using a standard Photothermal deflection spectroscopy (PDS) system. Further details concerning the experimental procedures are available elsewhere.[1,5,7]

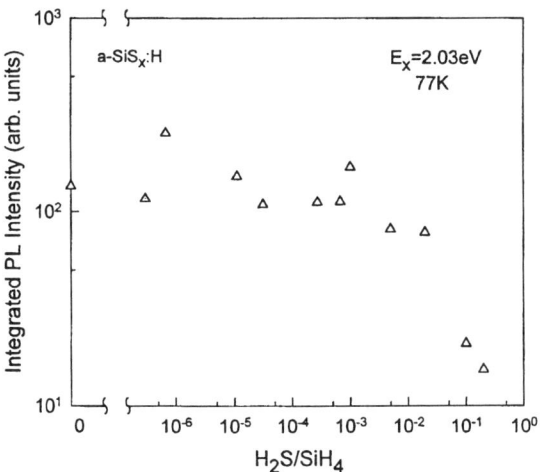

**FIGURE 1.** Integrated PL intensity at 77 K in a-SiS$_x$:H alloys as a function of sulfur concentration. The energy of the exciting light is 2.03 eV. The point at the far left is for a-Si:H (x = 0). For H$_2$S/SiH$_4$ ≤ $10^{-3}$ the PL lineshape is essentially independent of x.

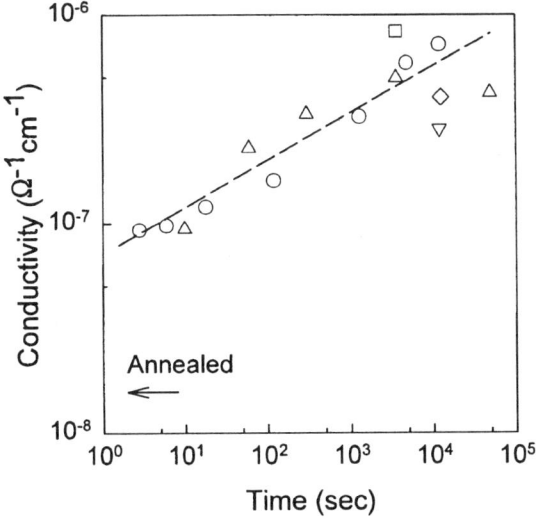

**FIGURE 2.** Dark conductivity in a-SiS$_x$:H for x = 3.3 × $10^{-2}$ after excitation with 100 mW/cm$^2$ of white light for the time indicated. Circles, triangles and squares represent different runs, each performed after annealing at ~200 C for 20 minutes. The arrow marks the level of the conductivity in the annealed samples.

## RESULTS

There are many experimental probes that track the quality of doped films of a-Si:H. One that is particularly useful is the intensity of the characteristic photoluminescence (PL) band. The intensity of the PL band tends to decrease on doping for normal n- and p-type dopants (group III acceptors and group V donors). In the case of sulfur double donors, the PL efficiency remains essentially constant throughout the doping range as shown in Fig. 1]. Above about 1 at. %S, which is beyond the doping range, the PL efficiency decreases.

For the purposes of this paper we concentrate on samples where the sulfur concentration is greater than approximately 1 at. % in order to emphasize the influence of the PPC effect over the influence of the ordinary Staebler-Wronski effect. In Fig. 2 we show the increase in the conductivity at 300 K of a sample of a-SiS$_x$:H where $x = 3.3 \times 10^{-2}$ as a function of irradiation time with 100 mW/cm$^2$ of white light. As can be seen from this figure, the conductivity increases from the annealed value of approximately $10^{-8}$ $\Omega^{-1}$cm$^{-1}$ to about $10^{-6}$ $\Omega^{-1}$cm$^{-1}$ after about $10^4$ s of irradiation with whit light. There may be a saturation of this PPC effect after about $10^5$ s but it is not certain from the data presented in Fig. 2.

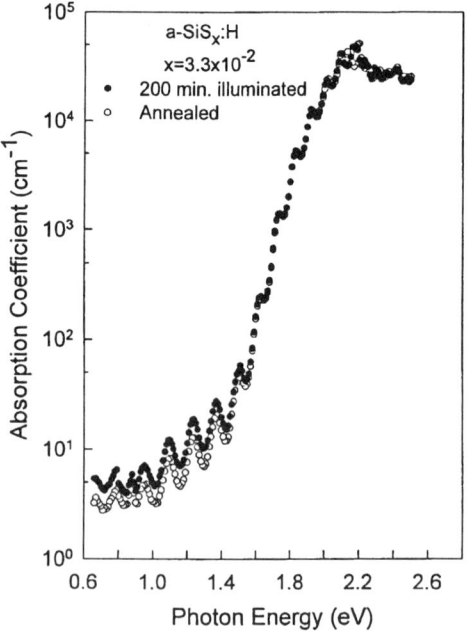

**FIGURE 3.** Absorption coefficient as a function of energy in a-SiS$_x$:H for x = 3.3 × 10$^{-2}$ as measured by PDS. The oscillations are fringes due to finite sample thickness. Open and filled circles are data for a sample annealed at ~200 C for 20 minutes and for a sample irradiated with 100 mW/cm$^2$ of white light for 200 minutes, respectively.

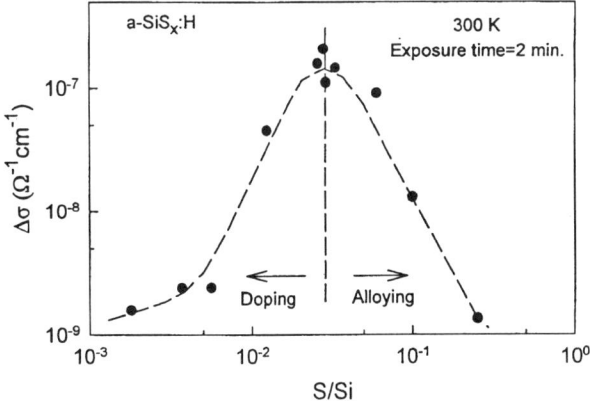

**FIGURE 4.** Excess conductivity $\Delta\sigma$ as defined in the text as a function of x in a-SiS$_x$:H. From the data it is clear that the PPC effect peaks near $x = 3 \times 10^{-2}$. The exposure time was 2 minutes in all cases.

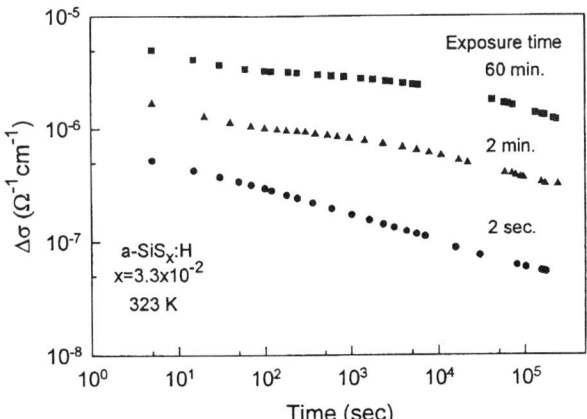

**FIGURE 5.** Decay of excess concuctivity $\Delta\sigma$ as a function of annealing time at 323 K in a-SiS$_x$:H for $x = 3.3 \times 10^{-2}$. The exposures were performed at 300 K. The solid squares, triangles and circles represent data for exposure times of 3600, 120 and 2 s, respectively.

The occurrence of the PPC effect is accompanied by a slight change in the optical absorption as measured by PDS. Figure 3 shows the optical absorption coefficient in a film of a-SiS$_x$:H where $x = 3.3 \times 10^{-2}$. It can be seen from this figure that the optical absorption doers not chang in the region of the optical band gap, but that well below the gap the absorption changes by a small amount after optical excitation.

The excess conductivity $\Delta\sigma$ (PPC effect) is shown as a function of sulfur concentration in Fig. 4. Up to about $x = 3 \times 10^{-2}$ the PPC effect increases, but after this composition the effect disappears. For this reason, we concentrate in this paper on a sample whose PPC effect is maximum ($x = 3.3 \times 10^{-2}$). At elevated temperatures the excess conductivity, which is generated by optical excitation at room temperature (300 K), decays. The decay rates depend on the temperature and also on the irradiation time at 300 K. Figure 5 shows the decays of the excess conductivity as a function of annealing time at 323 K. The decay rates depend strongly on the irradiation time at 300 K. It is clear from the data of Fig. 5 that the decay rates at 323 K decrease as the irradiation time at 300 K increases. In other words, the metastable states that are the hardest to produce at 300 K are also the hardest to anneal at 323 K. This behavior is often observed in the optical metastabilities introduced in a-Si:H (Staebler-Wronski effect).

Figure 6 shows the annealing of the optically-induced changes in the photo and dark conductivities in a-SiS$_x$:H where $x = 5.6 \times 10^{-3}$ after illumination with white light at 300 K for one hour. The annealed state is shown at the right-hand-side of the figure. It is clear for this figure that both the photo and dark conductivities increase upon annealing at progressively higher temperatures after irradiation at 300 K. By comparison with the data shown in Fig. 4 it is clear that the PPC effect is very small in this sample. This sample exhibits a complicated mixture of the Staebler-Wronski and PPC effects because the photoconductivity decreases after optical excitation at 300 K but the dark conductivity actually increases slightly. In addition, the dark conductivity actually increases even more on annealing at temperatures up to about 150 C. After annealing at about 200 C, however, the dark conductivity returns to its original lower value.

The activation energies for annealing of the PPC effect are shown in Fig. 7 for several values of x ($2.9 \times 10^{-2}$, $3.3 \times 10^{-2}$ and $6.0 \times 10^{-2}$). Over this range of sulfur concentrations the activation energy for annealing of the PPC effect is essentially independent of x and exhibits the value of approximately 0.6 eV. This activation energy is smaller than the activation energies typically observed for the annealing of the Staebler-Wronski effect in a-Si:H.

## DISCUSSION

It is clear from the previous section that the PPC effect becomes dominant in the a-SiS$_x$:H system when the sulfur concentration reaches the alloy range. As seen in Fig. 4, this metastable effect peaks at a sulfur concentration of about $3 \times 10^{-2}$. At this sulfur concentration the electrical properties, such as the sulfur doping efficiency, have deteriorated. The optical properties, such as the PL intensities shown in Fig. 1, have also changed.

We have previously suggested that the same PPC effect competes with the Staebler-Wronski effect at lower sulfur concentrations. We have also speculated that the PPC effect is the optical activation of sulfur donors that have been passivated by nearby hydrogen atoms.[2,3,7] Although this hypothesis remains unproved, the data at high sulfur concentrations presented in the previous section are consistent with this general picture. For example, the PDS data shown in Fig. 3 indicate a modest increase in the absorption after optical excitation, but no evidence for an onset to this absorption down to 0.6 eV. [The optically induced absorption associated with the deep-level silicon dangling bonds that have been implicated in the Staebler-Wronski effect show a distinct onset at about 0.8 eV.] Since the sulfur donors should be considerably shallower than 0.6 eV from the conduction band, this lack of an onset to the optically-induced absorption is consistent. Previous studies on compensation of boron, p-type dopants have shown that the inefficient n-type doping introduced by the sulfur continues up to sulfur atomic fractions of at least 0.05.[7] Since the PPC effect begins to disappear ab about this composition, this behavior is also consistent with the speculation that the PPC effect is the optical activation of sulfur donors that have been passivated by hydrogen.

The PPC effect exhibits complicated inducing and annealing behaviors, just as is the case for most metastabilities in amorphous semiconductors. In particular, there exists a distribution of inducing and annealing times just as occurs for the Staebler-Wronski effect in a-Si:H. Figure 5 shows the distribution of annealing times for the PPC effect - the longer the irradiation time the longer the decay at a given elevated temperature. The annealing behavior is even more complicated when there is a competition between the Staebler-Wronski and PPC effects as shown in Fig. 6.

The fact that the activation energy for annealing of the PPC effect is lower than that for annealing of the Staebler-Wronski effect (Fig. 7) is consistent with the previous results that show that the PPC effect anneals at a lower temperature than the Staebler-Wronski effect.[2,7] The lower annealing temperatures may be useful in device applications.

## SUMMARY

Measurements on alloys of a-SiS$_x$:H where x is on the order of 0.01 are consistent with the previous speculation that the important optically-induced metastability introduced by sulfur is the optical activation of sulfur donors that have been passivated by nearby hydrogen atoms.

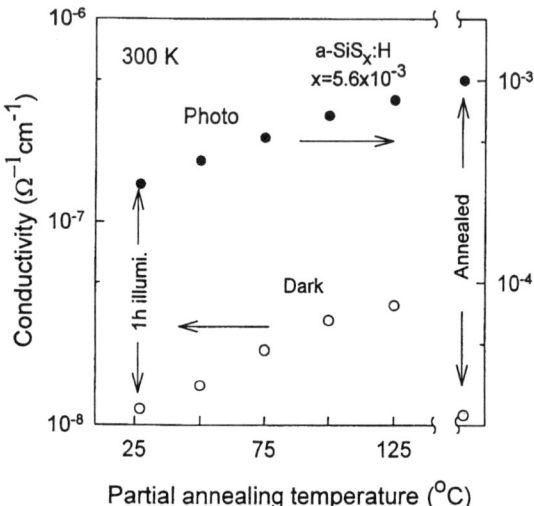

**FIGURE 6.** Dark and photoconductivity at 300 K in a-SiS$_x$:H for x = 5.6 × 10$^{-3}$ as a function of annealing at the temperatures indicated. The sample was illuminated for one hour at 300 K with 100 mW/cm$^2$ of white light. After illumination, the sample was successively annealed at the temperatures indicated for 20 minutes. The data for the sample annealed at ~200 C for 20 minutes are shown at the right.

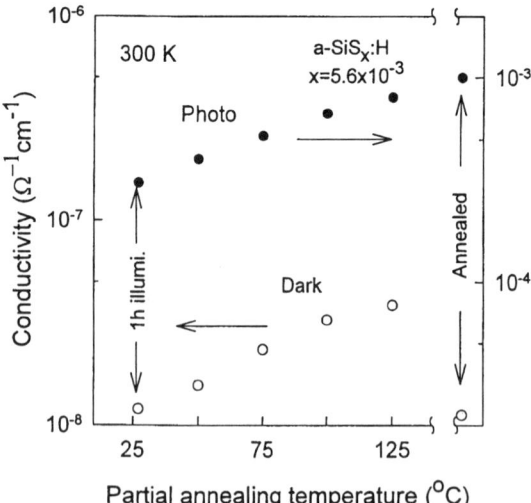

**FIGURE 7.** Temperature dependence of the relaxation time for annealing of the PPC effect in a-SiS$_x$:H for various values of x. The activation energy is approximately 0.6 eV for the three samples shown.

## ACKNOWLEDGMENTS

The research at the University of Utah was supported by the National Renewable Energy Laboratory (NREL) under subcontract number XAD-3121142.

## REFERENCES

1. S. L. Wang, Z. H. Lin, J. M. Viner and P. C. Taylor, MRS Symp. Proc. **336**, 559 (1994).

2. S. L. Wang and P. C. Taylor, Solid State Commun. **95**, 361 (1995).

3. S. L. Wang, J. M. Viner and P. C. Taylor, AIP Conf. Proc. **353**, 487 (1996).

4. S. L. Wang, J. M. Viner and P. C. Taylor, J. Non-Cryst. Solids **198-200**, 94 (1996).

5. S. L. Wang, J. M. Viner, M. Anani and P. C. Taylor, J. Non-Cryst. Solids **164-166**, 251 (1993).

6. S. L. Wang and P. C. Taylor, MRS Symp. Proc. **377**, 307 (1995).

7. P. C. Taylor and S. L. Wang, MRS Symp. Proc. **420**, 873 (1996).

8. M. K. Sheikman and A. Ya. Shik, Sov. Phys. Semicond. **10**, 128 (1976).

9. T. Itoh, S. Nitta, S. L. Wang and P. C. Taylor, J. Appl. Phys. (1996), in press.

# Pulse Duration and Wavelength Effects in Laser Scribing of Thin-Film Polycrystalline PV Materials

A. D. Compaan, I. Matulionis, S. Nakade, & U. Jayamaha

*Dept. of Physics and Astronomy, The U. of Toledo, Toledo, OH, 43606*

**Abstract.** This project is focussed on a study of wavelength-dependent effects and pulse-duration effects on laser scribing of polycrystalline thin-film PV materials. The materials studied here are CdTe, CI(G)S, $SnO_2$, ZnO, molybdenum and gold. This paper provides a summary of thresholds and optimum scribing energy densities for two types of Nd:YAG lasers, a 308 nm excimer laser, and a copper vapor laser. A comparison is presented of glass-side vs. film-side scribing. Discussion is also given of scribing of multilayer films such as ZnO/CIS/moly and gold/CdTe/$SnO_2$.

## Introduction

Polycrystalline thin-film photovoltaic materials present a challenge to the use of laser scribing for monolithic integration because they possess very different optical absorption lengths with very different wavelength dependences, different reflectivities, a wide range of thermal diffusivities, and a range of thicknesses as well. We are studying four qualitatively different laser systems: cw-lamp-pumped Nd:YAG and flashlamp-pumped-Nd:YAG ($\lambda$=532/1064 nm), copper-vapor (511/578 nm), and XeCl-excimer (308 nm). In addition to the broad range of wavelengths, these systems span a considerable range of pulse durations: 90-300, ~8, ~55, and ~15 ns, respectively.

## Vaporization Thresholds and Optimum Energy Densities

In the first part of this project, we identified threshold energy densities for the onset of ablation damage in each of the materials and for each of the laser systems (wavelength and pulse duration). Secondly, we identified the pulse energy density for the most efficient ablation rate. That is, above the threshold for vaporization, the amount of material vaporized increases linearly and then generally increases sub-linearly, as more of the pulse energy is absorbed into the vapor plume. Thus there is a maximum in the thickness vaporized per unit pulse energy density. These data are summarized in the two tables shown below. Most of these data were reported [1] at the 25th IEEE Photovoltaic Specialists Conference, however, additional data have been obtained and other data have been corrected after final calibrations. The additional work was done with the cw-lamp-pumped Nd:YAG with a pulse duration of 90 ns which is shown in bold in the tables. Table 1 shows the threshold energies for optically visible damage to the film.

### Table 1: Threshold for Vaporization of Thin-Film Materials

| λ(nm)@Δτ | $SnO_2$ (J/cm$^2$) | CdTe (J/cm$^2$) | ZnO (J/cm$^2$) | CIS (J/cm$^2$) | Moly (J/cm$^2$) |
|---|---|---|---|---|---|
| 1064 @10ns | 2.4 | 0.65 | 3.9 | 0.32 | 0.44 |
| 532 @10ns | 4.5 | 0.10 | 3.5 | 0.15 | 0.26 |
| **532@ 90ns** | **5** | **0.22** | **12*** | **0.23** | **0.5** |
| 308 @15ns | 0.44 | 0.08 | 0.24 | 0.13(CIGS) | 0.24 |
| 511, 578 @55ns | 3.4 | 0.1 | 0.6 | 0.1(CIGS) | 1.0 |

* large shot-to-shot variations

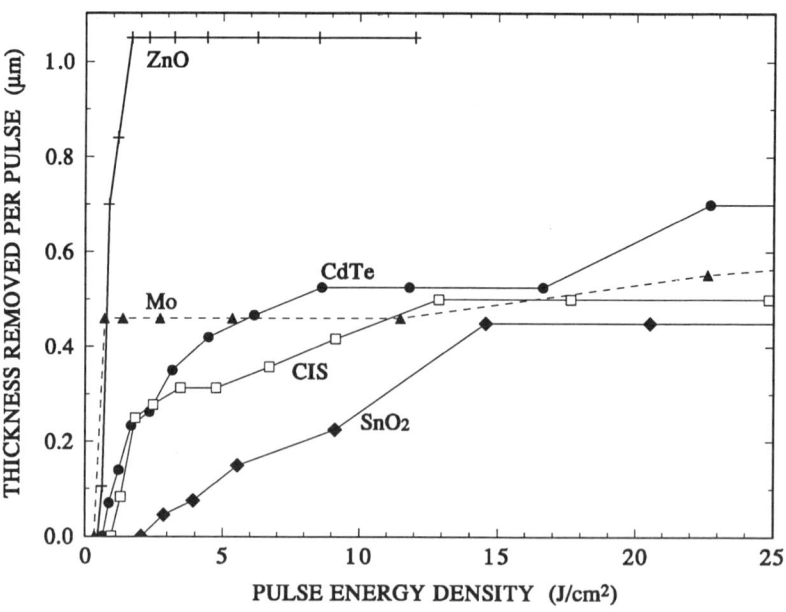

**Fig. 1.** Thickness vaporized per pulse vs. energy density, at 532 nm, 8 nsec, for five thin-film materials.

An example of the effect of increasing pulse energy density on the thickness of material removed is given in Fig. 1. These data show the depth of material vaporized per 8 nsec pulse of the flashlamp-pumped, frequency-doubled Nd:YAG laser for five different thin films. Data were calculated from the number of shots needed to remove the full film thickness. Typical film thickness was about 2 μm except for $SnO_2$ which was 0.5 μm. Note that, above a threshold value, the vaporization depth first rises approximately linearly but begins to show saturation at energy densities which typically are about twice the vaporization threshold. This effect is particularly strong with the short pulse lasers and arises from optical energy absorption into the vapor plume. The case of ZnO on glass is anomalous, for 532 nm pulses, in that either the film was undisturbed or the entire film chipped off under the laser focal spot. We believe this is due to defect-assisted and/or non-linear absorption in the transparent ZnO. At 308 nm, which is strongly absorbed, the ZnO vaporization depth was similar to other absorbing materials.

Fig 1 shows that, *e.g.*, one may remove a 2.1 μm layer of CdTe with 30 pulses at 0.8 $J/cm^2$ or with 3 pulses at 9 $J/cm^2$, but neither condition would represent the most efficient use of laser energy. The optimum efficiency occurs at 1.7 $J/cm^2$ at which point 8 pulses are needed. Table 2 provides a summary of the energy densities at which the most efficient vaporization of material occurs, for four lasers and five thin-film materials.

**TABLE 2: Optimum Efficiency for Laser Scribing of Thin-film PV Materials**

| λ(nm)@Δτ | $SnO_2$ (μm) | CdTe (μm) | ZnO (μm) | CIS (μm) | Mo (μm) |
|---|---|---|---|---|---|
| 1064 @10ns | 0.24 @ 5$J/cm^2$ | 0.47 @1.7$J/cm^2$ | 0.48 @1.7$J/cm^2$ | 0.40 @ 4$J/cm^2$ | 0.36 @ 1.7$J/cm^2$ |
| 532 @10ns | 0.27@9 $J/cm^2$ | 0.23 @1.7$J/cm^2$ | 0.72 @ 0.9$J/cm^2$ | 0.23 @1.8$J/cm^2$ | 0.47@ 0.7$J/cm^2$ |
| 532 @90ns | 0.49@7 $J/cm^2$ | 1.6@1.5$J/cm^2$ | 2.1*@12$J/cm^2$ | 0.86@3.6 $J/cm^2$ | 0.29@0.7$J/cm^2$ |
| 308 @15ns | 0.09@1.2 $J/cm^2$ | 0.11@0.35$J/cm^2$ | 0.19 @ 1$J/cm^2$ | 0.25 @2.5$J/cm^2$ | 0.52@0.4$J/cm^2$ |

\* highly variable

## Glass-side vs. Film-Side Scribing

For thin layers (≤3 μm) of semiconductors such as cadmium telluride and copper indium diselenide, the data of Figure 1 and Table 2 show that good scribing can take place from the film side with energy densities of 0.5 to 2 $J/cm^2$, with removal rates of about 0.2 to 0.5 μm per pulse. However, for thicker layers of 10 to 15 μm, and especially if there is thickness variation, uniform laser scribing from the film side is difficult. Figure 2a shows that a successful scribe can be obtained in 16 μm thick CdTe with 8 nsec pulses at 532 nm using very high energy densities (700 $J/cm^2$, with slightly overlapping pulses). In this case, glass-side scribing works better. Thus, we have used the same laser incident through the glass on the 16 μm thick CdTe/CdS/SnO$_2$/glass structures to obtain the scribe line shown in Fig. 2b. Again the pulses were slightly overlapping. Because of the lower energy density, the scribe line is narrower. Note that at this energy density of 1.3 J/cm$^2$ the SnO$_2$ is not damaged. The entire layer is removed in a single shot so that the efficiency is very high. In the case of this thick CdTe layer, glass-side scribing is more efficient than film-side scribing by a factor of about 700/1.3 ≈ 500. For back-side scribing the film removal mechanism is not vaporization of the full layer but vapor pressure build-up at the CdTe/CdS interface and subsequent rupture and ejection of the overlying layer. Note from Fig. 2 that the edges of the scribe can still be quite uniform even though the mechanism involves rupture of the film.

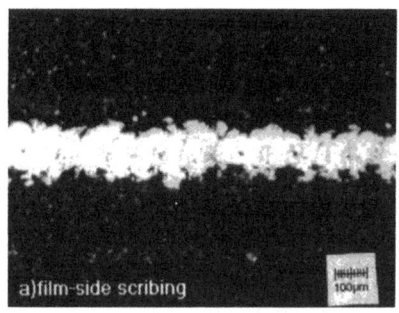 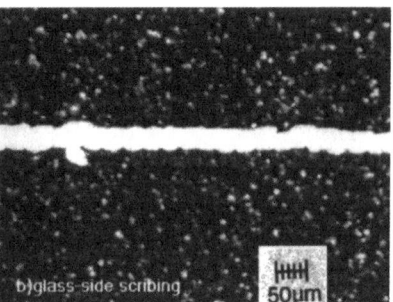

**Figure 2.** Optical micrographs of scribe lines in 16 μm thick CdTe on CdS/SnO$_2$/glass. a) film-side scribing with a pulse energy density 700 J/cm$^2$ at 10 Hz and scan speed 0.76 mm/sec; b) glass-side scribing with a pulse energy density of 1.3 J/cm$^2$ at 10 Hz and 0.27 mm/sec. There is a wide heat-affected zone in a) due to the high energy density.

In the case of Mo also, glass-side scribing works well at much lower energy densities. We find good clean scribes are obtained in 2μm thick Mo, with single pulses, at energy densities of about 4 J/cm$^2$ when the 532 nm, 10 nsec pulse is incident through the glass. If the pulse is incident on the film side, full film removal with a single pulse requires more than 60 J/cm$^2$.

## Scribing of Multilayers

Most of the work described above has been done on single layer films on glass. However, additional challenges arise when it is necessary to scribe through a thin-film layer on top of another. The case of CdTe on SnO$_2$ is relatively easy since the SnO$_2$

is transparent and has a high melting point. Among the more difficult multilayers are ZnO on CIS and metal layers on CdTe. In the case of ZnO on CIS we have recently used the 308 nm excimer laser to remove the ZnO successfully, apparently without forming a conductive copper-rich phase of CIS. The excimer laser was chosen because of its strong absorption in the wide bandgap ZnO. For the case of gold layers on CdTe we found that a thin layer of 20 nm Au is readily removed by laser scribing although a thicker layer of 150 nm was difficult to remove with 8 ns frequency-doubled Nd:YAG pulses. Work is continuing on these issues.

## Acknowledgments

This research was supported at UT by NREL (Bolko von Roedern) under contract no. ZAF-5-14142-08, which has also supported the contributions of materials, personnel, and/or equipment time from Solar Cells Inc (Toledo, OH), C J Laser Corp. (Dayton, OH), and ISET (Los Angeles, CA). The studies of scribing of thick CdTe were carried out on material supplied by Doug Rose of NREL. We are grateful also for helpful discussions with Doug Rose, with Gary Dorer and Rick Sasala of SCI, and Bulent Basol and Vijay Kapur of ISET.

## Reference

1. A.D. Compaan, I. Matulionis, M.J. Miller, and U.N. Jayamaha, "Optimization of Laser Scribing for Thin-Film Photovoltaics," *25th IEEE Photovoltaic Specialists Conference--1996*, (IEEE, Piscataway, N.J., 1996) pp. 769-772.

# FT-PL Analysis of CIGS/CdS/ZnO Interfaces

John D. Webb, Brian M. Keyes, Kannan Ramanathan, Patricia Dippo, David W. Niles, and Rommel Noufi

*National Renewable Energy Laboratory, Golden, Colorado 80401-3393*

**Abstract.** High-quality copper indium gallium diselenide (CIGS) films were subjected to a variety of surface treatments attendant to and including deposition of CdS and/or ZnO junctions or buffer layers. The resulting devices were analyzed at 87 K using Fourier transform photoluminescence (FT-PL) spectroscopy as part of a battery of analytical procedures, including surface analysis, ellipsometry, and I-V measurements, designed to elucidate the influences of the several interfaces on device performance. Our FT-PL system was upgraded with a miniature Joule-Thomson cryostat and a helium-neon laser excitation source to enable collection of highly-resolved, continuous PL spectra from 950-1750 nm. The PL intensity enhancements measured with the upgraded FT-PL system for devices fabricated using chemical bath deposition (CBD) of CdS, with or without a ZnO electrode, are much greater than for devices incorporating physical vapor deposited (PVD) CdS or ZnO/CIGS interfaces. Exposure of the CIGS films to components of the CBD solution alone, without deposition of CdS, also increases PL intensity, implying a reduction in the rate of non-radiative recombination in the films. Application of CBD CdS or a CBD background solution to the CIGS shifted its PL spectrum to shorter wavelengths, while application of PVD CdS or ZnO to the CIGS broadened its PL spectrum at longer wavelengths.

## INTRODUCTION

Thin films of cadmium sulfide (CdS) play an essential role in the creation of junctions or buffer layers in thin-film polycrystalline photovoltaic (PV) devices.[1,2] The best device performance has generally been achieved with solar cells incorporating CdS films grown by chemical bath deposition (CBD), rather than CdS films grown using other methods, such as physical vapor deposition (PVD).[2,3] Recently, considerable attention has focused upon the effects of components of the CBD solution on copper indium gallium diselenide (CIGS) PV films,[2] and upon side products of the CBD reaction. Typical CBD solutions contain a soluble cadmium salt, ammonia, and thiourea,[4] and sometimes utilize triethanolamine as a complexing agent for cadmium ion[1] or an ammonium salt as a pH buffer.[3] Sebastian et al[1] have observed an impurity phase apparently consisting of cadmium cyanamide (CdNCN) in CBD CdS films on glass substrates. This phase can be removed by rinsing the CdS film in dilute acetic acid.[1] Friedlmeier et al[5] have identified 30-50 nm-wide islands containing cadmium and oxygen on CIGS surfaces during the initial stages of CBD CdS film growth, which they postulate to consist of $Cd(OH)_2$. Urea[1] and cyanamide[5] have been identified as the primary products of the CBD reaction. In a recent study, Fourier transform infrared (FTIR) spectroscopy and X-ray photoelectron spectroscopy (XPS) were used to identify hydroxyl ions ($OH^-$), water ($H_2O$), cyanamides, and carbonates in CBD CdS films on a variety of substrates (none of these species were detected in CdS films deposited by evaporation).[6]

The variety of possible products and by-products of the CBD reaction, and the potential for aqueous ammonia to act as a complexing agent for both cadmium[1,7] and copper[7] ions, suggest that apparently minor variations in the CBD process chemistry may have a significant influence on the CdS/CIGS interface and on the performance of the resulting devices. While significant effects on

CIGS film properties resulting from exposure of the films to CBD solutions without thiourea have been cited recently,[2] other work incorporating slightly different chemical bath recipes[3] reveals no such effects. The effect of processing parameters on the ZnO/CdS interface, and on device performance, has also been investigated.[4]

The present study was conducted under the auspices of the DOE/NREL Thin-Film Partnership Program as part of the "junction team" activities.[2] These activities include elucidating the interactions between CBD and PVD CdS buffer layers and the CIGS absorber, determining the effects of CBD bath constituents on CIGS films, examining the interactions between ZnO and CdS, and characterizing ZnO/CIGS interfaces. This study emphasizes insights gained using Fourier transform photoluminescence (FT-PL) spectroscopy during team activities; results from the other analytical procedures utilized in these activities were distributed to team members and will be discussed briefly as they connect to this study.

## EXPERIMENTAL

As part of the team's activities, CIGS films having the composition 23.4 Cu, 19.3 In, 6.8 Ga, and 50.4 Se were prepared by elemental co-evaporation as referenced previously.[4] The baseline material data for these films were obtained using electron probe microanalysis (EPMA), Auger electron spectroscopy (AES), secondary-ion mass spectrometry (SIMS), XPS, scanning electron microscopy, ellipsometry, and FT-PL spectroscopy.

The CBD process for CdS film growth consisted of preparing a solution of 0.0015 M $CdSO_4$ and 1.5 M $NH_4OH$ at room temperature, inserting a CIGS film sample while stirring and heating the solution to 40° C, adding thiourea, and heating to 72° C over an interval of 4.5 minutes to yield a CdS film appr. 50 nm thick.[4] Background solution (BS) treatment of CIGS films consisted of carrying out the CBD process without thiourea. A "BS + thiourea" treatment, consisting of the BS procedure followed by a deionized water rinse, and a repetition of the CBD process without $CdSO_4$, was applied to other CIGS films. Thinner (5 nm) CBD CdS films on CIGS samples were prepared by abbreviated bath exposures. Films of PVD CdS, 5 or 50 nm in thickness, were applied to CIGS films by evaporation from undoped CdS chunks at a substrate temperature of appr. 150° C (monitor temperature of appr. 170° C). Intrinsic ZnO films, 50 nm in thickness, were applied to 50 nm films of PVD or CBD CdS on CIGS samples, or directly to CIGS samples, using rf sputtering as described previously.[4] Intrinsic ZnO films, 50 nm in thickness, were also applied directly to CIGS samples using chemical vapor deposition (CVD), as described previously.[8] Post-processing characterization of the samples was carried out using XPS, SIMS, AES, and FT-PL spectroscopy.

For FT-PL measurements, the Nicolet System 800 spectrophotometer with FT-Raman accessory and $LN_2$-cooled, Zn-doped Ge detector was operated at 4 $cm^{-1}$ resolution as described previously,[9] with some significant improvements. A Model R2205 Joule-Thomson (J-T) cryostat from MMR Technologies, Inc. (Mountain View, CA) was rebuilt to incorporate a vacuum shroud 2.5 cm in diameter, which provided the reflective FT-Raman optics with a clear aperture sufficient to collect the FT-PL signal. With argon supplied at a minimum pressure of 2000 psig, the J-T cryostat was able to chill the samples from room temperature to their measurement temperature of $87^0$ K within about ten minutes (with a custom gas blend, temperatures as low as $30^0$ K could be reached). One of two HeNe lasers equipped with a 632.8 nm bandpass filter and neutral density filters for power control was used in place of the Nd:YAG laser source provided with the FT-Raman accessory. This eliminated the need for 1064 nm Rayleigh line filters in the FT-Raman optical train, since the wavelength of the 632.8 nm HeNe emission is shorter than the operating range of the Ge detector (950-1750 nm, or 10500 to 5700 $cm^{-1}$), and allowed continuous FT-PL spectral measurements to be made over the entire detector range for the first time. Diameter of the HeNe beam convergent at the focus of the collection optics was measured as 100-120 μm for one of the two HeNe laser sources used (laser #2), and was apparently comparable but not necessarily identical for the other (laser #1). For FT-PL measurements, power incident on the samples was reduced to 0.15 mW to provide an excitation density of appr. $10^{14}$ $cm^{-3}$ (assuming 100 μm beam diameter) low enough to avoid band-filling shifts in the PL spectra, but high enough to give adequate signal-to-noise ratios in the spectra.

## RESULTS AND DISCUSSION

The upgraded FT-PL system allowed continuous, low-noise, well-resolved PL spectra of the treated and untreated CIGS samples to be measured. Photoluminescence intensity is directly related to the amount of radiative recombination within the material or device undergoing measurement. Consequently, an increase in this intensity implies a relative decrease in the rate of non-radiative recombination. In the case of chemically-treated surfaces, this reduction is described by a reduction in the surface/interface recombination velocity and is a result of the passivation of existing defects and/or band-bending at the treated surface.

Figure 1 shows a series of FT-PL spectra measured using laser #1 from a bare CIGS film, from a CIGS film exposed to the background solution, and from a CIGS film exposed sequentially to the background solution and to the thiourea solution. The measured increase in PL intensity resulting from either of the two chemical treatments implies a reduction in the surface combination velocity (the PL intensities measured after either treatment are comparable to each other within the $\pm 50\%$ variations observed as a function of excitation position on the samples). Figure 2 shows a series of FT-PL spectra measured using laser #1 from a bare CIGS film, CIGS films with 5 nm CBD or PVD CdS layers, and a CIGS film exposed to the background solution before PVD of a 5 nm CdS layer. As with the chemically-treated surfaces, deposition of thin CdS layers results in a reduction in the surface recombination velocity, with the 5 nm CBD CdS layer giving the most improvement.

The FT-PL data in Figure 3 were taken using laser #2 and so may not be directly comparable in terms of excitation density with the data in Figures 1 and 2. The FT-PL spectra were obtained from devices consisting of 50 nm CBD or PVD CdS films plus rf-sputtered ZnO layers, and from devices consisting of 50 nm ZnO layers applied to CIGS substrates using rf sputtering or CVD. Unlike the CIGS surface treatments just described, deposition of the thicker CdS and ZnO films will result in the formation of a p/n junction. Thus, the carrier distributions before and during the FT-PL measurements may be very different from those in treated or untreated CIGS films without junctions. This further complicates any direct comparison of the data in Figures 1 and 2 with those in Figure 3. Within the set of devices described in Figure 3, the CBD CdS/CIGS film yields by far the highest PL intensity.

The data in Figures 4-7 are the same FT-PL spectra shown in Figures 1-3, normalized to a common intensity to emphasize wavelength shifts. These data show that application of CBD CdS films or background solutions to CIGS shifts its PL spectrum to shorter wavelengths, while application of PVD CdS or intrinsic ZnO films to the CIGS broadened its PL spectrum towards longer wavelengths.

It is noteworthy that there is an insignificant difference in PL intensity or energy increases between the background solution and the background solution/thiourea solution treatments of the CIGS films. The components most responsible for the PL enhancements apparently reside in the background solution, which contains $CdSO_4$ and $NH_4OH$ only.

XPS analyses have shown numerous chemical distinctions between CBD and PVD-processed samples. As-deposited CIGS typically exhibits about 10 at. % of surface sodium which thermally diffuses from the Mo/glass substrates during CIGS film deposition. The chemical nature of the Na is not convincingly known, although it is known that the sodium compound oxidizes with exposure to air. Either the CBD process or rinsing with deionized water removes the sodium compound. The background solution contains $Cd^{+2}$ ions which adsorb to the CIGS surface prior to the addition of thiourea. During CBD, the CdS film forms only after $Cd^{+2}$ ions adsorb at the CIGS surface. Concurrent SIMS analyses confirms the presence of cadmium on the CIGS surface after background solution treatment, while both cadmium and sulfur are observed on the CIGS surface after background solution/thiourea solution treatment. In contrast, during PVD of CdS films, both cadmium and sulfur impinge simultaneously on the sodium-covered CIGS surface. The PVD process does not incorporate a step to remove the sodium on the CIGS surface. The point is that CdS/CIGS interfaces produced by the CBD and PVD processes for CdS deposition are different, and these chemical differences may have a substantial effect on PL intensity and hence on device performance.

The solution treatments confer more than half of the PL intensity enhancement achieved by application of a 5 nm CBD CdS film. The variability of PL intensities as a function of excitation position on the CBD samples could be caused by highly localized attack of the CBD solution, or deposition of reaction products or residues on the CIGS surface.[1,3,4,5,6] For example, dark spots up to 100 μm in diameter, observed to be prevalent on some CIGS devices, emitted significantly less PL than nearby, visually clean areas when analyzed using FT-PL micro-spectroscopy.[9] Subsequent FTIR microspectroscopic analyses of these spots showed that they contained residual thiourea, apparently left behind after the CBD process. Laboratory work is continuing to confirm the junction team's observations and to clarify the differences between the CBD and PVD CdS films.

## CONCLUSIONS

The FT-PL technique has been extended to allow sample temperature control, alternate laser excitation sources, and full coverage of PL spectra for CIGS samples containing up to 30% Ga. These advances, in concert with surface analysis and other activities of the junction team, have helped to identify the role of the CdS buffer layer, and of the CBD process itself, in reducing recombination velocity at CIGS surfaces and interfaces. Further improvements to the FT-PL system, including installation of a versatile diode laser excitation source, are underway and should contribute to the ongoing studies in this area.

## ACKNOWLEDGMENTS

The authors acknowledge the contributions of James Keane of NREL for CdS/CIGS device preparation, Larry Olsen of Washington State University for CVD of ZnO films, S. Asher of NREL for performing SIMS analyses, and John Perkins of NREL for contributions to the design of the J-T cryostat.

## REFERENCES

1. Sebastian, P. J., Campos, J., and Nair, P. K., *Thin Solid Films* **227**, 190-195 (1993).
2. Zweibel, K., Ullal, H. S., and von Roedern, B., "Progress and Issues in Thin-Film PV Technologies," in Proc. IEEE 25th Photovoltaic Specialists Conference, 1996, pp. 745-750.
3. Sheng, S. Li, Stanbery, B. J., Huang, C. H., Chang,, C. H., Chang, Y. S., and Anderson,T. J., "Effects of Buffer Layer Processing on CIGS Excess Carrier Lifetime," in Proc. IEEE 25th Photovoltaic Specialists Conference, 1996, pp. 821-824.
4. Ramanathan,K., Contreras, M. A., Tuttle, J. R., Keane, J., Webb,J. D., Asher,S., Niles,D., Dhere, R., Tennant, A. L., Hasoon, F. L., and Noufi, R., "Effect of Heat Treatments and Window Layer Processing on the Characteristics of CuInGaSe$_2$ Thin Film Solar Cells," in Proc. IEEE 25th Photovoltaic Specialists Conference, 1996, pp. 837-840.
5. Friedlmeier, T. M., Braunger, D., Hariskos, D., Kaiser, M., Wanka, H. N., and Schock, H. W., "Nucleation and Growth of the CdS Buffer Layer on Cu(In,Ga)Se$_2$ Thin Films," in Proc. IEEE 25th Photovoltaic Specialists Conference, 1996, pp. 845-848.
6. Kylner, A., Rockett, A., and Stolt, L., *Solid State Phenomena* **51-52**, 533-540 (1996).
7. Lide, D. R. (ed.), *CRC Handbook of Chemistry and Physics*, Boca Raton, Fla., CRC Press, 1995, Chapter 4, pp. 4-37, 4-38, 4-46, and 4-56.
8. Olsen, L. C., Agilar, H., Addis, F. W., Lei, W., and Li, J., "CIS Solar Cells with ZnO Buffer Layers," in Proc. IEEE 25th Photovoltaic Specialists Conference, 1996, pp. 997-1000.
9. Webb, J. D., Contreras, M., and Noufi, R., "Room-Temperature FT-Luminescence Analysis of Cu(In,Ga)Se$_2$ Films and Devices," in Proc. First IEEE/PVSC World Conference on Photovoltaic Energy Conversion, 1994, pp. 275-278.

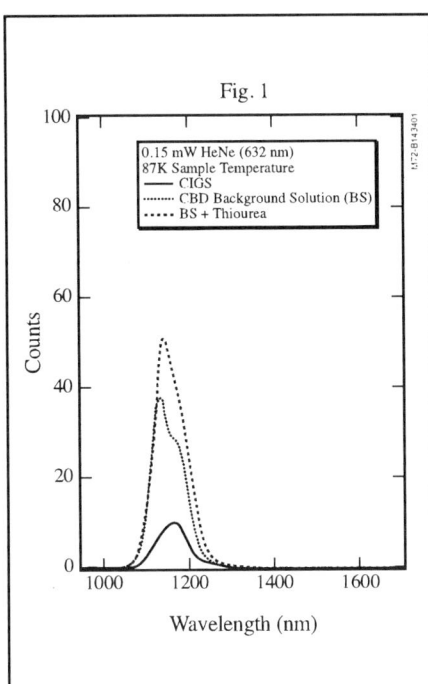

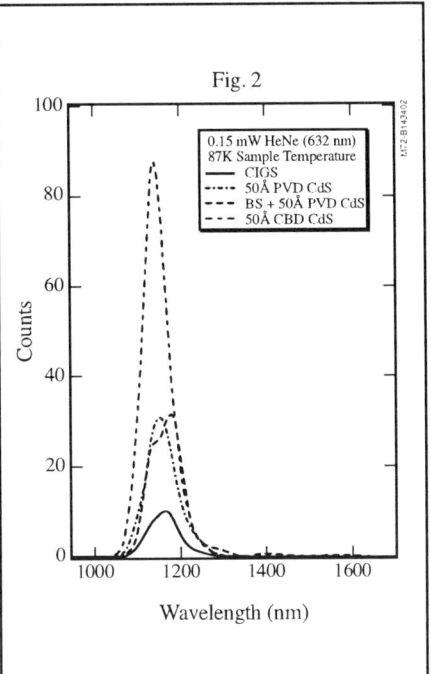

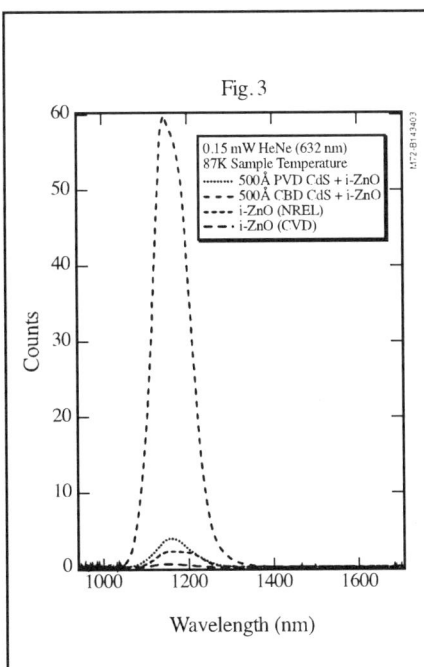

87K FT-PL Spectra:
(100μm spot size -> excitation density ≈ $1 \times 10^{14} cm^{-3}$)

PL intensity is directly related to the amount of radiative recombination. An increase in this intensity implies a decrease in the amount of nonradiative recombination.

**Fig. 1:** Increase in PL intensity with exposure to the CBD background solution (BS) and BS + Thiourea implies a passivation effect.

**Fig. 2:** Increase in PL intensity with deposition of CdS implies a passivation effect. Most notable is the improvement with the CBD CdS layer.

**Fig. 3:** These data were taken at a different time than those shown above and the magnitudes are not directly comparable with Figs. 1 and 2. Within this set, the CBD CdS shows far and away the greatest intensity.

Normalized 87K FT-PL Spectra: (100µm spot size -> excitation density ≈ 1X10$^{14}$cm$^{-3}$)
These plots show the peak positions for the various samples. No conclusions have been drawn from this data. The most notable feature is the presence of lower energy peaks for the PVD CdS samples.

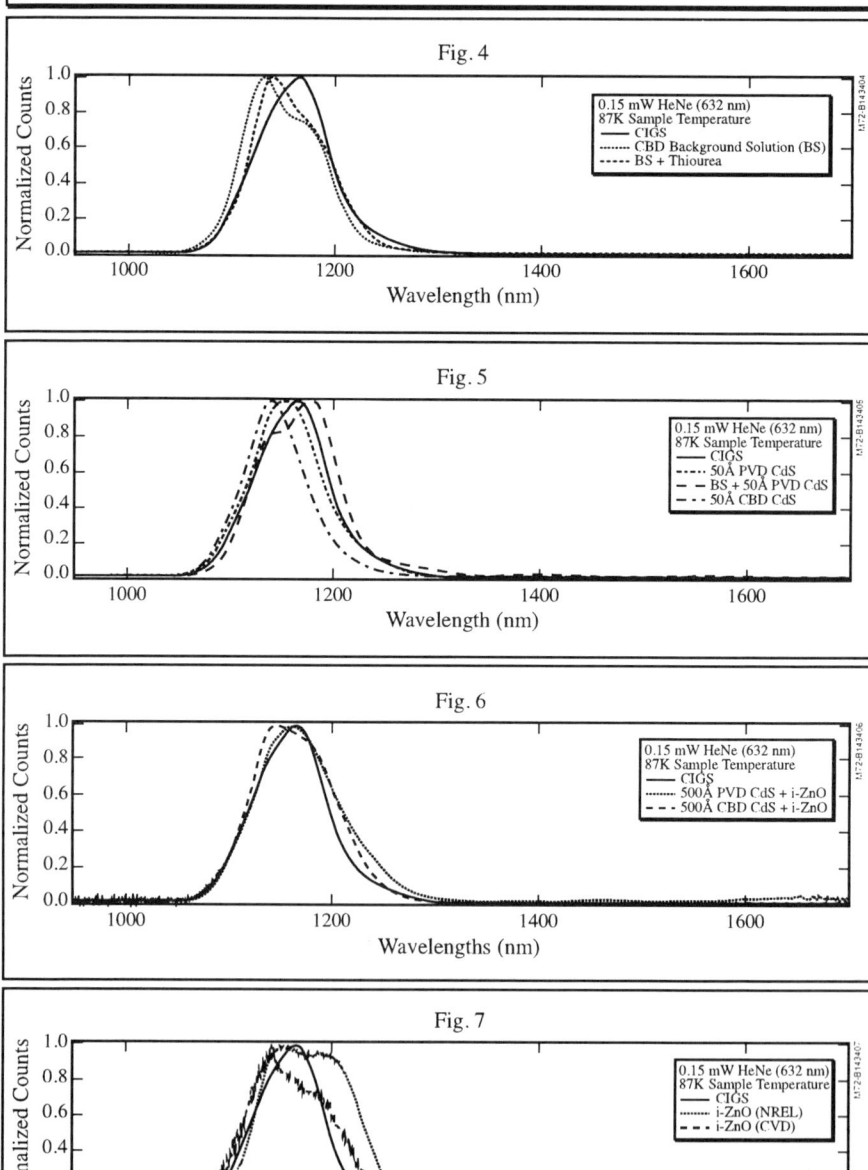

# Reaction Engineering and Precursor Film Deposition for CIS Synthesis

B.J. Stanbery, A. Davydov, C.H. Chang, and T.J. Anderson

*Department of Chemical Engineering
University of Florida
Gainesville, FL 32611-6005*

**Abstract.** We present an analysis of alternative reaction pathways for the synthesis of $CuInSe_2$-based films for photovoltaic applications based on our recent and ongoing investigations of the thermochemistry in the Cu-In-Se ternary, Na-Cu-In-Se quaternary, and constituent unary and binary systems. We also describe our efforts to determine the relationship between film growth conditions in our novel rotating-disc system and resultant phase constitution of precursor reactant films intended for subsequent *ex-situ* rapid thermal processing. A model for the phase chemistry of sodium in CIS films is presented.

## INTRODUCTION AND BACKGROUND

A variety of approaches have been used to fabricate thin films in the Cu–In–Se material system as absorbers for photovoltaic devices. Although the absorber films are often referred to as "$CuInSe_2$," detailed studies of their crystallographic structure show that the polycrystalline films contain secondary phases,[1] including metastable microstructures.[2] Recent comparisons of materials fabricated in different laboratories by different deposition methods[3] and by widely different post-deposition device optimization processes, demonstrate that these inhomogeneities play a significant role in carrier transport processes both within the absorber film, and across its interface with a buffer layer. In recognition of this variability in the actual phase constituents of polycrystalline films used for photovoltaic device fabrication in this material system, they are commonly referred to as CIS films (or CIGS when gallium is incorporated).

Sodium "contamination" of the films due to diffusion from soda lime glass substrates[4] or intentional incorporation during processing,[5] has been clearly shown beneficial to device performance when the sodium content is optimized.[6] Reported effects include increased grain size,[7] increased acceptor density,[8] higher open circuit voltage and fill factor,[9] and surface segregation of sodium.[10] The mechanism of sodium's influence on the kinetics of growth and resulting CIS film microstructure and phase constitution are unknown.

This research program is designed to elucidate these issues by combining experimental growth of thin films, synthesis of bulk samples for thermochemical and structural analysis, EMF measurements for determination of Gibbs energies, and a rigorous procedure of critical assessment for evaluation of the thermodynamic consistency of available thermochemical and phase diagram data. The results of these efforts are used to better understand the reaction chemistry of existing techniques for CIS synthesis and to engineer improved alternatives. This project is specifically directed toward development of a two-stage process for the synthesis of CIS absorber films. The first step is the low temperature deposition of binary precursor films and the second is *ex-situ* non-equilibrium Rapid Thermal Processing (RTP) to react them and form the desired absorber film structure.

## CONVENTIONAL AND NOVEL REACTION PATHWAYS

The approach of this study differs from that of previously reported RTP approaches which employ elemental precursor films[11,12] or attempt direct CIS recrystallization to increase grain size.[13] Conventional processes which synthesize CIS films under reaction conditions closer to equilibrium can be divided into selenization and recrystallization categories. Selenization approaches utilize precursor films predominately of Cu and In, and sometimes containing Ga[14] or minor Te[15] components. The precursor films react with a vapor produced by either the volatilization of condensed selenium or decomposition of hydrogen selenide to form the CIS absorber layer. The reaction chemistry of this approach is extremely complex due to the multiplicity of intermetallic Cu-In phases which can form.[16] Since the proposed approach of this study utilizes binary precursors, readers are referred elsewhere for discussions of the reaction chemistry of the selenization approaches.[17]

The earliest 'recipe' for the synthesis of CIS films for high efficiency photovoltaics (the Boeing bilayer process developed by Mickelsen and Chen[18,19]) is a recrystallization process in which a mixed-phase film layer of $CuInSe_2$ and $Cu_{2-x}Se$[20] deposited at low temperature reacts with a copper-deficient flux of coevaporated copper, indium, and selenium vapors at higher temperature to optimize the absorber film's stoichiometry during regrowth on the nucleation seeds within the initial layer of the film according to:

$Cu\,(v) + In\,(v) + Se_n(v) \Rightarrow CuInSe_2\,(s) + Cu_{2-x}Se\,(s)$ [precursor deposition]

$CuInSe_2\,(s) + Cu_{2-x}Se\,(s) + Cu\,(v) + In\,(v) + Se_n(v) \Rightarrow CuInSe_2\,(s)$ [regrowth]

This analysis is based on the assumption of near-equilibrium conditions, and to the extent that non-equilibrium components could be involved under some experimental conditions, a more complex description would be required.

For a final reaction temperature greater than the 523°C monotectic temperature in the Cu-Se binary system,[21] significantly increased grain size is observed in the final films.[22] This effect is explained as a consequence of melting of $Cu_{2-x}Se$ in the presence of excess selenium, resulting in a liquid phase assisted regrowth process.[23] Most evidence indicates that the $Cu_{2-x}Se$ phase is in fact not completely consumed during this recrystallization process, with small amounts of $Cu_{2-x}Se$ remaining on the $CuInSe_2$ grain boundaries or as inclusions.[24]

This approach has been continuously refined and adapted by many researchers, but an apparently different reaction chemistry yielding even higher efficiency devices was developed by Gabor et al.[25] Their three-stage process for growth is based on the reaction chemistry:

$In(v) + Se_n(v) \Rightarrow In_2Se_3 (s)$ [precursor deposition]

$In_2Se_3(s) + Cu(v) + Se_n(v) \Rightarrow CuInSe_2(s) + Cu_{1-x}Se_x(l)$ [regrowth]

$CuInSe_2(s) + Cu_{1-x}Se(l) + In(v) + Se_n(v) \Rightarrow CuInSe_2(s)$ ['titration']

where gallium is typically substituted in part for indium at each step for CIGS synthesis. A two-stage variant of this process omits the final step but carefully controls the total copper flux in the second stage to minimize the $Cu_{2-x}Se$ phase in the final film.[26] Both of these reaction sequences are in a sense "inverted" processes with respect to the Boeing bilayer recipe since they begin with an indium sesquiselenide layer and add copper during the second step. The Boeing bilayer process chemistry does not form the indium sesquiselenide at any time.

The approach proposed in this work is designed to realize the manufacturability benefits that RTP offers.[27] From the previous work described above, it appears that essential components of a successful alternative process are the initial formation of a nucleation seed layer and the subsequent formation of a coexisting liquid phase to achieve large grain size. It further appears that the the final growth should occur under indium-rich conditions.

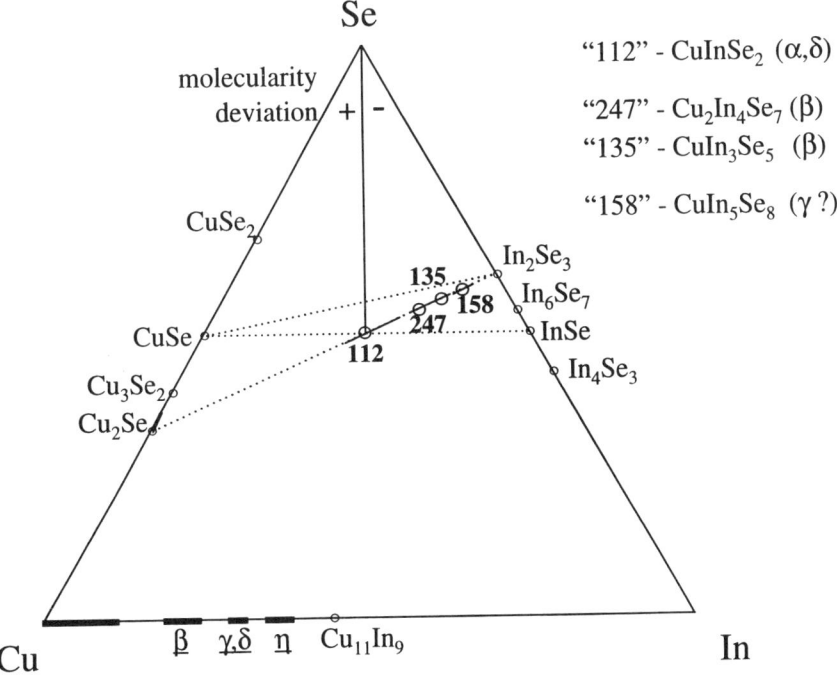

Figure 1. Cu-In-Se ternary composition diagram indicating compounds

Figure 1 is a composition diagram for the Cu–In–Se ternary system showing the accepted binary selenide compounds and several tie-lines connecting pairs of them, including the pseudobinary tie-line between the binary selenides with the highest melting temperature on each boundary ($Cu_2Se$ and $In_2Se_3$). In a thermodynamic system closed to mass transfer, the overall reaction product reulting from reacting two compound compositions must lie on that tie-line formed between the reactants, with the location determined by the initial reactant molar ratios.

Note that two of these tie-lines pass directly through the 112 phase composition ($CuInSe_2$), while the third, connecting the CuSe and $In_2Se_3$ phases, does not. This means that a reaction between the latter two binaries cannot yield only the 112 phase in a closed system. Nevertheless this combination is of interest, as will be described below.

First consider the following reaction between $Cu_2Se$ and $In_2Se_3$ along the pseudobinary tie-line to form the 112 phase:

$$Cu_2Se + In_2Se_3 \Rightarrow 2\ CuInSe_2$$

The ternary composition diagram of figure 1 can be extended in the third dimension to display temperature, where phase boundaries at various temperatures are represented as embedded two-dimensional manifolds. A cross-section along a tie-line through these surfaces yields the phase domains shown in a temperature-composition (T–x) diagram. The T–x diagram for the $Cu_2Se$–$In_2Se_3$ pseudobinary system was recently assessed[28] and is reproduced in figure 2 below.

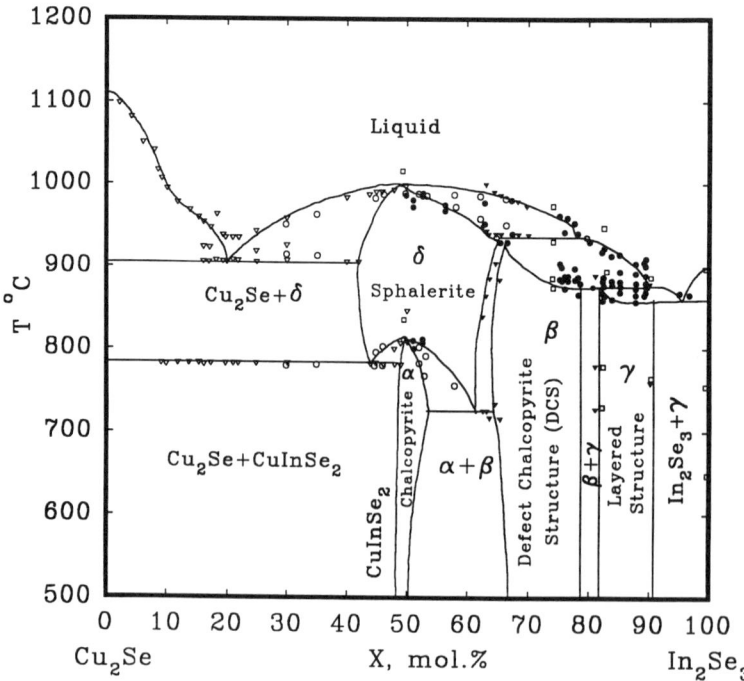

**Figure 2.** T-x diagram for the $Cu_2Se$-$In_2Se_3$ pseudobinary section

The lowest temperature at which a liquid phase coexists with a solid phase in equilibrium is ~870°C, corresponding to formation of $In_2Se_3(s)$ and the γ phase from the eutectic melt of composition $X_{In_2Se_3} \approx 0.96$. This temperature is slightly less than the 885°C melting temperature of pure $In_2Se_3$.

Figure 3 shows the T–x diagrams for the Cu–Se and In–Se systems[28] which comprise two of the three bounding surfaces along the edges of the ternary composition diagram in figure 1. Inspection of the Cu–Se diagram shows that only liquid phases persist in equilibrium above the 523°C monotectic temperature for overall compositions with more than 52.5 at% selenium. (In,Se) compositions with less selenium than that of the compound $In_6Se_7$ (<53.8 at% Se) will decompose into a liquid/solid mixture at temperatures above 156 to 600°C (depending on the overall composition), but will not form the high melting temperature compound $In_2Se_3$ at temperatures below the peritectic decomposition of $In_6Se_7$ at 660°C. Hence in separate closed systems at equilibrium, appropriately chosen selenium-rich copper binary precursors and selenium-poor indium binary precursors will each exist as liquids or liquid/solid mixtures at temperatures above 523°C to 600°C and any solid compounds would not be the highest melting-temperature ones found along each of these binary tie-lines so long as the temperature remains below 660°C.

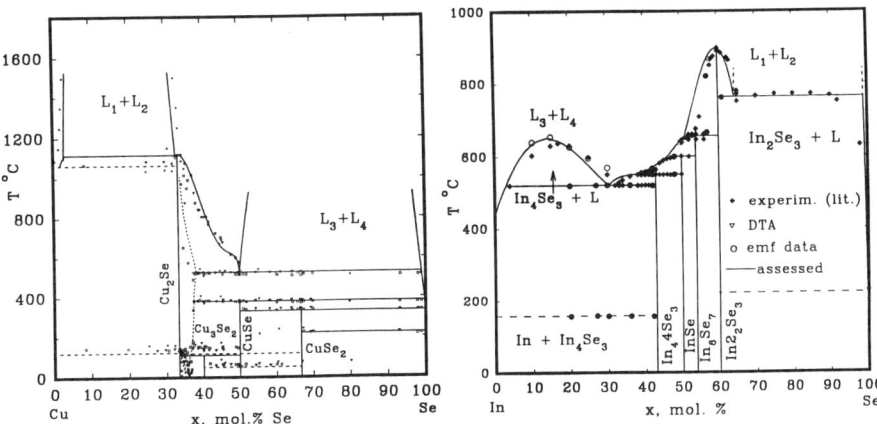

**Figure 3.** Cu-Se and In-Se binary T-x phase diagrams

These considerations alone do not answer the question of the reactions which would ensue upon liquid-phase mixing of such precursors. The equilibrium result of these reactions are found by reexamination of the T–x sections along the tie-lines in figure 1 between the precursor reactant compositions. Predicted phase diagrams along these other tie-lines suggest the existence of a very low temperature eutectic valley for those which cross the [Cu]/[In] = 1 meridian on the selenium-rich side of the 112 phase, for example the reaction:

$$2\,CuSe + In_2Se_3 \Rightarrow 2\,CuInSe_2 + Se.$$

This eutectic is absent for those tie-lines passing through the the stoichiometric 112 composition, reactions such as:

$$CuSe + InSe \Rightarrow CuInSe_2.$$

Rapid thermal processing is potentially a nonequilibrium process which provides an additional degree of freedom for process optimization. If the heating rate of the precursors is faster than the kinetic rate of a given reaction, that reaction may not proceed to its equilibrium extent if the temperature becomes high enough that a competing reaction pathway becomes more favorable. For example, the phase $Cu_3Se_2$ (see figure 3) undergoes a peritectoid decomposition into CuSe and $Cu_{2-x}Se$ at a temperature of 112°C. The rate of a solid-solid phase transformation at this low temperature is expected to be very low because substantial atomic rearrangement is required to effect the solid-solid transformation and the atomic transport mechanism is diffusion. Solid state diffusion is many orders of magnitude slower than liquid phase transport processes. Sufficiently rapid heating of $Cu_3Se_2$ to temperatures in excess of the CuSe peritectic decomposition at 377°C is expected to result in its direct decomposition into $Cu_{2-x}Se$ and selenium-rich liquid phase. Strategies such as this may be useful for circumventing the formation of undesirable reaction byproducts during the RTP synthesis of CIS films from alternative precursors.

## REACTANT ENGINEERING FOR PRECURSOR GROWTH

The underlying motivation for low temperature precursor growth is increased manufacturing equipment throughput, lower equipment cost, and thus a lower amortized capital cost for modules. The deposition of multilayer precursor films for subsequent RTP does not achieve these goals unless isothermal growth conditions can be found. Furthermore, the efficient incorporation of reactants into the precursor films is essential for economical manufacturing.

The fundamental parameters (temperature and flux ratio) for the growth of various phases in the In-Se binary system by means of vacuum coevaporation using elemental reactant fluxes has been thoroughly studied by Emery et al.[29] Their results show that under polymeric selenium flux-deficient growth conditions there exists a wide temperature domain of monophasic and diphasic stability for the compounds InSe and $In_4Se_3$, extending from 225 to at least 400°C. Their analysis[30] suggests that the low sticking coefficient found for selenium is most likely a consequence of the poor thermal accommodation of the larger selenium polymeric species characteristic of Langmuir evaporation from liquid selenium.[28]

No comparably thorough study of the growth of compounds in the Cu-Se binary system is known to these authors. Hence, we have undertaken to determine the parameter domain for the growth of selenium-rich precursor films in a rotating-disc growth system. This growth apparatus incorporates both a conventional double-oven thermal cracking source for selenium and a novel plasma cracker source compatible with both selenium and sulfur. The use of these advanced sources is expected to improve chalcogen sticking coefficients and increase the flux of dimeric and monomeric species. This in turn should reduce growth temperature and source depletion, as observed in the growth of other binary chalcogenides.[31]

Wavelength Dispersive Spectroscopy (WDS) analysis of samples from our preliminary experiments at a substrate temperature of ~200°C without cracked or ionized selenium reactants have demonstrated the growth of precursor films with 54 at% selenium which show only the CuSe phase and traces of an unidentified, possibly metastable phase in XRD analysis. Since the CuSe structure is a hexagonal layered one, an excess of selenium could reside either at grain boundaries or intercalated within the grains.

## A MODEL FOR THE ROLE OF SODIUM

The phase diagram for the binary Na–Se system shows 5 low temperature selenide compounds, the most stable of which is $Na_2Se$, with a melting temperature of 876°C. The phase diagram for the pseudobinary $Na_2Se$–$In_2Se_3$ section within the Na–In–Se ternary system has been determined in the composition range of 50–100% $In_2Se_3$.[32] The existence of two ternary compounds, the 112NIS ($NaInSe_2$) and 135NIS ($NaIn_3Se_5$) was reported but their crystallographic structures were not determined. Eutectics were found at 630°C and 645°C and 7.5 mole% and 42.5 mole% $Na_2Se$, respectively. These observations may be relevant to the three-layer process or others where $In_2Se_3$ is grown on sodium-containing substrates or intentionally codeposited with $Na_2Se$ or $Na_2S$.

No comparable phase study of the Na–Cu–In–Se quaternary system is known to these authors. Neumann et al[33] have, however, described a preliminary phase diagram for the system $Cu_{1-x}Li_xInTe_2$. Both of the alkali ternaries $LiInTe_2$ and $NaInSe_2$ differ from either of the copper ternaries $CuInTe_2$ or $CuInSe_2$, in that they do not undergo a chalcopyrite to sphælerite (solid-solid) phase transition at any temperature below their melting temperatures. The pseudobinary phase diagram in reference 33 shows an isothermal chalcopyrite to sphælerite phase transition for alloy compositions between 60 mole% and approximately 80 mole% $LiInTe_2$, and a maximum in the c lattice parameter in the same range. This data is evidence for the existence of a tetragonal quaternary compound $Li_3Cu_2In_5Te_{10}$ (corresponding to 60 mole% $LiInTe_2$). It is also possible that the analogous compound $Na_3Cu_2In_5Se_{10}$ may form in the Na–Cu–In–Se quaternary system by the reaction:

$$3\ NaInSe_2 + 2\ CuInSe_2 \Rightarrow Na_3Cu_2In_5Se_{10}$$

It is conjectured that their exists a wide solid miscibility gap in the Na–Cu–In–Se quaternary system along the $NaInSe_2$–$CuInSe_2$ tie-line in the psuedoternary composition diagram shown in figure 4, between $CuInSe_2$ and the presumed compound $Na_3Cu_2In_5Se_{10}$. This suggestion that sodium can incorporate into the $CuInSe_2$ lattice to an extent corresponding to the composition range of $CuInSe_2$'s phase stability (a few percent) and segregate to form the compound $Na_3Cu_2In_5Se_{10}$ is both plausible and consistent with numerous observations.

The ratio [In]/([Cu]+[In]) which would be found for a surface-segregated quaternary $Na_3Cu_2In_5Se_{10}$ compound is 5/7 or 71.4 %. This is very close to the ratio for the defect chalcopyrite structure $Cu_1In_3Se_5$, and within the range of values reported for the measured surface ratio[34] for CIS films grown on soda-lime glass.

Electronic transport measurements of CIGS films on Mo-coated soda-lime glass[6] show that sodium incorporation leads to a shallow acceptor state at about 75 meV above the valence band. At high levels of sodium, admittance spectrum indicate a high density of deeper trap states. This suggests a limited range of substitutional incorporation. Our model suggests that this range is determined by the molecularity deviation[35] of the copper chalcopyrite.

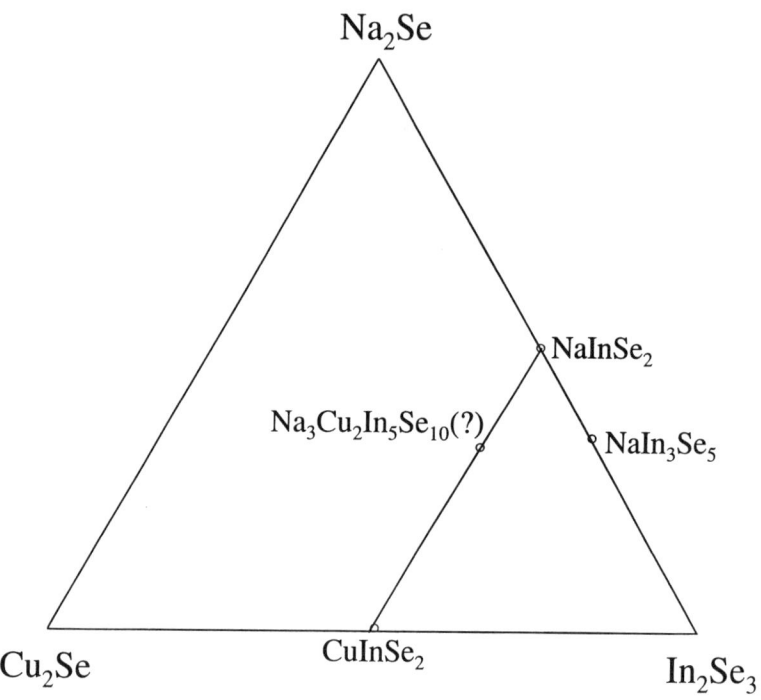

**Figure 4.** NaInSe$_2$-CuInSe$_2$ psuedoternary composition diagram

XPS and UPS analyses[36] of the chemical shifts and segregation of sodium in CIGS films on Mo-coated soda-lime glass show the presence of two different sodium species, one the consequence of a chemical reaction with environmental H$_2$O. It is relevant to note that both of the compounds NaInSe$_2$ and NaIn$_3$Se$_5$ were reported to be unstable in air at room temperature, as is Na$_2$Se itself.

## SUMMARY AND CONCLUSIONS

We have presented a review of some conventional reaction strategies for the synthesis of CIS photovoltaic absorber films from an equilibrium thermochemical perspective, emphasizing the importance of recrystallization and melting processes to their success. Application of these guidelines to non-equilibrium rapid thermal processing of less thermally stable chalcogenide precursor film couples is suggested as a means of further optimizing absorber film synthesis for manufacturability. Our experimental preparation of one such precursor in a novel rotating-disc reactor has been described.

Finally, we have proposed a model for the chemistry of sodium when present during high-temperature processing of CIS absorbers which is based on the reported existence of the ternary compound $NaInSe_2$ and published phase equilibrium data for the analagous Li–Cu–In–Te quaternary system. We believe that a model based only on a solid solution of $Na_2Se$ and $CuInSe_2$ is insufficient. Available information is inadequate to determine whether a solid solution of $NaInSe_2$ and $CuInSe_2$ ternaries, or a quaternary compound such as $Na_3Cu_2In_5Se_{10}$ forms in the Na–Cu–In–Se system as proposed here, but it is likely that an accurate model for the influence of sodium on the dynamics of CIS absorber film reaction chemistry will require an answer to this question.

## BIBLIOGRAPHY

1. J.R. Tuttle, D.S. Albin, J.P. Goral and R. Noufi, *The Conference Record of the 21st IEEE Photovoltaic Specialists Conference*, 748-754, Kissimee, FL (1990).
2. M.H. Bode, *Journal of Applied Physics* **76**, 159-162 (1994).
3. A. Rockett, F. Abou-Elfotouh, D. Albin, M. Bode, J. Ermer, R. Klenk, T. Lommasson, T.W.F. Russell, R.D. Tomlinson, J. Tuttle, L. Stolt, T. Walter and T.M. Peterson, *Thin Solid Films*, 1-11 (1994).
4. J. Hedström, H. Ohlsén, M. Bodegård, A. Kylner, L. Stolt, D. Hariskos, M. Ruckh and H.-W. Schock, *The Conference Record of the 23rd IEEE Photovoltaic Specialists Conference*, 364, Louisville, KY (1993).
5. M. Bodegård, J. Hedström, K. Granath, A. Rockett and L. Stolt, *Proceedings of the 13th European Photovoltaic Solar Energy Conference*, 2080, Nice, France, (1995).
6. U. Rau, M. Schmitt, D. Hilburger, F. Engelhardt, O. Seifert and J. Parisi, *The Conference Record of the 25th IEEE Photovoltaic Specialists Conference*, 1005-1008, Washington, DC (1996).
7. M. Bodegård, L. Stolt and J. Hedström, *Proceedings of the 12th European Community Photovoltaic Solar Energy Conference*, 1743 (1994).
8. M. Ruckh, D. Schmid, M. Kaiser, R. Schäffler, T. Walter and H.W. Schock, *Proceedings of the 1st World Conference on Photovoltaic Energy Conversion*, 156, Waikaloa, Hawaii (1994).
9. J.R. Tuttle, T.A. Berens, J. Keane, K.R. Ramanathan, J. Granata, R.N. Battacharya, H. Wiesner, M.A. Contreras and R. Noufi, *The Conference Record of the 25th IEEE Photovoltaic Specialists Conference*, Washington, DC (1996).
10. A. Rockett, M. Bodegård, K. Granath and L. Stolt, *The Conference Record of the 25th IEEE Photovoltaic Specialists Conference*, 985-987, Washington, D.C. (1996).
11. H. Oumous, A. Knowles, M.H. Badawi, M.J. Carter and R. Hill, *Proceedings of the 9th European Community Photovoltaic Solar Energy Conference*, 153, Freiburg (1989).
12. A. Knowles, H. Oumous, M.J. Carter and R. Hill, *The Conference Record of the 20th IEEE Photovoltaic Specialists Conference*, 1482, Las Vegas, NV (1988).
13. D.S. Albin, G.D. Mooney, A. Duda, J. Tuttle, R. Matson and R. Noufi, *Solar Cells* **30**, 47-52 (1991).
14. R. Gay, M. Dietrich, C. Fredric, C. Jensen, K. Knapp, D. Tarrant and D. Willett, *Proceedings of the 12th European Photovoltaic Solar Energy Conference*, 935-938, (1994).
15. B.M. Basol, V.K. Kapur and A. Halani, *The Conference Record of the 22nd IEEE Photovoltaic Specialists Conference*, 893-897, Las Vegas, NV (1991).
16. D.S. Albin, G.D. Mooney, J. Carapella, A. Duda, J. Tuttle, R. Matson and R. Noufi, *Solar Cells* **30**, 41-46 (1991).
17. N. Orbey, H. Hichri, R.W. Birkmire and T.W.F. Russell, *The Conference Record of the 25th IEEE Photovoltaic Specialists Conference*, 981-984, Washington, DC (1996).

18. R.A. Mickelsen and W.S. Chen, *Applied Physics Letters* **36**, 371-373 (1980).
19. W.S. Chen and R.A. Mickelsen, *Proceedings of the Society of Photo Optical Instrumentation Engineers*, 62, (1981).
20. E.R. Don, R. Hill and G.J. Russell, *Solar Cells* **16**, 131-142 (1986).
21. D.J. Chakrabarti and D.E. Laughlin, *Bulletin of Alloy Phase Diagrams* **2**, 305 (1981).
22. B.J. Stanbery, W.S. Chen, W.E. Devaney and J.M. Stewart, Phase 1 Annual Subcontract Report Report No. Contract ZH-1-19019-6, 1992.
23. J.R. Tuttle, M. Contreras, A. Tennant, D. Albin and R. Noufi, *The Conference Record of the 23$^{rd}$ IEEE Photovoltaic Specialists Conference*, 415-421, Louisville, KY (1993).
24. M.H. Bode, M.M. Al-Jassim, K.M. Jones, R. Ratson and F. Hasoon, *AIP Conference Proceedings 268: Photovoltaic Advanced Research and Development Review Meeting*, 140-148, Denver, CO (1992).
25. A.M. Gabor, J.R. Tuttle, D.S. Albin, M.A. Contreras, R. Noufi and A.M. Hermann, *Applied Physics Letters* **65**, 198-200 (1994).
26. A.M. Gabor, J.R. Tuttle, M. Contreras, D.S. Albin, A. Franz, D.W. Niles and R. Noufi, *Proceedings of the 12$^{th}$ European Photovoltaic Solar Energy Conference*, 939-943, Amsterdam, The Netherlands (1994).
27. F. Karg, V. Probst, H. Harms, J. Rimmasch, W. Riedl, J. Kotschy, J. Holz, R. Treichler, O. Eibl, A. Mitwalsky and A. Kiendl, *The Conference Record of the 23$^{rd}$ IEEE Photovoltaic Specialists Conference*, 441-446, Louisville, KY (1993).
28. C.H. Chang, A. Davydov, B.J. Stanbery and T.J. Anderson, *The Conference Record of the 25$^{th}$ IEEE Photovoltaic Specialists Conference*, 849-852, Washington, D.C. (1996).
29. J.-Y. Emery, L. Brahim-Ostmane, C. Herlemann and A. Cheny, *Journal of Applied Physics*, 3256 (1992).
30. C. Chatillon and J.-Y. Emery, *Journal of Crystal Growth*, 312-320 (1993).
31. D.A. Cammack, K. Shahzad and T. Marshall, *Applied Physics Letters* **56**, 845-847 (1990).
32. Z.Z. Kish, V.B. Lazarev, E.Y. Peresh, E.E. Semrad and I.S. Shaplygin, *Russian Journal of Inorganic Chemistry* **30**, 854-856 (1985).
33. H. Neumann, U.-C. Boehnke, G. Nolze, B. Schumann and G. Kühn, *Journal of Alloys and Compounds*, L11-L12 (1994).
34. D. Schmid, M. Ruckh, F. Grunwald and H.W. Schock, *Journal of Applied Physics* **73**, 2902-2909 (1992).
35. C. Rincón and S.M. Wasim, *Proceedings of the 7$^{th}$ International Conference on Ternary and Multinary Compounds*, 443-452, Snowmass, CO (1986).
36. C. Heske, R. Fink, E. Umback, W. Riedl and F. Karg, *Applied Physics Letters* **68**, 3431-3433 (1996).

# The Effect of Surface Processing Conditions on the Junction Properties of $CuIn_xGa_{1-x}Se_2$ Solar Cells

S. Zafar, J. D' Amico, S. Karthikeyan, R. Narayanaswamy, P. Panse,
H. Sankaranarayanan, C. S. Ferekides and D. L. Morel

*Department of Electrical Engineering*
*Center for Clean Energy and Vehicles*
*University of South Florida*
*Tampa, Florida 33620*

**Abstract.** Using a manufacturing friendly process we have developed effective techniques for incorporating Ga in CIGS films. In one technique we configure the Ga to form an effective BSF and to reduce point defects in the SC region while not increasing the band gap. This results in collection lengths of 2 µm and $J_{sc}$'s which are consistently in the 40 mA/cm² range. In a second technique we alloy Ga in the SC region and increase the band gap. This results in mild deterioration of both bulk and surface properties attributable to an unoptimized incorporation environment. The surface properties are shown to be dominated by recombination lifetime which varies systematically with the Se flux environment during surface formation. These insights are providing the foundation for ongoing advances in device performance and imply no fundamental limitations to performance for these processing techniques.

## INTRODUCTION

Although several groups have reported CIGS device efficiencies in excess of 15%(1), the technology has not yet been commercialized. There are many complex factors which enter the decision making process for commercialization, including those which are non-technical. However, technical factors are clearly at work here as well. We believe that the commercialization process would be hastened if improved processing options were developed. A primary thrust of this project is to contribute to this need by developing an easily manufacturable process for state-of-the-art CIGS devices. The issues which define our approach are avoidance of hydrogen selenide gas as the Se source and codeposition steps requiring tight control. This leads us to all-solid-state processing using elemental In, Cu, Ga and Se sources and delivery of these to the substrate in a primarily sequential manner. The details of our procedures have been discussed elsewhere(2), and in last year's edition of this paper we presented the results which we had attained for CIS(3). In this paper we focus on the results attained by introducing Ga into our process. The primary result of our success with Ga is that we have boosted our efficiencies to the 12% level while maintaining our manufacturing-friendly guidelines.

# BACKGROUND

Device properties are determined by two primary process variables, the precursor and the anneal profile. The precursor consists of layers of Cu, In, Ga, and Se which are deposited in various sequences. All of the precursor components are deposited at temperatures typically below 275° C. Fabrication is completed by annealing the precursor with a temperature ramp up to 550° C followed by a cooldown to room temperature. During the anneal profile Se flux is applied as an additional variable. There are virtually an infinite number of perturbations that can be made with these processing components. We have spent considerable effort exploring this deposition space, and from the understanding which we have developed based on this experience we have narrowed the field two primary approaches which we label PI and PII. As will be discussed in detail below, these offer two different options for the incorporation of Ga and the optimization of performance. The key difference between the two is that in PI the Ga is deposited before the In, while in PII the In is deposited first. Since to first order In and Ga are thought to be interchangeable in a homogeneous CIGS lattice, by reversing the order in which they enter the lattice we examine to what extent this is found to be the case.

An important component of our program is the use of a device model to help understand performance and to guide our fabrication efforts. It is generally accepted that CIGS device performance is dominated by recombination in the space charge layer. Consequently we have adopted the original Shockley-Read-Hall(SRH) recombination theory(4) as the primary component of our model. We have discussed the application of this model to CdTe as well as to CIGS previously(5)(6) and have found it to consistently match experimental observations and to provide significant insights to the underlying mechanisms. A key aspect of this model is that it does not follow a simple diode formula with single values of diode factor and $J_o$. The effective voltage dependent $J_o$ is primarily driven by the recombination lifetime, while an effective voltage dependent diode factor is found by taking the slope of the I-V curve which in turn is influenced by the location of the recombination centers in the gap. Although we use the model to guide our day-to-day activities, we recognize its limitations with respect to the advanced models developed at Penn State and Purdue. Those models contain the same physics but allow for layer-to-layer variations in properties as well as the integration of other mechanisms. When more complex analysis is required, we turn to those models for assistance.

We, as do others, also use XRD, EDS and other analytical tools to characterize our materials and devices. While we find them to be generally useful, their applicability usually falls short of the realm of point defects and ppm variations in composition in which much of device performance is determined. But this is the realm in which the SRH model is most useful. Recombination lifetimes are the primary variables for current flow, and these lifetimes are the direct result of microstructure and micro composition. By careful analysis of IV curves, especially those generated from $I_{sc}$-$V_{oc}$, we learn a great deal about device parameters and from those about the effects of process variations on material properties. A particular point of interest in this paper is the effect of Ga on these properties. Since our analysis suggests that the primary recombination zone is within about 700 Å of the metallurgical junction, process variations which affect this "surface" play a key role in determining $V_{oc}$ and thus are a focal point of our investigations.

## RESULTS AND DISCUSSION

### Type I

Type I devices are a direct link to our previous efforts with CIS in that the band gap

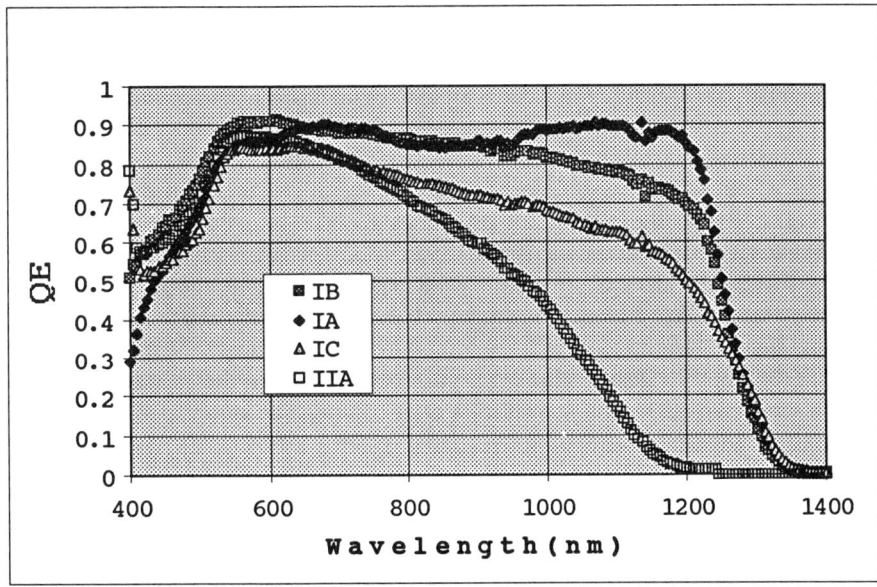

Figure 1. Quantum efficiency spectra for type I and Type II devices.

remains the same as CIS in spite of the fact that enough Ga is deposited to result in a Ga/(In + Ga) ratio of about 10 %, if all of the Ga were distributed uniformly in the film. The spectral responses of several type I devices(and a type II) are shown in figure 1. Using the wavelenth at 1/e of the peak quantum efficiency(QE) results in a band gap of .98 eV which is the same as our CIS devices whose spectral response is similar to device IC shown in figure 1. We note that these devices do not have AR coatings, and that the QE is calibrated with NREL calibrated Si and CIS reference cells.The integral of these responses results in the $J_{sc}$ values shown in table 1. The point to be made here is that the judicious incorporation of Ga has increased our current densities by about 5 mA/cm². We offer two perspectives for the role played by Ga. Since the unchanged band gap suggests that it is not present at alloy levels(or at least not chemically bonded) in the space charge(SC) layer, it must be located between the Mo contact and the space charge layer. Assuming that it has chemically incorporated in this region, the resulting larger band gap would form the equivalent

| Device | $J_{sc}$ mA/cm$^2$ | Depletion Width(μm) | Diffusion Length(μm) |
|---|---|---|---|
| IA | 39.7 | 1.0 | 2.0 |
| IB | 39.2 | 0.4 | 0.8 |
| IC | 34.5 | 0.3 | 0.3 |

**Table 1.** Device and material properties for type I devices.

of a back surface field(BSF) which would aid the collection of minority carriers. To investigate this possibility we used a spectral response model to fit the data of figure 1. The model is a straightforward adaptation of the standard formulas found in Sze(7) including the assumption that carriers generated in the SC region are collected. The biggest uncertainty in our procedures involved absorption constants. We used two sources(8) (9) which had differences between them, and one of the sources even demonstrated that the absorption properties especially in the rise portion were sensitive to Cu/ ( Cu + In) ratios. Within the limits of these uncertainties we were able to determine the values of SC width(W) and minority carrier diffusion lengths(L) shown in table 1. We see that for these devices increases in $J_{sc}$ result from increases in both W and L. The increase in L we attribute at least in part to the proposed BSF. The increase in W leads in to the other proposed role of Ga, that at trace levels it is able to favorably affect the point defect composition. The abrupt rise in QE at the absorption edge in device IA is testimony to the positive influence of Ga. The resulting overall collection length of 3 μm is remarkable, and the high values of $J_{sc}$ reflect that fact.

The range of performance exhibited in figure 1 is the result of variation in the Ga incorporation. We wish to note however, that the high end performance of devices IA and IB is reproducible and consistent. This solid and reliable current generation thus provides a firm foundation for studying and advancing $V_{oc}$ and FF performance. The power curve for device IB is shown in figure 2. With a FF of .66 and $V_{oc}$ of .463 its efficiency is 12.0%. These values of FF and $V_{oc}$ represent significant improvements over CIS as well. In this case we attribute the effect to reduction of recombination near the metallurgical junction. Thus trace levels of Ga are operative throughout the SC layer and are improving the microstructure. To examine this issue more closely we use the SRH model to fit $J_{sc}$- $V_{oc}$ data as shown in figure 3. The two sets of data points are measurements going up and down in intensity to look for hysteresis caused by trapping. The parameters used in the calculations are from direct measurements and from representative literature values for difficult parameters such as cross

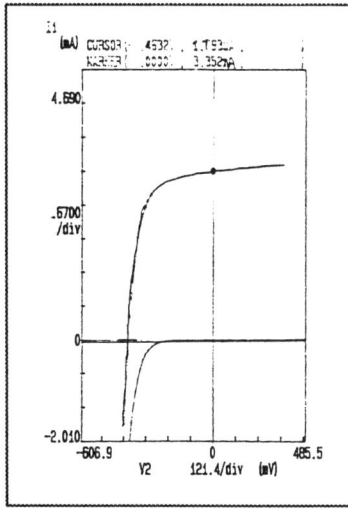
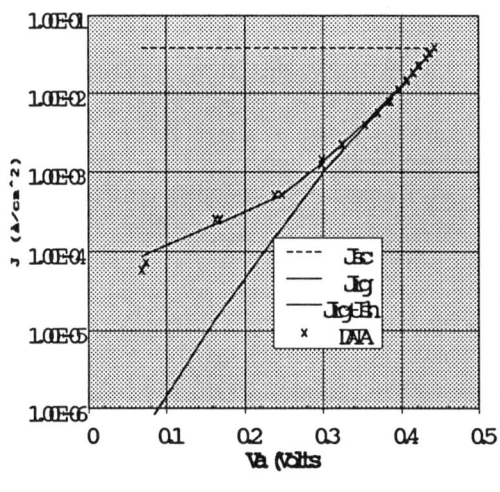

**Figure 2.** Power curve for device IB, a 12% type I device.

**Figure 3.** Fit to $J_{sc}$-$V_{oc}$ data for a type I device. The dashed line is $J_{sc}$ at one sun. The lower line is the theoretical curve without including shunting. The line through the data includes a shunt contribution to current.

sections. While these limit the accuracy of the calculations, relative comparisons are meaningful. For example, the recombination lifetime for the device in figure 3 is found to be 7.1 x $10^{-9}$ s from the fit. This is in good agreement with direct measurements of lifetime in companion devices by the DBOM technique(8). The effect of Ga on this lifetime will be discussed further below in reference to type II devices.

Previously we have reported our efforts to improve $V_{oc}$ 's in terms of reduction of recombination center density(RCD). As seen in figure 4, our best CIS devices had a RCD of 1 x $10^{17}$/cm$^3$ and $V_{oc}$ 's just over .4 volts. We had also demonstrated that by etching the surface of these devices to reduce the RCD $V_{oc}$ could be raised to .53 volts, and this became our goal. With type I devices we have now reached .49 volts which indicates significant progress toward our goal. Through judicious use of Ga to reduce the RCD we have also increased $J_{sc}$. With further refinements in surface processing, as we push the RCD below $10^{16}$/cm$^3$ we expect to realize improvements in FF to the mid .7 range as well while maintaining current densities of 40 mA/cm$^3$. This will result in device efficiencies above 15%.

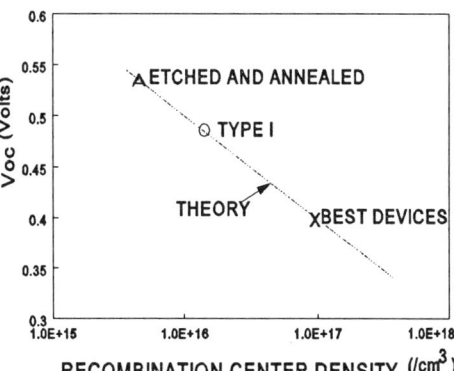

**Figure 4.** $V_{oc}$ versus recombination center density, data and SRH model.

## Type II

In terms of theoretical efficiency it is clear that Ga should be used to increase the band gap to the 1.4 eV range, and the highest reported efficiencies are for CIGS with band gaps in the 1.1 - 1.2 eV range. As seen in figure 1, for type II devices we are able to shift our band gap into this range. This results from depositing Ga after In in our process sequence. General observations suggest that Ga migrates towards the back of the device and In towards the front. Thus in order to incorporate alloy levels of Ga in the SC region it is necessary to place the Ga in that vicinity during the critical formation stage of that region. In addition, the details of the processing conditions during Ga incorporation determine its bonding effectiveness. We feel that the downward slope in the spectral response for device IIA is due to incomplete Ga incorporation. That is, poorly or unbonded Ga is deteriorating transport properties. The corresponding slope in device IC may be due to the same effect. Although type I devices only have trace quantities of Ga in the SC layer, as these are increased toward the alloy level we see the onset of the downward slope prior to band gap changes. Since codeposition processing does not reveal as strong an effect on the slope as we observe, at least at Ga levels below about 20%, this is not a fundamental problem but a process-dependent problem. We have recently made small modifications to our process and are observing improvements in these slopes. Details will be provided in a future publication.

The effect of Ga on surface properties of type II devices is also mixed at this point. Device efficiencies are in the 10 - 11% range due to a less than 1-1 increase in $V_{oc}$ with increasing band gap in addition to the loss in $J_{sc}$. Our highest $V_{oc}$ for a type II device is .57 volts. This is an increase of only 80 mV with respect to type I devices for a band gap increase of 150 mV. We attribute this shortfall to a deterioration of the interface caused by ineffective incorporation of Ga. As stated above Ga incorporation is a function of its availability and the immediate environment. In our process a major contributor to the environment at each step is the Se flux. To observe the influence of these parameters, in a single run we create deposition gradients across a matrix of up to 25 devices. The effect of gross relative Se/Ga ratios on recombination lifetimes for a typical type II device is shown in table 2. The Ga level increases by only 10% from left to right in the table, while the Se flux decreases by about 40%. The band

| Relative Se/Ga | | 1.6 | 1.4 | 1.2 | 1.0 |
|---|---|---|---|---|---|
| minority lifetime | $10^{-9}$ s | 2.8 | 2.0 | 0.71 | 0.42 |
| $V_{oc}$ | | .53 | .51 | .46 | .43 |

**Table 2.** Properties of a type II device versus relative Se/Ga flux.

gap of the devices with the lowest Se/Ga (and overall highest Ga content) is about .03 eV higher than the highest Se/Ga and on that basis are expected to generate a higher $V_{oc}$. As can be seen the opposite occurs, and it is clear that $V_{oc}$ for these devices is dominated by minority lifetime. The lifetime decreases with increasing Ga because the incorporation environment, as represented by the Se flux, is not proper. The region in which these effects are dominant is within 700 Å of the metallurgical junction. The primary influence is thus on dark current which directly affects $V_{oc}$. These mechanisms should also have some effect on light generated carriers in this region. We have preliminary data in support of this, but more detailed analysis is required because of difficulties with absorption constants.

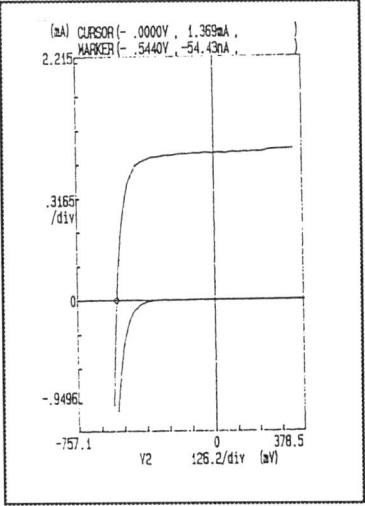

**Figure 5.** Power curve for a type II device with FF = .73.

In spite of the shortcomings of our type II surfaces we observe FF's in the low .7 region as seen in figure 5. Type I devices with superior surfaces have slightly lower FF's(up to .70) because they tend to be slightly more shunted. Thus their surfaces will allow FF's into the mid .7 range as stated above.

## CONCLUSIONS

The incorporation of Ga during the formation of CIGS devices although primarily driven by thermodynamics, can be significantly modified by manipulation of its delivery to the forming film and the environment during its critical incorporation period. These manipulations can be accomplished within the confines of a manufacturingly friendly process. By causing the bulk of Ga to remain in the back of the device in type I devices we have formed an effective BSF which results in large collection lengths and consistently produces $J_{sc}$'s in the 40 mA/cm² range. Trace quantities of Ga which remain in the SC region improve the point defect structure resulting in increases in both $V_{oc}$ and FF. These improvements in device properties result in device efficiencies in the 12% range. By altering the Ga incorporation procedure we can systematically increase the amount of Ga which is incorporated in the SC layer and raise the band gap accordingly. In doing so however, we find some deterioration in film properties. We attribute this to ineffective incorporation of Ga due to an unoptimized growth environment. In recent experiments we have observed improvements in $J_{sc}$'s for these structures reflecting modifications to the process aimed at bulk properties. Similar efforts will be directed toward surface properties. The progress which we have made in performance and the understanding upon which that progress is built provide a solid base for further progress. We are confident that these processing approaches can produce device efficiencies in excess of 15%.

## ACKNOWLEDGMENTS

These results were produced under contract from NREL.

## REFERENCES

1a. J. Hedstrom, H. Ohlsen, M. Bodegard, A. Kylner, L. Stolt, D. Hariskos, M. Ruckh, and. H. Schock, 1993, "ZnO/CdS/Cu(In,Ga)Se$_2$ Thin Film Solar Cells with Improved Performance", Proceedings of the XXIII rd IEEE PV Specialist Conference, Louisville, KY, pp. 364-371.

1b. D. Tarrant, and J. Ermer, 1993, "I-III-VI$_2$ Multinary Solar Cells Based on CuInSe$_2$", Proceedings of the XXIII rd IEEE PV Specialist Conference, Louisville, KY, pp. 372-378.

1c. M. A. Contreras, J. Tuttle, D. Albin, A. Tennant, K. Ramanathan, A. Gabor, J. Scofield, and R. Noufi, 1994, "High Efficiency Cu(In,Ga)Se$_2$ - Based Solar Cells: Processing of Novel Absorber Structures", Proceedings of the First World Conference on Photovoltaic Energy Conversion, Waikoloa, HA, to be published.

1d. W. N. Shafarman, R. Klenk, and B. E. McCandless, "Characterization of Cu(In,Ga)Se$_2$ Solar Cells with High Ga Content" Proceedings of the 25th IEEE PV Specialists Conference, Washington, D.C., May, 1996.

2. D. L. Morel, G. Attar, S. Karthikeyan, A. Muthaiah and S. Zafar, 1993, "Advanced Processing Technology for High-Efficiency Thn-Film CuInSe$_2$ Solar Cells", NREL Annual Subcontract Report NREL/TP-451-5653.

3. G. Attar, S. Karthikeyan, H. Natarajan, D. Nierman, S. Zafar, C. S. Ferekides and D. L. Morel, "The Effect of Interface States on CuInSe$_2$ Solar Cells", Proceedings of the 13th NREL Photovoltaics Program Review Meeting, November, 1995, Denver.

4a. W. Shockley and W. T. Read, Jr., Phys. Rev. B, **87,** no. 5, 835-842(1952).

4b. R. N. Hall, Phys. Rev., **83,** 228(1951) and **87,** 387(1952).

5. D. M. Oman, K.M Dugan, J.L Killian, V. Ceekala, C.S. Ferekides and D.L. Morel, *Appl. Phys. Lett.* 67(13), 1896 (1995).

6. D. L. Morel and C. S. Ferekides, 1994, "Heterojunction Development and Optimization in Thin-Film Compound Semiconductor Solar Cells", NREL Annual Subcontract Report NREL/XAD-3-12114-3.

7. S. M. Sze, "Physics of Semiconductor Devices", Wiley Interscience, New York, 1981.

8. J. R. Tuttle, R. Noufi and R. G. Dhere, "The Effect of Composition on the Optical Properties of CuInSe$_2$ Thin Films" Proceedings of the 19th IEEE PV Specialist Conference, New Orleans, 1987, p 1494.

9. IEC, private communication.

# CHALLENGE OF REPLACING CdS IN CuInSe$_2$-BASED SOLAR CELLS

Larry C. Olsen, F. William Addis, Wenhua Lei and Heriberto Aguilar
Washington State University at Tri-Cities
100 Sprout Rd. Richland, WA 99352

## ABSTRACT

This paper discusses some key issues concerning the replacement of CdS buffer layers in CIS solar cell structures, and describes investigations of alternative buffer layers deposited by MOCVD. One apparently unique property of CdS buffer layers grown by CBD is that a ZnO TCO can be deposited on top of a CdS/CIS structure without significantly degrading the photovoltaic properties of the CdS-CIS junction. Investigation of alternative buffer materials such as high resistance ZnO (i-ZnO), ZnSe and InSe have first identified MOCVD growth procedures that yield Al/X/CIS test structures (X = i-ZnO, ZnSe and InSe) with good properties, and then addressed the challenge of fabricating efficient, complete cells with conductive ZnO top contact layers. These studies have been conducted with Siemens CIS and CIGSS substrates, and with NREL CIGS substrates. A total area efficiency of 12.7 % and estimated active area efficiency of 13.4 % is reported for a CIGS cell with an i-ZnO buffer layer grown by MOCVD.

## INTRODUCTION

Replacement of CdS in the CIS solar cell structure could be important for these cells to be accepted in the marketplace, and/or may allow an all vacuum process to be possible for CIS cell fabrication. However, it is clear that CBD grown CdS films have some unique properties that have allowed the fabrication of CIGS cells with very high efficiencies. Understanding these unique properties may be essential to the development of alternative buffer layers. This paper considers the challenge of replacing CdS buffer layers in CIS solar cell structures, and describes investigations of alternative buffer layers deposited by MOCVD. In particular, studies have been carried out with buffer layers based on highly resistive ZnO (i-ZnO), ZnSe and InSe grown by MOCVD.

## SOME UNIQUE PROPERTIES OF CBD CdS BUFFER LAYERS

Many properties of CBD CdS films might be identified as particularly important to the fabrication of high efficiency CdS/CIGS cells, such as the 17.7 % result reported by the NREL group (1). The electron band structure of a CdS/CIS or CdS/CIGS structure is certainly appropriate for achieving good performance. Some researchers think the solution growth process leads to a well passivated interface. Based on simulation studies discussed below, the high resistance of CBD CdS buffer layers is beneficial for attaining high efficiency. One property of CBD CdS films that may be the most important of all is that conductive ZnO top contact layers, either RF sputtering or CVD, can be deposited on top of the CdS layer without any apparent degradation of the CdS/CIS junction. Thus, the use of CBD CdS buffer layers allows the completion of an efficient cell structure. We have found that deposition of conductive ZnO layers on structures based on alternative buffer layers often leads to low efficiency cells, despite the fact that test structures formed before the top contact layer was deposited indicated that good performance should be achieved with completed cells. After procedures for growth of alternative buffer films have been developed, determination of processes that allow successful completion of efficient cell structures becomes the focus of cell development. One of the major challenges of replacing the CdS buffer layer in CIS-based solar cells is to identify an approach to depositing a transparent conducting oxide that is compatible with the alternative buffer layer.

## ZnO BUFFER LAYER STUDIES

ZnO is one the leading candidates for a non-Cd buffer layer. Most efforts to deposit ZnO directly onto CIS or CIGS have resulted in low efficiency cells. An active area efficiency of 10.5 % was achieved by the European CIS group (2) using a sputtered ZnO layer, and we have previously reported a total area efficiency of 11.3 % (active area efficiency of 12.0 %) for a cell with a ZnO buffer layer grown by MOCVD (3). Further improvements in efficiency for cells based on ZnO buffer layers are presented in this paper. In the remainder of this section, ZnO film growth, optimization of ZnO buffer layers, modeling studies, physical characterization and solar cells are discussed.

### ZnO Film Growth

Chemical vapor deposition of ZnO is done by reacting a zinc adduct with tetrahydrofuran using the WSU MOCVD system which consists of a SPIRE 500XT reactor with added gas handling capabilities. The zinc adduct is formed by reacting dimethylzinc and triethylamine. The approach to CVD growth of ZnO is distinguished from other approaches in two important ways, namely: (1) The zinc and oxygen precursors are both relatively large molecules; (2) and no extra oxygen or ions are involved in the deposition process.

### Optimization Of ZnO Buffer Layers

Optimization studies have concentrated on investigating the effects of substrate temperature ($T_{sub}$), as-deposited film resistivity and film thickness on test cell performance. Measurements of ZnO film thickness and resistivity are made on a film grown on a glass witness to provide an estimate of these properties of the film grown on CIS. Illuminated characteristics of test cells are measured by illuminating the device such that $J_{sc} \approx 40$ mA/cm$^2$ for CIGSS substrates, and 36 mA/cm$^2$ for CIGS substrates. The cell efficiency may be regarded as an estimated active area, AM1.5G efficiency, assuming that one can add a top contact layer and collector grid without degrading the junction properties. Although aluminum is quite reactive, the test cells provide useful information regarding buffer layer properties since the ZnO films are typically ≥ 500 Å thick, which makes the ZnO-Al interface relatively far from the ZnO-CIS interface.

Best results for ZnO buffer layers have been achieved with a two-step growth process. First, a layer on the order of 100 Å to 150 Å is grown at 250°C, followed by growth of 500 Å to a 1000 Å at 100°C. It is also important for the as-deposited film resistivity to be > 10,000 ohm-cm. ZnO films grown at 250°C have a smoother morphology compared to films deposited at 100°C. Thus, the 250°C step may provide good coverage but must be limited in time to minimize interdiffusion effects. Further studies are required in order to understand the significance of the two step approach. The necessity of the as-deposited resistivity being relatively large may a basic requirement for direct ZnO/CIS cells. In particular, ZnO/CIS solar cells fabricated with low resistivity (.001 to .01 ohm-cm) ZnO in direct contact with CIS and related alloys generally lead to cells with low efficiencies. Based on modeling studies discussed below, one should be able to achieve efficiencies > 10 % with the resistivity > 1.0 ohm-cm. Therefore, it appears that the resistivity requirement is related to processing.

## Modeling Studies

Modeling calculations have been carried out using PC-1D, a one dimensional computer code based on a finite element numerical approach to solve the semiconductor equations. Up to three regions of different material parameters can be used to define a cell structure, each with its own doping profiles and electronic and optical properties. Recombination of the electron hole pairs can be defined in each region by SRH band to band transitions or through user-defined deep level transitions. Surface recombination at interfaces is also taken into account. Simulation studies were carried out for CIS cells with ZnO buffer layers, and with a low resistivity ZnO top contact layer. The main objective of the work was to determine if low efficiency n-ZnO/CIS cells could be understood by assuming a high surface recombination velocity (S), or by assuming a low value of recombination lifetime (tau) in the space charge region. Calculated efficiency verses buffer layer resistivity is plotted in Figure 1 for a range of values of S and tau. The simulation studies indicate that the low efficiencies observed for cells with low resistivity ZnO buffer layers may be caused by reduced values of tau resulting from processing.

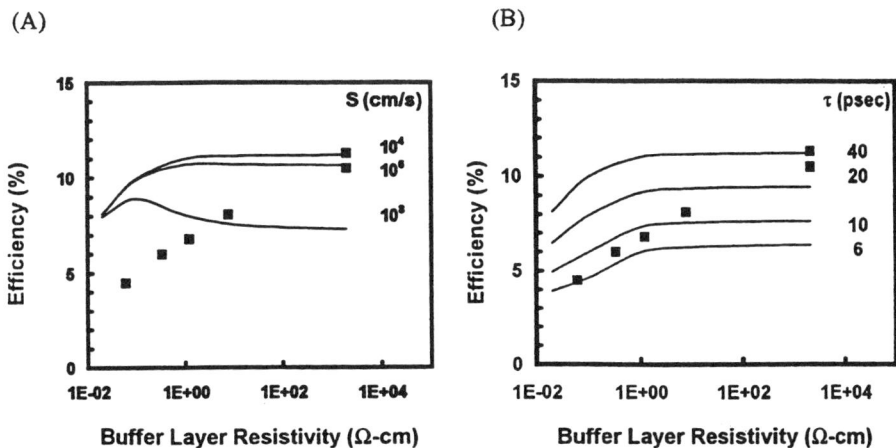

**FIGURE 1.** Calculated CIS cell efficiency vs ZnO buffer layer resistivity for:
(A) a range of values of surface recombination velocities and tau = 40 psec, and
(B) a range of lifetimes and S = 1E4 cm/sec. ZnO Thickness = 300 Å.

## Physical Characterization

CIS cells with ZnO buffer layers have been characterized by spectroscopic ellipsometry. In order to fit experimental data for pseudooptical constants of a ZnO film grown by the two-step process onto a Siemens CIS substrate, it was necessary to assume a layered structure depicted by Figure 2. The ZnO buffer layer grown with the two step method apparently results in two layers, the top layer being approximately 300 Å with a bandgap of 3.3 eV, and the bottom layer adjacent to the CIS material approximately 600Å with a bandgap of 2.6 eV. Whereas the top layer is apparently ZnO, the region adjacent to CIS is another compound with a bandgap intermediate between ZnO and CIS. A candidate material for the intermediate layer is $ZnInSe_2$. Additional ellipsometric measurements are planned in conjunction with low angle X-ray diffraction measurements to confirm these results.

## ZnO/CIS Solar Cells

ZnO/CIS solar cells have been fabricated by first depositing a ZnO buffer layer onto a CIS (or related alloy) substrate, followed by deposition of a low resistivity ZnO top contact layer and collector grid. A total area efficiency of 11.3 % was reported for a cell fabricated with a Siemens CIGSS substrate (3). Collaboration with NREL this past year has allowed fabrication of cells with NREL CIGS material. It was determined that the two-step growth method was appropriate for NREL substrates. Figure 3 describes results for a cell fabricated this year using a NREL substrate. After depositing a highly resistive ZnO buffer layer (referred to as i-ZnO) onto a NREL substrate, the i-ZnO/CIGS structure was sent to NREL for deposition of a low resistance ZnO top contact layer, deposition of a collector grid and AR coating. These last two steps are typically done at WSU, but cell completion was done at NREL in this case. The total area efficiency was 12.7 % and the estimated active area efficiency was 13.4 %.

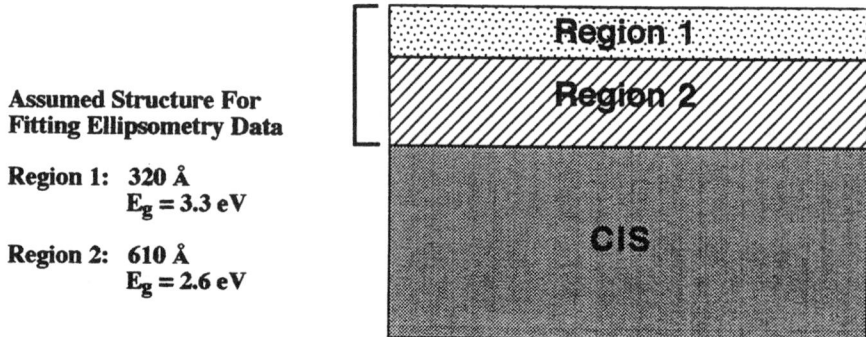

**FIGURE 2.**. Assumed layer structure that allows excellent fit of experimental data for pseudooptical constants measured with ellipsometry.

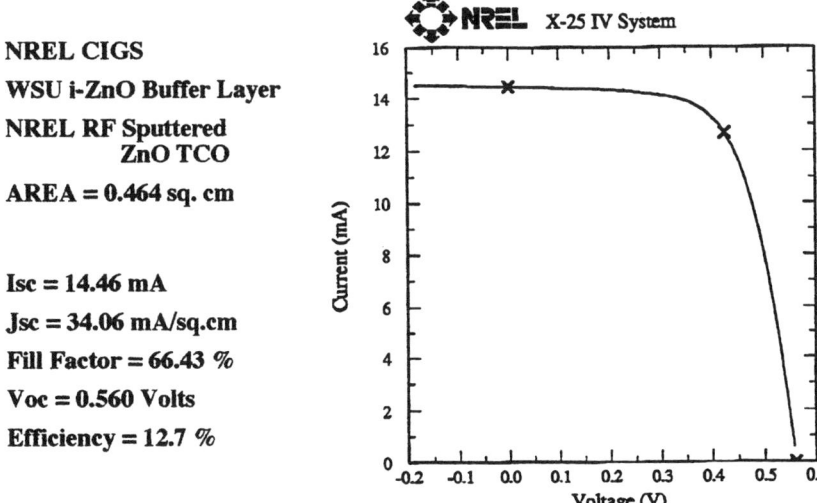

NREL CIGS
WSU i-ZnO Buffer Layer
NREL RF Sputtered ZnO TCO
AREA = 0.464 sq. cm

Isc = 14.46 mA
Jsc = 34.06 mA/sq.cm
Fill Factor = 66.43 %
Voc = 0.560 Volts
Efficiency = 12.7 %

**Figure 3.** Illuminated I-V characteristics as measured by NREL for a completed ZnO/CIGS solar cell with a high resistivity ZnO buffer layer.

## ZnSe BUFFER LAYER STUDIES

Investigations of CIS cells based on ZnSe buffer layers continued, but at a lower level than the studies of ZnO layers. Test cells fabricated using ZnSe buffer layers exhibit exceptional properties. In particular, Test cells based on Siemens CIS and CIGSS substrates, and NREL CIGS substrates all exhibit fill factors $\geq$ 0.7, and relatively large values of open circuit voltage. To date, a process for depositing a top contact layer that allows one to complete a cell with properties approaching those of test cells has not been developed. Recently, an encouraging result was achieved. A ZnSe/CIS(Siemens) structure was sent to NREL for deposition of a conducting ZnO top contact layer by RF sputtering, and then returned to WSU for deposition of $MgF_2$ and collector grids. Illuminated characteristics measured at WSU are shown in Figure 4. Although the resulting cell performance is significantly less than the 12 to 13 % suggested by test cell studies, the completed cell exhibited improved properties over those previously obtained with other ZnO TCO deposition processes.

## InSe BUFFER LAYER STUDIES

Investigation of InSe buffer layers was also initiated this past year. InSe films are grown by reacting $H_2Se$ and ethyldimethylindium with substrate temperatures in the range of 300 to 400 °C. Test cell studies have shown promise. Raman spectroscopic analyses indicate that the crystallinity of the films improve as $T_{sub}$ is increased from 300°C to 400°C. The Raman studies also suggest that the amount of the $\gamma$-$In_2Se_3$ phase increases with $T_{sub}$. This phase has a bandgap of approximately 2.0 eV.

Siemens CIS
WSU ZnSe Buffer Layer
NREL RF Sputtered ZnO TCO
AREA = 0.464 sq. cm

Isc = 16.5 mA
Jsc = 36.44 mA/sq.cm
Fill Factor = 62.15 %
Voc = 0.420 Volts
Efficiency = 9.51 %

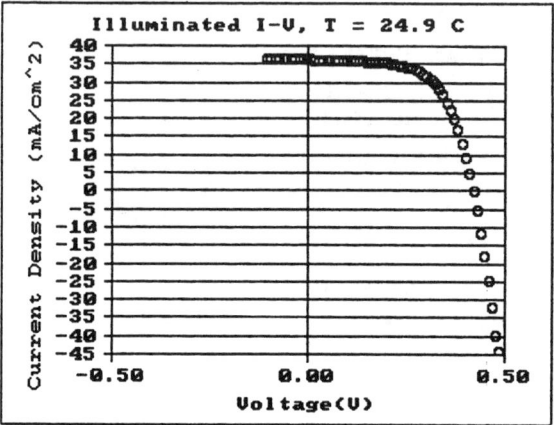

**FIGURE 4.** Illuminated I-V characteristics measured at WSU for a completed n-ZnO/ZnSe/CIS solar cell with the substrate being Siemens CIS material, and the conducting- ZnO TCO deposited by RF sputtering at NREL.

## ACKNOWLEDGMENTS

This work was supported by The National Renewable Energy Laboratory under Subcontract XG-2-11036-6. We are grateful for the guidance provided by our contract monitor Bolko Von Roedern, collaboration with Kannan Ramanathan and the measurement support by NREL. We are also very grateful to Siemens Solar and NREL for providing CIS substrates.

## REFERENCES

1. Ken Zweibel, Harin S. Ullal and Bolko von Roedern, "Progress And Issues In Polycrystalline Thin-Film PV Technologies," Proceedings 25th IEEE Photovoltaic Specialists Conf., 1996, pp 745 - 750.

2. J. Kessler. M. Ruckh, D. Hariskos, U. Ruhle, R. Menner and H.W. Schock, "Interface Engineering Between $CuInSe_2$ And ZnO," in Proceedings of the 23rd IEEE Photovoltaic Specialists Conference, 1993, pp. 447-452.

3. Larry C. Olsen, Heriberto Aguilar, F William Addis, Wenhua Lei and Jun Li, " CIS Solar Cells With ZnO Buffer Layers," Proceedings 25th IEEE Photovoltaic Specialists Conf., 1996, pp 997 - 1000.

# Monocrystalline CuInSe$_2$-CdO Cell

## Z.A. Shukri and C.H. Champness

*Electrical Engineering Department, McGill University, Montreal, Quebec, Canada H3A-2A7*

**Abstract** Photovoltaic cells of the form CuInSe$_2$ (p) - CdO (n) have been fabricated by sputtering a layer of CdO on the cleaved surface of a monocrystalline substrate obtained from a Bridgman-grown CuInSe$_2$ ingot. Despite an apparent 23% lattice mismatch between CdO and CuInSe$_2$, the cells gave conversion efficiencies of about 4% using unannealed substrates and about 6% with substrates annealed in argon at atmospheric pressure. The heat-treatment appears to reduce the acceptor concentration near the surface of the CuInSe$_2$ substrate, resulting in an increase of fill factor. The insensitivity of photovoltaic activity to lattice mismatch in the CuInSe$_2$ - CdO cells may be due to the seat of action lying below the substrate surface.

## 1. INTRODUCTION

Single crystal CuInSe$_2$ can provide information useful for the development of thin film photovoltaic solar cells because such bulk monocrystalline material is easier to handle, etch and polish and has controlled composition and precise crystallographic orientation. Test laboratory cells using a monocrystalline CuInSe$_2$ substrate with the structure CuInSe$_2$/CdS/ZnO have yielded a conversion efficiency of over 10% for an effective area of 8 mm$^2$ [1] [2] and with improvements in fabrication techniques, higher efficiencies are expected. In this case, the n-type partner was a thin layer of dip-deposited CdS, which has a lattice mismatch to CuInSe$_2$ of only 1%. The question is now raised how much degradation in performance arises with a significant lattice mismatch between the two semiconductors. However, this question has not been seriously examined experimentally. The present study is therefore a start in such a direction. Here a nominally heterojunction cell was fabricated of the form CuInSe$_2$(p)/CdO(n) (Fig. 1), where the CdO, a semiconductor with a direct energy gap of 2.3 eV [3], was reactively sputtered on a monocrystalline CuInSe$_2$ substrate. Much experience has been gained in this laboratory with the deposition of CdO [4]. This material has a rocksalt crystal structure and a lattice mismatch, in its a-parameter, of some 23 % with that of CuInSe$_2$. The resistivity and optical transparency of the CdO layer can be adjusted through its stoichiometry by control of the deposition rate through sputtering pressure, current, oxygen/argon mixture and gas flow rate. The deposition was made on a cleaved surface of a CuInSe$_2$ substrate obtained from a Bridgman-grown ingot [5].

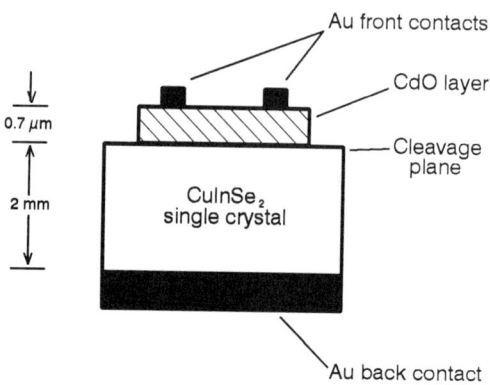

**FIGURE 1.** Schematic diagram of the CuInSe$_2$/CdO cell structure used in this study, showing the different device layers.

## 2. CuInSe$_2$ - CdO CELL FABRICATION

Starting with an ingot of p-type CuInSe$_2$, grown by a one-ampoule method [5], a sample was extracted by cleaving in either a {112} or {101} plane [6]. This sample was then shaped into the form of a platelet by abrasive grinding and polishing on a non-cleaved face to obtain a wafer of about 2 mm in thickness. During this polishing process, the cleaved surface was covered with a protective layer of Apiezon wax, which was afterwards removed. In some cases (see Table I), the wafer was annealed by heating it at about 340 °C in flowing argon for about 2 hours. Next a layer of gold, about 1 micrometer in thickness was evaporated on the back polished surface of the wafer, with the cleaved surface again protected with wax. With the wax removed, a thin copper wire was soldered to the gold back contact and the wafer was mounted into a mechanical holder. Then CdO was reactively sputtered on to the cleaved surface from a cadmium target using a metal mask to define a CdO area of up to 12 mm$^2$, with a thickness of about 0.7 micrometer. Details of the sputtering process are available from previous work [4]. Following this, two thin gold stripes were evaporated on to the CdO area through a metal mask and then fine copper wires were soldered to the stripes with Wood's metal. In Table I, columns 2 to 7, fabrication parameters are given for 7 fabricated CuInSe$_2$ - CdO cells.

## 3. MEASUREMENTS ON CELLS

The following measurements were made on the fabricated CuInSe$_2$ - CdO

Table 1   Characteristics of CuInSe$_2$/CdO Cells

| Cell No. | Cleavage plane | Annealing of CuInSe$_2$ Substrate [1] Time (hr) | Annealing of CuInSe$_2$ Substrate [1] Temp. (°C) | CdO Layer Characteristics Interference colour | CdO Layer Characteristics Thick [3] (μm) | CdO Layer Characteristics Area [4] (mm$^2$) | Illuminated Characteristics [2] J$_{sc}$ (mA/cm$^2$) | Illuminated Characteristics [2] V$_{oc}$ (volt) | Illuminated Characteristics [2] FF | Illuminated Characteristics [2] η (%) | Comments |
|---|---|---|---|---|---|---|---|---|---|---|---|
| 105-5 | {112} | - | - | 7th red | 0.91 | 10.0 | 21 | 0.43 | 0.42 | 4.0 | Un-annealed representative cell. |
| 125-1 | {112} | - | - | 5th green | 0.55 | 8.0 | 18 | 0.40 | 0.27 | 2.0 | Un-annealed cell with a thin CdO layer. |
| 83-5 | {112} | - | - | 6th green | 0.66 | 7.0 | 23 | 0.40 | 0.46 | 4.8 | Un-annealed cell degraded with time |
| 110-2 | {101} | - | - | 6th red | 0.78 | 6.0 | 27 | 0.40 | 0.39 | 3.8 | Un-annealed cell improved with forming. |
| 149-1 | {101} | 2 | 320 | 5th red | 0.65 | 11.0 | 24 | 0.41 | 0.50 | 5.4 | Annealed cell having largest active area. |
| 149-2 | {101} | 2.5 | 360 | 6th red | 0.78 | 8.0 | 28 | 0.42 | 0.51 | 5.7 | Annealed cell having highest L$_n$ estimate. |
| 125-3 | {112} | 2.5 | 360 | 6th red | 0.78 | 5.3 | 24 | 0.42 | 0.63 | 6.3 | Annealed cell having highest efficiency. |
| CIS-4 [5] | - | - | - | - | - | - | 36 | 0.48 | 0.68 | 12.0 | Active area = 0.54 cm$^2$. Window layer not CdO. |

1. Annealing of cleaved substrate prior to deposition of CdO layer and electrical contacts.
2. Illumination level was about 100 mW/cm$^2$ under simulated sunlight using a xenon arc lamp. Calibration carried out using a silicon reference cell tested at the National Renewable Energy Laboratory (NREL) of Golden, Colorado.
3. Estimate of CdO layer thickness derived from interference colour using the formula: 2nd=Nλ, where n is the refractive index (2.5), d is the film thickness, N is the interference order and λ is the interference colour wavelength.
4. Active area of the CdO layer.
5. Commercial polycrystalline thin film cell believed to have the structure Cu(InGa)Se$_2$/CdS/ZnO, supplied by Siemens Solar Industries of Camarillo, California.

cells : current density-voltage characteristics in darkness and under 100 mW/cm$^2$ of simulated solar irradiation, capacitance-voltage characteristics, photocurrent capacitance measurements and X-ray diffraction scans.

## 3.1 X-ray Diffraction

Fig. 2 compares X-ray diffraction scans, using copper K$\alpha$ radiation, for (a) CdO on {112} CuInSe$_2$, (b) CdO on glass and (c) powdered CdO and CuInSe$_2$ reported from JCPDS data. It is noted that in (a) the ratio of the intensities for the (111) and the (200) peaks is much higher than in (b) and the (220) and (311) peaks in (a) are hardly detectable but are clearly evident in (b). This suggests that some preferred orientation occurred with the (111) plane of the CdO partially parallel to the (112) surface of the CuInSe$_2$ substrate, despite the apparent large lattice mismatch between the two materials.

## 3.2 Current-Voltage Characteristics

Table I, columns 8 to 11, summarize the current-voltage characteristics obtained under 100 mW/cm$^2$ of simulated solar irradiation as measured on the CuInSe$_2$-CdO cells without antireflection coatings or optimized current collecting grids. The remarkable feature here is that such cells yielded efficiencies ($\eta$) generally more than 4% and short circuit current densities ($j_{sc}$) generally more than 20 mA/cm$^2$, despite the apparent large lattice mismatch of 23% between the substrate and its n-type window layer. The second feature to note is the higher efficiency of around 6% obtained on the devices using annealed substrates compared with those with unannealed substrates, having $\eta$ - values of around 4%. Fig. 3(a) brings out this difference clearly for cells 105-5 and 125-3. The latter annealed cell shows a higher fill factor and smaller series resistance than the former unannealed cell. Fig. 3(b) compares the dark currents for the same two devices, where it is noted that the annealed cell shows less shunt current at low voltage, lower series resistance at higher voltage and an ideality factor n nearer to 2.

## 3.3 Capacitance-Voltage Characteristics

Small signal parallel capacitance $C_p$ was measured on each of the fabricated cells as a function of reverse bias at a frequency of 10 kHz. Fig. 4 shows a Mott-Schottky plot of $(A/C_p)^2$ versus bias voltage for two cells, one annealed and the other unannealed. Here A is the CdO area. It is noted firstly that the experimental

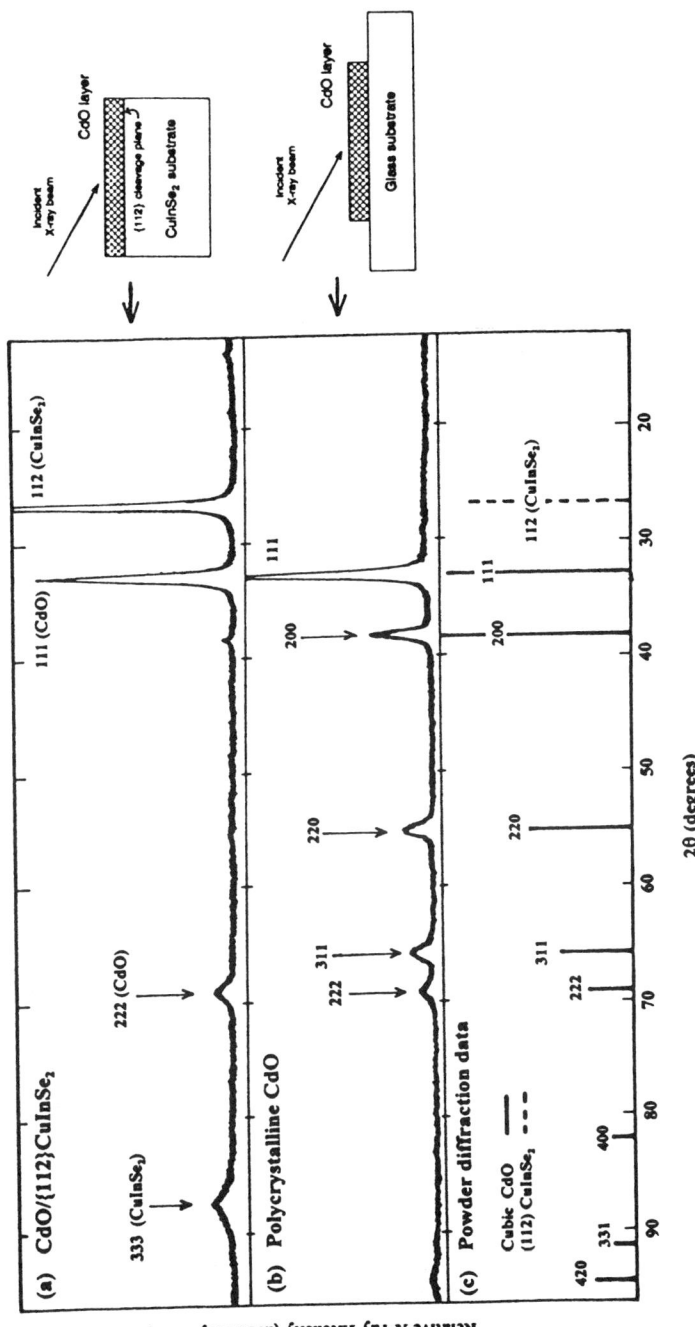

FIGURE 2. X-ray diffraction scans on: (a) a CdO film deposited directly onto a cleaved {112} CuInSe$_2$ surface, (b) a polycrystalline CdO film deposited onto a glass substrate and (c) powdered CuInSe$_2$ and CdO from JCPDS data.

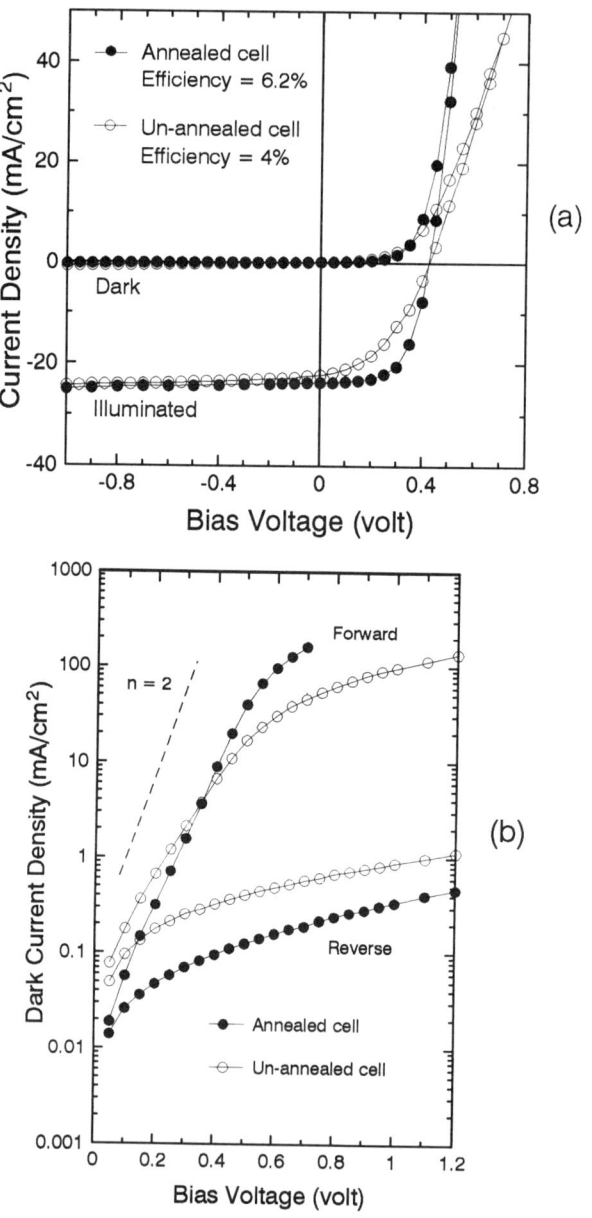

**FIGURE 3.** Illuminated and dark current-density-voltage characteristics of annealed cell 125-3 and unannealed cell 105-5, showing the effect of annealing (a) on characteristics under simulated solar illumination of 100 mW/cm$^2$, where the conversion efficiency increased from 4 to 6%, and (b) on dark characteristics, showing an improvement in the rectification ratio with annealing.

line for the annealed cell 149-2 shows a downward curvature on approaching zero bias, whereas the line for the non-annealed cell 125-1 does not. Secondly, it is observed that the slopes of the two curves at a larger reverse voltage of 3 or more volts are about the same. These features were consistently observed on many cells and indicate an apparent acceptor concentration of about $10^{17}$ cm$^{-3}$ deep within the CuInSe$_2$ substrate but a concentration of almost an order of magnitude lower near the active junction for those cells fabricated with annealed substrates. By contrast, the unannealed cells have the same acceptor concentration at the junction and deep within the CuInSe$_2$ substrate.

## 3.4 Diffusion Length Estimation

The photocurrent-capacitance method has been used extensively in this laboratory to estimate minority carrier diffusion lengths in Se-CdO photovoltaic cells [7]. It was therefore applied in the present case to the CuInSe$_2$-CdO cells but with appropriately longer monochromatic wavelengths. It was found however, that meaningful measurements could not be obtained on the unannealed cells, due to higher shunt current, which cannot be tolerated in this method. Nevertheless, measurements were made on the annealed cells and Fig 5(a) shows a plot for annealed cell 149-2 of the illuminated-to-dark photocurrent change $\Delta I$ against $1/C_p$ for fixed monochromatic slit widths and a wavelength of 1.2 $\mu$m and Fig 5(b) shows a similar plot for fixed values of wavelength ($\lambda$) and a slit width of 4 mm. Under appropriate conditions [8], extrapolation from a linear variation of the $\Delta I$ - $1/C_p$ plot to the abscissa should yield an intercept of $1/C_{pi} = -L_n/(\epsilon_o \epsilon_r A)$, where $\epsilon_o$ is the permittivity of a vacuum, $\epsilon_r$ is the relative dielectric constant of the absorber material and $L_n$ is the diffusion length of the electrons in the absorber layer. Applying this relation to Fig 5(a) for $\lambda = 1.2$ $\mu$m, the average $1/C_{pi}$ value yields $L_n \sim 3$ $\mu$m. Similar magnitudes were found on the other two annealed cells. A measurement made on a commercial C.I.S- based polycrystalline thin film cell with a 13% efficiency yielded an $L_n$ value of just over 1 $\mu$m.

## 3.5 Removal of CdO Layer

Arising from the result that considerable photovoltaic activity persisted, even with the apparent presence of a higher density of interface states from lattice mismatch, an experiment was carried out, where the CdO on cell 125-3 was completely removed. This was done by etching with dilute HCl, leaving the top gold contact stripes intact. Fig. 6 shows that after the CdO removal, significant photovoltaic activity still remained, suggesting that the seat of photovoltaic action was *within* the CuInSe$_2$ substrate itself.

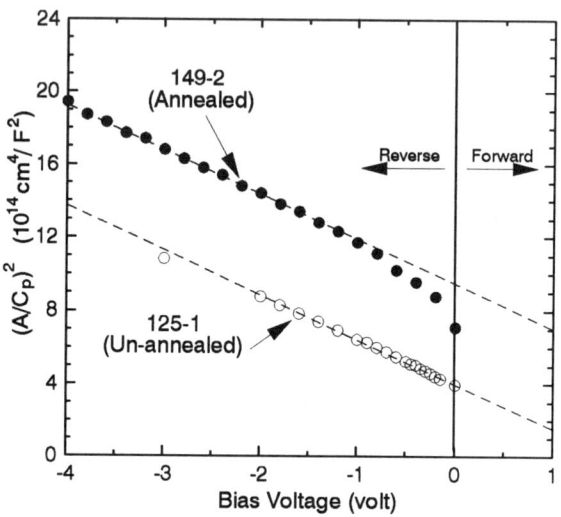

**FIGURE 4.** Mott-Schottky plot of $(A/C_p)^2$ against bias voltage showing a comparison between annealed cell 149-2 and un-annealed cell 125-1. The oscillator frequency in the capacitance measurement was fixed at 10 kHz.

## 4. DISCUSSION

The present work shows that, even with a lattice mismatch of as much as 23%, a cell can be made with $CuInSe_2$ as the p-type absorber layer with an efficiency of at least 6%. It is also evident that with optimization of the resistance and transparency of the CdO layer, the addition of an antireflection coating and a suitable current collecting grid, a higher efficiency can be attained in a $CuInSe_2$-CdO device. It is also to be noted that the diffusion length was some three times larger in the monocrystalline device than that in the 13% efficient polycrystalline cell. Thus, how can a cell, with more than 6% efficiency, be possible with the high density of interface states arising from a 23% lattice mismatch? The answer may be that the seat of photovoltaic action is not at the semiconductor hetero-interface but lies within the $CuInSe_2$ itself, as suggested by the experiment with persistence of photocurrent with the CdO layer removed. In this case, photovoltaic activity would be less sensitive to states located at the interface.

Apart from this, the present work confirms the beneficial effect of annealing the $CuInSe_2$ substrate in argon at atmospheric pressure for single crystal $CuInSe_2$ cells, principally by improving the fill factor. This was first reported in the fabrication of a $CuInSe_2$ - CdS - ZnO cell using a monocrystalline substrate [1] and the present work shows that the annealing has the same effect in the $CuInSe_2$ - CdO cells. The steeper slope of the Mott-Schottky plots near zero bias for cells made from annealed substrates indicates a lower acceptor concentration near the active junction of the device and the associated increase in resistance raises junction shunt

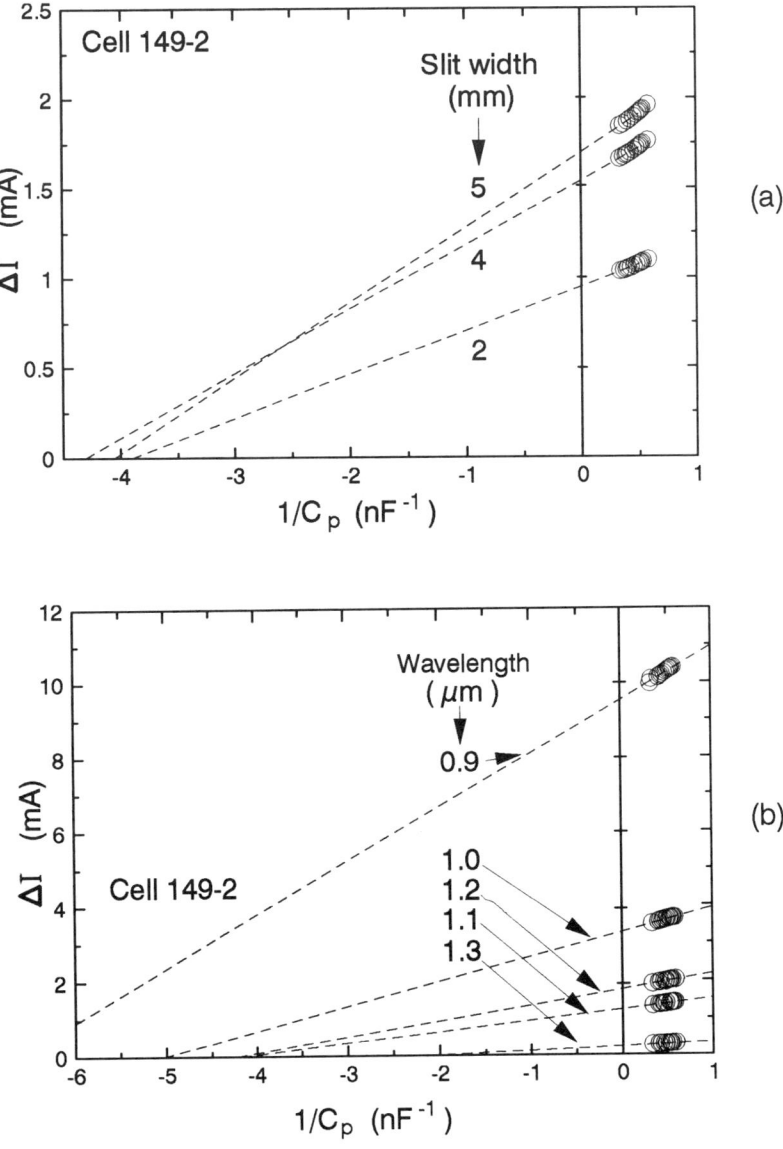

**FIGURE 5.** Plot of $\Delta I$ against $1/C_p$ for annealed sample 149-2 with variation of (a) monochromator slit width (b) incident wavelength, $\lambda$. The $1/C_p$ intercepts of the extrapolated broken lines yield an estimate of the diffusion length. In (a) $\lambda = 1.2$ μm and in (b) slit width = 4 mm.

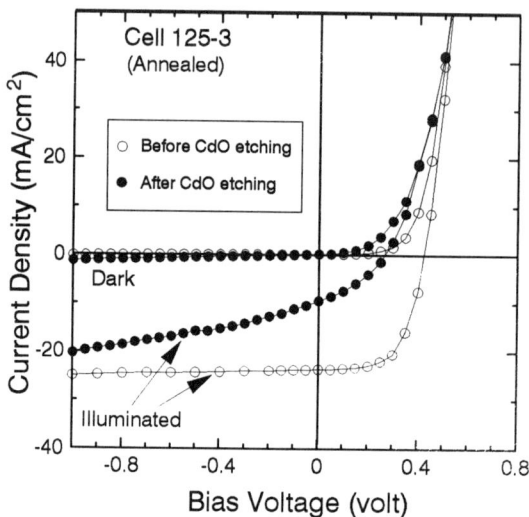

**FIGURE 6.** Illuminated and dark current-density-voltage characteristics of cell 125-3, measured before and after etching off the active CdO layer, showing evidence of some residual photovoltaic action, even after the CdO removal.

resistance, thereby increasing the fill factor. The reduction in acceptors in the $CuInSe_2$ is believed to be due to the partial out-diffusion of selenium during the heat-treatment.

## 5. ACKNOWLEDGMENTS

The authors wish to acknowledge the financial support of this work under the Strategic Grants Program of the National Science and Engineering Research Council of Canada.

## REFERENCES

[1] L.S. Yip and I. Shih. 1st World Conf. on Photovoltaic Energy Conversion, 210 (1995).
[2] L.S. Yip, Z.A. Shukri, I. Shih and C.H. Champness. Proc. 13th European Solar Energy Conf., Nice (1996).
[3] J.C. Boettger and A.B. Kunz. Phys. Rev. **B27**, 1359 (1983).
[4] C.H. Champness and C.H. Chan. Solar Energy Materials and Solar Cells **37**, 75-92 (1995).
[5] Z.A. Shukri, C.H. Champness and I. Shih. J. Crystal Growth **129**, 107-110 (1992).
[6] Z.A. Shukri and C.H. Champness. Surface Review and Letters, to be published (1996).
[7] C.H. Champness and C.H. Chan. Solar Energy Materials and Solar Cells, **30**, 6575 (1993).
[8] C.H. Champness and C.H. Chan. Proc. 6th Int. Photovoltaic Science and Engineering Conf., New Delhi, 827-832 (1992).

# Techniques for Increasing Ga Content in $CuIn_{1-x}Ga_xSe_2$ Thin Films Prepared by Two-Stage Selenization Process

Kevin Lynn and Neelkanth G. Dhere

*Florida Solar Energy Center
1679 Clearlake Road,, FL 32922-5703*

**Abstract.** A Cu-Ga(66 at. %) alloy target was employed for enhancing the gallium content in $CuIn_{1-x}Ga_xSe_2$ films prepared by two Se-vapor selenizations of metallic precursors. Combination with a Cu-Ga(22 at. %) sputtering target allowed preparation of $CuIn_{1-x}Ga_xSe_2$ films with a graded profile. Gallium content $Ga_x$ near the surface was raised to the range 0.28 - 0.32, while an even higher amount of gallium of up to 0.40 was obtained in the bulk of the films. Efficiency of solar cells prepared from $CuIn_{1-x}Ga_xSe_2$ films with moderately enhanced gallium content was 8.5%. Higher gallium proportions seem to be correlated with the formation of Cu-rich phases, surface inhomogeneities, and possibly a highly resistive phase. This combined with inferior crystallinity deteriorated solar cell efficiency.

## INTRODUCTION

$CuIn_{1-x}Ga_xSe_2$ (CIGS) thin-film photovoltaic solar cells have been studied by many groups because of their promise of large-scale economic manufacture [1-6]. The highest conversion efficiency of 17.7% is comparable to that of crystalline silicon cells [1]. PV Materials Laboratory of the Florida Solar Energy Center is carrying out studies with an aim to find a viable fabrication process to make CIGS solar cells that could be more easily transferred to a manufacturing process. The process uses two-stage selenization of magnetron sputtered metallic layers of Cu, In, and Ga in selenium vapor. Because sputtering of liquids can be problematic, CuGa alloy targets were fabricated. Initially the entire thicknesses of individual metallic layers were deposited sequentially and then the samples were selenized. This resulted in poor adhesion. Moreover, the process did not pass through a Cu-rich phase. Hence a two-stage selenization process was developed [5]. In this process, initially Cu, Ga, and In metallic layers were deposited followed by the first selenization to obtain a Cu-rich $CuIn_{1-x}Ga_xSe_2$ thin film, then the remaining metallic components were sputter-deposited followed by a second selenization to obtain an overall Cu-poor CuIn$_{1-x}$Ga$_x$Se$_2$ thin film. The best results using a CuGa target with 22% Ga was an efficiency of 9.02% corresponding to the highest Ga$_x$ content of x = 0.18.

Since the optimum Ga$_x$ content is known to be x=0.28-0.32, a second CuGa target with 66% Ga has now been added to achieve the higher gallium content. The goal was to obtain a high amount of Ga towards the back contact which decreases towards the p-n junction. At the junction, the Ga content will again rise slightly. This *double profiling* is supposed lead to several improvements. Electrons excited within the CIGS structure will "roll down" toward the p-n junction making it more likely to participate in the photovoltaic effect [6]. Also electrons excited near the back contact will be less likely to flow backwards because of the back-surface barrier. The slight rise in the gallium content at the p-n junction itself will raise the band gap in the vicinity of the junction thus allowing the absorption of only the higher energy photons near the junction. This has been shown to increase efficiencies. This paper presents results on the enhancement of gallium content in CuIn$_{1-x}$Ga$_x$Se$_2$ (CIGS) thin-films and its effect on the film properties and cell conversion efficiencies.

## EXPERIMENTAL TECHNIQUE

Five different deposition sequences have been chosen using the new CuGa [high 66% Ga] target. Figure 1 Shows the deposition and selenization sequences. The thickness of individual layers have been provided in Table 1. All selenizations were performed in the same manner, in a chamber under high vacuum ($10^{-6}$ Torr). Following the preheating of selenium (99.999% pure) in a Knudsen-type Mo boat for 15 minutes, the selenium evaporation rate was raised to 25-50 Å s$^{-1}$. The substrate temperature was then ramped up to 550° C in ~10 minutes. The samples remained at this temperature for approximately 30 minutes. The samples were then cooled to 300° C at a rate of about 12.5° C min$^{-1}$ while maintaining the selenium vapor incidence.

Structure and morphology of films were studied by x-ray diffraction (XRD) and scanning electron microscopy (SEM). Composition of films was analyzed by electron probe microanalysis (EPMA). Two electron-beam energies viz. 10 keV and 20 keV were utilized so as to obtain compositions of regions adjacent to the surface and in the bulk of the film respectively. After each selenization, electrical resistance of the films was measured. Cu-rich composition is indicated by a low resistance of a few ohms. Moderately Cu-poor CuIn$_{1-x}$Ga$_x$Se$_2$ thin-films of the desired composition have been found to give a resistance of few tens to few hundred kilo-ohms.

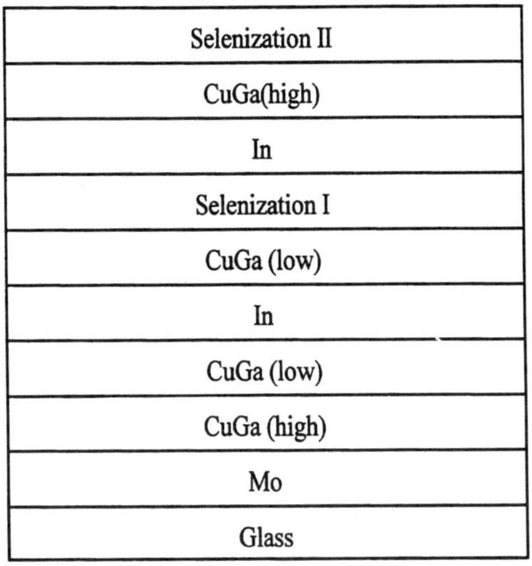

| Selenization II |
| --- |
| CuGa(high) |
| In |
| Selenization I |
| CuGa (low) |
| In |
| CuGa (low) |
| CuGa (high) |
| Mo |
| Glass |

**Figure 1.** Sequence of depositions and selenizations of metallic layers

**Table 1.** Thicknesses of CuGa (high), In, and CuGa (low) layers in Angstroms

| Series | Layer | | | | | | Totals | | |
| --- | --- | --- | --- | --- | --- | --- | --- | --- | --- |
| | CuGa (high) | CuGa (low) | In | CuGa (low) | In | CuGa (high) | CuGa (low) | CuGa (high) | In |
| NS#1 | 815 | 4475 | 4625 | 1119 | 2490 | 204 | 5594 | 1019 | 7115 |
| NS#2 | 1748 | 3832 | 3180 | 958 | 1713 | 437 | 4790 | 2185 | 4893 |
| NS#3 | 1748 | 3832 | 3499 | 958 | 1184 | 437 | 4790 | 2185 | 5383 |
| NS#4 | 1748 | 3066 | 3180 | 766 | 1713 | 437 | 3832 | 2185 | 4893 |
| NS#5 | 874 | 3066 | 4893 | 766 | 0 | 1311 | 3832 | 2185 | 4893 |

## RESULTS AND DISCUSSION

In the first experiment labeled New Series #1 (NS#1), the goal was to prepare a overall Cu-poor, 2.75 µm thick, thin film having the composition represented by

$Cu_{0.92}In_{0.72}Ga_{0.28}Se_2$. In practice, the deposited layer thicknesses corresponded to the composition $Cu_{0.92}In_{0.72}Ga_{0.28+0.1}Se_2$. The additional $Ga_{0.1}$ was included to compensate for gallium that evaporated from the film during selenizations. The sequence of deposition of metallized layers and their thicknesses have been provided in Figure 1 and Table I. After each selenization, ohmic resistance of the films was measured and the color of the films was noted. The film was uniform and dark grey after the first selenization. In this sample, the resistance measured after the first selenization was around 20Ω. After the second selenization, the resistance of the films was around 6.5 kΩ. This time the color of the film was a noticeably lighter grey. Average composition of samples from new series NS#1 as analyzed by EPMA is given in Table II. After analyzing a number of samples, it was determined that dark grey indicated a Cu-rich phase and the lighter shade represented a Cu-poor phase. As can be seen, the proportions Cu:In:Ga:Se near the surface correspond to 23.03:26.53:3.20:47.24. In the bulk of the film the proportions Cu:In:Ga:Se correspond to 27.96:20.78:6:11:45.13. $CuIn_{1-x}Ga_xSe_2$ thin films from series NS#1 had a $Ga_x$ content of x=0.244 in the bulk of the film. The best cell from this deposition sequence gave an efficiency of $\eta = 8.5\%$. This value is comparable to the earlier best efficiency of 9.02%.

From the first series NS#1, it was seen that Ga content was still too low. It was also decided to prepare a composite thin film with ~3000 Å thick $CuGaSe_2$ layer beneath a 2.7 μm thick $Cu_{0.92}In_{0.57}Ga_{0.43}Se_2$ layer. Hence in the second series NS#2 the layer thicknesses as shown in Figure 1 and Table 1 were chosen. These also included an additional amount of gallium equivalent $Ga_{0.1}$ to compensate for re-evaporation during selenization. As can be seen from EPMA results in Table 3, the proportions Cu:In:Ga:Se near the surface and the bulk of the film correspond to 27.85:19.79:6.40:45.96 and 27.99:17.45:8.90:45.65. Thus in these samples, $Ga_x$ content in the bulk of the film was significantly raised to x = 0.36. However, the copper content was excessive. The resistance values were low (~100Ω) after both the first and the second selenizations, indicating excessive proportion of copper content throughout the film.

Since copper content was too high, 10% more indium was added in the series NS#3 (Figure 1, Table 1). These films had a $Ga_x$ content of x = 0.35 and were more Cu-poor after the second selenization (~15 kΩ). Unfortunately, the surface morphology was inhomogeneous.

**Table 2**. Composition as analyzed by EPMA from depositions in new series NS#1

| Electron Beam Energy | Cu | In | Ga | Se |
|---|---|---|---|---|
| 10 keV | 23.03 | 26.53 | 3.20 | 47.24 |
| 20 keV | 27.96 | 20.78 | 6.12 | 45.13 |

**Table 3.** Average EPMA composition for samples in new series NS#2

| Electron Beam Energy | Cu | In | Ga | Se |
|---|---|---|---|---|
| 10 keV | 27.85 | 19.79 | 6.40 | 45.96 |
| 20 keV | 27.99 | 17.45 | 8.90 | 45.65 |

In the next series NS#4 (Figure 1, Table 1), the composition was adjusted by reducing the total amount of CuGa(22%) by 20% while maintaining the amount of indium as in the series NS#2. This gave promising results. Resistivity values indicated a Cu-rich phase after the first selenization and a Cu-poor phase after the second. Moreover, surface morphology was uniform after both selenizations. Results of EPMA analysis (Table 4) indicated $Ga_x$ content near the surface to be x ≈ 0.26 and x ≈ 0.33 in the bulk. However, the films were not very adherent. One of the samples delaminated partially during the chemical bath deposition of cadmium sulfide hetero-junction partner layer.

In the series NS#5 (Figure 1, Table 1), the entire amount of indium was included in the first metallized layers. Moreover, 40% CuGa(high) was transferred from the first metallized layers to the second. The idea was to prevent the formation of an undesirable InGa liquid phase during the second selenization. To obtain a Cu-rich phase after the first selenization, the amount of gallium in the first metallized layers was reduced to compensate for the additional indium. The results of composition analysis by EPMA for a representative sample are provided in Table 5. Although Ga content follows the profile model, there may be two problems. $Cu_y$ content did not fall between $0.86 \leq y \leq 0.96$ as it does in other high efficiency cells. Also resistivity values measured after the first selenization were high (~20 kΩ). The increased bandgap from the higher gallium proportion was indicated in the spectral response of the solar cells. However, the conversion efficiency of solar cells formed

**Table 4.** Average EPMA compositions for a samples in series NS#4

| Electron Beam Energy | Cu | In | Ga | Se |
|---|---|---|---|---|
| 10 keV | 22.24 | 23.25 | 5.20 | 49.31 |
| 20 keV | 25.31 | 18.62 | 8.25 | 47.82 |

**Table 5.** Average EPMA composition of samples in the new series NS#5

| Electron Beam Energy | Cu | In | Ga | Se |
|---|---|---|---|---|
| 10 keV | 26.94 | 15.81 | 8.92 | 48.33 |
| 20 keV | 24.33 | 15.33 | 11.01 | 49.34 |

from these films dropped to 4%. The efficiency reduction was mainly due to the lower fill factor arising from a very high series resistance. The higher gallium content thus seems to lead to a highly resistive gallium rich phase as well as poor crystallinity. The absence of a Cu-rich phase after the first selenization would impede the growth of large-grain and compact CIGS films with the benefit of fluxing action of $Cu_{2-x}Se$.

The most promising looking cells were from the series NS#4. By varying the sequence of depositions, it is expected to eliminate the highly resistive gallium rich phase. It should also improve the adhesion and crystallinity of $CuIn_{1-x}Ga_xSe_2$ thin films as well as the conversion efficiency of $CuIn_{1-x}Ga_xSe_2$ solar cells.

## CONCLUSIONS

Combination of Cu-Ga alloy targets with high and low gallium contents was employed for enhancing the gallium content and for preparing graded-bandgap $CuIn_{1-x}Ga_xSe_2$ thin films. The desired increase in the bandgap at high Ga content was accompanied by the formation of a copper-rich phase, surface inhomogeneity, and possibly a highly resistive phase. With the chosen layer sequences, especially with high Ga content layer near the surface, preparation of Cu-rich film after the first selenization became difficult. Reasonable conversion efficiency of 8.5% could be obtained in films with moderately high $Ga_x$ content of 0.20. However, higher $Ga_x$ values of 0.32 near the surface resulted in deterioration of crystalline quality of $CuIn_{1-x}Ga_xSe_2$ thin films and poor conversion efficiencies of solar cells.

## ACKNOWLEDGMENTS

This work was supported by National Renewable Energy Laboratory Contract # XG-2-11036-5. Authors are thankful to Alice Mason for EPMA analysis and to Dr. Kannan Ramanathan for completion of solar cells and many useful discussions.

# REFERENCES

1. Zweibel, K., Ullal H. S., and von Roedern, B., "Progress and Issues in Polycrystalline Thin-Film PV Technologies," Twenty-Fifth IEEE Photovoltaic Specialists Conference, Washington, D. C., May 13-17, 1996, pp. 745-750.
2. Niemi, E., Hedström, J., Martinsson, T., Granath, K., Stolt, L., Skarp, J., Hariskos, D., Ruckh, M., and Schock, H. W., "Small- and Large-Area CIGS Modules by Co-Evaporation," Twenty-Fifth IEEE Photovoltaic Specialists Conference, Washington, D. C., May 13-17, 1996, pp. 801-804.
3. Gay, R., Dietrich, M., Fredic, C., Jensen, C., Knapp, K., Tarrant, D., Willet, D., "Efficiency and Process Improvements in CuInSe2-Based Modules," Proc. 12th European Photovoltaic Solar Energy Conference (EPVSEC), Amsterdam, April 1994, pp. 935-938.
4. Delahoy, A., Britt, J., Butler, G., Faras, F., Sizemore, A., Ziobro, F., and Kiss Z., "Recent Advances in manufacturing Technology for CuInSe2-Based Power Modules," 12th EPVSEC, Amsterdam, April 1994, 1612-1615.
5. Dhere, N. G., Kuttath, S., Lynn, K. W., Birkmire, R. W., and Sharafam, W. N., Proc. IEEE First World Conference on Photovoltaic Energy Conversion, Waikoloa, Hawaii, December 5-9, 1994, pp. 190-193.
6. Contreras, M. A., Tuttle, J., Gabor, A., Tennant, A., Ramanathan, K., Asher, S., Franz, A., Keane, J., Wang, L., Noufi, R., "High Efficiency Graded Bandgap Thin-Film Polycrystalline Cu(In,Ga)Se2-Based Solar Cells," *Solar Energy Materials and Solar Cells*, v. 41/42, 1996, pp. 231-246.

# Sodium Dependence of Cu(In,Ga)Se$_2$ Junction Electronics

Jennifer E. Granata,[*] James R. Sites[*] and John R. Tuttle[†]

[*]*Department of Physics, Colorado State University, Fort Collins, CO 80523*
[†]*National Renewable Energy Laboratory, Golden, CO 80401*

**Abstract.** Cu(In,Ga)Se$_2$ photovoltaic devices were fabricated on five different substrates, with and without sodium-diffusion barriers and with and without the addition of sodium to the absorber. This was done to gain a better understanding of the effects of Na on junction electronics and device performance. Positive trends are observed in open-circuit voltage, fill factor, hole density and diode quality parameters. The presence of Na appears to have a positive effect on the junction and on the device performance as a whole.

## INTRODUCTION

The efficiency of Cu(In,Ga)Se$_2$ (CIGS)-based photovoltaic devices has been pushed to nearly 18% in the last year (1). The record cell, and all previous high-efficiency devices, was fabricated on soda-lime glass. The role played by the substrate has recently come under investigation. In the past few years, researchers have begun to find that impurities diffusing through the glass during back contact and absorber depositions may improve the performance of CIGS-based solar cells (2–4). This may in part explain why other substrates (alumina, stainless steel, flexible substrates) have generally failed to produce high efficiency devices.

Many authors have hypothesized that sodium is mainly responsible for the observed improvements (4–8). Here, we attempt to shed more light on the question of how sodium influences the performance of CIGS-based solar cells. This is the first step in achieving a more quantitative as well as qualitative understanding of the role of sodium.

To study the impact of sodium, we chose five different substrates, some containing sodium (three soda-lime glass substrates) and some which did not (smooth Al$_2$O$_3$ and stainless steel). We also chose to deposit a SiO$_2$ sodium-diffusion barrier before the molybdenum (Mo) back contact on some of the deposition runs and we added sodium to the absorber layer during deposition on half the runs. Device characterization reveals an improvement trend in open-circuit voltage (V$_{OC}$), fill factor

(FF), efficiency, diode quality factor (A), series resistance ($R_{series}$) and hole density as the sodium concentration increases.

## EXPERIMENTAL

ZnO/CdS/CIGS/Mo solar cells were fabricated on five different substrates using the two-stage process at the National Renewable Energy Laboratory (NREL) under four conditions. All other aspects of the cell fabrication were nominally identical. Details of the fabrication process can be found elsewhere (9). See Table 1 for the matrix of substrates and conditions employed. We assume that the sodium concentration is lowest in condition 1 and highest in condition 4. For all substrates except soda-lime glass C, a metal layer was deposited beneath the Mo to improve adhesion. When applicable, a $SiO_2$ sodium-diffusion barrier was then deposited (~ 550 Å), followed by a 1 $\mu$m layer of sputtered Mo. CIGS absorbers were then deposited. Na was added during two absorber runs in the form of $Na_2S$, evaporated during the precursor deposition. Devices were finished with 500 Å of chemical bath deposition CdS, followed by a bi-layer of radio-frequency sputtered ZnO (500 Å intrinsic, followed by 3500 Å Al-doped).

Finished devices were characterized at NREL and at Colorado State University (CSU). Device measurements include current-voltage (J-V) in the dark and under illuminated, one-sun conditions at 25°C, capacitance-voltage (C-V) in the dark and quantum efficiency (QE) without a white light bias.

## RESULTS

Figure 1a-d demonstrates the J-V results for all conditions, grouped according

**TABLE 1. Summary of Substrates and Experimental Conditions**

| Substrate | Soda-lime glass/Mo A | Soda-lime glass/Mo B | Soda-lime glass/Mo C | Smooth Alumina | Stainless Steel |
|---|---|---|---|---|---|
| Condition 1: | $SiO_2$ barrier, no Na | Metallic barrier, no Na | | $SiO_2$ barrier, no Na | $SiO_2$ barrier, no Na |
| 2: | No barrier, no Na | | No barrier, no Na | No barrier, no Na | No barrier, no Na |
| 3: | $SiO_2$ barrier, with Na | Metallic barrier, with Na | | $SiO_2$ barrier, with Na | $SiO_2$ barrier, with Na |
| 4: | No barrier, with Na | | No barrier, with Na | No barrier, with Na | No barrier, with Na |

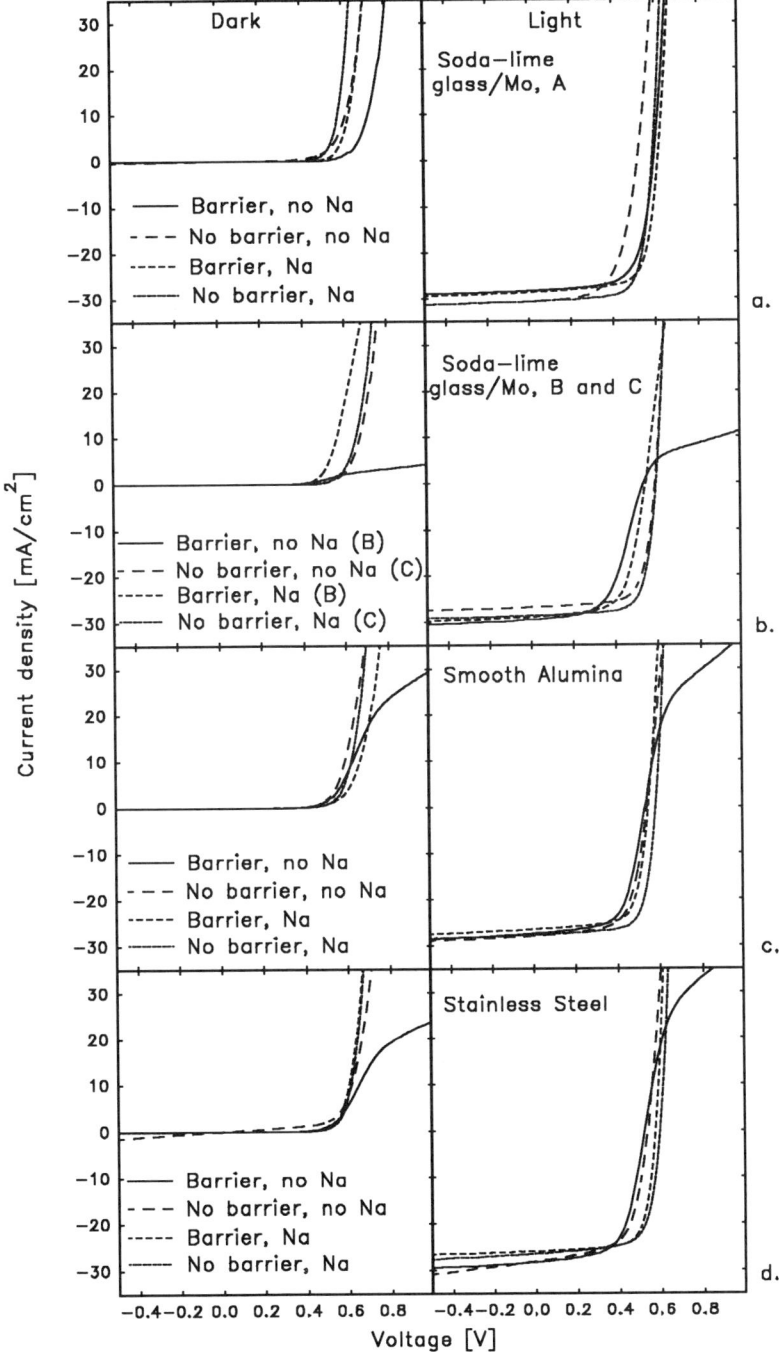

**FIGURE 1.** Current-voltage curves, grouped according to substrate.

to substrate. In all cases, $V_{OC}$ increases as sodium is added. FF and efficiency also increase with the addition of sodium. The increases in $V_{OC}$ and FF are obvious in Figure 1. Figure 2a-c illustrates the trends in $V_{OC}$, FF and efficiency, respectively.

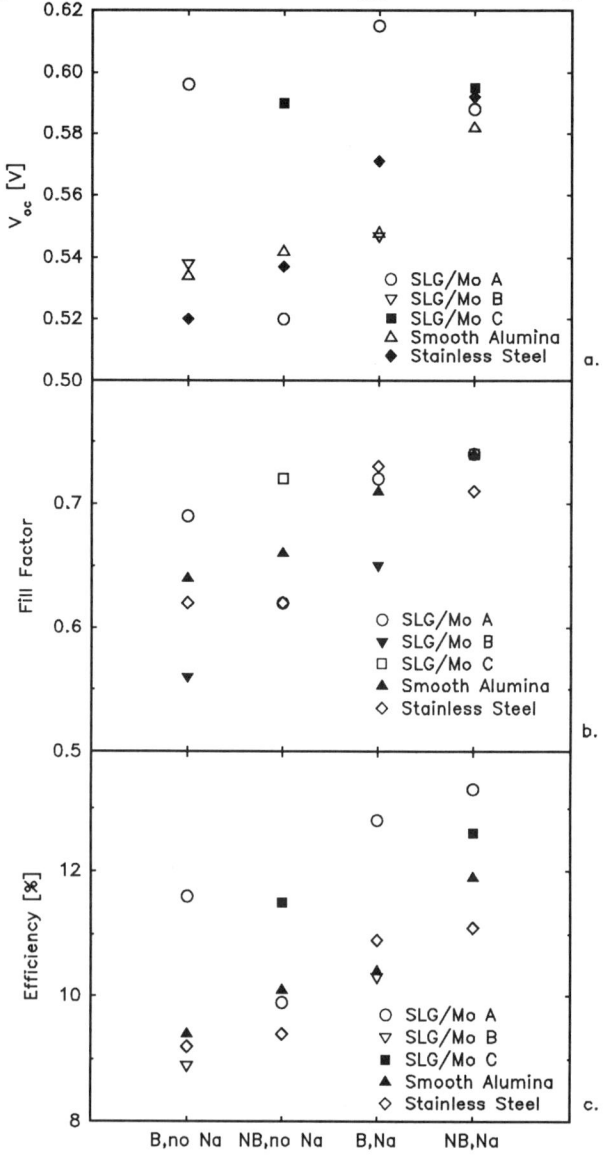

**FIGURE 2.** Trends observed in open-circuit voltage (2a), fill factor (2b) and efficiency (2c) as sodium concentration increases.

Diode quality factor and series resistance are extracted from the light J-V curves. See reference 10 for details of this analysis. In all cases, the diode quality factor decreases with an increase in sodium level, and the series resistance either decreases or remains constant. See Figure 3a-b.

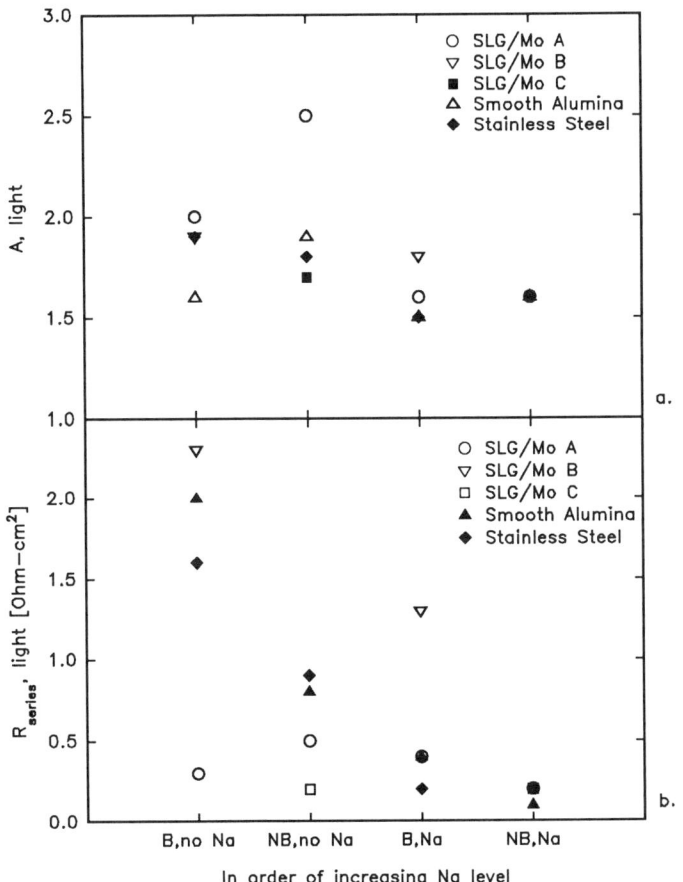

**FIGURE 3.** Trends observed in diode quality factor (3a) and series resistance (3b) as sodium concentration increases.

Figure 4 exhibits the quantum efficiency results for all conditions, grouped according to substrate. The J-V curves show a weak trend in $J_{sc}$ as the sodium concentration varies ($J_{sc}$ decreases slightly with increasing Na level, but not in every case). The same is true for the QE curves. A possible trend is observed in the collection beyond the bandgap wavelength, but if present, it is quite weak. As with $J_{sc}$, there may be a slight decrease in this collection. We also note that the sodium concentration does not appear to influence the position of the effective bandgap itself.

C-V data allow the hole density as a function of the distance from the metallurgical junction to be calculated. Details of this calculation can be found in reference 11. In Figure 5, the results of these calculations are shown, grouped as before. An increase in hole density (or carrier concentration) with increasing sodium level is consistently present. Figure 6 demonstrates this trend. In some cases (soda-lime glass/Mo B, stainless steel), the hole density increases by an order of magnitude with the addition of sodium.

## DISCUSSION

All of the above strongly implies that the presence of sodium has a beneficial effect on the performance of CIGS-based solar cells. The questions still remaining are how and why. The increases in $V_{OC}$ and hole density as well as the decreases in diode quality factor with the addition of sodium suggest improvements in the junction quality of these devices. Decreases in series resistance suggest improvements in overall electronic quality and a possible decrease in recombination as sodium concentration increases. Others have proposed that the sodium resides mainly in grain boundaries and on surfaces (3, 8). The decrease in series resistance supports this hypothesis, as does the increase in open-circuit voltage. Thus, if the Na resides in the grain boundaries, it may be acting to decrease grain-boundary recombination and hence increase $V_{OC}$. If Na is passivating or neutralizing defects at free surfaces, series resistance will decrease. However, the obvious improvements in junction quality lead us to believe that Na is having another effect as well. Some researchers have proposed that the presence of sodium improves the quality of the absorber during growth (3, 12). This may account in part for the observed improvements in junction quality. The increases in FF we believe are due mainly to the systematic decreases in A and $R_{series}$, pointing to a combination of improvements caused by the presence of sodium.

The J-V curves in Figure 1 also show a secondary trend. In three of four cases in which a direct ($SiO_2$) or indirect (metal) sodium-diffusion barrier was employed (condition 1), the J-V curves demonstrate a "rollover" effect in both the light and in the dark. It appears likely that the presence of a sodium-diffusion barrier induces a CIGS-Mo blocking barrier, as the equivalent run without the barrier (condition 2) shows no such signs of this rollover. When those same substrates were used and sodium was added to the absorber (conditions 3 and 4), the rollover is

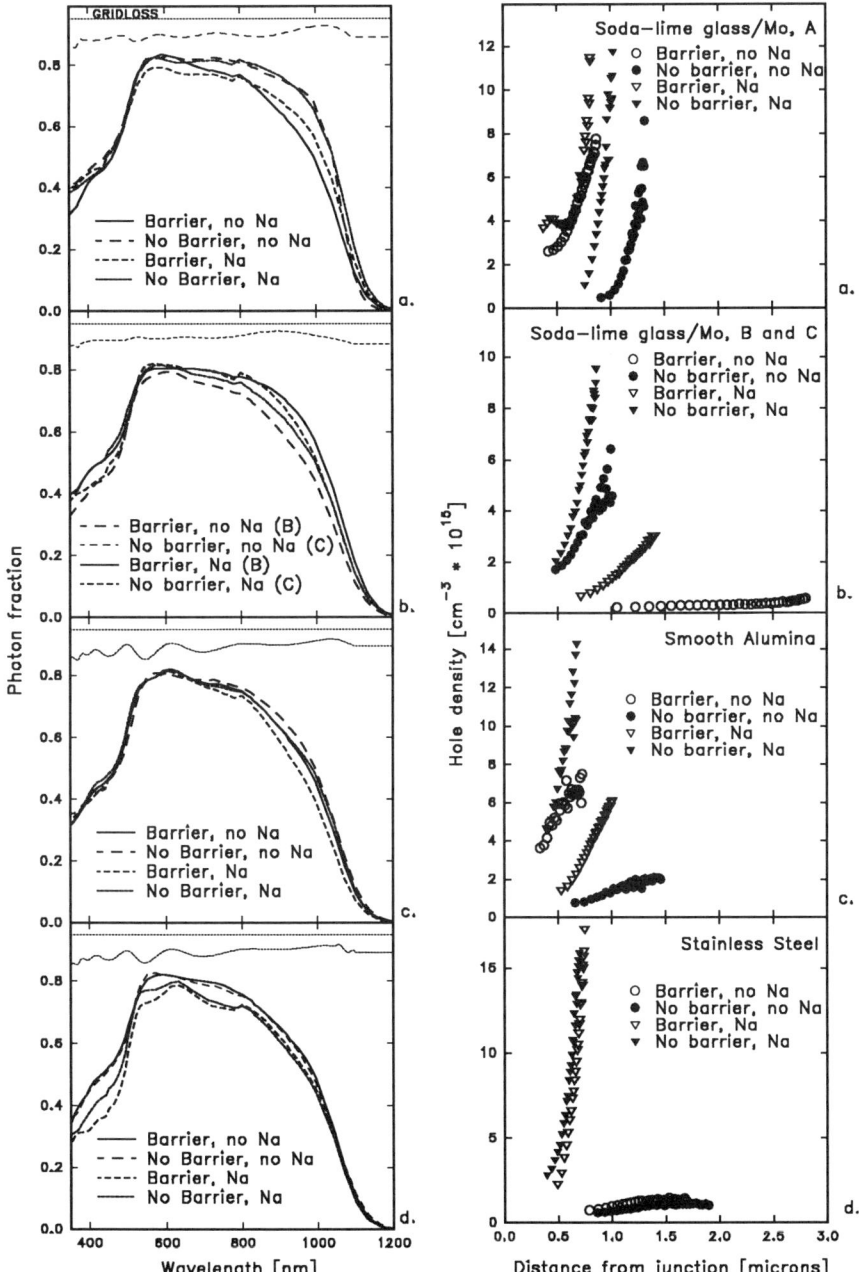

**FIGURE 4.** Quantum efficiency results, grouped according to substrate. 4a: soda-lime glass/Mo A; 4b: soda-lime glass/Mo B, C; 4c: smooth alumina; 4d: stainless steel.

**FIGURE 5.** Hole density Vs. distance from junction, grouped according to substrate. In 5b, open symbols refer to SLG/Mo B whereas filled symbols refer to SLG/Mo C.

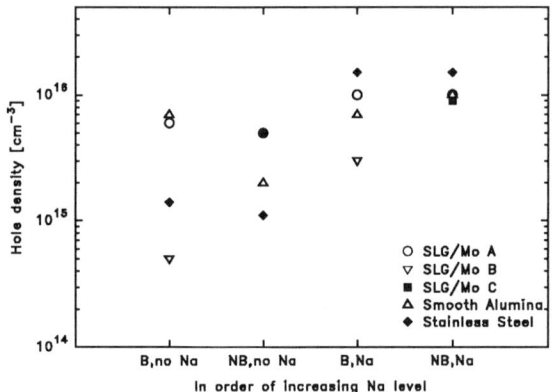

**FIGURE 6.** Trend observed in hole density as sodium concentration increases.

greatly decreased or disappears all together.

The immediate next step in this study is materials analysis of the same samples to measure the sodium levels in the Mo and absorber using secondary ion mass spectroscopy, to understand the effects of Na on the film quality using scanning electron microscopy, and to see how sodium influences the position of the junction using electron-beam-induced current techniques. A primary goal of continuing study is to identify the ideal Na level for optimal performance. In light of this goal, the next experimental steps will include varying the amount of Na introduced during deposition and possibly varying the Na source.

## CONCLUSIONS

By intentionally varying the amount of Na in the Mo and absorber layers of CIGS-based solar cells, we have shown to first order that the presence of Na has a beneficial effect on the performance of these devices. Positive trends are observed in open-circuit voltage, fill factor, efficiency, hole density, diode quality factor and series resistance with the addition of Na. These trends imply improvements in junction quality and overall electronic performance as sodium concentration increases. It is also observed that the presence of a sodium-diffusion barrier may induce a CIGS-Mo blocking barrier which is reduced or negated by the presence of sodium. Many questions remain, such as what is the ideal sodium level for best performance, where does the sodium reside, how does it get there, and how does Na influence the CIGS-Mo interface. These questions will be the subject of further study for these authors.

# ACKNOWLEDGEMENTS

The authors wish to thank colleagues at Siemens Solar, Inc. and at Solarex, Inc., for supplying substrate materials; James Keane for device processing; Alice Mason for EPMA measurements; and Sally Asher, Rick Matson, Kannan Ramanathan and Troy Berens for helpful discussions. This work was supported by NREL subcontract XAX-4-14000-01.

# REFERENCES

1. Tuttle, J. R., Ward, J. S., Duda, A., Berens, T. A., Contreras, M. A., Ramanathan, K. R., Tennant, A. L., Keane, J., Cole, E. D., Emery, K., and Noufi, R., "The Performance of Cu(In,Ga)Se$_2$-Based Solar Cells in Conventional and Concentrator Applications," in *MRS Spring 1996 Proceedings*, to be published.

2. Hedström, J., Ohlsen, H., Bodegård, M., Kylner, A., Stolt, L., Hariskos, D., Ruckh, M., and Schock, H. W., "ZnO/CdS/Cu(In,Ga)Se$_2$ Thin Film Solar Cells With Improved Performance," in *Proceedings of the 23rd IEEE Photovoltaics Specialists Conference*, 1993, pp. 364–371.

3. Bodegård, M., Stolt, L., and Hedström, J., "The Influence of Sodium on the Grain Structure of CuInSe$_2$ Films for Photovoltaic Applications," in *Proceedings of the 12th European Photovoltaic Solar Energy Conference*, 1994, pp. 1743–1746.

4. Ruckh, M., Schmid, D., Kaiser, M., Schäffler, R., Walter, T., and Schock, H. W., "Influence of Substrates on the Electrical Properties of Cu(In,Ga)Se$_2$ Films," in *Proceedings of the 1st World Conference on Photovoltaic Energy Conversion*, 1994, pp. 156–159.

5. Probst, V., Rimmasch, J., Riedl, W., Holz, J., Harms, H., Karg, F., and Schock, H. W., "The Impact of Controlled Sodium Incorporation on Rapid Thermal Processed Cu(In,Ga)Se$_2$-Thin Films and Devices," in *Proceedings of the 1st World Conference on Photovoltaic Energy Conversion*, 1994, pp. 144–147.

6. Rau, U., Schmitt, M., Hilburger, D., Engelhart, F., Seifert, O., Parisi, J., Riedl, W., Rimmasch, J., and Karg, F., "Influence of Na and S Incorporation on the Electronic Transport Properties on Cu(In,Ga)Se$_2$ Solar Cells," presented at the 25th IEEE Photovoltaics Specialists Conference, Washington, D. C., May 13–17, 1996.

7. Tuttle, J. R., Berens, T. A., Keane, J., Ramanathan, K. R., Granata, J., Bhattacharya, R. N., Wiesner, H., Contreras, M. A., and Noufi, R., "Investigations Into Alternative Substrate, Absorber, and Buffer Layer Processing for Cu(In,Ga)Se$_2$-Based Solar Cells," presented at the 25th IEEE Photovoltaics Specialists Conference, Washington, D. C., May 13–17, 1996.

8. Holz, J., Karg, F., and von Philipsborn, H., "The Effect of Substrate Impurities on the Electronic Conductivity in CIS Thin Films," in *Proceedings of the 12th European Photovoltaic Solar Energy Conference*, 1994, pp. 1592–1595.

9. Tuttle, J. R., Contreras, M. A., Gabor, A. M., Ramanathan, K. R., Tennant, A. L., Albin, D. S., Keane, J., and Noufi, R., *Prog. in Photovoltaics* 3, 383–391 (1995).

10. Sites, J. R., and Mauk, P. H., *Solar Cells* **27**, 411–417 (1989).

11. Mauk, P. H., Tavakolian, H., and Sites, J. R., *IEEE Transactions on Electron Devices* **37**, 422–427 (1990).

12. Rockett, A., Bodegård, M., Granath, K., and Stolt, L., "Na Incorporation and Diffusion in $CuIn_{1-x}Ga_xSe_2$," presented at the 25th IEEE Photovoltaics Specialists Conference, Washington, D. C., May 13–17, 1996.

# Effects of Processing Temperature on the Thickness of CdS and the Performance of CdTe Solar Cells

C. S. Ferekides, B. Tetali, D. Marinskiy, S. Marinskaya, and D. Morel

Department of Electrical Engineering
Center for Clean Energy and Vehicles
University of South Florida
Tampa, Florida 33620

**Abstract:** CdTe cells have been fabricated on soda lime glass substrates. The effect of the CdS thickness and CdTe deposition temperature on the spectral response (SR) and solar cell parameters has been studied. The CdTe deposition temperature has been found to be a key processing parameter in determining the extent of interdiffusion at the CdTe and CdS interface. When the deposition of CdTe is carried out at high temperatures a significant portion of the CdS films is "lost" due to interdiffusion which leads to enhancement of the blue response of the solar cells. Devices with identical blue response (400-500 nm) have been fabricated even though the starting CdS thicknesses were different; the cells for which the starting CdS thickness was greater exhibited higher open-circuit voltages and fill factors.

## INTRODUCTION

A new record efficiency for CdTe cells has been recently achieved by Golden Photon, Inc. (GPI)[1]. The small area device fabricated at GPI exhibited an efficiency of 16.1%. What makes this accomplishment remarkable is the fact that not only is this a new record but that this device was fabricated on inexpensive soda lime glass and yet exhibited a high short-circuit current density. Although, several CdTe groups have been consistently improving the performance of their CdTe solar cells, new record efficiencies have been difficult to realize for several reasons. Most groups have been able to fabricate devices with open-circuit voltages in the range of 800-850 mV and fill factors well above 70%[2]. However, the short-circuit currents have been in the range of 20-22 mA/cm$^2$ a fact that significantly limits cell efficiencies. The rather poor optical properties of soda lime glass and absorption in CdS for wavelengths below 500 nm have been the main reasons for the low $J_{sc}$'s. Due to cost restraints, soda lime glass is the only viable substrate for CdTe cells at this time. Therefore, emphasis has been placed on improving $J_{sc}$ and eventually cell efficiencies by utilizing thin CdS films. Even though CdTe cells prepared with thin CdS films typically exhibit higher $J_{sc}$'s as the blue response of the cells increases, this comes at the expense of lower ff's and $V_{oc}$'s. This trade off between $J_{sc}$ (blue SR) and $V_{oc}$/ff has been the main objective of one of the two CdTe Partnership teams.

The main focus of the CdTe research at the University of South Florida (USF) has been the utilization of soda lime glass substrates and low processing temperatures that would eventually meet manufacturing constraints. To that end CdTe cells are fabricated using a range of processing conditions in order to understand some of the efficiency limiting factors, and develop processes that would lead to further efficiency advancements.

## EXPERIMENTAL

All solar cell results discussed in this paper have been prepared on soda lime glass substrates obtained from Libbey Owens Ford (LOF); the LOF substrates are currently being used by the two leading CdTe companies Solar Cells, Inc., and Golden Photon, Inc. The CdS films are prepared by the chemical bath deposition (CBD) process and the CdTe films by close spaced sublimation (CSS). Based on previous studies the "threshold" CdS thickness for the USF CdTe cells - the thickness below which cell properties begin to degrade - has been found to be about 700-800Å. The optimum CdTe deposition temperature has been in the 600-620 °C range, although devices prepared at temperatures as low as 460 °C, indicated that high $V_{oc}$'s and ff's can also be attained at low temperatures(3).

The process conditions currently used for the deposition of CBD CdS limit the CdS thickness to a maximum value of about 1100Å. For these studies the thickness of the CdS films were varied in the range of 600-800Å (±50 Å) by maintaining all process conditions constant and simply removing the substrates from the solution at different time intervals. The CdTe depositions were carried out at two temperature ranges. A low temperature range of 540-560°C which was previously optimized for soda lime glass substrates, and a high temperature range of 580-620°C. When soda lime glass substrates are processed at high temperatures deformations in the glass are clearly visible, and it is believed that to some extent device performance is degraded. Nevertheless, solar cells are routinely fabricated at these conditions in order to understand the role of this important process parameter.

The following two sections discuss how device performance is affected by the CdS thickness and the CdTe deposition temperature. It will become apparent that the CdTe deposition conditions can have a dramatic effect on the final thickness of the CdS films. All thickness quoted in angstroms (Å) refer to the starting thickness of the CdS films, and are estimates based on deposition times. It should be noted that the CdS thickness of films prepared under identical conditions (different runs) could vary as much as 70 Å at times. Essentially no thickness variation was measured for CdS films from the same deposition run.

## RESULTS AND DISCUSSION

### CdS Thickness - Processing Temperature

Figure 1 shows the SR of a set of four cells. The response of these cells in the 400-500 nm range suggests that there is variation in the CdS thickness among the four devices. However, all four CdS films were produced during the same CBD run to ensure minimum thickness variation. The difference in processing (as noted on the figure) is the CdTe deposition temperature. Two of these devices were prepared at high temperatures and two at low temperatures. The formation of an interfacial $CdS_{1-x}Te_x$ layer has been verified by many(4-6) and therefore some thinning of the CdS films is always expected, depending on the extent of interdiffusion between the CdTe and CdS. The results of figure 1 suggest that the final thickness of the CdS films and the interface layer can be varied by adjusting the CdTe deposition temperature. The photovoltaic parameters for all twelve devices fabricated on the above substrates (3 cells per substrate) are shown in table I. The devices fabricated at higher temperatures exhibit slightly higher $J_{sc}$'s as one should expect based on the SR of these cells. Apparently variations in the

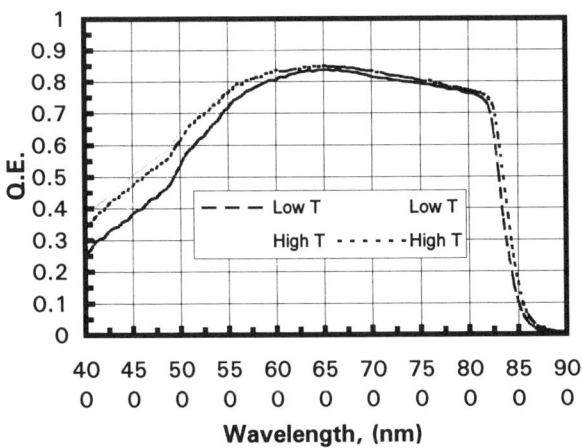

**Figure 1.** SR of four cells prepared with CdS of the same starting thickness.

thickness and/or composition of the interface layer have not affected the ff or $V_{OC}$ of these cells. Typically both the $V_{OC}$ and ff begin to degrade as the blue response of devices increases beyond about 55-60% at 450 nm(7).

In order to better understand the interrelation between the CdS thickness and CdTe deposition temperature, a number of substrates were coated with CdS films of different thicknesses. These substrates were subsequently used for fabricating cells at high and low temperatures. The target CdS thicknesses were 600, 700, and 800 Å, but as mentioned earlier these could vary up to 50 Å.

**TABLE I.** The photovoltaic parameters of the cells of figure 1.

| $T_{DEP}$ (CdTe) | $V_{OC}$ (mV) | | FF (%) | | $J_{SC}$ (mA/cm$^2$) | |
|---|---|---|---|---|---|---|
| Low  | 830 | 839 | 71 | 71 | 20.6 | 19.0 |
| Low  | 832 | 841 | 71 | 74 | 20.8 | 20.3 |
| Low  | 833 | 837 | 71 | 73 | 20.4 | 19.8 |
| High | 839 | 834 | 73 | 71 | 20.0 | 21.1 |
| High | 839 | 834 | 72 | 71 | 21.1 | 21.0 |
| High | 836 | 828 | 71 | 69 | 21.2 | 21.0 |

Figure 2 shows the SR of cells fabricated at low CdTe deposition temperatures. Since all these devices were prepared under the same conditions (but with different starting CdS thicknesses) the amount of CdS "lost" due to intermixing at the interface should be the same in all cases. The SR in the 400-500 nm range clearly indicates that the CdS thickness varies as expected.

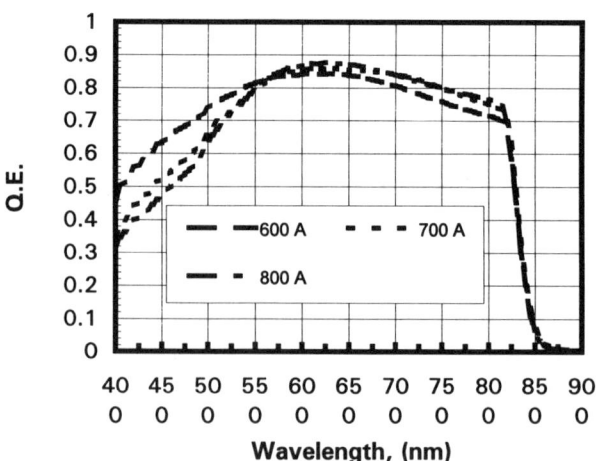

**Figure 2.** The SR of cells fabricated at low temperatures with CdS films of different starting thickness.

Figure 3 shows the SR of cells with CdS thicknesses similar to those in figure 2 but processed at high temperatures. The blue response of these devices is clearly higher than those in figure 2 suggesting that the CdS films are thinner. The two devices that have essentially identical SR were fabricated with CdS films with thicknesses of 600 and 700 Å.

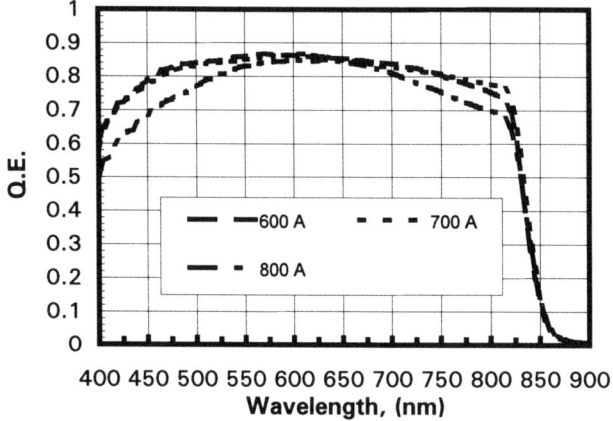

**Figure 3.** The SR of cells fabricated at high temperatures with CdS films of different starting thickness.

Table II lists the $V_{OC}$, ff, and $R_{SH}$ for the devices in figures 2 and 3. The variation in CdS thickness for the cells fabricated at low temperatures has

**TABLE II.** The photovoltaic parameters of the cells of figures 2 and 3.

| $T_{DEP}$ (CdTe) | CdS Thickness, (Å) | $V_{OC}$, (mV) | FF, (%) | $R_{SH}$, ($\Omega$-cm$^2$) |
|---|---|---|---|---|
| Low | 600 | 848 | 66.5 | 665 |
| Low | 700 | 846 | 67.7 | 1040 |
| Low | 800 | 841 | 68.9 | 800 |
| High | 600 | 756 | 52.6 | 500 |
| High | 700 | 791 | 60.8 | 800 |
| High | 800 | 802 | 65.7 | 650 |

essentially no effect on the photovoltaic properties (ff and $V_{OC}$) of these devices. Although the ff's of these cells are lower than the set in table I, this variation is most likely due to the back contact and is not believed to be associated with the CdS thickness. The set of devices fabricated at high temperatures exhibit what is considered a typical trade-off between the blue response and $V_{oc}$/ff. However, it is important to note that the $V_{oc}$ and ff of the two devices whose blue response is identical but the starting CdS thickness is different (600 and 700 Å) are not the same. The $V_{oc}$ and ff of the device for which the starting CdS thickness was 700 Å are higher by 45 mV and 8% respectively.

The above results suggest that the extent of interdiffusion at the CdTe/CdS interface and the final thickness of the CdS films are determined by the CdTe deposition temperature. The post-deposition heat treatment in the presence of $CdCl_2$ has been found to influence the CdTe/CdS interdiffusion. However, in this study this particular processing step was not varied and therefore all changes observed are attributed entirely to the CdTe deposition temperature. The devices fabricated at high temperatures (fig. 3) suggest that the thickness of CdS films can be considerably reduced even when this is in the 600-800 Å range. Most importantly the same enhanced blue response can be obtained even though the starting CdS thickness is different. In addition the devices for which the starting CdS was larger yielded higher $V_{oc}$'s and ff's. Although the overall efficiency of the high temperature devices (table II) was only in the range of 10-11%, these results may be important in the continuous efforts to achieve higher $J_{sc}$'s without degradation in any of the other device parameters.

## Shunt Resistance

Usually the shunt resistance (dark or light) of solar cells fabricated using CdS films of small thicknesses is a good indicator of device performance. The dark Ln I-V characteristics for two devices fabricated at low and high temperatures are shown in figure 4. These represent what is "typical" behavior for high and low temperature devices. The high temperature device exhibits a considerable amount of shunting as indicated by the low voltage section of the I-V, as well as higher $J_o$. The light $R_{SH}$ of a large number devices was calculated (by measuring the slope of the I-V at $J_{SC}$).

The shunt resistance for cells prepared at high temperatures was in the range of 300-800 $\Omega$ and consistently decreased with decreasing CdS thickness. The shunt resistance of devices prepared at low temperatures was in the range of 1000-2000 $\Omega$ and did not show any dependence on the CdS thickness. The ff of about thirty devices prepared using the low and high temperature deposition conditions and various CdS thicknesses is shown in figure 5. It is clear that for $R_{SH}$ values in excess of 800-1000 $\Omega$ the ff is above 70%, and it decreases rapidly as the value of $R_{SH}$ drops below 800 $\Omega$.

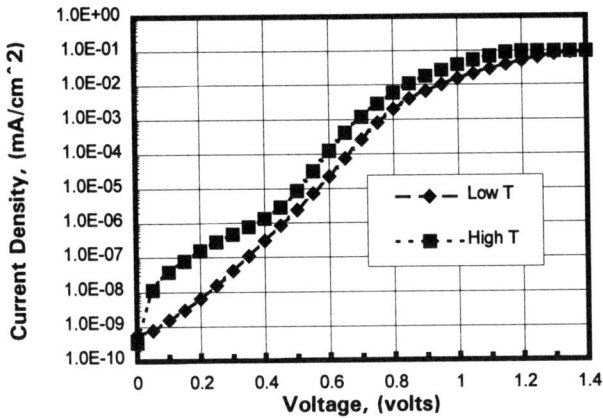

**Figure 4.** Typical dark I-V for cells fabricated at low and high temperatures.

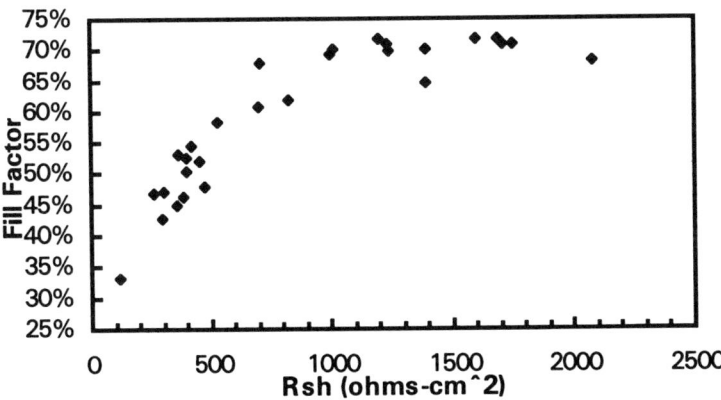

**Figure 5.** The dependence of the fill factor on $R_{SH}$ (light)

## SUMMARY

The effect of the CdS thickness and CdTe deposition temperature on the performance of CdTe cells has been studied. Based on the SR measurements of devices prepared at low and high temperatures for various CdS thickness, it has been shown that a considerable amount of CdS is used up during the CdTe deposition process due to intermixing at the CdTe/CdS interface. Although the final thickness of the CdS films may end up being the same after the devices are completely processed, the initial thickness is critical in improving the open-circuit voltage and fill factor. It is therefore suggested that improved performance can be achieved by fabricating cells with CdS films of larger starting thicknesses, and promoting interdiffusion and thinning of this layer using high temperatures during the fabrication process. Since one of the main objectives for this work is to develop low temperature processes, future emphasis will be placed on enhancing interdiffusion for processing temperatures in the range of 500-550°C.

## ACKNOWLEDGEMENTS

This work has been supported by the National Renewable Energy Laboratory Under subcontract XAF-5-14142-09.

## REFERENCES

1. Zweibel, K., private communication
2. Zweibel, K., H. Ullal, and B. von Roedern, "Progress and Issues in Polycrystalline Thin-Film PV Technologies", Proc. 25th IEEE Photovoltaic Specialists Conference, pp 745-750, (1996).
3. Atter, G., D. P. Bhethanabolta, K. Dugan, S. Karthikeyan, M. Kazi, J. L. Killian, A. B. Muthaiah, D. Niermen, D. M. Oman, R. Swaminathan, S. A. Zafar, C. S. Ferekides, and D. L. Morel, "CuInSe$_2$ and CdTe Tgin Films for Photovoltaic Applications", 12$^{th}$ NREL Photovoltaics Program Review AIP Conference Proceedings 306, pp 309-319, (1993).
4. Mao, L. H. Feng, Y. Zhu, J. Tang, W. Song, R. Collins, D. L. Williamson, and J. U Trefny, "Interdiffusion in Polycrystalline CdTe/CdS Solar Cells", 13$^{th}$ NREL Photovoltaics Program Review AIP Conference Proceedings 353, pp 352-359, (1995).
5. Birkmire, S. S. Hegedus, B. E. McCandless, J. E. Phillips, TWF Russell, W. N. Shafarman, S. Verma, and S. Yamanaka, "Polycrystalline Heterojunction Solar Cells: Processing Perspective", Photovoltaic Advanced Research and Development Project, AIP Conference Proceedings 268, pp 212-217, (1992).
6. Clemminck, M. Burgelman, M. Casteleyn, J. De Poorter, and A. Varvaet, Proc. 22$^{nd}$ IEEE Photovoltaic Specialists Conference, pp 1114-1119, (1991).
7. Advanced Processing Technology for High Efficiency Thin Film CuInSe$_2$ and CdTe Solar Cells, NREL Final Report, Contract # XG-2-11036-1

# Effect of Cu Doping on the Properties of ZnTe:Cu Thin Films and CdS/CdTe/ZnTe Solar Cells

J. Tang, D. Mao, and J. U. Trefny

*Department of Physics, Colorado School of Mines, Golden, Colorado 80401*

**Abstract.** The effects of Cu doping concentration and post-deposition annealing treatment on the properties of ZnTe thin films were investigated in an effort to decrease the Cu doping concentration and improve the long-term stability of CdS/CdTe/ZnTe solar cells. The structural, compositional, and electrical properties were studied systematically using x-ray diffraction (XRD), electron microprobe, Hall effect and conductivity measurements. XRD measurements indicated that the crystalline phase of as-deposited and low-temperature annealed ZnTe films is dependent on Cu doping concentration. Low-Cu-doped films exhibited zincblende phase, whereas high-Cu-doped films showed wurtzite phase. After annealing at high temperature ($\geq 350°C$), all films exhibited zincblende structure. Electron probe microanalysis revealed a deficiency of cations in low-Cu-doped films and an excess of cations in high-Cu-doped films. Hall effect measurements revealed a dependence of hole mobility on Cu doping concentration with the highest mobility (20 $cm^2/V \cdot s$) obtained at a low Cu concentration. Carrier concentrations higher than mid-$10^{18}$ $cm^{-3}$ were obtained at a Cu concentration of 2 at. % and relatively low annealing temperatures. Studies of the activation energy of dark conductivity suggested that intrinsic defects (e.g., Zn vacancies) are the dominant acceptors for Cu concentrations lower than 4.5 at.%. Finally, ZnTe films with Cu concentrations as low as 1 at. % were used successfully as a back contact layer in CdTe based solar cells. Fill factors over 0.70 were obtained using ZnTe films of low Cu concentrations.

## INTRODUCTION

The long-term stability of CdTe thin film solar panels places stringent requirements on the back contact formation techniques. One back contact material, thermally evaporated Cu-doped ZnTe, has yielded very low contact resistance (< 0.1 ohm.cm²) on electrodeposited CdTe thin films and high fill factors (>0.74) of the resulting cells (1). Because of the well known high diffusivity of Cu in polycrystalline thin films, it is desirable to bring the Cu concentration in these materials to as low a level as possible while still maintaining the performance of the contact. Our previous studies investigated mostly ZnTe films doped with relatively high levels of Cu, between 4 and 10 at. %, with the best devices obtained at a Cu concentration of 6 at. % (1,2). Hall effect measurements of the ZnTe films indicated that the acceptor concentration was in the mid $10^{20}$ cm$^{-3}$ range. This high carrier concentration is more than enough to narrow the ZnTe/metal interface barrier within tunneling distance. It also indicated that only a small fraction of the total Cu concentration was activated. The low doping efficiency of Cu is caused by the combination of several factors: (1) the low processing temperature that was used to limit the Cu diffusion into CdTe; (2) the possible formation of secondary phases between Cu and Te; (3) carrier compensation by structural defects in ZnTe.

We have extended our research on the back contact by studying ZnTe films doped with low concentrations of Cu. With a lower Cu concentration, the ZnTe:Cu post-deposition annealing temperature may be increased without excessive Cu diffusion into CdTe. The formation of secondary phases will also be reduced. These two factors may lead to an improved Cu doping efficiency in ZnTe, therefore maintaining the quality of the contact. The decreased amount of Cu will certainly be beneficial for the long-term stability of the final devices.

## EXPERIMENTAL PROCEDURES

ZnTe films doped with Cu were deposited by vacuum evaporation. The vacuum system had a base pressure of 1 x $10^{-6}$ torr and was partitioned into two sections: one for a ZnTe powder source (99.999%, Johnson-Matthey) and the other for a Cu source (99.9999%, Johnson-Matthey). The ZnTe and Cu deposition rates were measured by separate thickness monitors. The typical deposition rates of the ZnTe were 5-10 Å/s. The deposition rate of the Cu was adjusted from 0 to 0.2 Å/s to obtain different Cu atomic concentrations in the ZnTe, varying between 0 and 10%. Microscope glass slides were cleaned with Micro detergent and used as substrates. During vacuum deposition, the substrates were held at room temperature. Uniform, adherent ZnTe films were routinely obtained using the described procedure. Post-deposition annealing was performed in a $N_2$ environment. For most measurements, 0.8-1.0 μm-thick films were used.

The XRD measurements were performed with a Rigaku x-ray diffractometer using CuKα radiation. Electrical resistivity along the lateral direction was measured by depositing Au-stripes on the ZnTe films and performing 2-point or 4-point probe measurements. Hall effect measurements were made with a Bio-Rad HL5500 system. Wavelength-dispersive spectroscopy (WDS, Cameca model MBX electron microprobe) was used for compositional analysis.

Complete cells were fabricated using ZnTe:Cu back contacts. Light current-voltage measurements were performed using an ELH-type tungsten halogen lamp calibrated with a GaAs standard cell. For each sample, several cells (0.1 cm$^2$ area) were measured and the cell performance parameters were averaged.

# RESULTS AND DISCUSSION

## Structural Analysis

The structure of the ZnTe films was studied with x-ray diffraction. Figure 1 shows a set of XRD patterns of ZnTe films doped with various amounts of Cu. For as-deposited films and films annealed at 250°C, the crystalline phase of ZnTe showed a dependence on Cu doping level. Low Cu doped (≤3 at. %) films exhibited cubic (zincblende) phase with a preferred orientation of (111), whereas high Cu doped (6 at. % and 8 at. %) films showed hexagonal (wurtzite) phase. After annealing at a higher temperature (350°C), all films exhibited zincblende structure. Because of the close proximity of the lattice constants of ZnTe (6.1026 Å) and $Cu_2Te$ (6.11 Å), it was difficult to detect the formation of $Cu_2Te$ in the films by monitoring the changes in XRD patterns.

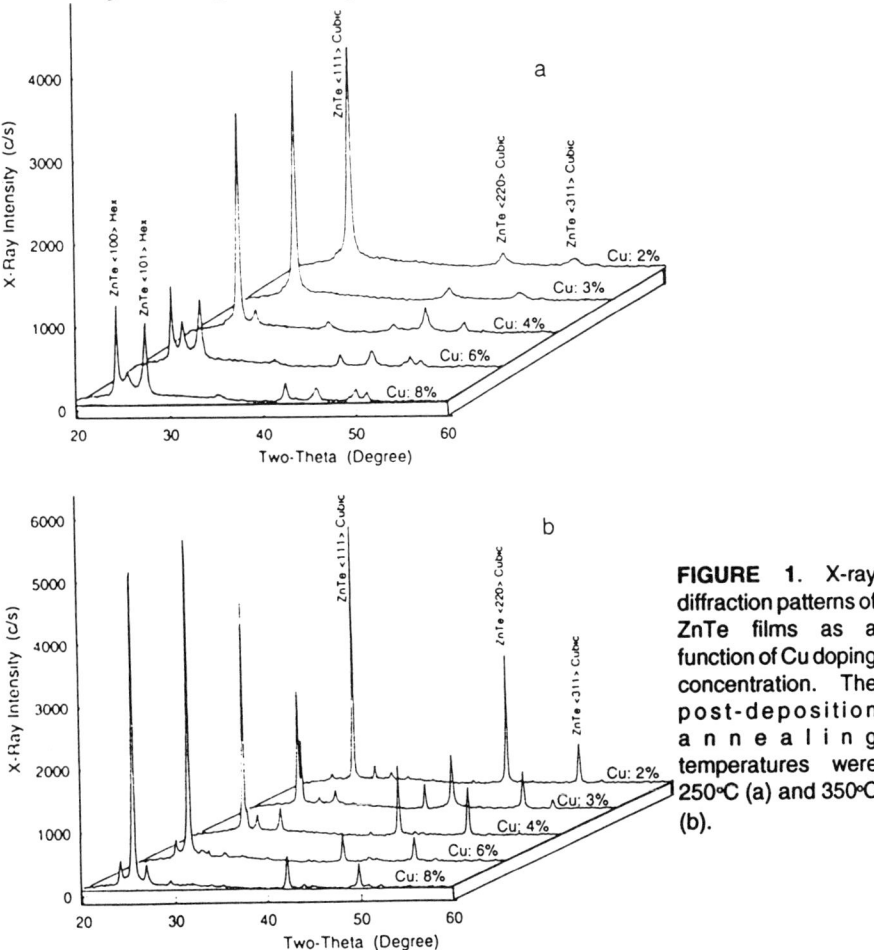

**FIGURE 1.** X-ray diffraction patterns of ZnTe films as a function of Cu doping concentration. The post-deposition annealing temperatures were 250°C (a) and 350°C (b).

## Compositional Analysis

Electron probe microanalysis was performed on a series of ZnTe:Cu films with nominal Cu concentrations ranging from 1 at. % to 8 at. %. A comparison of the concentrations of cations (Zn, Cu) and anions reveals some interesting features (Figure 2): at low Cu concentrations (< 6 at. %), there is a deficiency of cations; at Cu concentrations higher than 6 at. %, the cation concentration exceeds the anion concentration, suggesting that Cu in these films exists mostly in the $Cu_2Te$-like state. This has implications on the doping mechanism in these films. The cation deficiency at low Cu concentration suggests the existence of Zn vacancies which act as acceptors. The existence of $Cu_2Te$-like bonding, on the other hand, suggests that most of the Cu atoms are electrically inactive because all the valence electrons are fully coordinated. Finally, we notice that the trend we observed here is consistent with the results reported by Gessert et al. for rf-sputtered Cu-doped ZnTe (3,4), even though the deposition technique and substrate temperature used were rather different in these two studies.

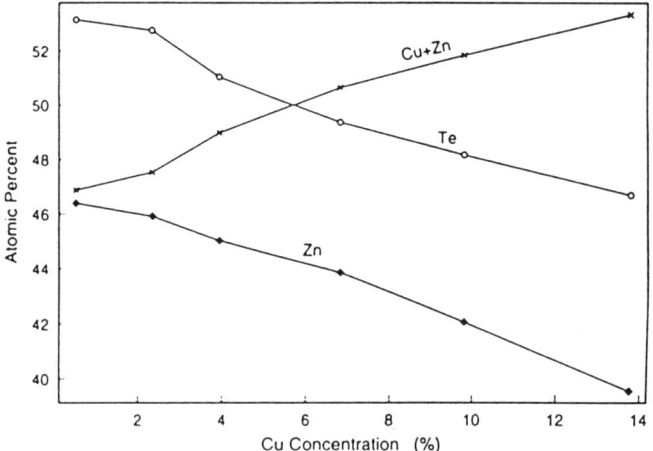

**FIGURE 2.** Atomic concentration of Zn, Te, and Zn+Cu as a function of Cu doping concentration in the ZnTe films.

## Hall Effect and Conductivity Measurements

Hall effect measurements were performed on ZnTe films containing various amounts of Cu (1 to 8 at. %) and annealed at different temperatures (250 - 450°C). The results are listed in Table 1. We can observe several features from this table:
1. The lowest resistivity was obtained at an annealing temperature of 350°C. At higher annealing temperatures, the resistivity increases again. The increase of resistivity at high annealing temperature was caused mostly by a decrease of carrier

concentration.

**Table 1.** The dependence of film resistivity, carrier concentration and mobility of ZnTe as a function of Cu concentration and post-deposition annealing temperature. The film thickness was 0.8 µm.

| Nominal Cu concentration | Annealing Temperature (°C) | ρ (Ω-cm) | Mobility (cm²/V·s) | Doping (cm⁻³) |
|---|---|---|---|---|
| 1% | 250 | $3.8 \times 10^4$ | 6.9 | $2.4 \times 10^{13}$ |
|    | 300 | 6.4 | 6.7 | $1.5 \times 10^{17}$ |
|    | 350 | 0.72 | 20 | $4.5 \times 10^{17}$ |
|    | 400 | 1.2 | 21 | $2.5 \times 10^{17}$ |
|    | 450 | 14.0 | 18 | $2.6 \times 10^{16}$ |
| 2% | 250 | 3.8 | 0.1 | $1.4 \times 10^{19}$ |
|    | 300 | 0.32 | 2.8 | $7.0 \times 10^{18}$ |
|    | 350 | 0.14 | 8.6 | $5.1 \times 10^{18}$ |
|    | 400 | 0.29 | 13 | $1.7 \times 10^{18}$ |
|    | 450 | 0.55 | 1.5 | $7.6 \times 10^{18}$ |
| 3% | 250 | 1.3 | 1.7 | $2.9 \times 10^{18}$ |
| 4% | 300 | 0.24 | 0.3 | $7.7 \times 10^{19}$ |
|    | 350 | 0.10 | 1.2 | $5.4 \times 10^{19}$ |
|    | 450 | 1.8 | 1.9 | $1.8 \times 10^{18}$ |
| 6% | 250 | 2.7 | 1.1 | $2.2 \times 10^{18}$ |
|    | 300 | 0.012 | 1.4 | $3.8 \times 10^{20}$ |
|    | 350 | 0.041 | 1.0 | $1.5 \times 10^{20}$ |
|    | 400 | 0.079 | 0.6 | $1.3 \times 10^{20}$ |
|    | 450 | 0.055 | 2.3 | $4.9 \times 10^{19}$ |
| 8% | 250 | 2.6 | 1.1 | $2.2 \times 10^{18}$ |

 2. Very high mobility (for polycrystalline p-type ZnTe) was obtained at low (1 to 2 at. %) Cu doping and relatively low annealing temperatures (350°C). With increase in Cu concentration, the mobility decreases, caused possibly by increased scattering by ionized impurities.

 3. 2.0 at. % seems to be the lowest Cu doping that can yield a p-type doping concentration of over $10^{18}$ cm⁻³. At even lower Cu concentration (1.0 at.%), charge compensation leads to a very low doping concentration.

 4. The highest Cu doping efficiency was obtained at a Cu concentration of 6 at. % for films annealed at 300°C or higher. For films annealed at a low temperature (250°C), however, the highest doping efficiency was obtained with a Cu concentration of 2 at. %. As shown in the table, the carrier concentration of both 2 at. % and 3 at. % doped films is higher than that of 6 at. % and 8 at. % doped films

for the annealing temperature of 250°C. Since low annealing temperature is desired for back contact formation on CdTe, this result suggests that a Cu doping of 2 at. % is preferable.

The conductivities of the films were measured as a function of temperature following annealing under different conditions. The activation energy of conductivity of ZnTe films was deduced from the temperature dependence of dark conductivity between 20–120°C. The results are shown in Table 2. We observe several features: first, the activation energy of conductivity is rather high (0.6 eV) for the undoped film. The ionization energy of the second level of a Zn vacancy is a possible interpretation for this activation energy (5); second, for Cu-doped films, the activation energy is smaller than that of the undoped film and decreases with increasing Cu concentration and annealing temperature. In general, the activation energy observed was between 0.2 and 0.5 eV. These values correspond well with the ionization energy reported for intrinsic defect levels in ZnTe (0.25 eV, and 0.5 eV) and are higher than the ionization energy of the $Cu_{Zn}$ acceptor level (0.12, 0.15 eV) (6, 7). This indicates that intrinsic defects (e.g., Zn vacancies) are the dominant acceptors. This can be correlated with the compositional analysis results. As discussed in a previous section, there is a high degree of cation deficiency in low Cu-doped films. This may lead to a high concentration of Zn vacancies. It is possible that the main effect of Cu is to create this cation deficiency rather than acting as an acceptor directly. If this is indeed the case, it may be possible to form Zn-deficient ZnTe films using other mechanisms, thereby eliminating Cu completely.

**TABLE 2.** Activation energy of dark conductivity of ZnTe films as a function of Cu concentration and post-deposition annealing temperature.

| Cu Concentration | Annealing Temperature (°C) | Activation Energy (eV) |
|---|---|---|
| 0% | 300 | 0.60 |
| 1.0% | 145 | 0.45 |
|  | 235 | 0.50 |
|  | 305 | 0.41 |
| 2.0% | 140 | 0.40 |
|  | 230 | 0.45 |
|  | 300 | 0.30 |
| 3.0% | 140 | 0.42 |
|  | 225 | 0.32 |
|  | 305 | 0.22 |
| 4.5% | 145 | 0.25 |
|  | 190 | 0.21 |
|  | 310 | 0.22 |

## Solar Cell Performances

CdS/CdTe solar cells were fabricated with ZnTe films containing low Cu concentrations. CdTe films prepared by Solar Cells, Inc. were used. Good ohmic contacts were obtained on all films using Cu concentrations as low as 1.0 at. %. Table 3 shows the results. The ZnTe:Cu post-deposition annealing temperature was varied according to the Cu concentration used. The highest annealing temperature was close to 300°C.

**TABLE 3.** Performance of solar cells fabricated with CdS/CdTe films provided by Solar Cells, Inc. and ZnTe back contacts containing low Cu concentrations. The values are averages obtained from several 0.1 cm² cells. Doping density of CdTe deduced from capacitance-voltage measurements is also shown. $J_{sc}$ values are not accurate because the light source used does not have a standard AM1.5 spectrum.

| Cu Concentration | η (%) | $V_{oc}$ (V) | $J_{sc}$ (mA/cm²) | FF | Doping Density in CdTe (cm⁻³) |
|---|---|---|---|---|---|
| 1.0% | 11.6 | 0.74 | 23.8 | 0.66 | $4.2 \times 10^{14}$ |
| 2.0% | 12.4 | 0.73 | 24.8 | 0.68 | $5.0 \times 10^{14}$ |
| 3.0% | 11.6 | 0.77 | 23.3 | 0.65 | $4.3 \times 10^{14}$ |
| 4.5% | 12.2 | 0.76 | 24.0 | 0.67 | $2.1 \times 10^{14}$ |

Table 4 shows the solar cell performance as a function of ZnTe:Cu post-deposition annealing temperature for a Cu concentration of 2 at. %. The best cell performance was obtained for an annealing temperature of 270°C. This is very different from the optimal annealing temperature for ZnTe films doped with 6 at. % Cu. In the latter case, 170°C annealing yielded the highest fill factors in the resulting cells and annealing at temperatures above 200°C always resulted in significantly degraded cell performance. The much higher processing temperature used for the 2 at. % Cu doped films implies that the long-term stability of the resulting cells can be much higher because most cell degradation mechanisms are thermally activated processes.

**TABLE 4.** Effect of ZnTe post-deposition annealing temperature on cell performance. The Cu concentration was 2 at.%, ZnTe film thickness was 500 Å. The cell area was 0.1 cm².

| Annealing Temp. (°C) | η (%) | $V_{oc}$ (mV) | $J_{sc}$ (mA/cm²) | FF | dV/dJ at $V_{oc}$ (Ω-cm²) |
|---|---|---|---|---|---|
| 200 | 1.8 | 532 | 9.4 | 0.36 | 20.1 |
| 270 | 13.3 | 796 | 23.9 | 0.70 | 4.50 |
| 280 | 12.1 | 763 | 23.2 | 0.68 | 4.75 |
| 350 | 7.08 | 678 | 20.7 | 0.50 | 18.8 |

In order to probe any possible Cu diffusion from the ZnTe:Cu film into the CdTe, we measured the doping density in the CdTe layer using capacitance-voltage measurements. If Cu does diffuse into the CdTe during the cell fabrication, it would act as an acceptor in the CdTe and higher doping in CdTe would be expected for cells made with ZnTe films containing higher concentrations of Cu. As shown in Table 3, we did not observe any correlation between CdTe doping density and the Cu doping level in ZnTe, indicating that Cu diffusion is negligible for as-prepared cells.

## CONCLUSIONS

We have investigated the effects of Cu doping concentration and post-deposition annealing temperature on the structural, compositional, and electrical properties of ZnTe using x-ray diffraction, electron microprobe, Hall effect and conductivity measurements. Our results indicate that low contact resistance and high fill factors can be obtained using low-Cu-doped ZnTe as a contact layer in CdS/CdTe solar cells. The decreased Cu doping concentration is expected to improve the long-term stability of resulting CdTe solar cells.

## ACKNOWLEDGMENTS

This work is supported by the National Renewable Energy Laboratory under Subcontract No. XAF-5-14142-11. We thank Rick Powell of Solar Cells, Inc. for providing the CdS/CdTe films that were used for cell fabrication. We also thank Alice Mason of NREL for performing electron probe microanalysis measurements.

## REFERENCES

1. J. Tang, D. Mao, L. Feng, W. Song, and J.U. Trefny, "The Properties and Optimization of ZnTe:Cu Back Contacts on CdTe/CdS Thin Film Solar cells," *in Proceedings of the 25th IEEE Photovoltaic Specialists Conference*, 1996, pp. 925-928.
2. L.H. Feng, D. Mao, J. Tang, R.T. Collins, and J.U. Trefny, *J. Electron. Materials* **25**, 1433 (1996).
3. T.A. Gessert, A.R. Mason, R.C. Reedy, R. Matson, T.J. Coutts, and P. Sheldon, *J. Electron. Materials* **24**, 1443 (1995).
4. T.A. Gessert, A.R. Mason, P. Sheldon, A.B. Swartzlander, D. Niles, and T.J. Coutts, *J. Vac. Sci. & Technol.* **A14**, 806 (1996).
5. T.L. Larsen, C.F. Varotto, and D.A. Stevenson, *J. Appl. Phys.* **43**, 172 (1972).
6. H. Tubota, *Jpn. J. Appl. Phys.* **2**, 259 (1963).
7. M. Aven and B. Segall, *Phys. Rev.* **130**, 81 (1963).

# Processing Issues for Thin Film CdTe/CdS Solar Cells

Brian E. McCandless, Robert W. Birkmire,
D. Garth Jensen, James E. Phillips, and Issakha Youm

*University Center of Excellence
For Photovoltaic Research and Education*

*Institute of Energy Conversion
University of Delaware
Newark, Delaware 19716 USA*

**Abstract.** Processing options addressing critical issues associated with fabrication of high efficiency thin film CdTe/CdS solar cells are presented. Particular focus is given to methods for: minimizing CdS loss during $CdCl_2$ heat treatment; obtaining high open circuit voltages; obtaining spatially uniform properties; and forming ohmic contacts. Data is presented for cells deposited by physical vapor deposition (PVD) which demonstrates that CdS loss can be overcome by use of $CdTe_{1-x}S_x$ absorber layers and by controlling the delivery of $CdCl_2$ vapor species during heat treatment. The $CdCl_2$ vapor treatments are shown to produce residue-free surfaces and uniform film properties which leads to uniform cell performance. State of the art open circuit voltages (>850 mV) are obtained with PVD cells by performing a short high temperature anneal prior to $CdCl_2$ treatment. Low resistance contacts using diffused Cu doping followed by surface etching are demonstrated on CdTe/CdS thin films deposited by five methods.

## INTRODUCTION

Processing high efficiency thin film (< 10 μm) CdTe/CdS solar cells consists of a film deposition and a post-deposition treatment which adjusts film properties in order to form an ohmic contact to the CdTe surface. The post-deposition treatment required to improve CdTe/CdS junction properties also results in detrimental thinning of the CdS layer due to interfacial diffusion, thereby imposing a lower limit to the thickness of the initial CdS buffer layer thickness that can be employed in maximizing current density [1]. Further, CdTe/CdS cells deposited at temperatures below ~500°C exhibit lower open circuit voltages ($V_{oc}$) than cells deposited at higher temperatures [2]. Finally, ohmic contact formation to the CdTe surface tends to be process-specific and depends on overly complicated processes to both achieve high surface doping density and minimize grain boundary leakage [3]. In this paper, processing options are presented which address these issues. $CdTe_{1-x}S_x$ absorber layers are shown to minimize CdS loss during $CdCl_2$ post-deposition treatment of CdTe/CdS cells deposited by physical vapor deposition (PVD). An all-vapor $CdCl_2$ post-deposition treatment is presented which also reduces the CdS loss and which results in a residue-free CdTe surface with good spatial uniformity and a $V_{oc}$ greater than 850 mV. Finally, the analysis of the electrical behavior of CdTe/CdS cells contacted using the diffused Cu plus etch process [3] is presented for CdTe films deposited by five different methods.

## CdTe/CdS CELL FABRICATION AND ANALYSES

The post-deposition process was first optimized on 1" x 1" CdTe/CdS structures deposited by physical vapor deposition described elsewhere [1]. The initial CdS thicknesses ranged from 0.1 to 0.2 μm. Contacts to the CdTe were made using the diffused Cu plus etch process described in reference [3]. The contact process was evaluated on CdTe films deposited by PVD, elemental vapor deposition (EVD), electrodeposition, sputtering, and close-spaced sublimation (CSS) using carbon contacts.

Scanning electron microscopy (SEM) and x-ray diffraction (XRD) were used to assess morphological and compositional changes in the CdTe/CdS structures caused by the treatments. Devices were analyzed by current-voltage (J-V) and quantum efficiency (QE) measurements. In a previous work, it has been shown that the QE response below 500 nm in a high efficiency device is limited by the parasitic absorption of the CdS layer [4]. Thus the QE measurement, when combined with XRD peak profiles, provides an analytical tool for measuring the redistribution of S in the final device.

## $CdTe_{1-x}S_x$ ABSORBER LAYERS

The diffusion of CdS into CdTe during $CdCl_2$ processing results in a net thinning of the CdS layer by as much as 80 nm. For PVD cells, it has been shown that the best $V_{oc}$ attained drops as CdS thickness is reduced below ~150 nm [5]. This phenomenon can be explained by formation of pinholes in the CdS layer as its thickness is reduced, resulting in the formation of $CdTe_{1-x}S_x$/ITO junctions in parallel with the desired $CdTe_{1-x}S_x$/CdS junctions. For the CdTe-CdS alloy system in equilibrium with $CdCl_2$ at 420°C, a miscibility gap exists from 5.8% S at the Te-rich side [6]. By incorporating S into the CdTe layer during deposition, it is expected that the driving force for additional S diffusion will be reduced.

Devices based on $CdTe_{1-x}S_x$ absorbers having S content near the solubility limit behave similar to conventional CdTe/CdS devices, achieving efficiencies ~11% [7].

Table 1 shows the tradeoff between $V_{oc}$ and $J_{sc}$ as the CdS thicknesses is reduced.

**TABLE 1.** Open circuit voltage ($V_{OC}$), short circuit current density ($J_{SC}$), fill factor (FF), and change in CdS thickness ($\Delta d$) for CdTe/CdS and CdTe$_{1-x}$S$_x$/CdS devices

| Cell Type | Initial d(CdS) (nm) | Final d(CdS) (nm) | $\Delta d$ (nm) | $V_{oc}$ (mV) | $J_{sc}$ (mA/cm$^2$) | FF (%) |
|---|---|---|---|---|---|---|
| CdTe/CdS | 105 | 25 | 80 | 572 | 24.2 | 56 |
| CdTe/CdS | 135 | 65 | 70 | 684 | 23.2 | 60 |
| CdTe/CdS | 170 | 115 | 55 | 775 | 20.5 | 67 |
| CdTe$_{95}$S$_5$/CdS | 105 | 75 | 30 | 769 | 22.6 | 60 |
| CdTe$_{95}$S$_5$/CdS | 135 | 85 | 50 | 782 | 21.6 | 62 |
| CdTe$_{95}$S$_5$/CdS | 170 | 130 | 40 | 769 | 20.3 | 56 |

Table 1 shows that the CdS lost to formation of CdTe$_{1-x}$S$_x$ in CdTe/CdS cells increases for decreasing initial CdS thickness. For example, a CdTe/CdS cell that initially had a 105 nm thick CdS layer, suffers a net loss of 80 nm. By comparison, cells made on the same CdS but with 5% S incorporated into the CdTe film as CdTe$_{1-x}$S$_x$ alloy exhibited similar performance over the entire range of CdS thickness and less CdS loss in each case. In fact, the equivalent CdS thickness lost is about the same in each case for the CdTe$_{1-x}$S$_x$/CdS cells. A comparison of the CdS optical transmission and the QE for both kinds of cells with a CdS thickness of 105 nm, is shown in Figure 1. In the figure, the QE of both devices at short wavelengths exceeds the initial CdS transmission, indicating a loss of CdS thickness during processing. The sample deposited with a uniform CdTe$_{1-x}$S$_x$ layer exhibited the least CdS loss and a longer wavelength QE falloff. It is expected that this approach, when combined with CdS films having initial thickness of ~50 nm, will result in state-of-the-art performance using a PVD deposition process.

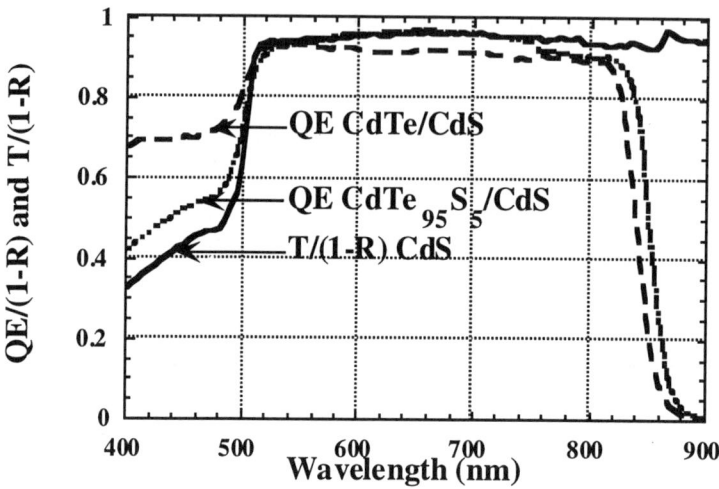

**FIGURE 1.** Quantum efficiency of CdTe/CdS and CdTe$_{1-x}$S$_x$/CdS cells and transmission of 105 nm thick CdS film prior to device fabrication. These data have been normalized for reflection.

## CdCl$_2$ VAPOR TREATMENT

All-vapor post deposition processing holds many advantages the over conventional coat-and-rinse techniques that are employed for CdTe cells. For example, the thermal separation of CdTe/CdS films from the chloride source allows independent control of both the reaction temperature and species concentration. This facilitates temperature-time configurations that: can reduce the CdS loss via interdiffusion; increase the $V_{oc}$; reduce the treatment time; and produce a residue-free CdTe surface. Figure 2 shows the CdTe (511) XRD peak profile in 2.5 µm CdTe/0.2 µm CdS structures after CdCl$_2$ vapor treatment. The bimodal profile obtained for the sample in which the CdCl$_2$ and CdTe/CdS structure were heated together indicates extensive CdTe-CdS mixing and corresponds to an equivalent CdS thickness of approximately 65 nm. Delaying the delivery of CdCl$_2$ vapor by two minutes after reaching reaction temperature results in a diffusion tail that corresponds to an equivalent CdS thickness of less than 20 nm.

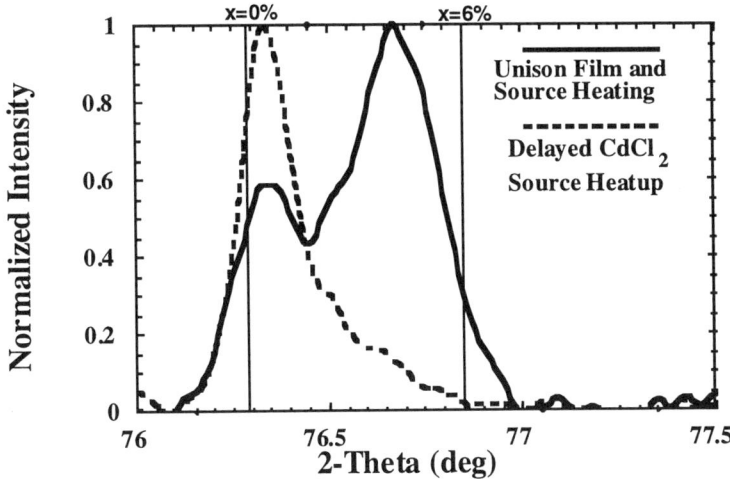

**FIGURE 2.** CdTe$_{1-x}$S$_x$ (511) XRD profile for CdTe/CdS structures heated in unison with (solid) and prior to (dotted) the CdCl$_2$ source.

The vapor chloride processing yields a spatially uniform grain size and a clean CdTe surface free of residual chlorides, oxides, and chlorates. This eliminates the necessity for rinsing or handling of rinsates prior to contact formation. From a device perspective, these benefits translate into spatially uniform properties and performance. Table 2 shows current-voltage parameters for four 0.2 cm$^2$ area cells fabricated on a 1" x 1" sample processed with CdCl$_2$ vapor treatment.

**TABLE 2.** Current-voltage parameters for CdTe/CdS cells demonstrating spatial uniformity of device performance with vapor chloride processing.

| Cell | $V_{oc}$ (mV) | $J_{sc}$ (mA/cm$^2$) | FF (%) | Eff (%) |
|---|---|---|---|---|
| 1 | 786 | 21.4 | 67 | 11.2 |
| 2 | 795 | 21.8 | 64 | 11.1 |
| 3 | 779 | 21.0 | 66 | 10.8 |
| 4 | 790 | 21.5 | 66 | 11.2 |

Until now, the highest $V_{oc}$ obtained with PVD CdTe has been typically less than 820 mV, compared to the greater than 840 mV obtained by close-spaced sublimation [8]. A possible explanation for this lower $V_{oc}$ with processes such as physical vapor deposition and electrodeposition (820 mV, ref. 9) is higher dark current is produced by defects in CdTe, that are not completely annealed out during the nominal 400°C CdCl$_2$ treatment. Therefore, higher temperature anneals may be expected to reduce the defect density, resulting in a lowered dark current and higher $V_{oc}$. Performing a short (10-30 minute) anneal in argon at

550°C prior to exposure to $CdCl_2$ has yielded enhanced $V_{oc}$ in PVD devices (Table 3).

**TABLE 3.** Current-voltage parameters for CdTe/CdS cells treated with and without a 550°C anneal prior to treatment with $CdCl_2$.

| Treatment Type | $V_{oc}$ (mV) | $J_{sc}$ (mA/cm$^2$) | FF (%) | Eff (%) |
|---|---|---|---|---|
| Argon anneal at 550°C for **30** min then $CdCl_2$ at 420°C for 20 min | 853 | 21.0 | 67 | 12.0 |
| Argon anneal at 550°C for **10** min then $CdCl_2$ at 420°C for 20 min | 834 | 20.7 | 66 | 11.4 |
| $CdCl_2$ at 420°C for 20 min | 806 | 21.4 | 65 | 11.2 |

## CURRENT VOLTAGE BEHAVIOR

The current voltage behavior under standard solar cell test conditions (AM1.5 Global @ 100 mW/cm$^2$ and 25° C) was measured for each kind of device that was contacted. Table 4 shows the basic J-V parameters of this test. The a.c resistance or slope of the J-V curve at open circuit voltage ($dV/dJ@V_{oc}$) is also included. Unusually large values of this open circuit a.c. resistance can be indicative of problems due to high series resistance. Figure 3 shows the plots of the entire current voltage behavior. In forward bias, at higher voltage, it appears that the Colorado School of Mines' device might be showing a non-ohmic current voltage behavior. By plotting this slope as a function of inverse total current, as shown in Figure 4, the small amount of non-ohmic behavior that could be due to the contacting process is highlighted. However, the figures show that the performance of these devices are not limited by the contact used.

**TABLE 4.** CdTe deposition technique and basic J-V parameters (measured at AM1.5 Global Illumination and 25° C) for the various groups whose devices were contacted.

| Group | CdTe Deposition Technique | Eff (%) | FF (%) | $V_{oc}$ (mV) | $(dV/dJ)@V_{oc}$ ($\Omega$-cm$^2$) |
|---|---|---|---|---|---|
| Col.Sch. of M. | Electro | 8.82 | 64.5 | 754 | 6.3 |
| IEC | Physical Vapor | 10.97 | 66.2 | 853 | 7.3 |
| S.C.Inc. | Elemental Vapor | 11.73 | 74.5 | 789 | 3.1 |
| U.of Tol. | Sputtering | 11.43 | 65.0 | 796 | 8.5 |
| U.of S.Fl. | Close-Spaced Sublimation | 14.68 | 71.2 | 845 | 3.7 |

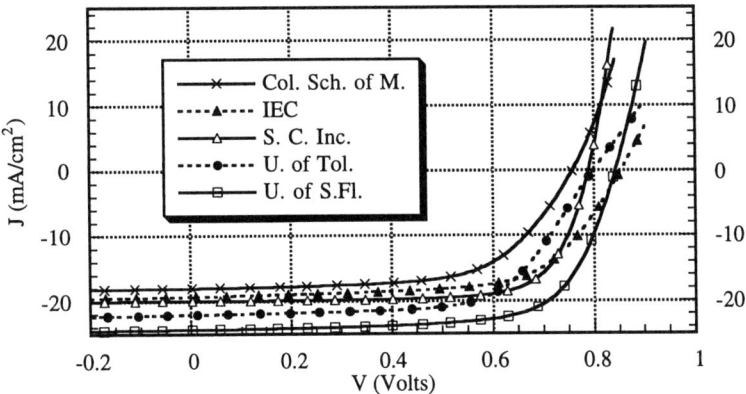

**FIGURE 3.** J-V plots (measured at AM1.5 Global Illumination and 25° C) for the various groups whose devices were contacted.

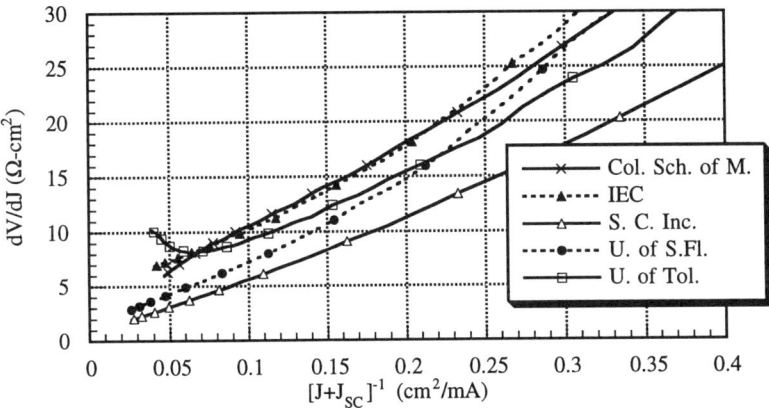

**FIGURE 4.** Slope of J-V curves shown in Figure 3 as a function of inverse total current for the various groups whose devices were contacted.

## SUMMARY AND CONCLUSIONS

The CdS loss that occurs during $CdCl_2$ processing of CdTe/CdS thin film cells can be minimized by the use of $CdTe_{1-x}S_x$ absorber layers and by vapor processing. The uniform action of vapor chloride treatments produces a chemically clean surface and result in spatially uniform cell performance. A short high temperature anneal step prior to $CdCl_2$ exposure enhances cell $V_{oc}$ and has yielded $V_{oc}$ greater than 850 mV with CdTe/CdS cells made by PVD, thereby demonstrating that the $V_{oc}$ in PVD deposited CdTe cells is comparable to cells with the CdTe deposited at higher temperatures. Analysis of the diffused Cu plus etch contacting process on CdTe/CdS cells made by five different methods indicates ohmic contact and similar device operation.

## ACKNOWLEDGMENTS

The authors acknowledge the skilled contributions of Shannon Fields, Kevin Hart, Rajesh Venugopal, and Johnny Yu. This work was supported by NREL subcontract number XAV-3-13170-01.

## REFERENCES

1. B. E. McCandless and R. W. Birkmire, Solar Cells, **31** (1991), pp. 527-534.
2. D. Bonnet, Int'l Journal Solar Energy, **12** (1992), pp. 1-14.
3. B.E. McCandless, Y. Qu, and R. W. Birkmire, Proc. 1st World Conf. on Photovoltaic Energy Conversion (1994), pp. 107-110.
4. R. W. Birkmire, B. E. McCandless, and S. S. Hegedus, Int'l Journal of Solar Energy, **12** (1992), pp. 145-154.
5. R. W. Birkmire, et. al., Annual Report to NREL, subcontract XAV-3-13170-01 (1994), pp. 29-31.
6. D. G. Jensen, B. E. McCandless, and R. W. Birkmire, Proceedings of the 1996 MRS Meeting, San Francisco (1996).
7. D. G. Jensen, B. E. McCandless, and R. W. Birkmire, Proc. 25th IEEE PVSC, (1996), pp.773-776.
8. C. Ferekides and J. Britt, Applied Physics Letters, **62** (22) (1993), pp.2851-2852.
9. A. K. Turner, J. M. Woodcock, M. E. Ozsan, et. al., Solar Energy Materials, **23** (1991), pp. 388-393.

# Raman and RBS Studies of Interdiffusion in RF-Sputtered CdS/CdTe Solar Cells

A. Fischer, U.N. Jayamaha, E. Bykov, D. Grecu, R.G. Bohn, and A.D. Compaan

Dept. of Physics and Astronomy, The University of Toledo, Toledo, OH 43606

**Abstract.** The performance of CdS/CdTe photovoltaic devices is strongly determined by the properties of the CdS/CdTe interface region which forms during the heat treatment of the solar cell. Due to interdiffusion of sulfur and tellurium across the original CdS/CdTe junction and the formation of $CdS_xTe_{1-x}$ at the interface, material properties such as the bandgap and the absorption coefficient of the newly formed material will be changed.

In order to improve our understanding of the interface and to be able to control it, near resonant Raman scattering on a series of single-phase $CdS_xTe_{1-x}$ alloys was performed and the Stokes shifts of the longitudinal optical (LO) phonons were measured over the entire composition x. The data have been fitted according to the modified random element isodisplacement (MREI) model. The results gained from the investigation of the alloys have then been applied to study the CdS/CdTe interface region of sputter-deposited solar cells. The formation of a two-phase $CdS_xTe_{1-x}$ alloy region at the CdS-CdTe solar cell interface has been confirmed and the composition of each phase was measured.

In addition, we have obtained Rutherford backscattering (RBS) spectra from thin bilayers of CdTe/CdS on fused silica, which provided information on interdiffusion with ~10 nm depth resolution.

## INTRODUCTION

For Raman scattering in semiconductors a resonant enhancement of the scattered intensity can be achieved if the energy of the incident photon from the laser matches approximately the bandgap of the semiconductor under investigation. In a structure such as the diffused interface, the resonance enhancement of Raman scattering provides an additional degree of flexibility for materials analysis. With "blue" excitation there is electronic resonance with a CdS-rich alloy whereas with "red" excitation resonance will be established only with a CdTe-rich alloy. This behavior allows one to selectively excite only one alloy at a time and to discriminate the Raman response from co-existing alloys with other compositions. This effect is exploited in the work described below. Four different excitation wavelengths were used in order to exploit resonance Raman enhancement: 457.9 nm and 514.5 nm for sulfur-rich films, 752.5 nm for tellurium-rich films, and 632.8 nm for intermediate compositions. The dependence of the bandgap upon the molar fraction of sulfur has been published by Jensen et al. (1).

Besides the possibility of exciting Raman scattering through the glass substrate, Raman scattering holds another advantage over x-ray analysis: Since the penetration depth of the exciting laser light is a function of its photon energy, one can obtain Raman data from different depths within the sample by using different laser wavelengths (e.g. the 457.9 nm line of an Ar-ion laser penetrates only 90 nm in CdTe at 10K whereas 752.5 nm from a Kr-ion laser probes down to a depth of 1500 nm.)

## RAMAN SCATTERING FROM $CdS_XTe_{1-X}$ ALLOYS

$CdS_XTe_{1-X}$ alloys were grown by laser physical vapor deposition (LPVD) as described previously (2). The actual film composition has been measured by wavelength dispersive x-ray spectroscopy (WDS) at NREL and shows generally some depletion of sulfur in the films relative to the original target mixture.

In a pseudo-binary alloy like $CdS_XTe_{1-X}$ each sublattice, that is either the CdS or CdTe lattice, respectively, oscillates at its characteristic frequency with Brillouin-zone-center LO phonons probed during the interaction with the radiation field in the Raman process. However, the Stokes shift of these phonons will depend upon the molar fraction of sulfur or tellurium in the film. From the known film composition the Stokes shift in the alloy films can be calibrated and used as a measure of the molar composition of alloys produced by interdiffusion across the CdS/CdTe interface in solar cells. Figure 1 contains the Raman frequencies of the entire set of samples.

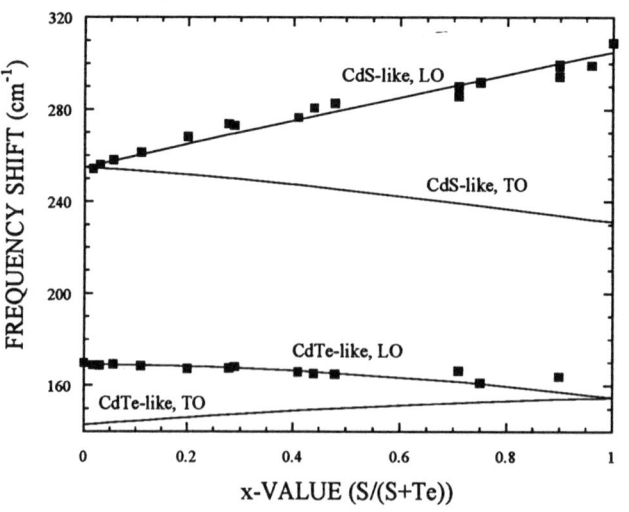

**Figure 1.** Phonon frequencies versus composition x of the $CdS_xTe_{1-x}$ system. The squares are measured data, the solid lines represent fits according to the MREI-model.

From Figure 1 it follows that for $CdS_XTe_{1-X}$ alloys the CdS-like LO-phonon shifts from 309 cm$^{-1}$ for pure CdS down to 254 cm$^{-1}$ for an alloy with trace amounts of sulfur present (x=0.017). On the other hand, the CdTe-like phonon frequencies show only a small decrease as x approaches one.

The solid lines in Figure 1 are fits according to the random-element isodisplacement (MREI) model (3). A more detailed description of these fits will be given in a subsequent paper. For the LO modes the experimental data and the MREI model are in good agreement. The CdS-like LO-phonon mode has been used in particular to probe phase compositions near the junction of CdS/CdTe solar cells.

A selection of alloys described in the previous paragraph has been heat treated at 400°C in air for 20 minutes. The effect of the presence of $CdCl_2$ during this treatment was studied by means of resonant Raman scattering.

In Figure 2 the 10 K Raman spectrum of the $CdS_xTe_{1-x}$ alloy with an as-grown composition of x=0.41 is compared with the Raman data of the same film after anneal. No $CdCl_2$ was applied to the sample in this case. The 752.5 nm line of the Kr-ion laser was used to excite the spectrum near resonance with the gap which is about 1.51 eV for x=0.41 at 10 K. The spectra before and after anneal are virtually the same. In particular, the CdS-like mode remains at 276.2 cm$^{-1}$ indicating an x-value of 0.41 even after the heat treatment.

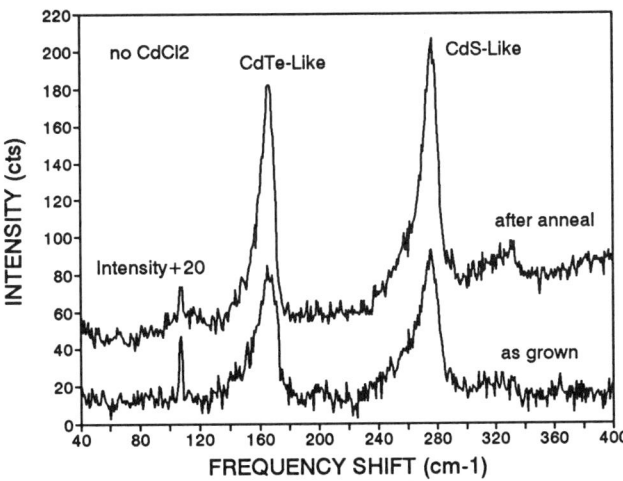

**Figure 2.** Raman spectrum of the $CdS_{0.41}Te_{0.59}$ alloy, as-grown (bottom) and after anneal at 400°C without $CdCl_2$ (top). The sample was excited with 752.5 nm in both cases.

To test if a sulfur-rich phase had formed after annealing the sample was excited with the 514.5 nm line of an Ar-ion laser, which would allow for a resonant enhancement of the Raman signal due to a possible sulfur-rich phase. In this case no Raman lines at all were observed indicating that a sulfur-rich phase had not been formed yet. The two results show that in case of a $CdCl_2$-free heat treatment no detectable phase separation had occurred within 20 min of annealing.

The same experiment was repeated; this time a thin layer (~200 nm) of $CdCl_2$ was deposited on top of the alloy film before the anneal step took place. In Figure 3 the Raman spectra taken after the heat treatment are shown for 752.5 nm excitation (a) and 514.5 nm excitation (b). Besides the x=0.41 alloy, each figure contains the Raman data of an additional alloy for which x=0.29. (In both cases the composition refers to the initial x-value before the heat treatment.)

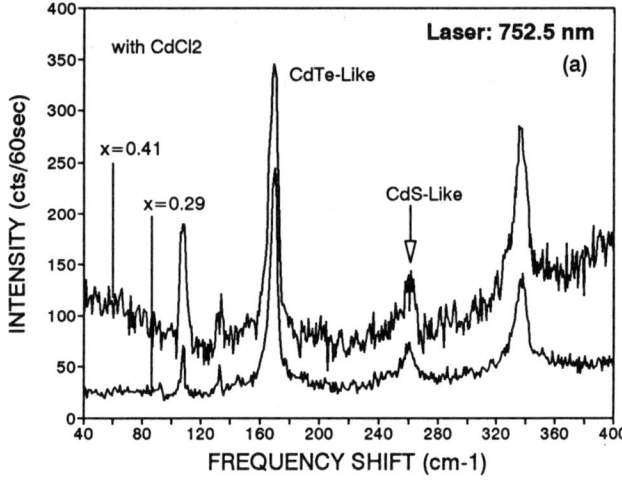

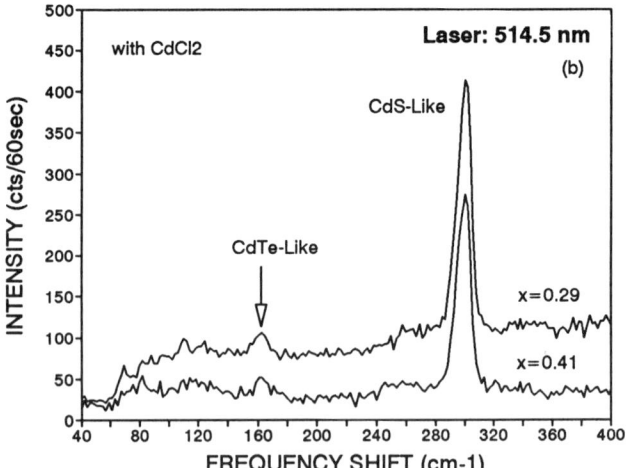

**Figure 3.** 10 K Raman data of $CdS_xTe_{1-x}$ alloys with two different initial composition x after anneal at 400°C with $CdCl_2$ present during the heat treatment, excited with 752.5 nm (a) and 514.5 nm (b). We infer the presence of two phases, one with x=0.08±0.02 for $\upsilon$=261 $cm^{-1}$ and a second phase with x=0.96±0.02 for $\upsilon$=301 $cm^{-1}$.

From the Raman spectra two important differences compared to the $CdCl_2$-free heat treatment can be observed:

1) The CdS-like mode shifts from its initial position of 276 $cm^{-1}$ to 261 $cm^{-1}$ if $CdCl_2$ is present during the anneal step.
2) For both x=0.41 and x=0.29 as-deposited alloys, after $CdCl_2$ treatment and anneal, the CdS-like LO peak occurs at 261 $cm^{-1}$ under red excitation and at 301 $cm^{-1}$ under green excitation.

From the calibration graph in Figure 1 one can readily gain the two x values corresponding to these Raman shifts: They are x=0.08±0.02 for $\upsilon$=261 cm$^{-1}$ and x=0.96±0.02 for $\upsilon$=301 cm$^{-1}$. This confirms that two separate phases co-exist within the material after the heat treatment under CdCl$_2$ presence.

The data show the importance of the presence of CdCl$_2$ during the anneal step. Besides its function as a defect passivator and catalyst for grain growth, CdCl$_2$ also seems to increase the diffusivity in a polycrystalline film such as CdS$_X$Te$_{1-X}$ and therefore considerably shortens the time it takes for the alloy to phase separate.

## RAMAN SCATTERING FROM THE CdS/CdTe INTERFACE OF SOLAR CELLS

The Raman results from the alloy films have been used to analyze interdiffusion across the CdS/CdTe junction. The solar cell structures studied here were grown by rf-sputtering in a standard fashion as described earlier (4), however, to produce a smooth bottom surface the cell structures have been grown on top of a 1 mm thick microscope slide. (In this particular case the solar cell structures lack their front and rear contacts.) Determined from scanning tunneling microscopy measurements (STM) the microscope slides used had a surface roughness of about 16 nm. After completion of the cell most of the CdTe was then removed by etching in a 0.05 vol% solution of bromine in methanol. After the etch the remaining film was polished leaving the film with a top surface roughness of about 70 nm (STM). Absorption measurements were performed with the 633 nm line of a HeNe laser to determine the remaining film thickness. (Since 633 nm photons are only absorbed in Te-rich material, the thicknesses given here correspond to the ones of the Te-rich alloy and of CdTe.) Raman data were then taken from the thinned samples as a function of remaining alloy/CdTe thickness.

In Figure 4 Raman data obtained at 10 K with blue excitation are shown for which the CdS-rich alloy regions are resonantly enhanced. Spectra are shown for three different thicknesses of CdTe remaining after thinning. Since the Raman lines are close in wavelength to the laser line, their absorption within the material occurs at approximately the same rate as the laser line so that the Raman probe depth is only about have the penetration depth of the laser which is about 90 nm for 457.9 nm excitation. Therefore the blue laser beam is a near-surface probe for Raman scattering, with an effective probe depth of about 45 nm. In Figure 4 the first order of the CdS-like phonon occurs at 307 cm$^{-1}$ which corresponds to an x-value of 0.97 ± 0.02. Second and third order of the CdS-like mode are found as well and are an indicator of reasonably high film quality. Comparing different alloy/CdTe thicknesses the LO peak position remains fixed at 307 cm$^{-1}$ indicating little variation of x-value in the region of CdS-rich material from 30 to 190 nm from the interface.

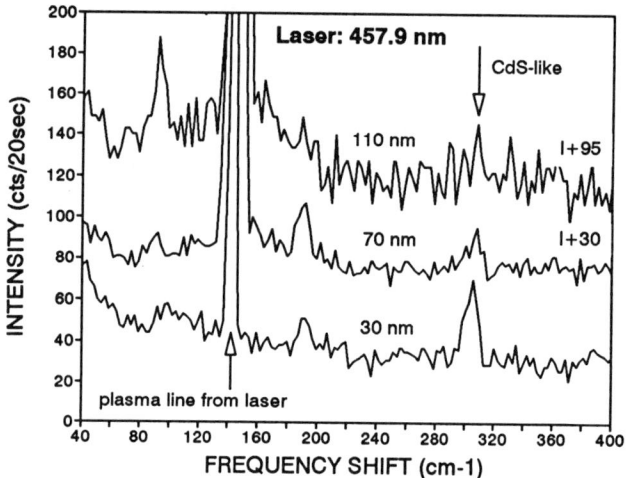

**Figure 4.** Raman data obtained at 10 K with 457.9 nm excitation from the alloy region of a sputtered solar cell structure for three different thicknesses of CdTe remaining after thinning. Note that the first order CdS-like phonon at 307 cm$^{-1}$ remains fixed indicating little variation of x-value in the region of CdS-rich material.

With red excitation, the CdTe-rich regions are resonantly enhanced and Figure 5 shows strong CdTe-like first and second order modes with a weak CdS-like mode at 261 cm$^{-1}$.

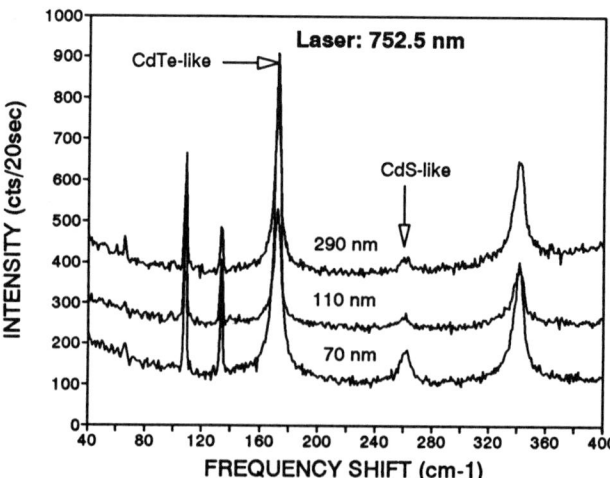

**Figure 5.** 10 K Raman data obtained with 752.5 nm excitation from the alloy region of a sputtered solar cell after CdTe thinning. Besides strong CdTe-like first and second order modes, a weak CdS-like local vibrational mode at 261 cm$^{-1}$ is seen. No shift nor broadening occurs for this mode with different alloy/CdTe thicknesses.

An important difference from blue excitation is that the optical absorption depth is 1500 nm at 752.5 nm so that the entire film is probed. Again this localized mode does not shift with different thicknesses nor does it show much broadening. Thus we conclude that the x-value of this material is nearly constant and, from the

$261 cm^{-1}$ frequency from Figure 1, that $x=0.08 \pm 0.02$. Again, we infer the existence of two separate phases that build during the anneal process. One phase is sulfur-rich with x close to 97 percent, a second Te-rich phase exists with x near 8 percent. These results agree with the findings from the $CdCl_2$ treated alloy films and correspond to the phase separation predicted by the phase diagram (5) of the $CdS_xTe_{1-x}$ system. Jensen et al. (6) and Özsan et al. (7) have suggested similar conclusions based on x-ray diffraction and Rutherford backscattering earlier.

## TIME EVOLUTION OF INTERFACIAL DIFFUSION

Raman spectra were taken from an etched and polished solar cell structure that was annealed before polishing for 5 min, 20 min, and 40 min, respectively. The remaining CdTe/alloy thickness was approximately 250 nm. With the 752.5 nm excitation the Te-rich phase is resonantly enhanced. As shown in Figure 6, the CdS-like mode arises at $261 cm^{-1}$ after 5 min anneal and does not shift as the anneal time increases. However, its peak height increases in time indicating that at a distance of 250 nm from the junction the alloy corresponding to about 8 percent sulfur in CdTe starts to form from the very beginning and builds up as the anneal time increases.

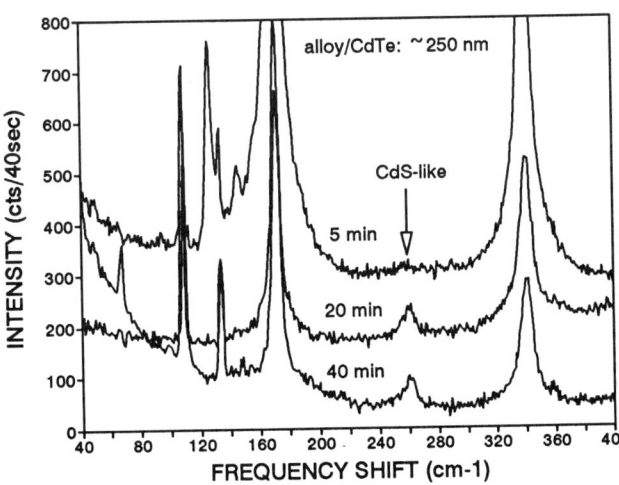

**Figure 6.** 10 K Raman data obtained with 752.5 nm excitation from the thinned alloy/CdTe region (~250 nm) of a sputtered solar cell versus anneal time. Peaks at 170 and 340 $cm^{-1}$ are first and second order CdTe-like LO-modes.

These data show that diffusion across the interface does not occur by a gradual change in molar composition from initially 100 percent CdS and CdTe on each side of the junction to a two-phase material corresponding to their thermal equilibrium concentrations. The two separate equilibrium phases form from the very beginning of the anneal process at the interface and expand deeper into the original materials (CdS and CdTe) as the sample continues to be exposed to elevated temperatures.

## RUTHERFORD BACKSCATTERING

At the IEEE Photovoltaic Specialists Conference-1996 (4) we introduced the first results of Rutherford Backscattering on CdS/CdTe solar-cell type structures. Because sulfur is light compared to cadmium and tellurium, it appears well isolated from other elements if the layers are thin. However, due to the interference of heavy elements in soda-lime and in borosilicate glasses, it is necessary to use pure silica substrates. Figure 7 illustrates the type of RBS spectra which can be obtained from as-deposited layers on silica.

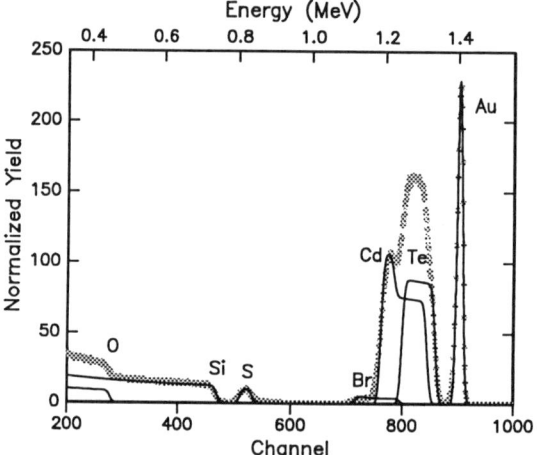

**Figure 7.** Rutherford backscattering spectrum obtained by Mark Stan at CWRU with 1.53 MeV He$^+$ from as-deposited bilayer structure: 7 nm Au / 100 nm / CdTe / 50 nm / CdS / silica. The sample was etched with Br:MeOH and residual Br at about the 5% level appears throughout the CdTe and CdS layers.

In this case the bilayer structure was etched briefly with a very weak Br:MeOH solution, following the same procedure as for the CdCl$_2$-treated and annealed samples described below. To avoid charging problems during RBS, the sample was coated with 7 nm of gold prior to analysis. The gray points in the figure are experimental data, the solid curve is a RUMP (8) simulation of the He$^+$ backscattered spectrum. Although the sample was extensively rinsed in clean MeOH, a relatively strong Br signal appears under both the CdS and CdTe at a relative intensity of about 5%!

Our objective in this study was to examine the interdiffusion of the sulfur across the CdS/CdTe interface. Thus additional samples were treated with CdCl$_2$ in the usual way (2) and annealed in air for 20 minutes. The results from a sample annealed at 385°C for 20 minutes is shown in Fig. 8. This spectrum shows only the low energy region near the sulfur peak. In comparison with the as-deposited sample, the S peak shows a distinct tail extending to higher energies. This is caused by S diffusing toward the surface into the CdTe or along grain boundaries. We have used the RUMP simulation package to model this diffusion with an error function form. In the figure the dashed curve is the simulation without diffusion; the solid curve is the simulated spectrum after diffusion with a diffusion coefficient of $D \cong 10^{-14}$ cm$^2$/sec.

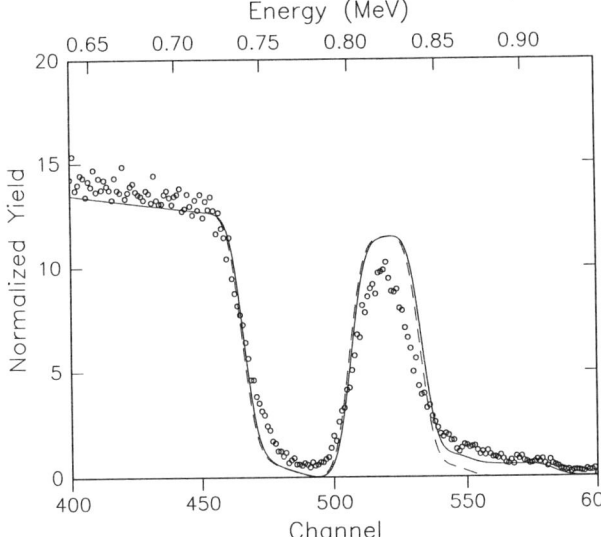

**Figure 8.** Low energy section of RBS spectrum from bilayer film after CdCl$_2$ treatment and anneal at 385 C for 20 min. Sulfur peak shows a tail to higher energies due to diffusion into the CdTe. Dashed curve shows RUMP simulation with no S diffusion; solid curve - simulation with diffusion.

Additional work with both RBS, Raman and photoluminescence is in progress to elucidate the interdiffusion mechanisms further.

## ACKNOWLEDGMENTS

We are grateful to Alice Mason from NREL for the WDS measurements and to Mark Stan from Case Western University for performing the RBS experiments. We are also thankful to Fray de Lande Castillo-Alvarado, Jaime Avendano, and Gerardo Contreras-Puente for performing the MREI fit. The financial support of NREL is gratefully acknowledged.

## REFERENCES

1. Jensen, D.G., McCandless, B.E., Birkmire, R.W., *Proceedings of the 25$^{th}$ IEEE Photovoltaic Specialists Conference*, p. 773 (1996).

2. Compaan, A.D., Feng, Z., Contreras-Puente, G., Narayanswami, C., and Fischer, A., presented at the Mat. Res. Soc. Symposium, Spring 1996, to be published.
3. Chen. Y.S., Shockley, W., and Pearson, G.L., *Phys. Rev.* **151**, 648 (1996).
4. Fischer, A., Narayanswamy, C., Grecu, D.C., Bykov, E., Nance, S.A., Jayamaha, U.N., Contreras-Puente, G., Compaan, A.D., Stan, M.A., and Mason, A.R., *Proceedings of the 25$^{th}$ IEEE Photovoltaic Specialists Conference*, p. 921 (1996).
5. Ohata, K., Saraie, J., and Tanaka, T., *Jpn. J. Appl. Phys.* **12**, 1198 (1973).
6. Jensen, D.G., McCandless, B.E., and Birkmire, R.W., presented at the Mat. Res. Soc. Symposium, Spring 1996, to be published.
7. Özsan, M.E., Johnson, D.R., Lane, D.W., and Rogers, K.D., *Proceedings of the 12$^{th}$ Euro. Photo. Sol. En. Conf.*, p. 1600 (1994).
8. *Computer Graphic Service, Ltd.*, Ithaca, N.Y. 14850-8716.

# Tin Oxide Stability Effects - Their Identification, Dependence on Processing and Impacts on CdTe/CdS Solar Cell Performance

Dave Albin, Doug Rose, Ramesh Dhere, Dave Niles, Amy Swartzlander, Alice Mason, Dean Levi, Helio Moutinho, and Peter Sheldon

*National Renewable Energy Laboratory*
*1617 Cole Boulevard*
*Golden, Colorado 80401*

**Abstract.** High efficiency polycrystalline thin film CdTe solar cells involve the growth of CdTe films on CdS/SnO$_2$/glass substrates. The CdS layer in such a structure is commonly reported to benefit from a brief hydrogen anneal prior to the deposition of the CdTe film. In this paper, we show that the SnO$_2$ layer can be susceptible to reduction in H$_2$ and that the degree of susceptibility is dependent on the type of SnO$_2$ used. Chemical vapor deposited (CVD) SnO$_2$/glass substrates (Solarex Corp.) show the most resistance to reduction while room-temperature sputtered SnO$_2$ films show the least resistance. When annealed under reducing conditions, Sn from the SnO$_2$ reacts with S-containing impurities and oxygen in as-grown chemical bath deposited (CBD) CdS films to form SnS. Cd-containing impurities are more volatile resulting in a loss of Cd relative to S in films annealed in H$_2$. These films appear dark due to the presence of SnS, a grayish-black impurity, in the CdS and possibly SnO in the SnO$_2$. In normal CSS CdTe deposition processes where H$_2$ annealing is followed by further heating to deposition temperatures in either He or He:O$_2$ ambient, S loss occurs at temperatures exceeding the H$_2$ anneal. If oxygen is absent, CdS films undergo loss of both Sn and S due to evaporation of the SnS. When O$_2$ is present, SnS converts to SnO$_2$ allowing for only the evaporation of sulfur. In this fashion, Sn levels on the CdS surface immediately prior to the deposition of CdTe, can be affected not only by the temperature of the H$_2$ anneal, but also by the oxygen present during the CdTe deposition step. Modifications to the CdS/CdTe device fabrication process including the use of more stable tin oxide layers (CVD-grown) and lower temperature H$_2$ anneals yield devices with higher open circuit voltage, fill-factors, and total-area efficiencies. Room-temperature sputtered tin oxide can be strengthened against reduction by annealing at 550°C in 400 torr O$_2$ prior to the CdS deposition step.

## INTRODUCTION

CdS/CdTe solar cells represent one of the leading polycrystalline thin film solar cell technologies present today [1]. The highest conversion efficiency for a laboratory-produced CdS/CdTe solar cell currently stands at 15.8% [2]. The CdTe layer in this cell was grown by close-space sublimation (CSS) on high-temperature Corning 7059 glass. The best efficiency to date by a commercial spraying technique using less expensive soda-lime glass is 14.8% [3]. CdTe films capable of device efficiencies exceeding 10% can readily be deposited by a variety of other techniques including evaporation, sputtering, electrodeposition, and powder sintering.

CdS/CdTe solar cells are typically configured in a superstrate structure with the following layer sequence: backcontact/p-CdTe/n-CdS/TCO/glass where the transparent conductor (TCO) is typically n-$SnO_2$. At NREL, the best CdTe solar cell achieved to date, 13.7% total-area efficiency, used a cadmium stannate TCO, a chemical-bath deposited (CBD) CdS layer, and a CSS-deposited CdTe layer [4]. In general however, $SnO_2$ remains the TCO commonly used. The $SnO_2$ layer is about 1 µm thick (or whatever thickness will yield a sheet resistance of approximately 5-10 $\Omega$/sq). At NREL, most devices are fabricated using chemical vapor deposited (CVD) $SnO_2$-coated 7059 glass substrates received from Solarex Corporation. These TCOs have a sheet resistance of about 9 $\Omega$/sq. These conducting substrates are used either "as-received" (hereafter referred to as c$SnO_2$/glass) or are subsequently coated with a undoped insulating sputter deposited $SnO_2$ layer (~1300 Å thick) prior to subsequent processing (hereafter referred to as i$SnO_2$/c$SnO_2$/glass). The purported intent of the i$SnO_2$ layer is to help maintain high open circuit voltages, $V_{oc}$'s, in the presence of suspected CdS pinholes which can shunt the absorber (CdTe) and front contact (c$SnO_2$) layers. The CdS layer is often deposited as thin as possible in order to improve short-circuit current, $J_{sc}$ [5] and is typically 60-100 nm thick in CSS-deposited CdTe solar cells. Due to the high absorption coefficient, the CdTe layer need not be much thicker than 2-3 µms though thicknesses of 10 µms are commonly used by us to minimize grain boundary shunting during the backcontact acid etch.

The exact location of the heterojunction can vary depending upon whether the cell behaves as a true heterojunction, or as a shallow buried homojunction. In either case, the chemistry of the $SnO_2$/CdS/CdTe interface undoubtedly dominates device performance. In addition, the surface chemistry of the CdS layer, at the point of CdTe growth affects the nucleation and growth characteristics of the CdTe layer.

When the CdS layer is grown by solution techniques like spraying or CBD, large amounts of impurities can be expected in the film. Oxygen is the primary contaminant in the case of CBD deposited films and may exist in either water or carbonate form [6,7,8]. Nitrogen is also a major contaminant and appears to result from the formation of cyanimide-containing compounds[7,8]. Understanding the affect of thermal anneals on CBD CdS films then becomes a matter of 1) determining what reaction byproducts exist after CBD, 2) knowing how these byproducts inter-react at elevated temperatures both with the annealing ambient and possible contaminants from the substrate, and finally, 3) determining the relative thermal stabilities of these phases, and therefore, which phases survive during high temperature anneals prior to the CdTe growth.

Hydrogen anneals of the CBD CdS layer are commonly used to reduce oxygen in these films. Hydrogen anneals have also long been known to reduce film resistivity [9-12] in CdS films by reducing grain boundary scattering associated with chemisorbed oxygen [13]. In CdS/CdTe solar cells, oxygen in the CdS may also provide detrimental hole recombination centers [14]. $H_2$ annealing can reduce the number of these defects and subsequently improve $V_{oc}$ and fill factor, FF, in CdS/CdTe cells. [15]. In the CSS CdTe process, $H_2$ anneals are also reported to reduce interface states by partial sublimation of the CdS surface before deposition of the CdTe layer [16]. Detailed analysis of the CdS stoichiometry before and after $H_2$ and $N_2$ anneals indicate that CdS sublimation can also be incongruent, with the preferential removal of sulfur atoms from the CdS resulting in Cd-rich surfaces, and likely, a high concentration of $V_S$ defects [17-21]. Anneals in $N_2$ and $H_2$ generally result in a slight decrease (60-100 meV) of the fundamental bandgap which may be associated with the presence of a large number of these $V_s$ defects [20-22]. Anneals in $N_2$ of CdS/SnO$_2$/glass substrates up to 550°C do not reportedly result in intermixing at the CdS/SnO$_2$ interface [19] though some indications exist that anneals in $H_2$ as low as 450 °C may cause reduction of the SnO$_2$ layer [2].

In this paper we present results which elaborate on TCO interactions with CBD grown CdS layers prior to CSS CdTe deposition which occur during $H_2$, He, and $O_2$ anneals. Microstructural changes in the CdS and SnO$_2$ layers of various CdS/SnO$_2$/glass and SnO$_2$/glass substrates as a function of various thermal anneals in $H_2$, $O_2$, and He were studied by a variety of characterization techniques. Devices were also fabricated to demonstrate the effect of microstructural and chemical changes in the CdS film on cell performance.

## EXPERIMENTAL PROCEDURE

Film Characterization

Characterization techniques used in this study include x-ray photoelectron spectroscopy (XPS), grazing-incidence x-ray diffraction (GIXRD), and both total optical reflection-transmission and photo-induced absorption measurements. XPS measurements of Cd, S, Sn, and O composition were performed in a Physical Electronic 5600 ESCA using a monochromated Al K$\alpha$ x-ray source operated at 300 W. The spot size for collecting photoelectrons was 0.8 mm. Depth-profiling was accomplished by sputtering with 1 kV Ar+ ions. The estimated accuracy of the measurements is within 1 at.%.

GIXRD measurements were performed using a Scintag Polycrystalline Texture Stress (PTS) x-ray diffractometer with a Cu-tube operated at 45 kV and 40 mA fitted with a thin film (parallel beam) attachment. Such measurements offer significant advantages over Bragg-Brentano XRD when analyzing very thin films. In this technique, the incident angle, $\Omega$, is fixed (rather than scanned at 1/2 the angular rate of the detector as in the regular 2θ/θ technique). When used in this configuration and with the proper parallel beam optics, the penetration depth of x-rays into the sample can be adjusted as a function of the incident angle, $\Omega$. Consequently, $2\theta$ scans at different values of $\Omega$ can be used to gather XRD data as a function of film depth. Using Huang's equation [23], the perpendicular penetration depth of $CuK_{\alpha 1}K_{\alpha 2}$ in CdTe, CdS, and SnO2 was calculated as a function of $\Omega$ and is shown in Fig. 1. In this calculation, we used x-ray mass absorption coefficients of 958.9, 1403.5, and 701 $cm^{-1}$ for $CuK_\alpha$ radiation in CdS, $SnO_2$, and CdTe, respectively.

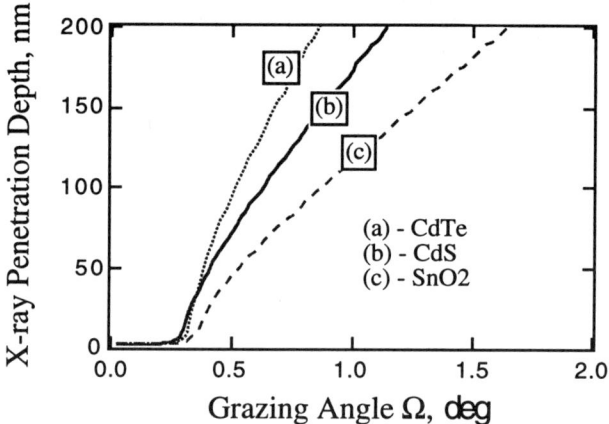

FIGURE 1. X-ray Penetration Depth normal to film surface as a function of Incident Grazing Angle, $\Omega$

Optical Reflection-Transmission (R-T) measurements were performed using a CARY 5G UV-VIS Spectrophotometer with total R-T integrating capability. Data was collected between 310 and 790 nm. Samples were oriented film-side towards the beam in both transmission, T, and reflection, R. R-T data was used to determine absorption (100%-T-R) as a function of wavelength in the various $SnO_2$/glass and $CdS/SnO_2$/glass substrates.

Photo-induced absorption measurements were used to characterize the electronic states occupied by photoexcited carriers. This modulation technique measures the changes in sample absorption induced by a 150 fs, 400 nm pump pulse which produces a carrier density of $2x10^{18}$ $cm^{-3}$. Changes in sample absorption are determined by measuring the transmission with and without the pump beam present [24]. Transmission is measured in the spectral range 480 to 700 nm using a low-intensity, broad-band white light continuum pulse delayed by 50 ps from the pump pulse. At this delay time, the photoexcited carriers have relaxed to the band edges and/or trap states. Due to the photoexcited carrier density needed to make these measurements, it is not expected that this technique is sensitive to defects with concentrations less than approximately $10^{18}$ $cm^{-3}$. Hence the wavelength of the photoinduced absorption peak is primarily determined by the product of the valence band to conduction band joint density of states with the electron and hole distribution functions. In essence, this peak will occur at an energy very near the bandgap, depending on the degree of band tailing due to defect states.

Conventional processing sequences use a $H_2$ anneal of 400°C, however, in this work, films were annealed at 500°C. This was done to increase the magnitude of various effects up to beyond the detectability limit of various characterization techniques. Where detection limits were more sensitive, as in the TRDA technique, lower anneal temperatures were used. All device fabrication results involve the more appropriate lower $H_2$ anneal temperatures.

Device Fabrication

The following process parameters were fixed in this study during solar cell fabrication: 1) the use of Corning 7059 glass; 2) a fixed CBD CdS process (Cd and ammonium actetate, amomonium hydroxide, thiourea chemistry); 3) CSS source-substrate temperature of 660 and 610°C respectively; 4) CdTe thickness of 10 µm; 5) a 400°C 30m air post CSS CdTe anneal; 6) a nitric-phosphoric acid etch prior to contacting; and 7) a HgTe doped carbon paste/Ag back contact. Parameters which were varied included: 1) tin oxide substrates with and without a room-temperature sputtered $iSnO_2$ buffer layer, 2) tin oxide substrates with and without a 550°C/10m/400 Torr $O_2$ anneal prior to CBD CdS growth, 3) the use of two different $H_2$ annealing steps: 400°C/10m/30 Torr, and 300°C/5m/30 Torr, 4) two CdS film thicknesses (~85 and 100 nm), and 5) three levels of $CdCl_2$ treatment after CdTe deposition (50% $CdCl_2$ in MeOH; 7, 15, and 22 min. soak time).

## RESULTS AND DISCUSSION

Various combinations of room-temperature sputtered $iSnO_2$ (1300 Å) and $cSnO_2$ (~ 1 µm) on Corning 7059 glass with and without the use of a 550°C $O_2$ treatment were annealed in $H_2$ at 500°C for 10 min. Optical R-T measurements were performed on these films both before and after the $H_2$ anneal. Total absorption losses associated with these substrates were calculated and are shown in Fig. 2.

As seen in this figure, it is clear that $H_2$ annealing of any TCO structure using the $iSnO_2$ layer results in considerable absorption (curves g and f). However, when $O_2$ anneals are performed on $iSnO_2$ containing TCO structures (curves b and e), absorption is reduced significantly. TCO structures which do not contain the $iSnO_2$ layer (curve d) do not darken with $H_2$ anneals.

GIXRD measurements at $\Omega$ = 0.2, 0.3, 0.5, and 1.0 degrees were performed on both $iSnO_2/cSnO_2$/glass (Fig. 3a) and $CdS/iSnO_2/cSnO_2$/glass (Fig. 3b) structures to determine the distribution of phases present as a function of depth from the surface. Normal penetration x-ray depths corresponding to both $SnO_2$ and CdS surfaces are indicated in these figures.

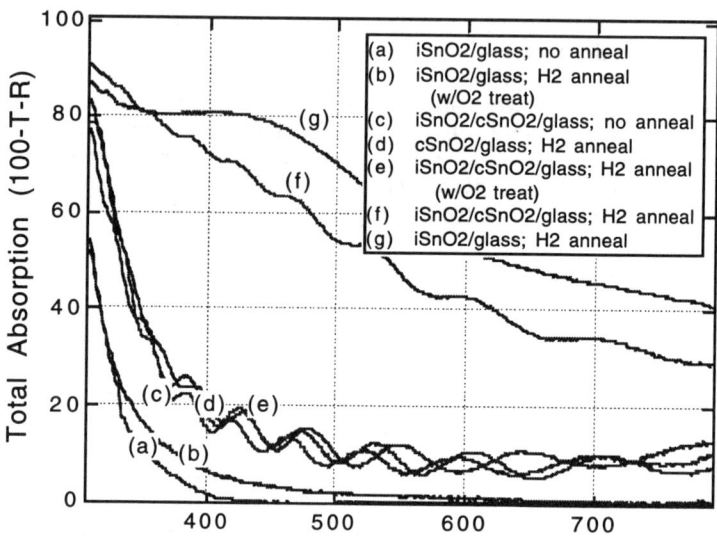

FIGURE 2. Total Absorption Curves showing absorption in tin oxide as a function of type (conducting vs. insulating) and pre-$H_2$ processing ($O_2$ anneals; 550°C/30m/400 Torr). All $H_2$ anneals performed at 500°C for 10m.

For the TCO case (Fig. 3a), we see that $H_2$ anneals clearly result in the formation of elemental tin (JCPDS 4-673) in addition to $SnO_2$ (JCPDS 41-1445). The concentration of tin appears to segregate somewhat away from the surface and this may indicate some Sn evaporation has occurred. GIXRD scans of Sn/glass substrates indicate that Sn at room temperature does not readily form $SnO_2$. The cause for the darkening in these films shown in Fig. 2 is believed to be associated with SnO formation though the presence of this phase by GIXRD was not verified.

For the CdS/TCO case (Fig. 3b), we observe $SnO_2$ as a phase decreasing towards the surface of the CdS and more importantly, peaks associated with SnS (JCPDS 39-354) present throughout the entire film. Due to convolution effects (i.e., data collected at higher values of $\Omega$ convolutes with data at lower values), GIXRD measurements indicate that the SnS phase segregates more at the surface of the CdS. The CdS films in this analysis were approximately 100 nm thick as-deposited and measured approximately 90 nm after the $H_2$ anneal (due to CdS sublimation). The presence of $SnO_2$ in these scans within 11 nm of the surface is not believed to indicate actual diffusion of $SnO_2$ from the substrate since molecular diffusion is believed energetically unfavorable. Rather, $SnO_2$ present in the CdS after $H_2$ anneals may be the result of elemental Sn reacting with oxygen present in these CBD-deposited films. The large exothermic standard heat of formation, $\Delta H_f^\circ$, of this oxide (-138.8 kcal/mole) would favor such reactions. The formation of SnS must involve either the reaction of elemental Sn with CdS or sulfur-containing

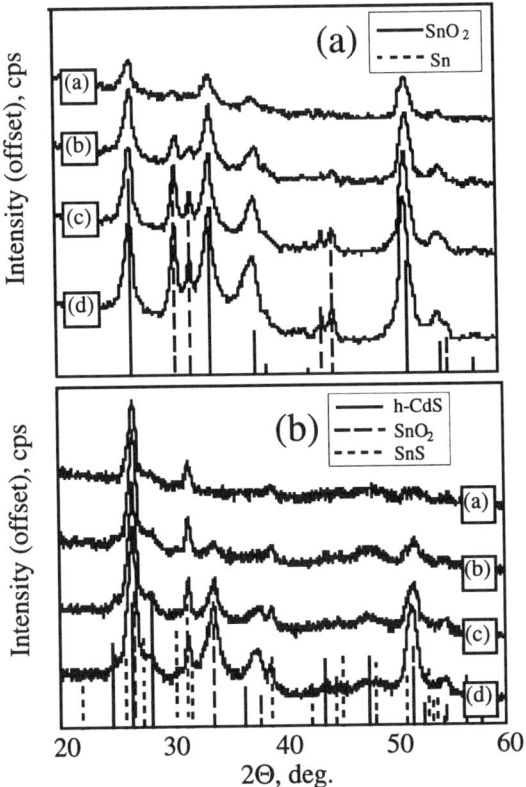

FIGURE 3. GIXRD data for (a) iSnO$_2$/cSnO$_2$/glass and (b) CdS/iSnO$_2$/cSnO$_2$/glass structure annealed at 500°C/20m in H$_2$. Calculated normal x-ray sampling depth: for Fig. 3(a), (a) 2.4 nm, (b) 3.6 nm, (c) 43.4 nm, (d) 116.1 nm, and Fig. 3(b), (a) 3.1 nm, (b) 11.3 nm, (c) 72.8 nm, and (d) 173.6 nm. Identified impurity phase in top figure: Sn metal; impurity phase in bottom figure: SnS.

impurities in the film. Since the formation of SnS is not as exothermic as CdS ($\Delta H_f°$ equals -25.8 vs. -35.7 kcal/moe), S-containing impurities in CBD grown CdS films are the more likely source of S for SnS formation at this point.

The presence of SnS compounds in CBD deposited CdS films is intriguing since these compounds are more thermally unstable than CdS. As shown in Figure 4, there are several possible Sn$_x$S compounds exhibiting higher vapor pressures than CdS in the region of 200 to 700 °C. $\Delta H_f°$ values associated with Sn$_x$S compounds decrease as sulfur increases. For example SnS (Sn/S = 1.0) has a higher $\Delta H_f°$ value than SnS$_2$ (Sn/S=0.5; $\Delta H_f°$ =-36.7), Sn$_2$S$_3$ (Sn/S=0.67; $\Delta H_f°$=-63.0), or Sn$_3$S$_4$ (Sn/S=0.75; $\Delta H_f°$ =-88.5). Consequently,

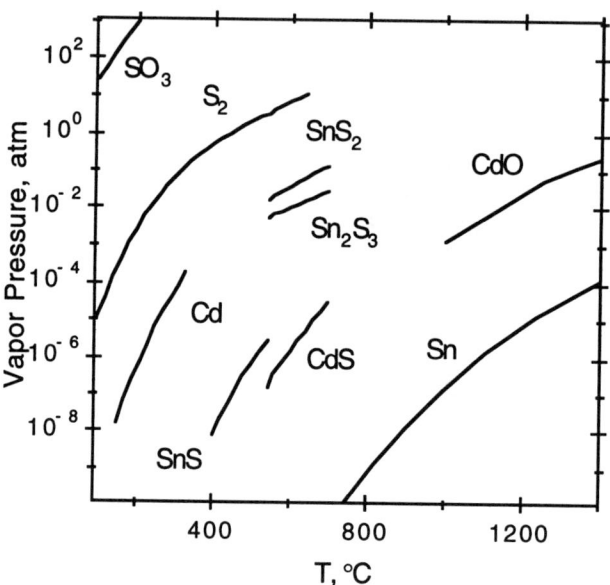

FIGURE 4. Vapor pressure data for various materials relevant in the formation of CBD-CdS films. Data from references 25-27.

*with no limitation regarding the activity of S*, SnS is the least likely $Sn_xS$ compound to form. However, none of the latter compounds dissociates to SnS. The decomposition of both $SnS_2$ and $Sn_2S_3$ should be evidenced by the presence of $Sn_2S_3(s)$ and $Sn_3S_4(s)$ respectively. Since SnS is clearly the phase identified by GIXRD above, it would appear that SnS is the reaction product of Sn in CBD CdS films and that this might be explained by a limited activity of S available for reaction.

To obtain a more instructive picture of what was happening in our CdS films, we monitored XPS compositional changes in Sn, O, Cd, and S as a function of depth after thermal anneals simulating actual device fabrication. Measurements were performed on $CdS/iSnO_2/SnO_2$/glass structures after rapid heating to a simulated deposition temperature of 610 °C in both He and He:$O_2$ ambients after a 500°C $H_2$ anneal, on a sample which had only been $H_2$ annealed, and a sample which had been untreated. These depth profiles are shown in Fig. 5. For the untreated case we find, as expected, no tin in the CdS film and a baseline oxygen level (i.e., defined as the average value at a depth of between 10 and 30 nm into the CdS) of about 11 at.% (much higher at the surface) associated with impurity by-products of the CBD process. The Cd/S ratio in this case was about 1.5 and was relatively independent of film thickness. (Note that due to preferential sputtering of the lighter O and S atoms during XPS depth profiling, all *uncalibrated* XPS ratios of Cd/S and Sn/O should be higher than the true stoichiometric value.)

To understand the effect of annealing the $CdS/iSnO_2/SnO_2$ substrate in $H_2$, we compare the compositional profiles of process B relative to process A in Figure 5. We observe three major effects. First, oxygen levels decrease significantly (bulk values of 4% down from 12%) as expected. Second, we observe a sharp increase in tin (surface and bulk values of 12.5 and 7-8 at.% up from zero). At this point, the measured Sn/O ratio in the films is extremely high (1 to 2 in the bulk). XPS measurements performed on $SnO_2$ films (as-received from Solarex without an $iSnO_2$ buffer layer with the $SnO_2$ structure verified by XRD) indicate that measured Sn/O ratios of 0.75 represent a true ratio of 0.5. With this calibration factor, our measured Sn/O ratios of 1.0 to 2.0 correspond to *actual* ratios of between 0.67 and 1.3. These latter values support the GIXRD results which indicate $SnO_2$ and SnS in the CdS layers. These films are severely darkened (curve f in Fig. 2.) as SnS is opaque in the visible region. Finally, we observe in this, and in many similar cases, a sharp drop in the Cd/S ratio. For the film shown in Fig. 5, Cd/S ratios at the surface and bulk for the as-fabricated case were measured to be 1.54 and 1.5 respectively. After the $H_2$ anneal, the Cd/S ratios decreased to 0.94 and 1.1 at the surface and in the bulk respectively. This indicates significant Cd-loss with $H_2$ annealing. Since sulfur is much more volatile than Cd, this was at first perplexing. However, the somewhat large ratio of Cd to S (1.5) in the as-deposited films does suggest excess Cd as a secondary phase to CdS. At 500°C these Cd-containing impurity phases are obviously not as stable as is SnS and consequently, the Cd/S ratio drops. Note that this change, ie., Cd-loss, does not directly imply the formation of $V_{Cd}$ defects; it simply indicates evaporation of Cd not bonded as CdS. Thermal anneals should rather result in the formation of either $V_S$ or $Cd_i$ defects in the CdS [18-20]. The CdS microstructure at this point consists of CdS grains (with undetermined intrinsic defect chemistry) with $SnO_2$ and SnS impurities probably segregated at CdS grain boundaries with very little Cd-containing impurities. The SnS phase (and possibly SnO in the $SnO_2$) is largely responsible for the visible darkening of the film.

To understand the effect of ramping the $CdS/iSnO_2/SnO_2$ substrate to deposition temperatures in He after the $H_2$ anneal we compare the compositional profiles of process C relative to process B in Figure 5. It is apparent that Sn levels drop significantly, to 6.6 at.% from 12.5 at.% at the surface and to 3 at.% from 7-8 at.% in the bulk. Oxygen levels are observed to increase slightly in the bulk to 8-10 at.% from about 4 at.%. These films no longer appear dark. The reason for the significant decrease in Sn is obviously associated with the evaporation of SnS present in the films after the $H_2$ anneal step. The sublimation of SnS was also substantiated by the increase in the Cd/S ratio both at the surface (1.1 from 0.94) and within the bulk (1.2 from 1.1). The films appear transparent again since SnS has been removed. The increase in bulk oxygen may be due to continued thermal outdiffusion of oxygen from reduced $SnO_2$.

Finally, the effect of ramping the $CdS/iSnO_2/SnO_2$ substrate to deposition temperatures in an $O_2$ containing He ambient after the $H_2$ anneal is understood by comparing the compositional profiles of process D relative to

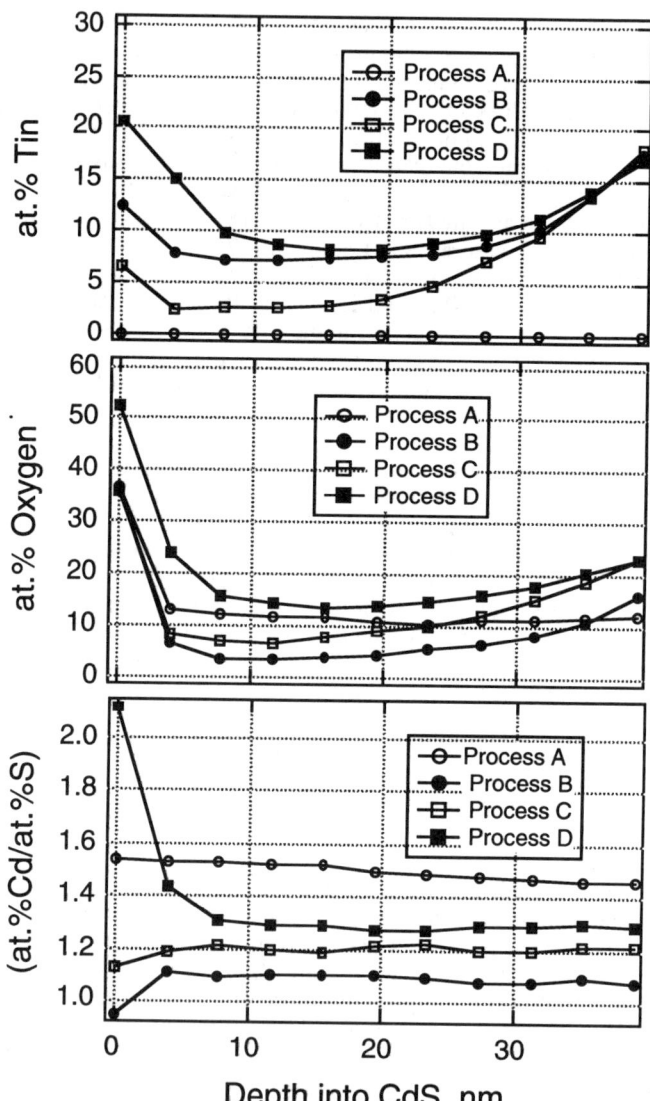

FIGURE 5. XPS determined measurments of Sn, O, and Cd/S vs. depth into CdS/iSnO$_2$/cSnO$_2$/glass substrates for four different processes. Process A - as fabricated; Process B - 500°C/10m H$_2$ (30 torr) anneal; Process C - 500°C/10m H$_2$ (30 torr) anneal, cool to 200°C, ramp to 610 °C in 16 Torr He (140°C/m); Process D - 500°C/10m H$_2$ (30 torr) anneal, cool to 200°C, ramp to 610°C in 15 Torr He:1 torr O$_2$ (140°C/m). Process C and D most representative of actual processing conditions (with exception of H2 anneal temp) *immediately prior* to CSS deposition of CdTe.

process B in Figure 5. Unlike the case of He alone, we see a tremendous *increase* in Sn both at the surface (up to 21 at.% from 12.5 at.%) and bulk (up to 9 at.% from 7-8 at.%). Oxygen has the effect of eliminating Sn sublimation by oxidation of SnS to $SnO_2$ which is much more thermally stable. If we measure the increase in both Sn and $O_2$ at the surface in Process D relative to Process B, we see that Sn increases by 8.5 at.% while $O_2$ increases by 17 at.%, a stoichiometric increase exactly equal to the formation of $SnO_2$ on the CdS surface. The loss of SnS results in the films becoming optically clear again. Note that in effect, the degree of oxygen incorporated onto the CdS surface during the ramp up to deposition temperatures in a He:$O_2$ ambient is exactly determined by the amount of Sn present on the CdS surface immediately prior to heating. The Sn level in turn is determined by the temperature and time of the $H_2$ anneal step. Note that the increase in Cd/S ratio associated with S loss in this case is much larger relative to the increase seen with He alone (Process C) up to 2.1 from 0.94 at.% at the surface and up 1.3 from 1.1 at.% in the bulk. Sulfur loss in Process C is associated with evaporation of *some* SnS while in Process D, *more* of the SnS is either removed by evaporation or reacts with $O_2$ to form $SnO_2$.

With this information we now propose a phenomenological model for how the CdS film develops prior to CSS CdTe deposition at temperatures of around 600 °C. CdS/i$SnO_2$/$SnO_2$/glass substrates undergo significant changes due to thermal processing before these higher temperatures. The tin oxide layer is quite susceptible to reduction when heated in hydrogen ambients depending upon how the tin oxide layer is grown. High $SnO_2$ fabrication temperatures (as in CVD) and the presence of oxygen during annealing of the tin oxide films can strengthen the film against reduction. When heated in $H_2$ at higher annealing temperatures (400-500°C), elemental Sn diffuses into the CdS and reacts with S and $O_2$ containing impurities to form SnS and $SnO_2$. The SnS darkens the film significantly. Subsequent heating of these films to actual CSS deposition temperatures (~600°C) in inert ambients (He) results in SnS loss with a corresponding improvement in film optical transmission. The CdS film surface consists of CdS grains, voids left behind associated with evaporated SnS, residual un-evaporated SnS, and some $SnO_2$ phases. When oxygen is present in the ambient during heating up to deposition temperatures, $SnO_2$ is formed at the expense of SnS and only sulfur is depleted. CdS film surfaces contain CdS and $SnO_2$. The loss of SnS again results in improved optical transmission relative to $H_2$ anneals alone.

In order to substantiate the plausible explanation of grain boundary segregation rather than the inclusion of SnS, and $SnO_2$ phases *intra*-granular, we made photo-induced absorption measurements on CdS films both with and without Sn present. Shown in Fig. 6 are measurements for CBD deposited CdS on glass and i$SnO_2$/$SnO_2$/glass substrates. Both are annealed under conditions exactly simulating device fabrication (i.e., lower $H_2$ anneal temperatures) in addition to a single 500°C anneal case. The CdS/glass sample, void of any Sn effects, was included to monitor changes in $E_g$. As can be seen in Fig. 6,

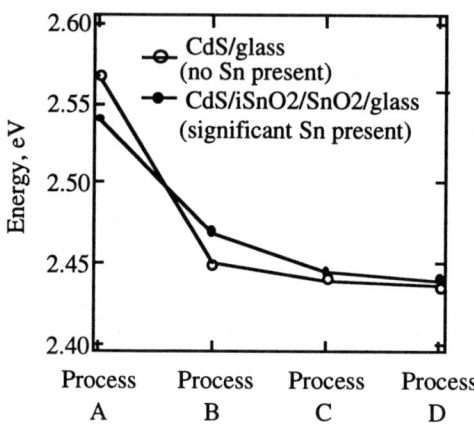

FIGURE 6. Room temperature photo-induced absorption determination of CdS bandgap for CdS/glass and CdS/iSnO$_2$/SnO$_2$/glass samples as a function of processing. Process A - as fabricated, Process B - 300°C/5m H$_2$ (30 torr) anneal, cool to 200°C, ramp to 610 °C in 15 Torr He: 1 Torr O$_2$; Process C - 400°C/10m H$_2$ (30 torr) anneal, cool to 200°C, ramp to 610°C in 15 Torr He: 1 Torr O$_2$; Process D - 500°C/10m H$_2$ (300 torr) anneal.

within the sensitivity limit of $10^{18}$ cm$^{-3}$ defects, the presence of Sn or SnS appears to have little effect on the density of states within the bandgap, i.e., no defect levels were observed within the gap (from 1.65 to 2.75 eV). Rather, we observed the commonly reported decrease in E$_g$ with thermal anneals. This decrease in E$_g$ may be indicative of the formation of V$_S$ defects within individual CdS grains. The possibility that defect states may be present outside the gap has not been investigated however.

Given the improved understanding regarding tin oxide reduction and resulting microstructural changes in the CdS film, we designed a device fabrication matrix to study process variables which we have shown affect Sn reduction and diffusion: 1) tin oxide substrates with and without the iSnO$_2$ buffer layer, 2) tin oxide substrates with and without a 550°C - 10m - 400 Torr O$_2$ anneal, and 3) the use of two different H$_2$ annealing steps: 400°C/10m and 300°C/5m. CSS CdTe depositions were performed in a 15 torr He, 1 torr O$_2$ ambient for all these devices. Other details of device processing are discussed in the introduction. An orthogonal matrix of 32 devices was planned. Of these, two were lost due to processing errors and five lost to adhesion failures. The latter adhesion failures were attributed to either problems with substrate preparation or excessive CdCl$_2$ treatments. Current-voltage, I-V, results under AM 1.5 illumination are shown for the remaining 25 devices in Table 1 below.

| with iSnO$_2$ | with O$_2$ anneal | CdS thick, nm | CdS H$_2$ anneal (°C/min) | CdCl$_2$ soak time (min) | V$_{oc}$ (volts) | FF (%) | Eff (%) |
|---|---|---|---|---|---|---|---|
| no | yes | 100 | 300/5 | 7 | .837 | 69.8 | 11.81 |
| no | yes | 85 | 400/10 | 7 | .812 | 69.8 | 12.13 |
| no | yes | 85 | 300/5 | 7 | .835 | 70.8 | 12.41 |
| yes | yes | 100 | 400/5 | 15 | .812 | 69.9 | 11.07 |
| yes | yes | 100 | 400/10 | 7 | .790 | 69.9 | 11.39 |
| yes | yes | 100 | 300/5 | 15 | .801 | 69.2 | 11.61 |
| yes | yes | 100 | 300/5 | 7 | .802 | 70.3 | 11.54 |
| yes | yes | 85 | 400/10 | 15 | .800 | 70.2 | 11.79 |
| yes | yes | 85 | 400/10 | 7 | .795 | 69.7 | 11.94 |
| yes | yes | 85 | 300/5 | 15 | .808 | 69.8 | 11.59 |
| yes | yes | 85 | 300/5 | 7 | .817 | 70.2 | 12.41 |
| no | no | 100 | 400/10 | 15 | .814 | 67.8 | 11.53 |
| no | no | 100 | 300/5 | 15 | .819 | 72.0 | 12.20 |
| no | no | 100 | 300/5 | 22 | .825 | 71.0 | 12.37 |
| no | no | 85 | 400/10 | 15 | .783 | 67.9 | 10.90 |
| no | no | 85 | 400/10 | 22 | .786 | 60.9 | 8.66 |
| no | no | 85 | 300/5 | 15 | .830 | 72.0 | 12.69 |
| yes | no | 100 | 400/10 | 15 | .721 | 68.0 | 10.09 |
| yes | no | 100 | 400/10 | 22 | .796 | 69.2 | 11.50 |
| yes | no | 100 | 300/5 | 15 | .809 | 72.9 | 12.55 |
| yes | no | 100 | 300/5 | 22 | .819 | 72.8 | 11.8 |
| yes | no | 85 | 400/10 | 15 | .720 | 68.6 | 10.28 |
| yes | no | 85 | 400/10 | 22 | .745 | 68.8 | 10.33 |
| yes | no | 85 | 300/5 | 15 | .809 | 71.8 | 12.17 |
| yes | no | 85 | 300/5 | 22 | .811 | 72.2 | 12.00 |

TABLE 1. Solar cell device results for series investigating the effect of SnO$_2$ (type and thermal treatment), CdS window thickness, H$_2$ anneal temperature/time, and CdCl$_2$ soak time. Device efficiencies shown above have been normalized to NREL-confirmed measurements performed on 3 cells in the set.

The orthogonal design of the above matrix allows one-to-one comparisons of only a single variable, all other variables held constant. With regards to the use or absence of the iSnO$_2$ layer for instance, we found out of 9 paired data sets, device efficiency improved 6 times, decreased 2 times, and remained unchanged once when the iSnO$_2$ layer was absent. The effect was even more dramatic in regards to V$_{oc}$. V$_{oc}$ increased an average of 34 mV in cases where the iSnO$_2$ layer was absent relative to when it was used. Equally impressive was the improvement realized by reducing H$_2$ anneal temperature and time. Out of 11 data pair sets, a decrease in H$_2$ anneal conditions led to an improvement in efficiency 10 times out of 11 largely due to improvements in V$_{oc}$ and Fill Factor, FF. The significant improvement in V$_{oc}$ and FF realized by avoiding the use of iSnO$_2$ layers and reducing the H$_2$ anneal temperature/time to 300 °C/5m in shown in Figs. 7 and 8.

It is also clear from Fig. 7 that variations in $V_{oc}$ due to different substrate types occur when using the higher temperature $H_2$ anneal. Using the 300 °C anneal downplays the effect of tin oxide stability to $H_2$ reduction. Also, the use of a 550°C $O_2$ anneal of the $iSnO_2$ layer can also alleviate problems

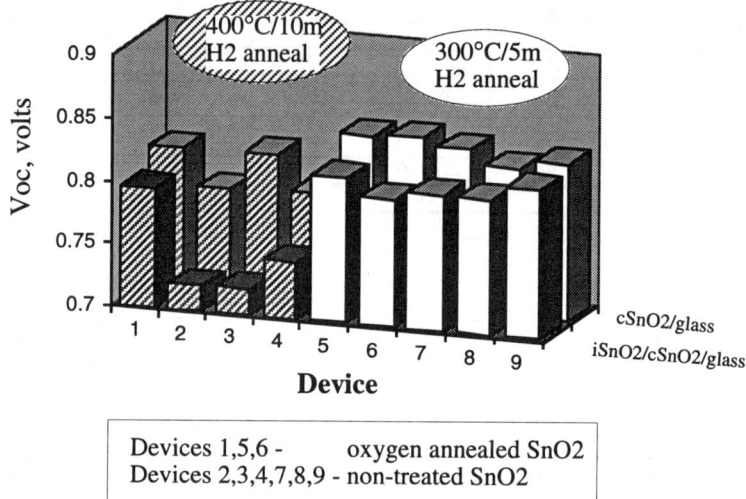

FIGURE 7. Variation of $V_{oc}$ as a function of tin oxide type (cond vs. insul), pre-treatment of tin oxide (w and w/o $O_2$ annealing), and $H_2$ anneal (400°C/10m and 300°C/5m).

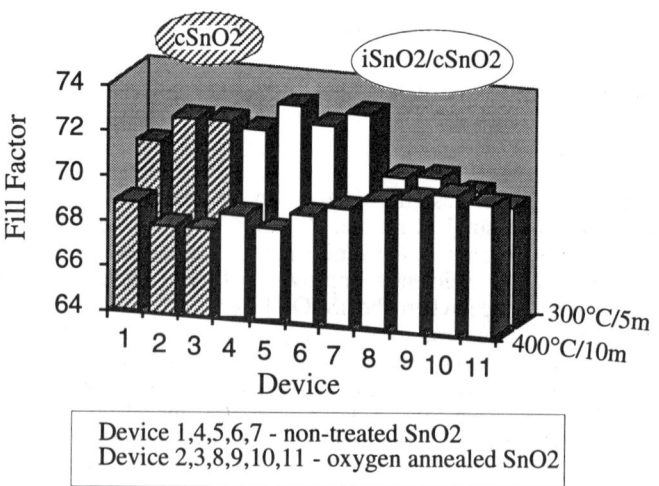

FIGURE 8. Variation of Fill Factor as a function of tin oxide type (cond vs. insul), pre-treatment of tin oxide (w and w/o $O_2$ annealing), and $H_2$ anneal (400°C/10m and 300°C/5m).

associated with using the $iSnO_2$ layer when the higher 400 °C anneals are used. $O_2$ anneals also appear to reduce variations in FF (albeit reducing them somewhat perhaps due to the extra handling of the substrates involved) among devices fabricated with the $iSnO_2/cSnO_2$/glass substrates (particularly when using the higher temperature 400 °C $H_2$ anneals).

As discussed previously, the CdS surface immediately prior to CSS deposition of CdTe in $He:O_2$ probably consists of CdS grains with inclusions of $SnO_2$ (the reaction product of SnS with $O_2$). These $SnO_2$ areas on the CdS surface probably have a negative effect on device performance. First, CdTe does not nucleate on $SnO_2$ as well as it does on CdS which may result in non CdS/CdTe junctions at the CdS/CdTe interface. For example, $SnO_2$/CdTe interfaces are reported to cause reductions in $V_{oc}$ and FF [28]. Though non-ideal, these latter interfaces may be desireable relative to CdS/CdTe interfaces resulting from CSS deposition in He only ambients. These latter interfaces are characterized by both SnS inclusions and possible voids left by evaporated SnS. Such defects could possibly explain the increased pinhole densities observed in CdTe films grown without oxygen. Additionally, since oxygen increases the nucleation site density of CdTe, irrespective of substrate [29], the use of oxygen may be able to force CdTe coverage in the presence of such voids.

To probe the effect these non-ideal interfaces might have on the junction, dark $J_o$ current from I-V data for a subset of the devices shown in Table 1 (the set involving the 15 min. $CdCl_2$ treatment) were determined after series resistance corrections. As shown in Table 2, we found out of 6 data pair sets, that the use of the 300°C $H_2$ anneal temperature relative to the 400 °C anneal resulted in a decrease in $J_o$ five times. This decrease in $J_o$ correlates nicely with the increased $V_{oc}$'s observed in Fig. 7.

| with $iSnO_2$ | with $O_2$ anneal | CdS thick, nm | CdS $H_2$ anneal (°C/min) | log $J_o$ dark (A/cm^2) |
|---|---|---|---|---|
| no | no | 850 | 300/5 | -10.4 |
| no | no | 850 | 400/10 | -9.32 |
| yes | no | 850 | 300/5 | -9.96 |
| yes | no | 850 | 400/5 | -8.48 |
| no | no | 1000 | 300/5 | -9.89 |
| no | no | 1000 | 400/10 | -9.00 |
| yes | no | 1000 | 300/5 | -9.83 |
| yes | no | 1000 | 400/10 | -9.02 |
| yes | yes | 850 | 300/5 | -8.78 |
| yes | yes | 850 | 400/10 | -9.51 |
| yes | yes | 1000 | 300/5 | -9.95 |
| yes | yes | 1000 | 400/10 | -9.68 |

TABLE 2. Paired data showing a decrease in dark $J_o$ when using a lower $H_2$ anneal temperature/time combination (5 out of 6 times).

# SUMMARY

Thermal treatments of the CdS layer prior to CSS CdTe deposition must be moderated, especially when $H_2$ is used, in order to prevent thermal reduction of the tin oxide film. When reduction occurs, tin reacts rigorously with impurities in CBD grown CdS to form SnS and $SnO_2$. The sulfide phase darkens the film and can affect CSS CdTe nucleation during subsequent higher temperature excursions in either inert or oxidizing ambients. With inert ambients like He, some of the SnS evaporates from the CdS, possibly leaving voids in addition to unevaporated SnS and $SnO_2$. Under oxidizing environments (i.e., $O_2$-containing ambients typically used during CSS deposition), SnS is oxidized to $SnO_2$ which maintains higher Sn levels at the surface (in the form of $SnO_2$), and may minimize voiding associated with SnS evaporation. In either case, the CdS surface is characterized by microstructural pertubations from the ideal CdS/CdTe interface. The presence of these microstructural defects is shown to have a demonstratable effect on device performance, in particular $V_{oc}$ and dark $J_o$. The use of more thermally resilient TCOs, lower $H_2$ anneal temperatures and times, and the possible use of post $SnO_2$ growth $O_2$ anneals at 550 °C are all shown to alleviate problems associated with Sn-related defects.

# ACKNOWLEDGMENTS

This work was supported by the U.S. Department of Energy under contract No. DE-AC36-83CH10093 to NREL.

# REFERENCES

1. K. Zweibel, H.S. Ullal, B.G. von Roedern, R. Noufi, T.J. Coutts, M.M. Al-Jassim, *Proc. 23rd IEEEE Photovoltaic Spec. Conf.*, (1993) p. 379.
2. C. Ferekides, J. Britt, Y. Ma, and L. Killian, *Proc. 23rd IEEE Photovoltaic Spec. Conf.*, (1993) p. 389.
3. See J.Kester, et al., *ibid*.
4. See X. Wu, et al., *ibid*.
5. J.E. Granata and J.R. Sites, *Proc. 25th IEEEE Photovoltaic Spec. Conf.*, (1996) p. 853.
6. D.W. Niles, G. Herdt, and M. Al-Jassim, accepted for publication in Feb. 15 issue, *J. of Appl. Phys.*, (1997).
7. A. Kylner, A. Rockett, and L. Stolt, *Solid State Phenomena*, 51-52 (1966) p. 533.
8. A. Kylner, J. Lindgren, and L.Stolt, *J. Electrochem. Soc.*, 143 [8] (1996) p. 2662.
9. Y.Y. Ma, A.L. Fahrenbruch and R.H. Bube, *Appl. Phys. Lett.*, 30 (1977) p. 423.
10. A.L. Fahrenbruch, V. Vasilchenko, F. Buch, K. Mitchell and R.H. Bube, *Appl. Phys. Lett.*, 25 (1974) p. 605.
11. H. Matsumoto, N. Nakayama and S. Ikegami, *Japan. J. Appl. Phys.*, 19 (1980) p. 129.
12. S.A. Tomas, O. Vigil, J.J. Alvarado-Gil, R. Lozada-Morales, O. Zelaya-Angel, H. Vargas, A. Ferreira da Silva, *J. Appl. Phys.*, 78 (1995) p.2204.

13. K.L. Chopra, R.C. Kainthla, D.K. Pandya and A.P. Thakoor, *Phys. Thin Films*, 12 (1982) p. 167.
14. I.J. Ferrer and P. Salvador, *J. Appl. Phys.*, 64 (1988), 1233.
15. A. Rohatgi, *Int. J. Solar Energy*, 12 (1992) p. 37.
16. T.L. Chu, S.S. Chu, and S.T. Ang, *J. Appl. Phys.*, 64 (1988) 1233.
17. R. Sundharsanan and A. Rohatgi, *Proc. 21st IEEE Photovoltaic Spec. Conf.*, (1990) p. 504.
18. D.W. Niles and F.S. Hasoon, *Prog. in Photovoltaics: Res. and Appl.*, 1 (1993) p. 279.
19. A.B. Swartzlander, D.W. Niles, F.S. Hasoon, and M. Al-Jassim, *Surface and Interface Analysis*, 21 (1994) p. 160.
20. H.R. Moutinho, R.G. Dhere, K. Ramanathan, P. Sheldon and L.L. Kazmerski, *Proc. 25th IEEEE Photovoltaic Spec. Conf.*, (1996) p. 945.
21. F. Goto, K. Shirai, and Masaya Ichimura, *Tech. Digest of the Intl. PVSEC-9*, Miyazaki, Japan, (1996) p. 435.
22. W.J. Danaher, L.E. Lyons, and G.C. Morris, *Solar Energy Materials*, 12 (1985) p. 137.
23. T.C. Huang, *Advances in X-ray Analysis*, ed. C.S. Barrett, J.V. Gilfrich, T.C. Huang, R. Jenkins, and P.K. Predecki, Plenum Press, 33 (1989) 91.
24. for a detailed explanation of this technique see B.D. Fluegel, Ph.D. Thesis, University of Arizona, 1992.
25. K. Mills, *Thermodynamic Data for Inorganic Sulphides, Selenides and Tellurides*, London: Butterworths, (1974).
26. N.A. Lange, *Lange's Handbook of Chemistry - 13th Edition*, ed. J.A. Dean, McGraw-Hill Book Company (1985).
27. R.E. Honig and D.A. Kramer, *RCA Review*, 30 [2] (1969) p. 292.
28. C.S. Ferekides, D. Marinskiy, S. Marinskaya, B. Tetali, D. Oman, and D.L. Morel, *Proc. 25th IEEEE Photovoltaic Spec. Conf.*, (1996) p. 751.
29. D.H. Rose, D.H. Levi, R.J. Matson, D.S. Albin, R.G. Dhere, and P. Sheldon, *Proc. 25th IEEEE Photovoltaic Spec. Conf.*, (1996) p. 777.

# Nanoparticle Colloids as Spray Deposition Precursors to CIGS Photovoltaic Materials

Douglas L. Schulz, Calvin J. Curtis, Rebecca A. Flitton,
Holm Wiesner, James Keane, Richard J. Matson,
Philip A. Parilla, Rommel Noufi, and David S. Ginley

*Center for Photovoltaic and Electronic Materials and Center for Basic Sciences, National Renewable Energy Laboratory, 1617 Cole Blvd., Golden, CO 80401-3393 (USA)*

**Abstract.** Cu-In-Ga-Se nanoparticle colloids have been used as precursors in the spray deposition of photovoltaic films. Precursor colloid was prepared by reaction of the metal iodides in pyridine with sodium selenide in methanol at reduced temperature according to one of two routes: synthesis of each of the component binary selenides (Type I) followed by physical mixing of the isolated particles; or a one-pot synthesis with all the metal iodides reacting together in one flask to form a mixed-metal Cu-In-Ga-Se colloid (Type II). The constituent nanoparticles in these colloids were analyzed by TEM and XRD and were determined to be amorphous as-synthesized. Crystalline phase formation of these nanoparticles was observed by XRD after a thermal treatment. These precursor colloids were sprayed onto Mo-coated glass substrates at elevated temperatures. The nanoparticle-derived Cu-In-Ga-Se films were characterized by SEM and XRD prior to being finished into CIGS solar cell devices according to standard NREL protocol. I-V characterization of these CIGS solar cells showed these devices are limited by a large series resistance.

## INTRODUCTION

Copper indium gallium diselenide (Cu(In,Ga)Se$_2$ or CIGS) is one of a few candidate materials presently being evaluated as an absorber layer in polycrystalline solar cell devices. CIGS devices possess the highest polycrystalline solar cell efficiencies to date with one small-area cell (0.41 cm$^2$) exhibiting a solar to electrical energy conversion efficiency of 17.7 %.(1) The fabrication of solar cell modules, however, requires compositional uniformity over square meter areas. This is non-trivial for CIGS technology as process control of four-element deposition, temperature, and pressure is often required. Furthermore, "band gap engineering" of the CIGS layer where the In/Ga ratio is intentionally graded to produce an absorber layer with optimal photovoltaic properties(2) demands further control over the deposition conditions. While industrial photovoltaic companies such as Siemens Solar, Showa Shell, and Solarex have demonstrated mini-module (areas < 100 cm$^2$) solar cell efficiencies exceeding 13 %, the translation of this technology to product-sized modules (area > 1 m$^2$) has yet to be realized. The typical CIGS solar cell device heterostructure is composed of a sodalime glass substrate, a molybdenum back contact, a p-type CIGS layer, an n-type CdS layer, a ZnO transparent conducting layer, a MgF$_2$ anti-reflection coating, and a Ni/Al grid top contact.

Spray deposition is a non-vacuum technique that is amenable to the manufacture of large area films with low processing costs. The value of this approach has recently been demonstrated for polycrystalline CdTe solar cells by industrial researchers at Golden Photon, Inc. who have successfully produced 29.3 W CdTe thin film modules (active area = 2,793 cm$^2$; 10.5 % active area efficiency; 7.8 % total area efficiency) using a spray deposition approach for the growth of the CdTe layer.(3) We have recently adopted a nanoparticle-based precursor spray deposition route to CdTe thin film materials.(4) This nanoparticle-based approach differs from conventional spray deposition in that the nanoparticles, prepared by solution synthesis, have approximate diameters around 10 nm while conventional particles, prepared by ball-milling of bulk, have diameters of about 1 μm. Nanoparticle spray deposition has several advantages over standard spray deposition including film deposition with no binder agent, enhanced kinetic reactivity owing to higher surface energies through larger surface areas, and a fixed chemical composition ratio within the precursor colloid. This paper reports the synthesis and characterization of two types of Cu-In-Ga-Se nanoparticle colloids, the spray deposition and characterization of CIGS precursor films from these nanoparticle colloids, and the subsequent fabrication of CIGS solar cells using the nanoparticle-derived CIGS precursor film.

## EXPERIMENTAL

Standard Schlenk techniques and a helium-filled Vacuum Atmospheres glove box were used in the isolation and handling of all Cu-In-Ga-Se nanoparticles. Methanol solvent was dried over magnesium methoxide and was distilled under nitrogen gas immediately prior to use. Pyridine solvent was dried over KOH pellets. These solvents were degassed with nitrogen gas for at least 15 minutes prior to experimentation. Glassware used in synthesis of CIGS nanoparticles was heated at 140 °C in a convection oven and was removed immediately into the Vacuum Atmospheres glove box antechamber. $Na_2Se$ was prepared by reaction of sodium metal with selenium metal in liquid ammonia(5) while CuI (99.999 %, Aldrich), $InI_3$ (99.999 %, Aldrich), and $GaI_3$ (Aldrich No. 39,911-6) were used as received. XRD data for the nanocrystalline colloid and CIGS films were collected using a Scintag X1 Diffraction System. CIGS thin film morphologies were observed by SEM using a JEOL JSM 840. The chemical composition of the CIGS films was determined by ICPAES using a Varian Liberty 150. The I-V response curves for CIGS solar cells were obtained using a measurement system consisting of an Optical Radiation Corporation Solar Simulator 1000, a temperature-regulated sample stage (25 °C), and computer-controlled I-V instrumentation. I-V characteristics are referenced to AM1.5 solar spectra approximating the global ASTM E 892-87 specification at 1000 W/m$^2$.

## Synthesis of Colloidal Copper Selenide

The metathesis reaction of copper (I) iodide with sodium selenide is given in Eq. 1. Accordingly, CuI (0.808 g, 4.24 mmol) was added to a 500 mL side-arm round bottom flask fitted with a magnetic stir bar. $Na_2Se$ (0.270 g, 2.16 mmol) was added to a 250 mL side-arm round bottom flask fitted with a magnetic stir bar. Approximately 250 mL dried, degassed pyridine and approximately 60 mL dried, degassed methanol were added via cannula to the CuI and $Na_2Se$ flasks, respectively. Both flasks were stirred vigorously until the reagents dissolved giving a bright yellow CuI solution and a dark red-brown $Na_2Se$ solution. The metal iodide and $Na_2Se$ flasks were cooled to -42 °C and -78 °C, respectively, by employing dry ice/isopropanol cooling baths. The $Na_2Se$ solution was added quickly to the CuI solution using a large bore cannula and the materials reacted instantly to give a dark brown mixture. The cooling baths were removed and the reaction mixture was allowed to warm with stirring. With the reaction mixture at ambient temperature, stirring was ceased and the precipitate was allowed to settle. Colorless NaI in pyridine/methanol supernatant was decanted via cannula and discarded. The remaining orange/brown slurry was transferred to two 50 mL centrifuge tubes and centrifuged at 4000 rpm for 5 minutes. Colorless supernatant was decanted via cannula and a pyridine/methanol mixture (1:3) was added. This slurry was sonicated for 10 minutes and then centrifuged for 15 minutes at 4000 rpm. A pale yellow supernatant was removed from the slurry via cannula and a pyridine/methanol mixture (1:3) was once again added to the brown slurry. After 15 minutes of sonication, the Cu-Se particles remained suspended in solution. Yields for this reaction typically exceed 90 % as determined in sacrificial samples by solvent evaporation.

$$2CuI + Na_2Se \xrightarrow{pyridine/methanol} Cu_2Se + 2NaI \quad (Eq. 1)$$

## Synthesis of Colloidal Indium Selenide

The metathesis reaction of indium (III) iodide with sodium selenide is given in Eq. 2. Accordingly, $InI_3$ (0.664 g, 1.34 mmol) was added to a 500 mL side-arm round bottom flask fitted with a magnetic stir bar. $Na_2Se$ (0.256 g, 2.05 mmol) was added to a 250 mL side-arm round bottom flask fitted with a magnetic stir bar. Approximately 225 mL dried, degassed tetrahydrofuran (THF) and approximately 60 mL dried, degassed methanol were added via cannula to the $InI_3$ and $Na_2Se$ flasks, respectively. Both flasks were stirred vigorously until the reagents dissolved giving a clear $InI_3$ solution and a dark red-brown $Na_2Se$ solution. The metal iodide and $Na_2Se$ flasks were cooled to -78 °C by employing dry ice/isopropanol cooling baths. The $Na_2Se$ solution was added quickly to the $InI_3$ solution using a large bore cannula and the materials reacted instantly to give an orange mixture. The cooling baths were removed and the reaction mixture was allowed to warm with stirring. With the reaction mixture at ambient temperature, stirring was ceased and the precipitate was allowed to settle. The colorless NaI in THF/methanol supernatant was decanted via cannula and discarded. The remaining orange/brown slurry was transferred to two 50 mL centrifuge tubes and centrifuged at 4000 rpm for 5 minutes. The clear supernatant was decanted via cannula and methanol was added. This slurry was sonicated for 10 minutes and then centrifuged for 15 minutes at 4000 rpm. A pale yellow supernatant was removed from the slurry via cannula and methanol was once again added to the brown slurry. After 15 minutes of sonication, the In-Se particles remained suspended in solution. Yields for this reaction typically exceed 90 % as determined in sacrificial samples by solvent evaporation.

$$2InI_3 + 3Na_2Se \xrightarrow{tetrahydrofuran/methanol} In_2Se_3 + 6NaI \quad (Eq. 2)$$

## Synthesis of Colloidal Gallium Selenide

The metathesis reaction of gallium (III) iodide with sodium selenide is given in Eq. 3. Gallium (III) selenide colloid was prepared analogously to indium (III) selenide starting from 0.603 g (1.34 mmol) $GaI_3$ and 0.264 g (2.11 mmol) $Na_2Se$. Yields for this reaction typically exceed 90 % as determined in sacrificial samples by solvent evaporation.

$$2GaI_3 + 3Na_2Se \xrightarrow{tetrahydrofuran/methanol} Ga_2Se_3 + 6NaI \quad (Eq. 3)$$

## Synthesis of Mixed Metal Cu-In-Ga Selenides

The metathesis reaction of copper (I) iodide, indium (III) iodide, and gallium (III) iodide with sodium selenide is given in Eq. 4. Mixed-metal Cu-In-Ga-Se colloid was prepared analogously to copper (I) selenide starting with 0.419 g (2.20 mmol) CuI, 0.670 g (1.35 mmol) InI$_3$, and 0.208 g (0.461 mmol) GaI$_3$ dissolved into approximately 250 mL dried, degassed pyridine and 0.477 g (3.82 mmol) Na$_2$Se dissolved into approximately 60 mL dried, degassed methanol. Yields for this reaction typically exceed 90 % as determined in sacrificial samples by solvent evaporation.

$$1.10\,CuI + 0.68\,InI_3 + 0.23\,GaI_3 + 1.91\,Na_2Se \xrightarrow{pyridine/methanol}$$
$$Cu_{1.10}In_{0.68}Ga_{0.23}Se_{1.91} + 3.82\,NaI \quad (Eq.\ 4)$$

## Preparation of Type I and II Nanoparticle CIGS Colloids

Type I CIGS colloid was prepared by physically mixing the constituent binary selenide powders after the solvent was removed from the colloids under a stream of nitrogen gas. Toward this end, 0.043 g (0.21 mmol) Cu$_2$Se, 0.080 g (0.17 mmol) In$_2$Se$_3$, and 0.028 g (0.07 mmol) Ga$_2$Se$_3$ powders were ground together in a mortar and pestle using THF as a slurrying agent. The slurry was allowed to dry and an additional 11.5 mL THF was added to the powder yielding a black mixture with the following composition: Cu$_{0.92}$In$_{0.76}$Ga$_{0.32}$Se$_{2.08}$.

The type II CIGS colloid was simply the product of the aforementioned synthesis of mixed metal Cu-In-Ga selenides which gave a dark red-brown colloidal suspension with the following composition: Cu$_{1.10}$In$_{0.68}$Ga$_{0.23}$Se$_{1.91}$.

## Spray Deposition of Nanoparticle CIGS Colloids

Spray deposition of the colloids to form precursor CIGS films was performed in a nitrogen-purged Plas-Lab 818-GB glove box fitted with an evacuable antechamber. Precursor CIGS films were deposited onto molybdenum-coated sodalime glass substrates (1"x2", Siemens Solar). Film thicknesses of 2.7 ± 0.2 μm were attained and reproducibility was good owing to the implementation of a Mac IIci computer-controlled Velmex XY stepper motor system used to raster the Vega 2000 aspiration-feed sprayer.

## CIGS Solar Cell Fabrication

Thermal processing of the CIGS precursor films was carried in a modified thermal evaporator at 2x10$^{-5}$ Torr under a selenium flux of 15 Å/sec. According to this treatment, the sample temperature was ramped up in ~15 min, held at the soak temperature for 10 min, raised to or held at 550 °C during the 13 min evaporation of In and Ga, and finally ramped to 300 °C prior to turning off the Se flux. Solar cell devices were fabricated by coating these CIGS films with CdS, ZnO, and Ni/Al grid according to standard NREL protocol.(2,6)

# RESULTS AND DISCUSSION

## Characterization of Type I CIGS Colloid: XRD of Cu$_2$Se, In$_2$Se$_3$, and Ga$_2$Se$_3$

A Cu-Se film for XRD characterization was prepared by evaporating the colloid onto a glass slide. XRD θ/2θ characterization of this Cu-Se sample gave only broad peaks which is consistent with an amorphous phase. The Cu-Se film was next subjected to an anneal at 400 °C for 100 min under flowing Ar. Figure 1a shows XRD θ/2θ data for the Cu-Se film after the anneal. Both the $Cu_{2-x}Se$ (PDF 6-680) and $Cu_2Se$ (PDF 29-575) phases are observed. It may be inferred that the original Cu-Se colloid was indeed $Cu_2Se$ and that the $Cu_{2-x}Se$ phase formed upon partial evolution of Se during the anneal required for phase identification.

An In-Se film for XRD characterization was prepared by evaporating the colloid onto a glass slide. XRD θ/2θ characterization of this In-Se sample gave only broad peaks which is consistent with an amorphous phase. The In-Se film was next subjected to an anneal at 400 °C for 100 min under flowing Ar. Figure 1b shows XRD θ/2θ data for the In-Se film after the anneal. The $β-In_2Se_3$ (PDF 20-494) phase is observed with some extraneous reflections. The extra XRD peaks could not be correlated to any In-Se phase published in the PDF although several polymorphs are known in this system. It is possible that the extraneous peaks are really from the $β-In_2Se_3$ phase owing to the poor figure of merit for the PDF data (i.e., $F_{30} = 1.6$). It may be inferred that the original In-Se colloid was amorphous $In_2Se_3$.

A Ga-Se film for XRD characterization was prepared by evaporating the colloid onto a glass slide. XRD θ/2θ characterization of this Ga-Se sample gave only broad peaks which is consistent with an amorphous phase. The Ga-Se film was next subjected to an anneal at 400 °C for 100 min under flowing Ar. Figure 1c shows XRD θ/2θ data for the Ga-Se film after the anneal. Cubic $α-Ga_2Se_3$ (PDF 5-724) phase is observed. It may be inferred that the original Ga-Se colloid was amorphous $Ga_2Se_3$.

## Characterization of Type II Nanoparticle CIGS Colloid

Chemical and phase composition of the Cu-In-Ga-Se nanoparticles within the colloid was probed using a number of techniques. TEM analysis showed the CIGS colloid exists as an amorphous phase with no lattice fringes apparent even at 300 kX magnification with electron diffraction showing only diffuse broad rings. A Cu-In-Ga-Se film for XRD characterization was prepared by evaporating the colloid onto a glass slide. XRD θ/2θ characterization of this sample gave only broad peaks which is consistent with an amorphous phase. This Cu-In-Ga-Se film was next subjected to an anneal at 400 °C for 100 min under flowing Ar. Figure 2a shows XRD θ/2θ data for this film after the anneal which shows pure CIGS phase formation (PDF No. 40-1487).(7) While the phase composition of the nanoparticles within the colloid remains unknown at present, it is likely either some amorphous mixture of the binaries (e.g., Cu-Se, In-Se, Ga-Se) or amorphous CIGS phase.

## Characterization of Cu-In-Ga-Se Films
### *Precursor Films Prepared from Type I Colloid*

The THF-based Type I colloid was sonicated for 1 h prior to being sprayed onto Mo-coated glass substrates heated above 35 °C. This ultrasonic treatment was an attempt to reduce the size of the Cu-Se, In-Se, and Ga-Se agglomerates that formed upon solvent removal. While particle adherence to the substrate was good, the film bulk region was characterized by a high density of pinholes. These pinholes are likely due to the presence of large, agglomerates within the colloids which led to non-optimal packing of the particles during spray deposition. While XRD of a thermally-treated (i.e., 550 °C/9 min anneal under a Se flux of 15 Å/s) film showed phase pure CIGS formation, the pinhole problem limits the applicability of Type I CIGS colloids. No solar cells were fabricated from the Type I colloid precursor mixture.

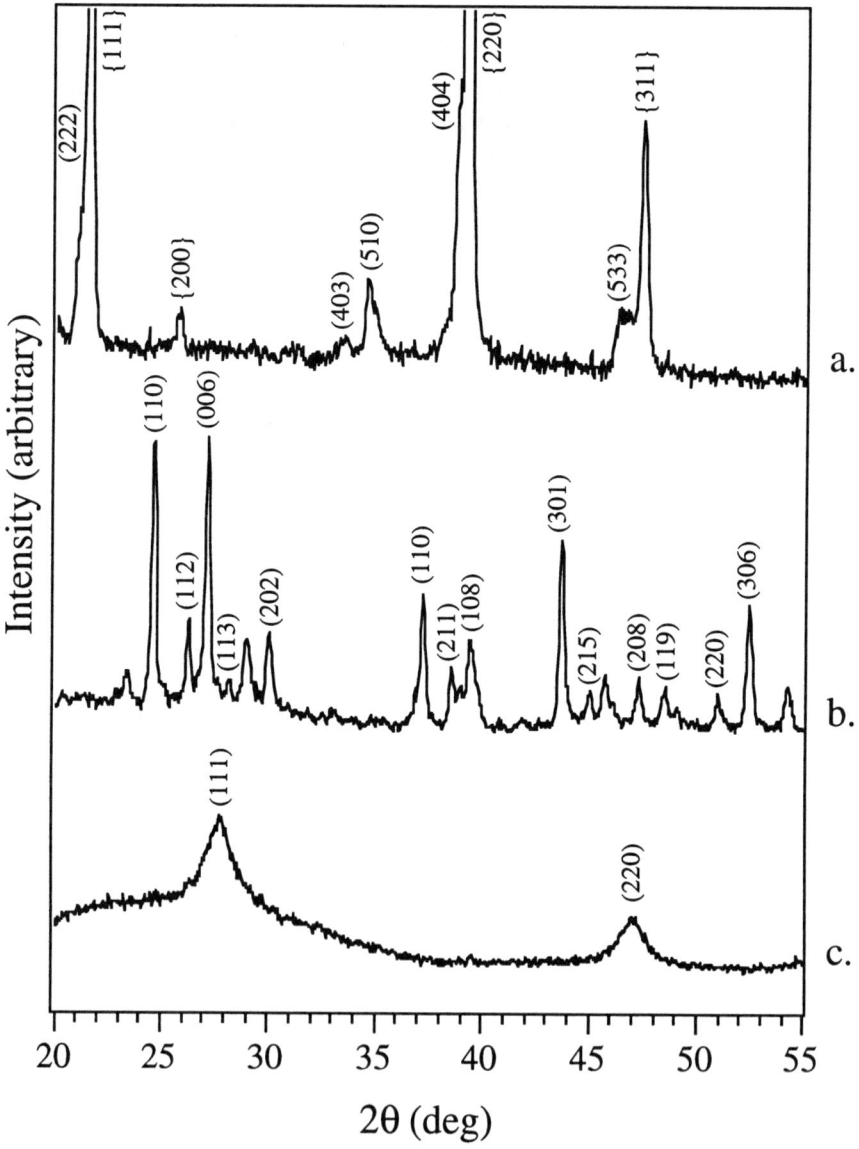

Figure 1. XRD θ/2θ data for post-annealed (i.e., 400 °C/100 min in Ar gas) [a] colloidal copper selenide with (hkl)s corresponding to $Cu_2Se$ and {hkl}s to $Cu_{2-x}Se$; [b] colloidal indium selenide with (hkl)s corresponding to $\beta$-$In_2Se_3$; and [c] colloidal gallium selenide with (hkl)s corresponding to $\alpha$-$Ga_2Se_3$.

## Precursor Films Prepared from Type II Colloid

The chemical composition of precursor films sprayed from Type II colloid was probed by ICPAES. It was determined that the chemical composition of the films was nearly identical to that of the precursor colloid. For example, a CIGS colloid prepared with the stoichiometry of $Cu_{1.10}In_{0.68}Ga_{0.23}$ produced a precursor film with a stoichiometry of $Cu_{1.13}In_{0.64}Ga_{0.23}$; a difference that is approximately equal to the error associated with the ICPAES measurement itself. The morphologies of the CIGS precursor films were measured by SEM. It was observed that the precursor film is composed of closely packed particles interspersed with voids. Similar porous morphologies have been observed in nanoparticle-derived films of $SnO_2$.(8)

Figure 2b shows XRD data for a CIGS film that was thermally processed according to the following treatment given a Se evaporation rate of 15 Å/s throughout: first, 15 min ramp to 550 °C; second, soak at 550 °C for 10 min; third, 13 min evaporation of In (~1700 Å) and Ga (~600 Å) at 550 °C; fourth, ramp to 300 °C in 15 min. A slight preferential orientation of the (112) family of planes is noted versus literature intensities (i.e., PDF No. 40-1487). Figure 3 is a cross-section SEM micrograph of this CIGS film. The SEM image clearly shows three layers; columnar molybdenum on the glass substrate, an intermediate smaller grained layer, and a top larger grained layer. The thickness of the top larger grained layer is similar to the total amount of In and Ga added during processing. Depth

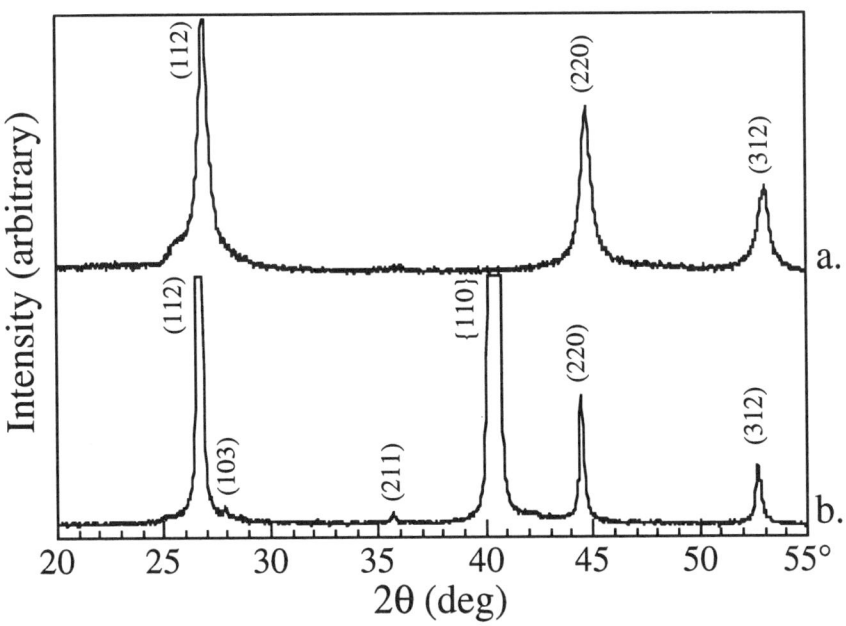

Figure 2. XRD θ/2θ data for [a] annealed colloidal copper-indium-gallium selenide with (hkl)s corresponding to CIGS; and [b] thermally-treated nanoparticle-derived CIGS film with (hkl)s corresponding to CIGS and {hkl}s corresponding to Mo.

profiled AES data of this film shows that the top surface is very Cu-deficient while the middle layer is In and Ga poor. It appears that the flux-assited grain growth occured only during In and Ga evaporation with incomplete conversion of the precursor layer into large grained materials. Furthermore, increasing the duration of the initial reaction period (i.e., soaking at 550 °C for 30 min) did not lead to enhanced grain growth within the intermediate layer. The porous nature of the precursor film (see Fig. 3) may limit the extent of grain growth owing to kinetic diffusion limitations. It is also possible that the atomically-mixed Cu-In-Ga-Se nanoparticles comprising the precursor film react prior to the formation of a Cu-rich Cu-Se liquid and form the thermodynamically stable CIGS phase.

## Characterization of CIGS Solar Cells

I-V characterization of CIGS solar cell devices showed the best cells exhibit efficiencies from 4.4 to 4.6%, open circuit voltages from 512 to 518 mV, short circuit current densities from 25.2 to 27.1 mA/cm$^2$, fill factors from 32.6 to 33.8%, and series resistance values from 13.1 to 13.5 $\Omega$·cm. By way of comparison, the solar cells fabricated in this study show markedly increased series resistance versus high efficiency cells (i.e., 13 vs 0.2 $\Omega$·cm). It has been shown that increased series resistance leads to decreased short circuit current which in turn degrades the maximum power point and ultimately, efficiency.(9) We expect the devices in the present study are limited by series resistance as a consequence of the porous, intermediate small-grained layer observed in Figure 3.

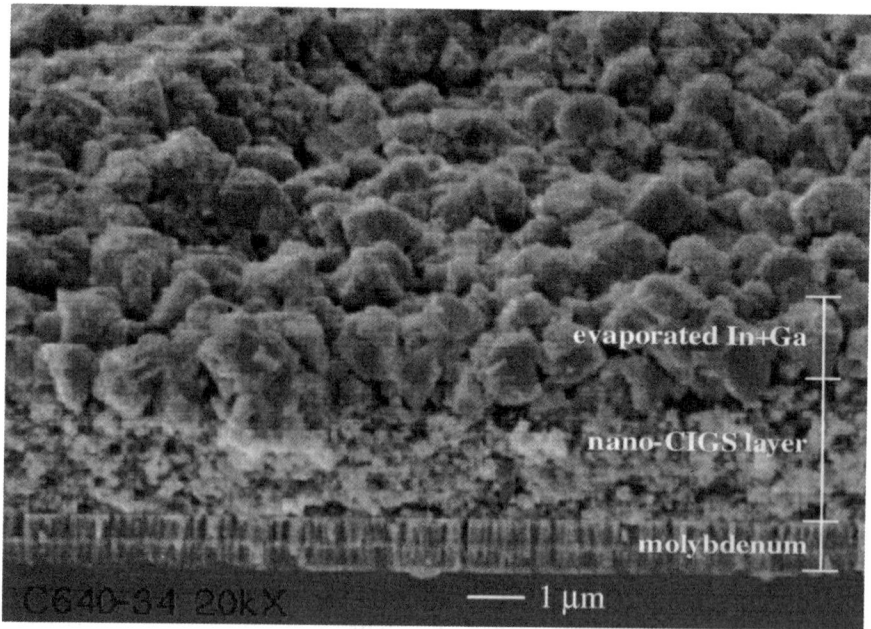

Figure 3. SEM image of a thermally-treated nanoparticle-derived CIGS film.

## SUMMARY

In summary, we have demonstrated a straightforward metathesis to methanolic Cu-In-Ga-Se nanoparticle colloids and employed these mixtures in the spray deposition of CIGS precursor films. This approach offers the advantage of fixed stoichiometry within the precursor particles, and therefore, guarantees homogeneous chemical composition throughout the film. It is apparent that the reaction kinetics of these precursor films differ markedly versus those prepared by thermal evaporation. As a consequence, these nanoparticle-derived solar cells possess an "unreacted" intermediate portion within the CIGS absorber layer. It is expected that the efficiencies of these solar cells are limited by a large series resistance that is likely a consequence of the unreacted layer.

## ACKNOWLEDGEMENTS

The authors gratefully acknowledge Amy Swartzlander and Kim Jones for the AES and TEM data, respectively. This research was funded by the U.S. Department of Energy, Office of Energy Research, Chemical Sciences and Materials Sciences Divisions and by the U.S. Department of Energy National Photovoltaic Program.

## REFERENCES

(1) Tuttle, J. R., Ward, J. S., Duda, A., Berens, T. A., Contreras, M. A., Ramanathan, K. R., Tennant, A. L., Keane, J., Cole, E. D., Emery, K., and Noufi, R., *Mat. Res. Soc. Symp. Proc.* **426**, 143-151 (1996).
(2) Gabor, A. M., Tuttle, J. R., Albin, D. S., Matson, R., Franz, A., Niles, D. W., Contreras, M. A., Hermann, A. M., and Noufi, R., *Mater. Res. Soc. Symp. Proc.* **343**, 143-148 (1994).
(3) Albright, S. P., personal communication (active area = 2,793 cm$^2$, 10.5 % active area efficiency 7.8 % total area efficiency) (1997).
(4) Pehnt, M., Schulz, D. L., Curtis, C. J., Jones, K. M., and Ginley, D. S., *Appl. Phys. Lett.* **67**, 2176-2178 (1995).
(5) Feher, F., *Sulfur, Selenium, Tellurium,* New York: Academic Press, 1963.
(6) Tuttle, J. R., Contreras, M., Bode, M. H., Niles, D., Albin, D. S., Matson, R., Gabor, A. M., Tennant, A., Duda, A., and Noufi, R., *J. Appl. Phys.* **77**, 153-161 (1995).
(7) Appropriate corrections were applied to the PDF No. 40-1487 (CuInSe$_2$) considering the reduction in cell volume consistent with the substitution of gallium ions for indium ions.
(8) Kamat, P. V., *Materials Technology* **9**, 147-149 (1994).
(9) Wolf, M., and Rauschenbach, H., *Advanced Energy Conversion* **3**, 455-479 (1963).

# CdS/CdTe Thin-Film Devices Using a $Cd_2SnO_4$ Transparent Conducting Oxide

X. Wu, P. Sheldon, T.J. Coutts, D.H. Rose, W.P. Mulligan, and H.R. Moutinho

*National Renewable Energy Laboratory, 1617 Cole Blvd., Golden, CO 80401*

## ABSTRACT

Transparent conducting oxide films of cadmium stannate ($Cd_2SnO_4$) have several significant advantages over conventional transparent conducting oxides. They are more conductive, more transparent, have lower surface roughness, are patternable, and are exceptionally stable. $Cd_2SnO_4$-based CdS/CdTe polycrystalline thin-film solar cells with efficiencies of 13.7% have been fabricated for the first time. Preliminary cell results have demonstrated that device performance can be enhanced by replacing the $SnO_2$ layer with a $Cd_2SnO_4$ transparent conductive oxide.

## INTRODUCTION

Cadmium telluride has long been recognized as a promising photovoltaic material for thin-film solar cells because of its near optimum bandgap of 1.5 eV and its high absorption coefficient. Small-area CdS/CdTe heterojunction solar cells with efficiencies of more than 12% and commercial-scale modules with efficiencies of 9% have been prepared by several techniques, including close-spaced sublimation (CSS), spray deposition, and electrodeposition. Conventional transparent conductive oxides, primarily $SnO_2$ films, have been used as the front contact current collector in CdS/CdTe cells and modules. However, $SnO_2$, with an inherent sheet resistivity of 10 $\Omega$/sq. (at thicknesses necessary for good optical transmission), is not an efficient current collector in CdS/CdTe cells of any appreciable size. In addition, $SnO_2$ thin films, deposited by chemical vapor deposition (CVD), have an average surface roughness of 100-250 Å depending on the chemistry used. It is well known that high efficiency CdS/CdTe solar cells use a thin CdS window layer with a thickness of about 600 Å. High $SnO_2$ surface roughness, coupled with a thin CdS layer, can significantly affect the uniformity of the CdS and the resulting $CdS_xTe_{1-x}$ intermixed layer, ultimately degrading device performance. $SnO_2$ films are also very difficult to pattern, which limits commercial applications for $SnO_2$-based devices including CdTe solar cells. Therefore, it is desirable to improve the qualities of the transparent conducting oxide film for CdS/CdTe solar cell applications. Nozik and later Haacke et al. were the first to report $Cd_2SnO_4$ transparent conducting oxide films deposited by r.f. sputtering (1,2). Recent improvements in r.f. sputter-deposited $Cd_2SnO_4$, produced by our laboratory, have yielded films with superior properties (3,4). In this work, we demonstrate that $Cd_2SnO_4$ films have several significant advantages over conventional $SnO_2$ films, including higher conductivity, better transmission, lower surface roughness, improved patternability, and exceptional stability. Thin-film CdS/CdTe devices using a $Cd_2SnO_4$ transparent conducting oxide have also been prepared. The preliminary cell results indicate that by replacing the $SnO_2$ layer with $Cd_2SnO_4$, the CdS/CdTe cell performance can be enhanced.

## EXPERIMENTAL

In this study, $Cd_2SnO_4$ films were prepared by r.f. magnetron sputtering. The sputtering was carried out in a modified CVC SC-3000 system, evacuated to a base pressure of $\sim 1 \times 10^{-6}$ Torr and then backfilled with high purity oxygen. Corning 7059 glass or soda-lime glass was placed on a water-cooled sample holder parallel to the target surface. The distance between the substrate and the target was varied from 6 to 9 cm. In this study, we used a commercial hot-pressed-oxide target with a composition of 33 mol% $SnO_2$ and 67 mol% CdO. X-ray diffraction showed that the target was single phase orthorhombic $Cd_2SnO_4$. Deposition was performed at an oxygen partial pressure of $10\text{-}20 \times 10^{-3}$ Torr with the r.f. power between 100 and 180 Watts, providing a deposition rate of about 100 Å/min. After deposition, all samples were annealed in a tube furnace in Ar or in an Ar/CdS atmosphere at an elevated temperature of 580° to 680°C for 10-30 minutes. The electrical, optical, and compositional properties of $Cd_2SnO_4$ films were characterized using Hall effect measurements, optical and infrared spectroscopy, x-ray diffraction (XRD), scanning electron microscopy (SEM), and atomic force microscopy (AFM).

The superstrate device structure in which the $Cd_2SnO_4$ layer replaces the $SnO_2$ layer as the front contact is shown in Figure 1. The properties of the $Cd_2SnO_4$ film are described in detail below. The CdS film was deposited by a chemical bath deposition (CBD) technique, using $CdAc_2$, $NH_4Ac$, $NH_4OH$ and thiourea in an aqueous solution. This technique is described elsewhere (5). Prior to CdTe deposition, the CdS films were annealed in $H_2$ at 400°C for 15 minutes. Following the CdS anneal, approximately 10 μm of CdTe was deposited by CSS. During deposition, the substrate and source temperatures were 600°C and 660°C, respectively. The space between the substrate and the source was 0.2 cm. Deposition was initiated at a total pressure of 15 Torr in a $He/O_2$ ambient (14.5 Torr He/0.5 Torr $O_2$). After CSS deposition, the substrates were soaked in a 25% to 75% saturated solution of $CdCl_2$ at $\sim 55$°C for 15 minutes; the saturated solution is 7.5 g cadmium chloride dissolved in 500cc methanol. The devices were then annealed at 400°C for 30 minutes in a tube-furnace with a continuously flowing $He/O_2$ mixture (100 sccm He and 25 sccm $O_2$). A HgTe-doped graphite paste back contact was then applied to the devices. A 1000 Å $MgF_2$ anti-reflection (AR) coating was deposited on some of the cells.

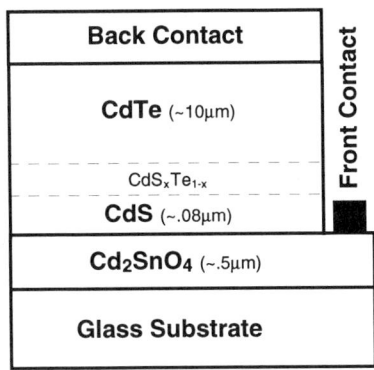

**FIGURE 1.** An improved CdS/CdTe superstrate cell structure.

## RESULTS and DISCUSSION

Single-phase $Cd_2SnO_4$ films have been successfully prepared using an unusual processing scheme developed in our laboratory. This process uses $Cd_2SnO_4$ films sputter deposited in pure oxygen at room temperature, followed by an Ar or Ar/CdS anneal at elevated temperature (3,4). Figure 2 shows the typical XRD pattern for an as-deposited and an Ar/CdS annealed $Cd_2SnO_4$ film. The as-deposited films were amorphous. After an Ar/CdS anneal, the film crystallized into a spinel crystal structure. There is no evidence that secondary phases, such

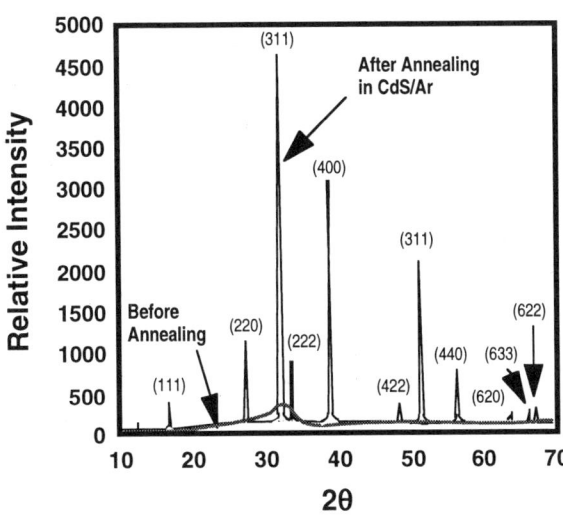

**FIGURE 2.** Typical x-ray diffraction pattern of $Cd_2SnO_4$ film.

as CdO, $SnO_2$, or $CdSnO_3$ form in these films after crystallization. We found that the electrical and optical properties of $Cd_2SnO_4$ films are well correlated to the microstructure of the $Cd_2SnO_4$ film. Under optimum processing conditions, $Cd_2SnO_4$ films exhibit higher conductivities, better transmission, lower surface roughness, improved patternability, and better stability than conventional transparent conducting oxides. As described below, these properties can be optimized to enhance CdS/CdTe device performance.

## $Cd_2SnO_4$ Resistivity

Transparent conducting oxide conductivity can be improved either by increasing the carrier concentration or by increasing the electron mobility. Through optimizing sputtering deposition and post-annealing conditions, $Cd_2SnO_4$ electron mobilities as high as 65 $cm^2$/Vs at a carrier concentration of $2 \times 10^{20}$ $cm^{-3}$ have been achieved. Even at a carrier concentration of $9 \times 10^{20}$ $cm^{-3}$, the mobility is still 55 $cm^2$/V-s. These mobilities are about 2 to 3 times higher than commercial $SnO_2$ films doped to similar levels. Free carriers are thought to result from oxygen deficiencies in the films, accommodated either as oxygen vacancies or cadmium interstitials or a combination of both. Table 1 shows a comparison of electrical properties between representative $Cd_2SnO_4$ and $SnO_2$ films prepared by NREL and others.

As shown in Table 1, the resistivity of $Cd_2SnO_4$ films is 2 to 6 times lower than that of $SnO_2$ films deposited on similar substrates. Low resistivity transparent

**TABLE 1.** Comparison of electrical properties between $Cd_2SnO_4$ and $SnO_2$ films.

| Sample | Thickness (Å) [a] | n ($cm^{-3}$) [b] | μ ($cm^2$/Vs) [b] | ρ (Ω cm) [b] | $R_s$ (Ω/sq.) [c] |
|---|---|---|---|---|---|
| $Cd_2SnO_4$ (NREL) | 5100 | $8.94 \times 10^{20}$ | 54.5 | $1.28 \times 10^{-4}$ | 2.58 |
| $Cd_2SnO_4$ (NREL) [d] | 5500 | $6.58 \times 10^{20}$ | 51.6 | $1.84 \times 10^{-4}$ | 3.2 |
| $SnO_2$ (mfg. 1) | ~10000 | $4.95 \times 10^{20}$ | 15.4 | $8.18 \times 10^{-4}$ | 8.6 |
| $SnO_2$ (mfg. 2) [d] | ~3400 | $4.19 \times 10^{20}$ | 33.0 | $4.53 \times 10^{-4}$ | 13.4 |
| $SnO_2$ (univ. 1) | ~10000 | $4.52 \times 10^{20}$ | 42.0 | $3.29 \times 10^{-4}$ | 3.3 |

[a] Determined from the position of neighboring interference maxima in optical transmittance curves, and cross checks were performed using a Dektak3 thickness profilometer.
[b] Measured by the Hall effect method.
[c] Measured by the Four-point probe technique.
[d] Soda-lime glass substrate.

conducting oxide films will be essential in reducing the series resistance and improving the fill factor and efficiency in CdS/CdTe small area cells and commercial modules. Reducing the series resistance in modules is of particular interest, because manufacturers will be able to increase the cell width between laser scribe lines, thereby reducing losses.

## Optical Transmission

Figure 3 shows the transmittance and absorbance of both a $Cd_2SnO_4$ and a $SnO_2$ thin film. The $Cd_2SnO_4$ film has superior transmission compared to the $SnO_2$ film even though the $Cd_2SnO_4$ film has a much lower resistivity. The electrical properties of these two samples ($Cd_2SnO_4$ - [NREL] and $SnO_2$ - [mfg. 1]) are listed in Table 1. It can also be seen in Figure 3 that the absorbance of the CTO film, in the visible range, is smaller than that of the $SnO_2$ film even at a carrier concentration of $9 \times 10^{20}$ cm$^{-3}$. This feature appears to be due to the unusually high electron mobility. The improved transmission and absorbance of the $Cd_2SnO_4$ film is ideal for superstrate CdS/CdTe devices and should yield improved short circuit currents.

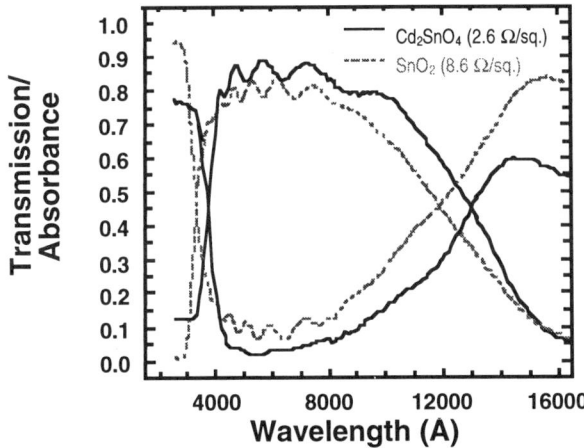

**FIGURE 3.** The comparison of the transmittance and absorbance between $Cd_2SnO_4$ and $SnO_2$ films.

## Surface Roughness

Figure 4 shows atomic force micrographs of the surfaces of both a $Cd_2SnO_4$ film and a $SnO_2$ film. The $SnO_2$ thin film (Figure 4a), was deposited by atmospheric pressure chemical-vapor deposition using a $SnCl_4$ chemistry and has an average surface of 212 Å. In contrast, the $Cd_2SnO_4$ thin film (Figure 4b) has a very smooth surface with an average surface roughness that is almost an order of magnitude lower than that of the $SnO_2$ film.

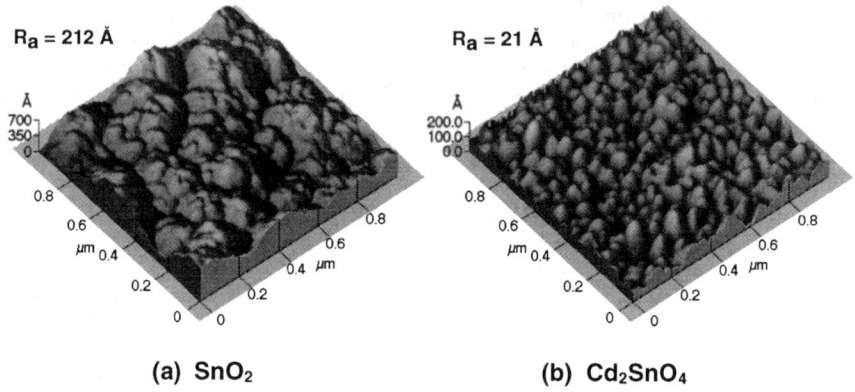

**FIGURE 4.** Comparison of $Cd_2SnO_4$ and $SnO_2$ film surfaces taken by AFM.

It is well known that in a heterojunction solar cell, higher short-circuit currents can be obtained by reducing the window layer absorption. In CdS/CdTe cells, this is achieved by reducing the CdS thickness (6). A CdS/CdTe cell with a thin CdS layer has much better spectral response in the blue portion of the spectrum. However, reducing the thickness of the CdS film to 600 Å - 700 Å can result in a reduction in open circuit voltage and fill factor. During CdTe deposition and subsequent $CdCl_2$ heat-treatment, the CdS film is either partially or completely consumed, forming a $CdS_xTe_{1-x}$ intermixed layer. In fact, McCandless et al. have reported that the CdS consumption increases as the CdS thickness decreases (7). As the CdS is thinned, pinholes can form yielding localized $CdTe/SnO_2$ junctions that have inferior open circuit voltages and fill factors. The probability of pinhole formation increases (especially for sputter-deposited or CSS-deposited CdS) when the $SnO_2$ surface roughness increases (8). Therefore, it is reasonable to expect that by reducing the transparent conducting oxide surface roughness, the probability of forming localized transparent conducting oxide/CdTe junctions will be reduced, thereby improving the device performance.

## Patternability

$SnO_2$ thin films are difficult to pattern, thus providing limitations for certain commercial applications. In contrast, $Cd_2SnO_4$ films can be patterned by etching them in either HCl or HF acid. This is clearly shown in Figure 5, which is a photolithographically patterned $Cd_2SnO_4$ film etched in dilute HCl for 2 minutes. The $Cd_2SnO_4$ film was completely removed in the exposed area, producing a pattern with excellent edge definition. This property may be useful in advanced commercial applications.

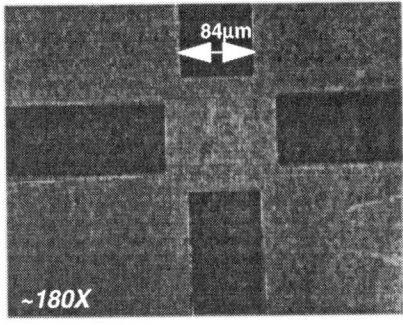

**FIGURE 5.** Optical micrograph of a $Cd_2SnO_4$ film etched in dilute HCl.

## Stability

$Cd_2SnO_4$ films have good mechanical properties. Films deposited on glass substrates have good adhesion, and are reasonably hard and scratch resistant. $Cd_2SnO_4$ films also show excellent thermal stability. As shown in Figure 6, the resistivity of a $SnO_2$ film degrades when annealed in an Ar ambient for 20 minutes at temperatures in excess of 500°C. In contrast, $Cd_2SnO_4$ thin films are stable up to temperatures as high as 650°C. This is particularly important since CSS deposition temperatures can exceed 600°C. This is highlighted in Table 2, which shows the sheet resistivity ($R_s$) before and after CSS CdTe deposition. In some cases, $SnO_2$ films showed significant degradation after 5 minutes of CdTe deposition. Only $Cd_2SnO_4$ films retained a sheet resistivity of < 3 Ω/sq.

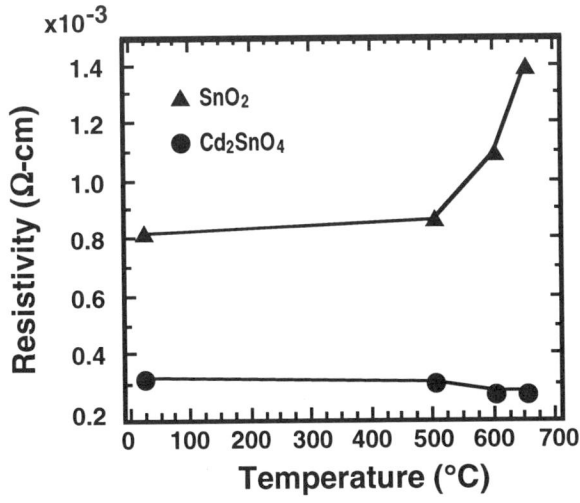

**FIGURE 6.** Thermal stability of $Cd_2SnO_4$ and $SnO_2$ thin films.

**TABLE 2.** $R_S$ stability after CdTe deposition.

| Material | Sample | $R_S$ (Ω/Sq.) Before CdTe Dep. [a] | $R_S$ (Ω/Sq.) After CdTe Dep. [a] |
|---|---|---|---|
| SnO$_2$ (mfg. 1) | S12 | 8.4 | 8.3 |
|  | S8 | 8.2 | 8.1 |
| SnO$_2$ (univ. 1) | U6 | 6.5 | 8.2 |
|  | U3 | 6.7 | 8.3 |
| Cd$_2$SnO$_4$ (NREL) | C4 | 2.6 | 2.5 |
|  | C5 | 2.6 | 2.7 |

[a] $R_S$ measured using a Tencor M-gauge.

Cadmium stannate is also a very stable chemical compound in vacuum or air. Using a mass spectrometer and heating a Cd$_2$SnO$_4$ film in vacuum to 400 K, we were unable to detect any cadmium-containing gaseous species. In addition, Cd$_2$SnO$_4$ films exposed to air ambient at room temperature showed no degradation in resistivity over a 1-year period.

## Cd$_2$SnO$_4$/CdS/CdTe Device Results

We have prepared a limited number of CdS/CdTe cells with a structure similar to that shown in Figure 1. In these cells, the Cd$_2$SnO$_4$ film thickness was 0.5-0.6 µm with a sheet resistivity of about 3 Ω/sq. The CdS thickness was between 800 and 1000 Å, and the CdTe film thickness was around 10 µm. Figure 7 shows the I-V curve of the best CdTe cell, which had an efficiency of 13.7% (measured

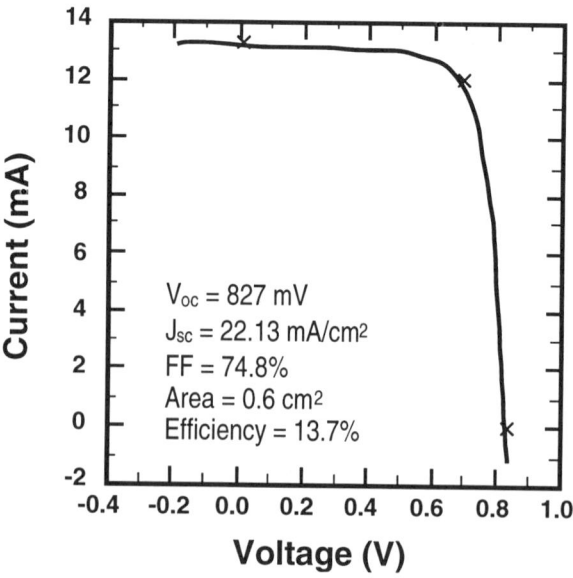

**FIGURE 7.** Current-voltage characteristic of a Cd$_2$SnO$_4$-based thin-film CdS/CdTe solar cell.

by NREL). We found that the fill-factor of most $Cd_2SnO_4$-based CdTe cells were higher than those of $SnO_2$-based devices. This is likely due to the low resistivity of $Cd_2SnO_4$ films. We believe that this advantage will be more significant when applied to large-area CdS/CdTe modules.

To compare the differences between $Cd_2SnO_4$- and $SnO_2$-based devices, two sets of CdS/CdTe cells were prepared using the same cell fabrication procedures. Figure 8 shows the $V_{oc}$ of these two sets of CdTe cells as a function of $CdCl_2$ concentration.

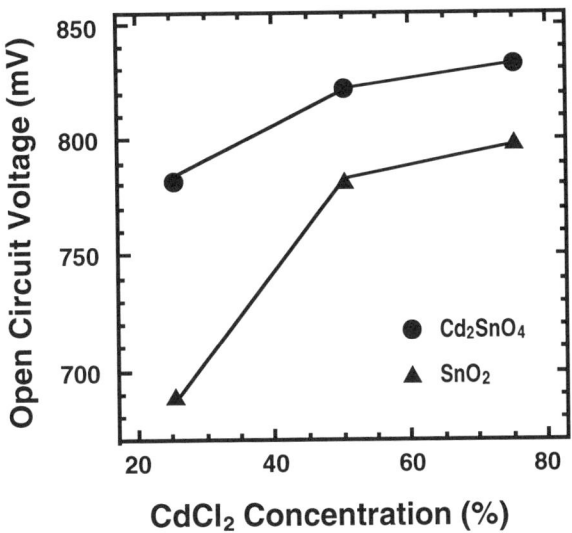

**FIGURE 8.** Open circuit voltage of two sets of CdS/CdTe devices as a function of $CdCl_2$ concentration.

As shown in this figure, the open circuit voltage of $Cd_2SnO_4$-based devices is always higher than that of $SnO_2$-based devices for a given $CdCl_2$ concentration. This result could, in part, be caused by the reduced surface roughness of the $Cd_2SnO_4$. However, it is likely that the chemistry at the $Cd_2SnO_4$/CdS interface is also an important factor. Figure 8 also indicates that the open circuit voltage of $Cd_2SnO_4$-based CdS/CdTe devices is less dependent on the $CdCl_2$ heat-treatment. This result indicates the potential for improved process reproducibility, thereby increasing yield.

## CONCLUSIONS

$Cd_2SnO_4$ transparent conducting oxides have several significant advantages over conventional tin oxides. They are more conductive, more transparent, have a lower surface roughness, are patternable, and are exceptionally stable.

$Cd_2SnO_4$-based CdS/CdTe polycrystalline thin-film solar cells with efficiencies of 13.7% have been prepared for the first time. The preliminary cell results have demonstrated that by replacing the conventional $SnO_2$ with a $Cd_2SnO_4$ film, CdS/CdTe device performance can be enhanced. $Cd_2SnO_4$-based CdS/CdTe cells have a higher fill factor and open-circuit voltage and are less dependent on the $CdCl_2$ heat-treatment than their $SnO_2$ counterparts. We believe that the performance of $Cd_2SnO_4$-based CdS/CdTe cells is far from optimized. Future work will concentrate on improving the efficiency of $Cd_2SnO_4$-based CdS/CdTe cells through the optimization of $Cd_2SnO_4$ film preparation and cell fabrication.

## ACKNOWLEDGMENT

The authors would like to thank Dr. D. Albin and Dr. R. Dhere for useful discussions regarding cell fabrication, and Dr. A. Dillon for residual gas analysis.

## REFERENCES

1. A.J. Nozik, "Optical and Electrical Properties of $Cd_2SnO_4$: A Defect Semiconductor," *Phys. Rev. B.*, Vol. 6, No. 2, pp. 453-459 (1972).
2. G. Haacke, W.E. Mealmaker, and L.A. Siegel, "Sputter Deposition and Characterization of $Cd_2SnO_4$ Films," *Thin Solid Films*, Vol. 55, pp. 67-81 (1978).
3. T.J. Coutts, X. Wu, W.P. Mulligan, and J.M. Webb, "High-performance, Transparent Conducting Oxides Based on Cadmium Stannate," *J of Electronic Materials*, Vol. 25, No. 6, pp. 935-943 (1996).
4. X. Wu, W.P. Mulligan, and T.J. Coutts, "Recent Developments in RF-sputtered Cadmium Stannate Films," *Thin Solid Films*, to be published.
5. T.L. Chu, S.S. Chu, N. Schultz, C. Wang, and C.Q. Wu, "Solution-Grown Cadmium Sulfide Films for Photovoltaic Devices," *J. Electrochem. Soc.*, Vol. 139, No. 9, pp. 2443-2446 (1992).
6. C. Ferekides, J. Britt, Y. Ma, and L. Killian, "High Efficiency CdTe Solar Cells by Close Spaced Sublimation," *Proc. 23rd IEEE Photovoltaic Spec. Conf.*, pp. 389-393 (1993).
7. B.E. McCandless and S.S. Hegedus, "Influence CdS Window Layers on Thin Film CdS/CdTe Solar Cell Performance," *Proc. 22nd IEEE Photovoltaic Spec. Conf.*, pp. 967-972 (1991).
8. A. Rohatgi, S.A. Ringel, R. Sudharsanan, and H.C. Chou, "An Improved Understanding of Efficiency Limiting Defects in Polycrystalline CdTe/CdS Solar Cells," *Proc. 22nd IEEE Photovoltaic Spec. Conf.*, pp. 962-966 (1991).

# Polycrystalline MBE-Grown GaAs for Solar Cells

D. J. Friedman, Sarah R. Kurtz, A. E. Kibbler, M. Al-Jassim,
K. Jones, B. Keyes, and R. Matson

*National Renewable Energy Laboratory, 1617 Cole Blvd, Golden, CO 80401*

**Abstract.** This paper will discuss initial studies of thin-film GaAs grown by molecular-beam epitaxy for use in developing a thin-film GaAs solar cell. Photocurrent and photoluminescence intensity are related to the material morphology as a function of growth conditions. Growth temperature and V/III ratio have a dramatic effect on the photocurrent. However, it seems likely that even after optimizing such growth parameters, it will be necessary to provide substrates that can provide templates to enhance grain size from the start of thin-film growth.

## INTRODUCTION

GaAs, with its 1.4eV band gap and related optical properties, is in many ways close to being an ideal material for solar cells. Indeed, several of the highest photovoltaic conversion efficiencies ever demonstrated have been for devices based on GaAs, for example the GaInP/GaAs tandem device which has achieved conversion efficiencies greater than 30% [1,2]. These (and all such high-efficiency devices) are based on epitaxial growth on single-crystal substrates. The resulting expense can be tolerated in certain applications, notably for high concentration and for space power; indeed, GaAs-based solar cells are widely used in the space photovoltaics (PV) industry. However, the expense of growing epitaxially onto III-V substrates is far too high for terrestrial flat-plate applications. In principle, a solution to the cost problem is to fabricate the device in thin-film form on an inexpensive substrate such as window glass. Some work on thin-film GaAs was done 15-20 years ago [3-5], and a vast amount of development work has been done on a-Si and II-VI thin films. Nonetheless, it seems appropriate to revisit GaAs thin films. The understanding of and the tools for the materials growth issues have become much more sophisticated since the previous thin-film-GaAs work was performed.

The key difference between crystalline and polycrystalline/thin-film material is the presence of grain boundaries in the latter. Grain boundaries can be expected to have various detrimental effects on device quality. For GaAs, with its relatively long minority-carrier diffusion lengths that can be many µm and its high free-surface recombination velocities ~$10^7$cm/sec, carrier recombination at the grain boundary is likely to be highly detrimental if not addressed. Additionally, grain boundaries may enhance dopant diffusion during growth, and may act as transport barriers in the finished device, increasing series resistance in key regions such as the emitter. Therefore, large grain sizes and good boundary passivation are likely to be necessary for successful thin-film GaAs solar cells.

Thin-film Ge substrates are being developed that may provide large-grain templates for the growth of large-grain GaAs [6], and efficiencies as high as ~20% have been demonstrated for GaAs grown on polycrystalline Ge with submillimeter grain sizes [7]. Thus, there seems to be the potential for thin-film GaAs with materials and device performance approaching that of epitaxial GaAs. If this goal is achieved, there is then a clear path to even higher performance by carrying over to the thin-film approach the multijunction structures such as GaInP/GaAs that have been successfully developed in the epitaxial form.

This paper will discuss preliminary results for materials characteristics of GaAs grown on molybdenum-coated glass. For the purposes of this preliminary exploration, no special grain-size-enhancing substrate preparation (such as, for instance, that of Ref. [6]) was used.

## MATERIALS FABRICATION

The substrates used for the sample fabrication were molybdenum-coated soda-lime glass (window glass). Such substrates represent the low end of the spectrum in both cost and sophistication; materials and device properties of GaAs grown on such substrates can be expected to represent a lower bound to what might be achieved with more sophisticated substrates. Other than an organics rinse for degreasing, no preparation of the substrates was done before they were loaded into the growth system.

Samples were grown using solid-source MBE (Molecular Beam Epitaxy — although "epitaxy" is a misnomer in this case). While MBE is generally thought of as a sophisticated and very expensive tool, it is essentially a very simple and unsophisticated evaporation technique, and it is reasonable to project that thin-film materials growth methods developed in a research MBE system will be transferable to a much less expensive evaporator suitable for manufacturing-scale production. If there are difficulties in transferring the technology to a manufacturing evaporator, they are likely to be related to either the difference in base pressures between the MBE and manufacturing evaporator chambers ($10^{-10}$ torr and $10^{-7}$ torr, respectively), or the generation of the arsenic flux. While the group-III materials are simply heated in a cone-shaped crucible, the arsenic is evolved from a valved cracker cell that permits precise control of the arsenic flux, and provides the capability to crack the evaporated arsenic molecules to $As_2$. All of the growths reported here, however, were grown with non-cracked $As_4$.

The Ga flux was kept at a beam-equivalent pressure of $1.4 \times 10^{-6}$ torr to provide a nominal growth rate of 2.5 μm/hour. Growth temperature and As flux were varied from sample to sample. A 0.2-μm-thick layer of InAs was grown on the substrate to act as a contacting layer to the molybdenum, followed by a nominal 5 μm of GaAs. No dopants were used. The V/III ratio referred to in this paper is the V/III beam flux ratio as measured by an ion gauge in the path of the beam.

## MEASUREMENT ISSUES

While the ultimate measure of device materials quality is the device performance, it is useful to have a lower-level measure of materials quality that reflects the properties of a single layer. For the initial materials development, we use the intensity of the spectral response at hv=1.6 eV, normalized to the corresponding value measured for high-quality epitaxially grown GaAs material, as the basic measure of photovoltaic materials quality. The spectral response of the samples was measured using an electrolyte junction to the front surface of the sample. As an additional measure of materials quality, the photoluminescence (PL) intensity of each sample was measured.

The grain structure of the material was examined by transmission electron microscopy (TEM) and cross-sectional scanning electron microscopy (X-SEM) with the aim of relating the grain morphology — as determined by the growth conditions — to the opto-electrical properties.

## EFFECT OF ARSENIC FLUX

The As flux proved to have a dramatic effect on the photocurrent (PC). Samples were grown with an As/Ga beam-equivalent-pressure (V/III BEP) ratio of 3, 6, and 14 at a constant growth temperature of 400°C. The PC's in percent of the PC for an epitaxial GaAs layer were PC=0.005%, 0.05%, and 2% for the 14, 3, and 6 V/III-BEP-ratio samples respectively. Figure 1 shows X-SEM micrographs of each of the samples. Examination of the X-SEM micrographs should be made with the caution that the pictures represent a convolution of the grain structure with the fracture pattern from the sample cleaving; nonetheless, some basic features of the grain structure are clear.

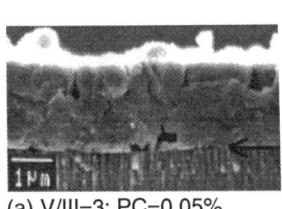
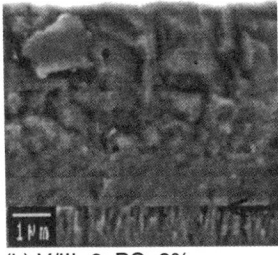

(a) V/III=3; PC=0.05%    (b) V/III=6; PC=2%    (c) V/III=14; PC=0.005%

**FIGURE 1.** X-SEM micrographs of GaAs layers showing the effect of arsenic flux on growth mode/grain size. The corresponding photocurrent (PC) at 1.6eV, in percent of the PC for an epitaxial GaAs layer, is also noted. The arrows denote the interface between the semiconductor layer and the molybdenum film coating the glass.

A comparison of this material grown with V/III=6 with the material grown at V/III=14 by X-SEM shows a marked difference in the grain structure in the two materials. The V/III=14 sample shows a columnar grain structure, with grain sizes/column diameters on the order of 0.1-0.3 µm. In contrast, the V/III=6 sample shows an evolution of the grain size away from the substrate surface, with grain sizes increasing to 1-2 µm near the surface of the sample. This difference can readily be understood in terms of the mobility of the Ga atoms on the growth surface: increasing the As flux decreases the time it takes to bind the Ga atom to the GaAs surface, thus decreasing the Ga mobility and favoring columnar growth rather than an increase of grain size as the growth proceeds.

While the greater thickness of the V/III=6 sample compared to the V/III=14 may have some effect on the photocurrent, the V/III=6 sample's much larger grain sizes, and hence the lesser effect of recombination at grain boundaries, is presumably responsible for its much better photocurrent.

## EFFECT OF GROWTH TEMPERATURE

Sample growth temperature $T_g$ was also found to have a strong effect on the PC of the resulting material. A series of samples were grown with V/III=6 and various growth temperatures. Figure 2 shows the resulting dependences of photocurrent and of PL intensity on $T_g$. The PC shows a very well-defined behavior as a function of $T_g$, rising two orders of magnitude as $T_g$ is raised from 400°C to 500°C. As with the effect of the V/III ratio described above, it is likely that the increase of PC with increasing $T_g$ is due to increasing mobility of adsorbed atoms on the growth surface with the increasing temperature. As the temperature rises above 500°C into the range at which the glass starts to soften, there is a decrease in the PC. This decrease is consistent with the morphology of the material as shown in the X-SEM pictures of Fig. 3. At $T_g$=500°C the GaAs film is a well-defined compact layer, while by $T_g$=540°C the near-surface region appears to have formed into a slightly porous, filamentary morphology. Such porosity would make the collection of photogenerated carriers more difficult and decrease the measured PC, as is observed. However, it is also possible that diffusion of material such as sodium from the glass may be affecting the electrooptical properties as well; more measurements are needed on this question.

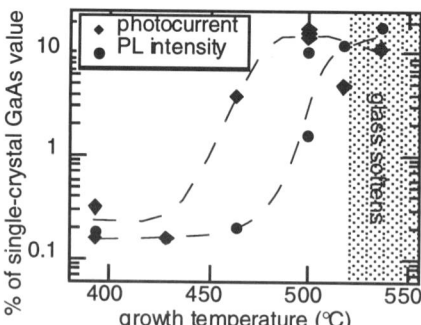

**FIGURE 2.** PC and PL intensity, normalized to their values as measured for a single-crystal GaAs epilayer grown on a GaAs substrate, for samples grown at various growth temperatures. Ga and As fluxes were held constant, as described in the text.

The PL intensity tracks the PC overall, but not in detail: the onset of the rise in the PL intensity is 50°C higher than for the PC, and the subsequent falloff of PC above the maximum is not observed in PL over the range of $T_g$ in which the samples were grown.

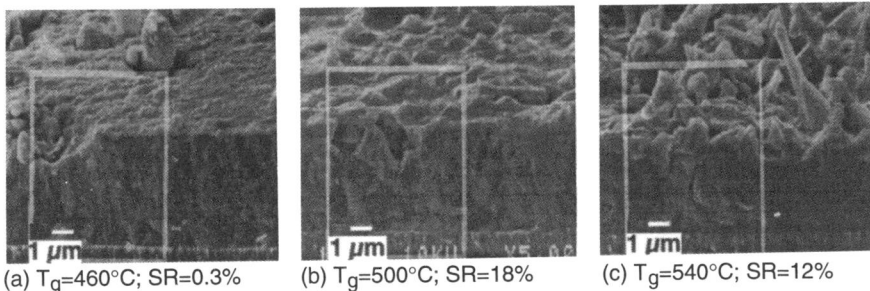

(a) $T_g$=460°C; SR=0.3%   (b) $T_g$=500°C; SR=18%   (c) $T_g$=540°C; SR=12%

**FIGURE 3.** X-SEM micrographs of GaAs layers showing the effect of growth temperature.

## LIQUID-PHASE GROWTH

While it is likely that specially-prepared large-grain substrates will be necessary to induce adequately large, oriented grains in the GaAs film, there are various methods to enhance the grain size during growth. The adjustment of growth temperature and V/III as discussed above are two simple methods. Another approach, which in a sense is comparable to going to extremely low values of V/III, is to provide a large excess of molten Ga for the growth, and let GaAs crystals grow out of the melt. One example of what this can provide is given by the sample of Fig. 1(a), which was grown at a very low V/III of 3. The as-grown sample surface was dotted with microscopic droplets of Ga because of the low As overpressure. The Ga was etched away with HCl.

**FIGURE 4.** SEM micrograph of surface of GaAs film grown with V/III so low that Ga droplets formed on the surface. This picture shows the area under one such droplet, after the Ga itself was etched away.

Figure 4 shows an SEM micrograph of an area on the surface of the sample where such a Ga droplet had been, after the Ga droplets were etched away.

Extremely well-formed crystallites of sizes up to 10 μm and more have formed. This promising picture suggests that it may be possible to nucleate relatively large domains, thus enhancing the materials properties, by depositing a thick film of Ga and then exposing it to arsenic [8]. We have performed a preliminary test of this by growing a film at $T_g$=500°C as in Fig. 3(b) but with an initial nucleation layer formed by first depositing 0.4μm of Ga, then exposing it to As for 10 minutes, and then proceeding with a more standard growth of coevaporating Ga and As. This film showed a PC of 27%, roughly twice the PC of the corresponding sample grown without the liquid-phase nucleation layer.

## CONCLUSIONS

The PC, which was taken for this study as the main measure of materials quality for solar cell applications, shows a very strong dependence on such growth parameters as growth temperature and V/III. For instance, changing the growth temperature by 100°C while holding all other parameters constant resulted in two orders of magnitude change in the PC. Additionally, techniques such as growing the film by arsenizing liquid Ga may further enhance materials quality. However, it seems likely that it will be necessary to grow on substrates more sophisticated than molybdenum-coated glass to get the grain structure needed for high efficiencies.

## ACKNOWLEDGMENTS

We thank J. Alleman, A. Mason, R. Noufi and K. Ramanathan for helpful discussions, measurements, and fabrication of the Mo-coated glass substrates.

## REFERENCES

[1] D.J. Friedman, S.R. Kurtz, K.A. Bertness, A.E. Kibbler, C. Kramer, and J.M. Olson, "30.2% efficient GaInP/GaAs monolithic two-terminal tandem concentrator cell", *Progr. Photovolt.* **3**, 1995, pp. 47-50.

[2] K.A. Bertness, S.R. Kurtz, D.J. Friedman, A.E. Kibbler, C. Kramer, and J.M. Olson, "29.5%-efficient GaInP/GaAs tandem solar cells", *Appl. Phys. Lett.* **65**, 1994, pp. 989-991.

[3] S.S. Chu, T.L. Chu, and H.T. Yang, "Thin-film gallium arsenide solar cells on tungsten/graphite substrates", *Appl. Phys. Lett.* **32**, 1978, pp. 557-559.

[4] S.S. Chu, T.L. Chu, Y.T. Lee, C.L. Jiang, and A.B. Kuper, "Improved polycrystalline thin film gallium arsenide MOS solar cells", *14th IEEE Photovoltaic Specialists Conference*, 1980, pp. 1306-1310.

[5] Y.C.M. Yeh, K.L. Wang, B.K. Shin, and R.J. Stirn, "Epitaxial and polycrystalline GaAs solar cells using OM-CVD techniques", *14th IEEE Photovoltaic Specialists Conference*, 1980, pp. 1338-1342.

[6] H. Atwater and e. al., *Proceedings of the 1996 DOE NREL/SNL Photovoltaic Program Review*, 1996, (this conference).

[7] R. Venkatasubramanian, B. O'Quinn, and E. Siivola, "High-efficiency GaAs solar cells on mm and sub-mm grain-size polycrystalline Ge substrates", *Proceedings of the 1996 DOE NREL/SNL Photovoltaic Program Review*, 1996, pp. (this conference).

[8] S.S. Chu, T.L. Chu, W.J. Chen, H. Firouzi, Y.X. Han, and Q.H. Wang, "Large grain gallium arsenide thin films", *17th IEEE Photovoltaic Specialists Conference*, 1984, pp. 896-899.

# O Impurity Chemistry in CdS Thin-Films Grown by Chemical Bath Deposition: An Investigation with X-ray Photoelectron Spectroscopy

David W. Niles,* Gregory Herdt,[†] and Mowafak Al-Jassim*

*National Renewable Energy Laboratory 1617 Cole Boulevard, Golden, CO 80401
[†]Fusion Semiconductor Corporation, 7600 Standish Place, Rockville, MD 20855

**Abstract.** We used x-ray photoelectron spectroscopy to investigate the chemistry of O impurity atoms in CdS thin-films grown for photovoltaic purposes by chemical-bath deposition (CBD). We compared the Cd 3d photoline, O 1s photoline, Cd MNN Auger line, and O KLL Auger line taken from a CBD CdS thin-film, CdS single-crystal reference, Cd metal reference, CdO reference, and Cd(OH)$_2$ reference. This comparison showed that the O present in thin-film CBD CdS is a manifestation of H$_2$O incorporated into the film during the CBD growth. Ar$^+$ ion sputtering – a technique frequently used in thin-film analyses – preferentially removed S from the CBD CdS thin-film and created CdS$_{1-x}$O$_x$ (x ~ 0.04) in the surface region from the incorporated O impurity.

## INTRODUCTION

CdS is a common window layer for both CdTe- and Cu(In,Ga)Se$_2$-based photovoltaic (PV) devices.(1-5) In the PV industry, CdS must be deposited by a suitable thin-film technique as part of a stacking sequence to make devices. Techniques for depositing CdS thin-films include chemical-bath deposition (CBD), physical-vapor deposition, close-spaced sublimation, sputtering, electrodeposition, and chemical vapor deposition. Researchers trying to promote efficiency records in the laboratory environment use CBD to deposit CdS thin-films, although details of the CBD recipe may vary from laboratory to laboratory.(6)

A concern in our laboratory is the economic feasibility of scaling up the CBD technique required for commercialization of high-efficiency CuInSe$_2$- and CdTe-based devices.(1) A vacuum-based evaporation technique such as physical-vapor deposition would appear more amenable to low-cost production of PV devices. In the event that CBD poses monetary problems with large-area production of PV modules, the thin-film PV community will need to explore alternative deposition techniques. The problem is that the highest-efficiency PV devices use CBD CdS, whereas other techniques have yielded inferior-quality devices.(5)

As a II-VI semiconducting material, thin-film CBD CdS is different from thin-film CdS made with other techniques. As other researchers have noted, CBD CdS is replete with impurities from the chemical bath, most notably O.(7-11) In all cases, the impurities concentrations may vary from trace amounts to several or even 10 a/o. Impurities incorporated from the chemical bath will not be an issue with other deposition techniques, although other deposition techniques may also be plagued with impurity-related issues. O appears to be the most prevalent impurity in CBD CdS thin-films. In our own laboratory, O is the most prevalent impurity typically at several atomic percent of the CdS thin-films. From simple arguments of reaction rates, it is difficult to understand how CdO, $Cd(OH)_2$, or $CdCO_3$ could be incorporated in the thin-films from the bath.(6)

The purpose of this manuscript is to describe experiments that use predominantly X-ray photoelectron spectroscopy (XPS) to illuminate the chemistry of O in our CBD CdS thin-films. We compare the photoemission signal from CBD CdS with the signal from single-crystal CdS, $Cd(OH)_2$, metallic Cd, and CdO to identify O and Cd chemistry. We positively identify the presence of Cd–O bonds in our CBD thin-films as an artifact of $Ar^+$ ion sputtering, and we conclude that the presence of $Cd(OH)_2$ is inconsistent with our data. In agreement with Kylner et al., and in contrast with most of the literature we found, the dominant chemical compound for the O in our CBD CdS thin-films is $H_2O$.(7-11)

## EXPERIMENT

CdS thin-films were grown by chemical-bath deposition on soda-lime glass substrates.(6) The substrates were sonicated and rinsed in 18 MΩ/cm deionized $H_2O$ to remove particles from the surface, dried with high-purity nitrogen, and etched for 5 minutes in a 5:1:1 solution of $H_2SO_4:H_2O_2:H_2O$ immediately prior to CBD CdS growth. A similar procedure was performed on InP[100] substrates for a separate experiment, but used here to confirm that the glass substrates did not present a unique surface that would affect the CdS thin-films. The results presented here are from the glass-based substrates; results from the InP substrates were identical.

The CBD CdS thin-films were grown in a covered temperature-controlled bath using a 200 mL solution of 1 mM $CdSO_2$, 20 mM ammonium acetate ($NH_4Ac$), 5 mM thiourea, and 0.4 M $NH_4OH$. The temperature was 85°C. The growth rate, determined by ellipsometry, was 8.0 ± 0.3 nm/min. The samples were stored in air between growth and XPS analysis. The samples were not annealed postgrowth.

A CdO standard for comparison was prepared by polishing a piece of metallic Cd, inserting into a ultraviolet box to expose the surface to ozone, and then heating to 150°C for 2 hours to remove deposited $H_2O$ from the surface. The resulting oxide layer was not pure CdO, but contained a sufficient CdO signature to use as a standard. A $Cd(OH)_2$ standard was prepared in a manner similar to that described by Sugimoto et al., except the $Cd(NO_3)_2$ was replaced with $CdCl_2$.(12)

A CdS[1010] single-crystal was also used as a standard reference.

XPS measurements were performed in a commercial Physical Electronic 5600 ESCA equipped with a monochromated Al Kα X-ray source operated at 300 W. The energy resolution of the hemispherical analyzer was set to ~0.18 eV for high-energy resolution scans. The spot size on the sample for collecting photoelectrons was 0.8 mm. The sputter gun was set to 1 kV Ar$^+$ ions to minimize the depth of sputter damage in the CBD thin-films.(13) The estimated sputter rate determined by sputtering through CBD CdS thin-films of known thickness was ~1 nm/min.

## RESULTS ANS DISCUSSION

Fig. 1 shows an XPS survey spectrum of a CBD CdS thin-film after sputtering ~3 nm. The dashed line is the Tougaard background for the CBD CdS thin-film.(14) Emission from the C 1s level at $E_B$ = 286 eV is below our detection and is not visible even on an expanded scale. One can also see a very small N 1s signal at $E_B$ = 398 eV for the CBD thin-film, but it is not visible in the survey spectrum or even on an expanded scale. We will not discuss in this manuscript the chemistry of N in the CBD thin-film.

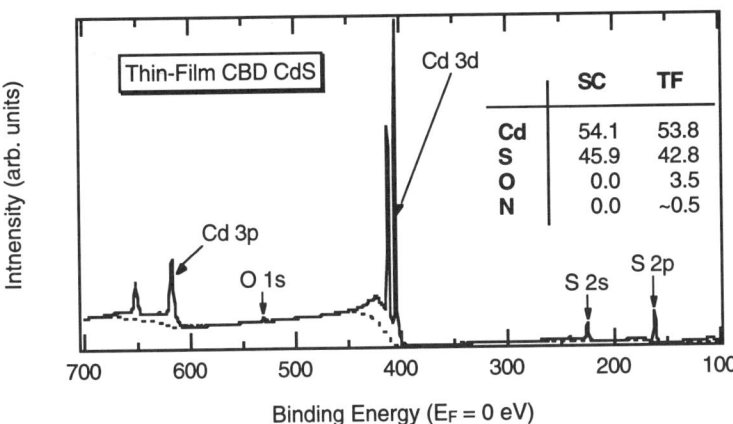

Fig. 1: XPS survey spectrum of thin-film CBD CdS.

The table incorporated into Fig. 1 lists the atomic compositions for the CBD thin-film (TF) and the single-crystal reference (SC). We used the standard sensitivity factors for our electron analyzer to determine the atomic concentrations.(15) The atomic concentrations of the CBD thin-film (ignoring the trace N signal) were Cd = 53.8 ± 1 a/o, S = 42.8 ± 1 a/o, and O = 3.5 ± 1 a/o. The corresponding atomic concentrations for the single crystal reference were Cd = 54.1 a/o and S = 45.9 a/o. The single-crystal surface is rich in Cd, indicating preferential removal of S by the Ar⁺ sputter beam.(13) To prove that the measured deviation from stoichiometry was at least partially indicates preferential sputtering, we increased the energy on the incoming Ar+ ions to 3 kV and consistently measured compositions of Cd = 52 ± 1 a/o and S = 48 ± 1 a/o. Extensive efforts to find Fermi level emission from metallic Cd were not productive. A lack of Fermi level emission led us to believe that the excess Cd did not form metallic islands. The surface of the CdS single-crystal is $CdS_{1-x}$, where x is ~0.15. The essential difference between the CBD CdS thin-film and the CdS single-crystal reference is that ~ 3.5 a/o of the S has been replaced by O.

Fig. 2 shows an XPS depth profile of the O content throughout a 70-nm-thick CBD CdS thin-film. A high-resolution scan of the O 1s lineshape showed no significant changes in lineshape with sputter depth. Detailed analysis of the composition at various depths yielded the O concentration of 3.8 ± 0.5 a/o. The Cd enrichment of the surface due to preferential removal of S stabilizes at depths of less than 10 nm. We do not see any evidence of further divergence of the CdS stoichiometry in either the single-crystal reference (not shown) or the CBD thin-film. The critical point for this analysis is that the O is not simply a surface contaminant, but is present throughout the bulk of the thin-film.

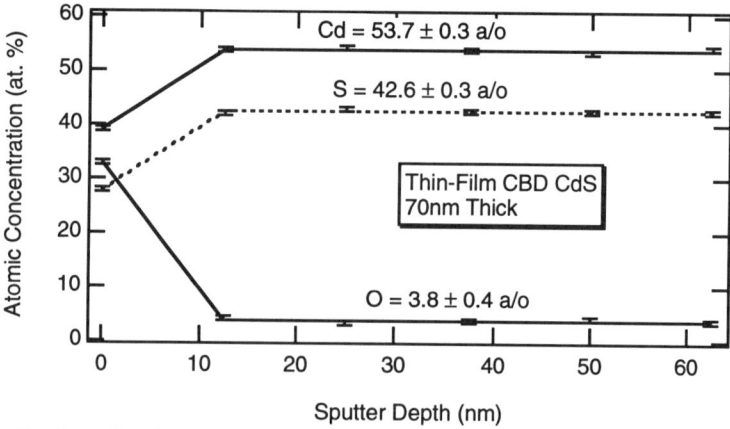

Fig. 2: Depth profile of a CBD CdS thin-film.

Fig. 3 shows emission from the O KLL Auger and O 1s photoemission lines for the CdO reference, the sputtered CBD CdS thin-film, unsputtered CBD CdS thin-film, and the $Cd(OH)_2$ reference.(17) The O 1s photoline from the CdO reference has a large signal at $E_B$ = 529.0 eV, representative of Cd–O bonds. The peak at 530.7 eV in the sputtered CBD thin-film sample (labeled "O Impurity") represents incorporated $H_2O$ from the bath, as we will show later. The O KLL spectrum for CdO has a clear peak at $E_B$ = 970.2 eV (labeled CdO in Fig. 6) that we will also use as a signature for Cd–O bonds. This peak at $E_B$ = 970.2 eV is not present in the O KLL spectrum from the unsputtered CBD CdS thin-film, the unsputtered single-crystal CdS (not shown in Fig. 3), or the Cd(OH)$_2$ reference, although a shoulder at the same energy is present. Both of these signatures of Cd–O bonds are present, however, in the sputtered CBD CdS thin-film. Their absence in the unsputtered CBD thin-film and presence in the sputtered CBD thin-film is proof that sputtered creates Cd–O bonds.

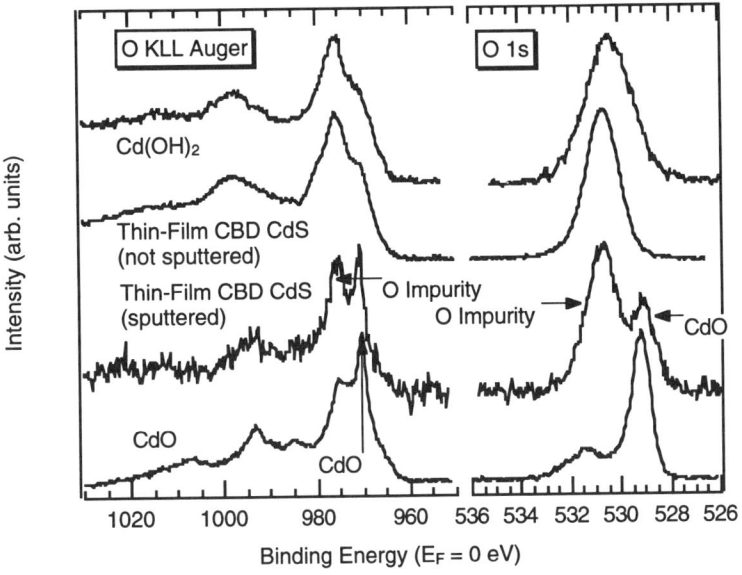

Fig. 3: O KLL Auger and O 1s photo-lines for relevant materials.

The O KLL Auger lines for the Cd(OH)$_2$ and unsputtered CBD CdS samples look identical, and they have a sharp peak at $E_B = 975.9$ eV. We have labeled this peak "O Impurity" in Fig. 3. The O 1s from the Cd(OH)$_2$ is significantly broader than the other spectra. This breadth may have two origins: the first is that the Cd(OH)$_2$ powder charged differentially and was extremely difficult to neutralize; the second is that the X-ray beam damaged the Cd(OH)$_2$ powder, turning the white powder brown.(18)

The spectra of Fig. 3 prove that the O 1s peak at $E_B = 529.0$ eV is from Cd–O bonds, but did not help us conclusively assign a bonding state for the peak at $E_B = 530.7$ eV. The O Auger parameter did not support an argument for Cd–(OH) bonds. Fig. 4 shows a set of Cd MNN and Cd 3d spectra for relevant Cd compounds, similar to the set of O KLL spectra shown in Fig. 3.(21) One can see quite clearly the remarkable similarity between the Cd MNN Auger spectra from the sputtered CBD thin-film and the single-crystal, and the marked differences among these two spectra and spectra from CdO, Cd(OH)$_2$, and Cd metal. The important point is that we do not see any evidence of the Cd(OH)$_2$ Cd MNN lineshape in the spectrum from CBD CdS. This leads us to conclude that the O must be bound to H in the form of $H_2O$ because there are not other elements present in the CBD CdS to which the O could bind. The dynamics of the sputtering process allows for the formation of Cd–O bonds from the damaged CdS and $H_2O$.

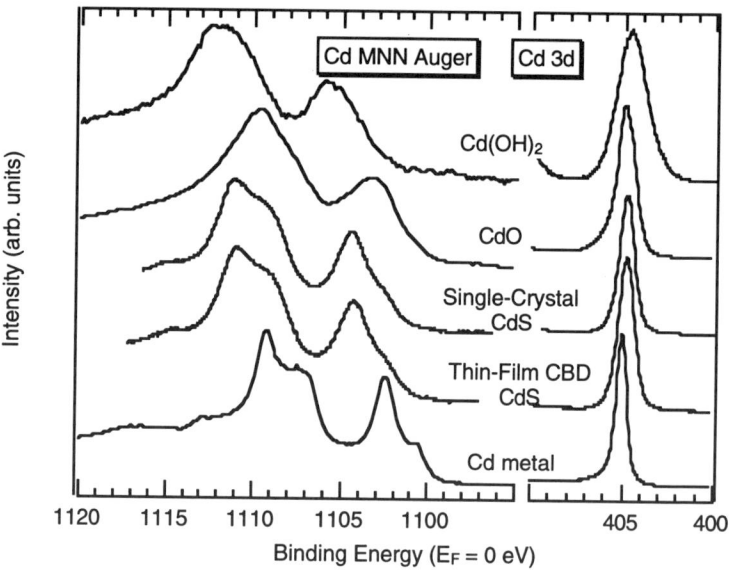

Fig. 4: Cd MNN Auger and Cd 3d photo-lines for relevant materials.

Fig. 5 shows high-energy resolution scans of S 2p and Cd 3d emission from CBD and single crystal CdS. The S 2p lineshapes for the CBD thin-film and single-crystal samples were identical, as highlighted by the difference spectrum. The Cd 3d lineshapes, however, are not identical. The FWHM for the single crystal is 1.008 eV, whereas it is 1.062 eV for the CBD thin-film. The binding energy changes from $E_B = 404.82$ eV for the single-crystal to $E_B = 404.76$ eV for the CBD thin-film. Although these changes are small, they are unmistakable. Furthermore, the shift toward lower binding energy and the increased width for the Cd 3d emission from the CBD thin-film compared to the single-crystal are consistent with the presence of Cd–O bonds.

By scaling the Cd 3d emission on the high-binding-energy side to equal the intensity on the high-binding-energy side of the Cd 3d emission from the CBD thin-film, one can generate a difference spectrum that itself looks like a Cd 3d 5/2 peak. The FWHM of the difference peak is 0.827 eV, and its binding energy is 404.52 eV. We measured a difference in binding energy between the Cd–S and Cd–O bonds in the Cd 3d photoline as 404.82 eV – 404.52 eV = 0.3 eV. This value is significantly smaller than the 1.1 eV difference between CdS and CdO, as measured by Gaarenstroom *et al.*(22) We do not believe that CdO exists as a separate phase on the surface, but rather, that the sputtering process promotes the formation of a $CdS_{1-x}O_x$ layer on the surface. In a $CdS_{1-x}O_x$ layer with only ~ 1.3 a/o O, only one or two out of every 50 anions is an O atom.(24) Because four anions surround every cation, one would not expect the S to respond to the presence of O. Four S atoms would surround ~ 90 % of the Cd atoms, 3 S atoms and 1 O atom would surround ~ 10 % of the Cd atoms, and in only very few cases would two or more O atoms contact a Cd. One would expect to observe ~ 90 % of the Cd appearing like Cd surrounded by 4 S atoms, and ~ 10 % of the Cd appearing like Cd surrounded by 3 S atoms and 1 O atoms. This is what we observe. Although x must vary with depth, x must be ~ 0.04 to account for the observed signal.

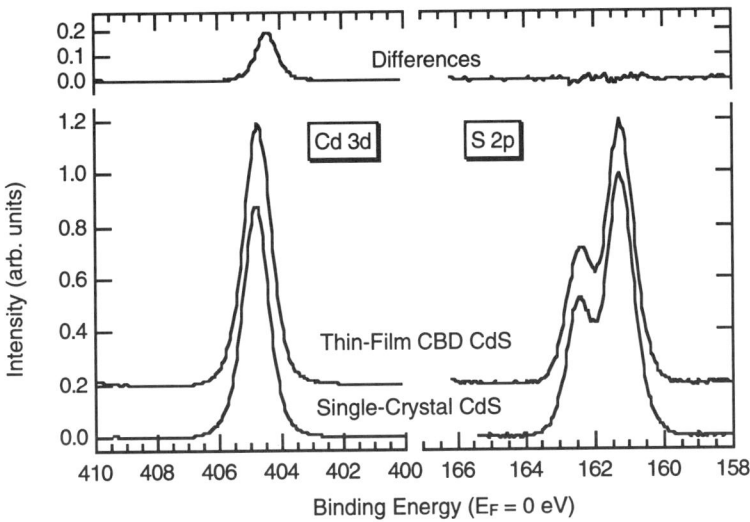

Fig. 5: Cd 3d and S 2p photolines for CBD thin-film and single-crystal CdS.

A critical issue for the PV community is to understand the relationship between impurities in CBD CdS and device performance. At this point, we are unable to correlate impurity concentration with device efficiency. Nevertheless, record performance CdS/CdTe and $CdS/Cu(In,Ga)Se_2$ devices employ CBS CdS, and a thorough understanding of this material is important to the understanding of PV devices.

## ACKNOWLEDGMENTS

The authors would like to thank A. Mason, K. Ramanathan, R. Bhattacharya, D. Schultz, and R. Noufi of the National Renewable Energy

Laboratory for assistance with the chemical-bath deposition technique and data acquisition, and A. Kylner of Uppsala University for a copy of Ref. 10 prior to its appearance in the literature.

# REFERENCES

1. J.R. Tuttle, M. Contreras, A.M. Gabor, K.R. Ramanathan, A.L. Tennant, D.S. Albin, J. Keane, R. Noufi, *Prog. Photovoltaics* **3**, 383 (1995), and references therein.
2. V. Nadanau, D. Braunger, D. Hariskos, M. Kaiser, Ch. Köble, A. Oberacker, M. Ruckh, U. Rühle, R. Schäffler, D. Schmid, T. Walter, S. Zweigart, H.W. Schock, *Prog. Photovoltaics* **3**, 363 (1995), and references therein.
3. P.V. Meyers, R.W. Birkmire, *Prog. Photovoltaics* **3**, 393 (1995), and references therein.
4. T. L. Chu, S. S. Chu, *Prog. Photovoltaics* **1**, 31, (1993).
5. M.A. Green, K. Emery, K. Bücher, D.L. King, *Prog. Photovoltaics* **4**, 59 (1996).
6. R. Ortega-Borges, D. Lincot, *J. Electrochem. Soc.* **140**, 3464 (1993).
7. Y. Hashimoto, T. Nakanishi, T. Andoh, K. Ito, *Jpn. J. Appl. Phys.* **34**, L382 (1995).
8. M.T.S.Nair, P.K. Nair, R.A. Zingaro, E.A. Meyers, *J. Appl. Phys.* **75**, 1557 (1994).
9. A. Kylner, J. Lindgren, L. Stolt, J. Electrochem. Soc. **143**, 2662 (1996).
10. A. Kylner, A. Rockett, L. Stolt, in *Polycrystalline Semiconductors IV-Physics, Chemistry and Technology*, edited by S. Pizzini, H.P. Strunk, and J.H. Werner, in the Series *Solid State Phenomena*, Trans. Tech. Publ., Zug, Switzerland (1995).
11. P.C. Rieke, S.B. Bentjen, *Chem. Mater.* **5**, 42 (1993).
12. T. Sugimoto, G. Dirige, A. Muramatsu, *J. Colloid Interface Sci.* **173**, 257 (1995).
13. *Practical Surface Analysis by Auger and X-Ray Photoelectron Spectroscopy*, edited by D. Briggs and M. P. Seah (Wiley, New York, 1984).
14. S. Tougaard, C. Jansson, *Surf. Inter. Anal.* **20**, 1013 (1993), and references therein.
15. *Handbook of Photoelectron Spectroscopy*, edited by Jill Christian (Physical Electronics, Eden Prairie, 1992).
16. Sputtering artifacts are described in more detail in Ref. 13.
17. The CdO reference was not a pure standard. Its composition was Cd = 60 a/o and O = 40 a/o. Emission from the peak at $E_B$ = 531.7 eV in the O 1s spectrum was ~ 5 times larger prior to a 130°C anneal, and its attenuation with heat is consistent with its assignment as $H_2O$. Nevertheless, it made suitable reference for the identication of Cd–O bonds.
18. B. Tielsch, J.E. Fulghum, *Surf. Inter. Anal.* **24**, 422 (1996).
19. C.D. Wagner, L.H. Gale, R.H. Raymond, *Anal. Chem.* **51**, 466 (1979).
20. C.D. Wagner, D.A. Zatko, R.H. Raymond, *Anal. Chem.* **52**, 1445 (1980).
21. L.C. Lynn, R. I. Opila, *Surf. Inter. Anal.* **15**, 180 (1990).
22. S.W. Gaarenstroom, N. Winograd, *J. Chem. Phys.* **67**, 3500 (1977).
23. J. S. Hammond, S. W. Gaarenstroom, N. Winograd, *Anal. Chem.* **47**, 2193 (1975).
24. The surface region has ~ 3.8 a/o O, but only about one-third of that is involved in Cd–O bonds. See Fig. 5.

# CRYSTALLINE MATERIALS (P)

# Thin Film GaAs Solar Cells On Glass Substrates By Epitaxial Liftoff

X. Y. Lee, Mark Goertemiller, Misha Boroditsky, Regina Ragan and Eli Yablonovitch

*Electrical Engineering Dept., University of California, Los Angeles, CA 90095-1594*

**ABSTRACT.** In this work, we describe the fabrication and operating characteristics of GaAs/AlGaAs thin film solar cells processed by the epitaxial liftoff (ELO) technique. This technique allows the transfer of these cells onto glass substrates. The performance of the lifted-off solar cell is demonstrated by means of electrical measurements under both dark and illuminated conditions.

We have also optimized the light trapping conditions in this direct-gap material. The results show that good solar absorption is possible in active layers as thin as 0.32 µm. In such a thin solar cell, the open circuit voltage would be enhanced. We believe that the combination of an epitaxial liftoff thin GaAs film, and nano-texturing can lead to record breaking performance.

## INTRODUCTION

In recent years, III-V solar cell technology has been actively pursued for use in space applications. While GaAs has been regarded as a high-efficiency high-cost option, new developments in MOVCD growth reactors, epitaxial liftoff processing and substrate re-use can lead to low cost III-V epitaxial thin films (1). As a result, III-V cells show significant potential for terrestrial use as well.

In this work, we describe the fabrication and operating characteristics of GaAs/AlGaAs thin film solar cells (2-4) using the epitaxial liftoff (ELO) technique (5). This technique allows the transfer of these cells onto glass substrates.

We also describe a way of enhancing light absorption by our solar cells through nanotexturing of the thin film after they have been transferred onto glass slides. This light trapping operation permits a much thinner than normal active layer which will allow a new higher voltage operating regime.

## DEVICE FABRICATION AND LIFTOFF

The structure used in our experiment was grown using MOCVD. It consists of an n-GaAs active region sandwiched between n-$In_{0.5}Ga_{0.2}Al_{0.3}P$ and p-$Al_{0.3}GaAs$

window/passivation layers, and capped by a p⁺-GaAs contact layer. The entire structure is grown on top of a 500 Å thick sacrificial layer of AlAs, which is subsequently etched to release the device from its substrate. A thin GaAs layer above the AlAs serves to protect device layers during liftoff. This thin protective layer is removed after the structure has been lifted off and before it is bonded to a glass slide.

Devices were fabricated prior to epitaxial liftoff by first depositing Cr/AuZn/Au p-type contacts onto the sample, followed by a mesa etch down to the n-type active region. After the evaporation of AuGe/Ni/Au contacts onto the n-type layer, the sample was annealed at 380°C, to make both contacts ohmic. The material and junction quality were monitored after each process step using nondestructive room temperature photoluminescence (6-8) and open-circuit voltage ($V_{oc}$) testing. No reduction in $V_{oc}$ was observed during the process sequence, indicating that the material properties did not degrade during processing.

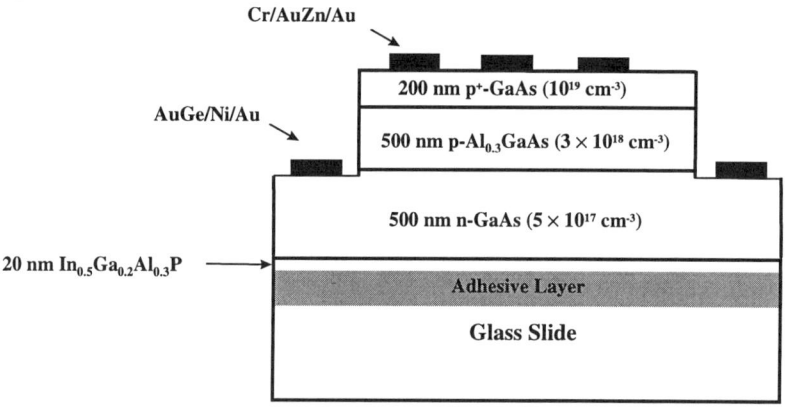

Figure 1. Structure of the solar cell after liftoff and bonding.

After device fabrication, the sample was covered with black wax. The thin-film GaAs solar cell was lifted off of its substrate by selectively laterally etching the buried AlAs layer in HF acid solution. Then the 100 nm thick GaAs protective layer between the InGaAlP window and AlAs epitaxial liftoff layer on the thin film is etched away using $H_2SO_4:H_2O_2:H_2O$ (1:8:500). Again, the device was monitored using absolute PL measurement before and after the removal of the GaAs layer to ensure the etching does not damage the cell. Finally the thin film is attached to a glass substrate with UV-curing polyurethane. The black wax was then dissolved and conductive copper wires were attached to the p- and n-type contacts be means of silver paint. The ELO process is summarized in Figure 2.

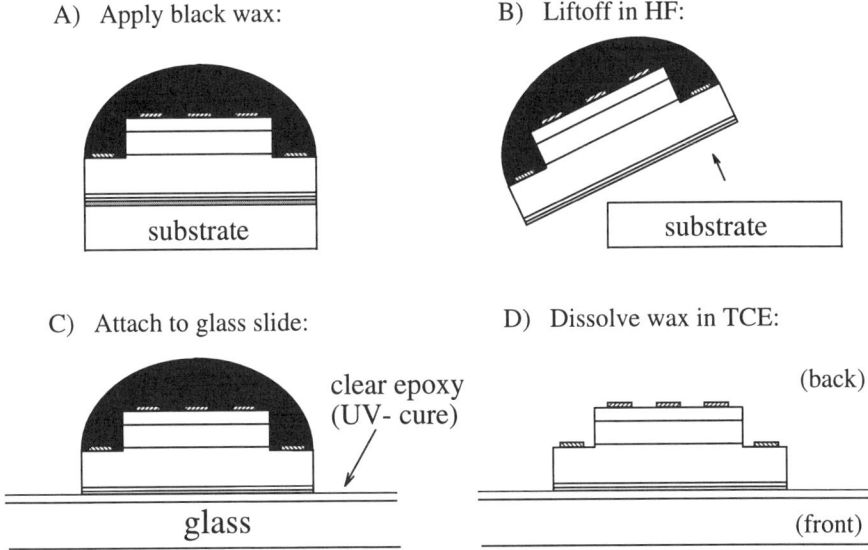

**Figure. 2**. Epitaxial liftoff process

## ELECTRICAL MEASUREMENTS

Electrical measurements were then performed on the lifted-off solar cell. The dark current-voltage (*I-V*) characteristic is shown in Figure 3. This plot exhibits a diode factor of $n = 1.93$ for voltages between 0.45 V and 0.9 V, which corresponds to nonradiative recombination current in the junction. The *I-V* curve is virtually identical to that obtained before liftoff, indicating little material degradation during that step. It is desirable for the operating point of the cell to occur in the $n = 1$ radiative recombination current regime, which is encountered at higher voltages. In this *I-V* curve, however, higher voltage regime is obscured by series resistance.

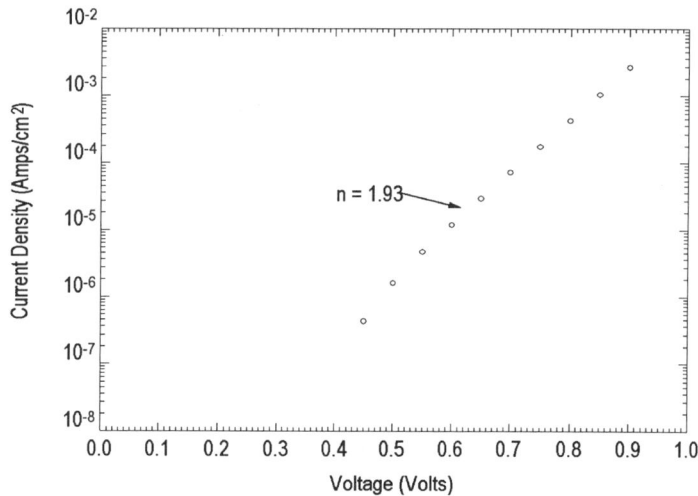

**Figure. 3.** Current-Voltage characteristic in the dark.

We also obtained $I$-$V$ curve under an illumination condition of $\approx$ 1 sun. The above solar cell is bifacial, in that it can be illuminated either through the glass slide (the "front" side) or from the contact ("back") side. The illumination was adjusted slightly above 1 sun to compensate for the absence of an anti-reflection coating between the InGaAlP window layer and the polyurethene adhesive layer. The result of the measurement is shown in Figure 4. Illumination from the front yields $V_{oc}$= 0.995 V and a short-circuit current $I_{sc}$ of 30.7 mA/cm$^2$. At an operating voltage of 0.85 V, the fill factor of this $I$-$V$ power curve is 77%. The superior performance in front illumination is of course due to the fact that incident light is not obstructed from the active region by metal contact pads. Moreover, reflection from the pads on the backside may enhance performance by recycling unabsorbed photons.

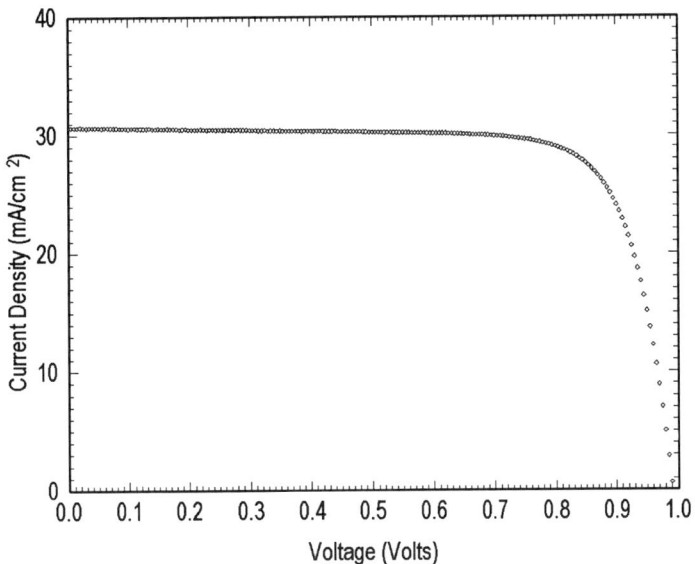

**Figure 4.** Current-Voltage characteristic under ≈1-sun illumination from the front side of the solar cell.

## DISCUSSION

Referring to Figure 3, it can be seen that the operating point lies in the undesirable $n = 2$ region of the *I-V* curve. This can be attributed edge leakage as has been thoroughly explored by the Purdue group (9). Another factor is the bandgap of the $In_{0.5}Ga_{0.2}Al_{0.3}P$ window layer. This layer was made with only 30% Al-content, in order to resist the HF acid. In principle a higher Al concentration can be used allowing more of the short wavelengths to be transmitted, but his is probably unnecessary since the InGaAlP window layer is only 20 nm thick. We believe that these results indicate good performance is possible in an epitaxial liftoff thin-film GaAs solar cell with a 500 nm active layer.

## NANOTEXTURING

Now we describe a process to enhance light absorption by our solar cells. Yablonovitch (10) showed that total randomization of light leads to an enhancement of absorption by a factor of $4n^2$. This result was confirmed experimentally by Deckman et al (11) by applying natural lithography technique to amorphous silicon solar cells. Our nano-texturing process applied to very thin GaAs/AlGaAs double heterostructure has achieved 90% of the enhanced absorption in the near band edge region. The structure of GaAs/AlGaAs sample used in our experiment is shown in Figure 5.

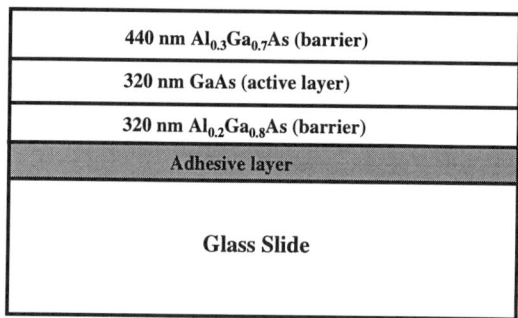

**Figure 5**. Structure used in nanotexturing experiment.

The double heterostructure sample is nano-textured by "Natural Lithography" (7) using commercially available carboxylate modified 0.953 µm polystrene spheres as ion etching mask. The spheres are diluted in 5.56% methanol such that we obtained 10.5% of spheres by volume. The diluted sphere solution is then spun onto the sample at 1700 rpm. The random close-paced array of polystrene spheres act as a mask on the sample surface. Then the sample is etched using reactive ion etching to impose the randomly closed packed sphere pattern onto the surface. The depth of the imposed pattern is 250 nm. After etching, the epitaxial liftoff is performed to separate the thin film from the substrate and then it is bonded to a glass slide with the untextured side against the glass.

The spectral dependence $R(\lambda)$ of reflectance of the textured and untextured films is then measured with respect to a white reference surface using a standard integrating sphere setup. Figure 6 shows the result of our measurement. The indicated 80% maximum attainable absorption is due to the reflection of the incident beam off the glass and the semiconductor adhesive interface.

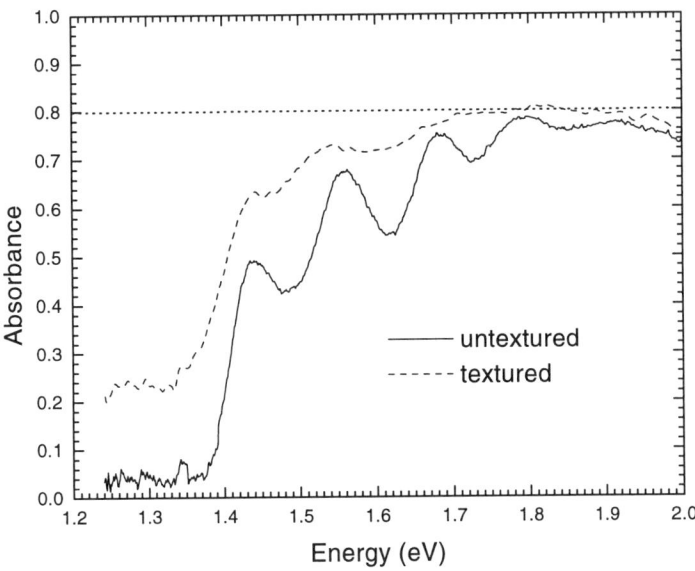

**Figure 6.** Absorption of a textured and untextured films.

Figure 7 shows the comparison between our data and theory. It indicates that our experimental data nearly reaches to 90% of the absorbance predicted by the theory based on the $4n^2$ expected absorption enhancement for light trapping (10). The absorbance oscillations which appear for energies above the band gap is due to Fabry-Perot oscillations which arise within the untextured sample. Residual Fabry-Perot oscillations occur also for nano-textured GaAs thin-films. This is the first time light trapping has been applied to a direct gap semiconductor of the order of one optical wavelength thick.

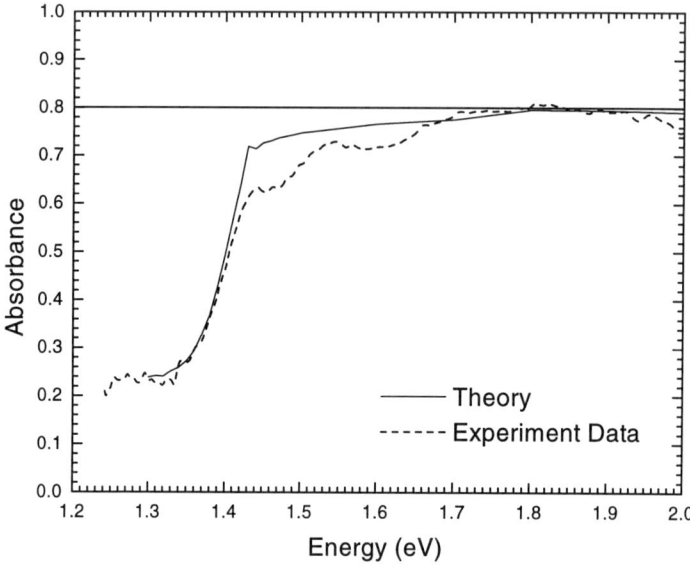

**Figure 7.** Comparison between experimental data and theory.

## CONCLUSION

In summary, we have integrated a working thin film GaAs solar cell onto a glass substrate, using the epitaxial liftoff technique. This configuration allows illumination through the non-contact side of the cell. The cell exhibits $V_{oc}$= 0.995 V and fill factor of 77%. We also present new light trapping parameters which are unique to an ultra thin direct bandgap solar cell. We believe that this combination of low cost through ELO, and high performance through light-trapping in ultra-thin films shows promise for terrestrial photovoltaics.

## REFERENCES

1. Yablonovitch, E., Stringfellow, G.B., and Greene, J.E., "Growth of Photovoltaic Semiconductors", *Journal of Electronic Materials*, **22**, 49-55, (1993)
2. Zahraman, K., Guillaume, J.-C., Nataf, G., Beaumont, B., Leroux, M., and Gibart, P., , "High-Efficiency $Al_{0.2}Ga_{0.8}As$ / Si Stacked Tandem Solar Cells using Epitaxial Lift-Off," *Japanese J. of Appl. Phys.*, **33**, 5807-5810, (1994).

3. G. B. Lush, M. P. Patkar, M. P. Young, M. R. Melloch, M. S. Lundstrom, S. M. Vernon, E. D. Gagnon, L. M. Geoffroy, and M. M. Sanfacon, "Thin film GaAs Solar Cells by Epitaxial Lift-off", presented at the *23rd IEEE Photovoltaic Specialists Conf.*, Louisville, KY, May 1993

4. Hageman, P.R., Bauhuis, G.J., van Geelen, A., van Rijsingen, P.C., and Giling, L.J., "Large Area Epitaxial Lift Off GaAs Solar Cells," presented at the *25rd IEEE Photovoltaic Specialists Conf.*, Washington, D. C., May 1996.

5. E. Yablonovitch, T. J. Gmitter, J. P. Harbison, and R. Bhatt, "Extremely Selectivity in the Lift-Off of Epitaxial GaAs Films," *Appl. Phys. Lett.*, **51**, 2222-2224, (1987).

6. Lee, X.Y., Wu, C. Q., Verma, A. K., Ranganathan, R., and Yablonovitch, E., "Non-Destructive Testing by Absolute Room Temperature Photoluminescence Quantum Efficiency of GaAs Solar Cells," presented at the *25rd IEEE Photovoltaic Specialists Conf.*, Washington, D. C., May 1996.

7. Sinton, R.A., Cuevas, A., "Contactless determination of current-voltage characteristics and minority-carrier lifetimes in semiconductors from quasi-steady-state photoconductance data," *Appl. Phys. Lett*, **69**, 2510-2512, (1996)

8. Rumyantsev, V.D., Rodriguez, J.A., "Luminescent characterization of contactless AlGaAs/GaAs heterostructure for solar cells and diodes", *Solar Energy Materials and Solar Cells*, **31**, 357-370, (1993)

9. Stellwag, T.B., Melloch, Mr.R., Lundstrom, M.S., Carpenter, M.S., Pierret R.F., "Orientation-dependent perimeter recombination in GaAs diodes"," *Appl. Phys. Lett.,* **56**, 1658-1660, (1990)

10. E. Yablonovitch, "Statistical Ray Optics", *J. Opt. Soc. Am*, **72**, 899-907 (1982)

11. Deckman, H.W., Wronski, C.R., Witzke, H., and Yablonovitch, E., *App. Phys. Lett.*, **42**, 968-970, (1983)

# MOLECULAR DYNAMICS MODELING OF HYDROGEN IN SILICON

Stefan K. Estreicher* and Peter A. Fedders[†]

*Physics Department, Texas Tech University, Lubbock, TX 79409
†Physics Department, Washington University, St. Louis, MO 63130

**Abstract.** Hydrogen-defect and hydrogen-impurity interactions determine the efficiency of passivation processes in solar cells. These interactions are investigated theoretically at (and near) the ab-initio level using a variety of tools. The hydrogen-vacancy studies started with H-monovacancy interactions. Ongoing work deals with the clustering of vacancies and will continue with the interactions between H and the most stable vacancy aggregates. The results discussed in the present paper deal with a remarkably stable and electrically/optically inactive defect, the ring-hexavacancy. As for hydrogen-impurity interactions, the current effort focuses on the role of O and H-enhanced diffusion of O. These interactions are the likely reason why the hydrogenation of O-rich material does not lead to an increase in the efficiency of the device.

## INTRODUCTION AND RESEARCH OBJECTIVES

Poly-Si solar cells often benefit substantially from the passivation by H of dangling bonds at vacancies, dislocations, grain boundaries, and other defects. However, the efficiency of O-rich samples is not improved by this process and, in all materials, some defects resist hydrogenation. Dislocations are sometimes referred to as 'good' (those that can be passivated) and 'bad' (those that cannot) dislocations. Such a primitive classification illustrates how poorly these defects are understood. Since they are one of key the efficiency-limiting factors, it is essential to better understand the fundamental interactions involved.

In addition to passivation reactions,(1,2) hydrogenation results in a number of other uncontrolled processes. One of them is the formation and growth of platelets.(3) Another is the enhancement by H of the diffusivity of interstitial oxygen. This process affects the electrical and optical properties of the material and may be the reason why the hydrogenation of CZ material does not lead to improved cell performance.

The research done at Texas Tech University addresses these issues from a theoretical (ab-initio) perspective. A variety of approaches are used to calculate a wide range of properties of defects and complexes. Static properties are calculated at or near the ab-initio Hartree-Fock (HF) level in molecular clusters: equilibrium geometries, electronic configurations, potential energy surfaces, dissociation energies, diffusion paths, activation energies for diffusion, etc. Other properties require the use of dynamic methods: the kinetics of complex formation and dissociation, vibrational frequencies, and diffusion properties are calculated with the ab-initio (tight-binding) molecular dynamics (MD) package developed by Sankey and co-workers.(4,5) In this approach, the host crystal is represented by periodic supercells of 64 atoms. For details about the theoretical techniques and the cluster and supercells approximations, see Ref. (1).

The results obtained in previous years dealt with the diffusion of interstitial H and of monovacancies, H-vacancy complex formation and diffusion, and the calculation of infra-red (IR) active vibrational modes of $\{V, H_n\}$ complexes for experimental identification.(6,7) Ongoing research focuses on vacancy-vacancy and hydrogen-oxygen interactions. This report discusses two of the most important results obtained so far.

First is the realization that vacancy-vacancy interactions lead to the formation of a very stable defect, the ring-hexavacancy, which is electrically and optically inactive.(8) But it may grow. It is planar and trigonal (the $C_3$ axis is the <111> axis of the crystal), resembles a tiny prismatic dislocation, and could be a precursor in the formation of a range of extended defects, including platelets. The large stability of this defect makes it unlikely that it would dissociate during thermal anneals. Further, it is a large void in the crystal and may be a strong gettering center.

Second is a MD simulation of H-enhanced diffusion of oxygen. These are the first successful MD simulations of this process.(9) The results provide an obvious explanation as to why hydrogen does not improve CZ-Si solar cell material. The full consequences of our results are still unclear, and many difficult open questions remain.

## VACANCY-VACANCY INTERACTIONS

We used MD simulations to find all the possible configurations of aggregates of N-vacancies ($V_N$), with $N = 1, \cdots, 6$. We removed all the combinations of symmetrically inequivalent sets of N silicon atoms from a 64-atoms periodic supercell and quenched. This procedure forces convergence toward a local minimum of the potential energy and allows a comparison of the (fully relaxed) structures and energies of inequivalent $V_N$ complexes.

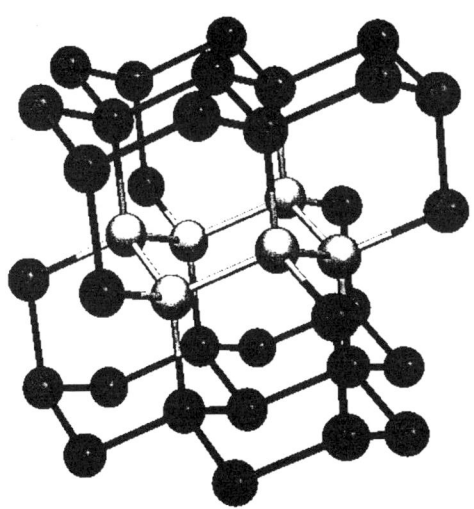

**FIGURE 1.** Hexagonal ring in the perfect crystal. The (fully relaxed) lowest-energy configurations of $V_N$, with $N = 1, \cdots, 6$ occur when Si atoms are successively removed from the ring.

The lowest-energy configuration of all the $V_N$'s occurs when Si atoms are removed successively from a hexagonal ring in the crystal (Fig. 1). $V_1$, $V_2$ and $V_3$ have only one possible configuration. $V_4$ and $V_5$ have several, but only one of the metastable structures is close in energy ($\sim +0.1$ eV) to the one where the atoms come off of a ring.(10) The lowest-energy configuration of $V_6$ (missing hexagonal ring) is much more stable than any other of the fifteen or so configurations of $V_6$ (the next one is at $\sim +0.9$ eV).(11)

The lowest-energy configurations of the $V_N$'s were re-optimized at the HF level using gradient techniques with no symmetry assumption. This provides a second set of geometries and energetics, which can be compared to the ones obtained independently at the MD level. Such comparisons allow theoretical predictions to be made with a much higher degree of confidence. Further, the ab-initio HF outputs also contain a lot of chemical details about the rebonding taking place: degrees of bonding and bond indices, overlap populations, Mulliken charges, etc.

The formation energies $\Delta E_N = \{E_N + E_0\} - \{E_{N-1} + E_1\}$ are shown in Fig. 2. $E_N$ is the energy of the most stable (fully reconstructed) $V_N$ aggregate. Thus, $\Delta E_N$ is the energy gained by adding an isolated monovacancy to an aggregate of N–1 vacancies. The energy of the heptavacancy was obtained by removing one Si atom adjacent to the missing hexagonal ring and allowing the crystal to reconstruct. The open circle shows the experimental binding energy of the divacancy (1.5 eV).(12)

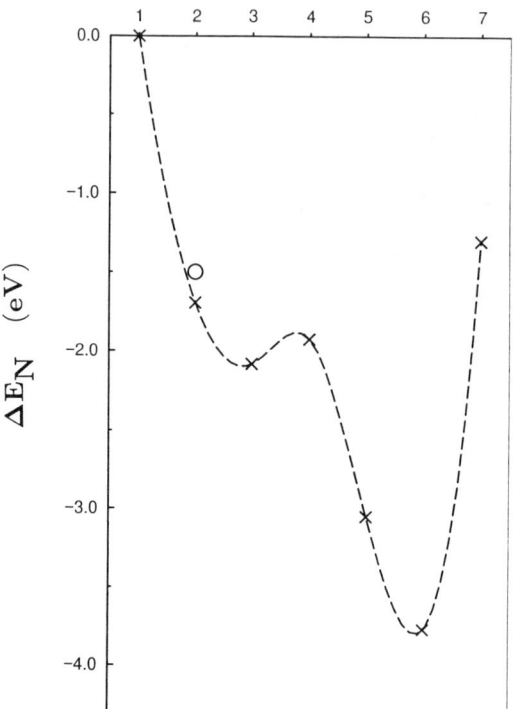

**FIGURE 2.** Energy difference between the lowest-energy configuration of $V_N$ and $V_1$ plus $V_{N-1}$. The hexavacancy is by far the most stable defect. The open circle is the experimental(12) binding energy of the divacancy.

$V_6$ is remarkable not only for its stability but also because the crystal is able to reconstruct almost perfectly around it. This results in no 'dangling-bond-like' character for this defect. The ring configuration of the hexavacancy is not an electron-hole recombination center.

A comparison of the energy eigenvalues of the perfect cluster to those of $V_N$, $N = 1, \cdots, 7$ (Fig. 3), shows that the top of the valence band remains unchanged and that there are no deep levels in the gap. Only the bottom of the conduction band is affected by $V_6$ ('band-tailing'). This band-tailing involves only unoccupied levels. The same result is obtained at the HF and MD levels, although the band tailing is most pronounced in the MD calculations. In contrast, all the other vacancy aggregates have deep levels clearly visible in the gap. While no *quantitative* predictions of the position of energy levels in the gap should be made, the *qualitative* features imply that $V_6$ is not an electron-hole recombination center and should be electrically inactive. The band-tailing suggests that a broad photoluminescence (but no sharp lines) is the most one could expect from this defect.

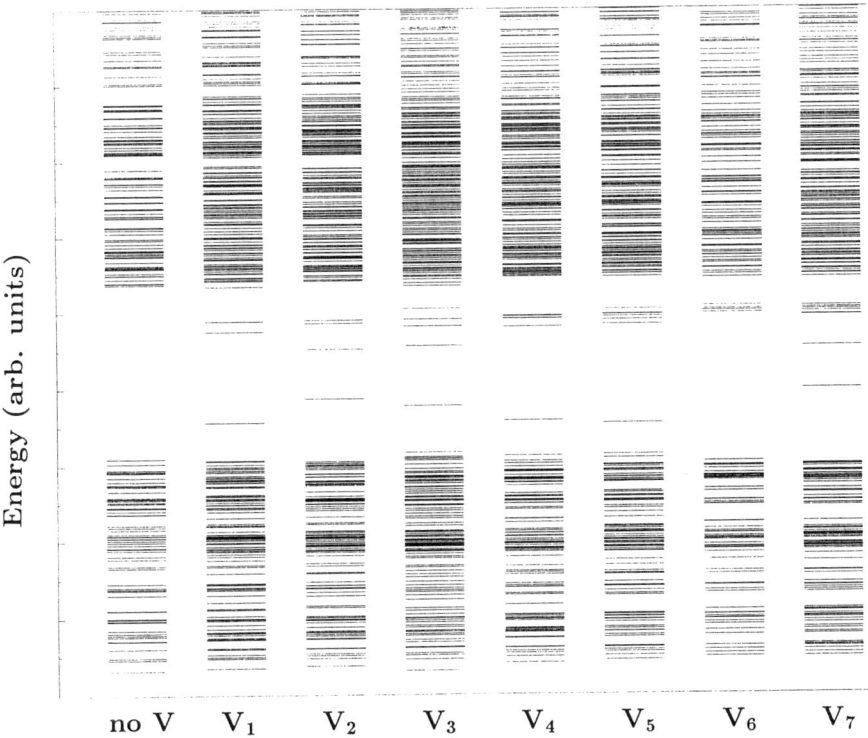

**FIGURE 3.** Energy eigenvalues near the band gap calculated at the approximate ab-initio HF level for the perfect cluster and the $V_N$'s, $N = 1, \cdots, 7$. The hexavacancy has no deep levels in the gap. It shows some band-tailing from the bottom of the conduction band (unoccupied levels).

The reconstructed hexavacancy is almost planar and has trigonal symmetry. The charge distribution calculated at the ab-initio HF level indicates that $V_6$ has the smallest dipole moment of all the vacancy aggregates we considered. The net dipole moment points along the trigonal axis but has negligible magnitude ($\sim 0.0007$ atomic units), implying that $V_6$ should not be IR active.

$V_6$ resembles a small prismatic dislocation loop.(13) It could also be the fundamental building block of hydrogen-related platelets.(3) Since it is a large void in the crystal, it could also serve as a gettering center for impurities such as H, O, or selected transition metals.

# HYDROGEN–ENHANCED DIFFUSION OF OXYGEN

Interstitial O is an abundant impurity in Si. It sits at a relaxed, puckered bond-centered (BC) site. The strong Si–O bonds result in a large activation energy for diffusion, 2.59 eV. Following thermal anneals above 450 °C, O precipitates into (unidentified) series of double donors called thermal donors (TDs). There is ample experimental evidence that H considerably enhances the formation rate of TDs. The earliest observation is in Ref. (14), recent work in Refs. (15) and (16), and a review in Ref. (17).

There is no direct experimental information on the nature H–O interactions (only the TD formation kinetics are seen). However, there are strong indications that atomic hydrogen is the key species, and that H concentrations as low as $10^{13}\,cm^{-3}$ are sufficient for the process to take place in samples containing some $10^{18}\,cm^{-3}$ interstitial O. Further, H acts as a catalyst: it normally does not participate in the final state, but lowers the reaction energies. The estimate is that the activation energy for diffusion of interstitial O drops from $2.59\,eV$ (no H) to about half that value.

There is also experimental evidence that O dimers are involved. Nobody really knows what the $\{O, O\}$ pairs are, but they diffuse much faster than isolated O does. It is likely that dimer formation and diffusion is the rate-limiting factor in TD formation. It is possible that the role of H is simply to enhance the formation rate of the dimers. Thus, H would interact with a single O and somehow force it to jump around at rather low temperatures.

IR (Ref. (18)) and muon spin rotation ($\mu$SR) (Ref. (19)) show that H is attracted to interstitial O. The IR work shows evidence of a correlation (97%) between the intensities of several Si–H vibrational lines and the stretching mode of interstitial O. This is interpreted as meaning that H often ends up in the immediate vicinity of O in the bulk.

The $\mu$SR work shows that the spectra are qualitatively different in FZ (O-poor) and CZ (O-rich) material. In particular, the tetrahedral interstitial (T) muonium is not seen in CZ material, but channeling data suggest that it is rapidly moving around interstitial O.

Finally, recent work in Ref. (20) shows unexpected (and unexplained) behavior of the concentration of TDs vs. depth under conditions of continuous hydrogenation for many hours at 350 °C. The profiles are 'box-like' with sharp edges. The concentration of thermal donors does not saturate, and the kinetics imply that the process is controlled by H (or D).

Three static ($0\,K$) studies(21-23) propose mechanisms by which H could lower the activation energy for diffusion of isolated interstitial O. While the three models differ on details, they all show that if H is near a Si atom adjacent to interstitial O, it can substantially lower the activation energy of O diffusion by saturating a Si dangling bond in the saddle point configuration.

There is only speculation as to *how* H gets near O, *why* it remains near O long enough for one jump to occur, and what happens *after* O has jumped.

Our MD simulations(9) imply that the situation is much more interesting. The calculations confirm that H is attracted to interstitial O in both the BC and the T configurations, and self-traps in the vicinity of O. Since our calculations simulate only a few picoseconds (real time, with a time step of 0.2 fs), we do not know at present if H needs longer times to get very close to O or if several H interstitials must reach the same O before a catalytic reaction is initiated. However, we observed H-enhanced diffusion of O by forcing the reaction to occur within computationally reasonable times. We provided a 'kick' to O while placing H at an adjacent T site. The kick consists in compressing the puckered Si–O–Si bond. Several steps in the resulting simulation are drawn in Fig. 4.

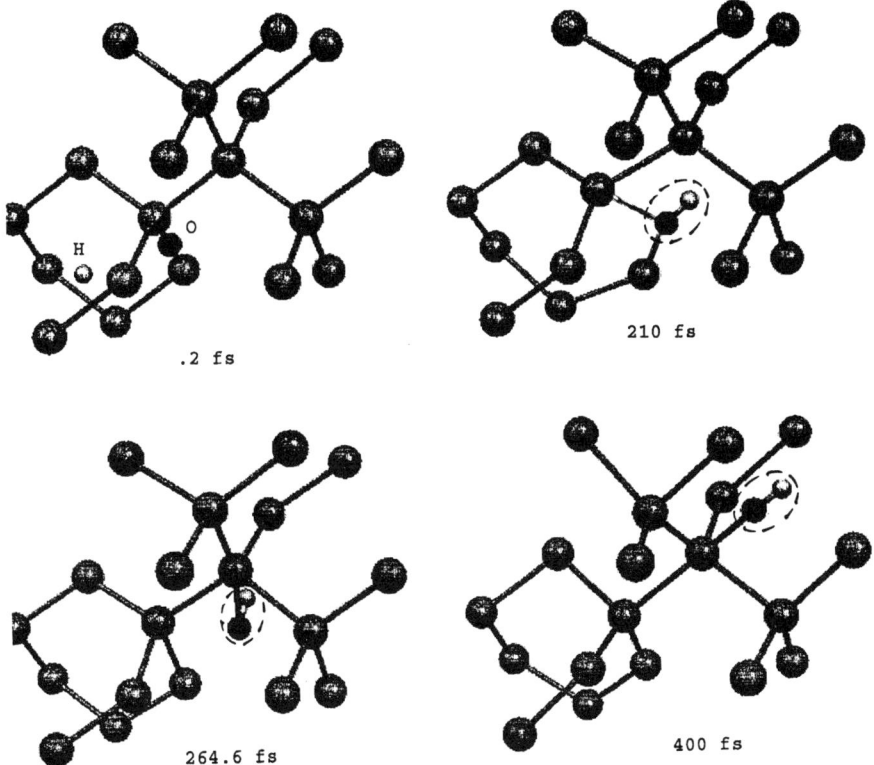

**FIGURE 4.** MD simulation of H-enhanced diffusion of oxygen. O starts at an unrelaxed BC site and H at a T site. H attaches to O, then the pair becomes mobile. The times are shown in (fs). Only a few of the 64 Si atoms are drawn and, except for the first figure, the OH pair is marked with a dashed oval.

If H is placed near O, it attaches to O almost immediately. This makes sense since the formation of an H–O bond (bond strength $\sim 5.2$ eV) should be favored over the formation of an H–Si bond (bond strength $\sim 3.7$ eV). The formation of the H–O bond results in the replacement of the configuration where O is attached to two (fixed) Si atoms by one where O is attached to one (fixed) Si atom and to H. At this point, the {O,H} pair begins to oscillate around the (fixed) Si atom. If given the kick described above, the {O,H} pair can be observe to jump through the crystal within a few picoseconds.

## SUMMARY AND CONCLUSIONS

Many calculations regarding vacancy aggregates are still in progress. However, the implications of the results we already have are as follows:
1. Vacancy-vacancy interactions lead to the growth of vacancy aggregates: One always gains energy by adding one vacancy to a cluster of vacancies.
2. The most stable vacancy aggregate is the ring-hexavacancy. Its stability suggests that it will not dissociate during thermal treatments. However, it grows by capturing additional vacancies (see Fig. 1).
3. The hexavacancy is electrically and optically inactive: it has no deep level in the gap and its dipole moment is almost zero.
4. The hexavacancy is planar and has trigonal symmetry, which reminds us of platelets and some types of dislocations. It is plausible that $V_6$ is a precursor or fundamental block of such extended defects.
5. The ring-hexavacancy is a large void in the crystal. It could be an efficient gettering center for H, O, or some transition metals.

The results of MD simulations obtained so far on hydrogen-oxygen interactions provide the following evidence.
1. There is a long-ranged H–O attraction,
2. if H gets close enough to bond-centered oxygen, {O,H} pair formation results.
3. the {O,H} pair undergoes much larger oscillations than O alone,
4. given a kick to speed up the reaction, the {O,H} pair diffuses rapidly.

None of these results could have been obtained experimentally. However, some consequences of the calculations may be verified in the laboratory provided that one is searching for the correct quantity. For example, theory predicts that $V_6$ is mostly invisible. However, it may become IR active under stress. Further, it may be a strong trap for interstitial H, with some characteristic stretching frequency, or H-induced electrical activity.(23) We intend to pursue our research along these lines. As concerning H–O interactions, much work remains to be done, especially regarding O dimers. This work involves very large amounts of CPU time and is of long-term nature.

## ACKNOWLEDGEMENTS

The work of SKE is supported in part by NREL under contract XAX-5-15230-01 and the grant D-1126 from the R.A. Welch Foundation. The work of PAF is supported in part by the NSF grant DMR 93-05344.

## REFERENCES

1. Estreicher S.K., Mat. Sci. Engr. R **14**, 319-412 (1995).
2. Sopori B.L., Deng X.J., Benner J.P., Rohatgi A., Sana P., Estreicher S.K., Park Y.K., and Roberson M.A., Solar En. Mat. and Solar Cells **41/42**, 159 (1996).
3. Sopori B.L., Jones K., and Deng X.J., Appl. Phys. Lett. **61**, 2560 (1992).
4. Sankey O.F. and Niklewski D.J., Phys. Rev. B **40**, 3979 (1989).
5. Sankey O.F., Niklewski D.J., Drabold D.A., and Dow J.D., Phys. Rev. B **41**, 12750 (1990).
6. Roberson M.A. and Estreicher S.K., Phys. Rev. B **49**, 17040 (1994).
7. Park Y.K., Estreicher S.K., Myles C.W., and Fedders P.A., Phys. Rev. B **52**, 1718 (1995).
8. Estreicher S.K., Hastings J.L., and Fedders P.A., submitted to Appl. Phys. Lett.
9. Estreicher S.K., Park Y.K., and Fedders P.A., NATO ARW # 950709, ed. Jones R., in print.
10. This was predicted using an empirical 'dangling-bond-counting' model by Chadi D.J. and Chang K.J., Phys. Rev. B **38**, 1523 (1988), and Oshiyama A., Saito M., and Sugino O., Appl. Surf. Sci. **85**, 239 (1995).
11. Estreicher S.K., Hastings J.L., and Fedders P.A., unpublished.
12. Corbett J.W., *'Electron Radiation damage in Semiconductors and Metals'*, Academic (New York, 1996) and Refs. therein.
13. Hirth J.P. and Lothe J., *Theory of Dislocations* (Krieger Publ., Malabar FL, 1992), p.614.
14. Fuller C.S. and Logan R.A., J. Appl. Phys. **28**, 1427 (1957).
15. Brown A.R., Claybourn M., Murray R., Nandhra P.S., Newman R.C., and Tucker J.H., Semic. Sci. Technol. **3**, 591 (1988).
16. Stein H.J. and Hahn S.K., Appl. Phys. Lett. **56**, 63 (1990).
17. Newman R.C. and Jones R. in *"Oxygen in Silicon"*, ed. Shimura F.,

Semic. and Semimet. **42** (Academic, NY, 1994).
18. Qi M.W., Bai G.R., Shi T.S., and Xie L.M., Mater. Lett. **3**, 467 (1985).
19. Patterson B.D., Rev. Mod. Phys. **60**, 69 (1986), see in particular pages 110 and 131).
20. Stein H.J. and Hahn S.K., J. Electrochem. Soc. **142**, 1242 (1995).
21. Estreicher S.K., Phys. Rev. B **41**, 9886 (1990).
22. Jones R., Öberg S., and Umerski A., Mat. Sci. Forum **83-87**, 551 (1992); Jones R., Phil. Trans. R. Soc. Lond. A **350**, 189 (1995).
22. Pantelides S.T. and Ramamoorthy M., unpublished.
23. Hartung J. and Weber J., Phys. Rev. B **48**, 14161 (1993); Appl. Phys. Lett. **77**, 118 (1995).

# High Concentration Low Wattage Solar Arrays and Their Applications

Robert Hoffmann,* Joseph O'Gallagher[†] and Roland Winston[†]

*Midway Labs, Inc., 350 N. Ogden Avenue, Chicago, Illinois 60607;
(312) 432-1796; Fax (312) 432-1797. [†]University of Chicago

**Category:** Research and Development; Photovoltaic Concentrators

**Abstract:** Midway Labs currently produces a 335x concentrator module that has reached as high as 19% active area efficiency in production. The current production module uses the single crystal silicon back contact SunPower cell. The National Renewable Energy Lab has developed a multi junction cell using GaInP/GaAs technologies. The high efficiency (>30%) and high cell voltage offer an opportunity for Midway Labs to develop a tracking concentrator module that will provide 24 volts in the 140 to 160 watt range. This voltage and wattage range is applicable to a range of small scale water pumping applications that make up the bulk of water pumping solar panel sales.

## INTRODUCTION

Midway Labs is currently manufacturing a solar photovoltaic concentrator module based on the Sandia Laboratories SBM III design. The MLB4316-115 module uses sheet radial fresnel lenses to focus the suns rays on small single crystal silicon solar cells. The major improvement to this design by Midway Labs is the addition of a glass secondary optical element to improve the module performance.

The secondary is a patented design by University of Chicago Professors Roland Winston and Joseph O'Gallagher. The secondary is a non-imaging total internal reflector secondary optical element (SOE) molded of optical lens glass. The SOE is used to increase the allowable tracking error of the module assembly by a factor of 7 to 8 over a module that does not use a secondary. Another characteristic of the SOE is the uniform distribution of the light flux onto the surface of the solar cell which partially compensates for the transmission loss through the glass of the SOE. This allows increased tracking error without sacrificing performance. (1).

The SOE in current use was specifically designed for the Midway Labs application to be effective and inexpensive to manufacture. This design greatly increases the tolerances of the module assembly and reduces the complexity and cost of the tracking system. The allowable tracking error of the current design is +2.0 degrees at 335x concentration. Differential thermal expansion of the module causes up to +1.0 degree misalignment over the ambient temperature range of -40 deg. C. to +40 deg. C. Manufacturing tolerances require +0.25 deg. and the tracking tolerance is +0.25 deg. This allows some increase in the concentration ratio without increasing tracker costs.

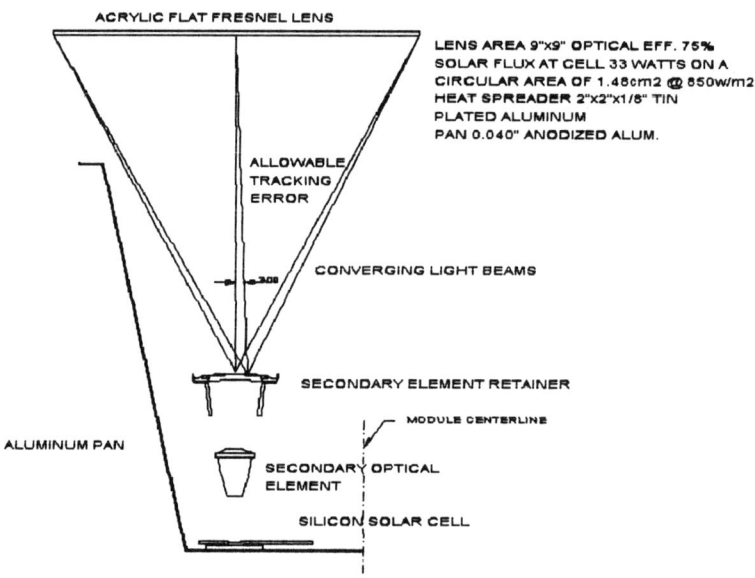

**FIGURE 1.** Optical schematic of current production MLB3416-115 (115 Watt Concentrator Module).

Due to the low voltage of single crystal cells the smallest array we can deliver economically is 230 watts at 12 volts nominal. This is 32 cells in series. The high current low voltage output of this array limits the ability to enter the small scale water pumping market. Because of the cost of high current wiring and controller components, it is natural to go to higher voltage and lower amperage in pump design. Systems commonly in service use flat panels and tracker operating at 24 volts nominal directly connected to a DC pump and no battery storage is used. Tracking flat panels are used to keep the pumps operating at their design input to maximize efficiency over a longer portion of the day than is possible for fixed flat panels. A tracking high voltage, high efficiency, high concentration module can offer significant efficiency and price advantage over the current technology.

There are some technical hurdles to be overcome before this GaInP/GaAs technology can be successfully used in a point focus application. The primary concern is the effect of chromatic aberration on the current matching of the two junctions as the light passes through the SOE. Sarah Kurtz of NREL determined that this effect must be evaluated using prototypes due to the difficulty in measuring the SOE output. The difficulties arise from the need of the SOE to be directly optically bonded to the cell and due to the small area of the light image from the SOE (2).

This paper reports on the initial evaluation of the cell and SOE configuration and an estimate of the market this product could serve.

## MODULE CONFIGURATION

The proposed module will build on Midway Labs' 10 years of developing concentrator arrays. With the latest generation of tracker and controller in service and performing well it has been determined that the concentration ratio of the proposed module can be increased from the present 335x to 560x without impacting reliability or ease of array assembly. With the SOE designed used, the increase in concentration ratio will result in the decrease in angle of acceptance from + 2.0 deg. to + 1.75 deg. The calculated output of the module is based on NREL testing and experience. The calculated I-V curve is shown below.

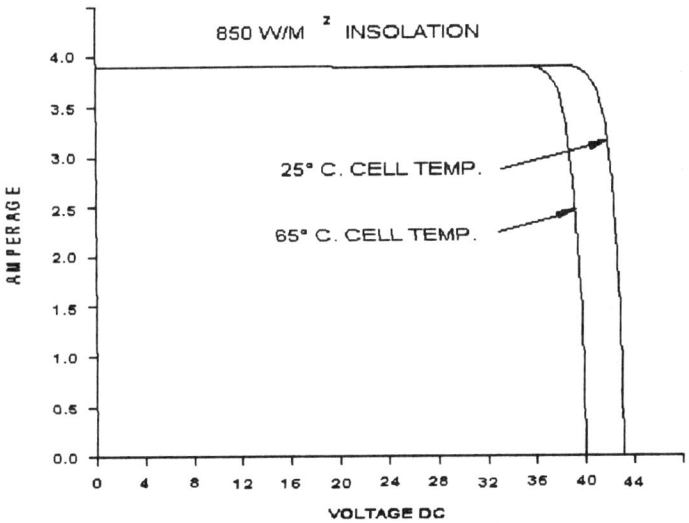

IV CURVE FOR 16 CELL 560x CONCENTRATOR MODULE

**FIGURE 2.** Calculated output of GaAS Cell Concentrator Module.

This module output is based on 16 cells in series and a 9"x9" primary fresnel lens that is used in the current production module. Test cells to evaluated this proposed configuration are being manufactured by NREL at this time and will be tested soon.

The module will require two-axis tracking and that tracking is shown in Figure 3.

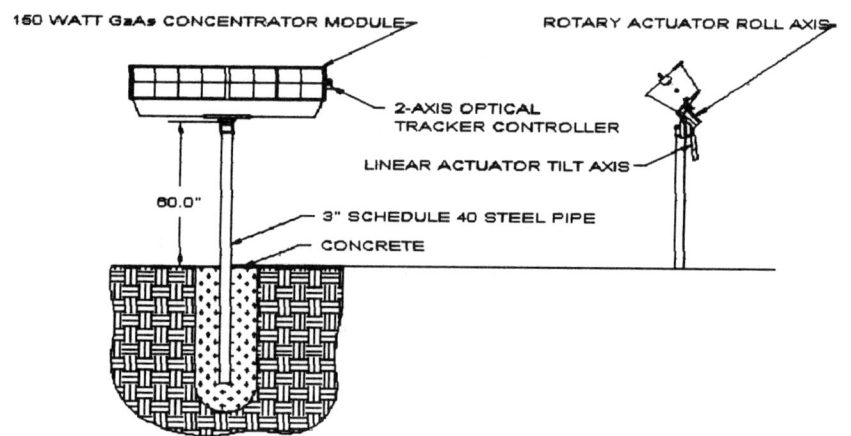

**FIGURE 3.** Proposed two-axis single module tracker configuration.

The high voltage output and relatively low amperage of this GaAs module relative to the current production module will allow this module to be used with the current production DC water pumps that are now on the market. As an example, if this single module array is mated with the Sunrise™ Model 5218 DC Submersible Water Pump the system would deliver about 2 gallons per minute of water from 150 feet of vertical lift. Based on the 6 hours of average available sunlight in Nebraska the system would deliver 720 gallons per day. This would supply water for 36 cow/calf pairs in a ranching application.

This type of application represents a large potential market for tracking concentrator arrays. Many other DC pump combinations are available for an array with the proposed output.

## POTENTIAL MARKET

The potential market for low cost photovoltaic power from concentrator modules in high volume production is well documented and generally accepted. The Utility Photovoltaic Group (UPVG) estimates there is a U.S. market for 9 Gigawatts of photovoltaic power if the installed price drops below $3.00/watt. Concentrator PV arrays can meet this pricing goals (3). The difficulty becomes in competing with conventional flat panels in lower volume production for market share which will allow the increased production needed to offer the low pricing that the large utility market requires.

A large potential market niche for concentrators is for off grid water pumping. The size of the off grid PV water pumping market is estimated by the Utility Photovoltaic Group to be 30 to 40 megawatts in the United States alone (4). A large percentage of PV water pumping applications currently use tracked flat panels so there is no market barrier to using tracked PV arrays. The centrifugal pumps used in these applications have a relatively small range of power input that allows them to operate at their rated efficiency. Operating outside of this required power greatly reduces their output. Trackers are used to keep the PV panels at their maximum output for longer hours per day than fixed panels thus increasing the pumping hours at optimum pump conditions. The markets experience with tracking in these applications reduces the reluctance to use concentrators which require tracking to function.

Low cost tandem junction cells are available from a satellite solar cell manufacturer using the scrap edge of current production 1 sun space cells. Performance may be lowered somewhat from the ideal GaAs configuration in terrestrial use due to the germanium substrate used in space cells. The cell manufacturer indicates this scrap material would be sufficient to meet 7 megawatts/year of concentrator module production. Cost per watt of 1 module arrays in production is estimated to be less than $5.00 per watt at production rates of 200 kilowatts/year. At production rates of 1 megawatt per year the tracking array cost per watt drops to less than $3.50. This pricing is very competitive with current flat panel and tracker requirements.

## CONCLUSION

The introduction of a high voltage concentrator array will allow the needed "creative marketing" necessary to carry the production of levels that will allow pricing to reach the <$3.00 per watt price. At this price the U.S. utility market is estimated to be 9 Gigawatts. By competing with the higher cost flat panels in growing market share much more realistic cost goals can be met at the production rate is increased. The world market is primarily low wattage applications and the lower the wattage a tracking concentrator array that can be produced at a competitive price the greater the market share available.

## REFERENCES

1. Ning, X.; Winston,R.; and O'Gallagher, J., Applied Optics **26**, p. 300 (1987).
2. Kurtz, S.; Friedman, D.J. and Olson, J.M., "The effect of chromatic aberrations on two-junction, two-terminal devices in a concentrator system," poster session 3P1 presented at 1st World Conference on PV Energy Conversion.
3. Friedman, D. J. and Kurtz, S., "Market and technical status and prospects for concentrator PV technology," presented at NREL PV Review Meeting, Nov. 4, 1996.
4. Utility Photovoltaic Group, SSA Opportunity Notice, p. 3, May 8, 1995.

# Plasma Processing Applied to Crystalline-Silicon Solar Cells

D. S. Ruby[1], C. B. Fleddermann[2], M. Roy[3] and S. Narayanan[3]

[1] *Sandia National Laboratories, Albuquerque, NM 87185-0752*
[2] *University of New Mexico, Albuquerque, NM 87131*
[3] *Solarex (a business unit of Amoco/Enron Solar), Frederick, MD 21701*

**Abstract.** We studied whether plasma-etching techniques can use standard screen-printed gridlines as etch masks to form self-aligned, patterned-emitter profiles on multicrystalline-silicon (mc-Si) cells from Solarex. We conducted an investigation of plasma deposition and etching processes on full-size mc-Si cells processed in commercial production lines, so that any improvements obtained would be immediately relevant to the PV industry. This investigation determined that reactive ion etching (RIE) is compatible with using standard, commercial, screen-printed gridlines as etch masks to form self-aligned, selectively doped emitter profiles. This process results in reduced gridline contact resistance when followed by plasma-enhanced chemical vapor deposition (PECVD) treatments, an undamaged emitter surface easily passivated by plasma-nitride, and a less heavily doped emitter between gridlines for reduced emitter recombination. This allows for heavier doping beneath the gridlines for even lower contact resistance, reduced contact recombination, and better bulk defect gettering. Our initial results found a statistically significant improvement of about half an absolute percentage point in cell efficiency when the self-aligned emitter etchback was combined with a PECVD-nitride surface passivation treatment. Some additional improvement in bulk diffusion length was observed when a hydrogen passivation treatment was used in the process. We attempted to gain additional benefits from using an extra-heavy phosphorus emitter diffusion before the gridlines were deposited. However, this required a higher plasma-etch power to etch back the deeper diffusion and keep the etch time reasonably short. The higher power etch may have damaged the surface and the gridlines so that improvement due to surface passivation and reduced gridline contact resistance was inhibited.

## INTRODUCTION

The use of plasma-enhanced chemical vapor deposition (PECVD) as a low-temperature surface passivation technique for silicon solar cells is a topic of increasing importance. PECVD is now widely recognized as a potentially cost-effective, performance-enhancing technique that can provide surface passivation

---

The work at Sandia National Laboratories was supported by the U.S. Department of Energy under contract DE-AC04-94AL85000.

and produce an effective antireflection coating layer at the same time [1]. For some solar-grade silicon materials, it has been observed that the PECVD process results in the improvement of bulk minority-carrier diffusion lengths as well, presumably due to bulk defect passivation [2].

In order to gain the full benefit from improved emitter surface passivation on cell performance, it is necessary to tailor the emitter doping profile so that the emitter is lightly doped between the gridlines, but heavily doped under them [3]. This is especially true for screen-printed gridlines, which require very heavy doping beneath them for acceptably low contact resistance. This selectively patterned emitter doping profile has historically been obtained by using expensive photolithographic or screen-printed alignment techniques and multiple high-temperature diffusion steps [3,4].

We have attempted to build on a self-aligned emitter etchback technique first described by Spectrolab [5]. In addition to the gridline-masked, plasma-etchback of the emitter they developed, we have included PECVD-nitride deposition because the low-recombination emitter produced by the etchback requires good surface passivation for improved cell performance. The nitride also provides a good antireflection coating and can be combined with plasma-hydrogenation treatments for bulk defect passivation.

## EXPERIMENTAL PROCEDURE

These cells were made using Solarex cast multicrystalline silicon and received Solarex's standard production line processing through the printing and firing of the gridlines, except for some cells which received an extra-heavy phosphorus emitter diffusion. Then, the cells underwent reactive ion etching (RIE) for 3 to 4 minutes to increase the sheet resistance of the emitters to 80-100 ohms/square. The self-aligned selective-emitter plasma-etchback and passivation process is shown in Figure 1. The cells were plasma-etched in a Technics PE II-A reactor using pure $SF_6$ at a power of 10W or 50W and a pressure of 100 mTorr. Then, the cells received either an ammonia-plasma hydrogenation (H-passivation) treatment or a silicon-nitride deposition (PECVD-nitride), both found to be effective for bulk and surface passivation in String Ribbon™ mc-Si [2]. Nitride-coated cells then received a forming gas anneal (FGA) at 300C for 30 minutes. The cells were then returned to the Solarex production-line for final cell processing.

The plasma-nitride depositions and H-passivation treatments were performed using a modified Pacific Western Coyote PECVD reactor. This is a commercial, RF parallel-plate reactor operating at 13.56 MHz with large batch-size and high-throughput potential. Reaction gases for nitride deposition were a 3% mixture of silane in nitrogen and pure ammonia. The H-passivation treatment consisted of an exposure to a pure ammonia plasma at 350C for 15 minutes.

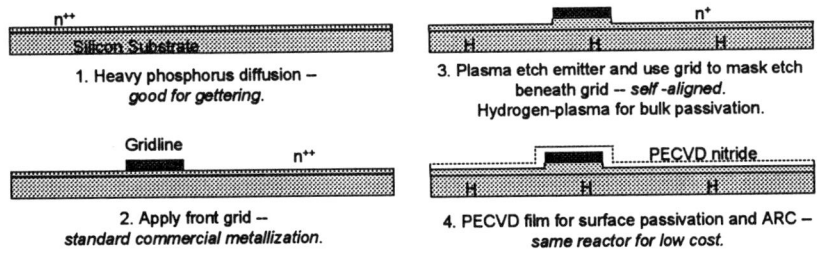

**FIGURE 1.** Process sequence for self-aligned emitter etchback. The emitter etchback can be performed after the hydrogen-plasma treatment to remove surface damage. However, in this work, the plasma-etching was performed before the plasma hydrogenation step.

## RESULTS

The initial experiment fabricated cells with four different process sequences (Table 1). One sequence consisted of the Solarex's standard process, while the other three sequences examined various aspects of the plasma-processing sequence depicted in Figure 1. This experiment used Solarex's standard emitter diffusion.

**TABLE 1.** Four processing sequences were applied to four groups of twelve 102.6-cm$^2$ Solarex mc-Si cells using matched material from the same ingot and in most cases with the same grain structure. The average illuminated IV parameters and their standard deviations are shown below.

| Eff. (%) | $I_{SC}$ (A) | $V_{OC}$ (mV) | FF (%) | $R_S$ (m$\Omega$) |
|---|---|---|---|---|
| Group 1. Control Cells: No emitter etchback, TiO$_2$ ARC | | | | |
| 12.6±0.1 | 2.91±.02 | 586±1 | 75.5±0.7 | 9.2±0.5 |
| Group 2. Plasma Etchback, TiO$_2$ ARC | | | | |
| 12.2±0.1 | 2.93±.02 | 580±1 | 73.4±0.4 | 15.0±1.0 |
| Group 3. Plasma Etchback, H-passivation, TiO$_2$ ARC | | | | |
| 12.8±0.3 | 2.97±.02 | 585±1 | 75.4±1.5 | 10.7±0.5 |
| Group 4. Plasma Etchback, PECVD-nitride ARC, FGA | | | | |
| 13.0±0.1 | 3.00±.01 | 587±1 | 75.3±0.2 | 10.7±0.5 |

The cells from Group 2 suffered an efficiency loss due primarily to loss of $V_{OC}$, which is expected since the etched-back emitter is now transparent to minority carriers and the front surface is not passivated. The loss in FF is due to the increase in series resistance, which is due to the higher sheet resistance of the etched-back emitter. In an optimized sequence, the cells would have more closely-spaced gridlines to compensate for the higher sheet resistance. In addition, an optimized sequence might use an extra-heavy emitter doping. The heavier doping could possibly result in additional gettering of bulk impurities, which could then be etched away. Also, heavier doping under the gridlines would better isolate them and reduce contact recombination. Finally, the heavier doping would also reduce the contact resistance that often limits screen-printed cell performance. The lack of current loss in these cells indicates that any increase in surface recombination is compensated for by reduced emitter recombination in the now lightly doped emitter.

The Group 3 cells have regained most of the $V_{OC}$ loss, probably due to the compensating effect of reduced bulk recombination from the hydrogenation treatment. Interestingly, this is accompanied by a reduction of the series resistance, which is in agreement with observations by Wenham et al., who attributed this to a decrease in the contact resistance of the screen-printed gridlines [7]. This, in combination with the benefits of heavier emitter doping mentioned above, would address many of the shortfalls that have been ascribed to the screen-printing process.

The cells from Group 4 have totally regained their initial $V_{OC}$ values and show a significant 3% gain in $I_{SC}$ now that the surface of the transparent emitter is passivated by the nitride film. The effect of the plasma-nitride deposition on reducing the gridline contact resistance is still apparent, resulting in an overall average increase in efficiency of almost half an absolute efficiency point. Even better results are expected when the nitride passivation is combined with bulk hydrogenation and the benefits of heavy emitter doping.

Internal quantum efficiency (IQE) curves of typical cells from Groups 1, 3, and 4 are shown in Figure 2. LBIC scans showed that the cells from Group 2 did not have the same grain structure as the others, and so it was not possible to find the same "median" grain from cells of Group 2 on which to measure the IQE.

The IQE curves show that while both plasma treatments increased the red-response relative to the control cell, the $NH_3$-hydrogenation treatment had the biggest effect. It is also clear that the nitride-ARC resulted in the best blue response due to its better passivation of the emitter surface. In fact, the IQE(400-nm) value (73%) is almost as high as that obtained previously on this material (78%) using a nitride coating optimized for low surface recombination [6]. This shows that the low-power RIE process may not have damaged the emitter surface significantly, if at all.

An additional 6 groups of cells were processed to compare the effect of an extra-heavy emitter diffusion. As previously mentioned, a more heavily diffused emitter is of interest because it may provide better gettering of impurities from the substrate and reduced contact resistance and recombination. These 6 groups of cells examined different process sequences similar to the experiment described in Table 1. The plasma etch used a higher power than the experiment of Table 1 in order to etch the more heavily diffused emitter in a reasonably short time. Unfortunately, this etch may have damaged the surface because none of the etched-back cells performed better than the control cells. There was no significant difference in the blue or red response, and the desired drop in series resistance was not observed, possibly due to gridline damage that caused noticeable gridline discoloration.

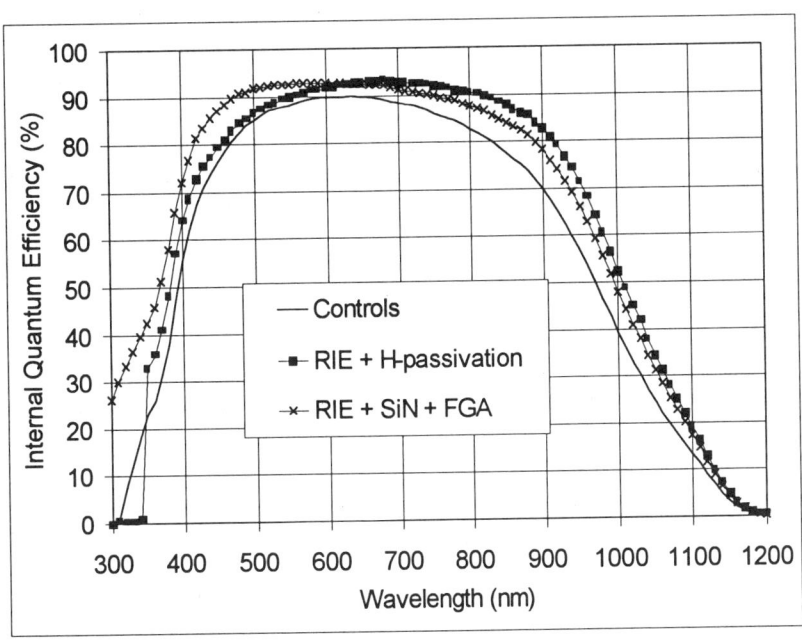

**FIGURE 2.** IQE curves of three Solarex cells representative of Groups 1, 3, and 4 described in Table 1.

## CONCLUSIONS

This investigation determined that low-power RIE is compatible with using standard, commercial, screen-printed gridlines as etch masks to form self-aligned, selectively-doped emitter profiles. This process results in reduced gridline contact resistance, an undamaged emitter surface easily passivated by plasma-nitride, and a less heavily doped emitter between gridlines for reduced emitter recombination. It allows for heavier doping beneath the gridlines for even lower contact resistance, reduced contact recombination, and better bulk defect gettering. Future work in this area will incorporate the heavier emitter doping as well as performing the bulk hydrogenation before a low-power RIE step so that surface damage from the bulk passivation step can be removed or reduced. This will be compared with the use of a protective nitride film before hydrogenation. Finally, all three plasma processes, the bulk passivation, emitter etchback, and nitride surface passivation, will be combined for the synergistic additive effect of their benefits.

## ACKNOWLEDGMENTS

The authors would like to thank J. M. Gee for initially suggesting the self-aligned emitter etchback concept, and W. K. Schubert for assistance in early experiments in this area. We are also grateful to the processing staff at Sandia's Photovoltaic Device Fabrication Laboratory (PDFL) and at Solarex's production line for the cell processing. Many thanks also go to L. W. Irwin, B. R. Hansen, and J. K. Snyder at Sandia's PDML for the cell measurements. The work presented in this paper was also presented at the 9$^{th}$ International PV Science and Engineering Conference, Miyazaki, Japan, 11-15 November 1996.

## REFERENCES

1. Z. Chen, P. Sana, J. Salami, and A. Rohatgi, "A Novel and Effective PECVD SiO2/SiN Antireflection Coating for Si Solar Cells," *IEEE Trans. Elect. Dev.*, 40, June 1993, pp. 1161-1165.
2. D.S. Ruby, W.L. Wilbanks, C.B. Fleddermann, and J.I. Hanoka, "The Effect of Hydrogen-Plasma and PECVD-Nitride Deposition on Bulk and Surface Passivation in String-Ribbon Silicon Solar Cells," *Proc. 13th EPSEC*, October 1995, pp. 1412-1414.
3. A. Blakers et al., "22.8%-Efficient Silicon Solar Cell," *Appl. Phys. Lett.* 55, 1989, pp. 1363-1395.
4. J. Coppye et al., "Non-Conventional Emitters for Polycrystalline Silicon Solar Cells," *Proc. 10$^{th}$ EPSEC*, April 1991, pp. 657-660.
5. N. Mardesich, "Solar Cell Efficiency Enhancement by Junction Etching and Conductive AR Coating Processes," *Proc. 15$^{th}$ PVSC*, May 1981, pp. 446-449.
6. D.S. Ruby, W.L. Wilbanks, and C.B. Fleddermann, "A Statistical Analysis of the Effect of PECVD Deposition Parameters on Surface and Bulk Recombination in Silicon Solar Cells," *Proc. 1$^{st}$ WCPEC*, Dec. 1994, pp. 1335-1338.
7. S.R. Wenham, M.R. Willison, S. Narayanan, and M.A. Green, "Efficiency Improvement in Screen-Printed Polycrystalline Silicon Solar Cells by Plasma Treatments," *Proc. 18$^{th}$ PVSC*, Oct. 1985, pp. 1008-1013.

# High-Flux Solar Furnace Processing of Crystalline Silicon Solar Cells

Y.S. Tsuo, J.R. Pitts, P. Menna*, M.D. Landry, J.M. Gee,**
and T.F. Ciszek

*National Renewable Energy Laboratory, Golden, Colorado 80401*
*\*ENEA-Centro Ricerche Fotovoltaiche, Portici 80055, Italy*
*\*\* Sandia National Laboratories, Albuquerque, New Mexico 87185*

**Abstract.** We studied the processing of crystalline-silicon solar cells using a 10-kW, high-flux solar furnace (HFSF). Major findings of this study include: (1) hydrogenated amorphous silicon films deposited on glass substrates can be converted to microcrystalline silicon by solid-phase crystallization in 5 seconds or less in the HFSF; (2) the presence of concentrated sunlight enhances the diffusion of phosphorus into silicon from a spin-on dopant source; (3) the combination of a porous-silicon surface layer and photo-enhanced impurity diffusion is very effective in gettering impurities from a metallurgical-grade silicon wafer or thin-layer silicon deposited using liquid-phase epitaxy; (4) a 14.1%-efficient crystalline-silicon solar cell with an area of 4.6 cm$^2$ was fabricated using the HFSF for simultaneous diffusion of front $n^+$-p and back p-$p^+$ junctions; and (5) we have shown that the HFSF can be used to texture crystalline-silicon surfaces and to anneal metal contacts printed on a silicon solar cell.

## INTRODUCTION

We believe the direct use of concentrated sunlight to process crystalline-silicon solar cells is an environmentally friendly manufacturing method that can provide a cost reduction for cell fabrication. The cost of photons for a high-flux solar furnace (HFSF) is orders of magnitude less than that of artificial light sources [1,2]. The cost reduction comes not only from the low-cost power source, but also from simpler manufacturing equipment, fewer processing steps, reduced wafer handling, and reduced pollution. Other advantages of HFSF processing over conventional furnace processing include: (1) it provides a cold-wall process, which reduces contamination of the samples under processing; (2) the temperature versus time profiles can be precisely controlled; (3) wavelength, intensity, and spatial distribution of the incident solar flux can be controlled and changed rapidly; (4) a number of high-temperature processing steps can be performed simultaneously; and (5) combined quantum and thermal effects may benefit overall cell performance.

## HIGH-FLUX SOLAR FURNACE

NREL's high-flux solar furnace uses an off-axis design, which allows the use of a variety of receivers/reactors. A flat, tracking heliostat 32 $m^2$ in area reflects sunlight onto a faceted primary concentrator of 25 curved, hexagonal mirrors with a 7-m focal length. Both the heliostat and primary concentrator are front-surface mirrors coated with enhanced-ultraviolet-reflectivity aluminum that reflects radiation from the entire solar spectrum (wavelengths from 300 to 2500 nm). The primary concentrator focuses the light to a position inside the target bay, where 94% of the energy falls inside a 10-cm circle. The peak flux can reach concentrations of 2500 suns (1 sun = 1 $kW/m^2$) without a secondary concentrator. Computer-controlled attenuator can be used to control the heating and cool-down rates. This is important for solar cell processing because fast cooling has been recognized as a drawback of rapid thermal processing using conventional heat sources [3].

The method of using light for silicon solar cell processing has been previously studied. Heating lamps, such as tungsten-halogen quartz lamps that are rich in infrared, and lasers, such as excimer lasers that are rich in ultraviolet, have been used with certain success in solar cell manufacturing [3-5]. However, the high-intensity, large-area light flux and the degree of control over the flux profile that can be achieved using a HFSF cannot be easily achieved using an artificial light source. The simplicity of the HFSF equipment also makes it easy to scale up for commercial manufacturing.

The cold-wall process, in which only the sample under treatment gets heated, not the reactor walls, also greatly reduces the possibility of impurity contamination from the chamber. Impurity contamination from the chamber walls has been a major factor in reducing the carrier lifetime of silicon in conventional high-temperature processes. This, in turn, reduces the performance of the solar cells manufactured. Periodic, sometimes daily, cleaning of the high-temperature furnace using hydrofluoric-acid-containing solutions is required. This increases the cost of manufacturing solar cells and also generates large quantities of toxic waste. In conventional hot-wall furnace processing, the need to heat the entire furnace is a waste of energy. In HFSF processing, on the other hand, the sample is heated directly by the concentrated sunlight.

Silicon cell processing steps that can be accomplished with the use of a HFSF include crystallization, gettering, surface texturing, doping, alloying and annealing metal contacts, hydrogen passivation, and oxide growth for surface passivation. The high heating and quenching rates and the flexible spectrum, intensity, and spatial distribution of HFSF processing may also be used to form novel solar cell structures.

## CELL FABRICATION

We have made solar cells using the HFSF to form the front $n^+$-p junction (using spin-on n-type dopants) and back-surface field (using either evaporated aluminum films or spin-on p-type dopants) at the same time. The junction formations were done with the samples heated to 850°C to 900°C in a 500-torr gas mixture of 2% oxygen and 98% nitrogen. A comparison of various heating conditions were reported in Ref. 6. The best cell conversion efficiency we have achieved so far, under standard test conditions, is 14.1% (open-circuit voltage = 0.592 V, short-circuit current density = 30.0 mA/cm$^2$, and fill factor = 79.5%) for a single-crystal silicon solar cell with an area of 4.6 cm$^2$. This cell has no surface texturing and no front- and back-surface passivation. This efficiency is similar to that of cells with the same structure but fabricated in a conventional diffusion furnace.

By comparing the junction depths of cells fabricated using our HFSF and those fabricated using a conventional furnace, we have shown that the presence of concentrated sunlight enhances the diffusion of phosphorus into silicon [7]. We believe that the light-enhancement in impurity diffusion is due to silicon interstitials forming in the surface layer, where strong light absorption causes intense electronic excitation. We have shown that a short annealing time of only 30 seconds at 1050°C is sufficient to produce a junction depth of 0.25 μm in a silicon wafer, which have been coated with a layer of n-type spin-on dopant.

## METALLIZATION

The use of screen-printed organometallic compounds for front metal-grid contacts has been previously demonstrated at NREL on ceramic substrates [8]. We have successfully deposited gold contacts on silicon using this method. In this experiment, a gold-containing organometallic compound was first screen printed on a polished silicon wafer surface, and then the HFSF was used to anneal the compound to form gold contacts.

## GETTERING

Gettering is a method of reducing impurities in a wafer by localizing the impurities in regions away from the active device regions. The effectiveness of gettering depends on the establishment of gettering sites for absorbing impurities, the diffusion coefficients of the impurities, and the segregation coefficient of the impurities at the gettering sites [9,10]. Popular gettering methods for Si solar cells include aluminum, phosphorus, and chlorine gettering [11]. Al gettering, for example, is usually a backside gettering method done in combination with back-electrode and back-surface-field formation. Deliberate surface damage, such as mechanical abrasion, laser scribing, ion bombardment, impact sound stressing, and reactive ion etching, is also widely used to introduce strains into the silicon lattice for the impurities to segregate. Deposited surface layers (polycrystalline Si, $Si_3N_4$, $Al_2O_3$, and Si-Ge alloys) have also been used to generate gettering sites [12]. For all of these gettering methods, an HFSF can be used to provide low-cost, high-temperature annealing. Variation of the wavelength and intensity of the light flux may be used to optimize the gettering process.

We also studied a novel porous silicon gettering method. In this method, a simple and low-cost chemical etching is used to generate the porous silicon layers on both sides of the silicon wafer to provide a large number of efficient gettering sites. Then, the HFSF is used to provide high-temperature annealing and the required injection of silicon interstitials. The gettering sites, along with the gettered impurities, can be easily removed at the end the process. The porous silicon removal process consists of oxidizing the porous silicon near the end the gettering process, followed by sample immersion in HF acid. Each porous silicon gettering process removes up to about 10 µm of wafer thickness. This gettering process can be repeated so that the desired purity level is obtained.

We used two types of samples for our gettering study: (1) metallurgical-grade silicon (MG-Si) prepared by directional solidification, and (2) thin-layer silicon deposited by liquid-phase epitaxy (LPE) on a MG-Si substrate or a high-purity cast polycrystalline silicon substrate. The impurity concentrations of the cast MG-Si, as well as the strong, visible-light emissions from porous silicon prepared from the MG-Si, were reported by Menna et al. [13]. The porous silicon layers were prepared using a $HNO_3$/HF (1:100) etching solution [14]. This chemical etching method (also known as stain etching) is very simple (no electrodes required), fast (takes 10 minutes or less), and produces porous silicon on both sides of the silicon wafer. The porous silicon etched silicon wafers were then annealed in an HFSF for 15 to 30 min at a sample temperature of about 1000°C.

We have shown that the significant increase of the surface area achieved with the porous Si (nearly 600 $m^2/cm^3$), as well as the larger lattice parameter of porous silicon compared with that of the bulk Si, greatly enhance the probability of tying up the contaminants during the annealing step. Strong gettering effects were observed for Al, B, Fe, Cu, and Cr impurities for both MG-Si samples [15] and thin-layer Si samples. We have also shown that intense incoherent light irradiation of the HFSF enhances the diffusion of metallic impurities [7].

## TEXTURE ETCHING

An HFSF can decompose fluorine-containing gases, such as hexafluoroethane, for texture etching of the silicon solar cell surface, which enhances light absorption. In an example of the HFSF etch, a wafer of n-type Si (100) was solvent-cleaned and mounted in the sample chamber fitted with a fused silica window. The chamber was pumped and purged repetitively with argon to remove air and provide an inert background. $C_2F_6$ gas was admitted into the system at a rate of 50 sccm, while the pressure was held at 1 torr. The solar beam was applied to the front surface of the wafer so that a heating rate of 100°C/min was maintained up to the reaction temperature of 775°C (applied flux of 373 kW/m$^2$). Etching was allowed to continue for 9 min and was then terminated by removing the solar beam. The resultant surface texture was a porous layer 1 µm thick, with typical filament sizes in the range of 0.05 µm. Surface texturing occurred only on the surface of the wafer exposed to the solar beam.

## SOLID-PHASE CRYSTALLIZATION

Polycrystalline or multicrystalline silicon (mc-Si) thin films are usually deposited on a substrate by a chemical vapor deposition (CVD), physical vapor deposition (PVD), or liquid-phase epitaxy (LPE) technique. Si films deposited by PVD or CVD on a low-cost, nonepitaxial substrate usually have an average grain size less than 1 µm and are also called microcrystalline silicon. Microcrystalline silicon (µc-Si) films deposited using plasma-enhanced CVD (PECVD) with strong hydrogen dilution of the Si-containing feedgas and high radio-frequency power densities are widely used as heavily doped window contact layers and tunnel-junction contact layers in a-Si:H solar cells [16]. However, larger Si grain sizes are needed for use as active semiconductor layers in such applications as thin-film transistors and photovoltaic solar cells. A post-deposition annealing of PVD or CVD-deposited a-Si:H or mc-Si is usually needed to obtain large crystalline grain sizes. PECVD is usually used to deposit the initial a-Si:H film because of its low-temperature deposition and high purity. Relatively high deposition rates can be used because the initial electronic property of the a-Si:H is not important. For conventional furnace annealing of PECVD-deposited a-Si:H, grain sizes as large as 4 µm and field-effect mobilities as high as 158 cm$^2$V$^{-1}$s$^{-1}$ have been reported [17]. (For comparison, the effective electron mobility in a-Si:H is only up to 1 cm$^2$V$^{-1}$s$^{-1}$; in single-crystal Si, it is up to 1400 cm$^2$V$^{-1}$s$^{-1}$.) However, such annealing usually takes more than 10 hours and cannot be used to selectively crystallize certain areas of an a-Si:H film.

For our HFSF crystallization experiments, we used 400-nm-thick a-Si:H films glow-discharge-deposited on Corning 7059 glass substrates using a radio frequency of 110 MHz [18]. With the concentration of the high-flux solar furnace at about 1500 suns, it took 5 seconds or less to crystallize a 1-in. x 1-in. area of a-Si:H, whereas it typically takes more than 10 hours to achieve the same result by conventional furnace annealing. The main reason we could achieve such a fast crystallization is because the HFSF can heat the a-Si:H film to very high temperatures while keeping the glass substrate at temperatures below 600°C to avoid softening the glass during solid-state crystallization. A comparison of the transmission properties of a film crystallized using an HFSF at 1500 kW/m$^2$ for 5 seconds, a similar film crystallized using a conventional quartz-tube furnace at 580°C for 48 hours, and an as-deposited a-Si:H film shows that the crystallization of the HFSF-annealed film is more complete than that of the furnace-annealed film, even though the treatment time of the former is nearly four orders of magnitude shorter [6]. The transmission property of the HFSF-crystallized mc-Si is very similar to a microcrystalline film deposited by PECVD using high hydrogen dilution and high radio-frequency power. We have also used lower solar fluxes, e.g., 1200 kW/cm$^2$, to crystallize a-Si:H for longer periods of time. On the other hand, a higher power level of 1600 kW/m$^2$ has been used to achieve solid-phase crystallization of a-Si:H in only 4 seconds.

In addition to high throughput resulting from the short crystallization time, advantages of an HFSF crystallization include low cost, excellent large-area capability, uniformity over a large area, and area-selectable crystallization. A major advantage of an HFSF over artificial light sources is that the long focal length of the HFSF optical train permits very flexible fixturing.

## SUMMARY

We have demonstrated that an HFSF can be used to provide all the high-temperature steps needed for the fabrication of a crystalline-silicon solar cell—from crystallization, gettering of impurities, surface texturing, and junction formation, to annealing metal contacts. We have also shown that HFSF processing provides the added benefits of optically enhanced crystallization and impurity diffusion. We conclude that HFSF processing of silicon solar cells has the potential to improve cell efficiency and reduce cell fabrication costs, and also to be an environmentally friendly manufacturing method.

## ACKNOWLEDGMENTS

This work was supported by the U.S. Department of Energy under Contract No. DE-AC36-83CH10093 through an NREL Director's Development Fund. The authors thank S. Asher for SIMS measurements, T.H. Wang for help with the cast MG-Si substrate preparations, and M. Al-Jassim for helpful discussions.

# REFERENCES

1. A. Lewandowski, *Environmentally Conscious Manufacturing Newsletter*, published by U.S. Dept. of Energy, **4**, No.1, March 1993.
2. A. Lewandowski, *Mat. Tech.*, **8**, 1993, pp. 237-249.
3. B. Hartiti, R. Schindler, A. Slaoui, B. Wagner, J.C. Muller, I. Reis, A. Eyer, and P. Siffert, *Progress in Photovoltaics: Research and Applications*, **2**, 1994, pp.129-142.
4. R.Z. Bachrach, K. Winer, J.B. Boyce, S.E. Ready, R.I. Johnson, and G.B. Anderson, *J. Electronic Materials*, **19**, 1990, 241.
5. A. Rohatgi, Z. Chen, P. Doshi, T. Pham, and D. Ruby, *Appl. Phys. Lett.*, **65**, 1994, pp. 2087-2089.
6. Y.S. Tsuo, J.R. Pitts, M.D. Landry, P. Menna, C.E. Bingham, A. Lewandowski, and T.F. Ciszek, "High-Flux Solar Furnace Processing of Silicon Solar Cells," *Solar Energy Materials and Solar Cells*, **41/42**, pp. 41-51, 1996.
7. Y.S. Tsuo, P. Menna, J.R. Pitts, M.M. Al-Jassim, S.E. Asher, and T.F. Ciszek, "Light-Enhanced Diffusion of Impurities in Silicon," submitted to *Journal of Applied Physics*, Sept. 1996.
8. J.R. Pitts, private communication.
9. Schroter, W., M. Seibt, and D. Gilles, *in Materials Science and Technology*, **Vol. 4**, VCH Publishers: NY (1991).
10. Wolf, S. and R.N. Tauber, *Silicon Processing for the VLSI Era*, **Vol.1**, Lattice Press: (1986).
11. A. Rohatgi, Sana, P., Ramanachalam, M.S., Salami, J., and Carter, W.B., *Proc. 23rd IEEE PV Specialists Conf.*, pp. 52-57, 1993.
12. M.A.Green, *High Efficiency Silicon Solar Cells*, Trans Tech Publications: Brookfield, VT, 1987.
13. P. Menna, Y.S. Tsuo, M. Al-Jassim, S. Asher, F.J. Pern, and T.F. Ciszek, "Light Emitting Porous Silicon from Cast Metallurgical Grade Silicon," *J. Electrochemical Society*, **143**, pp. L115-L117, 1996.
14. P. Menna, G. Di Francia, and V. La Ferrara, "Porous Silicon in Solar Cells - A Review and a Description of Its Applications as an Anti-Reflection Coating," *Solar Energy Materials and Solar Cells*, **37**, pp. 13-24, 1995.
15. P. Menna, Y.S. Tsuo, S.E. Asher, J.R. Pitts, M.M. Al-Jassim, and T.F. Ciszek, "Porous-Silicon Gettering of Impurities in Silicon", submitted to *Journal of Applied Physics*, Nov. 1996.
16. W. Luft and Y.S. Tsuo, "Hydrogenated Amorphous Silicon Alloy Deposition Processes", Chap.15, Marcel Dekker, Inc., New York; 1993.
17. K. Nakazawa and K. Tanaka, *J. Appl. Phys.* **68**, 1029 (1990).
18. R.E. Hollingsworth and P.K. Bhat, *Appl. Phys. Lett.*, **64**, pp. 616-618, 1994.

# Aluminum Gettering and Transition Metal Precipitates in PV Silicon

Henry Hieslmair*, Scott A. McHugo[†], Eicke R. Weber*

*Department of Materials Science, University of California, Berkeley,
Berkeley CA 94720
[†] Advanced Light Source, Lawrence Berkeley National Laboratory,
Berkeley, CA 94720

**Abstract.** Iron is used as a model impurity to study aluminum gettering and iron precipitation in silicon. Aluminum gettering was found to be effective in removing iron contamination from float zone silicon, greatly increasing minority carrier diffusion lengths from 30 ~ 50 μm to 150 ~ 170 μm. The same process does not significantly improve PV silicon materials, CZ, cast multicrystalline, and EFG. Finite difference modeling suggests that Al gettering in PV silicon is not diffusion limited. It is suggested that barriers to dissolution of impurity precipitates are the main cause of the insufficient gettering response in PV silicon. Initial results of a precipitation and dissolution study are presented, as well as new gettering simulation capabilities.

## ROLE OF TRANSITION METALS IN PV SILICON

Transition metals have been studied in silicon for many years. In the Integrated Circuit (I.C.) industry, transition metals have a detrimental effect on device performance and yield. Similarly, for good photovoltaic cell efficiencies, low transition metal impurity concentrations are required in order to have high minority carrier diffusion lengths and solar cell efficiencies. In photovoltaic (PV) grade silicon, commonly present metallic impurities are Fe, Cr, Ti, Mo and V in either dissolved or precipitated states (1). Our own studies indicate that the type and amount of impurity vary greatly from one PV silicon to the next, with up to $10^{14}$ Fe/cm$^3$ in some PV grade CZ silicon samples. The object of our continuing studies is gettering of transition metals and minimizing the deleterious effects of transition metals in PV grade silicon.

## Dissolved Impurities

Dissolved impurities can have an effect on minority carrier lifetimes if they create a deep trap level in the silicon bandgap. In p-type silicon, interstitial iron forms Fe-acceptor pairs, and is generally used as a model for other transition metals. Iron-boron pairs are easily detected with Deep Level Transient Spectroscopy (DLTS) with a deep level at Ev+ 0.1eV and Ec-0.29eV. A simplified S-R-H expression relating minority carrier diffusion length, $L$, and the dissolved impurity concentration as follows (2):

$$L_n = \sqrt{D_n \tau_n} \qquad \tau_n = \frac{1}{\sigma_n \langle v_{th} \rangle N_t} \cdot e^{(E_F - E_t)/kT} \qquad \text{Eq. 1}$$

where $\sigma$ is the minority carrier capture cross section, $v_{th}$ is the thermal velocity of the carriers and $N_t$ is the impurity concentration.

Figure 1a shows data from samples of various experiments. These FZ silicon samples ('As-quenched' and 'Pre-gettered') were annealed in the presence of iron and then quenched to room temperature. After room temperature storage, iron forms FeB pairs as confirmed by DLTS. Also shown are data from Zoth et al. (2) The two lines differ in the value of the effective carrier capture cross section.

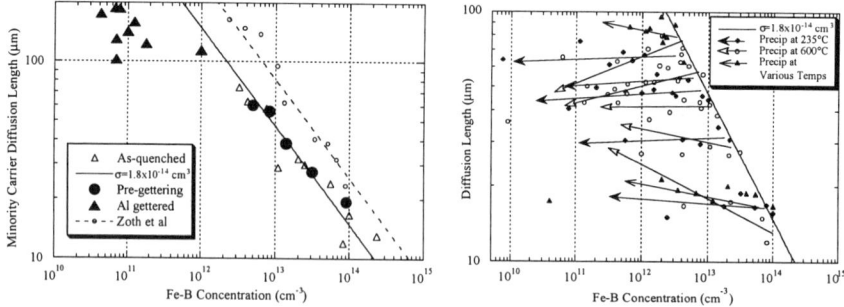

Figure 1a. As-quenched diffusion lengths of FZ silicon vs dissolved Fe concentration and Figure 1b. Diffusion length as iron precipitates, i.e. dissolved concentration drops as the iron forms clusters and precipitates.

## Precipitated Impurities

Iron and other transition metals can also precipitate from solution. Usually precipitation occurs at nucleation sites which include structural defects such as dislocations (3,4) found in most PV silicon materials. Precipitation at structural defects can greatly exacerbate the recombination activity of the structural defects.

However, the effects of impurity precipitation on overall diffusion lengths and cell efficiencies is uncertain. The types of defects present in the material as well as the cooling rates greatly affect the recombination activity of the defects. Studies with EBIC and metal impurities decorating various dislocations, performed by Shen et al.(5) and Sumino (6), hint at the complexity of the effects of impurity precipitation on recombination activity of defects and thus bulk minority carrier diffusion lengths. Our results, shown in Figure 1b, strongly suggest that Fe is as efficient as a recombination center in the dissolved state as in the precipitated state. Minority carrier diffusion lengths neither increased nor decreased by a significant amount as iron was precipitated, as indicated by the arrows in Figure 1b. Thus iron, as well as other transition metals, must be *removed* from the bulk in order to improve the minority carrier diffusion length.

## ALUMINUM GETTERING

### Demonstration of Aluminum Gettering in FZ Silicon

One of the most interesting approaches to removing iron or other transition metals from the bulk is aluminum backside gettering. A comprehensive study was begun to fully understand the nature of aluminum segregation gettering, both qualitatively and quantitatively (7). In this study, p-type float zoned silicon, with a boron concentration of approximately $1.15 \times 10^{15}$ cm$^{-3}$, was used to avoid possible complications arising from the interactions of the intentional iron contamination with dislocations, grain boundaries, oxygen precipitates, and other unintentional impurities. Additionally, the aluminum p$^+$ backsurface field, formed during the anneal at high temperatures, was removed with a strong etchant before any SPV measurements. Precipitation was also avoided as much as possible by performing the aluminum gettering at the same temperatures as the iron diffusion and by using a rapid thermal anneal and quench (RTAQ) to re-dissolve any remaining precipitates The FZ had an initial minority carrier diffusion length >400μm. Iron was used for our model impurity because its kinetics and solubility are well known and it is a common impurity element in I.C. and PV materials. Iron was diffused into the samples at the times and temperatures shown in Table 1.

Table 1: Iron diffusion and aluminum gettering temperatures and times

| Sample | 1 | 2 | 3 | 4 | 5 |
|---|---|---|---|---|---|
| Temp°C | 950 | 900 | 850 | 800 | 750 |
| Time (min) | 40 | 60 | 85 | 110 | 150 |

After the iron diffusion, the samples were quenched in ethylene glycol with a quenching rate of the order of 1000°C/s. This fast quench ensures that the iron remains in the dissolved state and does not precipitate. The iron diffusion times were chosen such that $\sqrt{Dt} \geq 700\mu m$ while our sample thickness was only 500μm. The samples were etched, removing any surface $FeSi_2$, and cleaned. Then, minority carrier diffusion measurements were measured with surface photovoltage (SPV) technique and iron concentrations were measured with deep level transient spectroscopy (DLTS).

Aluminum was then evaporated or sputtered onto the samples. All annealings were performed in a TCA cleaned furnace used for I.C. fabrication. The samples are slow cooled at the end of the gettering anneal, thus precipitating the remaining dissolved iron. In order to detect all the iron with DLTS, it must be redissolved. After a strong etch (to remove the Al-Si gettering layer) and clean, a rapid thermal anneal and quench (RTAQ) at 1000°C for 10 seconds is used to dissolve any residual precipitated iron in the bulk of the sample. After the RTAQ, the samples are etched again to avoid surface effects before DLTS and SPV measurements.

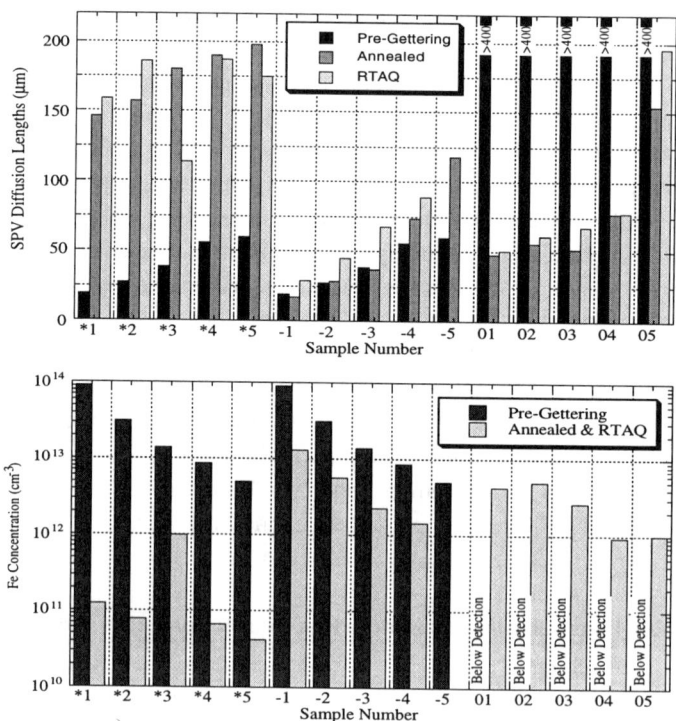

Figure 2a Minority carrier diffusion lengths and Figure 2b iron concentrations for various silicon samples, see Table 1.

Figure 2a shows the minority carrier diffusion lengths for each process. Figure 2b shows the initial and after RTAQ iron concentrations. In both figures, the sample numbers denote the iron diffusion and gettering temperatures as shown in Table 1. Additionally, the first group of five samples (*) had iron diffused and a back side aluminum layer, the middle group (-) had iron diffused and no back side aluminum layer, and the last group (0) had no iron and no aluminum back side layer. From these results we see significant diffusion length improvements via iron concentration reductions in the aluminum gettered samples to below solubility and background levels. For example, at a gettering temperature of 950°C, the aluminum gettering process removes iron to a level which is three orders of magnitude *below* the solubility of iron at that temperature.

The minority carrier diffusion length of the samples is plotted vs the iron concentration in Figure 1a. The aluminum gettered samples reflect the improvement by gettering, but they lie to the left of the dashed 'as-quenched' line. This indicates the presence of other recombination centers such as still undissolved iron (despite the RTAQ), other contaminants, or lifetime degradation resulting from the RTAQ. The iron detected in the samples with no intentional iron contamination (0), may have been initially present in the float-zone silicon.

## PV Silicon Materials Do not Respond Similarly

The experiment just discussed showed that Al gettering improved a 40 to 50 μm minority carrier diffusion length sample to 170 - 180 μm. Unfortunately, such improvements have not been observed in typical PV grade silicon materials such as low cost CZ or EFG. We have studied aluminum gettering of solar materials extensively, again using iron as a model impurity. A summary of gettering experiments is shown in Figure 3a and represents an average of numerous gettering results (8,9). These experiments were performed on various PV grade silicon

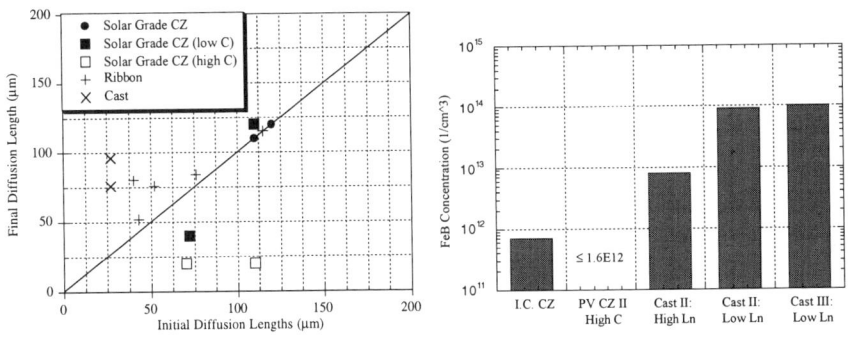

Figure 3a. Comprehensive summary of numerous gettering experiments. Each point is an average of experimental data. Figure 3b. As-grown iron impurities measured by DLTS after an RTAQ

materials without any intentional contamination and gettered at 850°C with various thicknesses of aluminum. The results for solar cell materials were not as positive as had been hoped for. Various gettering times and temperature treatments were explored and no significant minority carrier diffusion length improvements were observed. Generally PV silicon does not respond well to Al gettering processes.

## Reasons for Poor Gettering Response in PV Materials'?

At first glance, one might conclude that the segregation coefficient of aluminum gettering is too small. However, segregation coefficients, obtained from the previous aluminum gettering experiments involving FZ silicon, indicate that segregation coefficients range from $10^5$ to $10^6$ or higher, sufficiently high to improve PV materials. Thus, aluminum gettering should remove transition metals from the wafer silicon.

Perhaps aluminum gettering requires more time? Modeling of aluminum gettering indicates that, at least for dissolved iron, gettering performance should be better than it is. Figure 4 shows iron concentration profiles calculated using a finite difference technique which will be discussed later. For an initial iron concentration of $10^{13}$ Fe/cm$^3$, approximately three hours are necessary to reduce iron concentrations to well below DLTS detection limits. This is in the range of iron contamination in numerous as-grown PV silicon materials shown in Figure 3b. It is also possible that the PV grade silicon materials contain slow diffusing transition metals, although we have not observed other DLTS signals.

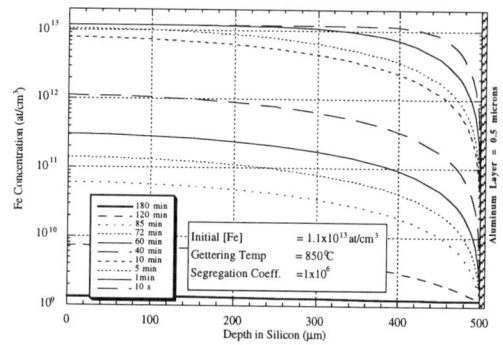

Figure 4. Iron concentration profiles during aluminum getter

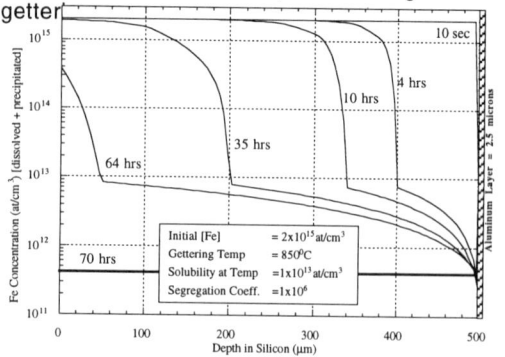

Figure 5. Iron concentration (dissolved and precipitated) vs wafer depth for increasing times during aluminum gettering.

Precipitated impurities could slow the gettering process by simple kinetics or with barriers to precipitate dissolution. The situation shown in Figure 5 is extreme, but serves to illustrate how high concentrations of precipitated impurities can greatly prolong the gettering process. While this may seem to explain the lack of gettering response of solar cell materials, gettering at higher temperatures, would greatly accelerate the gettering process. At 950°C this same process would only take a few hours. Yet experiments at higher gettering temperatures such as 950°C show no significant improvement.

Thus, it seems likely that there exist barriers to precipitate dissolution. Such barriers could arise from strain fields around dislocations or precipitate/lattice mismatch strains which would slow the gettering process. It is entirely possible that some barriers are so high as to prohibit gettering. Unfortunately, there is a poor understanding of transition metal dissolution in silicon.

## FUNDAMENTAL STUDY OF PRECIPITATION AND DISSOLUTION

Precipitation has been studied by many researchers, and is a prerequisite for studying dissolution of precipitates. The precipitation rate of an impurity depends on the number of precipitation sites and the effective precipitate radii. This was described by Ham with the following equation for spherical precipitates, (10)

$$4D\pi nr_o = 1/\tau \qquad \text{Eq. 2}$$

Here $nr_o$ refers to the precipitation site density and radius, $D$ is the diffusivity and $1/\tau$ is the precipitation rate. The $nr_o$ product can thus be measured by how fast the supersaturated impurity concentration drops. However, the number of precipitation sites depends on the degree of supersaturation. The least favorable precipitation sites will serve as nuclei only at high supersaturations and lower temperatures. Thus the effective $nr_o$ product seems to depend on the amount of supersaturation. Since the impurity must first be precipitated before dissolution can be studied, an understanding of precipitation becomes a prerequisite to studying dissolution.

### Iron Precipitation in FZ Silicon

Two groups of FZ silicon samples were intentionally contaminated with iron. The 'high' group had an initial concentration of $1 \times 10^{14}$ Fe/cm$^3$ and the 'low' group had a concentration of $4.2 \times 10^{12}$ Fe/cm$^3$. These samples were then annealed for various times at 700°C, 600°C, 475°C, 350°C or 235°C.

The precipitation results are shown in Figure 7. From these slopes effective $nr_o$ products were obtained and are shown in Figure 6a. They demonstrate a clear dependence on supersaturation. In Figure 6b, the precipitate site density, n, is plotted for various annealing temperatures. These densities were obtained by Teh Tan (11) by fitting the experimental precipitation data with a precipitation simulation. The results indicate that the density of sites for the 'high' and 'low' samples are similar. The difference in $nr_o$ product at the higher temperatures is thus the radius of the precipitate, $r_o$; the higher the impurity concentration, the larger $r_o$. As the annealing temperature drops, the supersaturation becomes so large in both samples that the difference in the 'high' and 'low' samples become negligible and both have the same high precipitate density and large $nr_o$ products.

Thus at the higher annealing temperatures, only the more preferable sites serve as precipitation sites. This is similar to observations by Shen[5] and McHugo[12] who observed copper decoration at certain sites only during a fast cool. This also implies that at *low* supersaturations, such as during a slow cool from melt or other high temperature processing, $nr_o$ and n will be small, and iron will only precipitate at the most

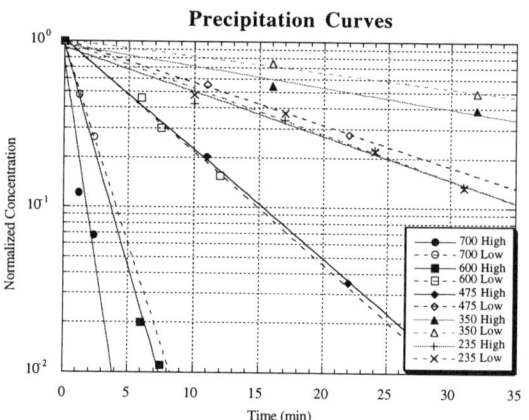

Figure 7. Dissolved iron concentration vs annealing time for different annealing temperatures.

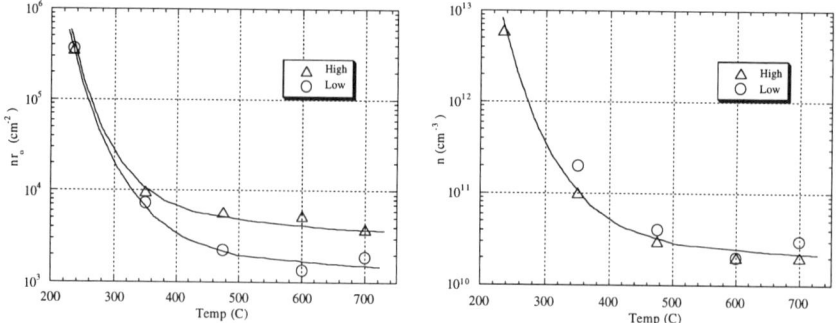

Figure 6 a&b. $nr_o$ products vs annealing temperatures and n vs annealing temperatures

favorable sites. We can speculate that these are probably sites on certain dislocations, at SiO$_2$ particles, or in grain boundaries.

Additionally, since at high temperatures iron diffuses quickly, it is likely that much of the iron will diffuse to these few favorable sites which become highly recombination active. The probable result of this is inhomogeneities in minority carrier diffusion lengths. Additionally, these few precipitates could also continuously emit the impurity when gettering at a lower temperature. This can significantly slow down the gettering process and would qualitatively explain our results.

Our future experiments will focus on precipitation and dissolution on CZ with well quantified oxygen precipitation sites, FZ with dislocations, and PV grade CZ and multicrystalline silicon materials.

## FINITE DIFFERENCE SIMULATIONS OF GETTERING

A finite differences simulation was developed to model the segregation of iron from the silicon to the aluminum. The segregation of iron from the silicon to the aluminum is a dynamic equilibrium situation. Thus the segregation boundary condition can be described by a modification of the flux at the interface, and is represented as:

$$J_{1\to 2} = \frac{D_{eff}}{\Delta x_{12}}\left(C_1^{Si} - \frac{C_2^{Al}}{m}\right) \quad \text{Eq. 3}$$

$$\text{where } D_{eff} = \frac{D_{Si}^{Fe} D_{Al}^{Fe}}{D_{Si}^{Fe} + D_{Al}^{Fe}}$$

This was derived through a control volume approach and is somewhat similar to the approach taken by Antoniadis (13). This treatment serves well for this type of problem, since it is better suited for very abrupt interfaces than the use of a chemical potential.

Ham's law was used for precipitation and dissolution as follows,

$$\frac{dC}{dt} = -\frac{1}{\tau}(C_o - C_{eq})e^{-t/\tau} \quad \text{where } 1/\tau = 4\pi D n r_o \quad \text{Eq. 4}$$

The radius of the precipitate was calculated by simply using,

$$\Delta C = \frac{4}{3}\pi(r_1^3 - r_o^3)n\Omega \qquad \text{Eq. 5}$$

where $\Omega$ is the impurity density in the precipitate. Most gettering techniques are segregation or relaxation phenomena. Others can often be described by an effective segregation coefficient and $nr_o$ product.

Models can illustrate the limitations of an experiment. The aluminum gettering experiments performed earlier were performed in order to find the segregation coefficient of the Al gettering process. The process times for these gettering experiments were thought to be adequate since the segregation coefficient was hypothesized by some researchers to be about $1 \times 10^4$. Our results indicate higher segregation coefficients. This means that the gettering process is not complete.

More importantly, as Figure 8 indicates, the experiment, as it was performed, is insensitive to segregation coefficients from $5 \times 10^5$ to $5 \times 10^7$ and higher. Since the process is diffusion limited, it is difficult to distinguish the higher segregation coefficient profiles at this stage in the process. At longer process times the profiles of the higher segregation coefficients will become distinguishable.

Simulations can also give us great insight into simultaneous processes and complicated temperature profiles. For example, Figure 9 is a demonstration of the capabilities of a gettering simulation. It shows the iron concentration (both precipitated and dissolved iron) in a wafer with a front side denuded zone and back side Al gettering layer as the wafer cools from 950°C to 500°C. Thus segregation and relaxation gettering are occurring simultaneously in this process and result in a very low iron concentration in the denuded zone as well as the backside.

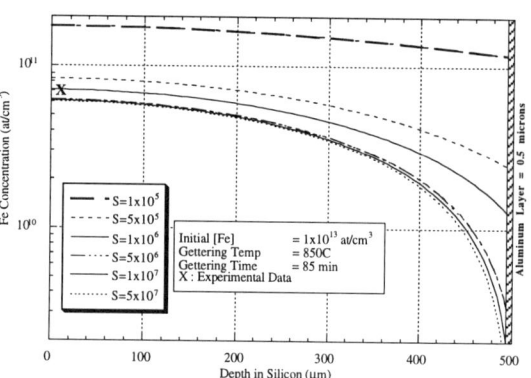

Figure 8. Iron concentration profiles for various aluminum gettering segregation coefficients

With these new capabilities we will explore new experimental procedures to design experiments which would yield good segregation coefficient information. We will also explore the effects of non-uniform precipitate densities on aluminum gettering kinetics.

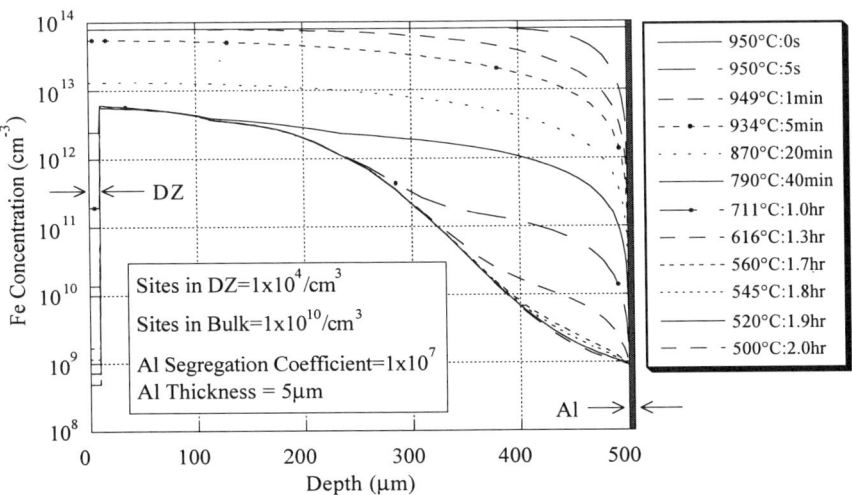

Figure 9. Iron concentration (dissolved + precipitated) profiles during a slow cool on a wafer with a front side 10μm denuded zone and a 5μm Al back side gettering layer.

## CONCLUSIONS

Aluminum gettering has been shown to work for iron dissolved in FZ silicon material. The aluminum gettering experiment allowed us to estimate a lower limit for the segregation coefficient at $1 \times 10^5$. An experiment more sensitive to higher segregation coefficients will be designed by modeling. The insensitivity of solar grade silicon materials to aluminum gettering is hypothesized to be due to barriers to dissolution of the impurity precipitates. Precipitate dissolution needs to be studied further. An iron precipitation experiment in FZ silicon was performed as a precursor to a dissolution study. It was observed that the number of precipitation sites increase with decreasing temperature and $r_o$ increases with initial concentration. Finite differences modeling has helped to understand the kinetics of aluminum gettering and precipitation of iron. More comprehensive simulations have been developed and will be used to help design experiments to better understand precipitate dissolution in solar grade silicon materials in order to design effective gettering procedures for this material.

## ACKNOWLEDGMENTS

The authors would like to thank Prof. Teh Tan for his contribution of the computer simulations of precipitation which yielded n values, and the useful discussions on computer simulations which greatly aided our own simulation capabilities. This work was funded under NREL subcontract #XD-2-11004-3.

---

1) J. P. Kalejs, B. R. Bathey, J. T. Borenstein and R. W. Stormont,"Effects of Transition Metal Impurities on Solar Cell Performance in Polycrystalline Silicon", in: 23rd IEEE Photovoltaic Specialists Conference, (1993) 184-89
2) G. Zoth, W. Bergholz, J. Appl. Phys, **67**, 6764, (1990)
3) C. Cabanel and J. Y. Laval: J. Appl. Phys. **67**, 1425 (1990)
4) V. Higgs and M. Kittler, Appl. Phys. Lett., **63**, 2085, (1993)
5) B. Shen, T. Sekiguchi, J. Jablonski and K. Sumino, J. Appl. Phys., **76**, 4540 (1994)
6) K. Sumino, Materials Science and Technology, **11**, 657 (1995)
7) H. Hieslmair, S. McHugo and E. R. Weber,"Aluminum Backside Segregation Gettering", in IEEE 25th Photovoltaics Specialists Conference, Washington D.C., May 13-17, 1996, IEEE, (1996), 441-4
8) H. Hieslmair, S. A. McHugo and E. R. Weber, "External Gettering of Silicon Materials Containing Various Efficiency-limiting Defects", 1995 NREL PV Review, (1995)
9) S. A. McHugo, H. Hieslmair and E. R. Weber, "Gettering of Metallic Impurities in Photovoltaic Silicon", Appl Phys A, **64**, 1 (1996)
10) F. S. Ham, *J. Phys. Chem. Solids*, **6**, 335 (1958)
11) This conference
12) S. A. McHugo and W. D. Sawyer, Appl. Phys. Lett., **62**, 2519, (1993)
13) D. A. Antoniadis and R. W. Dutton, IEEE J. Solid-State Circuits, SC-14, 412, (1979)

# INCORPORATION OF Cu AND Al IN THIN LAYER SILICON GROWN FROM Cu-Al-Si

T.H. Wang and T.F. Ciszek

National Renewable Energy Laboratory
1617 Cole Blvd., Golden, CO 80401, USA

## ABSTRACT

Cu and Al concentrations in silicon thin layers grown from Cu-Al-Si are determined by segregation at the solid-liquid interface, and for the fast diffusing Cu, also at the free silicon surface. Using the multicomponent regular solution model and experimental results, we found that Si-Al and Si-Cu interactions in the liquid solution are repulsive, and Al-Cu interaction is attractive. As a result, Al incorporation as a function of Cu and Al compositions in the growth solution is determined at about 900°C. Up to 0.2 $\Omega\cdot$cm P-type resistivities caused by Al doping are achieved because of suppression of Al incorporation by Cu, yet with a substantial amount of Al still present in the liquid for substrate surface-oxide removal. On the other hand, Cu concentration in the grown layers is reduced by Al in the liquid during growth and by surface segregation after growth. The surface segregation phenomenon can be conveniently used to getter Cu from the bulk of silicon layers so that its concentration (~$10^{16}$ cm$^{-3}$) is much lower than its solubility ($2.5\times10^{17}$ cm$^{-3}$) at the layer growth temperature and the reported $10^{17}$ cm$^{-3}$ degradation onset for solar-cell performance.

## INTRODUCTION

Growing silicon thin layers on metallurgical-grade silicon substrates by liquid-phase epitaxy offers the advantages of excellent crystallinity, high growth rate, perfect lattice match, and practicality over other current thin-silicon techniques for making low-cost silicon solar cells. Cu-Al has been found to be a good solvent system to grow macroscopically smooth Si layers with thicknesses in tens of microns at temperatures near 900°C [1]. This solvent system uses Al to ensure good wetting between the solution and substrate by removing silicon native oxides. Isotropic growth is achieved because of a high concentration of solute silicon in Cu-Al and the resulting microscopically rough interface. The incorporation of Al and Cu in the Si layers, however, needs to be controlled to obtain the desired electrical doping and to minimize the detrimental effect of Cu.

The solid solubility of Al and Cu in silicon (Al or Cu concentration in equilibrium with the liquid phase of Al-Si or Cu-Si) at 900°C is reported at

$1.5 \times 10^{19}$ cm$^{-3}$ and $2 \times 10^{17}$ cm$^{-3}$, respectively [2]. Therefore, if grown from a binary solution of Al-Si or Cu-Si, silicon crystals would have too low an electrical resistivity or too much Cu contamination for solar cells, although one study [3] shows that Cu nonetheless will not degrade solar-cell performance until above a level of $10^{17}$ cm$^{-3}$.

With ternary solutions of Al-Cu-Si at a near-constant growth temperature of 900°C, Al and Cu concentrations in grown silicon layers may be quite different from the binary case because of changes in free energy (by the third component, Cu or Al) in the liquid phase. We will simply treat the ternary solution with the multicomponent, regular solution model [4] and combine it with experimental results to determine the Al doping level with respect to liquid compositions and growth temperature. Because of its low diffusivity (~$10^{-13}$ cm$^2$ s$^{-1}$ [5]), Al redistribution after growth is negligible. However, segregation at the solid-liquid interface and at the free-silicon surface are both important for Cu concentration because of the high diffusivity of Cu (~$10^{-4}$ cm$^2$ s$^{-1}$ [5]), as we will see later. The driving force for the surface segregation of Cu arises from the difference between the chemical potential differentials (surface to bulk) of pure Si and Cu [6]. We will demonstrate the extent of the effect and use it as a mechanism to getter Cu from grown silicon layers.

## INTERACTION PARAMETERS FOR THE LIQUID

To use the regular solution model, we need to first determine the interactions between the elements. The activity coefficient for Si in a ternary system of Cu-Al-Si may be derived as

$$\gamma_{Si}^l = \exp\left\{\frac{1}{RT_e}\left[\Omega_{SiAl}^l\left(x_{Al}^l\right)^2 + \Omega_{SiCu}^l\left(x_{Cu}^l\right)^2 + \left(\Omega_{SiAl}^l + \Omega_{SiCu}^l - \Omega_{AlCu}^l\right)x_{Al}^l x_{Cu}^l\right]\right\}, \quad (1)$$

where $\Omega$'s are interaction parameters between the components denoted by subscripts, $R$ is the gas constant, and $T_e$ is the equilibrium temperature. Similar results are readily obtained for Cu and Al by permutation as all $\Omega$'s are symmetric.

Because the chemical potentials of respective elements in the liquid phase are equal to those in the solid phase at equilibrium, the activity of a given element in the solid phase must be equal to its counterpart in the liquid phase, which can be written as the product of its activity coefficient and composition. Because the silicon crystal is almost pure (to 99.9%), the Si activity is approximately unity and so is its activity coefficient. Therefore, the relationship between the compositions in the liquid when the mixture is at equilibrium (temperature $T_e$) with the solid phase may now be written as:

$$\frac{1}{x_{Si}^l} = \exp\left\{\frac{1}{RT_e}\left[\Omega_{SiAl}^l\left(x_{Al}^l\right)^2 + \Omega_{SiCu}^l\left(x_{Cu}^l\right)^2 + \left(\Omega_{SiAl}^l + \Omega_{SiCu}^l - \Omega_{AlCu}^l\right)x_{Al}^l x_{Cu}^l\right]\right\}. \quad (2)$$

From this equation, the interaction parameters can be determined by trying three mixtures of different compositions at the same growth temperature. A best set of solutions to equation (2), representing an average of four groups of experiments (each group has three different mixtures), is obtained as (at a common melting point of $T_e \approx 1173K$),

$\Omega^l_{SiAl} = 2.430RT_e$,
$\Omega^l_{SiCu} = 2.469RT_e$,
$\Omega^l_{AlCu} = -0.103RT_e$.

Both $\Omega^l_{SiAl}$ and $\Omega^l_{SiCu}$ are large positive numbers, so Si-Al and Si-Cu interactions are of a repulsive nature. $\Omega^l_{AlCu}$ is negative, however, implying an attractive interaction between Al and Cu. Therefore, Cu in the growth solution will not only dilute Al, but will also retain Al in the liquid, thus providing greater control for Al doping.

A binary Al-Si mixture of $x^l_{Al}/x^l_{Cu}/x^l_{Si} = 0.65/0.00/0.35$ that gives $\Omega^l_{SiAl} = 2.485RT_e$ does not fit into the above best set of solutions for interaction parameters. This indicates that the regular solution model is limited to the low-Al region (probably $x^l_{Al} < 0.4$), which is all we need in actual layer growth for enhancing wetting between the growth solution and substrate surface.

## SEGREGATION OF Al AT THE SOLID LIQUID INTERFACE

If it is grown out of an Al-Si binary solution at 900°C, a silicon crystal is expected to have $1.5 \times 10^{19}$ cm$^{-3}$ of Al, or a segregation coefficient of $8.6 \times 10^{-4}$. To find the new value when Cu is present, we first write the Al activity coefficient as:

$$\gamma^l_{Al} = \exp\left\{\frac{1}{RT_e}\left[\Omega^l_{SiAl}(x^l_{Si})^2 + \Omega^l_{AlCu}(x^l_{Cu})^2 + (\Omega^l_{SiAl} + \Omega^l_{AlCu} - \Omega^l_{SiCu})x^l_{Si}x^l_{Cu}\right]\right\} \quad (3)$$

$$= \exp\left\{2.43(x^l_{Si})^2 - 0.103(x^l_{Cu})^2 - 0.142 x^l_{Si}x^l_{Cu}\right\} .$$

Because aluminum is always very dilute (<0.04%) in the crystalline silicon matrix, the activity coefficient of Al in solid silicon can be treated as constant for all Al concentrations in silicon; this is known as Henry's law.

The solution $x^l_{Al}/x^l_{Cu}/x^l_{Si} = 0.28/0.49/0.23$ resulted in a silicon layer with $1.7 \times 10^{18}$ cm$^{-3}$ of Al, i.e., $x^s_{Al} = 0.34 \times 10^{-4}$; the constant Al activity coefficient $\gamma^s_{Al}$ in solid silicon at $T_e \approx 1173K$ is thus calculated:

$$\gamma^s_{Al} = \frac{x^l_{Al}\gamma^l_{Al}}{x^s_{Al}} = \frac{x^l_{Al}}{x^s_{Al}}\exp\left[2.43(x^l_{Si})^2 - 0.103(x^l_{Cu})^2 - 0.142 x^l_{Si}x^l_{Cu}\right] = 8991. \quad (4)$$

We thus get the segregation coefficient of Al for growth of silicon from the Al-Cu-Si mixtures at $T_e \approx 1173K$:

$$K_{Al} = \frac{1}{8991}\exp\left[2.43(x^l_{Si})^2 - 0.103(x^l_{Cu})^2 - 0.142 x^l_{Si}x^l_{Cu}\right]. \quad (5)$$

The concentration of Al as a function of liquid compositions at $T_e \approx 1173K$ is plotted in Fig.1, together with experimental results measured by secondary-ion mass spectroscopy (SIMS). This shows that the Al concentration in solid silicon is not only controlled by Al composition in the liquid, but by Cu as well. As a result, by adjusting both Cu and Al compositions to allow growth at about 900°C, resistivity of the thin-layer silicon can be easily controlled in a range of 0.01-0.2 $\Omega \cdot cm$, with a substantial amount of Al present in the liquid solution for substrate surface-oxide removal.

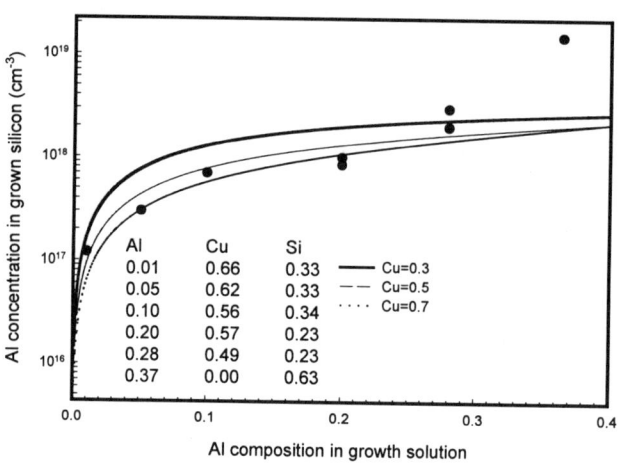

**FIGURE 1**. Multi-component regular solution model (lines) and experimental (dots) results of Al segregation at 900°C.

## Cu INCORPORATION

Segregation at the solid-liquid interface is considered first. When it is grown under a near-equilibrium condition from a Cu-Si melt, a silicon crystal is expected to be saturated with Cu to its solid solubility limit at the growth temperature, just like the case for Al. To verify that the reported solid solubility of Cu was not affected by its fast diffusion, a 2-cm-diameter ingot of single-crystal silicon was submerged in a Cu-Si melt equalized at 920°C for 8 hours (resulting in a Cu diffusion depth of ~ 1.7 cm from the ingot surface) to allow Cu to fully diffuse into the ingot. A 2-mm-thick slice was then cut and polished at room temperature shortly before SIMS measurement. Figure 2 is a SIMS map, taken after sputtering away the top surface, showing non-uniform Cu distribution with aggregations at swirl-defect sites. By averaging over 6 areas of 150 μm × 150 μm, the bulk concentration of Cu is calculated to be about $2.5 \times 10^{17}$ cm$^{-3}$, which is slightly higher than reported in the literature [7]. This concentration is dictated by segregation at the solid silicon and Cu-Si liquid interface.

When an Al-Cu-Si solution is used, the Cu segregation will be less than with Cu-Si solutions, similar to the case of Al,

$$K_{Cu} = \frac{1}{1.12 \times 10^5} \exp\left[2.469(x_{Si}^l)^2 - 0.103(x_{Al}^l)^2 - 0.064 x_{Si}^l x_{Al}^l\right], \quad (6)$$

where Cu activity coefficient in solid silicon, $\gamma_{Cu}^S = 1.12 \times 10^5$, has been determined from experiments as we did in section 3 for Al.

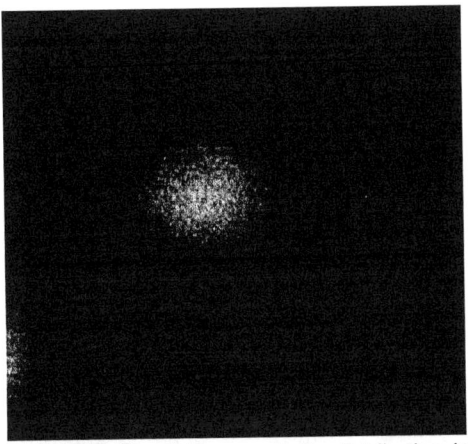

**FIGURE 2.** SIMS map of Cu distribution in an area of 250×250 μm² in Cu-diffused single crystal Si.

During post-growth cool-down, the Cu will become supersaturated and segregate to the surface or precipitate at the defect sites. The free-silicon surface is the preferred escape site for the supersaturated Cu atoms because the high free energy in the bulk will be spent on creating a new Cu-terminated surface if the Cu has lower surface energy than silicon. This indeed is the case: Cu segregates to the surface during cool-down after layer growth (see Fig. 3). We may incorporate this surface segregation phenomenon to effectively getter fast-diffusing Cu from the bulk of silicon without using a dedicated gettering procedure.

Fig. 4 is the Cu depth profile in a thicker sample grown at otherwise similar conditions as the one in Fig. 3. Because of the dynamic nature of SIMS measurements, the signals in the first 100 Å of the two samples are not accurate, but both of them show Cu enrichment in the 0.3-0.4-μm surface region. More surface-sensitive, ion-scattering spectroscopy (ISS) analysis reveals about 7% Cu at the top surface (about 50 Å deep) of a sample slowly cooled after growth. One can easily notice the difference in the bulk Cu concentrations between the two samples. A logical explanation is the difference in total Cu content caused by different substrate thicknesses. Both samples are expected to be saturated with Cu at the growth temperature of 900°C. The sample in Fig. 4 is thicker and would gather more Cu during growth from the in-diffusion of Cu than the sample in Fig. 3. During the sample cool-down period after growth, Cu out-diffuses to the surface and results in different levels of reduction in bulk Cu concentrations.

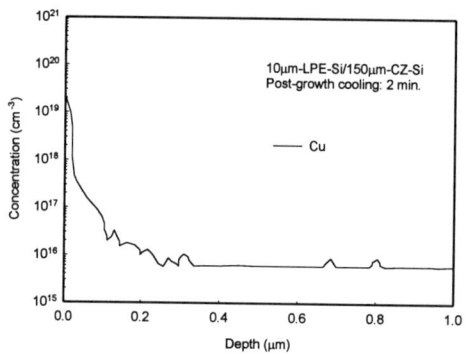

**FIGURE 3**. SIMS depth profile of Cu with a total sample thickness ≈ 160 μm.

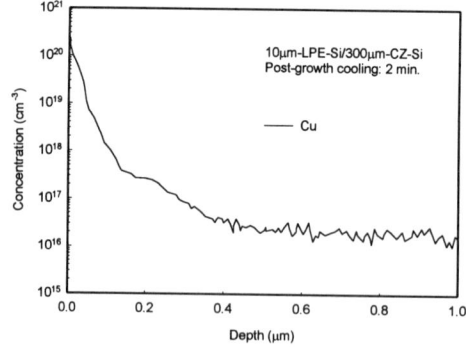

**FIGURE 4**. SIMS depth profile of Cu with a total sample thickness ≈ 310 μm.

After removing the top Cu-enriched surface region of about 0.5 μm by wet chemical etch, the bulk concentration of Cu is typically about $1 \times 10^{16}$ cm$^{-3}$, as seen in Fig. 5. Such a level of Cu in silicon is not expected to cause degradation effects for solar cells. This has a strong implication for using metallurgical-grade silicon as substrates for liquid phase epitaxial growth of high-quality silicon thin layers for solar cells, because the surface segregation will have a similar effect in gettering other fast-diffusing impurities like Ni or Fe. Slow-diffusing impurities in a low-purity substrate will not catch up to the epitaxial growth front, thus they will be of no concern.

For grain boundaries, the relative (to the bulk) chemical potentials of Si and Cu atoms at these locations are very likely to be lower than that of a free surface. This implies that the difference between the relative grain-boundary energy of Si and that of Cu is smaller than the difference between the relative surface energy of Si and Cu. The same argument may be made for other defects in silicon. Experimental evidence is abundant [8], including SIMS analysis in which no significant impurity enrichment at grain boundaries was observed, and, also, the necessity of a thick sample to Cu-decorate defects in silicon. Therefore, the free silicon surface is the preferred escape site for fast-diffusing impurities even for defected materials, provided a small silicon thickness and sufficient time are given.

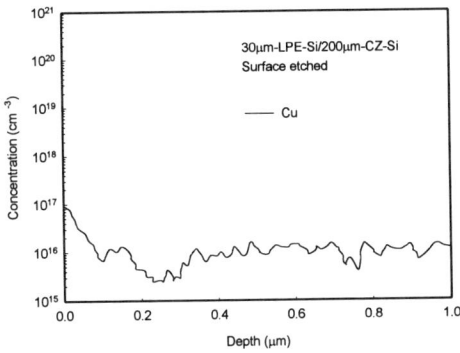

**FIGURE 5**. Cu depth profile after the original surface was removed by etching.

## CONCLUSIONS

The segregation of Al and Cu at the solid-liquid interface during growth and of Cu at the free silicon surface after growth in liquid-phase epitaxy of silicon at about 900°C has been studied. We found, by using the multicomponent regular solution model and experimental results, that Si-Al and Si-Cu interactions in the liquid solution are of a repulsive nature and Al-Cu interaction is of an attractive nature. Al concentration in solid silicon is controlled by both Al and Cu in the liquid. Consequently, resistivity of the thin-layer silicon can be easily controlled in a range of 0.01-0.2 Ω·cm, with a substantial amount of Al present in the liquid solution for substrate surface-oxide removal.

Surface segregation can be used to effectively getter Cu and other fast-diffusing impurities (assuming a lower surface energy than silicon) from bulk silicon when the impurity concentration exceeds its room-temperature solubility and when the silicon crystal is thin (a few hundred microns). When a free-silicon surface is available, this gettering process does not need special procedures, but only a prolonged cool-down step.

## ACKNOWLEDGMENTS

We thank Robert Reedy and Sally Asher for SIMS measurements and David King for ISS analysis. Support for this work was provided by the U.S. Department of Energy under contract No. DE-AC36-83CH10093 to the National Renewable Energy Laboratory.

## REFERENCES

[1] T.H. Wang and T.F. Ciszek, *Conference Record of the Twenty-Fourth IEEE Photovoltaic Specialists Conference*, Hawaii, 1994, p.1250.
[2] F. Trumbore, *Bell Sys. Tech. J.* **39**, 1960, p. 205.
[3] J.R. Davis, Jr., A. Rohatgi, R.H. Hopkins, P.D. Blais, P. Rai-Choudhury, J.R. McCormic, and H.C. Mollenkopf, *IEEE Trans. Electron. Devices*, **ED-27**, 1980, p. 677.
[4] A.S. Jordan, *J. Electrochem. Soc.*, Jan. 1972, p.123.
[5] W. Frank, U. Gosele, H. Mehrer, and A. Seeger, in: *Diffusion in Crystalline Solids*, ed. G.E. Murch and A.S. Nowick, Academic Press, Orlando, 1984, p. 90.
[6] J. Du Plessis and G.N. Van Wyk, *J. Phys. Chem. Solids*, **49**(12), 1988, p. 1441 and p. 1451.
[7] E.R. Weber, *Appl. Phys.* **A 30**, 1983, p.1.
[8] T.H. Wang, T.F. Ciszek, R. Reedy, S. Asher, and D. King, *Conference Record of the Twenty-Fifth IEEE Photovoltaic Specialists Conference*, Washington, D.C., 1996, p. 689.

# Current Status of HEM Grown Silicon Ingots

Chandra P. Khattak and Frederick Schmid

Crystal Systems, 27 Congress Street
Salem, Massachusetts 01970

**Abstract.** A fully automated HEM furnace has been set up to produce high performance multicrystalline silicon ingots for photovoltaic applications. Emphasis has been placed on design to achieve effective operation of these furnaces in developing countries. Silicon ingots of 58 cm square cross section, 200 kg have been produced with the labor requirement limited to loading and unloading of the furnace. A cycle time of 48 hours has been demonstrated and the material quality is very uniform.

## INTRODUCTION

The current mainstay of commercial photovoltaic systems in terrestrial applications is primarily crystalline silicon technology. Besides major producers there have been significant new plants set up worldwide and there are more plants in the planning stage. The developing countries have taken a leading role in this expansion and these plants are based upon crystalline silicon photovoltaic systems. While there has been major emphasis on module production and solar cell processing, there is interest in further integration and plans to set up silicon material production. This paper discusses the concerns of technical know-how and infrastructure required to support silicon material production in developing countries.

It is desirable that silicon material production in developing countries address at least the following requirements:
- high quality silicon is produced,
- the process requires minimum technical know how,
- high degree of automation to eliminate in-process decisions,
- equipment can operate with poor quality power with wide surges,
- damage to equipment is minimal in case of power outage,
- system is designed to minimize damage due to loss of coolant,
- equipment can operate under hot, humid and dusty environment.

If these requirements are not met, then developing countries will always be dependent on their material supply. Since major markets of photovoltaic systems are in developing countries, it is necessary to meet this challenge or the future growth of solar energy will be rather limited.

## INFRASTRUCTURE IN DEVELOPING COUNTRIES

Most developing countries have poor quality/unreliable power; this makes it difficult to operate high technology equipment with a high degree of automation. Other support facilities, such as consumables, maintenance, etc., are also limited. In some cases technical labor may be available but it is necessary to undertake pertinent training in the specific area of technology. Similarly, characterization facilities, if existing, need to be directed to meet the needs. There are usually strict regulations and long delays on procurement of spare parts. Therefore, it is a challenge to transfer technology and set up reliable production.

Silicon material technology involves production of high quality sheet for solar cell processing. This can be achieved by using ingot processes followed by slicing into wafers, or by using ribbon processes to produce silicon sheet directly from the melt. The requirements of precise temperature control and small melt volumes constrain the ribbon growing processes to operate effectively under poor quality input power. In comparison, ingot processes involve large melt volumes, insulated heat zones, tolerance of range of growth rates, etc. These features allow sufficient latitude in design of ingot growth furnaces to meet the needs of the developing countries.

Multicrystalline silicon ingots for photovoltaic applications have been produced using the Heat Exchanger Method (HEM). This material is ideally suited for high efficiency solar cells (1,2). Over 15% efficiency photovoltaic modules were produced using HEM silicon, the world's first 15% multicrystalline silicon modules (3). An HEM furnace was designed to meet the requirements of operation in developing countries without compromising the quality of silicon produced under those conditions.

## AUTOMATED HEM INGOT GROWTH

The design of the HEM furnace involved production of at least 55 cm square cross section ingots. This size was chosen as it gives optimum yield for nominal 100 mm x 100 mm solar cells (the standard of the industry), as well as nominal 125 mm x 125 mm solar cells (the perceived short term standard for the industry). Emphasis was placed on automation so minimum know-how of the process was necessary to operate under normal conditions; however, there was flexibility to make changes in the processing to optimize ingot growth under prevailing local conditions. While automation achieved the goal of easier technology transfer, it put additional constraints on operation under poor quality and unreliable power conditions. At all stages it was important to retain the low cost manufacturing features of the process.

A heat zone was designed to accommodate the desired size of the crucible and the relevant supports for the crucible. The interior section of the heat zone was fabricated using high density graphite and graphite felt insulation. The heat zone was designed with controlled heat extraction. During the melting there are no temperature gradients so that a rapid meltdown cycle can be carried out; however, during the growth stage gradual temperature gradients are introduced at the bottom of the heat zone so that the heat losses are primarily through the bottom of the crucible, while the sides and top portions remain insulated. With this approach large grains are nucleated and growth progresses with near vertical orientation of the grain boundaries. The HEM furnace uses a circular, resistive heating element and a square cross section crucible. The corners of the crucible are significantly closer to the heat source than the sides. Heat flow was designed to minimize this effect such that the growth interface was "squarish" rather than circular. The corners will still be the last material to solidify; however, the curvature of the solid-liquid interface near the corners will not be very sharp.

A saturable core reactor, 150 KVA power supply was built with an input voltage range of 350 to 480 V, 3 phase, 50/60 Hz AC. An instrument transformer converted the incoming power to nominal 120 V, single phase, to feed the motors, instruments and the control cabinet. An uninterruptible power supply (UPS) was incorporated between the power supply and the control instrumentation so that in the event of a power outage, the control instrumentation effects an orderly shutdown of the growth run in progress. Emphasis was placed on signal conditioning to compensate for poor quality and unreliable power. Once the power problems were addressed, it was possible to set up the entire control through a computer. This offered customization of programming, screens, data logging, process monitoring, alarms, etc. View screens were developed with color coordinated, schematic representations as well as detailed data reporting so that the operator could decipher any major deviations from the programmed course without reading the data points, even from a distance. The system was designed using multiple screens so that a range of information/data could be displayed; this range extended from limited information of interest to the operator, to the most detailed data of the process. The range of parameters was flexible and set by "master parameters"; any deviations set off visual and audible alarms which would immediately display the alarm screen.

The conventional HEM furnace used a helium cooled heat exchanger; this may present problems with availability of helium gas in some countries. The heat extraction system for the present furnace does not require helium gas and is designed so that after meltdown of the charge, heat losses were slowly increased from the bottom of the crucible. During this stage the furnace temperature was also decreased slowly from the melting temperature towards the growth temperature. Controlled nucleation of large grains was achieved at the bottom of the crucible, and thereafter, growth progressed with near vertical orientation of the grain boundaries. This approach insured growth with a slightly convex solid-liquid interface so that there were no impinging interfaces. At the growth temperature, uniform steady growth towards the top surface resulted from the temperature gradient imposed at the bottom of the crucible. When the solidification front approached the surface of the melt, circular solid could be observed at the center and this solid grew across the top surface. Rapid solidification was observed across the entire surface. After complete solidification the temperature gradient on the ingot was reduced in a controlled manner so that the ingot was annealed prior to a controlled cooldown.

## RESULTS

The HEM furnace was used to produce 55 cm square cross section, 155 kg multicrystalline ingot growth with a cycle time of 48 - 56 hours (4). Recently, it has been reported (5,6) that for multicrystalline ingots grown in silica crucibles the minority carrier lifetime was lower for silicon near the surface of the ingot. Silicon ingots of 58 cm square cross section, 200 kg were grown to yield high performance 25 silicon bars of 100 mm x 100 mm x 220 mm or 16 bars of 125 mm x 125 mm x 220 mm size so four of these bars can be used effectively to engage the full capacity of multiwire slicers. The total cycle time for production of a 200 kg HEM ingot was 48 hours; the profile of the various segments of ingot growth are shown in Figure 1. The furnace was loaded with charge, the chamber was closed and the process cycle initiated by the operator. The computer carried out pre-start checks, evacuated the chamber, determined that the rate-of-rise test was within the desired limits and started the process cycle via the program. During the entire cycle operator involvement was limited to loading and unloading of the furnace.

The 200 kg multicrystalline silicon ingot produced from the automated HEM furnace is shown in Figure 2. The ingot was sectioned parallel to one of the sides through the center and the structure of half such a section is shown in Figure 3. Another section was produced so that the cut was vertical along the diagonal of the square cross section ingot; this section is shown in Figure 4. A comparison of the grain size and orientation of the grain boundaries in Figures 3

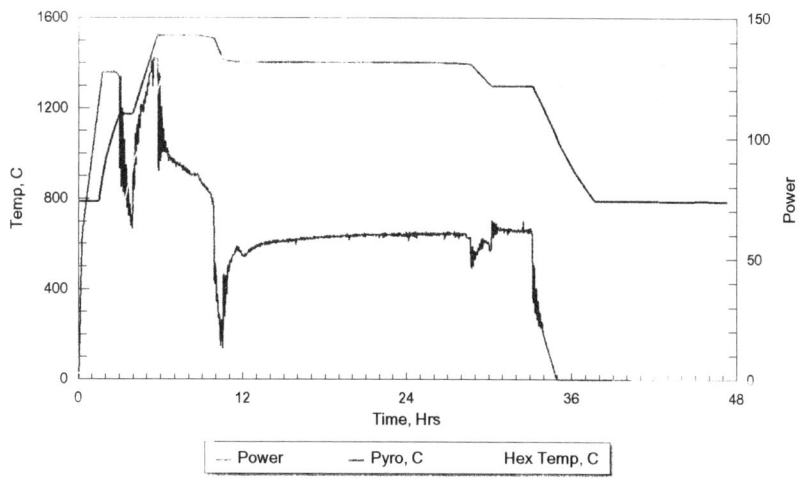

**FIGURE 1.** Process parameters for growth of 200 kg HEM multicrystalline silicon

and 4 shows that the structures are quite similar. This shows that the heat flow in the HEM furnace has been controlled so that uniform growth can be achieved even when the ingot cross section is square and the heat zone is circular.

Solar cells fabricated from HEM multicrystalline silicon ingots produced from the automated furnace have demonstrated high efficiencies (7-9), including the world record highest efficiency of 18.6% on multicrystalline silicon (9).

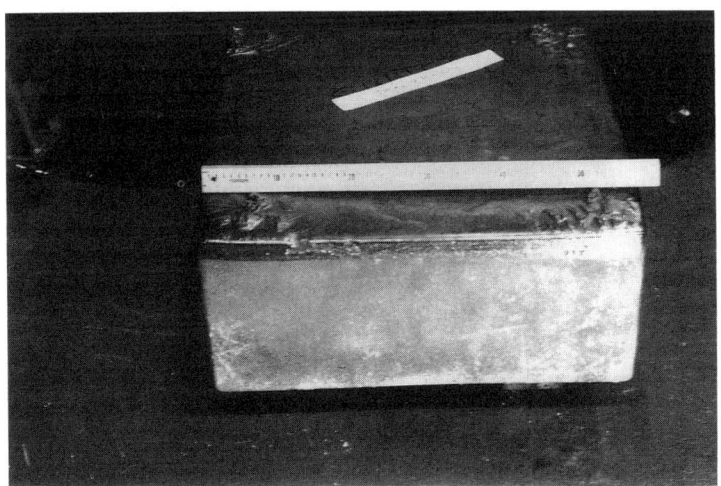

**FIGURE 2.** A 58 cm square cross section, 200 kg HEM multicrystalline silicon ingot

**Figure 3.** Structure of a 58 cm square cross section 200 kg HEM multicrystalline silicon ingot

**Figure 4.** Structure of the ingot shown in figure 3 cut along the diagonal of the square ingot

# REFERENCES

1. Khattak, C. P.; Schmid, F.; Schubert, W. K.; Cudzinovic, M.; Sopori, B. L. "Characteristics of HEM Silicon Produced in a Reusable Crucible." *The Conference Record of the Twenty-Third IEEE Photovoltaics Specialists Conference, Louisville, Kentucky, 1993, pp. 73-77.*

2. Khattak, C. P.; Schmid, F.; Schubert, W. K. "High-Efficiency Solar Cells Using HEM Silicon." *Proceedings of the 1994 IEEE First World Conference on Photovoltaic Energy Conversion, Waikoloa, Hawaii, 1994, pp. 1351-1355.*

3. King, D. L.; Schubert, W. K.; Hund, T. D. "World's First 15%-Efficient Multicrystalline Silicon Modules." *Proceedings of the 1994 IEEE First World Conference on Photovoltaic Energy Conversion, Waikoloa, Hawaii, 1994, pp. 1660-1662.*

4. Khattak, C. P.; Schmid, F. "Low Cost Multicrystalline Silicon for High Efficiency Solar Cells." *American Institute of Physics Proceedings 353, 13th NREL Photovoltaics Program Review, Lakewood, CO, 1995, pp. 118-125.*

5. Borne, E; Goaer, G.; Sarti, D.; Laugier, A. "3D Distribution Study of Impurities into a Polix Ingot." *Volume II of Proceedings of the Thirteenth European Photovoltaic Solar Energy Conference, Nice, France, 1995, pp. 1340-1343.*

6. Ferrazza, F.; Margadonna, D. "New Developments and Industrial Perspectives of Crystalline Silicon Technologies for PV." *Thirteenth European Photovoltaic Solar Energy Conference, Washington, D.C., 1995, pp 3-8.*

7. Khattak, C. P.; Schmid, F. "Automation in HEM Silicon Ingot Production." *Twenty Fifth IEEE Photovoltaic Specialists Conference, Washington, D.C., 1996, pp. 597-604.*

8. Narasimha, S.; Kamra, S; Rohatgi, A.; Khattak, C. P.; Ruby, D. "The Optimization and Fabrication of High Efficiency HEM Multicrystalline Silicon Solar Cells." *Twenty Fifth IEEE Photovoltaic Specialists Conference, Washington, D.C., 1996, pp. 449-452*

9. Rohatgi, A.; Narasimha, S.; Kamra, S.; Doshi, P.; Khattak, C. P.; Emery, K.; Field, H. "Record High 18.6% Efficient Solar Cell on HEM Multicrystalline Material." *Twenty Fifth IEEE Photovoltaic Specialists Conference, Washington, D.C., 1996, pp. 741-744.*

# Distributed Control and Process Monitoring for Photovoltaic Applications

M.D. Landry, Y.S. Tsuo, T.F. Ciszek, R. Roze,* and D. Hoegh**

*National Renewable Energy Laboratory, Golden, Colorado*
*\*Texas Instruments, Dallas, Texas*
*\*\*TMMI, Inc., Lakewood, Colorado*

**Abstract.** Distributed process control and monitoring in a multitasking personal computer environment improves the ease of use, reproducibility, and safe operation of photovoltaic solar cell processing systems. In this paper, we report the experience learned from our recent upgrading of a Tempress high-temperature furnace system at the National Renewable Energy Laboratory (NREL), which is used for processing crystalline-silicon solar cells in a clean-room environment. The furnace has eight quartz tubes with three temperature zones per tube, for a total of 24 temperature control loops. Customized programs and hardware interface were installed for automated control of the furnace and to provide extensive safety interlocks.

## INTRODUCTION

Photovoltaic (PV) solar cells are large-area semiconductor devices that need to be cost competitive with conventional energy sources. A successful PV solar cell processing line must have high throughput, high yield, and low cost. We believe distributed control and process monitoring based on multitasking personal computers (PCs) can help PV solar cell manufacturers achieve such goals. We recently upgraded a high-temperature furnace using a Pentium-processor-based PC running under the IBM OS-2 operating system (Fig. 1). In this paper, we discuss the design and lessons learned from this project.

Customized programs were written using Paragon TNT process monitoring and control software with a touch screen for user input. All PID (proportional-integral-derivative) temperature control, process gas valve sequencing, mass flow control, and process safety interlock functions are controlled via a multilevel, graphical user interface. The hardware interface for process control input and output (I/O) are done using commercially available Opto 22 I/O modules. The PC is networked to allow installations of an on-line database, remote process monitoring and control, and report functions. Additionally, over-temperature control and I/O watchdog functions are hardwired into the system for added safety.

## HIGH-TEMPERATURE FURNACE

Thermal dopant diffusion, oxide growth, and metal annealing and alloying are critical process steps in fabricating crystalline-silicon solar cell devices. Due to the completion of NREL's Solar Energy Research Facility (SERF), the existing crystalline-silicon diagnostic cell fabrication facility was to be moved to the new cleanroom user facility located in the SERF. To complete this move, new cleanroom-compatible equipment would have to be purchased and existing equipment would have to be made cleanroom compatible. A critical component of this laboratory is the high-temperature diffusion furnace. The furnace system is a Tempress brand, model 230, dual side-by-side vertical 4-tube stack arrangement (for a total of 8 process tubes, 3 temperature controlled heat zones per tube), capable of processing up to 6-inch-diameter silicon substrates. In essence, this furnace system is 8 individually controlled tube furnaces packaged together within a single frame. Originally purchased in 1980 for R&D use, this furnace system was the state of the art in production silicon wafer processing and allowed for a wide variety of processes to be initiated. Over time, many of the original microprocessor-based process and temperature controllers, as well as gas valves and MFCs (mass flow controllers), failed due to age, and as such, several of the furnace tubes were non-operational. Replacement electronic controllers were becoming difficult, if not impossible, to purchase. We decided to upgrade the existing furnace to make it cleanroom compatible by "bulkheading" the front face of the furnace into the cleanroom wall, thereby limiting access to the rear mechanical part of the furnace system. This necessitated updating the furnace control system with modern automatic process controllers and electropneumatic valving to allow remote operation from the cleanroom area. The heart of the Tempress 230 system, consisting of the 8 heater core elements, power transformers, SCR (silicon controlled rectifier) controlled firing circuits, stainless-steel gas scavenger system, and stainless-steel cleanroom-compatible front-loading access panels were in excellent condition, making this piece of equipment a good candidate for a control-system upgrade.

Because most of the furnace mechanical system would be bulkheaded behind the cleanroom wall in a service anteroom, access to the valves and electronic controls would be limited. The entire gas manifold system would have to be redesigned and reinstalled to allow for remote operation from within the cleanroom. A new gas distribution system, using new high-purity stainless-steel bellows valves, pneumatically actuated virgin Teflon diaphragm valves, new stainless-steel and Teflon tubing, low-voltage electropneumatic valve actuators, and electronic MFCs would be installed to ensure cleanliness and functionality. New automatic temperature-control and valve-control electronic hardware and software had to be installed. Safety interlocks consisting of software and hardwired redundant controls needed to be installed to prevent damage to the furnace and provide a safe operating environment. A cleanroom-compatible man/machine interface for furnace operation was also a consideration. Clarity and user friendliness to allow cleanroom staff to operate the furnace system safely and effectively was also a concern.

## Control System

After some research and outside consultation with process engineers involved in the integrated-circuit manufacturing industry, we decided on a distributed control system (DCS). This system, consisting of a host PC, analog and digital input/output electronic hardware, coupled with sophisticated instrumentation and control software, would be the most economical and flexible system-control option. We wanted a system that would allow us to eliminate the purchase of discrete computer-programmable automatic temperature controllers (because we needed control 24 individual heat zones), as well as discrete MFC readout controllers. Because this furnace was to be used in a research environment, off-the-shelf modular components would be used wherever possible to simplify maintenance and future modifications and upgrades. We ultimately decided to use an industrial automation and control software package from Intec controls, called Paragon TNT. This software package allowed us to design a furnace control program using multiple PID control loops. Operating in multitasking environment, it allows the simultaneous operation of multiple furnace tubes. The use of Paragon TNT also enabled us to design a multilayer graphical user interface system for furnace control and process programming. These user interface screens, coupled with an ELOgraphics touch screen monitor and mouse-emulation software, would facilitate furnace process programming and operation in a cleanroom environment. Paragon TNT was designed for operation within a multitasking environment, so we chose a Dell Pentium 120 MHz PC with 32 Mb of RAM to be used as the host PC, running under the IBM OS-2 Warp operating system. (Paragon TNT is now available for the Windows NT environment also.) All input/output signals from the furnace hardware to the host PC are accomplished using off-the-shelf Opto 22 hardware interface terminal boards. This architecture is modular in design, consisting of plug-in analog and digital I/O modules for various applications, such as thermocouple read-in, zero-cross firing power control output to the SCRs (back-to-back silicon controlled rectifiers, used to control the AC power to the heating elements), mass flow controller analog setpoint and readout, and electropneumatic valve control. With the Opto 22 I/O boards installed at the point of use (the rear of the furnace, where the gas manifold systems and SCR power control hardware is located), only one small instrumentation cable, consisting of two twisted shielded pairs, needs be run back to the host PC, thereby allowing easy installation of the host PC in the cleanroom area.

## Functions and Capabilities

The operator interface graphics are multilayered, consisting of a "main" page screen displaying real-time independent status of all eight individual tubes using the actual furnace front tube configuration of two side-by-side vertical 4-tube stacks. This main page allows the user random access to any individual tube at any time, regardless of current process status. Tubes not in use show a "green" tube number circular graphic. If a tube is in process, the number changes to "red" status. Selecting a tube (by merely touching the tube number graphic) brings up the next layer of screen, the process-tube control screen. This screen is a schematic representation of the selected individual three-zone furnace tube and gas manifold valve system. Displayed are individual zone temperatures, current temperature setpoint, valve status, gas flow graphics, MFC flow rates, and zone-specific PID time proportional output. Because this system is to be used as a research tool, the control screens were designed to be operated in either manual or automatic process control modes. The user can individually control the various gas valves and temperature setpoints manually, or select a variable process input screen, whereby the user inputs the multilevel process parameters (process time, ramp rate, gas flow, purge time, ramp down). Temperature and gas flow setpoints are entered by selecting the temperature setpoint or MFC flow button, calling up a large, easy-to-read numeric, calculator-style touchscreen keypad. A three-pen screen-displayed chart recorder can be called up using the "Trend" button, which displays time/temperature uniformity over the three zones of each tube.

## SAFETY INTERLOCKS

Safety interlocks include valve priority lockouts, preventing the accidental opening of inappropriate and/or hazardous valve combinations. This function can be overridden by authorized personnel for maintenance and repair purposes. The system is security locked by log-in passwords and operator-level authorization passwords. The system can be networked to allow remote monitoring and control from other PCs throughout the lab. These data can be stored and analyzed at a later date for statistical process control purposes. The multitasking software/PC combination replaces the tube-level individual PLCs (programmable logic controllers), simplifying the installation and reducing the overall cost.

Additional safety systems include a stand alone UPS (uninterruptable power supply) attached to the system-control PC to prevent loss of control in the event of a power failure. The OS-2 operating system and Paragon TNT software are extremely robust. But in the event of the PC itself or any component of the control software were to fail, the Opto 22 I/O interface boards have a built-in "watchdog" function, which can be configured to turn off all gases and SCR driver outputs to prevent damage to the furnace. To protect the furnace system from catastrophic meltdown, fail-safe hardware interlocks were installed, consisting of dedicated over-temperature controllers, each with an independent thermocouple for each furnace zone. This system will automatically open the shunt trip coil on the heater element circuit breaker for the specific tube encountering the over-temperature condition.

## COST ANALYSIS

After soliciting quotes for a new eight-tube furnace, as well as from various companies that specialize in vacuum and furnace system retrofits for the integrated-circuit industry, we determined that for our R&D application, designing and installing our own system would be the most cost effective. The cost of a new, similarly configured furnace was between $750,000 and $1.2 million. Quotes from commercial furnace rebuild companies ranged from $50,000 to $100,000 per tube.

Our cost breakdown is as follows: about $48,000 for hardware, including all Opto 22 modules and boards, all new valves, actuators and plumbing manifold hardware, all wiring and over-temperature controllers, the host PC and controller cards, touchscreen monitor and related software; and about $40,000 of in-house labor at about $50 per hour for 800 hours for work such as layout and design, software programming, hardware installation, and startup.

## ACKNOWLEDGMENTS

This work was supported by the U.S. Department of Energy under Contract No. DE-AC36-83CH10093.

Figure 1. Furnace System Diagram

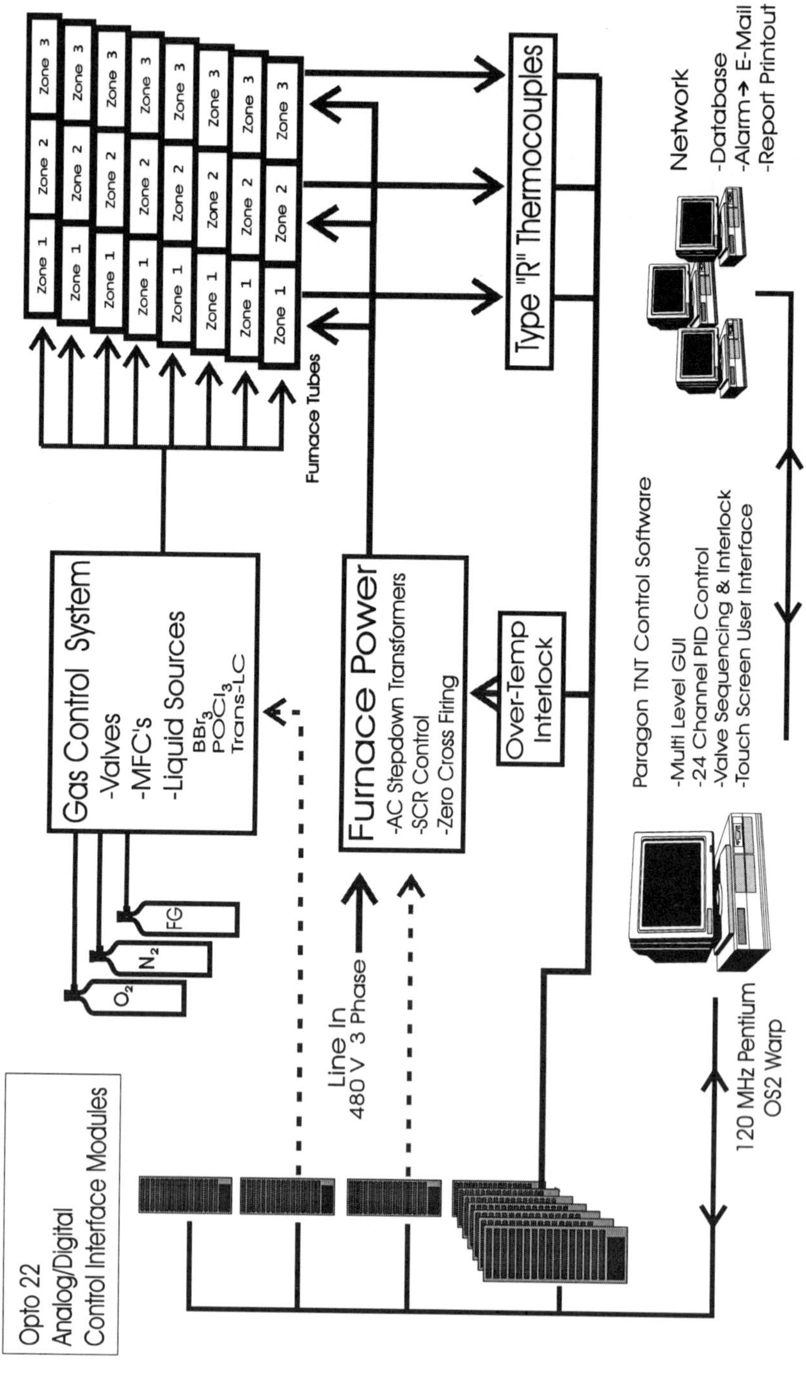

# COMPONENT AND SYSTEM EVALUATION AND RELIABILITY (P)

# Development of New EVA Formulations for Improved Performance at NREL

F. J. Pern

Center for Performance Engineering and Reliability
National Renewable Energy Laboratory (NREL)
1617 Cole Blvd., Golden, Colorado 80401, USA

**Abstract.** We review in chronological order the research stages and fundamental concepts involved in developing modified and new EVA formulations for improved performance against photo-induced degradation and discoloration. The new NREL EVA formulations use additives totally different from the present commercial formulations (EVA A9918 and EVA 15295). Validation of their long-term photostability and thermostability is presently under way. Together with UV-absorbing glass superstrates, they may offer better success in achieving a more reliable module performance and longer service life without significant EVA discoloration problems, which are commonly experienced with EVA A9918 and, at a lesser rate, EVA 15295.

## INTRODUCTION

Degradation and discoloration of photovoltaic (*PV*) module encapsulants made of ethylene-vinyl acetate (*EVA*) random copolymer have been observed mostly on crystalline silicon (*Si*) modules deployed in hot-dry and hot-humid areas (1-4). One of the most severe EVA browning cases was reported for the PV modules of the Carrisa Plains power plant in California, where more than 45% of the power loss in five years was associated with the browned EVA and other electrical degradation (4,5). Compositional analysis of the discolored EVA films removed from a number of modules shows that they lost significant transmittance in the visible spectrum because of yellowing or browning; lost most of the ultraviolet (*UV*) light absorber, Cyasorb UV 531, from their initial formulations; and gained a substantial increase in the gel content (degree of cross-linking) (1,2). In addition to the delamination observed on some modules, electron probe microanalysis (*EPMA*) and Auger depth-profiling analysis of the Pb-Sn coated Cu ribbons (tabs) removed from solar cells of Carrisa modules under the browned EVA layers revealed a great degree of oxidation and corrosion of the metallic ribbons (6). Attempts to determine by electron-beam-induced current (*EBIC*) analysis whether the Pb and Sn lost from the Cu ribbons diffused into and affected the cell's p-n junction were not successful because of the "electron-mirror" effect on the back side of the Si cells (7). A critical review on the issues of EVA discoloration was given by Czanderna and Pern (4).

This paper summarizes our research and development efforts conducted in the Module Encapsulation Research Laboratory (*MERL*) at NREL. Results from more than five years' research work have produced better insight into the fundamentals related to EVA degradation mechanisms, the effects of stabilizing additives used in the formulations on the EVA yellowing, and the factors that affect the EVA discoloration rate. These fundamental understandings were accomplished by extensive analytical measurements, specifically the fluorescence, transmittance, and color indices, and correlation of the results. The understandings have enabled us to modify the two commercial EVA formulations (see **Table 1**), developed jointly in the 1980's by the Jet Propulsion Laboratory and Springborn Laboratories (8,9), and create new EVA formulations. The NREL modified and new EVA formulations have shown significantly improved photostability against UV-induced discoloration in recent experiments.

**TABLE 1. Formulations of EVA Encapsulant and Chemical Additives[1]**

| Trade Name | Weight (pph) | Chemical Function | Manufacturer (Remark) |
|---|---|---|---|
| Elvax 150 | 100.0 | EVA base copolymer | Du Pont |
| Cyasorb UV 531 | 0.3 | UV absorber | American Cyanamid |
| Tinuvin 770 | 0.1 | UV light stabilizer | Ciba-Geigy |
| Naugard P | 0.2 | Antioxidant | Uniroyal |
| Lupersol 101 | 1.5 | Curing agent (slow cure) | Elf Atochem (for EVA A9918) |
| TBEC | 1.5 | (fast cure) | (for EVA 15295) |

| Trade Name | Chemical Name |
|---|---|
| Elvax 150 | Ethylene vinyl acetate random copolymer (EVA-33; 67 wt% ethylene and 33 wt% vinyl acetate) |
| Cyasorb UV 531 | 2-Hydroxy-4-n-octyloxybenzophenone |
| Tinuvin 770 | Bis(2,2,6,6-tetramethyl-4-piperidinyl)sebacate |
| Naugard P | Tris(mono-nonylphenyl)phosphite |
| Lupersol 101 | 2,5-Bis(tert-butyldioxy)-2,5-dimethylhexane |
| Lupersol TBEC | OO-t-Butyl-O-(2-ethylhexyl)-mono-peroxycarbonate |

1. A primer may be added to improve the adhesion to other materials in the module. Then, the formulations will be labeled as EVA A9918P and 15295P.

# EXPERIMENTAL

**Materials and sample preparations.** In earlier work, the samples were prepared and tested either in solutions of nonpolar cyclohexane or as thin films cast in cuvettes that allowed direct absorbance measurements. After a minipress with two heated platens became available, samples of laminated films between two borosilicate slides were then made in the ambient with controlled temperatures and heating times. In the last two years, laminated and cured samples were made in a custom-built double-bag vacuum laminator with a microprocessor controller, using programs as described elsewhere (10). Formulation additives of various UV absorbers (*UVA*), UV light stabilizers (*UVS*), and antioxidants (*AO*) were used as provided from several sources without further purification. Elvax 150,$^{TM}$ the plain EVA copolymer without any formulation additives, and other polymers were acquired from Du Pont and Aldrich Chemicals. The pellets of Elvax 150 were treated by soaking in two aliquots of methanol for two days, followed by drying in a vacuum oven at 35°-40°C before being used. Concentrations or quantities of additives (molar or weight ratios) used in the modified formulations were determined by detailed calculations prior to sample preparations. Thicker films of various formulations were cast and dried gradually in Petri dishes from HPLC-grade cyclohexane or tetrahydrofuran solutions before they were used in subsequent lamination and curing processes.

**Exposure and spectroscopic analysis.** Depending on the experiment design, exposures of the samples were selectively made with one of three light sources, which were an Oriel condensable 1-kW Xe arc lamp operated in a condensed mode (~1-in.-diameter light beam size and a ~17-sun intensity integrated over the 300-400 nm range), an Oriel 1-kW enhanced-UV solar simulator (~4.8 suns in the 300-400 nm range measured under a 1/8" borosilicate filter plate), and DSET Suntest CPS tabletop exposure systems (~1.1-1.2 suns in the 300-400 nm range). The conditions and length of exposure are indicated in each of the figures illustrated. The exposed samples were measured periodically with an HP-8452A UV-vis spectrophotometer, a HunterLab UltraScan spectrocolorimeter, and/or a SPEX Model FL112 fluorescence spectrophotometer.

# DEVELOPMENT STAGES AND RESULTS

## Photodecomposition and Stabilization of UV Absorbers

Results from accelerated exposure experiments show that the UVA, Cyasorb UV 531 (referred to as Cyasorb hereafter), used in the EVA formulations to protect the EVA from UV-induced photodegradation, does not have good photostability. As illustrated in Fig. 1a, its rapid decomposition in the nonpolar cyclohexane is further enhanced in the presence of Elvax 150. Subsequent additions of the UVS (Tinuvin 770) and AO (Naugard P) can only delay, but not prevent, its

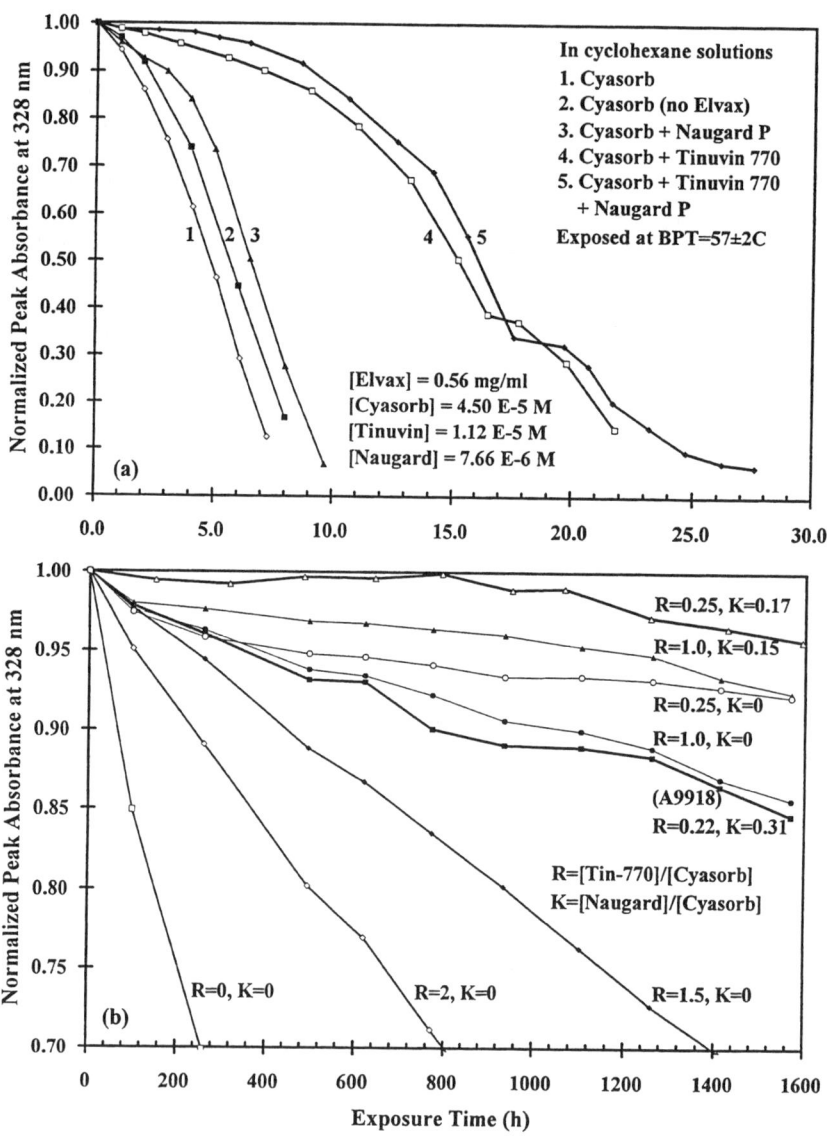

**Figure 1.** Photodecomposition of Cyasorb and its stabilization by the separate or joint presence of Tinuvin 770 and Naugard P in (a) cyclohexane solutions and (b) Elvax 150 (*EVX*) films exposed to condensed 1-kW Xe light filtered by an aqueous $CuSO_4$ solution and a long-pass 305-nm filter at a black panel temperature (BPT) of $57° \pm 2°C$. The cyclohexane solutions contained Elvax 150 and were used to cast the EVX films in cuvettes used in (b). The concentrations of each component or their ratios are indicated in the figures.

decomposition. Cyasorb's photodecomposition in Elvax 150 thin films cast in cuvettes behave similarly (Fig. 1b). Without Tinuvin 770 and Naugard P, the Elvax thin films become highly crosslinked, indicating that the Cyasorb and/or its decomposition products promoted the crosslinking. In fact, Pickett and Moore reported that Cyasorb UV 531 decomposes into free radical products (11). The free radicals could subsequently induce crosslinking reactions in the Elvax films.

The photostability of Cyasorb is also affected by the concentration ratio of Tinuvin 770 (UVS) and Naugard P (AO). A proper concentration ratio among the three produces a better stabilization of Cyasorb, as illustrated in Fig. 1b. The photostability of Cyasorb can also be improved by using other UV light stabilizers and antioxidants. Figure 2 demonstrates the effective stabilization of Cyasorb by some combinations of various UVS and AO additives (6). Similar studies were conducted using a number of other UV absorbers in combination with various UV light stabilizers and antioxidants. The accumulated results have established the foundation for the NREL EVA reformulations, because there are potentially better combinations of UV absorbers, UV light stabilizers, and antioxidants. In the absence of a laminator or even a minipress with heated platens, however, these early studies were limited to solutions and cast thin films in cuvettes.

## Comparison of Photochemical and Thermochemical Discoloration

To differentiate the discoloring effect of EVA by UV light and simple heating, two sets of solar cells laminated with EVA A9918 between a glass superstrate and a Tedlar substrate were exposed separately to heating in an oven and to UV irradiation from three 100-W RS4 lamps (General Electric) in a test chamber with a turntable. The tests were conducted at a controlled temperature of $85° \pm 2°C$ for 198 days, as reported previously (12). The results clearly show that UV light is far more important than thermal heating in inducing EVA browning (12).

## Comparison of Photostability of EVA 15295 and EVA A9918

The photostability of the two commercially available EVA formulations, regular-cure A9918 and fast-cure 15295 with/without a primer, was compared using films that were laminated and cured between two borosilicate slides using a minipress fitted with two heated platens. In all experiments of accelerated exposure with the Oriel enhanced-UV solar simulator, EVA 15295 consistently performed better with a significantly slower discoloration rate, as shown in Fig. 3 (13). Because the additives used in the two EVA formulations are identical, except for the curing agent, the earlier and faster discoloration rate for the EVA A9918 is attributed to a greater concentration of curing-generated, UV-excitable chromophores, as determined from fluorescence analysis, and a higher residual content of Lupersol 101, as calculated from the half life and activation energy of the Lupersol curing

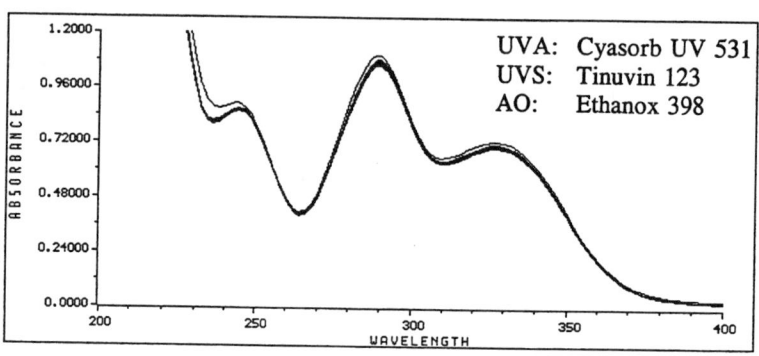
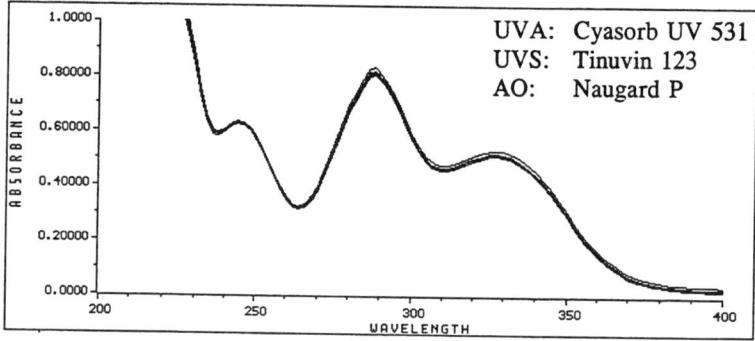
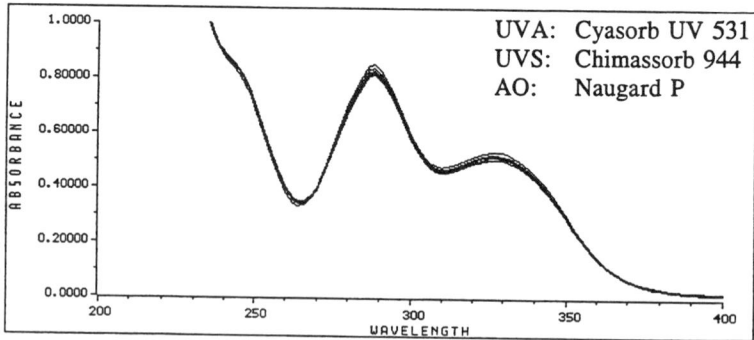

**Figure 2.** Photostabilization of Cyasorb UV 531, a benzophenone-type UV absorber, by different combinations of UV light stabilizers (Tinuvin 123, Chimassorb 944) and antioxidants (Ethanox 398, Naugard P) embedded in Elvax 150 thin films and cast in cuvettes. The samples were exposed to 1.8-kW Xe light for 1200 h in a DSET Suntest CPS tabletop system operated at 750 W/m$^2$ and a chamber BPT=57° ± 2°C. The absorbance spectra of the exposed films were measured periodically with an HP-8452A UV-vis spectrophotometer.

agents (13). Accordingly, we recommended to PV module manufacturers in 1994 that, until better EVA formulations or new polymer pottants become available, it is better to use the fast-cure EVA 15295 for crystalline-Si PV modules. The reasons include better photostability against UV-induced degradation and discoloration, greater economy in saving energy consumption, and higher production yield from a shorter processing time (13).

The NREL results indicate clearly that, to achieve a better photostability against a fast discoloration rate, it is important to reduce or minimize the concentration of curing-generated chromophores and the concentration of curing peroxide residue in the EVA laminates. As is well known by some PV manufacturers and also observed in our work, the fast-cure EVA may produce bubbles in the laminates if processing conditions are not well controlled. Therefore, we concluded that a better curing agent, which will cure fast, produce no bubbles, and generate less chromophores simultaneously, is highly desirable. We have identified better curing agents in our studies (7,10).

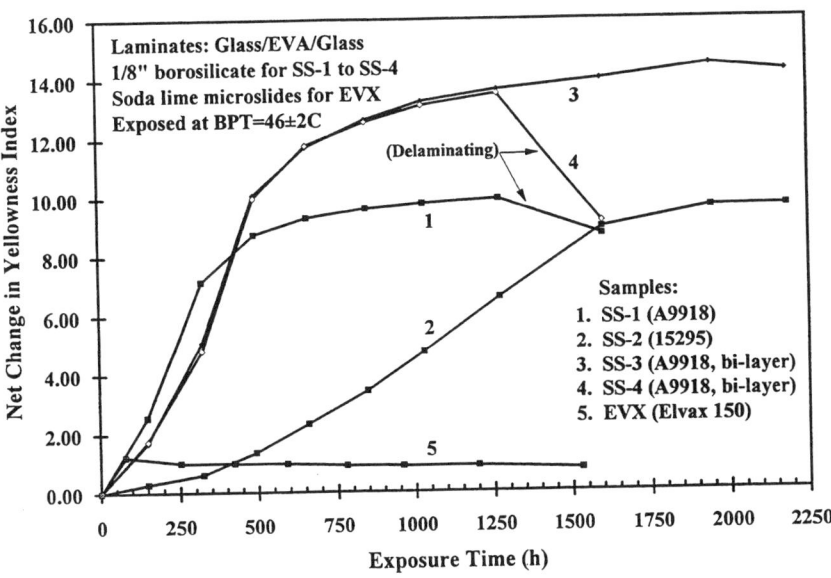

**Figure 3.** Net changes in yellowness index (*YI*) as a function of exposure time for four EVA film samples (curves 1, 2, 3, and 4) laminated between two 1/8"-thick borosilicate glass slides and cured, and a plain Elvax 150 film (curve 5) laminated between two soda lime microslides. Curve 2 is for EVA 15295, and curves 1, 3, and 4 are for EVA A9918 samples in which samples 3 and 4 used two layers of the EVA films in the lamination. Exposure to an Oriel enhanced-UV solar simulator's 1-kW Xe light was conducted simultaneously for all samples at BPT=46° ± 2°C.

## Discoloring Effects of Additives in Commercial EVA Formulations

The effect of the stabilizing additives presently used in the two commercial formulations on EVA discoloration was systematically investigated by using minipress-laminated and cured Elvax films. The results confirm that the (a) additives used in the EVA formulations, (b) curing conditions, (c) curing-generated, UV-excitable chromophores, and (d) type of peroxide curing agent promote *synergistically* the discoloration of EVA laminates. The results, which are important and provide further understanding of EVA degradation and discoloration mechanisms, facilitated our development of EVA reformulations and will be reported elsewhere (14).

## Effects of Curing-Generated Chromophores, Concentration Ratio, and Quantity of Additives

In the photodegradation of polymers, a common cause involves generation of free radicals in the presence of carbonyl groups and other UV-excitable chromophore moieties. Chemical reactions involving the propagation and termination of these free radicals may lead to the formation of double bonds, scission, crosslinking, and compounds of low molecular weight (15). Therefore, as pointed out earlier, it is highly desirable to reduce or minimize the concentration of the carbonyl and UV-excitable chromophore groups in laminated/cured EVA films. We found that the concentration of chromophores in cured EVA films is affected by curing temperature, curing time, and the type of curing agent, as discussed previously (16). The concentration of chromophores is found to be affected significantly by the antioxidant Naugard P, as shown in Fig. 4a. A thoughtful selection and combination of a new curing agent (L-X) and different antioxidants (AO-1, 2, 3, and 4) has resulted in greatly reduced concentration of chromophores in cured Elvax films (Fig. 4 [b and c]). The effect of the concentration of curing-generated chromophores on increasing the discoloration rate is demonstrated in Fig. 5 for a set of five Elvax film samples containing increased concentration of Naugard P in the presence of free-radical scavenger Tinuvin 770 (UVS) with correspondingly increased concentration (10). Other factors that have been identified to affect the discoloration rate of EVA were discussed previously (17).

In addition, we have concluded from a number of experiments that the actual amount of formulation additives (UVA, UVS, and AO) embedded in the Elvax is critical, even if their *concentration ratio* is close to the "optimal" as determined from previous studies. In general, using formulation additives in an amount greater than a "critical" point can be detrimental and can promote a relatively faster discoloration rate of laminated EVA films. The "critical" point, which is related to the relative concentration (molar or weight) ratios among the additives and Elvax, has to be derived experimentally for each combination of additives, however.

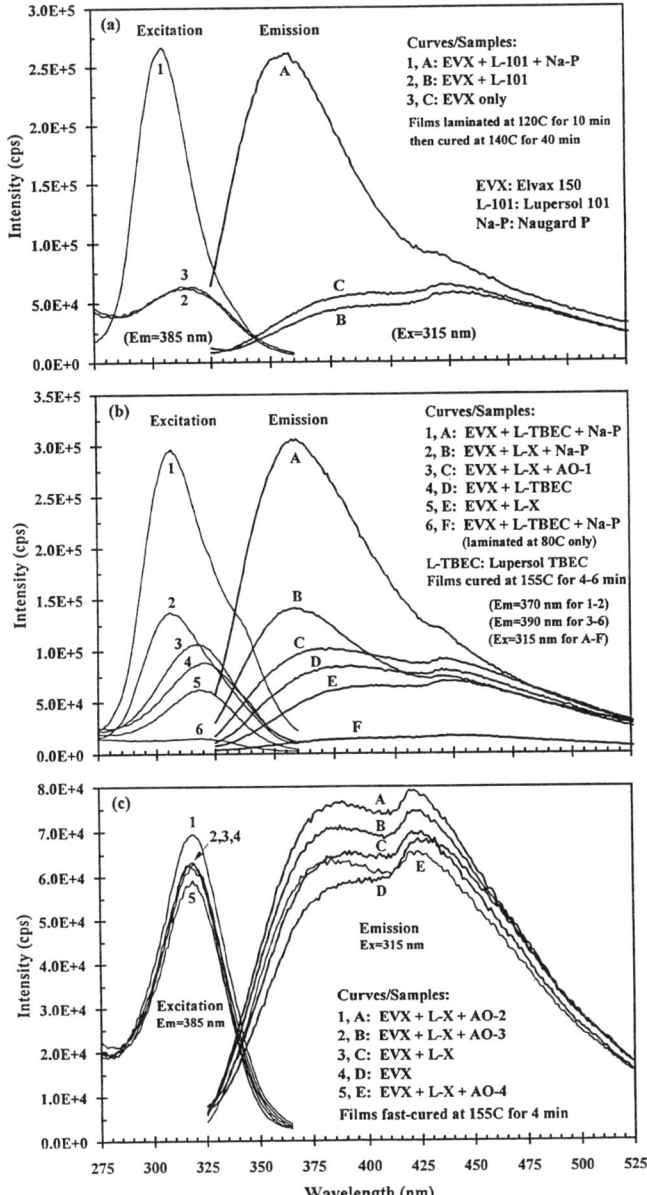

Figure 4. Fluorescence excitation and emission spectra (a) for cured EVX films that were prelaminated at 120°C for 10 min; (1, A): EVX added Naugard P (*Na-P*) and cured with L-101; (2, B): EVX cured with L-101; and (3, C): plain EVX but "pseudo-cured" as if with L-101; (b) laminated and cured EVX films impregnated with curing agents L-TBEC and L-X (see text) with or without antioxidant Na-P or AO-1, before and after curing at 155°C for 4 min to 6 min; and (c) laminated and cured EVX films with or without the curing agent L-X and antioxidants, AO-2, -3, and -4. The spectra are not subtracted from those of microslides. Note that the intensity scale in (c) is about one quarter of that used in (a) and (b).

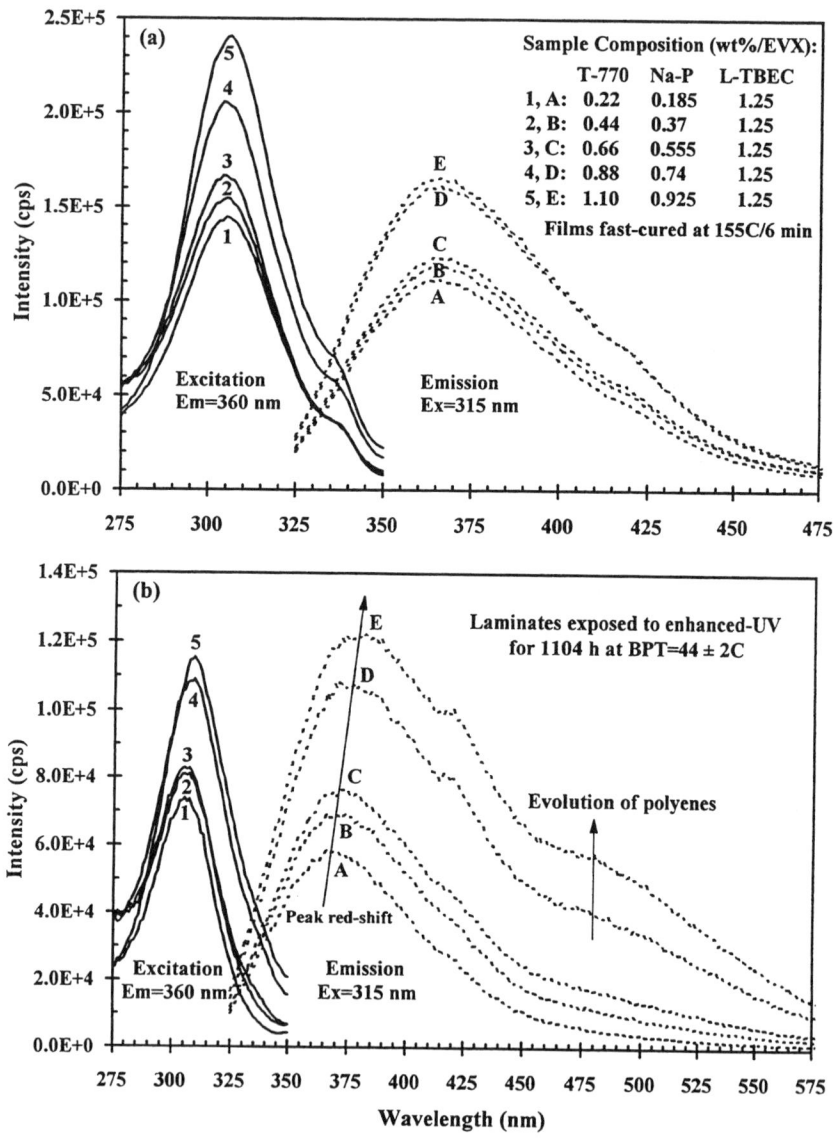

Figure 5. Fluorescence excitation and emission spectra for (a) as-cured Elvax 150 films laminated between two borosilicate slides and (b) UV-exposed at BPT=44° ± 2°C for 1104 h. The EVX films, which were cured with 1.25 wt% Lupersol TBEC, contained increasing concentrations of both Tinuvin 770 and Naugard P from sample 1 to 5 at a fixed molar ratio, as indicated in the figure.

## Modification of Two Commercial EVA Formulations

We have been able to modify with ease the present two commercial EVA formulations, A9918 and 15295, as given in Table 1 based on the results accumulated from our extensive experimental work and analysis, fundamental understanding of the critical factors that influence the discoloration rate of laminated/cured EVA films, and understanding of the chemical interplay of new UV light stabilizers, antioxidants, and UV absorbers. As an example, Fig. 6a demonstrates the significantly improved photostability of three modified EVA formulations compared with a discoloring sample of the EVA A9918 formulation (18). The samples are made of laminated and cured Elvax films cast from cyclohexane solutions. *Modifications of the commercial EVA formulations can be achieved by a number of ways and are the subject of a pending patent application.*

## New EVA Formulations

In addition to the modified EVA formulations above, several new stabilization additives, along with a new curing agent, have been identified experimentally that offer greater photostability to the laminated/cured EVA films. Hence, a number of new EVA formulations have been created. *The new EVA formulations do not use any of the stabilizers used in the A9918 and 15295 EVA formulations.* The advantages of using these new formulations are that they (1) produce significantly lower concentrations of chromophores upon curing (see Fig. 4 [b and c] for examples), (2) can use a shorter curing time without bubbling problems, and (3) have greater stability of the antioxidants against moisture and thermal decomposition (per product literature). These new EVA formulations are currently being subjected to accelerated UV exposure testing, and a preliminary result is shown in Fig. 6b. *An invention record has been registered at NREL, with a patent application pending for these new EVA formulations.*

## UV-Absorbing Glass and Gas-Permeable Polymer Superstrates

While modifying the commercial formulations and devising new formulations are a chemical means for improving the photostability of EVA encapsulants, we have also studied physical means since 1991 by testing various superstrates and substrates for PV modules. Most notably, cerium-containing thin-glass samples provided by OCLI in early 1991 and thicker (1/8") glass plates of Solarphire$^{TM}$ (formerly Airphire) from PPG in mid-1993 were found to be more effective than window and borosilicate glass of the same thickness in reducing substantially the EVA discoloration rate (13,18). A key factor is that the superstrate glass plate should be able to absorb (or filter) the UV light below ~350 nm to effectively block or reduce the rate of chromophore-induced photodegradation reactions. A

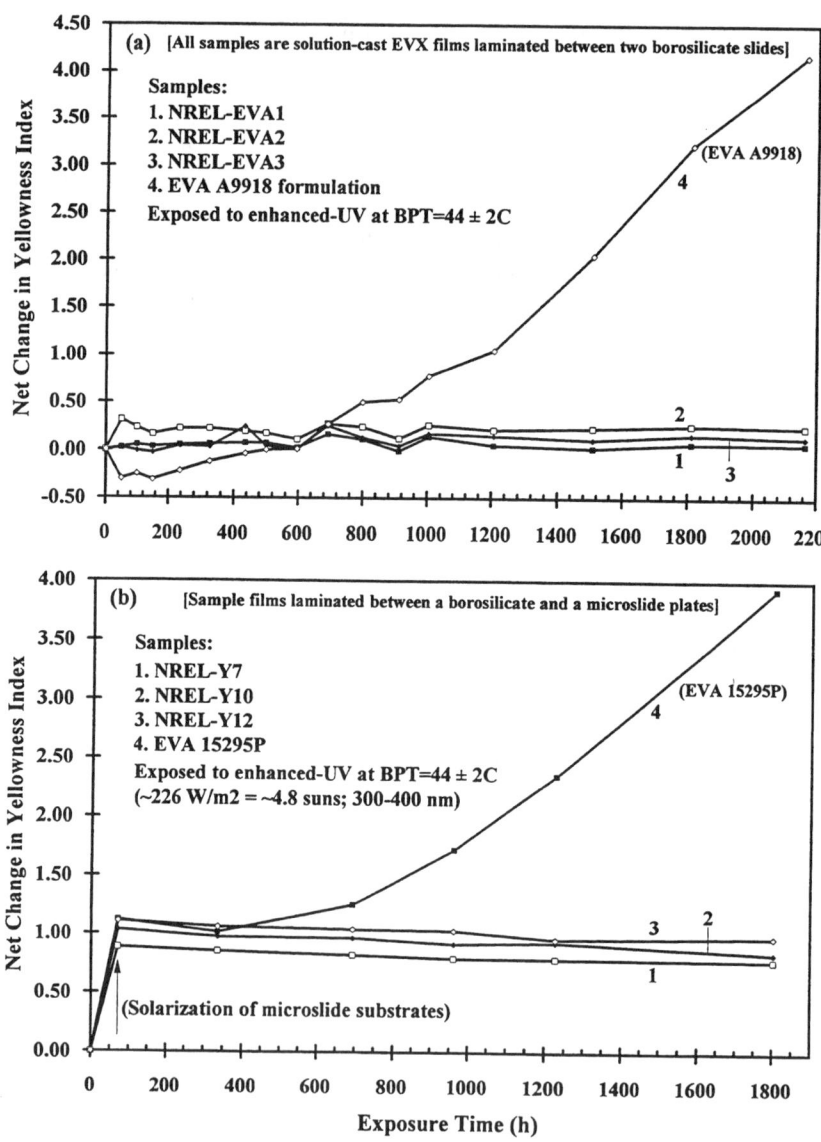

**Figure 6.** Net yellowness index changes obtained for (a) three non-discoloring EVA laminates of modified formulations (curves 1 to 3) and a discoloring film of EVA A9918 formulation (curve 4) laminated between two 1/8"-thick borosilicate slides; and (b) three non-discoloring new EVA laminates (curves 1 to 3) and a discoloring EVA 15295 film (curve 4) laminated between a 1/8"-thick borosilicate slide and a soda lime microslide. The samples were exposed to the Oriel enhanced-UV solar simulator's 1-kW Xe light at BPT=44° ± 2°C.

greatly improved photostability and longer service life may result from using both the UV-absorbing glasses as superstrates and the modified or new EVA formulations.

Another physical means to eliminate the discoloration problems of the commercial EVA is to use gas-permeable Tefzel films as superstrates. The gas permeability allows rapid destruction of curing-generated chromophores via the oxidative "photobleaching" reactions in the presence of air (oxygen) (13,18). However, long-term photochemical and mechanical stability of the Tefzel/EVA-encapsulated PV modules remains undetermined because the permeability of the Tefzel films to oxygen, moisture, and air pollutants is not insignificant. As an example, Hammond et al. recently reported corrosion of tinned Cu ribbons and delamination of the Tefzel superstrate films to a 20-80-in$^2$ size on Tefzel/EVA-laminated modules subjected to damp heat test at 85°C and 85 %RH for 1000 h (19). Thus, long-term effects of permeability of Tefzel or other polymer superstrate films on adhesion, oxidation/corrosion of metalization, and solar cells are research issues to be addressed in the future.

## CONCLUSIONS

After more than five years of extensive fundamental research and development efforts, modified and new EVA formulations have been created that show considerably improved photostability against UV-induced discoloration. These accomplishments are based on a detailed understanding of the fundamentals related to EVA degradation and discoloration mechanisms. Discoloration of laminated and cured EVA can be promoted or enhanced chemically and/or photochemically by (i) the additives in EVA formulations, (ii) the concentration ratio and quantity of additives used, (iii) the kind of peroxide curing agent, (iv) the thermal processing conditions (temperature, time, and pressure) employed for lamination and curing, and (v) a high concentration of curing-generated, UV-excitable chromophores. These factors affecting EVA discoloration rate can be suppressed, minimized, or eliminated by using new combinations of stabilization additives with desired chemical properties and stabilities. Results of performance validation of the modified and new formulations in the laminated form with crystalline-Si solar cells, with and without a UV-absorbing glass or gas-permeable polymer film superstrate, will be reported after accelerated weathering tests in weatherometers are completed in the near future.

## ACKNOWLEDGMENTS

The author is especially grateful for the technical assistance and contributions by S. H. Glick and the project support by A. W. Czanderna, R. DeBlasio, and R. Hulstrom of the Center for Performance Engineering and Reliability. The following NREL staff are thanked for their contributions: A. Mason for EPMA analysis; A. Swartzlander for Auger depth-profiling analysis; J. Webb and J. Johnson for FTIR and FTIR-ATR analysis; R. Matson for EBIC and OBIC analysis; K. Emery, H. Field, and D. Dunlavy for I-V and QE measurements of solar cells; and J. Folkvord for module and EVA analyses. L. Kazmerski is thanked for his encouragement. The following companies are gratefully acknowledged for providing samples of the stabilizers/additives (Ciba-Geigy), polymer pellets and adhesive films (Du Pont, Madico, 3M), Tefzel films (Du Pont, Clear Solutions), extruded EVA films with and without Tefzel overlay (Springborn Laboratories, Richmond Technology), solar cells with and without encapsulation (Siemens Solar and ASE Americas [formerly Mobil Solar]), and glass plates (OCLI, PPG). This work was performed with support of the U.S. Department of Energy under Contract No. DE-AC36-83CH-10093.

## REFERENCES

1. (a) Pern, F.J., "Recent Generic Studies of Ethylene Vinyl Acetate (EVA) Degradation," *Proc. of SERI 10th PV AR&D Review Meeting/Photovoltaic Module Reliability Workshop*, Oct. 23-26, 1990, Lakewood, CO, pp. 279-299. (b) Pern, F.J. and Czanderna, A.W., "Characterization of Ethylene Vinyl Acetate (EVA) Encapsulant: Effects of Thermal Processing and Weathering Degradation on Its Discoloration," *Solar Energy Materials and Solar Cells*, 25 (1992) 3-23.

2. Pern, F.J., "Luminescence and Absorption Characterization of The Structural Effects of Thermal Processing and Weathering Degradation on Ethylene Vinyl Acetate (EVA) Encapsulant for PV Modules," *Polym. Deg. and Stab.*, 41 (1993) 125-139. (NREL/TP-213-4605)

3. Pern, F.J., "Investigation of EVA Degradation Mechanisms," *Proc. of NREL 11th PV AR&D Review Meeting*, Denver, May 13-15, 1992, NREL/CP-413-4845, p.109. (Abstract). (b) Pern, F.J. and Czanderna, A.W., "EVA Degradation Mechanisms Simulating Those in PV Modules," *AIP Conf. Proc. of 11th PV AR&D Review Meeting*, Denver, CO, May 13-15, 1992. American Institute of Physics, Book no. 268, New York, 1992, pp. 445-452.

4. Czanderna, A.W. and Pern, F.J., "Encapsulation of PV Modules using Ethylene Vinyl Acetate Copolymer as a Pottant: A Critical Review," *Solar Energy Materials and Solar Cells*, 43 (1996), 101-183. (NREL/TP-412-7359)

5. (a) Gay, C. F. and Berman, E., "Performance of Large Photovoltaic Systems," *Chemtech*, March, (1990) 182-186. (b) Rosenthal, A. L. and Lane, C.G., "Field Test Results for the 6 MW Carrizo Solar Photovoltaic Power Plant," in *Solar Cells: Their Science, Technology, Applications and Economics*. Elsevier Sequoias, 30 (1991) 563-571. (c) Wenger, H. J., Schaefer, J., Rosenthal, A., Hammond R., and Schlueter, L., "Decline of the Carrisa Plains PV Power Plant: The Impact of Concentrating Sunlight on Flat Plates," in *Proc. of 22nd IEEE PVSC*, 1991, pp. 586-591. (d) A. L. Rosenthal and C. G. Lane, in: *Proc. PV Module Reliability Workshop*, ed. L. Mrig, Lakewood, CO, SERI/CP-4079, Oct. 25-26, (1990) 217-229. (e) V. J. Kusianovich, *ibid.*, pp. 241-5.

6. Pern, F.J., "Modification of EVA Formulation for Improved Stability," *Proc. PV Performance and Reliability Workshop*, Golden, CO, Sept. 8-10, 1993, NREL, pp. 358-74. (NREL/TP-410-6033)

7. Pern, F.J., unpublished results.

8. Cuddihy, E., Coulbert, C., Gupta, A., and Liang, R., "Flat-Plate Solar Array Project Final Report, Vol. VII--Module Encapsulation," *JPL Publication 86-31*, October 1986. (DOE/JPL-1012-125)

9. P. B. Willis, "Investigation of Materials and Process for Solar Cell Encapsulation," *Final Report of JPL Contract No. 954527, S/L Project 6072.1* by the Springborn Laboratories, Inc., to JPL, *JPL Publication*, DOE/JPL-954527-86/29, 1986.

10. Pern, F.J. and Glick, S.H., "Thermal Processing of EVA Encapsulants and Effects of Formulation Additives," *Proc. 25th IEEE PVSC*, May 13-17, 1996, Washington D.C., (1996) 1251-1254. (NREL-TP-412-20380)

11. Pickett, J.E. and Moore, J.E., "Photodegradation of UV Screeners," *Polym. Deg. and Stab.*, **42** (1993) 231-244.

12. Pern F.J., "A Comparative Study of Solar Cell Performance under Thermal and Photothermal Tests," *Proc. PV Performance and Reliability Workshop*, Golden, CO, Sept. 16-18, 1992, pp. 327-44. (SERI/CP-411-5184)

13. Pern, F.J., "Comparison of The Photostability for Two Common EVA Formulations: EVA A9918 and EVA 15295", *Proc. of 6th PV Performance and Reliability Workshop*, Sept. 21-23, 1994, Lakewood, CO, pp. 329-347. (NREL/CP-411-7414)

14. Pern, F.J. and Glick, S.H., "Effects of Additives on The Photostability of Thermally Cured EVA," in preparation.

15. (a) J. F. McKellar and N. S. Allen, *"Photochemistry of Man-Made Polymers,"* Applied Science, London, 1979. (b) N. M. Emanuuel and A. L. Bucgachenko, *"Chemical Physics of Polymer Degradation and Stabilization,"* VNU Science Press, Utrecht, The Netherlands, 1987.

16. Pern, F.J., and Glick, S.H., "Fluorescence Analysis as a Diagnostic Tool for Polymer Encapsulation Processing and Degradation," in Noufi, R. and Ullal, H.S. (Eds), *AIP Conf. Proc. for the 12th NREL PV Program Review Meeting*, book no. 306, American Institute of Physics, NY, 1994, pp.573-585. (NREL/TP-412-5996)

17. Pern, F.J., "Factors That Affect the EVA Encapsulant Discoloration Rate Upon Accelerated Exposure," *Solar Energy Materials and Solar Cells*, **41/42** (1996) 587-615. (NREL/TP-412-7700)

18. Pern, F.J., Glick, S.H., Czanderna, A.W., and DeBlasio, R., "Alternative Encapsulation Materials and Schemes," in Ullal, H.S. and Witt, C. E. (Eds), *AIP Conf. Proc. for the 13th PVAR&D Program Review Meeting*, May 16-19, 1995, Lakewood, CO. (NREL/TP-412-8013), Book no. 353, American Institute of Physics, NY, 1996, pp. 569-580.

19. Hammond, B., Whitfield, K., and Ji, L.-J., "PV Module Qulaification Test Experiences and Results," *Proc. PV Performance and Reliability Workshop*, B. Kroposki ed., Sept. 4-6, 1996, Lakewood, CO, pp. 273-280. (NREL/CP-411-21760)

# A Study of Various Encapsulation Schemes for c-Si Solar Cells with EVA Encapsulants

### F. J. Pern and S. H. Glick

Center for Performance Engineering and Reliability
National Renewable Energy Laboratory (NREL)
1617 Cole Blvd., Golden, Colorado 80401, USA

**Abstract.** Several encapsulation schemes for crystalline Si (c-Si) solar cells, grouped into three categories of superstrate/encapsulant/Si-cell/encapsulant/substrate, were studied using different superstrates and substrates with extruded EVA films as the main encapsulant materials. A number of technical problems were observed and practical solutions to the problems are presented. The results are useful for designing and fabricating various samples of encapsulated c-Si cells and mini-modules for accelerated weathering tests in our future work.

## INTRODUCTION

Encapsulation of photovoltaic (*PV*) modules with ethylene vinyl acetate copolymer (EVA) pottant has been very popular in PV manufacturing since the early 1980s. Two EVA formulations are commercially available: a regular (slow) cure EVA A9918 and a fast cure EVA 15295 with or without a primer, developed jointly by Jet Propulsion Laboratory and Springborn Laboratory (1,2). The encapsulation schemes for PV modules may vary significantly, depending on the type of solar cells/modules and their applications. For example, c-Si modules are commonly encapsulated with two EVA layers between a glass plate superstrate and a Tedlar film substrate; thin-film a-Si modules from Advanced Photovoltaic System (APS) deployed at the PV for Utility Scale Applications (*PVUSA*)--Davis site are laminated with one EVA layer between two glass plates (3); and some modules of a-Si thin film deposited on stainless steel foils from Sovonics are laminated with two EVA layers between a Tefzel film as superstrates and a polymer substrate (3). For CuInSe$_2$-based thin film modules, encapsulation with EVA results in a temporary 15%-25% loss of output power, which is mostly recoverable by light soaking, resulting from heating effects on the contact/series resistance during EVA encapsulation, as reported by Siemens Solar Industries (4,5). Different encapsulant materials may also be used; for example, ASE Americas uses a proprietary pottant (6) that is not EVA-based. Silicone resin and finite glass beads are used by Utility Power Group (UPG) for their glass/a-Si/glass modules (7).

The processing conditions for module encapsulation may vary considerably and, in most cases, are generally considered as company proprietary information, even if the same EVA is being used. The variations in processing may result in large difference in the long-term stability of the EVA because the EVA may undergo weathering-induced degradation. The discoloration rate is affected in part by the lamination/curing conditions in which UV-excitable chromophores are produced (8-10). Oxidation and corrosion of tinned Cu ribbons (11), degradation of solder bonds, and delamination of EVA may result from module aging in the field (12). Of the ~2%/yr system performance degradation for 10 systems (13), EVA degradation and discoloration may have accounted for about half of the loss. Formation of bubbles in modules during lamination/curing may also be a problem, especially when EVA 15295 is used and if the temperature-pressure-time parameters for the lamination-curing cycle are not well controlled. On the other hand, details of processing conditions become critical when Tefzel films are used as the superstrate because Tefzel films may shrink, wrinkle, and/or delaminate.

As a part of our research and development efforts to extend the service life of encapsulated PV modules, accelerated weathering tests in weatherometers have been planned. A number of alternative encapsulation schemes have been proposed by us (14) and the issues concerning module reliability were discussed (15). To produce statistically meaningful results for the performance reliability of different encapsulation schemes, a good understanding and control of processing conditions or parameters are critical for us to prepare a fairly large quantity of identical samples in a reproducible way. This paper summarizes our observations and the results of experiments for various encapsulation schemes using EVA films with various superstrates and substrates. Commercial products of acrylic adhesive films were also tested as alternative encapsulant materials.

## EXPERIMENTAL

**Materials.** The polymer films used for superstrates were Du Pont's clear Tedlar films of 1.5-mil (0.038-mm) thickness, and oriented $T^2$ Tefzel$^{TM}$ ZMC films of 1.5-mil, 2.5-mil (0.064-mm), and 5.0-mil (0.127-mm) thickness. Glass superstrates of various thickness from 1/8" to 1/32" (3.2-mm to 0.8-mm) were used: window glass, borosilicate from Ace Glass, Starphire$^{TM}$ from PPG, ITO glass from AFL, and Ce-containing Solarphire$^{TM}$ from PPG, Corning-213 (brown color), and OCLI-350 from OCLI. The encapsulants used are extruded EVA A9918P and 15295P films (P for *primed*) and were kindly provided by Springborn Laboratory and Richmond Technology. EVA films pre-attached with 1.5-mil Tefzel and $TiO_2$-blended EVA films were also provided by Richmond Technology. Scotch brand VHB (very high bond) double-coated acrylic adhesive films (F-9473PC, 10-mil

(0.25-mm) thick; #4905, 20-mil (0.5-mm) thick) from 3M were also tested as alternative encapsulant materials. Monocrystalline Si solar cells (Siemens Solar) of 1.2" x 1.2" (3-cm x 3-cm) or 1.6" x 1.6" (4-cm x 4-cm) size that were soldered manually with tinned Cu ribbons were purchased from Solar World. Polycrystalline Si solar cells of 4" x 4" (10-cm x 10-cm) size were kindly provided by Mobil Solar (presently ASE Americas). The materials used for substrates were borosilicate glass and 1" x 3" (2.5-cm x 7.6-cm) microslides from VWR, Tefzel, and various Tedlar films. The latter include trilaminates of 1.5-mil thick Tedlar provided by Madico: Tedlar/polyester/Tedlar (primed), Tedlar/polyester/EVA, and Tedlar/Al-foil/Tedlar. Pellets of EVA copolymers of 14%, 18%, 33%, and 40% vinyl acetate (VA) by weight and polyvinyl acetate (PVAc) were acquired from Aldrich Chemicals. Pellets of Elvax 150 or 150W, another EVA with 33% VA, was kindly provided by Du Pont. (*Note: English units will be used in the text and figures for convenience.*)

**Cleaning.** The Tedlar and Tefzel films were cleaned in detergent water for ~5 min in a ultrasonicator, rinsed with copious deionized (DI) water, and then cleaned again with high purity (Nanopure$^{TM}$) DI water for ~5 min in the ultrasonicator. A similar procedure was used to clean the various glass and microslide plates except that they were first soaked in an aqueous solution of 1N NaOH for 15 to 30 min before rinsing with DI water. Some Tefzel films were preshrunk thermally by subjecting them to the typical lamination/curing cycle for EVA 15295 in the laminator before being used.

**Lamination and Curing.** A custom-built double-bag vacuum laminator was used for the lamination and curing process. The system allows convenient programming, control and *in-situ* monitoring of temperature/time/pressure (vacuum). Typical processing conditions for both fast-cure EVA 15295P and slow-cure EVA A9918P were used (10). Pressing on the laminate samples was controlled by adjusting the bleed valve to the upper chamber to produce a pressure of ~0.5 atm or ~1 atm from the silicone rubber diaphragm. The timing to open the cover lid and remove laminated samples from the lower chamber was controlled by a preset temperature point during the chamber cool-down, after the lamination-curing cycle was completed. Samples with or without crystalline Si solar cells were prepared and studied. No edge sealing was made on the samples. In this work, most of the studies focused on using the fast-cure EVA 15295P because of a shorter processing time and greater photostability against discoloration than EVA A9918P (16). When Tefzel films were used, the corona-treated, cementable surface sides were used in contact with the EVA (or VHB). The features of laminated/cured samples were carefully observed, or tested if needed, for cracking, wrinkling, bubbling, shrinking, film popping, and/or delamination. The laminates were not sealed around the edges with epoxy. In the absence of a mechanical pull-strength test station, the adhesion strength for polymer film superstrates or substrates was tested and compared qualitatively by manual twisting and pulling.

## RESULTS AND DISCUSSION

As discussed below, the quality of laminates is affected directly by the configuration of the laminates, quality of the EVA, thickness of the pottant films, and the details of processing conditions.

### Effect of Laminate Configuration

The various encapsulation schemes can be categorized generally into three types in the order of *superstrate-encapsulant-solar cell-encapsulant-substrate*: Glass/P/Si/P/Polymer (or Polymer/P/Si/P/Glass), Glass/P/Si/P/Glass, and Polymer/P/Si/P/Polymer, where P stands for the polymeric encapsulant.

#### *Glass/P/Si/P/Polymer (or Polymer/P/Si/P/Glass) Configuration*

Results of lamination and curing for both configurations are typically satisfactory when the prescribed pressing pressure (~0.5 or ~1 atm) and sample removal temperature ($\leq 120°C$, the heater platen temperature in the lower chamber) were used. Thus, a solid superstrate or substrate support clearly is beneficial for producing good lamination of c-Si solar cells. All samples prepared in this configuration do not have bubbles or cracking, although the polymer/EVA layer around the edges of glass plate may appear curvy due to press-bending of the silicone rubber diaphragm from the upper chamber.

The glass/P/Si/P/polymer configuration is popular for crystalline Si modules in which P (EVA) and a Tedlar or Tefzel polymer film substrate are used. In this work, the encapsulation of c-Si cells was accomplished by placing the glass superstrate first towards the heater platen with the polymer film substrate on the top. The Tedlar film can be either a single layer or a trilaminate such as Tedlar/polyester/Tedlar or Tedlar/Al foil/Tedlar. The aluminum foil in the latter film is 0.7 mil-thick and is embedded to block moisture ingress from the ambient into the solar cells. Modules of this type may be suitable for high humidity regions. The $TiO_2$-blended EVA/Tedlar film can be used only as a substrate. White Tedlar film substrates have been shown to enhance total internal reflection and therefore increase the photocurrent in encapsulated solar cells (14). When Tefzel is used, the film thickness is not critical, although thinner films (1.5 mil or 2.5 mil-thick) have a better transmittance whereas 5.0 mil-thick, somewhat hazy film offers stronger protection when used as the substrate.

In the polymer/P/Si/P/glass configuration, transparent polymer films (1.5 mil-thick Tedlar or Tefzel) are used as superstrates to produce the laminates, with essentially identical processing conditions except that the front side of the solar cell is now facing upward. This laminate configuration will allow us to investigate, in future weathering tests, the permeability effects of air, moisture, and/or air pollutants through the polymer superstrate on metalizations and the front-side of solar cells. We have shown before that the air permeability of Tefzel films allows rapid photobleaching reactions to destroy the curing-generated chromophores in EVA (Fig. 1a) so that no discoloration is obtained, which is in contrast with that for glass/EVA/glass laminates shown in Fig. 1b (14,16). Even so, we suggest that cautions be taken for assessing the long-term performance of Tefzel-superstrated PV modules, such as a-Si thin films laminated with Tefzel films, because the long-term effects of moisture and air pollutants have not been determined (14). As supporting evidence to this concern, Hammond et al. recently reported corrosion of tinned Cu ribbons on Tefzel/EVA-laminated modules after a 1000-h damp heat test at 85°C and 85% RH (17).

**VHB film as alternative encapsulants.** When 3M's VHB acrylic adhesive films are used as the encapsulants, the entire lamination of all components can be achieved by a simple "finger-press" or "roll-press" method. Heating is optional, although a better laminate quality without bubbles is obtained if the stack of all components is processed as if an EVA 15295 pottant were being used. A 20 mil-thick VHB film is needed to reduce the "steps" and unfilled voids around the Si cell wafer that are produced when a 10 mil-thick film is used. In an accelerated exposure test, the 20-mil VHB films showed a blue shift with an increased transmittance (Fig. 2a) as the exposure time increased; meanwhile, the fluorescence emission peak intensity decreased (Fig. 2b), because of the photobleaching reactions resulted from air permeating through the VHB films. The Tefzel/VHB/glass and glass/VHB/glass laminates did not discolor as compared to other glass/EVA A9918/glass laminates (Figs. 2c, curves 1 and 2). The VHB adhesive films are made of acrylic polymer material with good weatherability but are more costly than EVA. Further study is required to determine whether these VHB films can provide greater long-term stability against degradation, discoloration, and delamination.

## *Glass/P/Si/P/Glass Configuration*

The glass/P/Si/P/glass configuration is not typically used in the manufacturing of c-Si modules (other than thin-film a-Si modules, for example). This configuration is needed in our work, which requires a large quantity (ca. 20) of good samples for accelerated testing in the Atlas Ci4000 Weather-Ometers™ in the near future. The glass/glass configuration will be useful to compare and contrast the glass/polymer configurations in assessing the degradative effect of the permeability of the polymer film, either as a superstrate or substrate, on the solar cell components and encapsulant. This glass/P/Si/P/glass configuration will allow exposed samples to be monitored for the performance degradation on the solar cell components by current-voltage (I-V) measurements and on the EVA (or VHB) encapsulant by fluorescence analysis, transmittance, and color indices measurements.

Two problems are associated with the lamination and curing of Si solar cells with two layers of EVA 15295P between two borosilicate glass plates: glass-cracking

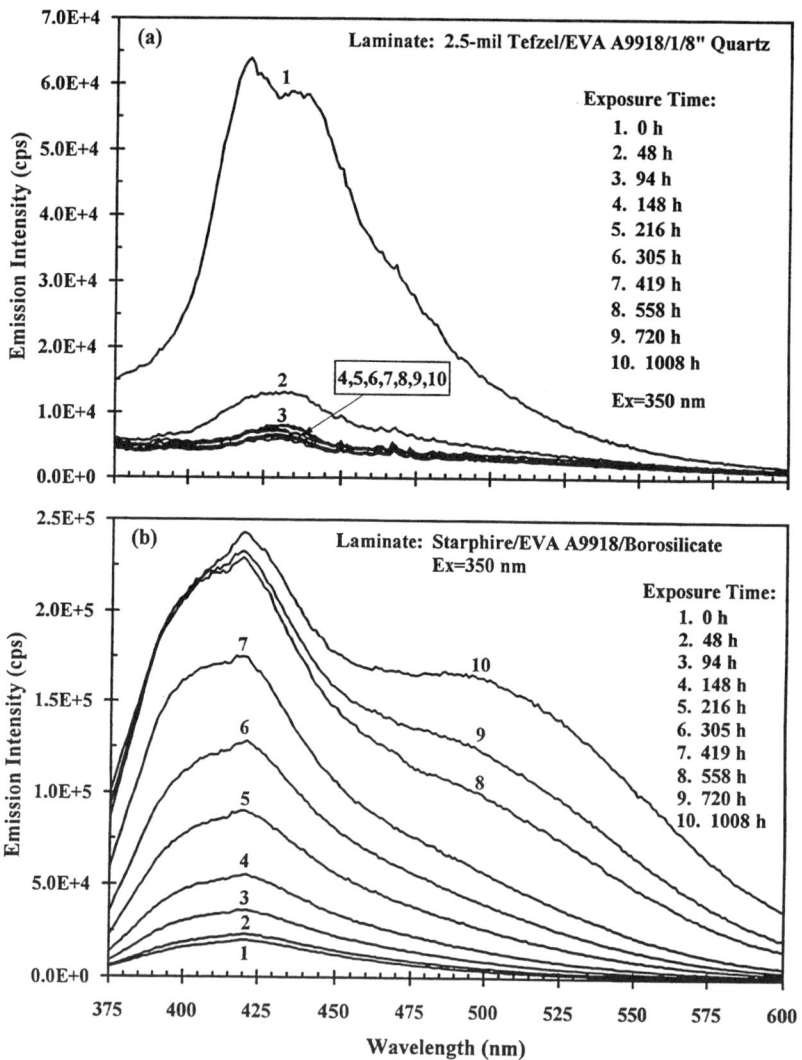

**Figure 1.** Fluorescence emission spectra at various exposure times to the enhanced-UV light from a 1-kW Xe, Oriel solar simulator at a black panel temperature (BPT) of 44° ± 2°C for (a) a laminate of 2.5-mil Tefzel/EVA A9918/quartz slide, showing the photobleaching of curing-generated chromophores, and (b) a PPG Starphire™/EVA A9918/borosilicate laminate, showing the evolution of an emission peak at ~500 nm originated by discoloring chromophores.

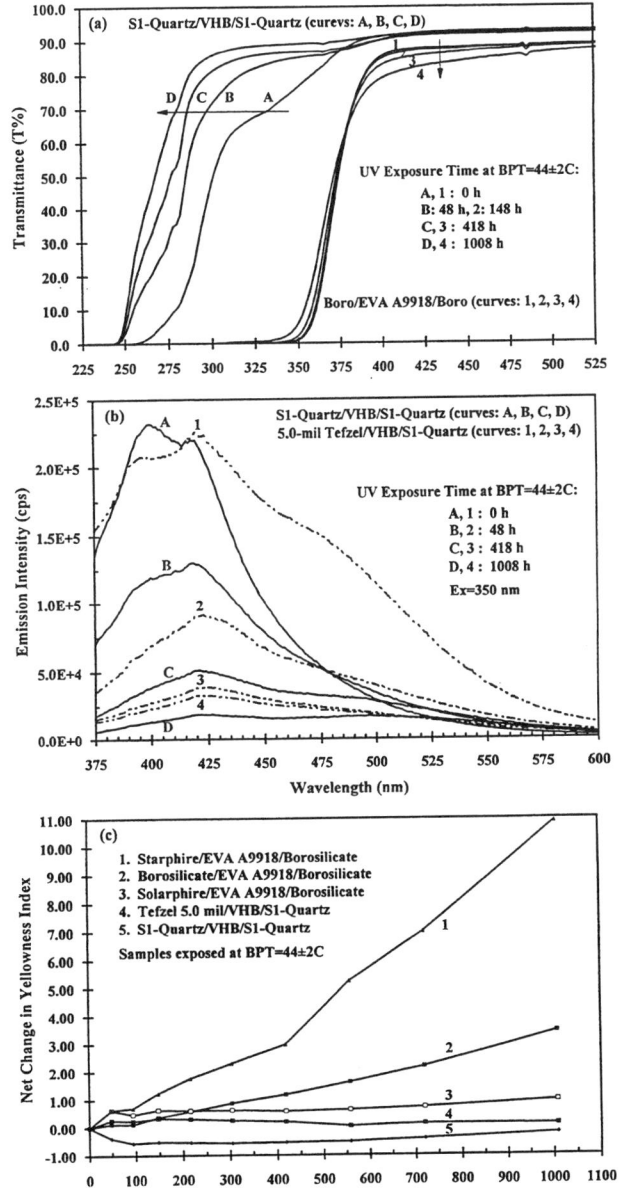

Figure 2. Simultaneous accelerated exposure of edge-unsealed laminates to the enhanced-UV light at a BPT= 44° ± 2°C for 1008 h. (a) Changes in transmittance for a non-discoloring 3M's 20-mil VHB film press-laminated between two quartz slides (curves A to D) and a discoloring A9918 EVA film laminated between two borosilicate slides (curves 1 to 4). (b) Photobleaching of existing chromophores in the 20-mil VHB films laminated between two quartz slides (curves A to D) and between a 5.0-mil Tefzel film and a quartz slide (curves 1 to 4). (c) Net changes in yellowness index (YI) for the A9918 EVA and VHB films laminated with various superstrates and glass substrates. Note the UV filtering effect of glass superstrates on the yellowing rate of EVA A9918: Starphire < borosilicate < Ce-containing Solarphire.

817

and bubbling in the laminates. In most cases, the cracking typically appears along the tinned Cu ribbon on the solar cell. The cracking is caused by the bending pressure from the silicone rubber diaphragm on the stack during the lamination/curing stages. The cracking may occur inside the laminator if the sample stacks are not surrounded (i.e., supported) in close proximity with spacers of similar height. If supporting spacers are used, the frequency of cracking is reduced. But cracking can still occur, *unpredictably*, on some samples much later—sometimes 30 min to 2 h after they have already cooled to room temperature—suggesting that a significant amount of stresses accumulated during the cooling. A solid aluminum spacer frame has been carefully designed and fabricated to solve the glass-cracking problems.

The small bubbles formed in the glass/EVA/Si/EVA/glass laminates during cooling is likely due to gradual release of the gaseous decomposition products of the EVA ingredients (primarily from the curing agent Lupersol TBEC) that were dissolved in the EVA melt when it was hot between the two glass plates. The problem of forming bubbles in the EVA may not be solvable because most of the bubbles are produced gradually when the laminates are being cooled to room temperature. In general, removing the hot laminate from the laminator at a higher temperature produces larger bubbles.

We also found that opening the laminator lid and removing samples at a higher temperature can increase the concentration of curing-generated chromophores, as shown in Fig. 3. From our earlier studies, a high concentration of curing-generated, UV-excitable chromophores may induce an earlier discoloration (8,10,16). Therefore, removing the samples at a lower temperature ($\leq 105°C$, but optimally $\leq 90°C$) is preferred.

## *Polymer/P/Si/P/Polymer Configuration*

The purposes of our study with the polymer/P/Si/P/polymer configuration are (1) to provide samples for future work to compare the weathering stability among the glass/glass, glass/polymer, and polymer/polymer laminates as affected by the gas permeability of polymer films; and (2) to examine if light-weight but strong laminates can be obtained without breaking of c-Si cells. Transparent Tedlar and Tefzel films of 1.5 mil-thick were used as the superstrate, and all varieties (type, color, and thickness) of Tedlar and Tefzel films available for this work were used as the substrate.

The results show that the quality of the laminates depends on several factors: thickness and pre-shrinking of Tefzel film, pressure of the diaphragm from upper chamber, chamber temperature to open the lid and remove the samples, and quality of the EVA encapsulant. The problems observed include bending, wrinkling, popping, bubbling, and haziness. The temperature when the lid is opened to remove the sample strongly affects bending and wrinkling of the Tedlar-Tefzel or

Tefzel-Tefzel laminates. The bending and wrinkling are greater if the laminates are removed at higher temperatures (>105°C). The degree of bending is aggravated when the substrate used is a 5.0-mil Tefzel film and is lessened if a pre-shrunk Tefzel film is used. If both the superstrate and substrate Tefzel films are of the same thickness, e.g., 1.5 mil, the bending is considerably less. For most samples, good, flat laminates are obtained if the samples are removed from the laminator chamber at a chamber temperature of ≤90°C and then cooled in the ambient. The wrinkling also occurs more often in parallel along the tinned Cu ribbons, apparently due to faster cooling along the metal ribbons. The bending and wrinkling obviously arise from thermal shrinking of the $T^2$ Tefzel films (18). The degree of thermal shrinking, which occurred during the entire period of lamination and fast-curing process up to a temperature of 157° ± 3°C, was somewhat irregular, depending on film thickness, and ranged from ~0.2% to ~4%, with an average of ~1.7%, as determined from many processing tests with single layers of Tefzel and 15295 EVA over a microslide or two Tefzel/EVA films with a c-Si solar cell.

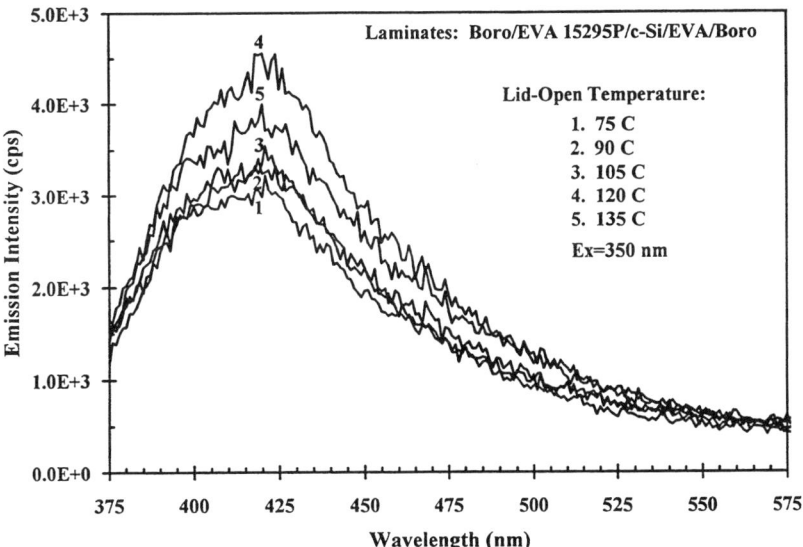

**Figure 3.** Fluorescence emission spectra for five EVA 15295P films laminated between two 2" x 2" x 1/8" borosilicate plates that were removed from the laminator at heater platen temperatures (lid-open temperatures) of (1) 75°C, (2) 90°C, (3) 105°C, (4) 120°C, and (5) 135°C, respectively, and cooled to room temperature in the ambient. The borosilicate plates for these samples were not cracked after cooling. The emission peak intensity for sample 5 (135°C) was less than that for sample 4 (120°C) and was attributed to a somewhat thinner film of sample 5. When removed at 135°C the EVA was still *hot and soft*, and it was possible the top borosilicate plate "sank" into the EVA, resulting in a thinner film.

The formation of bubbles in the laminates when using 15295P EVA was affected mostly by the pressing pressure and a high sample-removal temperature, e.g., at ≤135°C. A pressure of ~1 atm on the sample stacks from the silicone rubber diaphragm and a sample removal temperature of ≤105°C normally produced laminates without bubbles.

## Effects of EVA Quality and Contamination

White hazing problems were constantly and persistently observed when EVA and Tefzel/EVA films from one of the two EVA manufacturers, B, were used. A higher processing (curing) temperature would not solve the problem. By studying Elvax 3185 pellets, the raw copolymer used by the manufacturer B to make the formulated EVA films, it was found that the pellets would not produce a clear film upon curing at temperature of 145°C, 155°C, or even 165°C. The resultant films appear hazy white and are similar to the films from EVA-14 wt%VA and EVA-18 wt%VA. In contrast, clear films were obtained at curing temperature of 145°C from pellets of EVA-25 wt%VA, Elvax 150 (33 wt%VA), EVA-40 wt%VA, and PVAc.

Another issue related to the EVA films from the manufacturer B is the bubbling problem. Results of a careful investigation show that bubbles appeared more often on one side of the EVA film laminated to microslide plates. This particular side was identified to be the surface side of the extruded EVA films in direct contact with the waxed release paper used during extrusion. Thus, the hazing and bubbling problems are very likely associated with contamination from the waxed release paper when the EVA films, which could still be hot, were extruded, laid on, and rolled up with the release paper. Accordingly, the quality of the extruded EVA films has to be carefully controlled, not only by using proper raw polymer material but also by using good manufacturing practice.

## Adhesion of Tefzel Films on EVA

Although the wrinkling problems on the Tefzel films can be solved by applying a lamination pressure of ~1 atm and by removing the laminates at ≤90°C, the adhesion strength of Tefzel films on the cured EVA 15295P is far from satisfactory. In contrast, high adhesion strength is obtained for the Tedlar-to-EVA, Tedlar-to-VHB, and Tefzel-to-VHB laminates. The Tefzel/EVA/Tefzel laminates may appear visually to be fine; but by finger-twisting the laminates, the Tefzel films become loosened and can be peeled off with little difficulty. In some samples, the Tefzel films (either as superstrate or substrate) would pop off with a slight bending and produce a fairly large bubble between the Tefzel and EVA films. Hammond et al. recently reported that the Tefzel superstrate films were delaminated to a surface area size ranging from 20-in$^2$ to 80-in$^2$ on modules subjected to a damp heat test at 85°C and 85 %RH for 1000 h (17). In addition, the four new experimental EVA formulations, which were developed recently by Springborn Testing and Research (*STR*) (19), turned yellow to brown, which was attributed to the degrading effect of moisture induced during the damp heat test (17). We reported previously that UV-absorbing glass superstrates can greatly reduce the UV-induced discoloration rate of EVA as shown in Fig. 4 (14). STR also reported similar results with cerium-containing glass laminated with the four new experimental formulations (20). On the other hand, pinholes or punctures are also observed in Tefzel/EVA-laminated modules, resulting in rejection of the modules from acceptance in the PVUSA project (3). Improving substantially the adhesion strength of Tefzel film on EVA is a technical problem that remains to be solved.

## CONCLUSIONS

We have presented our results and observations for a variety of encapsulation schemes with different superstrates and substrates using EVA 15295P as the main encapsulant. The results show that the glass/P/Si/P/polymer configuration is mostly problem-free. Careful control of the lamination and processing conditions is required for the glass/P/Si/P/glass and polymer/P/Si/P/polymer configurations. Although Tefzel films are increasingly used as superstrates or substrates, their adhesion strength needs to be improved; besides, their ability to protect module components during long-term field exposure and accelerated weathering tests has not been studied. Similarly, the long-term weatherability of VHB films as encapsulants needs to be investigated. VHB films provide strong adhesion in the cases we studied and have potential for simple processing procedures for solar cell lamination.

## ACKNOWLEDGMENTS

The authors are grateful for the project support of A. W. Czanderna, R. DeBlasio, and R. Hulstrom. Detailed editing review of this paper by A. W. Czanderna is greatly appreciated. We also thank the two companies that provided the extruded EVA and Tefzel/EVA films, and the companies that provided the various glass samples. This work was supported by the U.S. Department of Energy under Contract No. DE-AC36-83CH-10093.

## REFERENCES

1. Cuddihy, E., Coulbert, C., Gupta, A., and Liang, R., "Flat-Plate Solar Array Project Final Report, Vol. VII--Module Encapsulation," *JPL Publication 86-31*, October 1986. (DOE/JPL-1012-125)

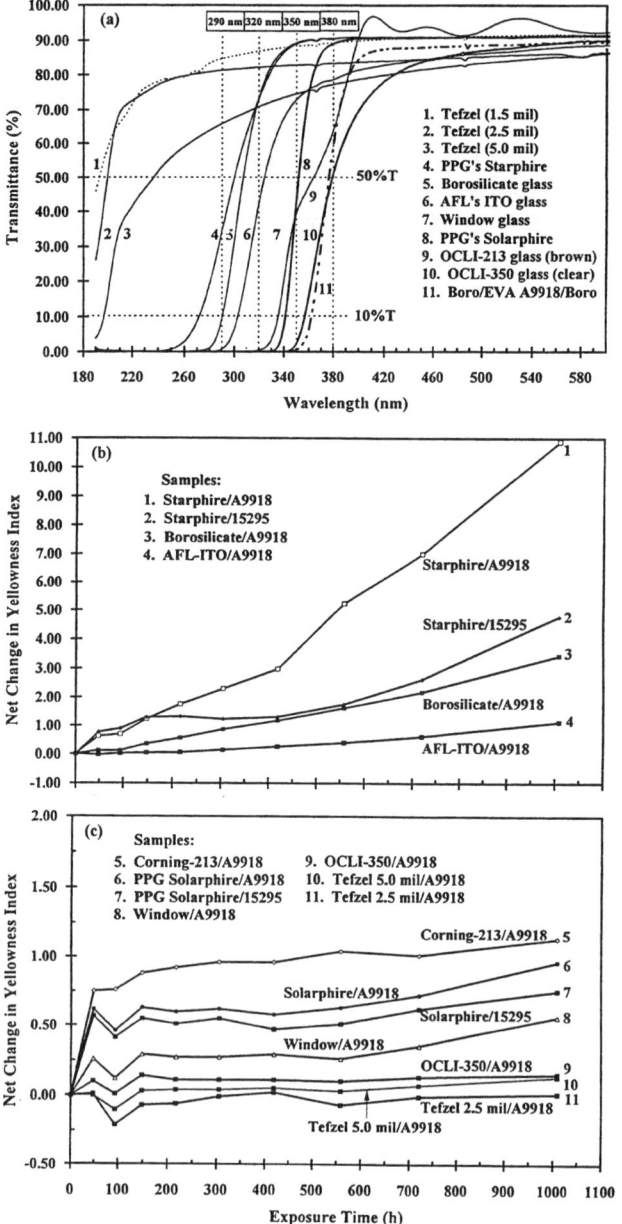

Figure 4. (a) Transmittance spectra of various glass slides and two Tefzel films. The transmittances (T) at 10% and 50% and wavelengths at 290, 320, 350, and 380 nm are indicated. (b, c) Net changes in the yellowness index as a function of exposure time obtained for the A9918 and 15295 EVA films laminated between various superstrates and a glass substrate (borosilicate or quartz) and exposed to the enhanced UV at a BPT = 44° ± 2°C for 1008 h.

2. Willis, B., "Investigation of Materials and Process for Solar Cell Encapsulation," Final Report of JPL Contract No. 954527, S/L Project 6072.1 by the Springborn Laboratory, Inc., to JPL, *JPL Publication*, DOE/JPL-954527-86/29, 1986.

3. Whitaker, C., "Pacific Gas and Electric Company's Perspective on Module Reliability," *Proc. PV Performance and Reliability Workshop*, L. Mrig ed., Sept. 16-18, 1992, Golden, CO, pp. 280-289. (NREL/CP-411-5184)

4. Willett, D., "Investigation of Lamination Induced Metastability in CIS-Based Modules," *Proc. PV Performance and Reliability Workshop*, L. Mrig ed., Sept. 8-10, 1993, Golden, CO, pp. 184-199. (NREL/CP-410-6033)

5. Willett, D., "Effects of Lamination Processes on CuInGaSeS Modules," *Proc. PV Performance and Reliability Workshop*, L. Mrig ed., Sept. 21-23, 1994, Lakewood, CO, pp. 179-190. (NREL/CP-411-7414)

6. Azzam, M., "Developmental Testing of ASE Americas New ASE-DG$^{TM}$ Product Line," *Proc. PV Performance and Reliability Workshop*, L. Mrig ed., Sept. 21-23, 1994, Lakewood, CO, pp. 349-360. (NREL/CP-411-7414)

7. Duran, G., "Progress in Thin-Film Module Encapsulation Design and Reliability," *Proc. PV Performance and Reliability Workshop*, L. Mrig ed., Sept. 21-23, 1994, Lakewood, CO, pp. 387-410. (NREL/CP-411-7414)

8. Pern, F.J., "Factors That Affect the EVA Encapsulant Discoloration Rate Upon Accelerated Exposure," *Solar Energy Materials and Solar Cells*, **41/42** (1996) 587-615.

9. Czanderna, A. W., "Overview if Current Issues: PV Cell and Module Performance and Reliability," *Proc. PV Performance and Reliability Workshop*, B. Kroposki ed., Sept. 4-6, 1996, Lakewood, CO, pp. 29-42. (NREL/CP-411-21760)

10. Pern, F.J. and Glick, S.H., "Thermal Processing of EVA Encapsulants and Effects of Formulation Additives," *Proc. 25th IEEE PVSC*, May 13-17, 1996, Washington D.C., IEEE, pp. 1251-1254. (NREL-TP-412-20380)

11. Quintana, M. A. and King, D. L., "Encapsulant Adhesion and Solder Bond Integrity in Field-Aged Modules," *Proc. PV Performance and Reliability Workshop*, B. Kroposki ed., Sept. 4-6, 1996, Lakewood, CO, pp. 245-257. (NREL/CP-411-21760)

12. Thomas, M. G., Rosenthal, A. L., Durand, S. J., and King, D. L., "A Ten Year Review of Performance of Photovoltaic Systems," *Proc. PV Performance and Reliability Workshop*, L. Mrig ed., Sept. 21-23, 1994, Lakewood, CO, pp. 279-285. (NREL/CP-411-7414)

13. Pern, F.J., "Modification of EVA Formulation for Improved Stability," *Proc. PV Performance and Reliability Workshop*, L. Mrig ed., Sept. 8-10, 1993, Golden, CO, pp. 358-374. (NREL/TP-410-6033)

14. Pern, F.J., Glick, S.H., Czanderna, A.W., and DeBlasio, R., "Alternative Encapsulation Materials and Schemes," in Ullal, H.S. and Witt, C. E. (Eds): *AIP Conf. Proc. for the 13th PVAR&D Program Review Meeting*, May 16-19, 1995, Lakewood, CO, American Institute of Physics, NY, Book no. 353, 1996, pp. 569-580. (NREL/TP-412-8013)

15. Czanderna, A.W. and Pern, F.J., "Encapsulation of PV Modules using Ethylene Vinyl Acetate Copolymer as a Pottant: A Critical Review," *Solar Energy Materials and Solar Cells*, **43** (1996), 101-183. (NREL/TP-412-7359)

16. Pern, F.J., "Comparison of The Photostability for Two Common EVA Formulations: EVA A9918 and EVA 15295", *Proc. of 6th PV Performance and Reliability Workshop*, L. Mrig ed., Sept. 21-23, 1994, Lakewood, CO, pp. 329-347. (NREL/CP-411-7414).

17. Hammond, B., Whitfield, K., and Ji, L.-J., "PV Module Qualification Test Experiences and Results," *Proc. PV Performance and Reliability Workshop*, B. Kroposki ed., Sept. 4-6, 1996, Lakewood, CO, pp. 273-280. (NREL/CP-411-21760)

18. (a) Levy, S. B., "Oriented Fluoropolymer Films" and (b) "T2 Films of Tefzel ETFE." Technical brochures, H-04311 and H-04310, for High Performance Films, Du Pont Electronics.

19. Holley, W. H., Agro, S. C., Galica, J. P., Thoma, L. A., Yorgensen, R. S., "UV Stability and Module Testing of Non-Browning Experimental PV Encapsulants," *Proc. 25th IEEE PVSC*, May 13-17, 1996, Washington D.C., IEEE, pp. 1259-1262.

20. Holley, W. H., Agro, S. C., Galica, J. P., Thoma, L. A., Yorgensen, R. S., "Effects of Glass Superstrates on Browning of Accelerated UV-Aged EVA Encapsulants," *Proc. PV Performance and Reliability Workshop*, L. Mrig ed., Sept. 21-23, 1994, Lakewood, CO, pp. 315-328. (NREL/CP-411-7414)

# PHOTOVOLTAIC MANUFACTURING (P)

# Development Of A Low Cost Integrated 15 kW A.C. Solar Tracking Sub-Array For Grid Connected PV Power System Applications

M. Stern, R. West, G. Fourer,
W. Whalen, M. Van Loo, and G. Duran

*Utility Power Group*
*9410 G De Soto Avenue*
*Chatsworth, California 91311*

**Abstract.** Utility Power Group has achieved a significant reduction in the installed cost of grid-connected PV systems. The two part technical approach focused on 1) The utilization of a large area factory assembled PV panel, and 2) The integration and packaging of all sub-array power conversion and control functions within a single factory produced enclosure. Eight engineering prototype 15kW ac single axis solar tracking sub-arrays were designed, fabricated, and installed at the Sacramento Municipal Utility District's Hedge Substation site in 1996 and are being evaluated for performance and reliability. A number of design enhancements will be implemented in 1997 and demonstrated by the field deployment and operation of over twenty advanced sub-array PV power systems.

## BACKGROUND

As a leading provider of PV system engineering, construction, and maintenance services since 1985, Utility Power Group ("UPG") is continuously striving to reduce the installed cost of utility scale grid connected photovoltaic ("PV") power systems. PV power systems generate electricity via the direct conversion of sunlight into electrical energy and can serve as an environmentally benign and domestically secure source of electricity to supply utility grid connected peak or intermediate loads. Considering that almost 700,000 MW of electricity generating capacity exists in the United States alone, and that less than 5MW of the total is derived from PV power systems, this technology has only negligibly penetrated the electricity generation market. As an inherently modular generation technology, PV provides additional value in transmission and distribution applications when located at the point of demand. The dominant factor limiting the use of PV power systems in grid connected applications today is the capital cost of the total installed system with respect to the annual kilowatt hours of energy generated.

The two primary PV power system capital cost groups are 1) PV Modules and 2) Everything Else, which is more commonly referred to by the acronym "BOS" (Balance-Of-System). Typically, 60% of the capital cost of new PV systems is associated with the PV modules while 40% of the costs are allocated to BOS. Both cost groups are highly affected by production and installation volume which permits standardization, automation, and integration.

Working with Siemens Solar Industries (Camarillo, California) to optimize the design of PV modules for power systems applications, UPG's primary focus has been on BOS component technology and manufacture within the context of an easily field deployed modular and integrated PV power system. This focus involves all engineering, manufacturing and construction tasks from the receipt of PV modules to the delivery of high quality and reliable grid connected electricity.

The need for an advanced integrated approach to PV power system design became apparent to UPG in 1994 during construction of a 100 kilowatt PV system in Fort Davis, Texas (Fig. 1) for Central and South West Services, Inc. which required a total of fifty one separate electrical enclosures to provide dc combining, conduit/wire routing, tracker control, disconnect, and inverter functions. Over 2200 PV modules were individually field installed on the array structures. A single inverter provided dc to ac power conversion, and a single tracker motor controller provided solar tracking commands.

**FIGURE 1.** Photograph Of 100 KW PV Power System In Fort Davis, Texas

## OBJECTIVE

The overall objective of Utility Power Group's two year sub-array development effort is to achieve a 20% reduction in the installed cost of PV power systems. Given UPG's focus on BOS costs and assuming such costs represent 40% of the total cost of an installed PV power system, UPG's overall objective could be re-stated as seeking to achieve a 50% reduction in BOS costs.

## TECHNICAL APPROACH

The technical approach employed by UPG towards achievement of the above objective is based upon the following lessons learned during the construction or installation of a number of PV power systems:

- Factory assembly provides cost savings over field assembly.
- Reducing the number of components reduces installation cost.
- Design simplicity reduces the cost of construction materials.
- Reproducibility of tasks reduces the cost of construction labor.
- Modularity reduces the cost of design and project management.

These lessons led UPG to select modularity and component integration as the two primary design criteria to be utilized. A modular alternating current ("ac") sub-array design simplifies the generation of specific and custom PV system design packages and reduce incremental administrative and project management costs. Redundancy and reliability are PV system operational cost factors which benefit from modularity and the reduced number of components associated with increased functional integration.

Based upon the above lessons learned and design criteria, UPG began development of an integrated and modular 15 kW ac sub-array in compliance with the 1996 National Electrical Code and applicable standards of Underwriters Laboratories.

Specifically, UPG has developed:

1. A large area Modular Panel (MP) which is factory assembled and tested by Utility Power Group utilizing production optimized and reliability tested PV laminates manufactured by Siemens Solar Industries.

2. An Integrated Power Processing Unit (IPPU) combining all sub-array electrical functions such as power collection, system protection, direct current ("dc") to ac conversion, and single axis solar tracking, into a single field deployable unit factory assembled and tested by Utility Power Group.

## Modular Panel Development

The factory assembled PV panels developed by UPG and Siemens Solar Industries (SSI) are specifically designed for high voltage power system applications and provides the following cost saving advantages:

1. Elimination of redundant structural frames on individual PV modules.
2. Simplification of the panel support structure.
3. Reduction of labor associated with field installation of panels.

UPG is currently marketing two POWERGLASS™ Modular Panels which are UL listed; the Model 11M55 and the Model 7SP75 (Fig. 2), which are rated at 606 Wdc and 525 Wdc respectively.

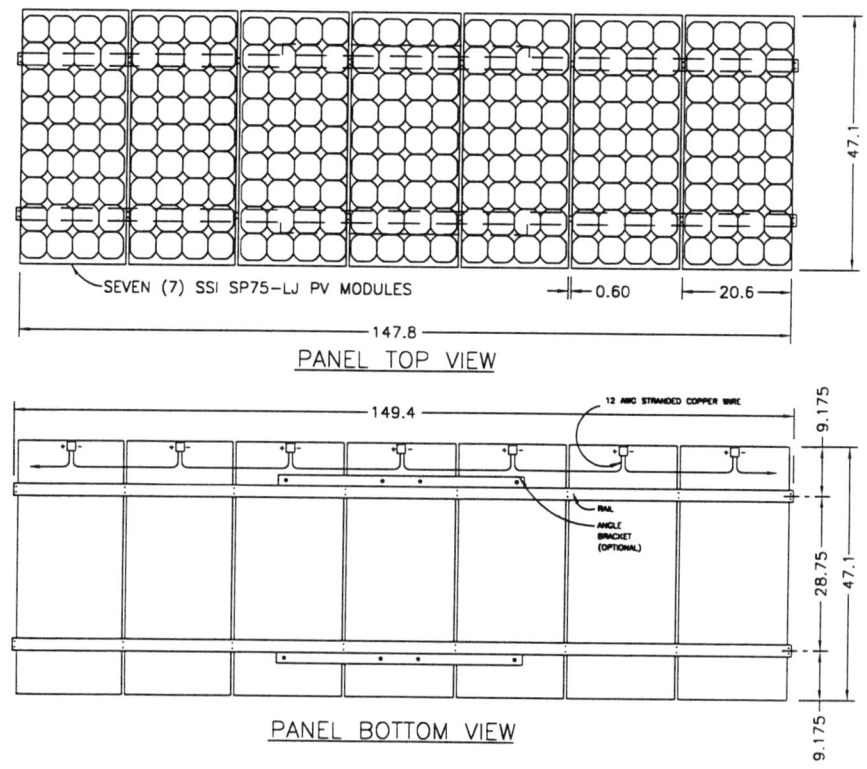

**FIGURE 2**. POWERGLASS™ 7SP75 Modular Panel.

# IPPU Development

The Integrated Power Processing Unit ("IPPU") developed by UPG provides significant BOS cost reduction through the integration and optimization of all electronic and electrical PV power system operations. To facilitate fabrication, testing, assembly, and field maintenance of the IPPU, UPG segmented the design into four sub-assemblies each utilizing high power printed circuit board (PCB) technology to perform a number of integrated functions. (Table 1).

**TABLE 1. IPPU Sub-Assemblies and Functions**

| **DC Interface PCB Assembly** | **Control PCB Assembly** |
|---|---|
| PV sub-circuit conductor termination | Start, wake-up, disable and stop logic |
| PV sub-circuit conductor protection | Machine state logic |
| Visible dc disconnect | Diagnostics and fault detection |
| Blocking (paralleling) diodes | Lowpower (nightly shutdown) |
| Lightning transient protection | Overpower, overvoltage bus |
| Crowbar function | Line synchronization error |
| DC contactor | Frequency out of tolerance |
| DC current sense, regulation and DAS | Utility voltage imbalance, blown fuse |
| DC ground fault current sense | Utility undervoltage |
| DC bus precharge and discharge | Utility overvoltage |
| | System fault, output current imbalance |
| **AC Interface PCB Assembly** | Control logic fault, undervoltage |
| 3-phase utility conductor interface | Overtemperature |
| Fused power distribution | DC ground fault |
| AC contactor (bridge isolation) | Tracker drive fault |
| Regulated low voltage control power supply | Disable, local or remote |
| Tracker drive motor relays and interface | Electrical max. power point tracking |
| Utility voltage sense | DC power calculator |
| Output inductor interface | LCD display driver |
| Output EMI filter | Tracker drive logic and user interface |
| DAS interface point | |
| | **DAS Board Option** |
| **Bridge / Regulator PCB Assembly** | Machine state information |
| 20khz 3-phase ultra fast IGBT bridge | Fault delineation |
| Opto isolated IGBT drivers | 3-phase AC power calculator |
| Digitally synthesized sinewave reference | Analog a/d conversion |
| Phase locked loop line sync circuit | AC line voltage |
| Output sinewave current sense | AC current by phase |
| DC voltage sense | AC power |
| Bus voltage regulation | DC voltage |
| Fast line-overcurrent detection | DC current |
| | DC power |
| | Remote enable/disable interface |
| | Isolated RS485 link to modem |

With this approach, UPG has not only integrated all sub-array functions into a single unit, but also has developed a d.c. to a.c. inverter with a higher conversion efficiency and a higher switching frequency than any other commercially available inverter.

# RESULTS

Table 2 compares the major field tasks required to construct a nominal 100 kW grid connected PV power system before and after the development of UPG's integrated sub-array. Modularity and integration are combined to significantly reduce the number of discrete field construction operations required, simplify design documentation, and permit standardization of key components.

TABLE 2. Before and After Comparison of Major Field Tasks.

| | Major Field Tasks: Before Integrated Sub-Array Development | | Major Field Tasks: After Integrated Sub-Array Development |
|---|---|---|---|
| 1 | Electrical Design | 1 | Modular Design Package |
| 2 | Mechanical and Structural Design | 2 | Site Mobilization |
| 3 | Site Mobilization | 3 | Site Preparation |
| 4 | Site Preparation | 4 | Support Pole Hole Augering |
| 5 | Support Pole Hole Augering | 5 | Support Pole Installation |
| 6 | Support Pole Installation | 6 | Drill Tracker Drive Assy Mtg Holes |
| 7 | Install Conduit Junction Boxes | 7 | Drill Torque Tube Bearing Mtg Holes |
| 8 | Install Tracker Motor Junction Boxes | 8 | Install Torque Tube Bearing Plates |
| 9 | Conduit Trenching | 9 | Install IPPU Assemblies |
| 10 | Dig & Form Electrical Equip Pad | 10 | Conduit Trenching |
| 11 | Electrical Equip Concrete Pad Form | 11 | Lay Conduit |
| 12 | Lay Conduit | 12 | Set Pre-Cast Xfrmr Pad and Xfmr |
| 13 | Backfill Conduit Trenches | 13 | Pull Conduit and Torque Tube Wires |
| 14 | Pour Concrete Electrical Equip Pad | 14 | Backfill Conduit Trenches |
| 15 | Drill Torque Tube Bearing Mtg Holes | 15 | Install Torque Tubes |
| 16 | Drill Tracker Drive Assy Mtg Holes | 16 | Install PV Panels |
| 17 | Install Torque Tube Bearing Plates | 17 | Connect All Wires |
| 18 | Install Tracker Drive Assemblies | 18 | Test IPPU's |
| 19 | Install Torque Tubes | 19 | Sub-Array Start-Up |
| 20 | Install Struts | 20 | |
| 21 | Install PV Module Rails | 21 | |
| 22 | Install PV Modules | 22 | |
| 23 | Install Row Junction Boxes | 23 | |
| 24 | Install Tracker Limit Junction Boxes | 24 | |
| 25 | Install Inverter | 25 | |
| 26 | Install Transformer | 26 | |
| 27 | Install DC Interface Enclosures | 27 | |
| 28 | Install Tracker Control Enclosure | 28 | |
| 29 | Wire PV Modules | 29 | |
| 30 | Pull Conduit and Torque Tube Wires | 30 | |
| 31 | Connect All Wires | 31 | |
| 32 | Test Source Circuits | 32 | |
| 33 | Test Sub-Arrays | 33 | |
| 34 | Test Inverter | 34 | |
| 35 | Test Tracker Controller and Motors | 35 | |
| 36 | System Start-Up | 36 | |

# CONCLUSION

UPG successfully completed the design, fabrication, and installation of a modular and integrated 15 kW a.c. solar tracking PV power system sub-array which exceeded the 20% cost reduction goal. Eight (8) sub-arrays were supplied to the Sacramento Municipal Utility District in 1996 for evaluation (Figure 2) and UPG has received orders for an additional forty-two (42) units to be constructed in 1997.

# ACKNOWLEDGMENT

This work was supported in part by the National Renewable Energy Laboratory (NREL) in Golden, Colorado under the Photovoltaic Manufacturing Technology (PVMaT) Program (Phase 4A1 Subcontract No.: ZAF-5-14271-06). The authors wish to acknowledge the role of the U.S. Department of Energy in support of NREL's PVMaT Program, Sandia's BOS Program, and PV COMPACT's UPVG TEAM-UP Program without which the development, demonstration, and commercialization of UPG's low-cost integrated PV power sub-array would not have been possible.

**FIGURE 3**. Prototype Sub-Arrays In Sacramento, California

# PVMat Improvements for Commercial Production of Thin-Film CdTe Modules

### Jeffrey Phillips and Terry Brog, Ph.D.

*Golden Photon, Inc.*
*4545 McIntyre Street*
*Golden, CO 80403*

**Abstract.** Golden Photon, Inc., of Golden, CO, has developed and refined a process for fabricating thin-film Cadmium Telluride based photovoltaic devices. Golden Photon has invested a considerable amount of effort in 1996 to implement processing improvements related to process scaling as part of PVMat. These processing improvements have brought about increases in both product stability and wattage, while other process improvements have resulted in a sizable increase in through process yields. This paper contains a summary of these activities and the gains made at Golden Photon, Inc., in 1996 for PVMat.

Golden Photon, Inc. (GPI), a subsidiary of ACX Technologies, has been investing in both Cadmium Telluride (CdTe) technology and production of Photovoltaic (PV) modules for approximately five years. These modules are fabricated via a proprietary thin-film spray deposition process onto LOF glass substrates. As part of the PVMat contract with the National Renewable Energy Laboratory, GPI has attempted to scale this process from the prototype manufacturing facility located at 4545 McIntyre St., Golden, CO, to a 10 MW facility. GPI has identified various obstacles, however, that must be overcome before any process scaling will occur, and spent considerable time and effort in 1996 to correct these problems.

GPI has compiled multiple successes in calendar year 1996 related to the manufacture and scale-up of thin-film CdS/CdTe photovoltaic modules. These successes were the result of process improvements being identified and implemented into the standard production process. These process improvements have resulted in 1) reduced module degradation over time; 2) higher average wattage output; and 3) increased production process yields. The combination of these factors help to drive down the costs associated with the manufacturing of these thin-film photovoltaic modules while providing a reliable product to the solar industry. Discoveries which led to these improvements are discussed below.

In 1996, GPI began scaled its production process from approximately 1200 panels per month to 1900 panels per month. This was done even though a number of modules produced during this time showed considerable degradation ($\approx 30\%$) from the initial wattage after an eight week outdoor exposure. Other panels, however, showed only moderate losses in wattage (refer to Figures 1 and 2). Because of this inconsistency in batch performance, R&D and Process Engineering were tasked to identify variables in the process which were the source for module degradation. This needed to be corrected before any continuation in process scaling.

The results from this study determined that a major contributor to the batchwise variation in degradation mechanism was a non-homogeneous CdTe microstructure across individual panels and between batches. This was determined from scanning electron microscopy characterization with samples analyzed after the CdTe re-crystallization process. Some batches had a consistent microstructure across a panel and from panel to panel whereas other batches had a wide range of microstructural inhomogeneity across a panel. GPI evaluated and implemented an improved recrystallization thermal profile which has yielded a more reproducible CdTe microstructure and reduced module degradation (refer to Figures 1 and 2). This improvement has also led to a World Record Small Cell Efficiency for CdS/CdTe thin-films on soda-lime-glass of 14.7%.

The second area targeted at GPI for PV manufacturing was to increase module wattage output. At the end of 1995, the average wattage by batch was approximately 21 watts. This represented considerable improvement from the beginning of the year, but fell well short of the wattage goals established by GPI Operations Management. The targeted goals were to demonstrate a continual wattage increase with time, but this had not been realized in production. In 1996, Operations and R&D were tasked by GPI management to identify mechanisms responsible for the lower wattage observed and to implement a pathway for improving the wattage to the targeted levels. Modest gains in wattage output were made throughout 1996 (refer to Figure 3), but a primary breakthrough came with the discovery of the non-uniformity and small grained nature of the CdTe microstructure. The process improvement discussed above that led to improved module stability also resulted in an increase of approximately 1.5 Watts in the average module output.

The improved CdTe morphology alone, however, was not the sole contributor to improvement in the maximum module output (refer to Figure 4). The maximum wattage panel produced each month was observed to fluctuate between 28 Watts and 31 Watts (measurement based on GPI internal testing). The reduced wattage observed in February, 1996, was attributed to material contamination, and Operations has implemented an improved characterization plan to screen raw materials prior to use. The wattage output, however, showed considerable increase from 1995 where the maximum wattage panels rarely exceeded 26 Watts. This sizable increase has been partly attributed to better control over film deposition (film quality and thickness) and the testing and implementation of a superior CdTe dopant incorporation technique. GPI believes that implementing process controls at other key stages in the process will result in a continued increase in the average and maximum wattage values.

The third area of interest at GPI towards PVMat was increasing the total throughput process yields. In 1995, throughput process yields ranged from approximately 35% to 60%. Most of the yield losses were associated with glass breakage with a secondary issue being low wattage modules. Operations supervisors, consultants, and engineers targeted two key stages for improving the total process yields. These were tin oxide deposition and cadmium sulfide deposition. These films are deposited onto a soda lime glass substrate using a spray deposition process which can lead to lower yields from glass breakage. In 1995, these individual processes demonstrated yields that ranged from 50% to 95%, but could not be controlled to acceptable levels. Routine furnace maintenance showed improvements, but these could not be maintained over time.

From analyzing the problem, it became apparent that ambient atmospheric conditions outside these furnaces played an integral role in the individual process yields. For example, air drafts into the production zone could create disturbances in these film deposition furnaces causing the glass to fracture. The glass breakage would throw off the heat balance from zone to zone leading to greater temperature instability. GPI has since installed an air heating system for both furnaces that minimize the impact of outside interference and has increased the yields for these furnaces to greater than 95%. This has resulted in throughput process yields typically exceeding 70% and sometimes achieving 80%. The 80% average yield is now the target set by Operations Management.

In January, 1996, GPI was producing PV modules with an average output of 20 watts and at a rate of approximately 1200 modules per month. The total process yield of completed modules was approximately 50%.
- Implementation of a Statistical Process Control plan identified numerous areas of process variability. Through more rigorous process measurement, specification analysis and subsequent specification changes, machine modifications and process changes directly led to yield improvements in GPI's cell interconnection operation as well as improvements in module output and total process yield.
- A complete re-design of GPI's encapsulation scheme resulted in a significant raw material and labor savings.
- An Incoming Raw Material and Pre-Qualification plan was implemented to screen raw materials prior to use by Operations.

By June, 1996, the average module output had increased to almost 26 watts, and GPI's productivity had increased to over 1900 modules per month. The final process yield had also increased to greater than 75% with costs decreased by approximately 20%. Although these milestones were impressive, module degradation continued to be a major obstacle for product liability and continued production. In July, GPI shifted its principle focus from scaling the process towards gaining a better understanding of mechanisms that led to module degradation. The team assembled by GPI has shown significant successes as evidenced by a marked decrease in module degradation and a significant increase in module output.

GPI has demonstrated multiple successes in 1996 in the area of manufacturing photovoltaic modules in accordance with PVMat guidelines. GPI scientists, engineers, operators, and management have devoted efforts towards identifying mechanisms responsible for panel degradation, limiting wattage output, and limiting process yields. Initial results indicate that GPI is making considerable progress as evidenced by higher wattage modules, more stable modules, and higher process yields. GPI has implemented key process improvements into its standard production practices and increased initial wattage by approximately 4 to 5 watts per module. These modules are also more stable based on internal testing at GPI. Finally, variables that contributed to reducing the overall process yield were studied, identified, and corrected in GPI's manufacturing process. This has resulted in a significant increase in the overall process yields and help to lower the costs associated with manufacturing scale up. GPI continues to investigate factors to improve wattage and stability so that GPI can continually successfully scale its thin-film photovoltaic production line.

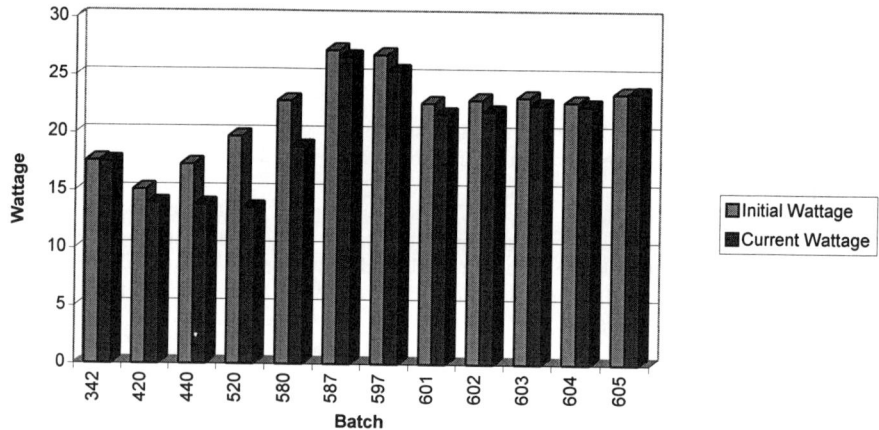

**Figure 1.** Golden Photon Batch to Batch Degradation

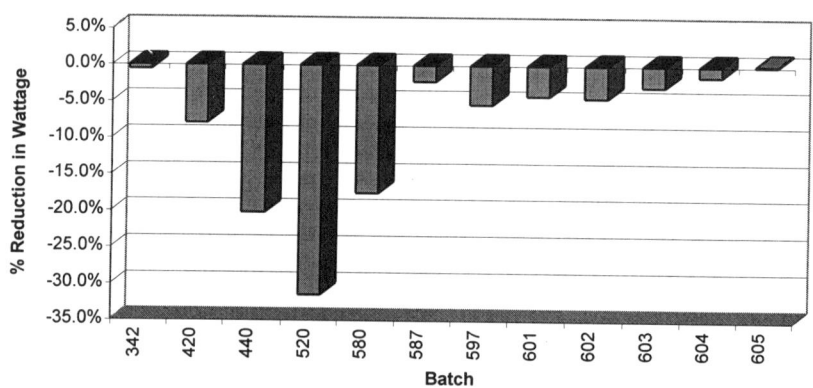

**Figure 2.** Percentage of Wattage Loss for Numerous Production Batches

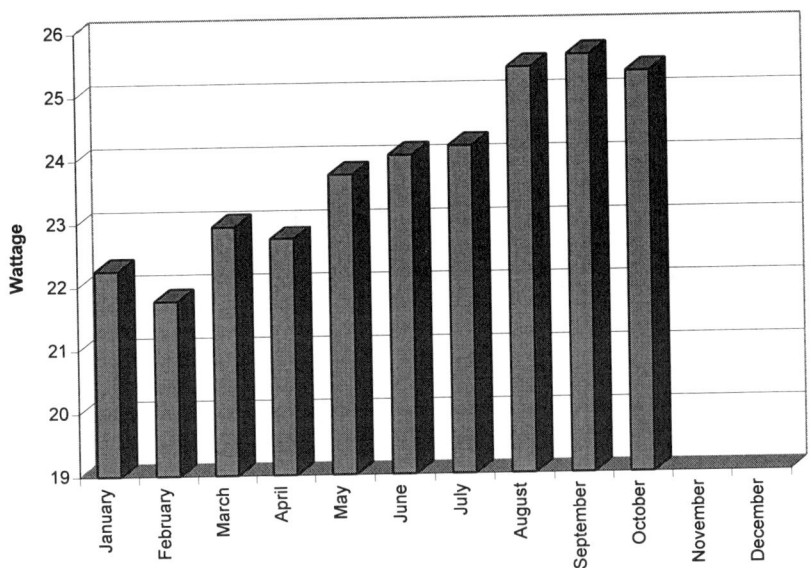

**Figure 3.** Golden Photon's Average Monthly Wattage

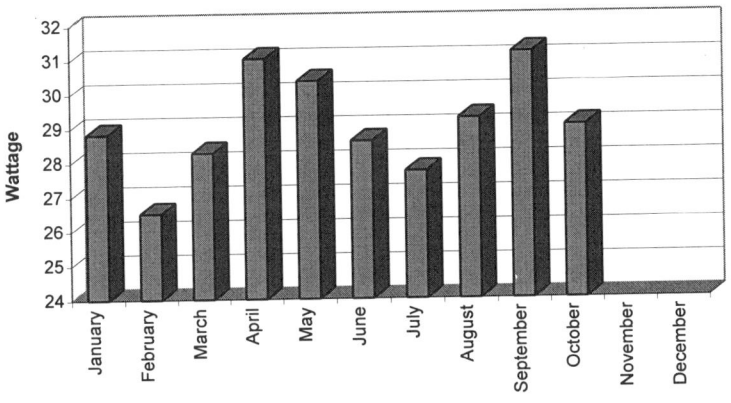

**Figure 4.** Golden Photon's Maximum Wattage Panel by Month

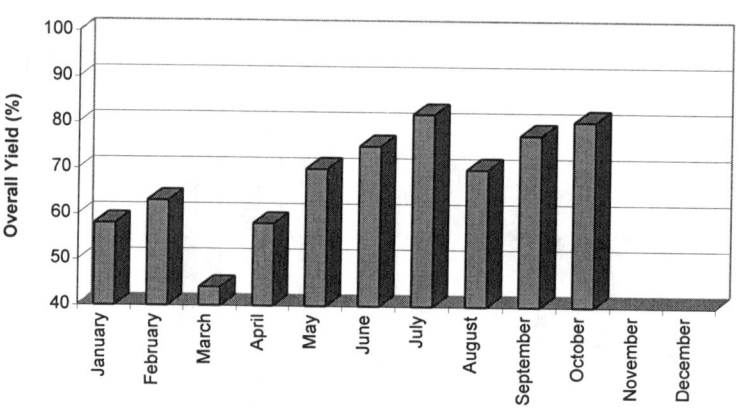

**Figure 5.** Golden Photon Average Production Yield by Month

# PVMaT Improvements in the Manufacturing of the PVI Powergrid™

Neil Kaminar

*Photovoltaics International, LLC (PVI)*
*171 Commercial Street*
*Sunnyvale, CA 94086*
*Phone (408) 746-3062*
*Fax (408) 746-3890*
*Email neil.kaminar@pvintl.com*

**Abstract.** This paper describes the results of the first phase of a three phase program to develop manufacturing technologies that will enable sales of the PVI Powergrid at very competitive prices. Key to this goal is low-cost manufacturing along with high output. The program focuses on development of a state-of-the-art Fresnel lens and module side extrusion system, development of a second generation automated receiver assembly station, development of low-cost, roll-formed panel frame members, and significant reduction of volatile organic compound (VOC) emissions during module assembly. The first phase has been successfully completed with design, specification, and initiation of fabrication of the various pieces of machinery. A new die for a 20 inch wide Fresnel lens was also developed.

## BACKGROUND

The PVI Powergrid is a linear focus concentrator which uses low-cost components and manufacturing techniques intended to reduce the price of a photovoltaics system to its absolute lowest possible point. The Powergrid uses a linear-focus Fresnel lens made by a plastic extrusion process, the lowest cost method of manufacturing. The plastic module sides are also extruded. The Powergrid uses solar cells manufactured using the low-cost methods used for one-sun cells. Twelve modules are mounted on a stationary panel frame to move in unison for single-axis tracking, see Fig. 1.

**Figure 1**. PVI Powergrid™ Panel

PVI developed the Powergrid under a Concentrator Initiative program with Sandia National Laboratories. Under the Sandia program, and a contract with the California Energy Commission, the company installed pilot production equipment and was able to demonstrate a production rate of 830 kilowatts per year. The pilot production demonstration program determined several manufacturing areas that needed improvement:

- Production extrusion of the lenses
- Production extrusion of the module sides
- Automated receiver assembly
- Panel frame manufacturing
- Module assembly process with VOC's

Our program under NREL PVMaT is designed to address these areas.

## LENS AND MODULE SIDE EXTRUSION

Accurate extrusion of the linear Fresnel lens and the module sides is key to high output and thus low cost per Watt for the Powergrid. Subcontracting the extrusion development at an outside company has not proved satisfactory, primarily due to the lack of control sophistication available with tradition commercial extrusion. Performance of production lenses has been low and the development time has been excessive. Accuracy of the module sides has also been poor. Under the PVMaT program, PVI is bringing the extrusion process in-house by developing a state-of-the-art extrusion system. This task has two areas of effort: development of a computer controlled extrusion system that will be far more advanced than any system commercially available, and development of a new set of extrusion tooling for a 20-inch wide lens.

## Development of a State-of-the-art Extrusion System

An advanced extrusion system has been designed and much of the fabrication has been completed. Installation at PVI and initial checkout is planned to be completed in January of 1997. The PVI extrusion system borrows heavily from the medical industry which must be able to extrude medical devices, such as catheter tubes very accurately.

Normally, extrusion machines are controlled using analog devices which the operator sets based upon various analog instruments and "feel". This leads to wide variation in the many extrusion parameters and inconsistent results.

The PVI extrusion system will be monitored and controlled using a computer. It will have various feedback loops that will automatically adjust the extrusion parameters based upon sensors, such as part width or lens transmission, see Fig. 2. The operator will be able to call up and analyze any trends or parameter relationships using the computer.

A number of other advanced features will make the PVI extrusion system capable of making high-quality lenses. It will include a flying cut-off saw that will accurately cut the parts to their final length with minimum waste. Post-extrusion tooling will include vacuum sizers and roll forming that will assure correct width and facet form. PVI has chosen the best possible components to assemble this extruder.

All of the components have been or are being fabricated. Some components have been delivered and the remaining are expected by the end of 1996. Initial system startup and checkout is scheduled for January of 1997 and production is scheduled for April of 1997.

**Figure 2.** In-line Transmission Sensor, Proof-of-concept Test

## Development of Advanced Extrusion Tooling

The extrusion tooling consists of a manifold to distribute the plastic from the extruder to the die, the die, and the after-forming tooling including vacuum-formers, an air rack to cool the plastic, and hardware necessary to support the soft plastic while it sets up. All of this equipment is important, but the die is the most critical.

In order to save time, PVI has been developing the extrusion tooling by leasing equipment from an extrusion company. The testing is done at the extrusion company using PVI tooling and some of the equipment designated for the PVI extrusion system, such as the puller. A puller is needed to pull the extrudate from the die and through the after-forming equipment. It is a critical piece of equipment and needs to provide a consistent speed if accurate parts are to be made. The puller designed by PVI provides the necessary accuracy while the pullers generally available from the extrusion company do not. PVI supplies the temperature controllers, pressure sensors, and other equipment for the same reason.

Although the equipment at the extrusion company is not as advanced as our own system will be, the early testing of the die and other components has been very valuable and saved many months of development time. It has allowed us to proof the PVI designed tooling and make changes to improve lens quality. With the improved 20-inch wide lenses, and other changes to the module, we have been able to improve the output of the PVI panel to approximately 1,400 Watts.

The 20-inch lens die represents a substantial improvement in lens extrusion technology. It uses a flat manifold and die to produce a flat lens which is later curved, see Fig. 3. Our previous dies, developed at an outside subcontractor, were curved. The flat die allows us to use well-developed sheet-die technology. It also allows us to use a die plate which can easily be removed and replaced, a feature that allows development of the die to take place in a matter of days as compared to months using the previous technology.

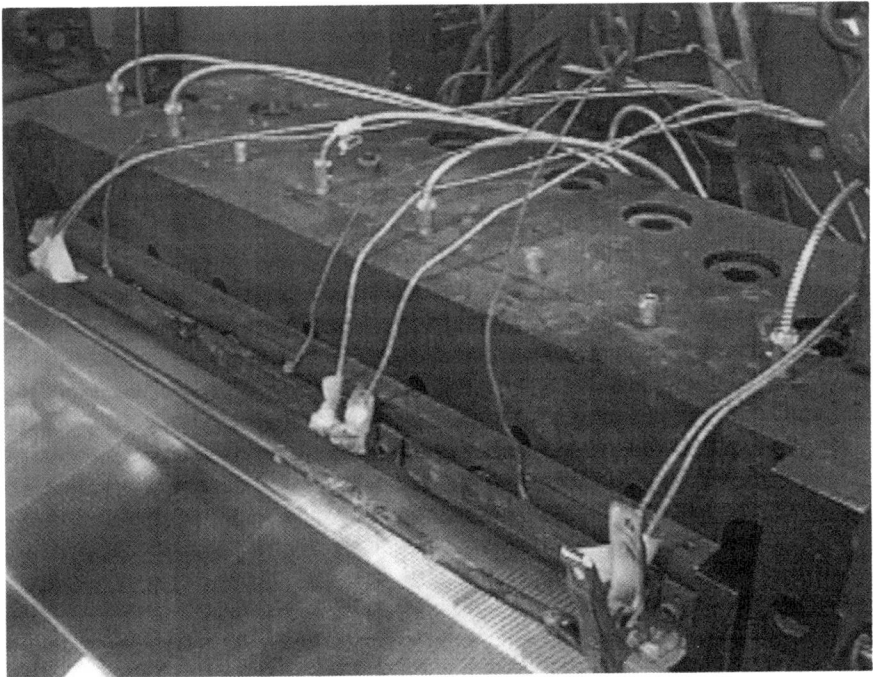

**Figure 3.** 20-inch Lens Die Being Tested

Additional die development will be required when our own system is on-line. Once development is completed, we expect mid-80's percent optical transmission with a 11.5 to 1 geometric concentration ratio.

## AUTOMATED RECEIVER ASSEMBLY STATION

Under the Sandia and CEC contracts, PVI developed an automated receiver assembly station. This station did not produce quality receivers and was taken out of service. Under the NREL PVMaT contract, we are developing a second generation station that addresses all of the problems associated with the first station, see Fig. 4. An automated receiver assembly station is crucial to PVI meeting the cost goals because of the high labor content of manual assembly.

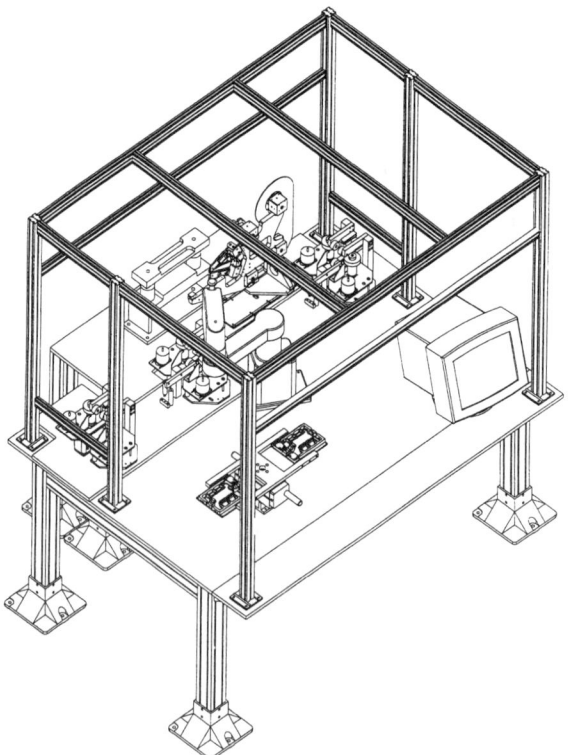

**Figure 4.** Second Generation Automated Receiver Assembly Station Design

The first station attempted to solder the leads to the cells in situ on the heat sinks. This proved extremely difficult and unreliable because of the heat dissipating properties of the heat sink. In the second generation station, the leads will be soldered to the cells before they are placed on the heat sink. Higher quality solder joints are possible and 100% inspection can be done.

We plan to use IR lamp soldering to join the leads to the cells. We have experimented with this technique and have found that, if done properly, it can be very reliable. We have also experimented with laser soldering to join the cells in series on the heat sink and plan to use this technique for that purpose. The laser gives intense heat only at the solder tabs on the leads.

We will use an off-the-shelf adept robot to perform a majority of the various tasks necessary for automated assembly. The robot will place the punched leads and cells in the soldering fixture. It will also move the completed cell assembly to the heat sink.

PVI has completed the design and detailing of the second generation station. All aspects of the station, such as the IR lamp soldering and laser soldering, have been verified in individual proof-of-concept tests. Some parts of the station have been fabricated and others are now being quoted. We expect the station to be fully operational by October of 1997.

## PANEL FRAME MANUFACTURING

The Powergrid panel employs a stationary frame that doubles as an electrical conduit. The frame is designed to resist twisting by using closed square-tube sections. It is supported at four points which makes it easy to mount on roof tops, similar to flat-panel frames. The present manufacturing technique is to use extruded aluminum sections that are machined after forming for wires and bearing mounts. This makes the frames expensive. A lower cost substitute is needed.

Under the NREL PVMaT contract, PVI is developing a light-weight roll-formed steel frame. The steel is environmentally protected by galvanizing. The frame members will be formed after the holes are punched. A lock seam will be employed to close the section and provide the necessary stiffness.

In Phase 1, PVI built prototypes of the panel frame and identified a source that can punch and roll-form the frame members. The roll tooling is expensive and should not be ordered before the design is proven through the prototypes. We plan to order the tooling in 1997. The roll-formed steel frames represent a sizable savings in cost.

## LOW VOC MODULE ASSEMBLY

The module is presently glued together using an adhesive which has a high VOC content. Also, the receivers are encapsulated using a very high VOC content liquid silicone that also is a hazardous material. Under the NREL PVMaT contract, PVI is developing techniques that will reduce the VOC's and hazardous materials while automating the module assembly process as much as practical.

The silicone encapsulant is the worse component by far in terms of both VOC's and hazard. We are developing techniques to replacing it with EVA.

The use of EVA is made possible by recent developments that make the material far less UV sensitive.[1] Another factor which makes the use of EVA possible in the PVI Powergrid concentrator is the use of lens material which restricts transmission of UV light that the EVA is sensitive to, about 350 µm and shorter wave length.

PVI is developing techniques to use the EVA in our receivers. We have produced a short test oven that is used to make test samples. We have also produced a larger production oven that can handle full size receivers. Several samples have been made that have undergone thermal cycling in the PVI test chamber. Other samples have been tested at Springborn Labs for UV resistance. Sample testing is continuing. No insurmountable problems have been encountered.

The EVA ovens use a diaphragm between a top vacuum chamber and a bottom vacuum chamber. The diaphragm operates similarly to the diaphragm used in flat-panel module production. Higher throughput ovens are being designed for larger scale production.

The adhesive used to join the plastic parts is difficult to eliminate. It does a very good job and no viable alternatives have been discovered. PVI has investigated ultrasonic and thermal bonding, but these processes are too expensive and problematic for our product. We are investigating alternative adhesives with 100% solids, but so far have not discovered one that can pass the qualification testing. Our solution so far has been to re-design the module to use the minimum amount of adhesive. During 1997, PVI plans to continue to search for an alternative.

## SUMMARY AND CONCLUSION

The first phase of the contract was basically to perform design and specification tasks. These tasks have been completed and fabrication has started. During 1997, the improvements will be put in place on the production line. PVI plans to demonstrate the new capability in 1998.

The completion of the NREL PVMaT contract will result in an estimated saving of almost three dollars per Watt for the PVI Powergrid. During the first five years of full production, the savings to the public are estimated to be over 600 million dollars. The PVI Powergrid is a unique concentrator technology which has the potential of being very low cost in the short term.

## REFERENCES

[1] Holley, W.A., *Advanced Development of PV Encapsulants, Semiannual Technical Progress Report 30 June 1995 - 31 December 1995*, National Renewable Energy Laboratory, June 1996, NREL/TP-411-21280

# Market-Driven Improvements in the Manufacturing of EFG Modules

Michael Kardauskas, Juris Kalejs, Jeff Cao, Eric Tornstrom, Ronald Gonsiorawski, Colleen O'Brien, and Mert Prince

*ASE Americas, Inc., 4 Suburban Park Drive, Billerica, MA 01821-3980*

**Abstract.** Most of the subtasks scheduled for the first year of the current PVMaT Phase 4A2 subcontract at ASE Americas have been completed on schedule. Improvements have been made in EFG wafer technology, cell efficiency, and module manufacturing methods, which have reduced module costs by 7%. Further improvements will be developed during the coming year. Full implementation of the new technologies in manufacturing, during the third and final year of the subcontract, is expected to reduce module production costs by 25%.

## INTRODUCTION

The current PVMaT Phase 4A2 subcontract at ASE Americas is designed to reduce the cost of manufacturing photovoltaic modules by 25% over the three-year course of the subcontract. The program aims to accomplish this objective by reducing costs at each of the three stages of manufacture of the module: in wafer manufacturing, cell production, and module assembly. The improvements in each of these areas are designed to complement one another so that, for example, improvements in wafer manufacturing will have their greatest impact on the performance of cells, and changes being made in the design of cells will reduce the cost of assembling them into modules.

The technology employed is based on wafers produced from octagonal silicon tubes grown by the Edge-defined Film-fed Growth (EFG) process. Those wafers are cut from the tubes by high speed lasers, and then processed into cells. Details of the production process have been published previously (1). Major subtasks in the current program include the reduction of the thickness of EFG wafers from 300 to 250 µm, improvement of average solar cell efficiency to over 15%, and the simplification of processes and reduction of costs in cell interconnect and module manufacturing.

This report describes work carried out under the first phase of this subcontract. Most of this effort was exploratory in nature, to determine the best approaches to improving various processes and materials. The second phase will focus on implementing those approaches in the form of prototype equipment and test runs in production. During the third and final phase of the program, the new equipment and processes will be introduced into the manufacturing line, providing a 25% reduction in module manufacturing costs.

# THE PHASE I PROGRAM

## Task 1: EFG Wafer Improvements

### Task 1 Objectives

There were several major objectives in Task 1. The first of these was to improve wafer quality by reducing levels of impurities and defects and by optimizing crystal growth variables. The second was to reduce wafer impurity levels by designing an enclosure for EFG growth furnaces which would exclude contaminants in the ambient air. The third objective was to develop an automated silicon feedstock sorter to reduce both labor and feedstock costs. Finally, average wafer thickness was to be reduced from 300 µm to 250 µm with no reduction in manufacturing yield. (Wafer thickness distribution at the beginning of the subcontract is illustrated in Figure 1.) The attainment of this last objective would both reduce cost, by reducing silicon consumption, and raise cell efficiency, by improving minority carrier collection in the base region. One of the subtasks related to this objective was to quantify the efficiency increase to be expected from a given reduction in wafer thickness.

### Task 1 Results

The objective of improving wafer quality was approached on three fronts. The first was to evaluate different processes for purifying the graphite components of EFG growth furnaces. The comparison was performed on the basis of the

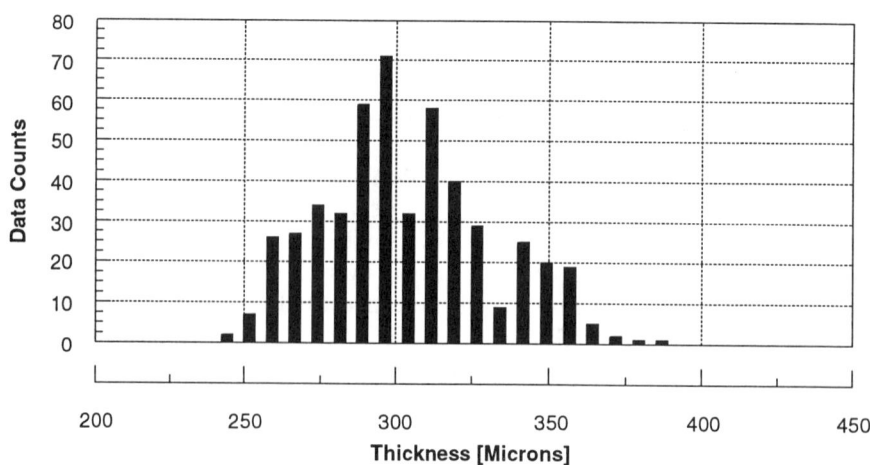

**FIGURE 1.** Thickness distribution of EFG wafers at the beginning of the subcontract. Average thickness is about 300 µm. A goal of the program is to reduce this value to 250 µm.

efficiencies of solar cells produced from wafers grown in furnaces containing graphite purified using the various processes. The best of those processes was then introduced into manufacturing by converting the EFG furnaces one-by-one to the use of the better purified materials, until the entire production line was changed over. Many fewer instances of silicon contamination during crystal growth have been observed since the conversion to the new purification process, as compared to an equal period in the previous year.

Other work designed to improved wafer quality included a study on the effect of crystal growth speed on wafer quality. The results of this work indicated that a reduction of growth speed by about 5% would improve the yield of flat EFG wafers, a finding that has been verified in production. Experiments were also carried out for the purpose of optimizing the level of oxygen, in the form of carbon monoxide (CO) added to the silicon growth ambient to improve bulk minority carrier lifetime. These experiments produced unexpected results indicating that CO additions produced little improvement, which conflicted with earlier findings (1). Studies on this effect are continuing, in an effort to resolve the two sets of results. However, the new results have already led to a reduction in production costs, as most EFG furnaces are now being operated without CO. The savings arise from increased life of EFG dies and graphite furnace parts, which become coated with silicon oxide deposits when CO is used during crystal growth. These savings, combined with those resulting from improved wafer quality, reduced ASE module production costs by 3.3%.

The design for the new enclosure designed to exclude ambient air from EFG growth furnaces has been completed, and construction of the enclosure is underway at a subcontractor's facility. It is expected to be installed at ASE Americas during the first quarter of Phase II of this PVMaT project. Silicon feedstock sorting equipment has also been designed and tested successfully.

Experiments were carried out to determine the effect of wafer thickness on cell efficiency, which generated the data shown in Figure 2. This data shows that a wafer thickness reduction from 300 µm to 250 µm will result in a cell efficiency increase of about 0.3-0.4% absolute. To date, average wafer thickness has been successfully reduced from 300 µm to 275 µm on full production runs of 500-600 wafers each, with no reduction in yield. However, further thickness reduction to 250 µm causes a significant increase in losses during laser cutting. Based on these results, the effort to reduce wafer thickness is now being directed toward improvements in laser cutting. Experiments using a new generation of $CO_2$ laser showed some promise for achieving smoother edges on the entrance side of the cut, but the new laser had a tendency to redeposit the cut material on the opposite side of the wafer. Further work now in progress is focussed on developing a cutting technology based on copper vapor lasers. These lasers cut cleanly and produce much less heating of the bulk silicon during the cutting process than do YAG and $CO_2$ lasers, reducing the length of microcracks formed during the cutting process. However, the copper vapor lasers available today have lower power capacities than those of the other types, requiring the development of new optics to enable cutting at required production rates.

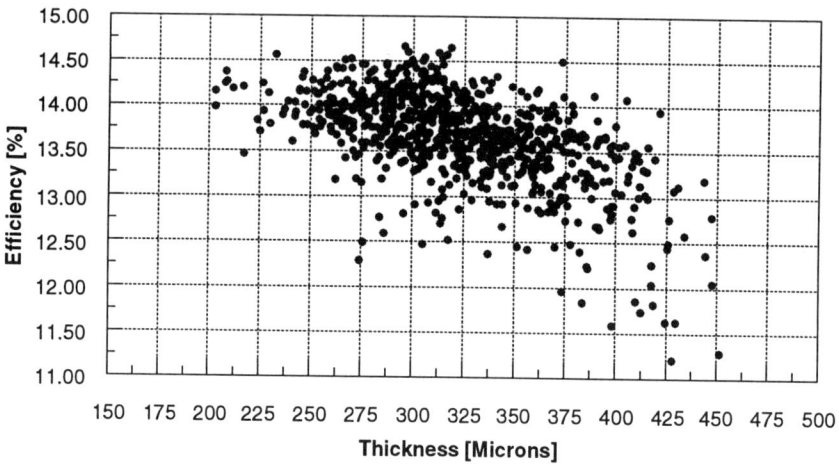

**FIGURE 2.** Cell efficiency as a function of EFG wafer thickness. A decrease in thickness of 50 μm is expected to raise efficiency by 0.3-0.4% absolute.

## Task 2: EFG Cell Improvements

### Task 2 Objectives

There were four major objectives of the cell improvement program during Phase I. The first of these was to optimize the front and rearside metallization patterns and processes to improve cell efficiency. The second, related to the first, was the achievement of average cell efficiency levels of 14.25% on the ASE production line by the end of Phase I. The third was the investigation of the use of textured coatings to produce light-trapping effects. If successful, this technology would be implemented in the ASE production line during Phases II and III. Finally, a new process would be developed for the removal of the glass present on wafers after phosphorus diffusion. This new process would be designed to reduce both chemical consumption and wastewater treatment requirements compared with the industry standard method of immersing the wafers in hydrofluoric acid, and would also be compatible with highly automated processing.

### Task 2 Results

Studies were carried out to determine the optimum thickness of aluminum to apply to the rear side of EFG cells to produce an effective Back Surface Field (BSF). This work was then followed by optimization of the heating and cooling profiles during the firing of the thick film metallization patterns. Cell efficiency was found

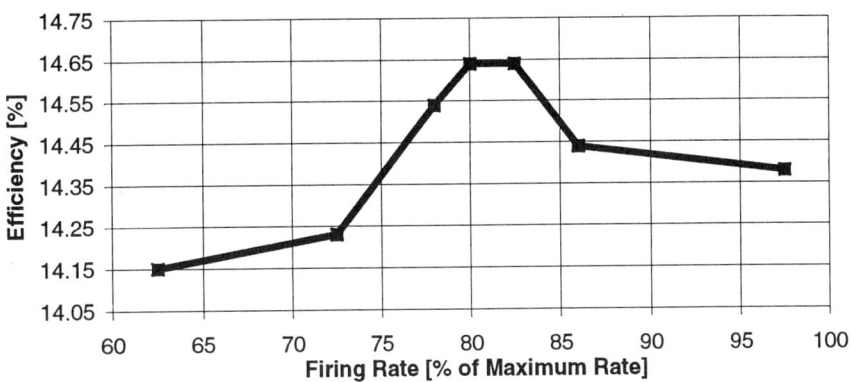

**FIGURE 3.** The effect of metallization firing rate on cell efficiency. Optimizing this parameter produced test batches of cells with efficiencies as high as 14.6%.

to be rather sensitive to firing conditions, as can be seen in Figure 3. Combining the results of the two optimizations resulted in the production of small test batches of cells with efficiencies as high as 14.6%. When the optimized processes were adopted on the production line, a significant number of cell lots were produced with average efficiencies over 14% (Figure 4), with a few lots having efficiencies as high as 14.2%. The increase in average efficiency reduced module costs by about 3%.

Textured coatings of zinc oxide (ZnO) were applied to EFG cells at Harvard University in an attempt to produce light trapping in the cells. Although coatings

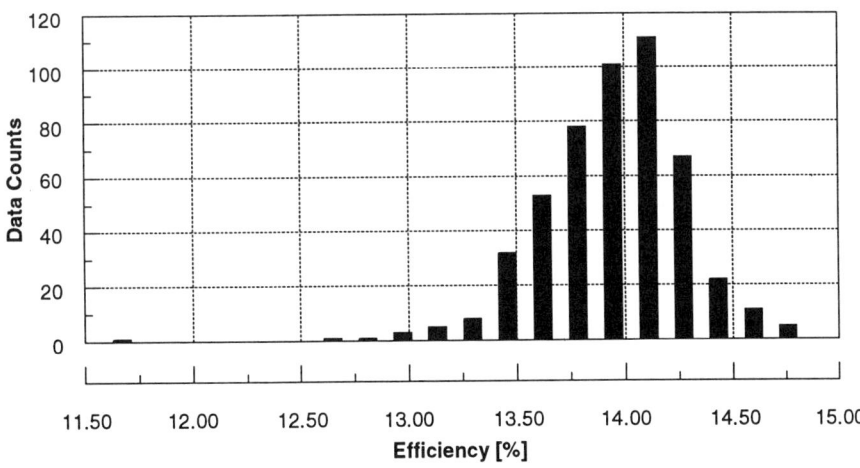

**FIGURE 4.** The distribution of cell efficiencies in a production lot of EFG cells with an average of nearly 14.0%. Average efficiencies as high as 14.2% have been achieved.

were produced which appeared matte black, which was an objective of this task, the coatings did not significantly improve cell performance. Further analysis of the coatings is now being carried out at both Harvard and Sandia Laboratories to determine the cause of the shortfall. One possible cause is light absorption in the films, which might be solved by a change in ZnO deposition conditions.

The process under development for the purpose of reducing the use of hydrofluoric acid (HF) during diffusion glass removal was successfully proven to be effective in laboratory trials. The new process was demonstrated to remove the glass layer cleanly using 95-98% less HF than the standard process. Plans for prototype etching equipment using the new process have been drawn up, and those plans incorporate the automatic control features originally envisioned for the new process. During Phase II, the prototype equipment will be built and tested. If those tests also prove successful, a production scale unit capable of etching sufficient wafers to produce 4 MW of cells per year will be ordered for installation early in Phase III of the program.

## Task 3: EFG Module Improvements

### Task 3 Objectives

The first goal of the module development program during Phase I was to simplify the task of interconnecting EFG cells by changing the cell design to permit all electrical connections to be made from one side. The second objective was to evaluate new materials and processes which would permit the integration of the cell interconnect, lamination, and module framing processes, using the new cell design. The goal of this work would be to reduce cost as much as possible. The third objective was to reduce the cost of the cell lamination process. This was expected to require the development of a new laminator design, and possibly new encapsulant materials. Finally, an effort would be made to further reduce module costs by other design changes, which would be developed during the course of Phase I.

### Task 3 Results

Three different approaches were tested to bring all electrical contacts to one side of the cell. All of them suffered from increased resistance losses compared to the standard interconnect method using solder bonding of copper tabs. The approach that shows the most promise is that of wrapping the busbar from the front grid around the edge of the cell to the cell back. This redesign will require changes in the shape of the busbars on EFG cells, and a new deposition technique to bring the contacts to the rear side. These changes will be tested in development trials during Phase II.

New module materials that have been evaluated include electrically conducting epoxy formulations, which may be used to replace solder for cell interconnection. The epoxies have higher electrical resistivities than solder, and are more expensive, but may be useful for implementing the single-sided interconnect approach. Another

**FIGURE 5.** New encapsulants are under development that would permit a significant reduction in lamination time. One concept being studied is the coextrusion of two different polymers, combining the adhesive properties of one with the higher melt flow of the other.

promising material for this purpose is a tin-plated copper strip product which is coated with a pressure sensitive conductive adhesive, which could replace both copper tabs and solder in a new interconnect process. New encapsulant materials were also investigated. The most promising of these is a coextrusion of two different polymers, designed to provide maximum adhesion to glass while reducing the time currently required for lamination using ASE's standard encapsulant (Figure 5). Additionally, a new module diode housing has been designed, which will reduce

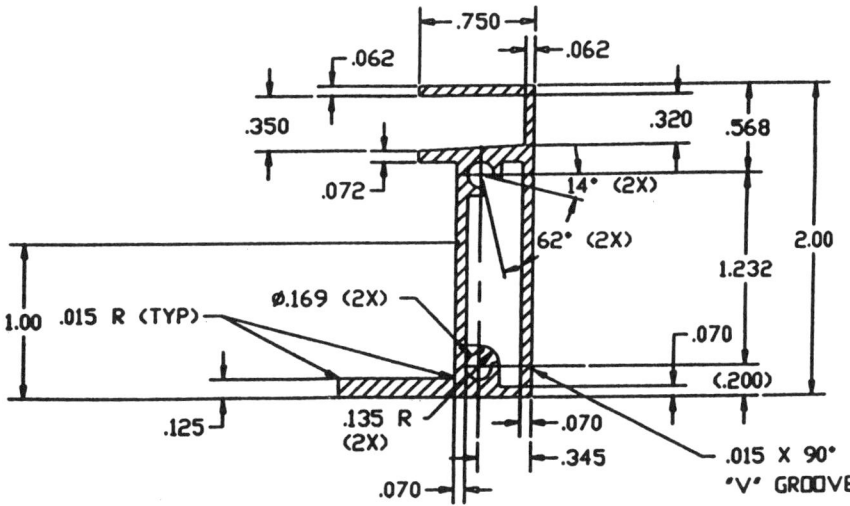

**FIGURE 6.** Cross-section of the extruded aluminum frame now used by ASE. Engineering work is being done to replace this frame with a less expensive one made of roll formed metal.

costs by eliminating unnecessary components. New frames have also been designed which would replace ASE's current design constructed from four lengths of extruded aluminum (Figure 6). The new frames would be formed from a single length of roll-formed aluminum, which would be wrapped around a module and fastened at only one location, eliminating several operations during frame assembly. The new frame designs are currently being assessed for structural strength and cost.

Minor changes in module assembly during Phase I provided about a 1% reduction in module cost. The total of this savings and those achieved in the wafer and cell manufacturing areas amounted to about 7% during Phase I.

## THE PHASE II PROGRAM

During Phase II, the exploratory work conducted in Phase I will be extended to fully demonstrate manufacturable applications of new materials and technologies. This will permit introduction of those technologies into manufacturing production during Phase III. The major objectives of the second phase are:

1) Reoptimization of EFG crystal growth processes using the new furnace enclosure, leading to further improvements in wafer quality. 250 μm thick EFG tubes will be grown and processed under production conditions. Work on the development of new laser cutting technology will continue.

2) Improvements in wafer quality, grid design, anti-reflection coatings, and cell processing which will raise average cell efficiency to 14.5%.

3) Integration of the new cell design, interconnection method, encapsulant, and frame to produce prototypes of ASE's future production module. This design will enter manufacturing during Phase III of this program.

## ACKNOWLEDGMENTS

The work described in this report includes both work carried out by the authors at ASE Americas, Inc., and work performed by a number of subcontractors. Those subcontractors include: Amerimax Specialty Products, Inc., Dr. R.O. Bell, BPP, Bright Technology, Inc., CQB Associates, Prof. S. Danyluk, Georgia Institute of Technology, Prof. T. Gross, GT Equipment, Inc., and Harvard University.

This work has been supported in part by DOE/NREL under PVMaT subcontract number ZAF-6-14271-13.

## REFERENCE

1. Kardauskas, M.J., Rosenblum, M.D., Mackintosh, B.H., and Kalejs, J.P, "The Coming of Age of a New PV Wafer Technology - Some Aspects of EFG Polycrystalline Silicon Sheet Manufacture," in *Proceedings of the 25th IEEE Photovoltaic Specialists Conference,* New York: IEEE Press, 1996, pp. 383-388.

# Advanced Polymer PV System

J. I. Hanoka, P. M. Kane, R. G. Chleboski, and M. A. Farber

*Evergreen Solar, Inc., Waltham, MA 02154*

**Abstract.** During this first year of a two year sub-contract from NREL under PVMat, a project to lower module manufacturing and related systems costs has been pursued. A novel backskin material has been developed which allows for a frameless module with a wide variety of simpler and lower cost mounting possibilities. Using rapid prototyping techniques and extensive feedback from potential customers, a new junction box has been designed. Preliminary work has been done towards the development of a new encapsulant. This new encapsulant, in turn, will enable the development of a continuous non-vacuum lamination method and the foundation for this was also established. The net result of all of this, when successfully deployed, will be a cost reduction of at least $0.50/watt for both module making and systems installation cost.

## INTRODUCTION

Evergreen Solar, Inc. (Evergreen) is a new solar cell company which was started in the fall of 1994. In order to reduce the costs of making crystalline silicon solar cells and modules, Evergreen is developing novel technology in each of the three principal areas involved in PV manufacture: (1) wafers or substrates; (2) cell making; and (3) modules and systems. For (1) Evergreen is developing String Ribbon [1,2,3], a method for producing silicon sheet in the form of continuous ribbon from the melt. For cell making, Evergreen is developing what we believe will be the lowest cost and simplest process in the industry for crystalline silicon. The third area, modules and systems, is the subject of this paper. In September of 1995, Evergreen began a two year PVMat subcontract (#ZAF-5-14271-09) which had the following objectives:

(1) Development of a novel backskin material which would then result in a frameless module.

(2) Reduced systems cost. This included development of novel mounting methods utilizing such a frameless module and the design of a better junction box.

(3) Development of a new encapsulant material which could be laminated in air.

(4) Development of a non-vacuum, continuous lamination method utilizing this new encapsulant.

In this paper we report on progress made in this project during the first year of this sub-contract. We will focus on work done under objectives (1) and (2) above.

## DEVELOPMENT OF A NOVEL BACKSKIN MATERIAL

The theme underlying Evergreen's PVMat contract is that the explosive growth of low cost but high performance polymers fueled by the huge markets these materials serve, offers interesting opportunities for photovoltaics. These are possibilities which PV, by itself, would not generate given the present size of the PV market vis-a-vis typical polymer markets and quantities. Of course these materials are not usually exactly in a form or precise composition which makes them immediately usable for PV. Thus, one thrust of Evergreen's PVMat contract has been to develop the necessary modifications to be able to utilize such materials.

For the development of a novel backskin material, the following criteria were used. It had to be a material of lower cost than the conventional Tedlar laminate which is about $1/sq. ft., but without sacrificing any of the desirable properties of Tedlar backskin. To avoid any possibility of puncture, it should be thicker than the approximately 9 mils thickness of the Tedlar backskin. It had to be a material which could be formed and molded in the lamination process in such a way that a frameless module design would be viable, especially regarding bonding to all adjacent surfaces and protecting the glass edges of the module. And finally, it should be a material which would be expected to pass all the PV module tests as well as the UL requirements.

After an extensive search of possible resin suppliers, a material based on modified polyolefins was chosen. This material basically satisfied all the above criteria. Following its selection, 40 mil sheet was prepared from it.

One of the earliest concerns with this novel backskin material was its resistance to thermal creep. In architectural applications, for example, it is possible for modules to reach temperatures as high as 90° C. Accordingly, a number of tests were used to assure that the thermal creep resistance of this material was satisfactory.

In the first such test, 1" x 6" strips of the material were subjected to 85°C under a nominal load for 45 days. The nominal load was that estimated for typical field operating conditions. In the second set of more stringent tests, 1" x 6" strips with nominal loading were heated to temperatures exceeding 250°C for periods of several hours. In both cases, no significant thermal creep was measured.

# FRAMELESS MODULE

The development of the novel backskin material paved the way for a frameless module in which the backskin material is also employed as an edge protector for the glass superstrate and then wraps around the edge to form a frameless module. In this case, the sealing of the module and the edge seal of the backskin material, as well as the wraparound edge, are all formed in the lamination process itself. So, at the end of the lamination step, a frameless module with edge seal and wraparound edging emerges.

The backskin material which was eventually chosen also exhibited excellent bonding capabilities to glass, metals, and other polymers. Fig. 1 shows the top portion of an Evergreen module using string ribbon solar cells and the new backskin. The edge protection of the front glass and the wraparound features of the backskin can be seen.

Several tests were employed to determine the integrity of the bond between the backskin material and the glass superstrate. The principal test was done in an environmental chamber. It was decided to use humidity freeze, since this was viewed as one of the most stringent of the test protocols for PV modules. The humidity freeze cycle is 85°C at 85% R.H. for 20 hours, followed by cooling down to -40°C. The standard protocol calls for 10 such cycles. We decided to see if we could test the glass/backskin bond to failure and found that even after 100 such cycles the bond integrity showed no weakening.

**FIGURE 1.** A String Ribbon frameless module

The strength of this bond, we did find, was contingent on the lamination temperature. Too low a temperature could produce a weaker bond.

The major advantage of the frameless module was that "slide bars" of metal or polymer could readily be bonded to the backskin in such a way as to allow the module to then be slid into place along two pieces of "C" channel. With such a possibility, the need for any mounting screws or nuts and bolts could be eliminated. Fig. 2 shows aluminum slide bars on the back of an Evergreen module. In the case of the aluminum/backskin material bond integrity, both extensive humidity freeze and creep tests under load were performed to test bond strength and reliability. Fig. 3a shows the set-up used for thermal creep testing. Samples were tested for 45 days at 85°C with a load estimated at more than twice that for any actual in practice loads and demonstrated no measurable creep. Fig. 3b shows how static load testing was done. In this case the weights on the sample were such as to approximate twice the maximum estimated loading a module would ever experience with this mounting configuration. The results in this case were also excellent and indicated clearly that the bond strength was high and likely to remain so under use.

**FIGURE 2.** Edgewise view of a frameless Evergreen module showing the new junction box and, just below it, the aluminum slide bar.

## REDUCED SYSTEMS COSTS

In addition to fostering the capability for a frameless module, the novel backskin also permitted a large variety of lower cost mounting possibilities. In particular, it is possible to attach either metal or plastic mounting components of many possible configurations. Fig. 3 shows the back of a module with aluminum mounting "slide bars." These slide bars allow the module to be mounted onto "C" channel by simply sliding the module onto two pieces of "C" channel. Such a mounting configuration requires no screws, is rapid, and consequently low in labor cost. Fig. 4 shows several modules mounted this way in an outdoor suburban Boston location.

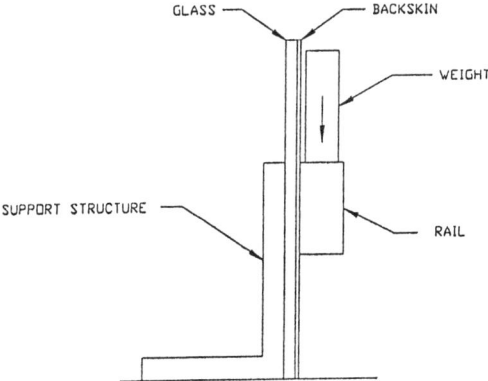

**FIGURE 3A.** Thermal creep test arrangement for the new backskin material.

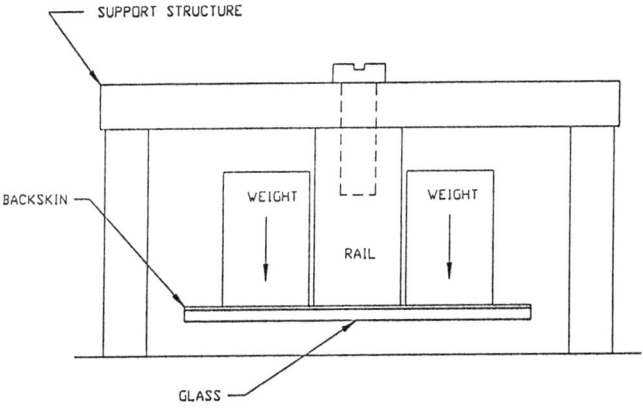

**FIGURE 3B.** Set up for static load testing of the aluminum slide rail for the frameless module.

**FIGURE 4.** Four Evergreen frameless modules mounted onto a simple C Channel and aluminum tubing structure.

## IMPROVED JUNCTION BOX

A portion of the PVMat program was devoted to marketing studies of the novel frameless module and mounting arrangements. In the course of this, Evergreen also obtained feedback concerning junction boxes from a number of potential customers and integrators. As a result, Evergreen decided to design a new junction box which would incorporate the most desired features which emerged from this marketing study.

Using the techniques now widely practiced for injection molded polymer parts, Evergreen embarked on a rapid prototyping method which used 3-dimensional modeling, continual feedback for several design iterations, and then stereolithography and a laser sensitive polymeric material to form prototypes. The prototypes were then used in a final feedback loop and then an injection mold was ordered. Fig. 5 shows the junction box and lid. Here the junction box has an extension to mount onto an aluminum frame. The j-box can also be made without this extension if the j-box is mounted away from the module edges.

This new junction box had the following features: a size large enough for comfortable field wiring; a hinged lid which stays open in any position and a single captive screw; field and factory wiring clearly separated; a molded in terminal strip, 600 V d.c. rated; conduit capable with 4 knock-outs; built-in fuse capability; multiple module attachment possibilities—onto or away from an aluminum frame when there is such a frame; and spare terminals for multiple module wiring configurations.

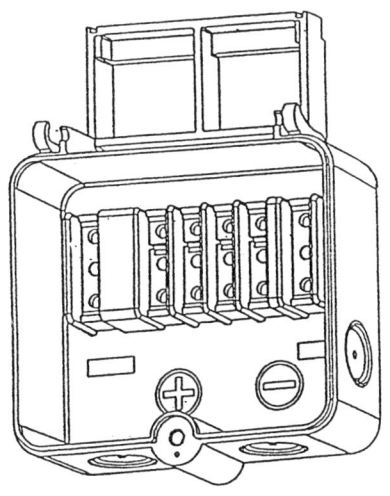 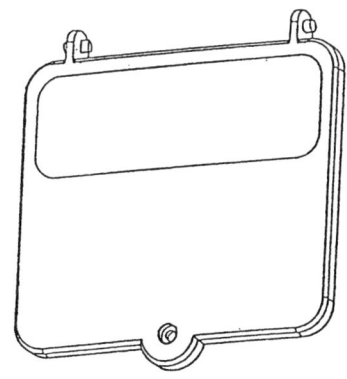

**FIGURE 5.** Evergreen's new junction box.

## NEW ENCAPSULANT AND LAMINATION PROCESS

As of this writing, Evergreen has just begun testing of the new encapsulant material which can be laminated in air. We are also in the final stages of a prototype continuous lamination machine which will utilize such an encapsulant.

## SUMMARY

The initial reception to Evergreen's frameless module, mounting system, and junction box has been very positive both among potential customers, integrators and even thin-film PV makers. In this brief report, we have summarized some of the work which led up to this during the first year of a two year PVMat contract. Given the progress made so far, the prognosis for reaching the cost goal reduction of $0.50/watt seems very positive.

## ACKNOWLEDGMENTS

We wish to thank Holly Thomas of NREL for her continued support and encouragement in this work which was done under PVMat contract #ZAF-5-14271-09. The assistance of Joseph Fava and Jennifer Martz at Evergreen is also gratefully acknowledged.

# REFERENCES

1. E. M. Sachs, D. Ely, and J. Serdy, Journal of Crystal Growth, (82) 1987, 117-121.

2. J. I. Hanoka, B. Behnin, J. Michel, B. Sapori and M. Symko, 5$^{th}$ NREL Workshop on the Role of Impurities and Defects in Silicon Device Processing, Copper Mountain, CO 1995.

3. R. Wallace, J. I. Hanoka, A. Rohatgi, and G. Crotty, PVSEC-9, Miyazaki, Japan, 1996.

# The AC Photovoltaic Module is Here!

Steven J. Strong*, John H. Wohlgemuth†, and Robert H. Wills‡

*Solar Design Associates, Inc., Harvard, MA 01451-0242
†Solarex, 630 Solarex Court, Frederick, MD 21701
‡Advanced Energy Systems, Inc., P.O. Box 262, Wilton, NH 03086

**ABSTRACT**. This paper describes the design, development, and performance results of a large-area photovoltaic module whose electrical output is ac power suitable for direct connection to the utility grid. The large-area ac PV module features a dedicated, integrally mounted, high-efficiency dc-to-ac power inverter with a nominal output of 250 watts (STC) at 120 Vac, 60 H, that is fully compatible with utility power. The module's output is connected directly to the building's conventional ac distribution system without need for any dc wiring, string combiners, dc ground-fault protection or additional power-conditioning equipment. With its advantages, the ac photovoltaic module promises to become a universal building block for use in all utility-interactive PV systems. This paper discusses AC Module design aspects and utility interface issues (including islanding).

## INTRODUCTION

The AC Module has the potential to become the first photovoltaic power "building product," as it can be marketed as a complete, packaged solution for the emerging residential rooftop and commercial demand-side management markets.

The advantages of the AC Module include

- The minimum system size of one AC Module provides a low barrier to market entry
- The minimum array increment of one AC Module and the elimination of balance-of-system equipment allows for maximum system flexibility in initial sizing and simple future array expansion
- There are no constraints on module/array orientation, shading or solar exposure—each module has its own dedicated maximum-power-point tracker
- Module and string mismatch losses from long series strings are eliminated
- No special string combiners, dc wiring or ground-fault protection are needed
- A dramatic reduction in manufacturing costs is possible with mass production of the inverters
- System design and installation costs are reduced through product standardization

- AC Modules are utility interactive and, thus, no energy is wasted, even when the load is less than the PV generating capacity.
- AC Modules are inherently safer than conventional, high-voltage dc PV systems
- AC Module systems improve system availability because of distributed hardware (i.e., redundancy), testability (via communications links) and simplicity high voltage dc components or wiring)

## INVERTER DESIGN ISSUES

There were four key considerations which influenced the design of the micro-inverter: cost, efficiency, communications, and safety. By using state of the art power metal-oxide semiconductor field-effect transistors (MOSFETs) and soft switching techniques, a no-load loss of 2.4 Watts and 90% input-output efficiency over most of the operating range was achieved (Figure 1).

Regarding safety, the ac module inverter is fully isolated from dc input to ac output and has four anti-islanding schemes (over-and-under frequency detection, over-and-under voltage detection, active unstable frequency, and output power shifting). It is also internally protected against dc reverse polarity, dc overvoltage, over-temperature, and ac line transients.

The inverter uses a novel and highly-accurate control scheme that keeps output current total harmonic distortion (THD) to less than 3%, while maintaining the power factor at > 0.99. All relevant setpoints (temperature, voltage and frequency limits, communications addresses, etc.) are stored in an EPROM, which can be changed easily in the field.

Based on their modular design and their potential for mass production, the manufacturing cost for these units is expected to fall as production quantities increase.

## AC MODULE TESTING

To date, more than one hundred analog modular inverters have been fielded for testing. In addition to units under test and evaluation at Advanced Energy Systems (AES) and Solarex. Installations have been fielded at the Sacramento Municipal Utility District (two 4-kWp roof-top systems in March of 1996), the 1996 Summer Olympics at Georgia Tech (18 ac modules integrated in natatorium entry canopy, April 1996), the Pagent of Peace at the National Christmas Tree display (2 kWp, November 1996), and the State University of New York (20 kWp, January 1997). In addition, two prototype digital inverters were delivered with modules (one to NREL and one to Sandia) in January 1997 for testing and evaluation.

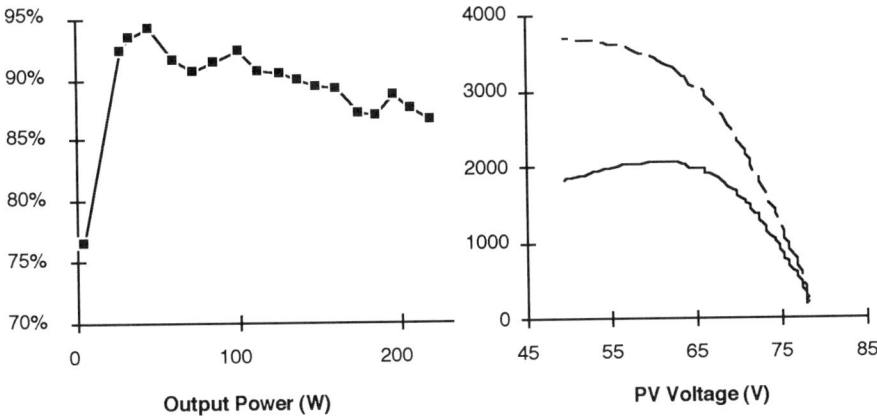

FIGURE 1. Measured Inverter Efficiency (S/N # 25)

FIGURE 2. I-V Curve (top) and Output Power (Wx10)

The Solarex tests have been in a curtain wall structure developed in conjunction with the Department of Energy (DOE), Kawneer, and Solar Design Associates. It has two large-area modules facing south (tilted), two vertical south, one east vertical and one west vertical module. The results in the next section are from this test house. One of the most exciting features of the AC Module is its ability to communicate using power-line-carrier (PLC) spread-spectrum communication. Without needing extra wiring, a central monitoring computer can read values of dc input voltage and current, ac output voltage and current (and hence power), line frequency, inverter temperature, run time, accumulated kWh, and current operating or fault status. It is also possible to command inverters on and off, to flash an identifying LED, and to perform automatic diagnostic I-V curve traces—the PV module output current is increased from zero to the maximum power point and above and for each step the PV voltage is recorded. This information is very valuable for early diagnosis of module problems, and can be automated for large array maintenance testing (Figure 2).

## TEST RESULTS

Figure 3 shows AC Module output power on May 1, 1996. The most striking feature is that individual modules act independently—east-facing S/N 79 starts up in the morning and drops to diffuse-only at noon. West-facing #84 starts to see direct beam after noon and #86, being vertical, sees less intensity than the tilted modules. Clearly, only an independent module-integrated inverter approach could work with so many orientations.

The maximum power tracking (MPPT) capability of the inverters can be seen in Figure 4. Note that, as expected, the PV voltage drops as module temperature rises. Lower and shaded modules run at higher voltages.

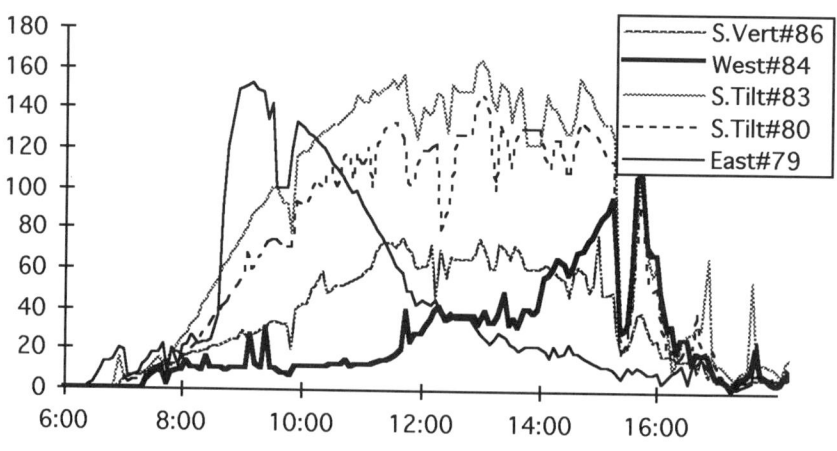

**FIGURE 3.** AC Module Output Power (W) vs. Time of Day

Finally, inverter shows warmer temperatures while operating at higher powers, as expected (east is warmer in the morning and west is warmer in the afternoon). Also note that south-tilted #83 is just starting to current limit its output due to a

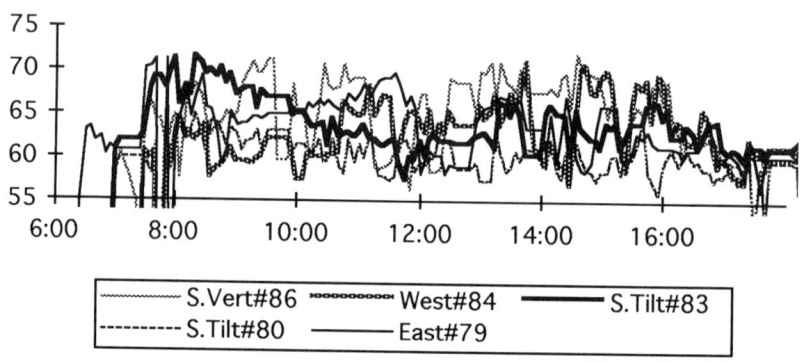

**FIGURE 4.** PV Module MPPT Voltage

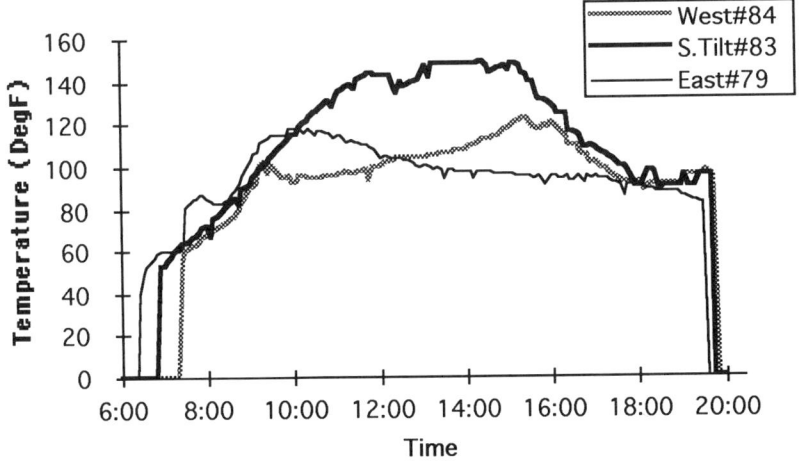

**FIGURE 5.** Inverter Temperature

65°C (150°F) maximum temperature setpoint—if this setpoint is reached, the inverter backs off on current and the operating point rises above MPP (Figure 5).

## FUTURE DIRECTIONS

We are now finalizing the microinverter for mass production. The unit is at Underwriters' Laboratories (UL) for listing under UL 1741. A number of high-visibility initial demonstration projects are currently underway, including a 20 kWp skylight canopy for the international terminal of Baltimore/Washington airport.

## ACKNOWLEDGMENTS

This work is supported by cost-shared agreements with the U.S. Department of Energy (DOE) under the PV:Bonus Program, for the development of the analog microinverter, ac module, and building-integrated PV applications; and, with the National Renewable Energy Laboratory and Sandia National Laboratories, under the DOE-sponsored PVMaT Program for the development of the digital microinverter and standardized ac PV systems. The authors wish to thank the program managers at DOE and the laboratories for their interest in and support of our work.

# Photovoltaic Manufacturing Technology (PVMaT) Improvements for ENTECH's Fourth-Generation Concentrator Systems

## M.J. O'Neill and A.J. McDanal

*ENTECH, Inc.*
*1077 Chisolm Trail, Keller, TX 76248*

**Abstract.** This paper describes recent improvements in manufacturing technology for fourth-generation photovoltaic concentrator systems. The fourth-generation systems are firmly based on prior generations of a field-proven, high-efficiency, stable photovoltaic technology. The fourth-generation manufacturing process has been streamlined and validated through pilot runs and field deployments. Future plans include a 1.5 MW installation in 1998, as part of the Solar Enterprise Zone (SEZ) program in Nevada.

## INTRODUCTION & BACKGROUND

The ENTECH technical team has been developing, field testing, refining, and commercializing photovoltaic concentrator systems since 1978 (1-10). These systems have been based on a patented arched Fresnel lens optical concentrator, which provides maximal optical efficiency coupled with exceptional real-world error tolerance. First-generation line-focus concentrator systems were deployed in the early 1980's, and provided the highest system performance levels of that era. Second-generation systems were deployed in the mid-1980's, and provided better performance at lower cost. Third-generation systems were deployed in the early 1990's, and were the first to use low-cost silicon cells made by one-sun module manufacturers (e.g., 11-12), with patented prismatic cell covers (13) to boost cell performance levels. Fourth-generation systems have been deployed in 1995 and 1996, and are the first to use a patented dry-film cell laydown and encapsulation approach (14). For utility-scale applications, these latest systems use a 25 kW building block, called a *SolarRow*, which can be replicated in cookie-cutter fashion to comprise virtually any size of power plant (15). For smaller, off-grid applications, an 800 W autonomous array, called *SunLine*, is available for village power, water pumping, etc. Each *SolarRow* uses 72 concentrator modules, while each *SunLine* uses 2 concentrator modules. Both array types use solid-state, microprocessor-based controllers (e.g., 16) and 12 V DC motor-driven linear actuators for both tilt (north/south) and roll (east/west) sun-tracking. The following paragraph describes the long-term performance and stability of this line-focus photovoltaic concentrator technology.

## SYSTEM PERFORMANCE & STABILITY

The field-measured performance and stability of the ENTECH line-focus photovoltaic technology has been excellent through four generations of systems dating back to the early 1980's. The first-generation technology was first deployed in a 25 kW system at DFW Airport, Texas, in 1982. Each year, from 1982 to 1987, Sandia National Labs thoroughly tested each source circuit in this system for performance. No measurable degradation was ever detected by Sandia for this first-generation system (17-18).

More recently, a third-generation system was deployed in early 1991 at the PVUSA test site in Davis, California. This array has been independently tested side-by-side with other leading photovoltaic technologies in array sizes of at least 20 kW (19). Figure 1 shows the long-term results of this independent testing. Note that the ENTECH array has been a PVUSA performance leader for the past 5½ years. Note that the other performance leaders, Siemens and Amonix, are, like the ENTECH array, based on the use of crystalline silicon cells. The arrays with mid-range performance are based on various polycrystalline materials, while the arrays with the lowest performance are based on amorphous silicon.

PVUSA has also rated the power output of this third-generation ENTECH system under identical meteorological conditions at the beginning of operation (April 1991), after 2½ years of operation (October 1993), and after 5 years of operation (April 1996). Figure 2 shows the results of these ratings, all normalized to the initial power rating. Note that the ratings are all the same within the error bars on the measurements. Thus, there has been no measurable degradation in the ENTECH array after five years of continuous operation. For the economic viability of any photovoltaic technology, power stability is just as critical as initial performance. Thus, the stability results in Figure 2 are just as important as the initial performance results in Figure 1.

From 1991-96, ENTECH has developed fourth-generation systems, with substantial assistance from Sandia National Labs (Sandia) under the Concentrator Initiative Program and from the National Renewable Energy Lab (NREL) under the Photovoltaic Manufacturing Technology (PVMaT) Program. In 1995, two 100 kW power plants were deployed, using the fourth-generation technology, at CSW Solar Park near Ft. Davis, Texas, and at TUE Energy Park in Dallas, Texas. In 1995-96, several smaller 800 W arrays, also using fourth-generation technology, were deployed at various locations around the country. The fourth-generation modules and arrays are firmly based on the successful modules and arrays in the earlier generations of ENTECH equipment. The fourth-generation equipment is described in more detail in the following paragraphs.

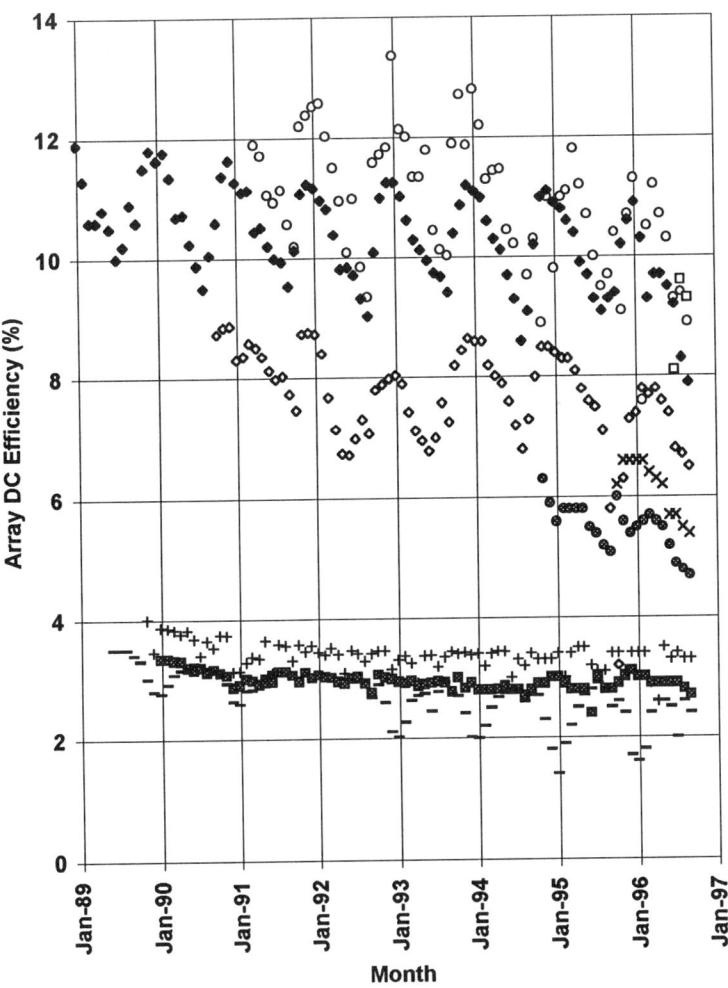

FIGURE 1. PVUSA Monthly Average Efficiency Measurements for Emerging Technology (EMT) Arrays

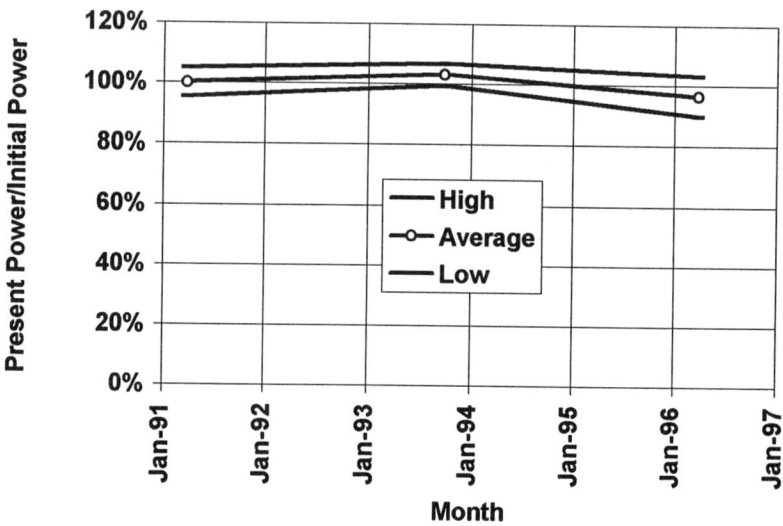

FIGURE 2. PVUSA Power Ratings for the ENTECH Array

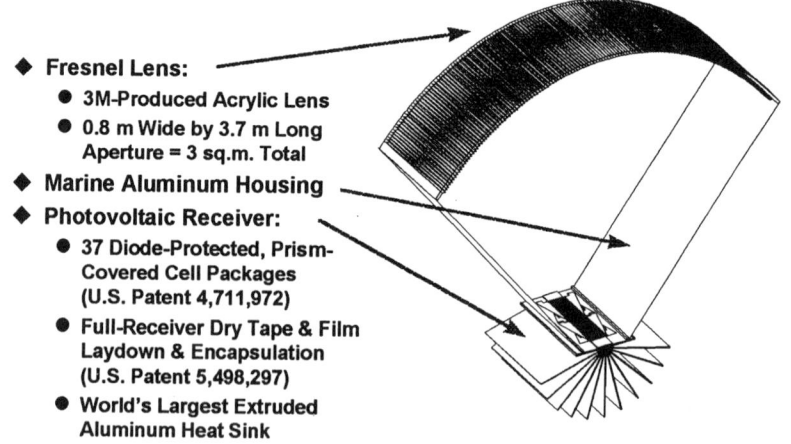

- ◆ Fresnel Lens:
  - 3M-Produced Acrylic Lens
  - 0.8 m Wide by 3.7 m Long Aperture = 3 sq.m. Total
- ◆ Marine Aluminum Housing
- ◆ Photovoltaic Receiver:
  - 37 Diode-Protected, Prism-Covered Cell Packages (U.S. Patent 4,711,972)
  - Full-Receiver Dry Tape & Film Laydown & Encapsulation (U.S. Patent 5,498,297)
  - World's Largest Extruded Aluminum Heat Sink

FIGURE 3. Fourth-Generation Concentrator Module Schematic

# MODULE DESCRIPTION

Figure 3 shows a cross-sectional schematic of the fourth-generation concentrator module. There are four key functional elements in the module. The Fresnel lens gathers and focuses the direct portion of the available solar irradiance into a focal line. The solar cell packages are arranged along the focal line produced by the lens to convert the sunlight to electrical power. The heat sink convectively dissipates waste heat from the solar cell packages to the surrounding atmosphere. The housing structure supports the lens and heat sink, and provides an environmental enclosure for the internal surfaces of the module. Module assembly is accomplished by snapping together six mating parts: the four sheet aluminum housing parts (two sidewalls and two endplates), the receiver assembly (one heat sink with cell packages attached), and the lens.

The fourth-generation module is presently the world's largest photovoltaic module, with an aperture area of 3 square meters and peak power output of 430 W at standard test conditions or 390 W at peak operational conditions, as measured by Sandia.

# ARRAY DESCRIPTIONS

Figure 4 shows a photograph of four *SolarRows* at CSW Solar Park near Ft. Davis, Texas. Each *SolarRow* is an integrated, autonomous two-axis tracking array containing 72 concentrator modules with an operational power output of 25 kW. The structural posts and frames are all hot-dipped galvanized steel for a 30+ year maintenance-free field lifetime. Sun-tracking is implemented with a microprocessor controller and two motor-driven linear actuators, one for north/south tilt tracking, and the other for east/west roll tracking. The controller and DC drive motors are powered by 12 V deep-discharge batteries. The batteries are trickle-charged by small (1 A max. output) "smart" chargers, which prevent overcharging of the batteries by automatically reducing output current as the state of charge increases. The array power is collected in a DC combiner box at the end of the row, where two wires exit the box carrying 63 A at 400 V to the nearest inverter. The *SolarRow* is presently the world's largest two-axis sun-tracking array, with an aperture area of 220 square meters and an operational power output of 25 kW. The *SolarRow* can be replicated in cookie-cutter fashion to provide virtually any size of solar power plant.

Figure 5 shows a photograph of a *SunLine* array, which is a small two-module cousin to the *SolarRow*. The *SunLine* system has been developed for the remote, off-grid market, including village power and water pumping applications. Each *SunLine* provides a power output of about 800 W (DC) under peak conditions.

- Four *SolarRows* at CSW Solar Park, Ft. Davis, TX
- Each 25 kW *SolarRow* Comprises:
  - Integrated, Galvanized Steel, Dual-Axis Sun-Tracking Array
  - 72 Concentrator Modules with 220 sq.m. Total Aperture
  - Microprocessor Controls, Roll/Tilt Drives, & Battery Storage
  - DC Power Interface Box per NEC Code with 2-Wire Output

FIGURE 4. *SolarRow* Array for Utility-Scale Applications

- Two Fourth-Generation Concentrator Modules
- Solar-Charged 12 VDC Battery Powers the Microprocessor Controller and the Sun-Tracking Drives for Both Tilt and Roll Axes
- All-Galvanized Structure Can Be Fully Assembled in Factory for Quick and Easy Field Installation

FIGURE 5. *SunLine* Array for Remote, Off-Grid Applications

# MANUFACTURING TECHNOLOGY

The streamlined manufacturing technology employed to produce the fourth-generation module has been described in detail in previous publications (8-10). The same basic manufacturing technology approach has also been applied to the new fourth-generation arrays (*SolarRow* and *SunLine*):

- a team of "best-qualified" organizations produces the parts
- the parts count is minimized
- all tolerances are expanded to their allowable limits
- factory assembly replaces field assembly wherever possible
- already mass-produced components are used wherever possible.

*SolarRow:* Fourteen conventional concrete piers are used as the foundation. The structural steel posts and frames have been designed to be compatible with conventional low-cost structural steel fabrication techniques, and hot-dip zinc galvanizing. The structural parts count has been minimized by making individual structural elements as large as practical. Each *SolarRow* has 24 east-west tubes, 24 north-south tubes, and 13 posts. The tubes bolt together to form the interconnected frame, which sits in bushings on top of the posts. Using a forklift, a crew of four individuals and a supervisor can install the structure for a *SolarRow* in about a day. Each module is installed within the frame using 8 rivets. The same 5-man crew can mechanically install all 72 modules in the row in about a day. Factory-assembled motor/actuator drives are installed at the center of the row. A factory-assembled junction box with military-style plug connectors houses the microprocessor-based controller. This box is bolted onto the center post, and factory-assembled cables plug in to the box to make the connections with drive motors and batteries. Another factory-assembled junction box houses the DC power combining circuitry and switches. This box is bolted onto the end post nearest the inverter, with 6 incoming wires and 2 outgoing wires carrying the power produced by the *SolarRow*. A roll bar is mounted along the south edge of the *SolarRow* to effect the east/west rotation of the modules in a Venetian blind fashion. Module wiring is completed using crimp-style connections, and long wire runs are made inside the structural tube members, precluding the need for conduit. The completed array has a total aperture area of 220 square meters, about the same as the floor area of a nice house. The east-to-west extent of the *SolarRow* is over 100 meters, a little longer than a football field.

*SunLine:* A 5-inch (12 cm) diameter steel pipe is embedded in concrete to provide the mount for the array, which can be factory-assembled for rapid field installation. The structure, drives, and controls are smaller versions of those used in *SolarRow*, described above. The modules are identical to those in *SolarRow*.

# FUTURE PLANS

Nevada Power, ENTECH, SAIC and Energy Unlimited have been selected to jointly provide 1.5 MW of solar power, coupled with 18.5 MW of wind power, as part of the Solar Enterprise Zone (SEZ) program in Nevada. SEZ is administered by the Corporation for Solar Technology and Renewable Resources (CSTRR). Installation of this 20 MW renewable energy plant is presently scheduled for 1998.

# REFERENCES

1. M.J. O'Neill et al., "Photovoltaic Manufacturing Technology Improvements for ENTECH's Concentrator Module: Phase 1 Final Technical Report," NREL/TP-214-4486, Denver, 1991.
2. M.J. O'Neill et al., "ENTECH's Fourth-Generation Linear Concentrator Module," 1992 DOE/Sandia Crystalline Photovoltaic Technology Project Review Meeting, SAND92-1454, Albuquerque, 1992.
3. M.J. O'Neill, "Fourth-Generation, Line-Focus, Fresnel Lens Photovoltaic Concentrator," 4th Sunshine Workshop on Crystalline Silicon Solar Cells, Tokyo, Japan, 1992.
4. M.J. O'Neill and A.J. McDanal, "Manufacturing Technology Improvements for a Line-Focus Concentrator Module," 23rd IEEE-PVSC, Louisville, 1993.
5. M.J. O'Neill, "Photovoltaic Manufacturing Technology (PVMaT) Improvements for ENTECH's Concentrator Module," 12th NREL PV Program Review Meeting, Denver, 1993.
6. M.J. O'Neill and A.J. McDanal, "Fourth-Generation Concentrator System: From the Lab to the Factory to the Field," 1st WCPEC, Hawaii, 1994.
7. M.J. O'Neill, "Chapter 10, Silicon Low-Concentration, Line-Focus, Terrestrial Modules," in the book, *Solar Cells and Their Applications*, ed. by Larry Partain, John Wiley, 1995.
8. M.J. O'Neill and A.J. McDanal, "Manufacturing Technology Improvements for ENTECH's Photovoltaic Concentrator Module," 14th NREL PV Program Review Meeting, Denver, 1995.
9. M.J. O'Neill and A.J. McDanal, "Fourth-Generation Photovoltaic Concentrator System Development," Sandia Labs Contractor Report No. SAND95-1553, Albuquerque, 1995.
10. M.J. O'Neill and A.J. McDanal, "Photovoltaic Manufacturing Technology (PVMaT) Improvements for ENTECH's Concentrator Module - Final Subcontract Report," National Renewable Energy Lab Report No. NREL/TP-411-20277, Golden, 1995.
11. T. Bruton et al., "Recent Developments in Concentrator Cells and Modules Using Silicon Laser Grooved Buried Grid Cells," 11th European PVSEC, Montreaux, Switzerland, 1992.
12. R. King et al., "Silicon Concentrator Solar Cells Using Mass-Produced, Flat-Plate Cell Fabrication Technology," 23rd IEEE-PVSC, Louisville, May 1993.
13. M.J. O'Neill, "Photovoltaic Cell Cover for Use with a Primary Optical Concentrator in a Solar Energy Collector," U.S. Patent No. 4,711,972, 1987.
14. M.J. O'Neill and A.J. McDanal, "Photovoltaic Receiver," U.S. Patent No. 5,498,297, 1996.
15. M.J. O'Neill and A.J. McDanal, "The 25 kW SolarRow - A Building Block for Utility-Scale Concentrator Systems," 25th IEEE-PVSC, Washington, 1996.
16. A.B. Maish et al., "SolarTrak Controller Development for Today's Applications," 25th IEEE PVSC, Washington, May 1996.
17. M.J. O'Neill, "Five-Year Performance Results for the Dallas-Fort Worth (DFW) Airport Solar Total Energy System," ASHRAE Transactions, Volume 94, Part 1, 1988.
18. E.C. Boes and A.B. Maish, "Advances in Concentrator Technology," 19th IEEE-PVSC, New Orleans, 1987.
19. Christina Jennings et al., "PVUSA - The First Decade of Experience," 25th IEEE-PVSC, Washington, 1996.

# PV Mat Manufacturing Improvements for Continuous Roll-to-Roll Amorphous Silicon Module Production

M. Izu, H.C. Ovshinsky and S.R. Ovshinsky

Energy Conversion Devices, Inc.
1675 West Maple Road
Troy, Michigan 48084

**ABSTRACT.** Under the PVMat 2A Program, Energy Conversion Devices, Inc. (ECD) has performed manufacturing technology development work utilizing its proprietary continuous roll-to-roll triple-junction a-Si alloy solar cell production line. Among the accomplishments achieved under this program, ECD demonstrated the production of the world's first 4 ft$^2$ PV modules utilizing triple-junction two-bandgap solar cells manufactured in a commercial, continuous roll-to-roll production line. These 4 ft$^2$ modules had 9.5% initial efficiency and 8% stable module efficiency. ECD has recently designed and constructed a 5 MW continuous roll-to-roll a-Si solar cell processor for its U.S. joint venture, United Solar Systems Corp. (United Solar). The state-of-the-art processor incorporates major advances in solar cell design and manufacturing processes achieved by United Solar and ECD, with support from DOE/NREL. The advanced continuous roll-to-roll triple-junction a-Si module production technology will reduce module production costs, increase stabilized module efficiency, and increase commercial production capacity.

## INTRODUCTION

Continuous roll-to-roll manufacturing technology, utilizing a thin, flexible stainless steel substrate developed by Energy Conversion Devices, Inc. (ECD) offers a number of advantages in a fully automated, high-throughput PV module production plant (1-21). The stainless steel substrate is rugged, and this improves production yield by eliminating substrate breakage - a significant problem in many glass substrate a-Si alloy PV module manufacturing plants. The transport mechanism on a thin, flexible stainless steel substrate is also mechanically simpler and significantly less expensive than the transport mechanism for conventional glass substrates. Heating and cooling of the substrate during solar cell deposition can also be accomplished quickly, compared to glass substrate manufacturing. As a result, the capital equipment cost for a large-volume plant utilizing continuous roll-to-roll technology will be relatively low. Under the PVMat 2A Program,

Subcontract No. 2M-2-11040-7, Energy Conversion Devices, Inc. has performed manufacturing technology development work utilizing ECD's advanced continuous roll-to-roll triple-junction a-Si alloy solar cell production line, which was engineered and manufactured by ECD. The production line consists of:

(1) continuous roll-to-roll substrate washing machine;
(2) continuous roll-to-roll back-reflector deposition machine;
(3) continuous roll-to-roll a-Si alloy deposition machine;
(4) continuous roll-to-roll top clear conductor deposition machine; and
(5) continuous step and repeat cell cutting machine.

The production line produces triple-junction two-bandgap a-Si alloy solar cells consisting of Si/Si/Si-Ge structure on a 5 mil. thick, 14 in. wide, 2500 ft. stainless steel roll at a speed of 1 ft./min. Major accomplishments are listed below.

- Successful incorporation of a high-performance Ag/ZnO back-reflector system into our continuous roll-to-roll commercial production operation.
- Incorporation of high-quality a-SiGe narrow bandgap solar cells into a commercial continuous roll-to-roll manufacturing process.
- Demonstration that the continuous roll-to-roll production of high-efficiency triple-junction two-bandgap solar cells is consistent and uniform throughout a 2500 ft. run with high yield. The average initial subcell efficiency and yield were 10.21% and 99.7%, respectively.
- Achievement of 11.1% initial subcell efficiency of triple-junction two-bandgap a-Si alloy solar cells manufactured in continuous roll-to-roll production.
- Production of the world's first 4 ft$^2$ PV modules utilizing triple-junction two-bandgap solar cells manufactured in a commercial, continuous roll-to-roll production line. These 4 ft$^2$ modules had 9.5% initial efficiency and 8% stable module efficiency.
- Demonstration of the long-term stability of ECD's 4 ft$^2$ production module. The stable module efficiency after 2,380 hours of sunlight soaking at 50°C showed 15% degradation after 600 hours. ECD modules have passed heat and humidity/freeze cycles of NREL recommended module reliability testing procedure.
- Process optimization to reduce the layer thickness and to improve gas utilization resulted in a 77% material cost reduction for germane, and a 58% reduction for disilane has been achieved.
- Design and construction, at ECD's expense, and completion of initial optimization of a 200 kW multi-purpose continuous roll-to-roll a-Si alloy solar cell deposition machine having upgraded machine and construction specifications.
- Demonstration of triple-junction solar cells with 9.5% initial efficiency with top a-Si cells deposited in the serpentine deposition chamber.

- Development of a new back-reflector evaluation technique using PDS to effectively analyze the optical losses of textured back-reflector and developed an improved, textured Ag/ZnO back-reflector system demonstrating 26% gain in $J_{sc}$ over previous textured Al back-reflector systems.
- Development of a concept design for an automated high-volume PV manufacturing plant for producing a low-cost, large-are PV module with an expected material cost reduction of 71%. The module manufacturing cost is estimated to be $1.00 per peak watt.

ECD has recently designed and constructed a 5 MW continuous roll-to-roll a-Si alloy solar cell processor for United Solar Systems Corp. (United Solar), an American joint venture between ECD and Canon Corporation. The state-of-the-art production equipment incorporates major advances in solar cell design and manufacturing processes achieved by United Solar and ECD, with support from DOE/NREL. The advanced continuous roll-to-roll triple-junction a-Si module production technologies will reduce the module production costs, increase the stabilized module efficiency, and expand the commercial production capacity. Products from the plant will be available for a variety of applications including rooftop solar products developed by ECD and United Solar under DOE/NREL's PV Bonus Program, such as solar shingles for residential rooftops and solar metal roofing products for commercial and residential rooftops. These products, which blend aesthetically into conventional roofing can be installed by commercial roofers without additional supporting structures and have been showcased in an energy-efficiency house in Atlanta, Georgia, and at a National Association of Home Builders townhouse in Maryland.

In this paper, we describe new design features that have been incorporated into the 5 MW a-Si processor.

## Improved Solar Cell Structure

The 5 MW continuous roll-to-roll a-Si alloy solar cell processor is designed to produce triple-junction, triple-bandgap a-Si/a-SiGe/a-SiGe solar cells in a single pass in a continuous roll-to-roll process. Figure 1 shows the solar cell structure.

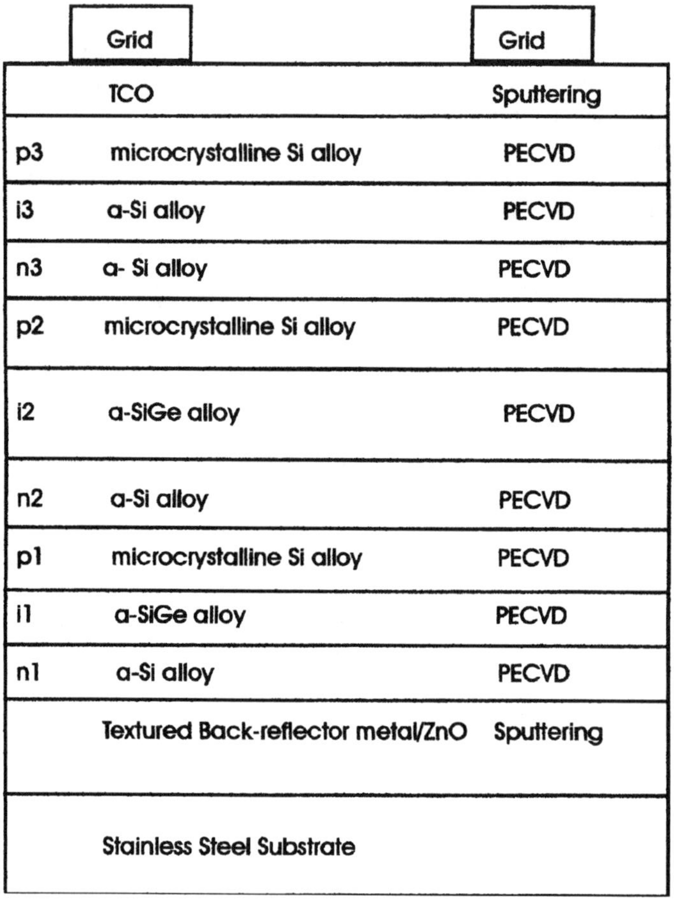

| | | |
|---|---|---|
| | Grid | Grid |
| | TCO | Sputtering |
| p3 | microcrystalline Si alloy | PECVD |
| i3 | a-Si alloy | PECVD |
| n3 | a-Si alloy | PECVD |
| p2 | microcrystalline Si alloy | PECVD |
| i2 | a-SiGe alloy | PECVD |
| n2 | a-Si alloy | PECVD |
| p1 | microcrystalline Si alloy | PECVD |
| i1 | a-SiGe alloy | PECVD |
| n1 | a-Si alloy | PECVD |
| | Textured Back-reflector metal/ZnO | Sputtering |
| | Stainless Steel Substrate | |

**FIGURE 1.** Structure of a triple-junction triple-bandgap solar cell

United Solar recently achieved a new world record efficiency of 14.5% initial efficiency utilizing the solar cell structure shown in Figure 1. The research that led to this world record was performed under DOE/NREL's Thin Film Partnership Program. Previously, in 1994, United Solar/ECD and the Department of Energy (DOE) jointly announced United Solar's achievement of 10.2% stable module efficiency, a critical milestone in the drive to develop a low-cost, practical solar technology. That achievement was based on a 13% initial cell efficiency. Earlier this year, United Solar and DOE also announced the achievement of stable cell efficiency of 11.8% (13.2% initial efficiency). The new record of 14.5% is a significant increase and will form the basis for substantially improved final-product performance and cost-reduction.

## 5 MW a-Si Alloy Solar Cell Processor

As shown in Figure 2, the 5 MW Processor is a roll-to-roll processing system which produces sequentially deposited thin films of doped and undoped a-Si alloy semiconductor on the back-reflector coated surface of stainless steel web. The resulting cell structure will be triple-junction in the form of NIP/NIP/NIP, as shown in Figure 1.

**FIGURE 2.** 5 MW Amorphous Silicon Alloy Solar Cell Processor

The input to the process is a roll of back-reflector coated stainless steel. Web rolls are approximately 2,500 feet in length, 14 inches in width, and weighing up to 1,000 pounds. The beginning of the input roll will be welded to the end of the preceding roll in situ. Web transport through the system will be single pass at the rate of 2 ft/min.

The triple-junction cell structure will consist of doped and undoped thin films of a-Si alloy semiconductor deposited sequentially and simultaneously in proper order as the web is transported through each successive chamber section. Deposition is achieved by low-pressure RF plasma chemical vapor deposition.

The multi-section a-Si alloy deposition machine consists of a pay-off chamber section, 9 process sections for depositing the triple-junction cell and terminating with a take-up chamber. Each section is separated by a "gas gate." The gas gates are designed so that the migration of dopants between chambers is eliminated. The position, tension and speed of the web through the processing chambers are computer-controlled. Web passage through the process chambers is such that deposition takes place on the underside, and front facing in the web transport system is not allowed.

Pay-off and take-up chambers contain all of the necessary web-drive components, auxiliary rolls, tension sensors and steering mechanisms. Pumping for these chamber sections is done by a Roots blower and mechanical displacement-type vacuum pumps.

In each of the process chambers, transport of reactant and dilutent gases to the reaction zone is from mixing manifolds designed to provide the specified mixture to be appropriately distributed in each individual reaction zone. Reaction zone pressures are controlled by a gas-pumping system using variable conductance valves, flow controllers, and pressure transducers used together in a closed-loop computer-controlled feedback system for reach process section.

Each of the layers making up the cell structure is deposited in sequence and simultaneously. Heat is provided to maintain the web at the specified temperature by an array of quartz-type lamps located above the web, and radiating downward controlled by thermocouples in individual closed-loop circuits.

In this advanced solar cell processing machine, we have incorporated several important new features:

1. The machine is designed to minimize incorporation of contaminants into deposited material. The machine has been fabricated, constructed and assembled under a strict procedure to minimize contaminants.
2. The machine operation is computer-controlled. The improved software developed by ECD enables an operator, monitoring interactive CRT displays, to start, operate, and shut-down the machine using computer keyboards that control all the machine functions including pumping, gas recipes, flow meters, heaters, valves, and web drive systems. The computerized system incorporates automatic sequence of operations with many safety interlocks, and other special design features that make the production operation safe and reliable. The computer system also provides a historical data acquisition system which monitors and stores all the processing data in computer discs.

3. The machine is equipped with a maintenance-free effluent scrubbing exhaust system consisting of a burn-box and scrubber.
4. The web transport system is improved, and the web pass line is horizontal and flat, as seen in Figure 2. Previously built machines utilized a catenary web pass line.
5. The gas-feed and cathode systems for depositing bottom and middle intrinsic layers incorporate diffusion-controlled gas-feed/deposition mechanism which allows adjustment of profile of Ge and Si in the layers.
6. The cathode configuration and deposition conditions have been optimized to deposit microcrystalline P layers.
7. Improved mechanical parts design for internal components of the process chambers has significantly reduced component heat distortion during processing.
8. Serviceability and accessibility have been improved with new non-catenary design. Using powered lid lifts for all chambers, all internal chamber components are accessible for servicing at waist-high levels. Further, the use of separate above-the-ground troughs for wiring and piping has improved the appearance as well as the accessibility of the external machine components.

United solar and ECD are currently in the process of starting-up production. All the machines and systems have been tested and functioning as designed. The machine is producing high-performance triple-junction triple-bandgap solar cells. The production line, scheduled to come on line in early 1997, will produce high-performance PV modules at an annual capacity of 5 MW.

*New Rooftop PV Modules.* The 5 MW line will be used to manufacture a variety of PV modules including new rooftop PV modules recently developed by ECD and United Solar, under the DOE PV Bonus Program, such as solar shingles and solar metal roofing products for building integrated PV applications. These rooftop PV modules, which blend aesthetically into conventional roofing can be installed by commercial roofers without additional supporting structures and special training. These modules have been showcased in an energy-efficient house in Atlanta, Georgia, and at a National Association of Home Builders (NAHB) 21st Century Townhouse in Maryland, as shown in Figures 3 and 4. New rooftop PV modules are receiving positive public reactions. Very recently, United Solar's PV shingle modules have received a Grand Award from Popular Science.

**FIGURE 3.** Energy-efficient house in Atlanta, Georgia, displaying United Solar's solar shingles

**FIGURE 4.** ECD's PV Metal Roofing Modules on the NAHB Research Center 21st Century Townhouse Project

## ACKNOWLEDGMENTS

The authors would like to thank Drs. S. Guha, J. Yang and H. Fritzsche for important discussions and suggestions. Also, the authors would like to thank other project members who contributed to this work.

## REFERENCES

1. Izu, M. and Ovshinsky, S.R., Production of Tandem Amorphous Silicon Alloy Cells in a Continuous Roll-to-Roll Process, *SPIE Proc.*, 407, **42** (1983).

2. Izu, M. and Ovshinsky, S.R., Roll-to-Roll Plasma Deposition Machine for the Production of Tandem Amorphous Silicon Alloy Solar Cells, *Thin Solid Films* 119, **55** (1984).

3. Morimoto, H. and Izu, M., edited by Hamakawa, Y., Amorphous Semiconductor Technology & Devices, *JARECT, 16* (1984)

4. Ovshinsky, S.R., Roll-to-Roll Mass Production Process for Amorphous Silicon Solar Cell Fabrication, *Proc. International PVSEC-1*, 577 (1988).

5. Yang, J., Ross, R., Glatfelter, T., Mohr, R., Hammond, G., Bernotaitis, C., Chen, E., Burdick, J., Hopson, M. and Guha, S., High Efficiency Multiple-Junction Solar Cells Using Amorphous Silicon and amorphous Silicon-Germanium Alloys, *Proc. 20th IEEE PV Spec. Conf.*, 241 (1988).

6. Yang, J., Ross, R., Mohr, R. and Fournier, J.P., Physics of High-Efficiency Multiple-Junction Solar Cells, *Proc. MRS Symp.*, **95**, 517 (1987).

7. Nath, P. and Izu, M., Performance of Large-Area Amorphous Silicon Based Single and Multiple-Junction Solar Cells, *Proc. of the 18th IEEE Photovoltaic Specialists Conf.*, Las Vegas, Nevada, 939 (1985).

8. Nath, P., Hoffman, K., Call, H., Vogeli, C., Izu, M. and Ovshinsky, S.R., 1 MW Amorphous Silicon Thin-Film PV Manufacturing Plant, *Proc. of the 3rd International Photovoltaic Science and Engineering Conf.*, Tokyo, Japan, 395 (1987).

9. Nath, P., Hoffman, K., Vogeli, C. and Ovshinsky, S.R., Conversion Process for Passivation Current Shunting Paths in Amorphous Silicon Alloy Solar Cells, *Appl. Phys. Lett.* 53, **11**, 986 (1988).

10. Nath, P., Hoffman, K., Call, J., DiDio, G., Vogeli, C. and Ovshinsky, S.R., Yield and Performance of Amorphous Silicon Based Solar Cells Using Roll-to-Roll Deposition, *Proc. 20th IEEE PV Spec. Conf.*, 293 (1988).

11. Nath, P., Hoffman, K., Vogeli, C., Whelan, K. and Ovshinsky, S.R., A New Inexpensive Thin Film Power Module, *Proc. 20th IEEE PV Spec. Conf.*, 1315 (1988).

12. Nath, P., Hoffman, K. and Ovshinsky, S.R., Fabrication and Performance of Amorphous Silicon Based Tandem Photovoltaic Devices and Modules, *4th International PV Science and Engineering Conf.*, Sydney, Australia (1989).

13. Guha, S., Advances in High-Efficiency, Multiple-Bandgap, Multiple-Junction Amorphous Silicon Based Alloy Thin Film Solar Cells, *A Paper Presented at MRS Spring Meeting, San Diego*, April (1989).

14. Yang, J., Ross, R., Glatfelter, T., Mohr, R. and Guha, S., Amorphous Silicon-Germanium Alloy Solar Cells with Profiled Bandgaps, *MRS Symposium Proc.*, **149**, 435 (1989).

15. Izu, M., Deng, X., Krisko, A., Whelan, K., Young, R., Ovshinsky, H.C., Narasimhan, L. and Ovshinsky, S.R., Manufacturing of Triple-Junction 4 ft$^2$ a-Si Alloy PV Modules, *Proc. of 23rd IEEE PVSC*, 919 (1993).

16. Izu, M., Ovshinsky, S.R., Deng, X., Krisko, A., Ovshinsky, H.C., Narasimhan, K.L. and Young, R., Continuous Roll-to-Roll Amorphous Silicon Photovoltaic Manufacturing Technology, *AIP Conf. Proc. 306, 12th NREL PV Program Rev.*, 198 (1993).

17. Guha, S., Yang, J., Banerjee, A., Glatfelter, T., Hoffman, K., Ovshinsky, S.R., Izu, M., Ovshinsky, H.C. and Deng, X., Amorphous Silicon Alloy Photovoltaic Technology - From R&D to Production, *Proc. of MRS Spring Meeting* (1994).

18. Deng, X., Izu, M., Narasimhan, K.L. and Ovshinsky, S.R., Stability Test of 4 ft$^2$ Triple-Junction a-Si Alloy PV Production Modules, *Proc. of MRS Spring Meeting* (1994).

19. Izu, M., Ovshinsky, H.C., Deng, X., Krisko, A.J., Narasimhan, L., Crucet, R., Laarman, T., Myatt, A. and Ovshinsky, S.R., Continuous Roll-to-Roll Serpentine Deposition for High Throughput a-Si PV Manufacturing, *Proc. of 24th IEEE PVSC*, 820, (1994)

20. Izu, M., Ovshinsky, H.C., Whelan, K., Fatalski, L., Ovshinsky, S.R., Glatfelter, T., Younan, K., Hoffman, K., Banerjee, A., Yang, J. and Guha, S., Lightweight Flexible Rooftop PV Module, *Proc. of 24th IEEE PVSC*, 990 (1994).

21. Izu, M., Deng, X., Ovshinsky, H.C., Crucet, R., Ovshinsky, S.R. and Polisan, A., Production Start-Up of 2 MW a-Si PV Manufacturing Line at Sovlux Plant, *Proc. of 25th IEEE PVSC*, 1327 (1996).

# SMALL BUSINESS
# INNOVATIVE RESEARCH (P)

# Photovoltaic Research in the Small Business Innovative Research Program

Ward I. Bower* and Alec Bulawka[†]

*Sandia National Laboratories and [†]U.S. Department of Energy
Albuquerque, NM 87185 and Washington, DC 20585

**Abstract:** The Small Business Innovative Research Program (SBIR) is currently authorized to be funded through September 30, 2000. The National Photovoltaics Program is a contributor to the Department of Energy (DOE) SBIR program. The small business photovoltaic industry has been benefiting from the SBIR program through awards that have funded basic research, new processes and products that have PV and other commercial applications. This paper provides information on SBIR opportunities, selected details of the SBIR program, statistics from the 1995 and 1996 DOE SBIR program, and methods for improving PV industry participation and success in the SBIR program.

## INTRODUCTION

The PV industry is comprised of a very high percentage of small businesses that perform vital research and development for the advancement of commercializing the PV technology. Virtually all of the balance-of-system manufacturers for PV systems belong to the small business community. Many of the manufacturers of PV component processing equipment such as module laminators, wire saws, and even robotics used to improve assembly of PV products are also small businesses. Many cell and module manufacturers are also classified as small businesses.

The DOE Photovoltaics Program, as with most federally funded programs, has seen significant budget cuts in the last year, making it more difficult to fund critical research and development work to improve or develop new PV systems and components. Project awards through the SBIR programs offered by different Federal agencies are excellent funding mechanisms for advancing the PV-related technologies through innovative small business research and development.

Competition is fierce for this funding, as this paper will show; however, the small business PV community has been very successful in obtaining awards. The small business must demonstrate that they are doing exceptional work, that the work is vital to the overall effort, that commercialization is planned and that their business plans are sound. Additionally, and very critical to winning an award, is the ability to write an exceptional proposal that is extremely clear and concise, while it shows the near-term and long-term value of the proposed work, and a demonstrated plan for success. Plans for commercialization are critical, but the small business does not have to be a large firm. Companies with fewer than 10 employees have won more than one-third of the SBIR awards. The SBIR program is not an assistance program for small businesses. It is merit-based, the competition is fierce, and the small business must have good research capabilities and project plans in order to score high enough to win an award.

# SBIR OPPORTUNITIES FOR THE PV SMALL BUSINESS

## Overview Of The SBIR Program

The SBIR program was created in 1982 by the Small Business Innovation Development Act, Public Law 97-219, and reauthorized in 1992 until the year 2000 by reauthorization legislation, Public Law 99-443, and the Small Business Research and Development Act, Public Law 102-564. The purposes of the new law were to expand and improve the SBIR program, emphasize the program's goal of increasing private sector commercialization of technology developed through Federal R&D, increase small business participation in Federal R&D, and improve the Federal Government's dissemination of information concerning the SBIR program (2).

Federal agencies with extra-mural R&D budgets of over $100 million are required to establish an SBIR program using a stated percentage of their budget. The percentage began with a required 0.2% of the annual budget in (FY) 1983 through 1985. It increased to 1.25% in FY's 1986-1992 for the civilian agencies. Public Law 102-564 then increased the percentage gradually, starting with 1.5% in FY's 1993 and 1994 and reaching a maximum of 2.5% in FY 1997 and extending through FY 2000 or through September 30, 2000 (3).

The Federal agencies that have SBIR programs include the Department of Defense, Agriculture, Commerce, Education, Energy, Health and Human Services (National Institutes of Health), Transportation, the National Aeronautics and Space Administration (NASA), the National Science Foundation (NSF), the Nuclear Regulator Commission (NRC) and the Environmental Protection Agency (EPA).

The SBIR program consists of three phases, but only two of those phases are funded by SBIR dollars. The SBIR program does not fund private commercial development. The Phase III part of the program is outlined below and is the commercialization phase, but it does not use SBIR dollars.

Phase I solicitations are issued once a year by each sponsoring agency except for the Department of Defense (two times per year) and the Department of Health and Human Services (three times per year). Phase I is a competition in which each small business competes for awards for published topics. Each small business submits its best proposal for solutions to the published topic. The proposals are then reviewed by expert peer groups outside the government. The winner is chosen based on the proposal's technology merit, the promise of success of the proposal, and the competence of the people who will do the work. Phase I objectives test the feasibility of the proposed solution. Phase I is limited to a cap of $100,000 and approximately six months of work. Some agencies have caps of less than $100,000. The DOE SBIR cap for 1996 was $75,000. There is no limit on how many proposals a company makes or how many awards it can win. The successful completion of Phase I objectives is a prerequisite for further SBIR support through a Phase II application.

Phase II is the full scale research phase of the program. Awards are made to winners of the Phase II competition, but the competition is different in that each small business is competing on different subjects. The Phase II proposals must be more detailed and must be based on a continuation of the Phase I work. The decision to award Phase II is based more on the relative importance of the technology, the commercial market place, the needs of a government agency, the likelihood of success and the likelihood of significance. Commitment of the small business to take the work to commercialization, or Phase III, is also weighed heavily in the review of Phase II proposals.

Under Phase III, it is intended that non-SBIR capital be used by the small business to pursue commercialization of the research results. The small business, financial institutions, capital pools, large companies, or wealthy individuals should fund Phase III to assure product or service commercialization. Federal agencies may also contract with the small business as part of Phase III for products or processes that meet the mission needs of those agencies or for further research or R&D on the product or process.

## SBIR Program Goals, Objectives and Awards

One of the SBIR program's goals is to improve the opportunities for small business in Federal R&D, while boosting the nation's competitiveness. The objectives of the SBIR program also include strengthening the role of small business in meeting Federal research and R&D needs. These objectives are met by increasing private sector commercialization of technology developed through DOE-supported R&D, stimulating technological innovation in the private, small business, sector and improving the return for investment from federally funded research (1).

SBIR award opportunities are available to PV-related small businesses through each of the participating agency's programs; however, note that five of the eleven agencies make more than 90% of the SBIR awards. The Department of Defense makes about one-half of the total awards. The Department of Energy, Health and Human Services, NASA and NSF makes approximately 40% of the total awards. Further, in order for small business to participate in a particular SBIR program, the topics chosen by that agency must be a technological match for the small business proposal. Generally, topics are selected within the agency's programs to meet their needs, but suggestions and communications with program directors may be helpful in defining a topic that opens opportunities for both the agency and the small business.

The DOE SBIR program has included PV-related topics for the last five years. In Fiscal Year (FY) 1995, a total of 38 PV-related proposals was submitted and eleven of those proposals were granted Phase I awards. That amounted to a 29% success rate for the PV-related proposals. SBIR awards to PV-related small businesses have also been made through the NSF, NASA, and the Department of Defense in recent years.

## Qualification of "Small Businesses"

Companies with fewer than 500 employees, including its affiliates, are considered "small businesses." Further, the business must be independently owned and operated, must not be dominant in the field of operation in which it is proposing, have its principal place of business in the United States, and be organized for profit. The business must also be at least 51% owned by United States citizens or lawfully admitted permanent resident aliens. Details of the qualifications are given in each of the Program Solicitation documents. The rules listed here are general rules. There are sometimes fine points of ownership or organization that must be decided on an individual basis.

## General SBIR Schedules

Each agency sets its own schedule for its SBIR program solicitation dates and the closing dates for proposals. These dates vary from year to year, but are generally in the same time frame. Table 1 lists the latest schedule for most of the SBIR agencies. This schedule is updated quarterly and is readily available through the SBIR and STTR Resource Center and is on the World Wide Web at http:/www.seeport.com/SBIR/resources.htm.

TABLE 1. Schedule of SBIR solicitations by agency FY 1997

| AGENCY | ISSUE DATE | CLOSING DATE |
|---|---|---|
| Department of Agriculture | June 1, 1997 | September 4, 1997 |
| Department of Commerce | October 1, 1996 | January 15, 1997 |
| Department of Defense (twice per year) | | |
| 97.1 Solicitation | October 1, 1996 | January 8, 1997 |
| 92.2 Solicitation | May 1, 1997 | July 16, 1997 |
| Department of Education | January 25, 1997 | March 28, 1997 |
| Department of Energy | December 2, 1996 | March 3, 1997 |
| Department of Health & Human Services | | |
| 1. Public Health Services | September 6, 1996 | November 5, 1996 |
| 2. Public Health Services | August 15.1996 | December 15, 1996 |
| 3. Public Health Services | January 15, 1997 | April 15, 1997 |
| Department of Transportation | February 14, 1997 | May 1, 1997 |
| EPA | November 13, 1997 | January 15, 1997 |
| NASA | May 13, 1997 | July 24, 1997 |
| National Science Foundation | April 1, 1997 | June 12, 1997 |

Note: The above dates are subject to change by the Federal participating agencies.

## The DOE SBIR Program

The cumulative funding for the DOE SBIR program is $384 million over the first twelve years (1). The program's budget for FY 1997 is about $78 million, based on a 2.5% contribution. These funds are used to support an annual competition for Phase I awards of up to $75,000 for about 6 months to explore the feasibility of innovative concepts and for Phase II awards. Phase II is the principal research or R&D effort, and the awards are capped at $750,000 for a two-year period. To date, DOE has funded an average of about 145 Phase I applications and about 55 Phase II applications per year. Success ratios for applicants have been about 12% in Phase I and 45% in Phase II.

Each agency issues at least one annual solicitation for Phase I grant applications. The DOE's solicitation contains topics in technical areas such as the following: Basic Energy Sciences, Health and Environmental Research, High Energy and Nuclear Physics, Magnetic Fusion Energy, Computational and Technology Research, Energy Efficiency and Renewable Energy, Nuclear Energy, Fossil Energy, Environmental Management, and Nonproliferation and National Security. Each year about 45 topics are allocated among the technical areas in proportion to their contributions to the budget; the funds are placed in a common pool, and applicants are selected competitively for award on scientific and technical merit.

The DOE's SBIR program has two features that are unique. First, it provides a mechanism for uninterrupted funding between Phases I and II for those awardees that choose to submit their Phase II applications six weeks before the end of their Phase I awards. Funding continuity has been provided to these awardees for thirteen consecutive years. Second, the DOE program provides aid to awardees in seeking follow-on funding for Phase III. The DOE has sponsored a Commercialization Assistance Project for the past six years that has provided individual assistance in developing business plans and in preparation of presentations to potential investment sponsors. In the 1993 project, awardees made presentations to about 55 sponsors from venture capital firms and large corporations. One-half of the companies that completed the 1991 and 1993 projects have already received a total of $30 million for commercialization of their SBIR research. In addition, these companies have increased sales from their SBIR funded research and development by more than $40 million.

In the last few years, more than one-half of the successful applicants to the DOE SBIR program have arranged for support from a DOE national laboratory or a university. The DOE national laboratories' technical expertise and equipment may be available to the applicant under existing DOE approved mechanisms for providing services and materials. The small business must have the necessary expertise to direct the entire project, because the laboratory cannot perform tasks beyond the small business's understanding. A minimum of two-thirds of the funded research of analytical effort during Phase I must be performed by the small business organization. Similar support may be obtained from a university.

The FY 1997 DOE SBIR solicitation is scheduled to be issued on December 2, 1996, with a closing date of March 3, 1997. The solicitation will be available electronically on the World Wide Web at http://sbir.er.doe.gov/sbir.htm. Those who do not have access to the World Wide Web may telephone (301) 903-5707, write to the following address, or telephone the information office:

SBIR Program Manager, ER-32<br>
U.S. Department of Energy<br>
19901 Germantown Road<br>
Germantown, MD 20874-1290

## DOE SBIR Summary for FY 1995

The DOE SBIR program received a total of 1569 submitted Phase I proposals for the 1995 program. Approximately 12.8% of those submittals were selected for 201 Phase I awards. During the same time, the-PV related topic received 38 proposals. Of those, approximately 29% were selected for a total of 11 Phase I awards. This percentage was well above the average of 12% for Phase I awards reported for the entire DOE SBIR program. This was the fourth consecutive year that a PV-related topic appeared in the DOE SBIR program.

## DOE SBIR Summary for FY 1996

The DOE SBIR program received a total of 1437 submitted Phase I proposals for the 1996 program. Approximately 12% of those submittals were selected for 173 Phase I awards. For the same time, the PV related topic received 62 proposals. That was a 163% increase in submitted proposals over 1995. On the down side of the statistics, only three (4.8%) of the PV proposals were selected for Phase I awards. This was the fifth consecutive year that a PV-related topic appeared in the DOE SBIR Program Solicitation. The topics for 1996 included:

a) Innovative Deposition and Manufacturing Processes for PV Cells and/or Modules.
b) Innovative Device Design and Improved Device Efficiency.
c) Innovative, Modular, Smart 3-Phase Inverters.
d) Innovative, Modular, Smart PV Charge Controllers.

The Phase I winners of awards for the 1996 competition included the following titles:

a) "Module Integrated Maximum Power Point Tracking Charge Controllers - The Missing Link in Large Scale Photovoltaic Systems."
b) "Novel Use of Gas Jet Plasma to Prepare Amorphous Silicon Alloy."
c) "High Rate Deposition of Transparent Conducting Zinc Oxide Using Activated Oxygen for Photovoltaic Manufacturing Cost Reduction."

Phase II for 1996 was more successful than Phase I with four (approximately 36%) of the 1995 Phase I awards continuing into Phase II. This, however, was below the 45% average for Phase II awards for the entire program.

## SUMMARY

The SBIR Program offers excellent funding opportunities for R&D for small businesses associated with the National Photovoltaic Program. The DOE SBIR program has offered PV-related topics for the last five years. Other SBIR programs, not specifically addressed in this paper, have also solicited numerous PV-related topics. The overall success rated for PV-related small business has been good, with a high of 29% seen in 1995. Further, Phase II success rates have been excellent with approximately 36% seen in 1996. The success rates for awards for the PV-related small businesses have been lower than the overall DOE SBIR program averages. The PV small business industry can supplement their R&D dollars by actively participating in the SBIR programs as offered by the appropriate agencies. Proposals must be well written and concise. The proposals must address the topic with respect to applicability to the agencies needs, the small business's commercialization plans, and a demonstrated well-developed R&D capability within the small business. Funding awards, with a maximum of $100,000 for Phase I and with up to $750,000 for Phase II are possible. The DOE SBIR program caps Phase I awards at $75,000. Support from national laboratories or universities may be included in SBIR proposals with prior arrangements for using DOE approved mechanisms to use Federal expertise or equipment.

## REFERENCES

1. U.S. Department of Energy, SBIR Fact Sheet, Available on the Internet at http://sbir.er.doe.gov/sbir96/sbirfact.htm or SBIR Program Manager, Germantown, MD, Sept. 20, 1996.
2. U.S. Department of Energy, Program Solicitation, DOE/ER-0653, Small Business Innovation Research, Germantown, MD, 1996.
3. "The SBIR Program: 1993 - 2000," guidelines on Public Law 102-564: Small Business R&D Enhancement Act of 1992, available at http://www.dsu.edu:8000/text/sbir2000.html.

# ENVIRONMENT, SAFETY, AND HEALTH IN PHOTOVOLTAICS

# Emerging Photovoltaic Technologies: Environmental and Health Issues Update

## Vasilis M. Fthenakis and Paul D. Moskowitz

*Biomedical and Environmental Assessment Group, Department of Applied Science, Brookhaven National Laboratory, Upton, New York 11973*

ABSTRACT. New photovoltaic (PV) technologies promise low-cost, reliable PV modules and have the potential for significant PV penetration into the energy market. These prospects for commercialization have attracted renewed interest in the advantageous environmental impact of using PV and also in the potential environmental, health and safety (EHS) burdens in PV manufacturing and decommissioning. In this paper, we highlight recent studies on EHS issues: a) An integrated energy-environmental-economic analysis which shows that large-scale use of PV can significantly contribute to alleviating the greenhouse effect; in the United States alone, it could displace 450 million tons of carbon emissions by the year 2030, b) Recycling of the spent modules and scrap is economically feasible; current research centers on improving the efficiency and economics of recycling CdTe and CIS modules, c) Toxicological studies conducted by the National Institute of Environmental Health Sciences (NIEHS) compared the acute toxicity of CdTe, CIS, and CGS; CdTe was the most toxic, and CGS the least toxic of the three. Additional studies are now comparing the systemic toxicity of these compounds with the toxicity of their precursors.

## INTRODUCTION

Recently, interest in thin-film photovoltaic (PV) cells, e.g., cadmium telluride (CdTe), copper indium diselenide (CIS), copper gallium diselenide (CGS), and silicon (Si), has been growing. These cells have demonstrated good performance, can be produced by various methods, appear to be environmentally stable, and have prospects for low manufacturing cost. As the commercial potential of these technologies becomes more apparent, concerns about the environmental, health and safety (EHS) issues associated with their manufacture, use, and disposal have also arisen. EHS issues related to manufacturing were discussed in earlier studies [e.g., 1]. In this paper, our discussion is limited to recent work on: a) the potential impact of large-scale PV implementation on reducing $CO_2$ emissions, b) the feasibility of recycling PV scrap and spent modules, and c) toxicology of new thin-film materials.

## THE POTENTIAL TO DECREASE $CO_2$ EMISSIONS

The basic tool for this analysis was MARKAL-MACRO, an integrated planning model for energy-environment-economic systems analysis. The model was developed at Brookhaven National Laboratory (BNL) in a collaborative effort under the auspices of the International Energy Agency (IEA). The model has been applied around the world at national, regional, and local levels. It is in active use in 25 countries as a means to examine strategies to mitigate the production of greenhouse gases. It is used by the U.S. Department of Energy in developing a least-cost energy strategy for the United States. MARKAL is a dynamic, linear programming model which describes all possible flows of energy, from resource extraction to energy transformation and end-use devices. The model predicts the evolution over a specified period, e.g., 30 years, of a specific energy system. The supply and demand sides are integrated, and the model selects that combination of technologies that minimize the total cost of energy systems. More then 200 technologies and options are described, including conventional sources of energy, and renewables. Thus, the model is able to evaluate the potential of PV in a competitive environment (2).

## Projections of PV Capacity in the United States

We used the National Renewable Energy Laboratory's (NREL) projections on the efficiency modules and their cost as inputs into MARKAL (Figure 1). Between 1995 and 2030, module efficiency was assumed to increase from 7% to 16%, and the system's cost assumed to decrease from $7/W to $0.64/W. Our predictions are based on these projections, as well as cost/efficiency projections for all competitive technologies. We studied three scenarios. A Base Scenario, labeled "BASE, with market penetration for PV fixed at current (1996) levels. The second scenario, called "AVG", assumes NREL's expectations of improvements in PV technology and costs (and average U.S. solar insolation of 1800 kWh/m$^2$-y). The third scenario, "SW", assumes that, instead of average solar insolation, all PV power plants would be built in the U.S. Southwest,

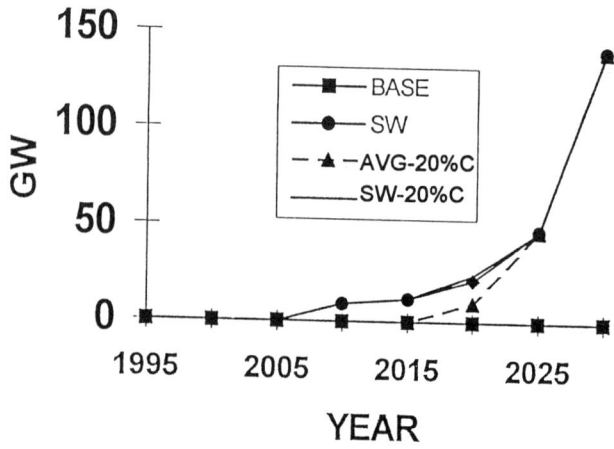

**FIGURE 1.** Projections of PV Energy Market penetration in the U. S.

where solar insolation is 2200 kWh/m$^2$-y. Solar insolation is the only difference between the two scenarios. Each scenario was simulated without constraint on carbon emissions, and then with carbon emissions in 2010 through 2030 constrained at 20% less than 1990 levels, which is one of the proposed $CO_2$ emission reduction targets under discussion in the Framework Convection on Climate Change. Figure 1 shows the projected PV market penetration in these different cases as a function of time. For the "SW" case PV becomes competitive and enters the market on a large scale (from 100 MW in 2005, to 9 GW in 2010, to 140 GW in 2030. Carbon constraints do not additionally boost PV penetration into the market. In the "AVG" case, PV does not enter the market in any significant way until 2030. However, when constraints on carbon emissions are imposed, PV technologies start to penetrate in 2015 and reach their full market potential in 2030.

## Reduction of Carbon Dioxide Emissions Due to PV Penetration into the Market

Figure 2 shows that for the "SW" case, PV has the potential to displace a cumulative 450 million tons of carbon emissions from 1995 to 2030 on an economic basis alone. At the projected capacity of 140 GW in 2030, PV could displace 60 million tons of carbon emissions a year, a very significant reduction by any single technology. Although this scenario brings PV in the GW capacity-scale in about 2010, it does not produce large reduction of $CO_2$ emissions before 2020 because initially PV displace expensive, but efficient gas turbines rather than coal-burning plants. The reduction in $CO_2$ would be greater if constraints on carbon emissions are imposed.

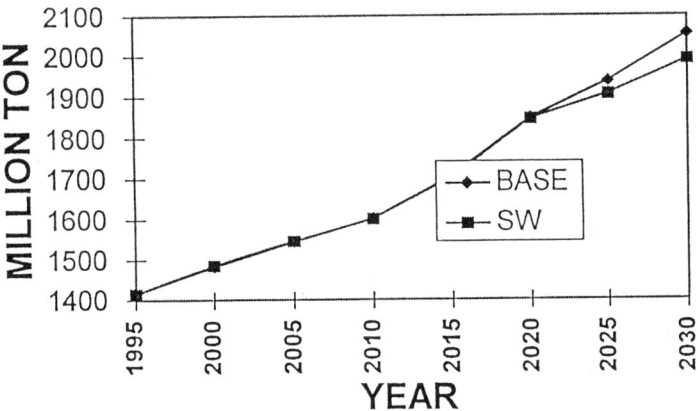

**FIGURE 2.** Projections of Carbon Emissions due to PV Market Penetration in the U. S.

# RECYCLING OF SCRAP AND SPENT MODULES

## Regulatory Requirements

Disposal of modules and production wastes containing Cd, Se, Pb, Cu, or Ag are regulated by Federal and State statutes and regulations. The Resource and Conservation Recovery Act (RCRA) is the key regulatory framework which will affect the wastes generated by the PV industry in the United States. The RCRA relates to solid wastes and defines them as any "discarded material" that is not specifically excluded. PV materials and modules are not excluded from the general definition of a solid waste; they are likely to be categorized as solid wastes unless handled in a manner that qualifies them for exclusion.

A solid waste can be classified as hazardous or non-hazardous, depending on the extent of risk it could pose to human health and safety, property, and the environment. Classification as a hazardous waste implies much more stringent requirements for all activities associated with its generation, collection, consolidation, transportation, processing, and disposal. Pursuant to RCRA, the US-EPA publishes regulations defining hazardous wastes by inclusions in specific lists, and by characteristics of ignitability, corrosivity, reactivity, and toxicity. The key determinant of hazardous waste applicable to PV materials is the Toxicity Characteristic Leaching Procedure (TCLP), a weak-acid leaching test on shredded material, considered to simulate leaching in landfills. TCLP limits are defined for thirty-nine materials. A solid waste that yields a soluble concentration of any of these materials in excess of their TCLP limits is characterized as hazardous. Certain wastes are not subject to RCRA control; these include reclaimed and reused secondary materials from production processes.

The California Hazardous Waste Control Law (HWCL) incorporates and expands upon RCRA in its definition of hazardous waste. HWCL incorporates RCRA's TCLP test as part of its toxicity characterization to identify RCRA wastes, and adds the Waste Extraction Test (WET) to identify non-RCRA wastes that exhibit toxicity characteristics. The threshold limits of the WET are given as the Soluble Threshold Limit Concentration (STLC), and they are similar to the TCLP limits. In addition, California considers the total concentration of listed materials, in any form, within a waste matrix. Total concentrations are determined by a strong-acid extraction test and compared to the limits of the Total Threshold Limit Concentration (TTLC) (3).

There are regulatory issues related to specific photovoltaic materials and technologies, and issues related to specific compounds across several technologies. The first category includes the photovoltaic alloys CdTe, CIS and CGS whose toxicological properties are not known, but their precursors present either well established health risks (i.e., Cd), or perceived risks (e.g., Se). The second category includes Pb from soldering, Ag from Ag paste, and Cu from Cu ribbon, which when used in quantities above certain thresholds, raise regulatory concerns about leaching into the environment.

Limited TCLP testing that was conducted for CIS and CdTe modules, gave the following findings:

a. CIS modules pass the TCLP for Cu, Se, and In;
b. Some CdTe modules failed the TCLP for Cd.

According to the Waste Classification of California's EPA, the following conclusions were reached:

a. CdTe modules are likely to be characterized as hazardous, based on TTLC and STLC for Cd;
b. CIS modules are likely to be characterized as hazardous, based on TTLC for Se and STLC for Se and Cd;
c. Some polycrystalline-Si modules may be characterized as hazardous based on TTLC and STLC for Ag or Pb.

The following additional observations were made:
a. Some polycrystalline-Si modules may fail TCLP for Ag and Pb;
b. Excess Pb from solder and Cu from ribbon can be problematic for several technologies;
c. Amorphous-Si modules do not seem to present any hazardous-waste issues.

## Experiences of Comparable Industries

The experiences and practices of the industries of printed-circuit boards, electronics, telecommunications, and fluorescence light tubes in managing hazardous wastes were described in detail elsewhere (3). The strategy of the printed-circuit board industry is especially instructive. A typical printed-circuit (PC) board contains about 2% PbSn solder and it would fail TCLP for Pb by a wide margin. However, by working closely with the EPA, the electronics industry managed to get several important RCRA rulings which allow exceptions from the hazardous-waste classification: a) unused defective boards sent for reclamation are exempted as "commercial products"; b) trimmings from board manufacturing and PbSn solder dross sent for reclamation are excepted as "by-products"; c) processed products (e.g., shredded boards) generally are considered solid waste, but some manufacturers have received EPA variances to retain the status of scrap metal for boards they shred to prevent unauthorized re-use (3).

From these experiences, we conclude that unused defective PV panels are likely to be exempt from RCRA regulation if reclaimed as an unlisted commercial product. Various processing byproducts (e.g. flakes, dusts) may also be excepted from RCRA if reclaimed as unlisted by-products. Thus, PV panels, modules, and manufacturing by-products intended for reclamation may not require handling as hazardous wastes, regardless of their TCLP test results, if changes to the underlying law are shaped, or if the EPA and State regulatory agencies grant classifications of given PV products or a family of products as non-hazardous. To take advantage of such potential exceptions, cost-effective recycling practices should be available. In the following section, we discuss the feasibility of PV recycling.

## PV Recycling Feasibility

A typical PV facility will generate a significant amount of scrap at the start of its operation and, within six months to a year, will reach a steady-state level of production generating relatively little waste. Then, twenty-five to thirty years later, modules at the end of their useful life will have to be decommissioned. Figure 3 shows this trend for a hypothetical 10 MWp facility, assuming there is 20% total scrap during the first year, and 5% scrap from the first to the last year of operation. Under the same assumptions, Figure 4 depicts the quantities of

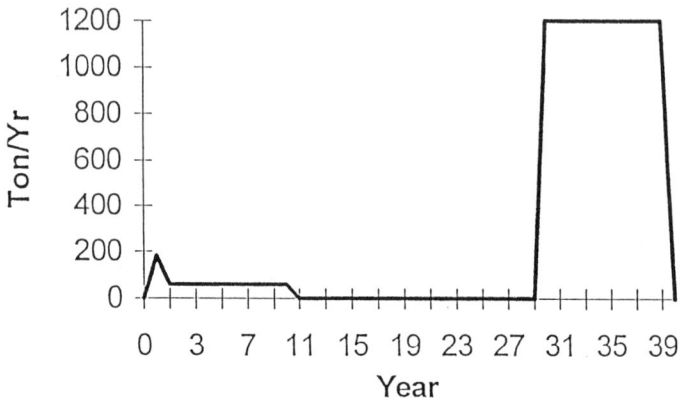

**FIGURE 3.** Quantity of Scrap and Decommissioned Modules Produced by a 10 MW/yr Manufacturing Facility.

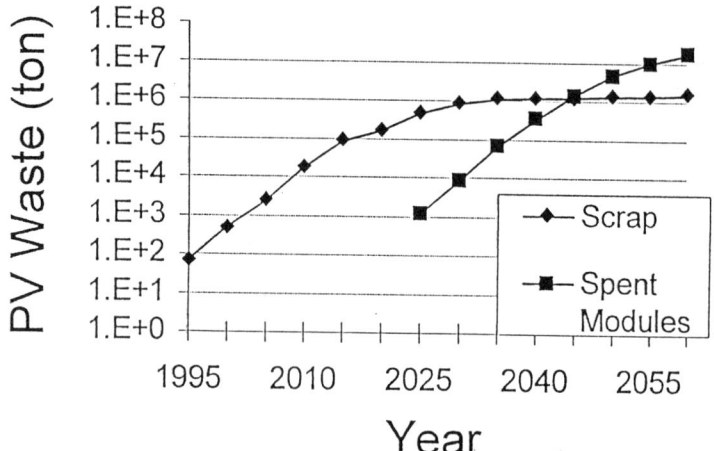

**FIGURE 4.** Projections of Quantities of Scrap and Decommissioned Modules in the U. S.

scrap and spent modules corresponding to the MARKAL-PV capacity projections.

In discussing PV recycling, one should distinguish between near-term and future needs and capabilities because of the long lapse between the start of manufacturing and decommissioning, and the corresponding differences in scale and technology. Near-term needs can be met by either centralized or de-centralized approaches, whereas future, large-scale needs would be more economically served by centralized strategies. Currently, pyro-metalurgical processes are more suitable for centralized recycling, whereas hydro-metalurgical processes are more likely feasible for small de-centralized operations, as shown in Figure 5.

## Centralized Strategies

Large smelters (e.g., Noranda, ASARCO), routinely recycle circuit boards, computer monitors, consumer electronics and telecommunication equipment to recover metals for their value. The recycling of PV materials might be

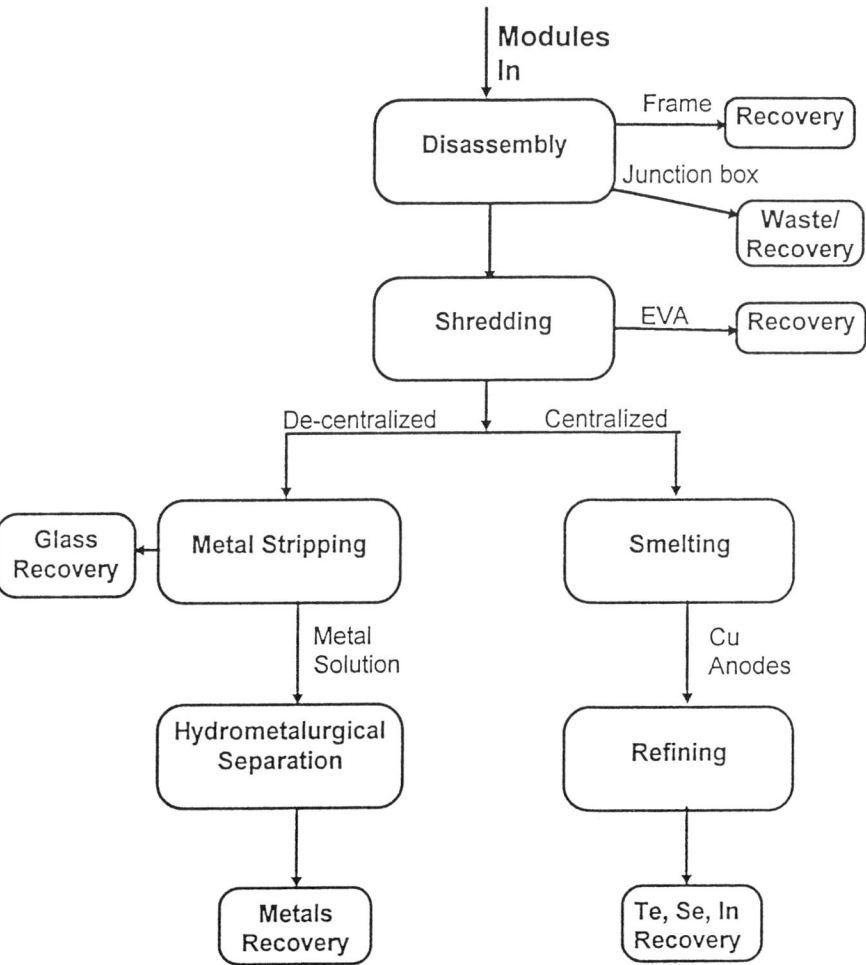

**FIGURE 5**. Scrap and Module Recycling Options.

incorporated into such commercial facilities. The low density of metals in PV modules and scrap does not generate any significant recycling value, but their glass content has a certain value to smelter operators who buy silica for their fluxing operation. The glass credit allows, therefore, for the costs of treatment to be reduced, which presently is estimated to be ~$200-$400/ton ($0.02 to $0.05 per W) for large deliveries (e.g., 20 ton containers). Transportation cost is an additional $200/ton. An important advantage of this option is that PV materials (other than sludge) intended for smelting are not subject to hazardous-waste regulations, since they are used as an ingredient in an industrial process to make a product.

Cadmium, tellurium, selenium, and contact metals (e.g., Ni) can be treated in copper smelters where the shredded material is processed through a liquid metal bath reactor, converters and anode furnaces. The glass content of the shred is used up in the fluxing operation of the smelter; the EVA and plastic decompose at the high temperatures of the smelter (e.g., 1000 C to 1400 C) into carbon dioxide and vapors of the monomers. In the furnaces, the anodes collect molten copper and the metals dissolved in it. These anodes are processed at the copper refinery, where metallurgical grade of tellurium and (if the supply is sufficient) of selenium are recovered electrolytically. Back metals (e.g., Ni) accumulate in the solution and are removed in the purification and acid-recovery phases. Cd does not dissolve in the Cu solution but remains in the waste stream of the copper smelter. A zinc smelter can recover cadmium, but the process is sensitive to tellurium and other metals, and so cannot accept CdTe scrap.

A problem affecting the economics of recycling for the PV industry is the large amount of glass or plastic in PV panels which makes the recovery of the expensive metals (e.g., In) uneconomical. On the other hand, these metals contaminate the glass so that large quantities of scrap and spent modules could be classified as hazardous, with all the associated cost implications. It would make sense to separate these metals from the glass if it can be done economically. The de-centralized recycling strategies discussed below follow this approach.

## De-centralized Recycling Strategies

In small-scale operations, metals can be stripped off glass and plastic by physical (e.g., hammer-mill, sand blasting, pyrolysis), or chemical methods. Chemical stripping using appropriate solvents (e.g., acids, oxidizers) is the most effective way to facilitate subsequent metal recovery. The metal-containing liquor can be treated in one of the following ways:

a. By the traditional method of precipitating metals as hydroxides, collecting them as (hazardous) sludge, and disposing of a hazardous waste site or sending it to a smelting plant.
b. By methods for concentrating metals in solution and recycling the solution within the plant. This can be accomplished by processes such as ion exchange, reverse osmosis, dialysis, and solvent extraction.
c. By methods for recovering metals directly by electrochemical means (e.g., electrodeposition).

The most common method is hydroxide precipitation, since most metals can be precipitated at alkaline pH. However, the process must be optimized for complete and selective precipitation. Amongst the solution-concentration techniques, ion exchange is widely practiced and can be very effective, but is more expensive than precipitation. Electrodeposition is occasionally practiced for cadmium, but is not normal practice. Both chemical precipitation and electrodeposition techniques are potentially adequate for separating cadmium/tellurium and copper/indium/selenium but they need to be optimized to be fully effective. On-going research, sponsored by the DOE Small Business Administration (SBIR) program, aims in such optimization. Three SBIR award recipients study the recycling of CdTe and CIS modules: Solar Cells Inc. of Toledo, Ohio; Drinkard Metalox of Charlotte, North Carolina; and Interphases Research of Thousand Oaks, California. Their work is summarized below.

Solar Cells' efforts focus on CdTe recycling. They tested various physical methods for separating the metals and plastic from the glass, including high-pressure water blasting, and crushing in a hammer-mill. The latter appears to be more advantageous since it creates a dry mixture of crushed glass and EVA flakes from which the flakes are readily separated by screening. The crushed glass then is stripped of the metals in successive steps of chemical dissolution, centrifugal separation, and precipitation or electrodeposition (Fig. 6). From this process, glass and Te are effectively separated and recovered while Cd remains in the sludge and has to be sent to a smelter or to controlled disposal. Solar Cells currently is in Phase II of the SBIR work, trying to accomplish total Cd separation and improve the overall material recovery and economics. The estimated cost for this process is approximately 0.04 $/W, excluding transportation and labor costs (4).

Drinkard Metalox is testing both CdTe and CIS recycling. Their process includes chemical stripping of the metals and EVA, skimming off the EVA from solution and successive steps of electrodeposition, precipitation, and evaporation to separate and recover the metals. In this process, substrates can potentially used again for PV deposition. Drinkard is now working in Phase II of SBIR work to improve process efficiency and economics. They project a recycling cost in the range of $0.01 to $0.04/W.

Interphases Research, studied CIS dissolution and deposition mechanisms and proposed a closed-loop electrochemical recycling scheme in which PV material from a defected panel is dissolved and re-deposited on a new substrate. This strategy is innovative and has the potential to be the most economical in facilities that use electrochemical deposition techniques, but it does not meet all the needs of a facility since it only applies to unbroken modules. Furthermore, although a PV layer can readily be dissolved from a PV/glass panel, dissolving it from an encapsulated module will be extremely difficult and slow. Funding of this SBIR project stopped after Phase I was competed.

In-situ recycling of dissolved metals via any of the above technologies requires metal stripping using a liquid solvent. Therefore, it is a logical option for facilities that use electrochemical deposition processes and create metal-containing liquid waste (e.g., $CdSO_4$, $CdCl_2$). In such facilities, scrap recycling and process-waste treatment can be integrated, consistent with the waste minimization objectives of RCRA; this should be more economic than external treatment.

# TOXICOLOGY

No clinical information is available on the potential human health effects from exposure to CdTe or CIS. In this section, we summarize human toxicological data on the precursor compounds Cd, Te, and Se and some newly developed animal data for CdTe, CIS, and CGS.

## Cadmium, Tellurium, and Selenium

The primary route of human exposure to cadmium has been by inhalation of Cd particles and fumes. Cd has a relatively high solubility, and is readily transported and bound into the blood. The acute health effects range from pneumonitis and pulmonary edema to death. Death has been caused by 1 hr exposure to concentrations of 40-50 mg/m$^3$ and a 5 hr exposure to 9 mg/m$^3$ (5a). The Immediately Dangerous to Life or Health (IDLH) level for Cd dust and fumes (Cd/CdO) is 40mg/m$^3$ as Cd (5b). The TWA TLV is 0.05 mg/m$^3$ as Cd (6). The primary chronic adverse effects of Cd inhalation are lung cancer and kidney damage, as shown in epidemiological studies of workers in nickel-cadmium battery, arsenic smelting, and other cadmium-bearing facilities. The National Institute for Occupational Safety and Health (NIOSH) has determined that there is a statistically significant increase in lung cancer deaths in workers exposed to a 20-year time-weighted average of 21-40 ug/m$^3$ of cadmium dust or fumes. Cadmium chloride, cadmium oxide, and cadmium sulfide produce lung tumors in rats. However, the potency of these compounds varies with the route of exposure and the compound, indicating differences in bio-availability of Cd among different compounds of different solubility.

The toxicity of tellurium metal is relatively low, but the toxicity of sodium telluride and sodium tellurite is high. Tellurium dioxide vapor has caused cough, shivering, fever and weakness, all symptoms of metal-fume fever. There are no reports of serious occupational illness from exposure to Te or its compounds. Acute exposure, however, of laboratory animals to Te and its compounds has caused bleeding and damage to the liver, kidneys, nervous system, lungs, gastrointestinal tract, and the heart ( 7).

There is a debate on what are the major effects of selenium intake in humans. In older studies, high doses were toxic to animals (8), whereas other studies suggest that selenium potentially has a cancer-preventing role. A very recent study showed that selenium may prevent lung, colon, rectal, and prostate cancers (9).

## Cadmium Telluride, Copper Indium Diselenide, and Copper Gallium Diselenide

Preliminary animal studies by Harris et al. (10), suggest that CdTe has little acute or sub-chronic toxicity. The investigators conducted a short-term (28 day) study of the systemic and reproductive toxicity of CdTe on laboratory rats. The compounds in particle form (mean diameter ~ 2 mm) were administered orally at doses ranging from 100 to 1000 mg/kg/day. All animals showed decrease in weight gain or weight loss, while the kidney weight increased. Despite these symptoms, there were no effects on the reproductive performance of females, including the number of live pups delivered nor in their subsequent survival and rates of weight gain. Based on their evaluation of developmental and reproductive factors, Harris et al. concluded that CdTe does not produce the toxic effects of the precursor Te. We are cautioned, however, on the difficulty of obtaining information on human exposure from oral administration, whereas such exposure would occur primarily through inhalation.

In a follow-up study designed to simulate the inhalation pathway, Morgan et al. (11), examined the acute toxicity of CdTe, CIS and CGS by intratracheal instillation of particulates in rats. Doses of 12, 25, 50, and 100 mg/kg body weight were administered. Body weight was significantly decreased in the rats, while lung weight, and brochoalveolar lavage fluid (BALF) increased, along with histopathological changes in the lung. Based on these parameters, CdTe was the most toxic, and CGS was the least toxic of the three compounds. The *in vitro* solubility of CdTe was considerably greater than that of CGS and CIS and likely contributed to its greater toxicity. There is no apparent reason for the observed difference in the toxicity of CGS and CIS since both compounds have similar solubilities. Additional studies are underway to evaluate pulmonary adsorption, systemic distribution, and systemic toxicity of the three compounds. The precursor compounds will also be evaluated in the same study to use the known characteristics of cadmium and tellurium as a reference for the extrapolating of the animal data to human occupational exposures.

## CONCLUSION

In addition to being a renewable and clean technology, PV energy can have a significant contribution into alleviating the Greenhouse effect.

Thin-film modules can be recycled at a reasonable cost, but a workable recycling program will require careful attention to the regulatory framework, materials economics, and to the experiences of comparable industries. Recycling used PV modules would be greatly simplified if they could be sent to a metal smelting/refining facility to be used as an ingredient in their operations. However, small quantities and high transportation costs make this option relatively expensive. Separating the PV materials from the glass can reduce the amount of potentially hazardous waste generated and the corresponding recycling or disposal costs. DOE SBIR funded research is in progress to create effective, economical means of recycling that can be used in both small- and large-scale operations.

There is a need to develop toxicological data for new compounds used in PV manufacturing. Recent studies comparing the acute toxicity of CdTe, CIS, and CGS showed that CdTe was the most toxic, and CGS was the least toxic of the three. Additional studies are underway to evaluate their pulmonary adsorption, systemic distribution, and systemic toxicity and to compare their toxicities with those of their precursors.

## ACKNOWLEDGMENTS

This work was supported by the Photovoltaics Division, Office of Photovoltaic and Wind Technologies, under Contract DE-AC02-76CH00016 with the U. S. Department of Energy. The animal studies reported herein were performed by NIEHS.

## REFERENCES

1. Fthenakis V. M. and Moskowitz P. D., Thin-Film Photovoltaic Cells: Health and Environmental Issues in their Manufacture, Use and Disposal, Progress in Photovoltaics: Research and Applications, Vol. 3, 295-306, 1995.
2. Hamilton L. D., Goldstein G. A., Lee J., Manne A. S., Marcuse W., Morris S. C., and Wene C-0. 1992. MARKAL-MACRO: an overview, BNL 48377. Brookhaven National Laboratory, Upton, NY.
3. Eberspacher C., Fthenakis V. M. and Moskowitz P. D., Environmental, Health and Safety Issues Related to Commercializing CIS-based Photovoltaics, BNL-63334, Brookhaven National Laboratory, Upton, NY.
4. Sasala R. A., Bohland J., and Smigielski K., Physical and Chemical Pathways for Economic Recycling of Cadmium Telluride Thin-film Photovoltaic Modules, 25th PVSC, pp. 865-868, IEEE, 1996.
5a. Documentation of the TLV Values, American Conference of Governmental Industrial Hygienists, (ACGIH), Cincinnati, OH, 4th ed., 1980-1981.
5b. NIOSH/OSHA Pocket Guide to Chemical Hazards, Publ. 78-210, September 1978.
6. Threshold Limit Values, American Conference of Governmental Industrial Hygienists (ACGIH), ISBN:0-936712-99-6, Cincinnati, Ohio, 1992.
7. Friberg L., Nordberg G. F., Kessler E., and Vouk V. B. (eds), Handbook of the Toxicology of Metals, 2nd ed., Vols. I, II, Elserier Science Publishers, Amsterdam, 1996.
8. Moxon A. L., Alkali Disease, or Selenium Poisoning, Bulletin 311, South Dakota Agr. Exp. Sta. Brookings p. 91, cited in The Bulletin of Selenium-Tellurium Development Association, June 1996, ISSN 1024-4204.
9. Journal American Medical Association, Dec. 25, 1996.
10. Harris M. W. The General and Reproductive Toxicity of the Photovoltaic Material Cadmium Telluride, Presented at the Society of Toxicology 1994 Annual Meeting, Dallas, Texas, March 13-17, 1994.
11. Morgan L. M., Shines C. J., Jeter S. P., Wilson R .E., Elwell P. E., Price H. C., and Moskowitz P. D., Acute Pulmonary Toxicity of Copper Gallium Diselenide, Copper Indium Diselenide, and Cadmium Telluride Intratracheally Instilled into Rats, Environmental Research 71, 16-24, 1995.

# Author Index

## A

Addis, F. W., 597
Aguilar, H., 597
Ahrenkiel, R. K., 225
Albin, D., 665
Albright, S., 162
Al-Jassim, M., 703, 709
Anderson, T. J., 579
Anna Selvan, J. A., 271
Arya, R. R., 49, 133, 479
Asher, S. E., 83

## B

Banerjee, A., 13
Barnett, A. M., 445
Basol, B. M., 107
Bauer, J., 143
Beck, N., 271
Berens, T. A., 83
Bhattacharya, R., 83
Birkmire, R. W., 123, 647
Boerner, A., 33
Bohn, R. G., 655
Bonn, R. H., 335
Boroditsky, M., 719
Bower, W. I., 893
Bradley, D., 49
Braunstein, R., 537
Brayman, S., 451
Britt, J. S., 115
Brog, T., 835
Buchanan, W. A., 547
Bulawka, A., 893
Bykov, E., 655

## C

Cai, L., 199
Cannon, T. W., 395
Cao, J., 851
Carlson, D. E., 49, 479
Champness, C. H., 603
Chang, C. H., 579
Chapman, R. N., 491
Checchi, J. C., 445
Chen, L. F., 49, 479
Chen, W., 451
Chleboski, R. G., 859
Ciszek, T. F., 189, 751, 771, 787
Cohen, J. D., 3
Compaan, A. D., 567, 655
Contreras, M. A., 83
Coutts, T. J., 693
Crandall, R. S., 27
Culik, J. S., 445
Cuperus, J., 271
Curtis, C. J., 683
Czanderna, A. W., 295

## D

Dalal, V. L., 33
D'Amico, J., 589
Davydov, A., 579
Dearmore, R., 143
DeBlasio, R., 369
Delahoy, A. E., 115
Dhere, N. G., 613
Dhere, R., 665
Dietrich, M. E., 143
Dippo, P., 573
Dong, S., 537
Dorer, G. L., 171
Dubail, S., 271
Duran, G., 827
Durand, S., 323

## E

Eser, E., 547
Estreicher, S. K., 729

## F

Farber, M. A., 859
Fedders, P. A., 729
Ferekides, C. S., 589, 631

Fernandez, G. T., 143
Fischer, A., 655
Fischer, D., 271
Fleddermann, C. B., 745
Flitton, R. A., 683
Fogleboch, J., 133
Ford, D. H., 445
Fourer, G., 827
Frausto, O. D., 143
Fredric, C. V., 143
Friedman, D. J., 247, 703
Fthenakis, V. M., 913

# G

Gafiteanu, R., 215
Gay, R. R., 143
Gee, J. M., 189, 751
Ginley, D. S., 683
Ginn, J. W., 335
Girvan, R., 33
Glick, S. H., 811
Goerlitzer, M., 271
Goertemiller, M., 719
Goetz, M., 271
Gonsiorawski, R., 851
Gordon, R. G., 39
Gösele, U. M., 215
Granata, J. E., 621
Grecu, D., 655
Grimmer, D. P., 451
Guha, S., 13, 27

# H

Halani, A., 107
Hall, R. B., 445
Handleman, C. K. P., 463
Hanley, C. J., 513
Hanoka, J. I., 859
Hegedus, S. S., 3, 547
Herdt, G., 709
Hieslmair, H., 759
Hoegh, D., 787
Hof, Ch., 271
Hoffmann, R., 739
Hund, T., 379

# I

Itoh, T., 557
Iwaniczko, E., 27
Izu, M., 881

# J

Jackson, E. L., 445
Jansen, K., 49
Jayamaha, U., 567
Jayamaha, U. N., 655
Jeffrey, F. R., 451
Jensen, C. L., 143
Jensen, D. G., 647
Jester, T. L., 433
Jones, K., 703
Jorgensen, G. J., 295

# K

Kalejs, J., 851
Kaminar, N., 841
Kane, P. M., 859
Kapur, V. K., 107
Kardauskas, M., 851
Karthikeyan, S., 589
Kaushal, S., 33
Kaydanov, V., 162
Keane, J., 83, 683
Kendall, C. L., 445
Keppner, H., 271
Kern, G. A., 463
Kessler, J., 133
Kester, J. J., 162
Keyes, B. M., 573, 703
Khattak, C. P., 779
Kibbler, A. E., 703
King, D. L., 347
King, R. R., 433
Kiss, Z. J., 115
Kroll, U., 271
Kroposki, B., 285, 313
Kurtz, S. R., 247, 703

# L

Lambarski, T. J., 471
Landry, M. D., 751, 787
Lee, X. Y., 719

Lei, W., 597
Leidholm, C. R., 107
Levi, D., 665
Liebe, J., 537
Lynn, K., 613

## M

Mahan, A. H., 27
Mao, D., 639
Marinskaya, S., 631
Marinskiy, D., 631
Martens, S. A., 451
Marudachalam, M., 123
Mason, A., 665
Matson, R. J., 683, 703
Matulionis, I., 567
Maxson, T., 33
McCandless, B. E., 123, 647
McDanal, A. J., 873
McHugo, S. A., 759
Meier, J., 271
Menna, P., 751
Meyers, P. V., 153
Minyard, G. E., 471
Mitchell, K. W., 433
Mitchell, R. L., 407
Morel, D. L., 589, 631
Moskowitz, P. D., 913
Moutinho, H. R., 665, 693
Mulligan, W. P., 693
Myers, D. R., 395

## N

Nakade, S., 567
Narashimha, S., 199
Narayanan, S., 745
Narayanaswamy, R., 589
Nelson, B. P., 27
Newton, J., 479
Niles, D. W., 573, 665, 709
Nitta, S., 557
Noak, M., 451
Norsworthy, G., 107
Noufi, R., 83, 573, 683

## O

O'Brien, C., 851
O'Gallagher, J., 739
Olsen, L. C., 597
O'Neill, M. J., 873
O'Quinn, B., 259
Oswald, R., 479
Ovshinsky, H. C., 881
Ovshinsky, S. R., 881

## P

Panse, P., 589
Parilla, P. A., 683
Pellaton Vaucher, N., 271
Pern, F. J., 795, 811
Pernet, P., 271
Peterson, T., 3
Phillips, J. A., 162, 835
Phillips, J. E., 547, 647
Pitts, J. R., 751
Platz, R., 271
Pohl, J., 271
Poplawski, C., 49
Powell, R. C., 171
Prince, M., 851

## R

Ragan, R., 719
Rajan, K., 49, 479
Ramanathan, K. R., 83, 573
Ramos, A. R., 143
Rand, J. A., 445
Reedy, R. C., Jr., 27
Reiter, N., 171
Ribelin, R., 162
Roe, R., 107
Rohatgi, A., 199
Romero, R., 479
Rose, D. H., 665, 693
Rosenthal, A., 323
Roy, M., 745
Roze, R., 787
Ruby, D. S., 745
Russell, L., 133
Russell, M. C., 463

## S

Sandwisch, D. W., 425
Sankaranarayanan, H., 589
Sasala, R. A., 171
Scandrett, B., 451
Schiff, E., 3
Schmid, F., 779
Schmitzberger, J. A., 143
Schultz, J. M., 123
Schulz, D. L., 683
Shafarman, W. N., 123, 547
Shah, A., 271
Sheldon, P., 665, 693
Shukri, Z. A., 603
Siivola, E., 259
Sites, J. R., 621
Sittler, G., 335
Skibo, S., 133
Stanbery, B. J., 579
Stern, M., 827
Stone, J. L., 521
Strong, S. J., 867
Swartzlander, A., 665

## T

Tan, T. Y., 215
Tang, J., 639
Tang, Y., 537
Tarrant, D. E., 143
Taylor, P. C., 557
Tetali, B., 631
Thomas, H. P., 407
Thomas, M., 323, 451
Tornstrom, E., 851
Torres, P., 271
Trefny, J. U., 639
Tscharner, R., 271
Tsuo, Y. S., 529, 751, 787
Tuttle, J. R., 83, 621

## U

Ullal, H. S., 75, 521
Unold, T., 27

## V

Van Loo, M., 827
Venkatasubramanian, R., 259
Viner, J. M., 557
von Roedern, B., 3, 313, 537
Vuille, J., 271

## W

Wagner, S., 3
Wallace, W. L., 529
Wang, S. L., 557
Wang, T. H., 771
Webb, J. D., 573
Weber, E. R., 759
Wei, S.-H., 63
West, R., 827
Whalen, W., 827
Wiedeman, S., 133
Wiesner, H., 683
Wieting, R. D., 143
Willet, D., 143
Willing, F., 49, 479
Wills, R. H., 867
Winston, R., 739
Witt, C. E., 407
Wohlgemuth, J. H., 415, 867
Wood, G., 49
Woods, L. M., 162
Wu, X., 693
Wyrsch, N., 271

## Y

Yablonovitch, E., 719
Yang, J., 13, 27
Yang, L., 49, 479
Youm, I., 647

## Z

Zafar, S., 589
Zhang, S. B., 63
Ziegler, Y., 271
Zunger, A., 63
Zweibel, K., 3

# AIP Conference Proceedings

| | Title | L.C. Number | ISBN |
|---|---|---|---|
| No. 331 | Non-Neutral Plasma Physics II<br>(Berkeley, CA 1994) | 95-79630 | 1-56396-441-4 |
| No. 332 | X-Ray Lasers 1994<br>Fourth International Colloquium<br>(Williamsburg, VA 1994) | 95-76067 | 1-56396-375-2 |
| No. 333 | Beam Instrumentation Workshop<br>(Vancouver, B. C., Canada 1994) | 95-79635 | 1-56396-352-3 |
| No. 334 | Few-Body Problems in Physics<br>(Williamsburg, VA 1994) | 95-76481 | 1-56396-325-6 |
| No. 335 | Advanced Accelerator Concepts<br>(Fontana, WI 1994) | 95-78225 | 1-56396-476-7 (Set)<br>1-56396-474-0 (Book)<br>1-56396-475-9 (CD-Rom) |
| No. 336 | Dark Matter<br>(College Park, MD 1994) | 95-76538 | 1-56396-438-4 |
| No. 337 | Pulsed RF Sources for Linear Colliders<br>(Montauk, NY 1994) | 95-76814 | 1-56396-408-2 |
| No. 338 | Intersections Between Particle and<br>Nuclear Physics 5th Conference<br>(St. Petersburg, FL 1994) | 95-77076 | 1-56396-335-3 |
| No. 339 | Polarization Phenomena in Nuclear Physics<br>Eighth International Symposium<br>(Bloomington, IN 1994) | 95-77216 | 1-56396-482-1 |
| No. 340 | Strangeness in Hadronic Matter<br>(Tucson, AZ 1995) | 95-77477 | 1-56396-489-9 |
| No. 341 | Volatiles in the Earth and Solar System<br>(Pasadena, CA 1994) | 95-77911 | 1-56396-409-0 |
| No. 342 | CAM -94 Physics Meeting<br>(Cacun, Mexico 1994) | 95-77851 | 1-56396-491-0 |
| No. 343 | High Energy Spin Physics<br>Eleventh International Symposium<br>(Bloomington, IN 1994) | 95-78431 | 1-56396-374-4 |
| No. 344 | Nonlinear Dynamics in Particle Accelerators:<br>Theory and Experiments<br>(Arcidosso, Italy 1994) | 95-78135 | 1-56396-446-5 |
| No. 345 | International Conference on Plasma Physics<br>ICPP 1994<br>(Foz do Iguaçu, Brazil 1994) | 95-78438 | 1-56396-496-1 |

| | Title | L.C. Number | ISBN |
|---|---|---|---|
| No. 346 | International Conference on Accelerator-Driven Transmutation Technologies and Applications (Las Vegas, NV 1994) | 95-78691 | 1-56396-505-4 |
| No. 347 | Atomic Collisions: A Symposium in Honor of Christopher Bottcher (1945-1993) (Oak Ridge, TN 1994) | 95-78689 | 1-56396-322-1 |
| No. 348 | Unveiling the Cosmic Infrared Background (College Park, MD, 1995) | 95-83477 | 1-56396-508-9 |
| No. 349 | Workshop on the Tau/Charm Factory (Argonne, IL, 1995) | 95-81467 | 1-56396-523-2 |
| No. 350 | International Symposium on Vector Boson Self-Interactions (Los Angeles, CA 1995) | 95-79865 | 1-56396-520-8 |
| No. 351 | The Physics of Beams Andrew Sessler Symposium (Los Angeles, CA 1993) | 95-80479 | 1-56396-376-0 |
| No. 352 | Physics Potential and Development of $\mu^+\mu^-$ Colliders: Second Workshop (Sausalito, CA 1994) | 95-81413 | 1-56396-506-2 |
| No. 353 | 13th NREL Photovoltaic Program Review (Lakewood, CO 1995) | 95-80662 | 1-56396-510-0 |
| No. 354 | Organic Coatings (Paris, France, 1995) | 96-83019 | 1-56396-535-6 |
| No. 355 | Eleventh Topical Conference on Radio Frequency Power in Plasmas (Palm Springs, CA 1995) | 95-80867 | 1-56396-536-4 |
| No. 356 | The Future of Accelerator Physics (Austin, TX 1994) | 96-83292 | 1-56396-541-0 |
| No. 357 | 10th Topical Workshop on Proton-Antiproton Collider Physics (Batavia, IL 1995) | 95-83078 | 1-56396-543-7 |
| No. 358 | The Second NREL Conference on Thermophotovoltaic Generation of Electricity | 95-83335 | 1-56396-509-7 |
| No. 359 | Workshops and Particles and Fields and Phenomenology of Fundamental Interactions (Puebla, Mexico 1995) | 96-85996 | 1-56396-548-8 |
| No. 360 | The Physics of Electronic and Atomic Collisions XIX International Conference (Whistler, Canada, 1995) | 95-83671 | 1-56396-440-6 |
| No. 361 | Space Technology and Applications International Forum (Albuquerque, NM 1996) | 95-83440 | 1-56396-568-2 |
| No. 362 | Two-Center Effects in Ion-Atom Collisions (Lincoln, NE 1994) | 96-83379 | 1-56396-342-6 |

| | Title | L.C. Number | ISBN |
|---|---|---|---|
| No. 363 | Phenomena in Ionized Gases<br>XXII ICPIG<br>(Hoboken, NJ, 1995) | 96-83294 | 1-56396-550-X |
| No. 364 | Fast Elementary Processes in<br>Chemical and Biological Systems<br>(Villeneuve d'Ascq, France, 1995) | 96-83624 | 1-56396-564-X |
| No. 365 | Latin-American School of Physics<br>XXX ELAF Group Theory and Its Applications<br>(México City, México, 1995) | 96-83489 | 1-56396-567-4 |
| No. 366 | High Velocity Neutron Stars<br>and Gamma-Ray Bursts<br>(La Jolla, CA 1995) | 96-84067 | 1-56396-593-3 |
| No. 367 | Micro Bunches Workshop<br>(Upton, NY, 1995) | 96-83482 | 1-56396-555-0 |
| No. 368 | Acoustic Particle Velocity Sensors:<br>Design, Performance and Applications<br>(Mystic, CT, 1995) | 96-83548 | 1-56396-549-6 |
| No. 369 | Laser Interaction and Related Plasma Phenomena<br>(Osaka, Japan 1995) | 96-85009 | 1-56396-445-7 |
| No. 370 | Shock Compression of Condensed Matter-1995<br>(Seattle, WA 1995) | 96-84595 | 1-56396-566-6 |
| No. 371 | Sixth Quantum 1/f Noise and Other Low<br>Frequency Fluctuations in<br>Electronic Devices Symposium<br>(St. Louis, MO, 1994) | 96-84200 | 1-56396-410-4 |
| No. 372 | Beam Dynamics and Technology Issues for<br>+ - Colliders 9th Advanced ICFA<br>Beam Dynamics Workshop<br>(Montauk, NY, 1995) | 96-84189 | 1-56396-554-2 |
| No. 373 | Stress-Induced Phenomena in Metallization<br>(Palo Alto, CA 1995) | 96-84949 | 1-56396-439-2 |
| No. 374 | High Energy Solar Physics<br>(Greenbelt, MD 1995) | 96-84513 | 1-56396-542-9 |
| No. 375 | Chaotic, Fractal, and Nonlinear Signal Processing<br>(Mystic, CT 1995) | 96-85356 | 1-56396-443-0 |
| No. 376 | Chaos and the Changing Nature of<br>Science and Medicine: An Introduction<br>(Mobile, AL 1995) | 96-85220 | 1-56396-442-2 |
| No. 377 | Space Charge Dominated Beams and<br>Applications of High Brightness Beams<br>(Bloomington, IN 1995) | 96-85165 | 1-56396-625-7 |
| No. 378 | Surfaces, Vacuum, and Their Applications<br>(Cancun, Mexico 1994) | 96-85594 | 1-56396-418-X |
| No. 379 | Physical Origin of Homochirality in Life<br>(Santa Monica, CA 1995) | 96-86631 | 1-56396-507-0 |

|  | Title | L.C. Number | ISBN |
|---|---|---|---|
| No. 380 | Production and Neutralization of Negative Ions and Beams / Production and Application of Light Negative Ions (Upton, NY 1995) | 96-86435 | 1-56396-565-8 |
| No. 381 | Atomic Processes in Plasmas (San Francisco, CA 1996) | 96-86304 | 1-56396-552-6 |
| No. 382 | Solar Wind Eight (Dana Point, CA 1995) | 96-86447 | 1-56396-551-8 |
| No. 383 | Workshop on the Earth's Trapped Particle Environment (Taos, NM 1994) | 96-86619 | 1-56396-540-2 |
| No. 384 | Gamma-Ray Bursts (Huntsville, AL 1995) | 96-79458 | 1-56396-685-9 |
| No. 385 | Robotic Exploration Close to the Sun: Scientific Basis (Marlboro, MA 1996) | 96-79560 | 1-56396-618-2 |
| No. 386 | Spectral Line Shapes, Volume 9 13th ICSLS (Firenze, Italy 1996) |  | 1-56396-656-5 |
| No. 387 | Space Technology and Applications International Forum (Albuquerque, NM 1997) | 96-80254 | 1-56396-679-4 (Case set) 1-56396-691-3 (Paper set) |
| No. 388 | Resonance Ionization Spectroscopy 1996 Eighth International Symposium (State College, PA 1996) | 96-80324 | 1-56396-611-5 |
| No. 389 | X-Ray and Inner-Shell Processes 17th International Conference (Hamburg, Germany 1996) | 96-80388 | 1-56396-563-1 |
| No. 390 | Beam Instrumentation Proceedings of the Seventh Workshop (Argonne, IL 1996) | 97-70568 | 1-56396-612-3 |
| No. 391 | Computational Accelerator Physics (Williamsburg, VA 1996) | 97-70181 | 1-56396-671-9 |
| No. 392 | Applications of Accelerators in Research and Industry: Proceedings of the Fourteenth International Conference (Denton, TX 1996) | 97-71846 | 1-56396-652-2 |
| No. 393 | Star Formation Near and Far Seventh Astrophysics Conference (College Park, MD 1996) | 97-71978 | 1-56396-678-6 |
| No. 394 | NREL/SNL Photovoltaics Program Review Proceedings of the 14th Conference— A Joint Meeting (Lakewood, CO 1996) | 97-72645 | 1-56396-687-5 |